# *Electrical Engineering*

# Newnes Know It All Series

**PIC Microcontrollers: Know It All**
Lucio Di Jasio, Tim Wilmshurst, Dogan Ibrahim, John Morton, Martin Bates, Jack Smith, D.W. Smith, and Chuck Hellebuyck
ISBN: 978-0-7506-8615-0

**Embedded Software: Know It All**
Jean Labrosse, Jack Ganssle, Tammy Noergaard, Robert Oshana, Colin Walls, Keith Curtis, Jason Andrews, David J. Katz, Rick Gentile, Kamal Hyder, and Bob Perrin
ISBN: 978-0-7506-8583-2

**Embedded Hardware: Know It All**
Jack Ganssle, Tammy Noergaard, Fred Eady, Lewin Edwards, David J. Katz, Rick Gentile, Ken Arnold, Kamal Hyder, and Bob Perrin
ISBN: 978-0-7506-8584-9

**Wireless Networking: Know It All**
Praphul Chandra, Daniel M. Dobkin, Alan Bensky, Ron Olexa, David Lide, and Farid Dowla
ISBN: 978-0-7506-8582-5

**RF & Wireless Technologies: Know It All**
Bruce Fette, Roberto Aiello, Praphul Chandra, Daniel Dobkin, Alan Bensky, Douglas Miron, David Lide, Farid Dowla, and Ron Olexa
ISBN: 978-0-7506-8581-8

**Electrical Engineering: Know It All**
Clive Maxfield, Alan Bensky, John Bird, W. Bolton, Izzat Darwazeh, Walt Kester, M.A. Laughton, Andrew Leven, Luis Moura, Ron Schmitt, Keith Sueker, Mike Tooley, DF Warne, Tim Williams
ISBN: 978-1-85617-528-9

**Audio Engineering: Know It All**
Douglas Self, Richard Brice, Don Davis, Ben Duncan, John Linsely Hood, Morgan Jones, Eugene Patronis, Ian Sinclair, Andrew Singmin, John Watkinson
ISBN: 978-1-85617-526-5

**Circuit Design: Know It All**
Darren Ashby, Bonnie Baker, Stuart Ball, John Crowe, Barrie Hayes-Gill, Ian Grout, Ian Hickman, Walt Kester, Ron Mancini, Robert A. Pease, Mike Tooley, Tim Williams, Peter Wilson, Bob Zeidman
ISBN: 978-1-85617-527-2

**Test and Measurement: Know It All**
Jon Wilson, Stuart Ball, GMS de Silva, Tony Fischer-Cripps, Dogan Ibrahim, Kevin James, Walt Kester, M A Laughton, Chris Nadovich, Alex Porter, Edward Ramsden, Stephen Scheiber, Mike Tooley, D. F. Warne, Tim Williams
ISBN: 978-1-85617-530-2

**Mobile Wireless Security: Know It All**
Praphul Chandra, Alan Bensky, Tony Bradley, Chris Hurley, Steve Rackley, John Rittinghouse, James Ransome, Timothy Stapko, George Stefanek, Frank Thornton, Chris Lanthem, John Wilson
ISBN: 978-1-85617-529-6

For more information on these and other Newnes titles visit: www.newnespress.com

# *Electrical Engineering*

Clive Maxfield
John Bird
M. A. Laughton
W. Bolton
Andrew Leven
Ron Schmitt
Keith Sueker
Tim Williams
Mike Tooley
Luis Moura
Izzat Darwazeh
Walt Kester
Alan Bensky
DF Warne

AMSTERDAM · BOSTON · HEIDELBERG · LONDON
NEW YORK · OXFORD · PARIS · SAN DIEGO
SAN FRANCISCO · SINGAPORE · SYDNEY · TOKYO

Newnes is an imprint of Elsevier

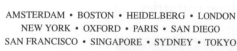
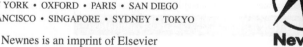

Newnes is an imprint of Elsevier
30 Corporate Drive, Suite 400, Burlington, MA 01803, USA
Linacre House, Jordan Hill, Oxford OX2 8DP, UK

**Library of Congress Cataloging-in-Publication Data**
Application submitted

**British Library Cataloguing-in-Publication Data**
A catalogue record for this book is available from the British Library.

ISBN: 978-1-85617-528-9

For information on all Newnes publications
visit our Web site at www.elsevierdirect.com

Typeset by Charon Tec Ltd., A Macmillan Company. (www.macmillansolutions.com)

Printed and bound in the United Kingdom

Transferred to Digital Printing, 2011

Working together to grow
libraries in developing countries

www.elsevier.com | www.bookaid.org | www.sabre.org

ELSEVIER    BOOK AID International    Sabre Foundation

# Contents

*Note from the Publisher: The authors of this book are from around the world and as such symbols vary between US and UK styles.*

# About the Authors

**Alan Bensky** MScEE (Chapter 19) is an electronics engineering consultant with over 25 years of experience in analog and digital design, management, and marketing. Specializing in wireless circuits and systems, Bensky has carried out projects for varied military and consumer applications. He is the author of *Short-range Wireless Communication, Second Edition*, published by Elsevier, 2004, and has written several articles in international and local publications. He has taught courses and gives lectures on radio engineering topics. Bensky is a senior member of IEEE.

**John Bird** BSc (Hons), CEng, CMath, CSci, FIET, MIEE, FIIE, FIMA, FCollT Royal Naval School of Marine Engineering, HMS Sultan, Gosport; formerly University of Portsmouth and Highbury College, Portsmouth, U.K., (Chapters 1, 2, 3, 4, 5, 6, 7, 8, Appendix A) is the author of *Electrical Circuit Theory and Technology,* and over 120 textbooks on engineering and mathematical subjects, is the former Head of Applied Electronics in the Faculty of Technology at Highbury College, Portsmouth, U.K.

More recently, he has combined freelance lecturing at the University of Portsmouth, with technical writing and Chief Examiner responsibilities for City and Guilds Telecommunication Principles and Mathematics, and examining for the International Baccalaureate Organisation.

John Bird is currently a Senior Training Provider at the Royal Naval School of Marine Engineering in the Defence College of Marine and Air Engineering at H.M.S. Sultan, Gosport, Hampshire, U.K. The school, which serves the Royal Navy, is one of Europe's largest engineering training establishments.

**Bill Bolton** (Chapter 18, Appendix B.) is the author of *Control Systems*, and many engineering textbooks, including the best-selling books *Programmable Logic Controllers* (Newnes) and *Mechatronics* (Pearson—Prentice-Hall), and has formerly been a senior lecturer in a College of Technology, Head of Research, Development and Monitoring at the Business and Technician Education Council, a member of the Nuffield Advanced Physics Project, and a consultant on a British Government Technician Education Project in Brazil and on Unesco projects in Argentina and Thailand.

**Izzat Darwazeh** (Chapter 9) is the author of *Introduction to Linear Circuit Analysis and Modelling*. He holds the University of London Chair of Communications Engineering in the Department of Electronic and Electrical at UCL. He obtained his first degree in Electrical Engineering from the University of Jordan in 1984 and the MSc and PhD degrees, from the University of Manchester Institute of Science and Technology (UMIST), in 1986 and 1991, respectively. He worked as a research Fellow at the University of Wales-Bangor—U.K. from 1990 till 1993, researching very high speed optical systems and circuits. He was a Senior Lecturer in Optoelectronic Circuits and Systems in the Department at Electrical Engineering and Electronics at UMIST. He moved to UCL in October 2001 where he is currently the Head of Communications and Information System (CIS) group and the Director of UCL Telecommunications for Industry Programme. He is a Fellow of the IET and a Senior Member of the IEEE.

His teaching covers aspects of wireless and optical fibre communications, telecommunication networks, electronic circuits and high speed integrated circuits and MMICs. He lectures widely in the U.K. and overseas. His research interests are mainly in the areas of wireless system design and implementation, high speed optical communication systems and networks, microwave circuits and MMICs for optical fibre applications and in mobile and wireless communication circuits and systems. He has authored/co-authored more than 120 research papers. He has co-authored (with Luis Moura) a book on Linear Circuit Analysis and Modelling (Elsevier 2005) and is the co-editor of the IEE book on Analogue Optical Communications (IEE 1995). He collaborates with various telecommunications and electronic industries in the U.K. and overseas and has acted as a consultant to various academic, industrial, financial and government organisations.

**Walt Kester** (Chapters 16, 17) is the author of *Mixed-Signal and DSP Design Techniques*. He is a corporate staff applications engineer at Analog Devices. For over 35 years at Analog Devices, he has designed, developed, and given applications support for high-speed ADCs, DACs, SHAs, op amps, and analog multiplexers. Besides writing many papers and articles, he prepared and edited eleven major applications books which form the basis for the Analog Devices world-wide technical seminar series including the topics of op amps, data conversion, power management, sensor signal conditioning, mixed-signal, and practical analog design techniques. He also is the editor of *The Data Conversion Handbook*, a 900+ page comprehensive book on data conversion published in 2005 by Elsevier. Walt has a BSEE from NC State University and MSEE from Duke University.

**Michael Laughton** BASc, (Toronto), PhD (London), DSc (Eng.) (London), FREng, FIEE, CEng (Chapters 25) is the editor of *Electrical Engineer's Reference Book, 16th Edition*. He is the Emeritus Professor of Electrical Engineering of the University of London and former Dean of Engineering of the University and Pro-Principal of Queen Mary and Westfield College, and is currently the U.K. representative on the Energy Committee of the European National Academies of Engineering, a member of energy and environment policy advisory groups of the Royal Academy of Engineering, the Royal Society and the Institution of Electrical Engineers as well as the Power Industry Division Board of the Institution of Mechanical Engineers. He has acted as Specialist Adviser to U.K. Parliamentary Committees in both upper and lower Houses on alternative and renewable energy technologies and on energy efficiency. He was awarded The Institution of Electrical Engineers Achievement Medal in 2002 for sustained contributions to electrical power engineering.

**Andrew Leven** (Chapter 17, 19) is the author of *Telecommunications Circuits and Technology*. He holds a diploma in Radio Technology, HNC, BSc (Hons) Electronics, MSc Astronomy, C. Eng M.I.E.E, Teaching Diploma, M.I.P., International Education and Training Consultant (Formerly Senior Lecturer in Telecommunications, Electronics and Fibre Optics at James Watt College of Higher Education, U.K.)

**A. Maddocks** (Chapter 25) was a contributor to *Electrical Engineer's Reference Book, 16th Edition*.

**Clive "Max" Maxfield** (Chapter 10) is the author of *Bebop to the Boolean Boogie*. He is six feet tall, outrageously handsome, English and proud of it. In addition to being a hero, trendsetter, and leader of fashion, he is widely regarded as an expert in all aspects of electronics and computing (at least by his mother).

After receiving his B.Sc. in Control Engineering in 1980 from Sheffield Polytechnic (now Sheffield Hallam University), England, Max began his career as a designer of central processing units for mainframe computers. During his career, he has designed everything from ASICs to PCBs and has meandered his way through most aspects of Electronics Design Automation (EDA). To cut a long story short, Max now finds himself President of TechBites Interactive (www.techbites.com). A marketing consultancy, TechBites specializes in communicating the value of its clients' technical products and services to non-technical audiences through a variety of media, including websites, advertising, technical documents, brochures, collaterals, books, and multimedia.

In addition to numerous technical articles and papers appearing in magazines and at conferences around the world, Max is also the author and co-author of a number of books, including *Bebop to the Boolean Boogie (An Unconventional Guide to Electronics)*, *Designus Maximus Unleashed (Banned in Alabama)*, *Bebop BYTES Back (An Unconventional Guide to Computers)*, *EDA: Where Electronics Begins*, *The Design Warrior's Guide to FPGAs*, and *How Computers Do Math* (www.diycalculator.com).

In his spare time (Ha!), Max is co-editor and co-publisher of the web-delivered electronics and computing hobbyist magazine *EPE Online* (www.epemag.com). Max also acts as editor for the Programmable Logic DesignLine website (www.pldesignline.com) and for the iDESIGN section of the Chip Design Magazine website (www.chipdesignmag.com).

On the off-chance that you're still not impressed, Max was once referred to as an *"industry notable"* and a *"semiconductor design expert"* by someone famous who wasn't prompted, coerced, or remunerated in any way!

**Luis Moura** (Chapter 9) is the author of *Introduction to Linear Circuit Analysis and Modelling*. He received the diploma degree in electronics and telecommunications from the University of Aveiro, Portugal, in 1991, and the PhD degree in electronic engineering from the University of North Wales, Bangor, U.K. in 1995. From 1995 to 1997 he worked as a research Fellow in the Telecommunications Research Group at University College London, U.K. He is currently a Lecturer in Electronics at the University of Algarve, Portugal. In 2007 he took one year leave of absence to work in the company Lime Microsystems U.K. as Senior Design Engineer. He was designing frequency synthesisers for multi-mode/multi-standard wireless transceivers.

**Ron Schmitt** (Chapters 20, 21) is the author of *Electromagnetics Explained*. He is the former Director of Electrical Engineering, Sensor Research and Development Corp. Orono, Maine.

**Keith H. Sueker** (Chapters 15, 22, 23) is the author of *Power Electronics Design*. Sueker received his BEE with High Distinction from the University of Minnesota, he continued his education at Illinois Institute of Technology where he received his MSEE, he also completed his course work for his PhD. He spent many years working for Westinghouse Electric Corporation in various positions. He then moved on to Robicon Corporation as a consulting engineer, he retired in 1993. His responsibilities included analytical techniques and equipment design for power factor correction and harmonic mitigation. Sueker has written a number of IEEE papers and several articles for trade publications. Also, he has prepared a monograph and 90 minute video tape on these subjects. He and Mr. R. P. Stratford have presented tutorial sessions on power factor and harmonics at IEEE-IAS annual meetings, and he has presented additional tutorials in other cities. He also presented a tutorial on transformers for the local IEEE-IAS in the spring of 1999 and repeated it in the fall of 2003. Sueker delivered a tutorial on power electronics for the local IEEE-IAS/PES in the spring of 2005. He was also pleased to serve on the IEEE committee for awarding the "IEEE Medal for Engineering Excellence" for four years. He is currently a Life Senior Member of the IEEE and also a registered Professional Engineer in the Commonwealth of Pennsylvania.

**Mike Tooley** (Chapters 11, 12, 14, 24) is the author of *Electronics Circuits*. He is the former Director of Learning Technology at Brooklands College, Surrey, U.K.

**Douglas Warne** (Chapters 25) is the editor of *Electrical Engineers Reference book, 16th Edition*. Warne graduated from Imperial College London in 1967 with a 1st class honours degree in electrical engineering, during this time he had a student apprenticeship with AEI Heavy Plant Division, Rugby, 1963–1968. He is currently self-employed, and has taken on such projects as Co-ordinated LINK PEDDS programme for DTI, and the electrical engineering, electrical machines and drives and ERCOS programmes for EPSRC. Initiated and manage the NETCORDE university-industry network for identifying and launching new R&D projects. He has acted as co-ordinator for the industry-academic funded ESR Network, held the part-time position of Research Contract Co-ordinator for the High Voltage and Energy Systems group at University of Cardiff and monitored several projects funded through the DTI Technology Programme.

**Tim Williams** (Chapters 11, 13, 15) is the author of *The Circuit Designer's Companion*. He is employed with Elmac Services, Chichester, U.K.

# CHAPTER 1

# An Introduction to Electric Circuits

John Bird

## 1.1 SI Units

The system of units used in engineering and science is the Système International d'Unités (International system of units), usually abbreviated to SI units, and is based on the metric system. This was introduced in 1960 and is now adopted by the majority of countries as the official system of measurement.

The basic units in the SI system are listed with their symbols, in Table 1.1.

*Derived SI units* use combinations of basic units and there are many of them. Two examples are:

- Velocity—meters per second (m/s)
- Acceleration—meters per second squared (m/s$^2$)

**Table 1.1: Basic SI units**

| Quantity | Unit |
|---|---|
| length | meter, m |
| mass | kilogram, kg |
| time | second, s |
| electric current | ampere, A |
| thermodynamic temperature | kelvin, K |
| luminous intensity | candela, cd |
| amount of substance | mole, mol |

**Table 1.2: Six most common multiples**

| Prefix | Name | Meaning | |
|--------|------|---------|--|
| M | mega | multiply by 1,000,000 | (i.e., $\times 10^6$) |
| k | kilo | multiply by 1,000 | (i.e., $\times 10^3$) |
| m | milli | divide by 1,000 | (i.e., $\times 10^{-3}$) |
| μ | micro | divide by 1,000,000 | (i.e., $\times 10^{-6}$) |
| n | nano | divide by 1,000,000,000 | (i.e., $\times 10^{-9}$) |
| p | pico | divide by 1,000,000,000,000 | (i.e., $\times 10^{-12}$) |

SI units may be made larger or smaller by using prefixes that denote multiplication or division by a particular amount. The six most common multiples, with their meaning, are listed in Table 1.2.

## 1.2 Charge

The *unit of charge* is the coulomb (C) where one coulomb is one ampere second. (1 coulomb = $6.24 \times 10^{18}$ electrons). The coulomb is defined as the quantity of electricity that flows past a given point in an electric circuit when a current of one ampere is maintained for one second. Thus,

charge, in coulombs $Q = It$

where $I$ is the current in amperes and $t$ is the time in seconds.

### Example 1.1
If a current of 5 A flows for 2 minutes, find the quantity of electricity transferred.

### Solution
Quantity of electricity $Q = It$ coulombs

$I = 5\,A, t = 2 \times 60 = 120\,s$

Hence, $Q = 5 \times 120 = 600\,C$

## 1.3 Force

The *unit of force* is the newton (N) where one newton is one kilogram meter per second squared. The newton is defined as the force which, when applied to a mass of one kilogram, gives it an acceleration of one meter per second squared. Thus,

force, in newtons $F = ma$

where $m$ is the mass in kilograms and $a$ is the acceleration in meters per second squared. Gravitational force, or weight, is mg, where $g = 9.81$ m/s$^2$.

### Example 1.2

A mass of 5000 g is accelerated at 2 m/s$^2$ by a force. Determine the force needed.

### Solution

Force = mass × acceleration

$$= 5 \, \text{kg} \times 2 \, \text{m/s}^2 = 10 \, \frac{\text{kg m}}{\text{s}^2} = \mathbf{10 \, N}$$

### Example 1.3

Find the force acting vertically downwards on a mass of 200 g attached to a wire.

### Solution

Mass = 200 g = 0.2 kg and acceleration due to gravity, g = 9.81 m/s$^2$

Force acting downwards = weight = mass × acceleration

$$= 0.2 \, \text{kg} \times 9.81 \, \text{m/s}^2$$
$$= \mathbf{1.962 \, N}$$

## 1.4 Work

The *unit of work or energy* is the *joule* (J) where one joule is one Newton meter. The joule is defined as the work done or energy transferred when a force of one newton is exerted through a distance of one meter in the direction of the force. Thus,

work done on a body, in joules $W = Fs$

where $F$ is the force in Newtons and $s$ is the distance in meters moved by the body in the direction of the force. Energy is the capacity for doing work.

## 1.5 Power

The *unit of power* is the watt (W) where one watt is one joule per second. Power is defined as the rate of doing work or transferring energy. Thus,

power in watts, $P = \dfrac{W}{t}$

where $W$ is the work done or energy transferred in joules and $t$ is the time in seconds. Thus,

energy, in joules, $W = Pt$

### Example 1.4

A portable machine requires a force of 200 N to move it. How much work is done if the machine is moved 20 m and what average power is utilized if the movement takes 25 s?

### Solution

Work done = force × distance

$\qquad$ = 200 N × 20 m

$\qquad$ = 4000 Nm or 4 kJ

Power = $\dfrac{\text{work done}}{\text{time taken}}$

$\qquad = \dfrac{4000 \text{ J}}{25 \text{ s}} =$ **160 J/s = 160 W**

### Example 1.5

A mass of 1000 kg is raised through a height of 10 m in 20 s. What is (a) the work done and (b) the power developed?

### Solution

(a)  Work done = force × distance and

$\qquad$ force = mass × acceleration

$\quad$ Hence, work done = (1000 kg × 9.81 m/s$^2$) × (10 m)

$\qquad\qquad\qquad$ = 98 100 Nm

$\qquad\qquad\qquad$ = **98.1 kNm or 98.1 kJ**

(b)  Power = $\dfrac{\text{work done}}{\text{time taken}} = \dfrac{98\,100 \text{ J}}{20 \text{ s}} = 4905$ J/s

$\qquad\qquad\qquad$ = **4905 W or 4.905 kW**

## 1.6 Electrical Potential and e.m.f.

The *unit of electric potential* is the *volt* (V) where one volt is one joule per coulomb. One volt is defined as the difference in potential between two points in a conductor which, when carrying a current of one ampere, dissipates a power of one watt, i.e.,

$$\text{volts} = \frac{\text{watts}}{\text{amperes}} = \frac{\text{joules/second}}{\text{amperes}}$$
$$= \frac{\text{joules}}{\text{ampere seconds}} = \frac{\text{joules}}{\text{coulombs}}$$

A change in electric potential between two points in an electric circuit is called a *potential difference*. The *electromotive force* (e.m.f.) provided by a source of energy such as a battery or a generator is measured in volts.

## 1.7 Resistance and Conductance

The *unit of electric resistance* is the *ohm* ($\Omega$) where one ohm is one volt per ampere. It is defined as the resistance between two points in a conductor when a constant electric potential of one volt applied at the two points produces a current flow of one ampere in the conductor. Thus,

$$\text{resistance, in ohms } R = \frac{V}{I}$$

where $V$ is the potential difference across the two points in volts and $I$ is the current flowing between the two points in amperes.

The reciprocal of resistance is called *conductance* and is measured in siemens (S). Thus,

$$\text{conductance, in siemens } G = \frac{1}{R}$$

where $R$ is the resistance in ohms.

### Example 1.6
Find the conductance of a conductor of resistance (a) $10\,\Omega$, (b) $5\,k\Omega$ and (c) $100\,m\Omega$.

### Solution
(a) Conductance $G = \dfrac{1}{R} = \dfrac{1}{10}$ siemen $= \mathbf{0.1\,S}$

(b) $G = \dfrac{1}{R} = \dfrac{1}{5 \times 10^3} \, S = 0.2 \times 10^{-3} \, S = \mathbf{0.2 \, mS}$

(c) $G = \dfrac{1}{R} = \dfrac{1}{100 \times 10^{-3}} \, S = \dfrac{10^3}{100} \, S = \mathbf{10 \, S}$

## 1.8 Electrical Power and Energy

When a direct current of $I$ amperes is flowing in an electric circuit and the voltage across the circuit is V volts, then,

power, in watts $P = VI$

Electrical energy = Power × time

$= VIt$ joules

Although the unit of energy is the joule, when dealing with large amounts of energy, the unit used is the *kilowatt hour* (kWh) where

$1 \, kWh = 1000$ watt hour

$= 1000 \times 3600$ watt seconds or joules

$= 3{,}600{,}000 \, J$

### Example 1.7

A source e.m.f. of 5 V supplies a current of 3 A for 10 minutes. How much energy is provided in this time?

### Solution

Energy = power × time and power = voltage × current.

Hence,

Energy $= VIt = 5 \times 3 \times (10 \times 60)$

$= 9000 \, Ws$ or $J$

$= 9 \, kJ$

### Example 1.8

An electric heater consumes 1.8 MJ when connected to a 250 V supply for 30 minutes. Find the power rating of the heater and the current taken from the supply.

*Solution*

Energy = power × time,

$$\text{power} = \frac{\text{energy}}{\text{time}}$$

$$= \frac{1.8 \times 10^6 \, \text{J}}{30 \times 60 \, \text{s}} = 1000 \, \text{J/s} = 1000 \, \text{W}$$

i.e., Power rating of heater = 1 kW

Power $P = VI$, thus, $I = \dfrac{P}{V} = \dfrac{1000}{250} = 4 \, \text{A}$

Hence, the current taken from the supply is 4 A.

## 1.9  Summary of Terms, Units and Their Symbols

### Table 1.3: Electrical terms, units, and symbols

| Quantity | Quantity Symbol | Unit | Unit symbol |
|---|---|---|---|
| Length | $l$ | meter | m |
| Mass | $m$ | kilogram | kg |
| Time | $t$ | second | s |
| Velocity | $v$ | meters per second | m/s or $\text{m s}^{-1}$ |
| Acceleration | $a$ | meters per second squared | $\text{m/s}^2$ or $\text{m s}^{-2}$ |
| Force | $F$ | newton | N |
| Electrical charge or quantity | $Q$ | coulomb | C |
| Electric current | $I$ | ampere | A |
| Resistance | $R$ | ohm | $\Omega$ |
| Conductance | $G$ | siemen | S |
| Electromotive force | $E$ | volt | V |
| Potential difference | $V$ | volt | V |
| Work | $W$ | joule | J |
| Energy | $E$ (or $W$) | joule | J |
| Power | $P$ | watt | W |

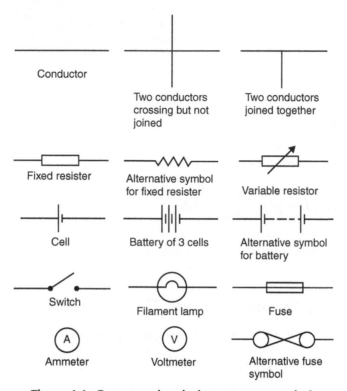

**Figure 1.1: Common electrical component symbols**

## 1.10  Standard Symbols for Electrical Components

Symbols are used for components in electrical circuit diagrams and some of the more common ones are shown in Figure 1.1.

## 1.11  Electric Current and Quantity of Electricity

All *atoms* consist of protons, neutrons and electrons. The protons, which have positive electrical charges, and the neutrons, which have no electrical charge, are contained within the nucleus. Removed from the nucleus are minute negatively charged particles called *electrons*. Atoms of different materials differ from one another by having different numbers of protons, neutrons and electrons. An equal number of protons and electrons exist within an atom and it is said to be electrically balanced, as the positive and negative charges cancel each other out. When there are more than two electrons in an atom the electrons are arranged into shells at various distances from the nucleus.

All atoms are bound together by powerful forces of attraction existing between the nucleus and its electrons. Electrons in the outer shell of an atom, however, are attracted to their nucleus less powerfully than are electrons whose shells are nearer the nucleus.

It is possible for an atom to lose an electron; the atom, which is now called an *ion*, is not now electrically balanced, but is positively charged and is thus able to attract an electron to itself from another atom. Electrons that move from one atom to another are called free electrons and such random motion can continue indefinitely. However, if an electric pressure or voltage is applied across any material there is a tendency for electrons to move in a particular direction. This movement of free electrons, known as *drift*, constitutes an electric current flow. Thus current is the rate of movement of charge.

*Conductors* are materials that contain electrons that are loosely connected to the nucleus and can easily move through the material from one atom to another.

*Insulators* are materials whose electrons are held firmly to their nucleus.

The unit used to measure the *quantity of electrical charge Q* is called the *coulomb* C (where 1 coulomb = $6.24 \times 10^{18}$ electrons).

If the drift of electrons in a conductor takes place at the rate of one coulomb per second the resulting current is said to be a current of one ampere.

Thus, 1 ampere = 1 coulomb per second or $1\,A = 1\,C/s$. Hence, 1 coulomb = 1 ampere second or $1\,C = 1\,As$. Generally, if $I$ is the current in amperes and $t$ the time in seconds during which the current flows, then $I \times t$ represents the quantity of electrical charge in coulombs, i.e., quantity of electrical charge transferred,

$$Q = I \times t \text{ coulombs}$$

### Example 1.9
What current must flow if 0.24 coulombs is to be transferred in 15 ms?

**Solution**

Since the quantity of electricity, $Q = It$, then

$$I = \frac{Q}{t} = \frac{0.24}{15 \times 10^{-3}} = \frac{0.24 \times 10^3}{15} = \frac{240}{15} = \mathbf{16\,A}$$

**Example 1.10**

If a current of 10 A flows for 4 minutes, find the quantity of electricity transferred.

**Solution**

Quantity of electricity, $Q = It$ coulombs
$I = 10\,A; t = 4 \times 60 = 240\,s$
Hence, $Q = 10 \times 240 = \mathbf{2400\,C}$

## 1.12  Potential Difference and Resistance

For a continuous current to flow between two points in a circuit a *potential difference* or *voltage*, V, is required between them; a complete conducting path is necessary to and from the source of electrical energy. The unit of voltage is the volt, V.

Figure 1.2 shows a cell connected across a filament lamp. Current flow, by convention, is considered as flowing from the positive terminal of the cell, around the circuit to the negative terminal.

The flow of electric current is subject to friction. This friction, or opposition, is called *resistance R* and is the property of a conductor that limits current. The unit of resistance

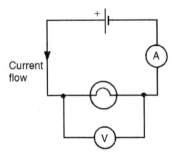

**Figure 1.2: Current flow**

is the *ohm*; 1 ohm is defined as the resistance which will have a current of 1 ampere flowing through it when 1 volt is connected across it, i.e.,

$$\text{resistance } R = \frac{\text{potential difference}}{\text{current}}$$

## 1.13  Basic Electrical Measuring Instruments

An *ammeter* is an instrument used to measure current and must be connected in series with the circuit. Figure 1.2 shows an ammeter connected in series with the lamp to measure the current flowing through it. Since all the current in the circuit passes through the ammeter it must have a very low resistance.

A *voltmeter* is an instrument used to measure voltage and must be connected in parallel with the part of the circuit whose voltage is required. In Figure 1.2, a voltmeter is connected in parallel with the lamp to measure the voltage across it. To avoid a significant current flowing through it, a voltmeter must have a very high resistance.

An *ohmmeter* is an instrument for measuring resistance.

A *multimeter*, or universal instrument, may be used to measure voltage, current and resistance. The *oscilloscope* may be used to observe waveforms and to measure voltages and currents. The display of an oscilloscope involves a spot of light moving across a screen. The amount by which the spot is deflected from its initial position depends on the voltage applied to the terminals of the oscilloscope and the range selected. The displacement is calibrated in volts per cm. For example, if the spot is deflected 3 cm and the volts/cm switch is on 10 V/cm, then the magnitude of the voltage is 3 cm × 10 V/cm, i.e., 30 V.

## 1.14  Linear and Nonlinear Devices

Figure 1.3 shows a circuit in which current $I$ can be varied by the variable resistor $R_2$. For various settings of $R_2$, the current flowing in resistor $R_1$, displayed on the ammeter, and the p.d. across $R_1$, displayed on the voltmeter, are noted and a graph is plotted of p.d. against current. The result is shown in Figure 1.4(a) where the straight line graph passing through the origin indicates that current is directly proportional to the voltage. Since the

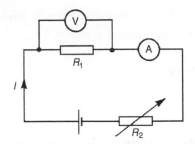

**Figure 1.3: Circuit in which current can be varied**

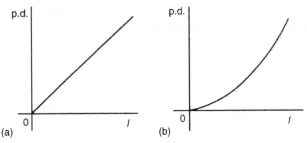

**Figure 1.4: Graphs of voltage vs. current: (a) linear device (b) nonlinear device**

gradient, i.e., (voltage/current), is constant, resistance $R_1$ is constant. A resistor is thus an example of a linear device.

If the resistor $R_1$ in Figure 1.3 is replaced by a component such as a lamp, then the graph shown in Figure 1.4(b) results when values of voltage are noted for various current readings. Since the gradient is changing, the lamp is an example of a *nonlinear device*.

## 1.15 Ohm's Law

*Ohm's law* states that the current $I$ flowing in a circuit is directly proportional to the applied voltage $V$ and inversely proportional to the resistance $R$, provided the temperature remains constant. Thus,

$$I = \frac{V}{R} \quad \text{or} \quad V = IR \quad \text{or} \quad R = \frac{V}{I}$$

*Example 1.11*

The current flowing through a resistor is 0.8 A when a voltage of 20 V is applied. Determine the value of the resistance.

*Solution*

From Ohm's law,

$$\text{resistance } R = \frac{V}{I} = \frac{20}{0.8} = \frac{200}{8} = 25\,\Omega$$

## 1.16 Multiples and Submultiples

Currents, voltages and resistances can often be very large or very small. Thus multiples and submultiples of units are often used. The most common ones, with an example of each, are listed in Table 1.4.

*Example 1.12*

Determine the voltage which must be applied to a 2 kΩ resistor in order that a current of 10 mA may flow.

*Solution*

Resistance $R = 2\,k\Omega = 2 \times 10^3 = 2000\,\Omega$

**Table 1.4: Common multiples and submultiples of units**

| Prefix | Name | Meaning | Example |
|---|---|---|---|
| M | mega | multiply by 1,000,000 (i.e., $\times 10^6$) | 2 MΩ = 2,000,000 ohms |
| k | kilo | multiply by 1000 (i.e., $\times 10^3$) | 10 kV = 10,000 volts |
| m | milli | divide by 1000 (i.e., $\times 10^{-3}$) | $25\text{ mA} = \dfrac{25}{1000}\text{ A}$ $= 0.025\text{ amperes}$ |
| μ | micro | divide by 1,000,000 (i.e., $\times 10^{-6}$) | $50\,\mu\text{V} = \dfrac{50}{1000\,000}\text{ V}$ $= 0.00005\text{ volts}$ |

Current $I = 10\,mA$

$$= 10 \times 10^{-3}\,A \quad \text{or} \quad \frac{10}{10^3} \quad \text{or} \quad \frac{10}{1000}\,A$$

$$= 0.01\,A$$

From Ohm's law, potential difference,

$$V = IR = (0.01)\,(2000) = \textbf{20 V}$$

### Example 1.13

A coil has a current of 50 mA flowing through it when the applied voltage is 12 V. What is the resistance of the coil?

### Solution

$$\text{Resistance } R = \frac{V}{I} = \frac{12}{50 \times 10^{-3}} = \frac{12 \times 10^3}{50}$$

$$= \frac{12\,000}{50} = \textbf{240 } \boldsymbol{\Omega}$$

### Example 1.14

A 100 V battery is connected across a resistor and causes a current of 5 mA to flow. Determine the resistance of the resistor. If the voltage is now reduced to 25 V, what will be the new value of the current flowing?

### Solution

$$\text{Resistance } R = \frac{V}{I} = \frac{100}{5 \times 10^{-3}} = \frac{100 \times 10^3}{5}$$

$$= 20 \times 10^3 = \textbf{20 k}\boldsymbol{\Omega}$$

Current when voltage is reduced to 25 V,

$$I = \frac{V}{R} = \frac{25}{20 \times 10^3} = \frac{25}{20} \times 10^{-3} = \textbf{1.25 mA}$$

### Example 1.15

What is the resistance of a coil that draws a current of (a) 50 mA and (b) 200 μA from a 120 V supply?

### Solution

(a) Resistance $R = \dfrac{V}{I} = \dfrac{120}{50 \times 10^{-3}}$

$$= \dfrac{120}{0.05} = \dfrac{12\,000}{5} = \mathbf{2400\,\Omega} \text{ or } \mathbf{2.4\,k\Omega}$$

(b) Resistance $R = \dfrac{120}{200 \times 10^{-6}} = \dfrac{120}{0.0002}$

$$= \dfrac{1\,200\,000}{2} = \mathbf{600\,000\,\Omega} \text{ or } \mathbf{600\,k\Omega}$$

$$\text{or } \mathbf{0.6\,M\Omega}$$

### Example 1.16

The current/voltage relationship for two resistors A and B is as shown in Figure 1.5. Determine the value of the resistance of each resistor.

### Solution

For resistor A,

$$R = \dfrac{V}{I} = \dfrac{20\,\text{A}}{20\,\text{mA}} = \dfrac{20}{0.02} = \dfrac{2000}{2} = \mathbf{1000\,\Omega} \text{ or } \mathbf{1\,k\Omega}$$

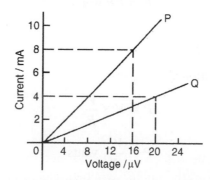

**Figure 1.5: Current/voltage for two resistors A and B**

For resistor B,

$$R = \frac{V}{I} = \frac{16\,\text{V}}{5\,\text{mA}} = \frac{16}{0.005} = \frac{16\,000}{5} = 3200\,\Omega \text{ or } 3.2\,\text{k}\Omega$$

## 1.17 Conductors and Insulators

A *conductor* is a material having a low resistance which allows electric current to flow in it. All metals are conductors and some examples include copper, aluminium, brass, platinum, silver, gold and carbon.

An *insulator* is a material having a high resistance which does not allow electric current to flow in it. Some examples of insulators include plastic, rubber, glass, porcelain, air, paper, cork, mica, ceramics and certain oils.

## 1.18 Electrical Power and Energy

### 1.18.1 Electrical Power

*Power P* in an electrical circuit is given by the product of potential difference *V* and current *I*. The unit of power is the *watt*, *W*. Hence,

$$P = V \times I \text{ watts}$$

From Ohm's law, $V = IR$.

Substituting for *V* in equation (1.1) gives:

$$P = (IR) \times I$$

i.e.,

$$P = I^2R \text{ watts}$$

Also, from Ohm's law, $I = \dfrac{V}{R}$

Substituting for *I* in the equation above gives:

$$P = V \times \frac{V}{R}$$

i.e.,

$$P = \frac{V^2}{R} \text{ watts}$$

There are three possible formulas that may be used for calculating power.

*Example 1.17*

A 100 W electric light bulb is connected to a 250 V supply. Determine (a) the current flowing in the bulb, and (b) the resistance of the bulb.

*Solution*

Power $P = V \times I$, from which, current $I = \dfrac{P}{V}$

(a) Current $I = \dfrac{100}{250} = \dfrac{10}{25} = \dfrac{2}{5} = \mathbf{0.4\,A}$

(b) Resistance $R = \dfrac{V}{I} = \dfrac{250}{0.4} = \dfrac{2500}{4} = \mathbf{625\,\Omega}$

*Example 1.18*

Calculate the power dissipated when a current of 4 mA flows through a resistance of 5 kΩ.

*Solution*

Power $P = I^2R = (4 \times 10^{-3})^{-2}(5 \times 10^3)$

$\qquad\qquad = 16 \times 10^{-6} \times 5 \times 10^3 = 80 \times 10^{-3}$

$\qquad\qquad = \mathbf{0.08\,W\ or\ 80\,mW}$

Alternatively, since $I = 4 \times 10^{-3}$ and $R = 5 \times 10^3$ then from Ohm's law,

voltage $V = IR = 4 \times 10^{-3} \times 5 \times 10^{-3} = 20\,V$

Hence, power $P = V \times I = 20 \times 4 \times 10^{-3} = \mathbf{80\,mW}$

*Example 1.19*

An electric kettle has a resistance of 30 Ω. What current will flow when it is connected to a 240 V supply? Find also the power rating of the kettle.

*Solution*

Current, $I = \dfrac{V}{R} = \dfrac{240}{30} = \mathbf{8\,A}$

Power, $P = VI = 240 \times 8 = 1920\,W$

$\qquad\qquad\qquad = \mathbf{1.95\,kW}$

$\qquad\qquad\qquad = $ power rating of kettle

## Example 1.20

A current of 5 A flows in the winding of an electric motor, the resistance of the winding being 100 Ω. Determine (a) the voltage across the winding, and (b) the power dissipated by the coil.

### Solution

Potential difference across winding, $V = IR = 5 \times 100 = \textbf{500 V}$

Power dissipated by coil, $P = I^2R = 5^2 \times 100$

$$= \textbf{2500 W or 2.5 kW}$$

(Alternatively, $P = V \times I = 500 \times 5 = \textbf{2500 W or 2.5 kW}$)

## Example 1.21

The hot resistance of a 240 V filament lamp is 960 Ω. Find the current taken by the lamp and its power rating.

### Solution

From Ohm's law,

current $I = \dfrac{V}{R} = \dfrac{240}{960} = \dfrac{24}{96} = \dfrac{1}{4}$ **A or 0.25 A**

Power rating $P = VI = (240)\left(\dfrac{1}{4}\right) = \textbf{60 W}$

### 1.18.2  Electrical Energy

$$\text{Electrical energy} = \text{power} \times \text{time}$$

If the power is measured in watts and the time in seconds then the unit of energy is watt-seconds or *joules*. If the power is measured in kilowatts and the time in hours then the unit of energy is *kilowatt-hours*, often called the *unit of electricity*. The electricity meter in the home records the number of kilowatt-hours used and is thus an energy meter.

## Example 1.22

A 12 V battery is connected across a load having a resistance of 40 Ω. Determine the current flowing in the load, the power consumed and the energy dissipated in 2 minutes.

*Solution*

Current $I = \dfrac{V}{R} = \dfrac{12}{40} = \mathbf{0.3\,A}$

Power consumed, $P = VI = (12)(0.3) = \mathbf{3.6\,W}$

Energy dissipated $=$ power $\times$ time $= (3.6\,\text{W})(2 \times 60\,\text{s}) = \mathbf{432\,J}$ (since $1\,\text{J} = 1\,\text{Ws}$)

### Example 1.23

A source of e.m.f. of 15 V supplies a current of 2 A for 6 minutes. How much energy is provided in this time?

*Solution*

Energy $=$ power $\times$ time, and power $=$ voltage $\times$ current

Hence, energy $= Vt = 15 \times 2 \times (6 \times 60)$

$\qquad\qquad\quad = 10\,800\,\text{Ws or J} = \mathbf{10.8\,kJ}$

### Example 1.24

An electric heater consumes 3.6 MJ when connected to a 250 V supply for 40 minutes. Find the power rating of the heater and the current taken from the supply.

*Solution*

Power $= \dfrac{\text{energy}}{\text{time}} = \dfrac{3.6 \times 10^6}{40 \times 60}\,\dfrac{\text{J}}{\text{s}}\,(\text{or W}) = 1500\,\text{W}$

i.e., power rating of heater $= \mathbf{1.5\,kW}$

Power $P = VI$, thus $I = \dfrac{P}{V} = \dfrac{1500}{250} = 6\,\text{A}$

Hence, the current taken from the supply $= \mathbf{6\,A}$

## 1.19 Main effects of electric current

The three main effects of an electric current are:

(a)  magnetic effect

(b)  chemical effect

(c)  heating effect

Some practical applications of the effects of an electric current include:

*Magnetic effect:*    bells, relays, motors, generators, transformers, telephones, car ignition, and lifting magnets

*Chemical effect:*    primary and secondary cells, and electroplating

*Heating effect:*    cookers, water heaters, electric fires, irons, furnaces, kettles, and soldering irons

# CHAPTER 2

# Resistance and Resistivity

John Boyd

## 2.1 Resistance and Resistivity

The resistance of an electrical conductor depends on four factors, these being: (a) the length of the conductor, (b) the cross-sectional area of the conductor, (c) the type of material and (d) the temperature of the material.

Resistance, $R$, is directly proportional to length, $l$, of a conductor. For example, if the length of a piece of wire is doubled, then the resistance is doubled.

Resistance, $R$, is inversely proportional to cross-sectional area, $a$, of a conductor, i.e., $R$ is proportional to $1/a$. Thus, for example, if the cross-sectional area of a piece of wire is doubled, then the resistance is halved.

Since $R$ is proportional to $l$ and $R$ is proportional to $1/a$, then $R$ is proportional to $l/a$. By inserting a constant of proportionality into this relationship, the type of material used may be taken into account. The constant of proportionality is known as the *resistivity* of the material and is given the symbol $\rho$ (Greek rho). Thus,

$$\text{resistance } R = \frac{\rho l}{a} \text{ ohms}$$

$\rho$ is measured in ohm meters ($\Omega$m).

The value of the resistivity is the resistance of a unit cube of the material measured between opposite faces of the cube.

Resistivity varies with temperature and some typical values of resistivities measured at about room temperature are given in Table 2.1.

Note that good conductors of electricity have a low value of resistivity and good insulators have a high value of resistivity.

### Example 2.1

The resistance of a 5 m length of wire is 600 Ω. Determine (a) the resistance of an 8 m length of the same wire, and (b) the length of the same wire when the resistance is 420 Ω.

### Solution

Resistance, $R$, is directly proportional to length, $l$, i.e., $R \propto l$. Hence, 600 Ω $\propto$ 5 m or $600 = (k)(5)$, where $k$ is the coefficient of proportionality. Hence,

$$k = \frac{600}{5} = 120$$

When the length $l$ is 8 m, then resistance

$$R = kl = (120)(8) = 960 \ \Omega$$

When the resistance is 420 Ω, $420 = kl$, from which

$$\text{length } l = \frac{420}{k} = \frac{420}{120} = \textbf{3.5 m}$$

### Example 2.2

A piece of wire of cross-sectional area $2 \, mm^2$ has a resistance of 300 Ω. Find (a) the resistance of a wire of the same length and material if the cross-sectional area is $5 \, mm^2$, and (b) the cross-sectional area of a wire of the same length and material of resistance 750 Ω.

**Table 2.1: Typical resistivity values**

| Copper | $1.7 \times 10^{-8}$ Ωm | (or 0.017 μΩm) |
|---|---|---|
| Aluminum | $2.6 \times 10^{-8}$ Ωm | (or 0.026 μΩm) |
| Carbon (graphite) | $10 \times 10^{-8}$ Ωm | (or 0.10 μΩm) |
| Glass | $1 \times 10^{8}$ Ωm | (or $10^{4}$ μΩm) |
| Mica | $1 \times 10^{-13}$ Ωm | (or $10^{7}$ μΩm) |

*Solution*

Resistance $R$ is inversely proportional to cross-sectional area, $a$, i.e., $R \propto (1/a)$

So $300\,\Omega \propto (1/2\,\text{mm}^2)$ or $300 = (k)(1/2)$ from which the coefficient of proportionality,

$k = 300 \times 2 = 600$

(a)  When the cross-sectional area $a = 5\,\text{mm}^2$

then $R = (k)(1/5) = (600)(1/5) - \textbf{120}\,\boldsymbol{\Omega}$

(Note that resistance has decreased as the cross-sectional area is increased.)

(b)  When the resistance is $750\,\Omega$ then $750 = (k)(1/a)$, from which cross-sectional area,

$$a = \frac{k}{750} = \frac{600}{750} = \textbf{0.8 mm}^2$$

### Example 2.3

A wire of length 8 m and cross-sectional area $3\,\text{mm}^2$ has a resistance of $0.16\,\Omega$. If the wire is drawn out until its cross-sectional area is $1\,\text{mm}^2$, determine the resistance of the wire.

*Solution*

Resistance $R$ is directly proportional to length $l$, and inversely proportional to the cross-sectional area, $a$, i.e., $R \propto (l/a)$ or $R = k(l/a)$, where $k$ is the coefficient of proportionality.

Since $R = 0.16$, $l = 8$ and $a = 3$, then $0.16 = (k)(8/3)$ from which

$k = 0.16 \times (3/8) = 0.06$

If the cross-sectional area is reduced to $\{1/3\}$ of its original area, then the length must be tripled to $3 \times 8$, i.e., 24 m.

New resistance $R = k\,(l/a) = 0.06\,(24/1) = \textbf{1.44}\,\boldsymbol{\Omega}$

### Example 2.4

Calculate the resistance of a 2 km length of aluminum overhead power cable if the cross-sectional area of the cable is $100\,\text{mm}^2$. Take the resistivity of aluminum to be $0.03 \times 10^{-6}\,\Omega\text{m}$.

*Solution*

Length $l = 2\,\text{km} = 2000\,\text{m}$; area, $a = 100\,\text{mm}^2 = 100 \times 10^{-6}\,\text{m}^2$; resistivity $\rho = 0.03 \times 10^{-6}\,\Omega\text{m}$

$$\text{Resistance } R = \frac{\rho l}{a} = \frac{(0.03 \times 10^{-6}\,\Omega\text{m})(2000\,\text{m})}{(100 \times 10^{-6}\,\text{m}^2)}$$

$$= \frac{0.03 \times 2000}{100}\,\Omega$$

$$= \mathbf{0.6\,\Omega}$$

*Example 2.5*

Calculate the cross-sectional area, in $\text{mm}^2$, of a piece of copper wire, 40 m in length and having a resistance of $0.25\,\Omega$. Take the resistivity of copper as $0.02 \times 10^{-6}\,\Omega\text{m}$.

*Solution*

$$\text{Resistance } R = \frac{\rho l}{a} \text{ so cross-sectional area } a = \frac{\rho l}{R}$$

$$= \frac{(0.02 \times 10^{-6}\,\Omega\text{m})(40\,\text{m})}{0.25\,\Omega}$$

$$= 3.2 \times 10^{-6}\,\text{m}^2$$

$$= (3.2 \times 10^{-6}) \times 10^{-6}\,\text{mm}^2 = \mathbf{3.2\,mm^2}$$

*Example 2.6*

The resistance of 1.5 km of wire of cross-sectional area $0.17\,\text{mm}^2$ is $150\,\Omega$. Determine the resistivity of the wire.

*Solution*

$$\text{Resistance } R = \frac{\rho l}{a}$$

$$\text{so resistivity } \rho = \frac{Ra}{l} = \frac{(150\,\Omega)(0.17 \times 10^{-6}\,\text{m}^2)}{(1500\,\text{m})}$$

$$= \mathbf{0.017 \times 10^{-6}\,\Omega m \text{ or } 0.017\,\mu\Omega m}$$

*Example 2.7*

Determine the resistance of 1200 m of copper cable having a diameter of 12 mm if the resistivity of copper is $1.7 \times 10^{-8}\,\Omega\text{m}$.

*Solution*

Cross-sectional area of cable, $a = \pi r^2 = \pi \left(\frac{12}{2}\right)^2 = 36\pi \text{ mm}^2 = 36\pi \times 10^{-6} \text{ m}^2$

$$\text{Resistance } R = \frac{\rho l}{a} = \frac{(1.7 \times 10^{-8} \text{ }\Omega\text{m})(1200 \text{ m})}{(36\pi \times 10^{-6} \text{ m}^2)}$$

$$= \frac{1.7 \times 1200 \times 10^6}{10^8 \times 36\pi} \text{ }\Omega = \frac{1.7 \times 12}{36\pi} \text{ }\Omega$$

$$= \textbf{0.180 }\boldsymbol{\Omega}$$

## 2.2 Temperature Coefficient of Resistance

In general, as the temperature of a material increases, most conductors increase in resistance, insulators decrease in resistance, while the resistance of some special alloys remains almost constant.

The *temperature coefficient of resistance* of a material is the increase in the resistance of a 1Ω resistor of that material when it is subjected to a rise of temperature of 1°C. The symbol used for the temperature coefficient of resistance is $\alpha$ (Greek alpha). Thus, if some copper wire of resistance 1Ω is heated through 1°C and its resistance is then measured as 1.0043 12 then $\alpha = 0.0043$ Ω/Ω°C for copper. The units are usually expressed only as "per °C." So, $\alpha = 0.0043$/°C for copper. If the 1Ω resistor of copper is heated through 100°C then the resistance at 100°C would be $1 + 100 \times 0.0043 = 1.43$ Ω.

Some typical values of temperature coefficient of resistance measured at 0°C are given in Table 2.2.

(Note that the negative sign for carbon indicates that its resistance falls with increase of temperature.)

**Table 2.2: Typical values of temperature coefficient of resistance**

| Copper | 0.0043/°C | Aluminum | 0.0038/°C |
|--------|-----------|----------|-----------|
| Nickel | 0.0062/°C | Carbon | −0.00048/°C |
| Constantan | 0 | Eureka | 0.00001/°C |

If the resistance of a material at 0°C is known, the resistance at any other temperature can be determined from:

$$R_\theta = R_0(1 + \alpha_0\theta)$$

where $R_\theta$ = resistance at 0°C

$R_\theta$ = resistance at temperature $\theta$°C

$\alpha_0$ = temperature coefficient of resistance at 0°C

### Example 2.8

A coil of copper wire has a resistance of 100 $\Omega$ when its temperature is 0°C. Determine its resistance at 70°C if the temperature coefficient of resistance of copper at 0°C is 0.0043/°C.

### Solution

Resistance $R_\theta = R_0 (1 + \alpha_0\theta)$

So resistance at 70°C, $R_{70} = 100[1 + (0.0043)(70)]$

$$= 100[1 + 0.301]$$
$$= 100(1.301)$$
$$= \mathbf{130.1\,\Omega}$$

### Example 2.9

An aluminum cable has a resistance of 27 $\Omega$ at a temperature of 35°C. Determine its resistance at 0°C. Take the temperature coefficient of resistance at 0°C to be 0.0038/°C.

### Solution

Resistance at $\theta$°C, $R_\theta = R_0(1 + \alpha_0\theta)$

Hence resistance at 0°C, $R_0 = \dfrac{R_\theta}{(1 + \alpha_0\theta)}$

$$= \dfrac{27}{[1 + (0.0038)(35)]}$$

$$= \dfrac{27}{1 + 0.133} = \dfrac{27}{1.133}$$

$$= \mathbf{23.83\,\Omega}$$

*Example 2.10*

A carbon resistor has a resistance of 1 k$\Omega$ at 0°C. Determine its resistance at 80°C. Assume that the temperature coefficient of resistance for carbon at 0°C is $-0.0005$/°C.

*Solution*

Resistance at temperature $\theta$°C, $R_\theta = R_0(1 + \alpha_0\theta)$

i.e., $R_\theta = 1000[1 + (-0.0005)(80)]$

$\qquad = 1000[1 - 0.040] = 1000(0.96)$

$\qquad = \mathbf{960\ \Omega}$

If the resistance of a material at room temperature (approximately 20°C), $R_{20}$, and the temperature coefficient of resistance at 20°C, $\alpha_{20}$, are known then the resistance $R_\theta$ at temperature $\theta$°C is given by:

$$R_\theta = R_{20}[1 + \alpha_{20}(\theta - 20)]$$

*Example 2.11*

A coil of copper wire has a resistance of 10 $\Omega$ at 20°C. If the temperature coefficient of resistance of copper at 20°C is 0.004/°C, determine the resistance of the coil when the temperature rises to 100°C.

*Solution*

Resistance at temperature $\theta$°C, $R = R_{20}[1 + \alpha_{20}(\theta - 20)]$

Hence resistance at 100°C,

$R_{100} = 10[1 + (0.004)(100 - 20)]$

$\qquad = 10[1 + (0.004)(80)]$

$\qquad = 10[1 + 0.32]$

$\qquad = 10(1.32)$

$\qquad = \mathbf{13.2\ \Omega}$

*Example 2.12*

The resistance of a coil of aluminum wire at 18°C is 200 Ω. The temperature of the wire is increased and the resistance rises to 240 Ω. If the temperature coefficient of resistance of aluminum is 0.0039/°C at 18°C determine the temperature to which the coil has risen.

*Solution*

Let the temperature rise to θ°

Resistance at θ°C, $R_\theta = R_{18}[1 + \alpha_{18}(\theta - 18)]$

i.e.,         $240 = 200[1 + (0.0039)(\theta - 18)]$

$240 = 200 + (200)(0.0039)(\theta - 18)$

$240 - 200 = 0.78(\theta - 18)$

$40 = 0.78(\theta - 18)$

$\dfrac{40}{0.78} = \theta - 18$

$51.28 = \theta - 18$, from which,

$\theta = 51.28 + 18 = 69.28°C$

Hence the temperature of the coil increases to 69.28°C.

If the resistance at 0°C is not known, but is known at some other temperature $\theta_1$, then the resistance at any temperature can be found as follows:

$R_1 = R_0(1 + \alpha_0\theta_1)$   and   $R_2 = R_0(1 + \alpha_0\theta_2)$

Dividing one equation by the other gives:

$$\frac{R_1}{R_2} = \frac{1 + \alpha_0\theta_1}{1 + \alpha_0\theta_2}$$

where $R_2$ = resistance at temperature $\theta_2$.

*Example 2.13*

Some copper wire has a resistance of 200 Ω at 20°C. A current is passed through the wire and the temperature rises to 90°C. Determine the resistance of the wire at 90°C, correct to the nearest ohm, assuming that the temperature coefficient of resistance is 0.004/°C at 0°C.

### Solution

$R_{20} = 200\,\Omega$, $\alpha_0 = 0.004/°C$

$$\frac{R_{20}}{R_{90}} = \frac{[1 + \alpha_0(20)]}{[1 + \alpha_0(90)]}$$

$$\text{Hence, } R_{90} = \frac{R_{20}[1 + 90\alpha_0]}{[1 + 20\alpha_0]}$$

$$= \frac{200[1 + 90(0.004)]}{[1 + 20(0.004)]}$$

$$= \frac{200[1 + 0.36]}{[1 + 0.08]}$$

$$= \frac{200(1.36)}{(1.08)} = \mathbf{251.85\ \Omega}$$

So, the resistance of the wire at 90°C is 252 $\Omega$.

# CHAPTER 3

# Series and Parallel Networks

John Bird

## 3.1 Series Circuits

Figure 3.1 shows three resistors $R_1$, $R_2$ and $R_3$ connected end to end, i.e., in series, with a battery source of $V$ volts. Since the circuit is closed, a current $I$ will flow and the voltage across each resistor may be determined from the voltmeter readings $V_1$, $V_2$ and $V_3$.

In a series circuit:

(a) the current $I$ is the same in all parts of the circuit; therefore, the same reading is found on each of the two ammeters shown, and,

(b) the sum of the voltages $V_1$, $V_2$ and $V_3$ is equal to the total applied voltage, $V$, i.e.,

$$V = V_1 + V_2 + V_3$$

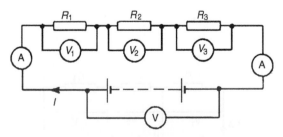

Figure 3.1: Series circuit

From Ohm's law:

$V_1 = IR_1$, $V_2 = IR_2$, $V_3 = IR_3$ and $V = IR$

where $R$ is the total circuit resistance.

Since $V = V_1 + V_2 + V_3$

then $IR = IR_1 + IR_2 + IR_3$

Dividing throughout by $I$ gives:

$$R = R_1 + R_2 + R_3$$

So, for a series circuit, the total resistance is obtained by adding together the values of the separate resistances.

### Example 3.1

For the circuit shown in Figure 3.2, determine (a) the battery voltage $V$, (b) the total resistance of the circuit, and (c) the values of resistance of resistors $R_1$, $R_2$ and $R_3$, given that the voltages across $R_1$, $R_2$ and $R_3$ are 5 V, 2 V and 6 V, respectively.

### Solution

(a)  Battery voltage $V = V_1 + V_2 + V_3$

$$= 5 + 2 + 6 = \mathbf{13\,V}$$

(b)  Total circuit resistance $R = \dfrac{V}{I} = \dfrac{13}{4} = \mathbf{3.25\,\Omega}$

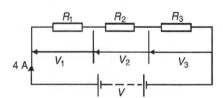

**Figure 3.2: Circuit for Example 3.1**

(c)  Resistance $R_1 = \dfrac{V_1}{I} = \dfrac{5}{4} = \mathbf{1.25\,\Omega}$

Resistance $R_2 = \dfrac{V_2}{I} = \dfrac{2}{4} = \mathbf{0.5\,\Omega}$

Resistance $R_3 = \dfrac{V_3}{I} = \dfrac{6}{4} = \mathbf{1.5\,\Omega}$

(Check: $R_1 + R_2 + R_3 = 1.25 + 0.5 + 1.5 = 3.25\,\Omega = R$)

### Example 3.2

For the circuit shown in Figure 3.3, determine the voltage across resistor $R_3$. If the total resistance of the circuit is $100\,\Omega$, determine the current flowing through resistor $R_1$. Find also the value of resistor $R_2$.

### Solution

Voltage across $R_3$, $V_3 = 25 - 10 - 4 = \mathbf{11\,V}$

Current $I = \dfrac{V}{R} = \dfrac{25}{100} = \mathbf{0.25\,A}$, which is the current flowing in each resistor

Resistance $R_2 = \dfrac{V_2}{I} = \dfrac{4}{0.25} = \mathbf{16\,\Omega}$

### Example 3.3

A 12 V battery is connected in a circuit having three series-connected resistors having resistances of $4\,\Omega$, $9\,\Omega$ and $11\,\Omega$. Determine the current flowing through, and the voltage across the $9\,\Omega$ resistor. Find also the power dissipated in the $11\,\Omega$ resistor.

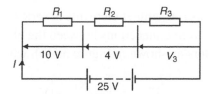

**Figure 3.3: Circuit for Example 3.2**

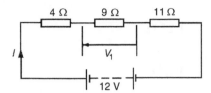

**Figure 3.4: Circuit for Example 3.3**

*Solution*

The circuit diagram is shown in Figure 3.4.

Total resistance $R = 4 + 9 + 11 = 24\,\Omega$

Current $I = \dfrac{V}{R} = \dfrac{12}{24} = \textbf{0.5\,A}$, which is the current in the $9\,\Omega$ resistor.

Voltage across the $9\,\Omega$ resistor, $V_1 = I \times 9 = 0.5 \times 9$

$$= \textbf{4.5\,V}$$

Power dissipated in the $11\,\Omega$ resistor, $P = I^2R = 0.5^2(11)$

$$= 0.25(11)$$
$$= \textbf{2.75\,W}$$

## 3.2 Potential Divider

The voltage distribution for the circuit shown in Figure 3.5(a) is given by:

$$V_1 = \left(\frac{R_1}{R_1 + R_2}\right) V$$

$$V_2 = \left(\frac{R_2}{R_1 + R_1}\right) V$$

The circuit shown in Figure 3.5(b) is often referred to as a *potential divider* circuit. Such a circuit can consist of a number of similar elements in series connected across a voltage source, voltages being taken from connections between the elements. Frequently the divider consists of two resistors as shown in Figure 3.5(b), where:

$$V_{\text{OUT}} = \left(\frac{R_2}{R_1 + R_2}\right) V_{\text{IN}}$$

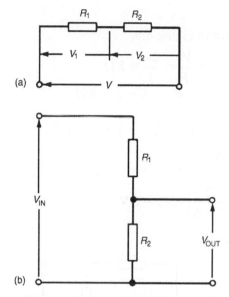

**Figure 3.5: Potential divider circuit**

A potential divider is the simplest way of producing a source of lower e.m.f. from a source of higher e.m.f., and is the basic operating mechanism of the *potentiometer*, a measuring device for accurately measuring potential differences.

### Example 3.4
Determine the value of voltage $V$ shown in Figure 3.6.

### Solution
Figure 3.6 may be redrawn as shown in Figure 3.7, and voltage

$$V = \left(\frac{6}{6+4}\right)(50) = \mathbf{30\ V}$$

### Example 3.5
Two resistors are connected in series across a 24 V supply and a current of 3 A flows in the circuit. If one of the resistors has a resistance of 2 Ω determine (a) the value of the other resistor, and (b) the voltage across the 2 Ω resistor. If the circuit is connected for 50 hours, how much energy is used?

*Solution*

The circuit diagram is shown in Figure 3.8.

(a) Total circuit resistance $R = \dfrac{V}{I} = \dfrac{24}{3} = 8\,\Omega$

Value of unknown resistance, $R_x = 8 - 2 = \mathbf{6\,\Omega}$

(b) Voltage across $2\,\Omega$ resistor, $V_1 = IR_1 = 3 \times 2 = \mathbf{6\,V}$

Alternatively, from above,

$$V_1 = \left(\frac{R_1}{R_1 + R_x}\right)V = \left(\frac{2}{2 + 6}\right)(24) = 6\,\text{V}$$

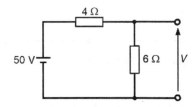

Figure 3.6: Circuit for Example 3.4

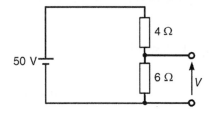

Figure 3.7: Redrawn version of Figure 3.6

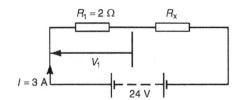

Figure 3.8: Circuit for Example 3.5

Energy used = power × time

$$= V \times I \times t$$

$$= (24 \times 3\,\text{W})\,(50\,\text{h})$$

$$= 3600\,\text{Wh} = \mathbf{3.6\,kWh}$$

## 3.3 Parallel Networks

Figure 3.9 shows three resistors, $R_1$, $R_2$ and $R_3$ connected across each other, i.e., in parallel, across a battery source of $V$ volts.

In a parallel circuit:

(a) the sum of the currents $I_1$, $I_2$ and $I_3$ is equal to the total circuit current, $I$, i.e., $I = I_1 + I_2 + I_3$, and

(b) the source voltage, $V$ volts, is the same across each of the resistors.

From Ohm's law:

$$I_1 = \frac{V}{R_1}, \quad I_2 = \frac{V}{R_2}, \quad I_3 = \frac{V}{R_3} \text{ and } I = \frac{V}{R}$$

where $R$ is the total circuit resistance.

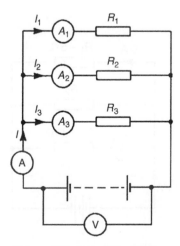

**Figure 3.9: Parallel resistors**

Since $I = I_1 + I_2 + I_3$

then $\dfrac{V}{R} = \dfrac{V}{R_1} + \dfrac{V}{R_2} + \dfrac{V}{R_3}$

Dividing throughout by $V$ gives:

$$\frac{1}{R} = \frac{1}{R_1} + \frac{1}{R_2} + \frac{1}{R_3}$$

This equation must be used when finding the total resistance $R$ of a parallel circuit. For the special case of *two resistors in parallel*:

$$\frac{1}{R} = \frac{1}{R_1} + \frac{1}{R_2} = \frac{R_2 + R_1}{R_1 R_2}$$

Hence, $\qquad\qquad R = \dfrac{R_1 R_2}{R_1 + R_2} \qquad \left( \text{i.e., } \dfrac{\text{product}}{\text{sum}} \right)$

### Example 3.6

For the circuit shown in Figure 3.10, determine (a) the reading on the ammeter, and (b) the value of resistor $R_2$.

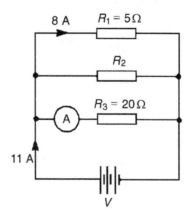

**Figure 3.10: Circuit for Example 3.6**

**Solution**

Voltage across $R_1$ is the same as the supply voltage $V$. Hence, supply voltage $V = 8 \times 5 = 40 \, \text{V}$.

(a) Reading on ammeter, $I = \dfrac{V}{R_3} = \dfrac{40}{20} = \mathbf{2\,A}$

(b) Current flowing through $R_2 = 11 - 8 - 2 = 1 \, \text{A}$

$\quad$ Hence, $R_2 = \dfrac{V}{I_2} = \dfrac{40}{1} = \mathbf{40 \, \Omega}$

**Example 3.7**

Two resistors, of resistance $3 \, \Omega$ and $6 \, \Omega$, are connected in parallel across a battery having a voltage of $12 \, \text{V}$. Determine (a) the total circuit resistance and (b) the current flowing in the $3 \, \Omega$ resistor.

**Solution**

The circuit diagram is shown in Figure 3.11.

(a) The total circuit resistance R is given by:

$$\frac{1}{R} = \frac{1}{R_1} + \frac{1}{R_2} = \frac{1}{3} + \frac{1}{6}$$

$$\frac{1}{R} = \frac{2+1}{6} = \frac{3}{6}$$

$\quad$ Hence, $R = \dfrac{6}{3} = \mathbf{2 \, \Omega}$

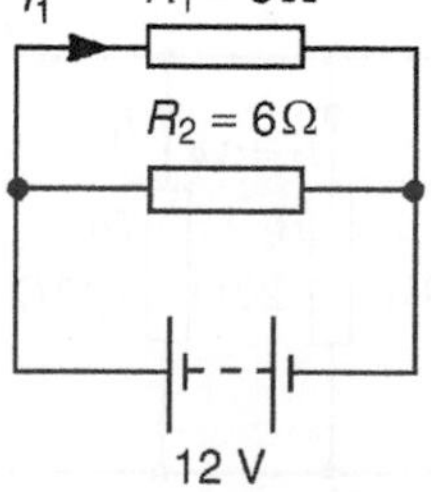

**Figure 3.11: Circuit for Example 3.7**

$$\left[ \text{Alternatively, } R = \frac{R_1 R_2}{R_1 + R_2} = \frac{3 \times 6}{3 + 6} = \frac{18}{9} = 2\,\Omega \right]$$

(b) Current in the $3\,\Omega$ resistance, $I_1 = \dfrac{V}{R_1} = \dfrac{12}{3} = \mathbf{4\,A}$

### Example 3.8

For the circuit shown in Figure 3.12, find (a) the value of the supply voltage $V$ and (b) the value of current $I$.

### Solution

(a) Voltage across $20\,\Omega$ resistor $= I_2 R_2 = 3 \times 20 = 60\,\text{V}$; hence, supply voltage $V = \mathbf{60\,V}$ since the circuit is connected in parallel.

(b) Current $I_1 = \dfrac{V}{R_1} = \dfrac{60}{10} = 6\,\text{A}; I_2 = 3\,\text{A}$

$$I_3 = \frac{V}{R_3} = \frac{60}{60} = 1\,\text{A}$$

Current $I = I_1 + I_2 + I_3$ and hence, $I = 6 + 3 + 1 = \mathbf{10\,A}$

Alternatively, $\dfrac{1}{R} = \dfrac{1}{60} + \dfrac{1}{20} + \dfrac{1}{10} = \dfrac{1 + 3 + 6}{60} = \dfrac{10}{60}$

Hence, total resistance $R = \dfrac{60}{10} = 6\,\Omega$

Current $I = \dfrac{V}{R} = \dfrac{60}{6} = \mathbf{10\,A}$

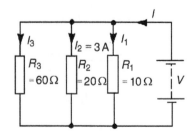

**Figure 3.12: Circuit for Example 3.8**

*Example 3.9*

Given four $1\,\Omega$ resistors, state how they must be connected to give an overall resistance of
(a) $1/4\,\Omega$ (b) $1\,\Omega$ (c) $1\tfrac{1}{3}\,\Omega$ (d) $2\tfrac{1}{2}\,\Omega$

*Solution*

(a) *All four in parallel* (see Figure 3.13),

$$\text{Since } \frac{1}{R} = \frac{1}{1} + \frac{1}{1} + \frac{1}{1} + \frac{1}{1} = \frac{4}{1}, \text{ i.e., } R = \frac{1}{4}\,\Omega$$

(b) *Two in series, in parallel with another two in series* (see Figure 3.14), since $1\,\Omega$ and
$1\,\Omega$ in series gives $2\,\Omega$, and $2\,\Omega$ in parallel with $2\,\Omega$ gives:

$$\frac{2 \times 2}{2 + 2} = \frac{4}{4} = 1\,\Omega$$

(c) *Three in parallel, in series with one* (see Figure 3.15), since for the three in parallel,

$$\frac{1}{R} = \frac{1}{1} + \frac{1}{1} + \frac{1}{1} = \frac{3}{1}, \text{ i.e., } R = \frac{1}{3}\,\Omega \text{ and } \frac{1}{3}\,\Omega \text{ in series with } 1\,\Omega \text{ gives } 1\frac{1}{3}\,\Omega$$

(d) *Two in parallel, in series with two in series* (see Figure 3.16), since for the two in parallel

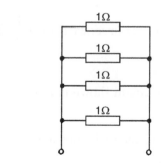

Figure 3.13: Circuit for Example 3.9(a)

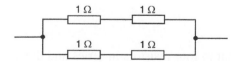

Figure 3.14: Circuit for Example 3.9(b)

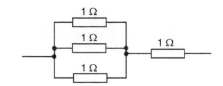

**Figure 3.15: Circuit for Example 3.9(c)**

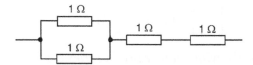

**Figure 3.16: Circuit for Example 3.9(d)**

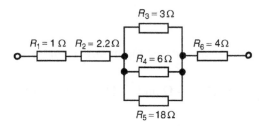

**Figure 3.17: Circuit for Example 3.10**

$$R = \frac{1 \times 1}{1 + 1} = \frac{1}{2}\ \Omega, \text{ and } \frac{1}{2}\ \Omega,\ 1\,\Omega \text{ and } 1\,\Omega \text{ in series gives } 2\frac{1}{2}\ \Omega$$

### Example 3.10
Find the equivalent resistance for the circuit shown in Figure 3.17.

### Solution
$R_3$, $R_4$ and $R_5$ are connected in parallel and their equivalent resistance $R$ is given by:

$$\frac{1}{R} = \frac{1}{3} + \frac{1}{6} + \frac{1}{18} = \frac{6 + 3 + 1}{18} = \frac{10}{18}$$

Hence, $R = \dfrac{18}{10} = 1.8\,\Omega$

The circuit is now equivalent to four resistors in series and the equivalent circuit resistance = $1 + 2.2 + 1.8 + 4 = \mathbf{9\,\Omega}$.

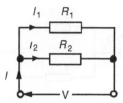

**Figure 3.18: Current division circuit**

## 3.4 Current Division

For the circuit shown in Figure 3.18, the total circuit resistance $R_T$ is given by:

$$R_T = \frac{R_1 R_2}{R_1 + R_2}$$

and $V = IR_T = I\left(\dfrac{R_1 R_2}{R_1 + R_2}\right)$

Current $I_1 = \dfrac{V}{R_1} = \dfrac{I}{R_1}\left(\dfrac{R_1 R_2}{R_1 + R_2}\right) = \left(\dfrac{R_2}{R_1 + R_2}\right)(I)$

Similarly,

current $I_2 = \dfrac{V}{R_2} = \dfrac{I}{R_2}\left(\dfrac{R_1 R_2}{R_1 + R_2}\right) = \left(\dfrac{R_1}{R_1 + R_2}\right)(I)$

Summarizing, with reference to Figure 3.18:

$$I_1 = \left(\frac{R_2}{R_1 + R_2}\right)(I) \quad \text{and} \quad I_2 = \left(\frac{R_1}{R_1 + R_2}\right)(I)$$

**Example 3.11**

For the series-parallel arrangement shown in Figure 3.19, find (a) the supply current, (b) the current flowing through each resistor and (c) the voltage across each resistor.

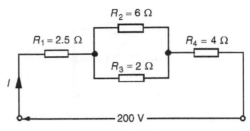

**Figure 3.19: Circuit for Example 3.11**

*Solution*

(a) The equivalent resistance $R_x$ of $R_2$ and $R_3$ in parallel is:

$$R_x = \frac{6 \times 2}{6 + 2} = \frac{12}{8} = 1.5\,\Omega$$

The equivalent resistance $R_T$ of $R_1$, $R_x$ and $R_4$ in series is:

$$R_T = 2.5 + 1.5 + 4 = 8\,\Omega$$

Supply current $I = \dfrac{V}{R_T} = \dfrac{200}{8} = \mathbf{25\,A}$

(b) The current flowing through $R_1$ and $R_4$ is 25 A. The current flowing through $R_2$

$$= \left(\frac{R_3}{R_2 + R_3}\right) I = \left(\frac{2}{6 + 2}\right) 25$$
$$= \mathbf{6.25\,A}$$

The current flowing through $R_3$

$$= \left(\frac{R_2}{R_2 + R_3}\right) I = \left(\frac{6}{6 + 2}\right) 25$$
$$= \mathbf{18.75\,A}$$

(Note that the currents flowing through $R_2$ and $R_3$ must add up to the total current flowing into the parallel arrangement, i.e., 25 A.)

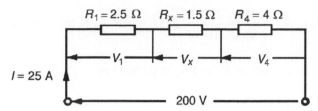

**Figure 3.20: Equivalent circuit of Figure 3.19**

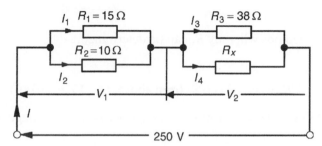

**Figure 3.21: Circuit for Example 3.12**

(c)  The equivalent circuit of Figure 3.19 is shown in Figure 3.20.

voltage across $R_1$, i.e., $V_1 = IR_1 = (25)(2.5) =$ **62.5 V**

voltage across $R_x$, i.e., $V_x = IR_x = (25)(1.5) =$ **37.5 V**

voltage across $R_4$, i.e., $V_4 = IR_4 = (25)(4) =$ **100 V**

Hence, the voltage across $R_2 =$ voltage across $R_3 =$ **37.5 V**

### Example 3.12

For the circuit shown in Figure 3.21 calculate (a) the value of resistor $R_x$ such that the total power dissipated in the circuit is 2.5 kW, and (b) the current flowing in each of the four resistors.

### Solution

(a)  Power dissipated $P = VI$ watts, hence, $2500 = (250)(I)$

i.e., $I = \dfrac{2500}{250} = 10\,\text{A}$

From Ohm's law, $R_T = \dfrac{V}{I} = \dfrac{250}{10} = 25\,\Omega$, where $R_T$ is the equivalent circuit resistance.

The equivalent resistance of $R_1$ and $R_2$ in parallel is:

$$\frac{15 \times 10}{15 + 10} = \frac{150}{25} = 6\,\Omega$$

The equivalent resistance of resistors $R_3$ and $R_x$ in parallel is equal to $25\,\Omega - 6\,\Omega$, i.e., $19\,\Omega$.

There are three methods whereby $R_x$ can be determined.

### Method 1
The voltage $V_1 = IR$, where $R$ is $6\,\Omega$, from above, i.e., $V_1 = (10)(6) = 60\,\text{V}$

Hence, $V_2 = 250\,\text{V} - 60\,\text{V} = 190\,\text{V} = $ voltage across $R_3 = $ voltage across $R_x$

$I_3 = \dfrac{V_2}{R_3} = \dfrac{190}{38} = 5\,\text{A}$. Thus, $I_4 = 5\,\text{A}$ also,

since $I = 10\,\text{A}$

Thus, $R_x = \dfrac{V_2}{I_4} = \dfrac{190}{5} = \mathbf{38\,\Omega}$

### Method 2
Since the equivalent resistance of $R_3$ and $R_x$ in parallel is $19\,\Omega$,

$$19 = \frac{38R_x}{38 + R_x} \quad \left(\text{i.e.,} \ \frac{\text{product}}{\text{sum}}\right)$$

Hence, $19(38 + R_x) = 38R_x$

$$722 + 19R_x = 38R_x$$

$$722 = 38R_x - 19R_x = 19R_x$$

Thus, $R_x = \dfrac{722}{19} = \mathbf{38\,\Omega}$

*Method 3*
When two resistors having the same value are connected in parallel, the equivalent resistance is always half the value of one of the resistors. In this case, since $R_T = 19\,\Omega$ and $R_3 = 38\,\Omega$, then $\boldsymbol{R_x = 38\,\Omega}$ could have been deduced on sight.

(b) Current $I_1 = \left(\dfrac{R_2}{R_1 + R_2}\right) I = \left(\dfrac{10}{15 + 10}\right)(10)$

$$= \left(\dfrac{2}{5}\right)(10) = \mathbf{4\,A}$$

Current $I_2 = \left(\dfrac{R_1}{R_1 + R_2}\right) I = \left(\dfrac{15}{15 + 10}\right)(10)$

$$= \left(\dfrac{3}{5}\right)(10) = \mathbf{6\,A}$$

From part (a), method 1, $\boldsymbol{I_3 = I_4 = 5\,A}$

### Example 3.13
For the arrangement shown in Figure 3.22, find the current $I_x$.

### Solution
Commencing at the right-hand side of the arrangement shown in Figure 3.24, the circuit is gradually reduced in stages as shown in Figures 3.23(a)–(d).

From Figure 3.23(d), $I = \dfrac{17}{4.25} = 4\,A$

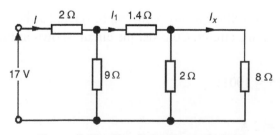

**Figure 3.22: Circuit for Example 3.13**

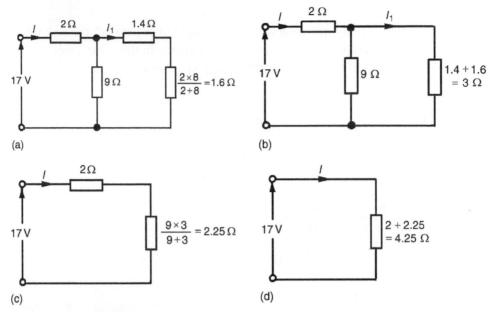

**Figure 3.23: Solution to Example 3.13, in four stages**

From Figure 3.23(b), $I_1 = \left(\dfrac{9}{9+3}\right)(I) = \left(\dfrac{9}{12}\right)(4) = 3\,\text{A}$

From Figure 3.22, $I_x = \left(\dfrac{2}{2+8}\right)(I_1) = \left(\dfrac{2}{10}\right)(3) = \mathbf{0.6\,A}$

## 3.5  Relative and Absolute Voltages

In an electrical circuit, the voltage at any point can be quoted as being "with reference to" (w.r.t.) any other point in the circuit. Consider the circuit shown in Figure 3.24. The total resistance,

$$R_T = 30 + 50 + 5 + 15 = 100\,\Omega$$

and current, $I = \dfrac{200}{100} = 2\,\text{A}$

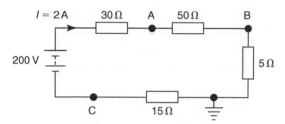

**Figure 3.24: Relative voltage**

If a voltage at point A is quoted with reference to point B then the voltage is written as $V_{AB}$. This is known as a *relative voltage*. In the circuit shown in Figure 3.24, the voltage at A w.r.t. B is I × 50, i.e., 2 × 50 = 100 V and is written as $V_{AB} = 100$ V.

It must also be indicated whether the voltage at A w.r.t. B is closer to the positive terminal or the negative terminal of the supply source. Point A is nearer to the positive terminal than B so is written as $V_{AB} = 100$ V or $V_{AB} = +100$ V or $V_{AB} = 100$ V +ve.

If no positive or negative is included, then the voltage is always taken to be positive.

If the voltage at B w.r.t. A is required, then $V_{BA}$ is negative and is written as $V_{BA} = -100$ V or $V_{BA} = 100$ V −ve.

If the reference point is changed to the *earth point* then any voltage taken w.r.t. the earth is known as an *absolute potential*. If the absolute voltage of A in Figure 3.24 is required, then this will be the sum of the voltages across the 50 Ω and 5 Ω resistors, i.e., 100 + 10 = 110 V and is written as $V_A = 110$ V or $V_A = +110$ V or $V_A = 110$ V +ve, positive since moving from the earth point to point A is moving towards the positive terminal of the source. If the voltage is negative w.r.t. earth then this must be indicated; for example, $V_C = 30$ V negative w.r.t. earth, and is written as $V_C = -30$ V or $V_C = 30$ V −ve.

### Example 3.14
For the circuit shown in Figure 3.25, calculate (a) the voltage drop across the 4 kΩ resistor, (b) the current through the 5 kΩ resistor, (c) the power developed in the 1.5 kΩ resistor, (d) the voltage at point X w.r.t. earth, and (e) the absolute voltage at point X.

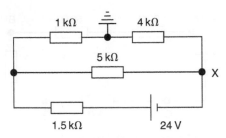

**Figure 3.25: Circuit for Example 3.14**

*Solution*

(a)  Total circuit resistance, $R_T = [(1 + 4)\,\text{k}\Omega$ in parallel with $5\,\text{k}\Omega]$ in series with $1.5\,\text{k}\Omega$

i.e., $R_T = \dfrac{5 \times 5}{5 + 5} + 1.5 = 4\,\text{k}\Omega$

Total circuit current, $I_T = \dfrac{V}{R_T} = \dfrac{24}{4 \times 10^3} = 6\,\text{mA}$

By current division, current in top branch

$= \left(\dfrac{5}{5 + 1 + 4}\right) \times 6 = 3\,\text{mA}$

Hence, **volt drop across 4 kΩ resistor**

$= 3 \times 10^{-3} \times 4 \times 10^3 = \mathbf{12\,V}$

(b)  **Current through the 5 kΩ resistor**

$= \left(\dfrac{1 + 4}{5 + 1 + 4}\right) \times 6 = \mathbf{3\,mA}$

(c)  **Power in the 1.5 kΩ resistor**

$= I_T^2 R = (6 \times 10^{-3})^2 (1.5 \times 10^3) = \mathbf{54\,mW}$

(d) The voltage at the earth point is 0 volts. The volt drop across the $4\,k\Omega$ is $12\,V$, from part (a). Since moving from the earth point to point X is moving towards the negative terminal of the voltage source, the voltage at point X w.r.t. earth is $-\mathbf{12\,V}.$

(e) The *absolute voltage at point X* means *the voltage at point X w.r.t. earth*; therefore, the absolute voltage at point X is $-12\,V$. Questions (d) and (e) mean the same thing.

# CHAPTER 4

# Capacitors and Inductors

John Bird

## 4.1 Introduction to Capacitors

A capacitor is an electrical device that is used to store electrical energy. Next to the resistor, the capacitor is the most commonly encountered component in electrical circuits. For example, capacitors are used to smooth rectified AC outputs, they are used in telecommunication equipment—such as radio receivers—for tuning to the required frequency, they are used in time delay circuits, in electrical filters, in oscillator circuits, and in magnetic resonance imaging (MRI) in medical body scanners, to name but a few practical applications.

## 4.2 Electrostatic Field

Figure 4.1 represents two parallel metal plates, A and B, charged to different potentials. If an electron that has a negative charge is placed between the plates, a force will act on the electron tending to push it away from the negative plate B towards the positive plate, A. Similarly, a positive charge would be acted on by a force tending to move it toward the negative plate. Any region such as that shown between the plates in Figure 4.1, in which an electric charge experiences a force, is called an *electrostatic field*. The direction of the field is defined as that of the force acting on a positive charge placed in the field. In Figure 4.1, the direction of the force is from the positive plate to the negative plate.

Such a field may be represented in magnitude and direction by *lines of electric force* drawn between the charged surfaces. The closeness of the lines is an indication of the field strength. Whenever a voltage is established between two points, an electric field will always exist. Figure 4.2(a) shows a typical field pattern for an isolated point charge, and

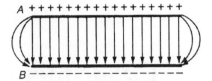

**Figure 4.1: Electrostatic field**

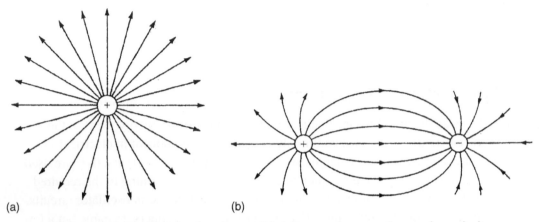

(a)                              (b)

**Figure 4.2: (a) Isolated point charge (b) Adjacent charges of opposite polarity**

Figure 4.2(b) shows the field pattern for adjacent charges of opposite polarity. Electric lines of force (often called *electric flux lines*) are continuous and start and finish on point charges. Also, the lines cannot cross each other. When a charged body is placed close to an uncharged body, an induced charge of opposite sign appears on the surface of the uncharged body. This is because lines of force from the charged body terminate on its surface.

The concept of field lines or lines of force is used to illustrate the properties of an electric field. However, it should be remembered that they are only aids to the imagination.

The *force of attraction or repulsion* between two electrically charged bodies is proportional to the magnitude of their charges and inversely proportional to the square of the distance separating them,

i.e., force $\propto \dfrac{q_1 q_2}{d^2}$   or   force $= k\, \dfrac{q_1 q_2}{d^2}$

where constant $k \approx 9 \times 10^9$ in air.

This is known as *Coulomb's law*.

Hence, the force between two charged spheres in air with their centers 16 mm apart and each carrying a charge of $+1.6\,\mu\text{C}$ is given by:

$$\text{force} = k\frac{q_1 q_2}{d^2} \approx (9 \times 10^9)\frac{(1.6 \times 10^{-6})^2}{(16 \times 10^{-3})^2}$$
$$= \textbf{90 newtons}$$

## 4.3 Electric Field Strength

Figure 4.3 shows two parallel conducting plates separated from each other by air. They are connected to opposite terminals of a battery of voltage $V$ volts.

Therefore an electric field is in the space between the plates. If the plates are close together, the electric lines of force will be straight and parallel and equally spaced, except near the edge where fringing will occur (see Figure 4.1). Over the area in which there is negligible fringing,

Electric field strength, $E = \dfrac{V}{d}$ volts/meter

where $d$ is the distance between the plates. Electric field strength is also called *potential gradient*.

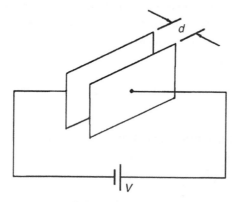

**Figure 4.3: Two parallel conducting plates separated by air**

## 4.4 Capacitance

Static electric fields arise from electric charges, electric field lines beginning and ending on electric charges. Thus, the presence of the field indicates the presence of equal positive and negative electric charges on the two plates of Figure 4.3. Let the charge be $+Q$ coulombs on one plate and $-Q$ coulombs on the other. The property of this pair of plates that determines how much charge corresponds to a given voltage between the plates is called their *capacitance*:

$$\text{capacitance } C = \frac{Q}{V}$$

The *unit of capacitance* is the *farad F* (or more usually $\mu F = 10^{-6}$ F or $pF = 10^{-12}$ F), which is defined as the capacitance when a voltage of one volt appears across the plates when charged with one coulomb.

## 4.5 Capacitors

Every system of electrical conductors possesses capacitance. For example, there is capacitance between the conductors of overhead transmission lines and also between the wires of a telephone cable. In these examples, the capacitance is undesirable but has to be accepted, minimized or compensated for. There are other situations where capacitance is a desirable property.

Devices specially constructed to possess capacitance are called *capacitors* (or condensers, as they used to be called). In its simplest form, a capacitor consists of two plates that are separated by an insulating material known as a *dielectric*. A capacitor has the ability to store a quantity of static electricity.

The symbols for a fixed capacitor and a variable capacitor used in electrical circuit diagrams are shown in Figure 4.4.

The *charge Q* stored in a capacitor is given by:

$$Q = I \times t \text{ coulombs}$$

where $I$ is the current in amperes and $t$ the time in seconds.

Fixed capacitor

Variable capacitor

**Figure 4.4: Symbols for a fixed capacitor and a variable capacitor**

### Example 4.1

(a) Determine the voltage across a 4 μF capacitor when charged with 5 mC.

(b) Find the charge on a 50 pF capacitor when the voltage applied to it is 2 kV.

### Solution

(a) $C = 4 \mu F = 4 \times 10^{-6}\,F; Q = 5\,mC = 5 \times 10^{-3}\,C$

$$\text{Since } C = \frac{Q}{V} \text{ then } V = \frac{Q}{C} = \frac{5 \times 10^{-3}}{4 \times 10^{-6}} = \frac{5 \times 10^{6}}{4 \times 10^{3}}$$

$$= \frac{5000}{4}$$

Hence, voltage $= 250\,V$ or $1.25\,kV$

(b) $C = 50\,pF = 50 \times 10^{-12}\,F; V = 2\,kV = 2000\,V$

$$Q = CV = 50 \times 10^{-12} \times 2000 = \frac{5 \times 2}{10^{8}}$$

$$= 0.1 \times 10^{-6}$$

So, charge $= 0.1\,\mu C$

### Example 4.2

A direct current of 4 A flows into a previously uncharged 20 μF capacitor for 3 ms. Determine the voltage between the plates.

### Solution

$I = 4\,A; C = 20\,\mu F = 20 \times 10^{-6}\,F;$

$t = 3\,ms = 3 \times 10^{-3}\,s$

$Q = It = 4 \times 3 \times 10^{-3}\,\text{C}$

$$V = \frac{Q}{C} = \frac{4 \times 3 \times 10^{-3}}{20 \times 10^{-6}} = \frac{12 \times 10^{6}}{20 \times 10^{3}} = 0.6 \times 10^{3}$$
$$= 600\,\text{V}$$

So, the voltage between the plates is 600 V.

### Example 4.3
A 5 μF capacitor is charged so that the voltage between its plates is 800 V. Calculate how long the capacitor can provide an average discharge current of 2 mA.

### Solution
$C = 5\,\mu\text{F} = 5 \times 10^{-6}\,\text{F};\ V = 800\,\text{V};$

$I = 2\,\text{mA} = 2 \times 10^{-3}\,\text{A}$

$Q = CV = 5 \times 10^{-6} \times 800 = 4 \times 10^{-3}\,\text{C}$

Also, $Q = It$. Thus, $t = \dfrac{Q}{I} = \dfrac{4 \times 10^{-3}}{2 \times 10^{-3}} = 2\,\text{s}$

Therefore, the capacitor can provide an average discharge current of 2 mA for 2 s.

## 4.6 Electric Flux Density

Unit flux is defined as emanating from a positive charge of 1 coulomb. Thus electric flux $\Psi$ is measured in coulombs, and for a charge of $Q$ coulombs, the flux $\Psi = Q$ coulombs.

Electric flux density $D$ is the amount of flux passing through a defined area $A$ that is perpendicular to the direction of the flux:

$$\text{electric flux density, } D = \frac{Q}{A} \text{ coulombs/meter}^2$$

Electric flux density is also called *charge density*, σ.

## 4.7 Permittivity

At any point in an electric field, the electric field strength $E$ maintains the electric flux and produces a particular value of electric flux density $D$ at that point. For a field established in *vacuum* (or for practical purposes in air), the ratio $D/E$ is a constant $\varepsilon_0$, i.e.,

$$\frac{D}{E} = \varepsilon_0$$

where $\varepsilon_0$ is called the *permittivity of free space* or the free space constant. The value of $\varepsilon_0$ is $8.85 \times 10^{-12}$ F/m.

When an insulating medium, such as mica, paper, plastic, or ceramic, is introduced into the region of an electric field the ratio of $D/E$ is modified:

$$\frac{D}{E} = \varepsilon_0 \varepsilon_r$$

where $\varepsilon_r$, the *relative permittivity* of the insulating material, indicates its insulating power compared with that of vacuum:

$$\text{relative permittivity } \varepsilon_r = \frac{\text{flux density in material}}{\text{flux density in vacuum}}$$

Here, $\varepsilon_r$ has no unit. Typical values of $\varepsilon_r$ include: air, 1.00; polythene, 2.3; mica, 3–7; glass, 5–10; water, 80; ceramics, 6–1000.

The product $\varepsilon_0 \varepsilon_r$ is called the *absolute permittivity*, $\varepsilon$.

$$\varepsilon = \varepsilon_0 \varepsilon_r$$

The insulating medium separating charged surfaces is called a *dielectric*. Compared with conductors, dielectric materials have very high resistivities. Therefore, they are used to separate conductors at different potentials, such as capacitor plates or electric power lines.

### Example 4.4

Two parallel rectangular plates measuring 20 cm by 40 cm carry an electric charge of 0.2 μC. Calculate the electric flux density. If the plates are spaced 5 mm apart and the voltage between them is 0.25 kV determine the electric field strength.

*Solution*

Charge $Q = 0.2 \, \mu\text{C} = 0.2 \times 10^{-6}\text{C}$;

Area $A = 20\,\text{cm} \times 40\,\text{cm} = 800\,\text{cm}^2 = 800 \times 10^{-4}\,\text{m}^2$

**Electric flux density** $D = \dfrac{Q}{A} = \dfrac{0.2 \times 10^{-6}}{800 \times 10^{-4}} = \dfrac{0.2 \times 10^4}{800 \times 10^6}$

$$= \dfrac{2000}{800} \times 10^{-6} = \mathbf{2.5 \, \mu C/m^2}$$

Voltage $V = 0.25\,\text{kV} = 250\,\text{V}$; Plate spacing, $d = 5\,\text{mm} = 5 \times 10^{-3}\,\text{m}$

**Electric field strength** $E = \dfrac{V}{d} = \dfrac{250}{5 \times 10^{-3}} = \mathbf{50 \, kV/m}$

## Example 4.5

The flux density between two plates separated by mica of relative permittivity 5 is $2\,\mu\text{C/m}^2$. Find the voltage gradient between the plates.

*Solution*

Flux density $D = 2\,\mu\text{C/m}^2 = 2 \times 10^{-6}\,\text{C/m}^2$;

$\varepsilon_0 = 8.85 \times 10^{-12}\,\text{F/m}$; $\varepsilon_r = 5$.

$$\dfrac{D}{E} = \varepsilon_0 \varepsilon_r,$$

hence, **voltage gradient** $E = \dfrac{D}{\varepsilon_0 \varepsilon_r}$

$$= \dfrac{2 \times 10^{-6}}{8.85 \times 10^{-12} \times 5} \, \text{V/m}$$

$$= \mathbf{45.2 \, kV/m}$$

## Example 4.6

Two parallel plates having a voltage of 200 V between them are spaced 0.8 mm apart. What is the electric field strength? Find also the flux density when the dielectric between the plates is (a) air, and (b) polythene of relative permittivity 2.3.

*Solution*

**Electric field strength** $E = \dfrac{V}{D} = \dfrac{200}{0.8 \times 10^{-3}} = $ **250 kV/m**

(a) For air: $\varepsilon_r = 1$

$\dfrac{D}{E} = \varepsilon_0 \varepsilon_r$. Hence,

**Electric flux density** $D = E \varepsilon_0 \varepsilon_r$

$$= (250 \times 10^3 \times 8.85 \times 10^{-12} \times 1)\, \text{C/m}^2$$

$$= \textbf{2.213 } \boldsymbol{\mu}\textbf{C/m}^2$$

(b) For polythene, $\varepsilon_r = 2.3$

**Electric flux density** $D = E \varepsilon_0 \varepsilon_r$

$$= (250 \times 10^3 \times 8.85 \times 10^{-12} \times 2.3)\, \text{C/m}^2$$

$$= \textbf{5.089 } \boldsymbol{\mu}\textbf{C/m}^2$$

## 4.8 The Parallel Plate Capacitor

For a parallel plate capacitor, as shown in Figure 4.5(a), experiments show that capacitance $C$ is proportional to the area $A$ of a plate, inversely proportional to the plate spacing $d$ (i.e., the dielectric thickness) and depends on the nature of the dielectric:

Capacitance, $C = \dfrac{\varepsilon_0 \varepsilon_r A}{d}$ farads

where $\varepsilon_0 = 8.85 \times 10^{-12}\,\text{F/m}$ (constant)

$\varepsilon_r$ = relative permittivity

$A$ = area of one of the plates, in m$^2$, and

$d$ = thickness of dielectric in m

Another method used to increase the capacitance is to interleave several plates as shown in Figure 4.5(b). Ten plates are shown, forming nine capacitors with a capacitance nine times that of one pair of plates.

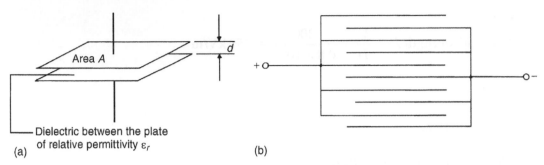

**Figure 4.5: Parallel plate capacitor**

If such an arrangement has $n$ plates, then capacitance $C \propto (n - 1)$.

Thus, capacitance $C = \dfrac{\varepsilon_0 \varepsilon_r A(n - 1)}{d}$ farads

### Example 4.7
(a) A ceramic capacitor has an effective plate area of $4\,cm^2$ separated by $0.1\,mm$ of ceramic of relative permittivity 100. Calculate the capacitance of the capacitor in picofarads. (b) If the capacitor in part (a) is given a charge of $1.2\,\mu C$, what will be the voltage between the plates?

### Solution
(a)  Area $A = 4\,cm^2 = 4 \times 10^{-4}\,m^2$;

$d = 0.1\,mm = 0.1 \times 10^{-3}\,m$;

$\varepsilon_0 = 8.85 \times 10^{-12}\,F/m$; $\varepsilon_r = 100$

Capacitance $C = \dfrac{\varepsilon_0 \varepsilon_r}{d}$ farads

$= \dfrac{8.85 \times 10^{-12} \times 100 \times 4 \times 10^{-4}}{0.1 \times 10^{-3}}\,F$

$= \dfrac{8.85 \times 4}{10^{10}}\,F = \dfrac{8.85 \times 4 \times 10^{12}}{10^{10}}\,pF$

$= \mathbf{3540\,pF}$

(b) $Q = CV$ thus, $V = \dfrac{Q}{C} = \dfrac{1.2 \times 10^{-6}}{3540 \times 10^{-12}}$ V $= 339$ V

## Example 4.8

A waxed paper capacitor has two parallel plates, each of effective area $800\,cm^2$. If the capacitance of the capacitor is $4425\,pF$, determine the effective thickness of the paper if its relative permittivity is 2.5.

### Solution

$A = 800\,cm^2 = 800 \times 10^{-4}\,m^2 = 0.08\,m^2$;

$C = 4425\,pF = 4425 \times 10^{-12}\,F$;

$\varepsilon_0 = 8.85 \times 10^{-12}\,F/m$; $\varepsilon_r = 2.5$

Since $C = \dfrac{\varepsilon_0 \varepsilon_r A}{d}$    then    $d = \dfrac{\varepsilon_0 \varepsilon_r A}{C}$

Hence, $d = \dfrac{8.85 \times 10^{-12} \times 2.5 \times 0.08}{4425 \times 10^{-12}} = 0.0004$ m

**So, the thickness of the paper is 0.4 mm.**

## Example 4.9

A parallel plate capacitor has nineteen interleaved plates each $75\,mm \times 75\,mm$ separated by mica sheets $0.2\,mm$ thick. Assuming the relative permittivity of the mica is 5, calculate the capacitance of the capacitor.

### Solution

$n = 19$; $n - 1 = 18$;

$A = 75 \times 75 = 5625\,mm^2 = 5625 \times 10^{-6}\,m^2$;

$\varepsilon_r = 5$; $\varepsilon_0 = 8.85 \times 10^{-12}\,F/m$;

$d = 0.2$ mm $= 0.2 \times 10^{-3}\,m$

Capacitance $C = \dfrac{\varepsilon_0 \varepsilon_r A(n-1)}{d}$

$$= \frac{8.85 \times 10^{-12} \times 5 \times 5625 \times 10^{-6} \times 18}{0.2 \times 10^{-3}}\, \text{F}$$

$$= 0.0224\,\mu\text{F} \quad \text{or} \quad 22.4\,\text{nF}$$

## 4.9 Capacitors Connected in Parallel and Series

### 4.9.1 Capacitors Connected in Parallel

Figure 4.6 shows three capacitors, $C_1$, $C_2$ and $C_3$, connected in parallel with a supply voltage $V$ applied across the arrangement.

When the charging current $I$ reaches point $A$ it divides, some flowing into $C_1$, some flowing into $C_2$ and some into $C_3$. Therefore, the total charge $QT$ $(=I \times t)$ is divided between the three capacitors. The capacitors each store a charge and these are shown as $Q_1$, $Q_2$ and $Q_3$, respectively. Hence:

$$Q_T = Q_1 + Q_2 + Q_3$$

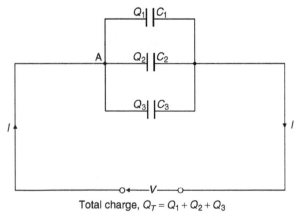

Total charge, $Q_T = Q_1 + Q_2 + Q_3$

**Figure 4.6: Three capacitors connected in parallel**

But $Q_T = CV$, $Q_1 = C_1V$, $Q_2 = C_2V$ and $Q_3 = C_3V$. Therefore, $CV = C_1V + C_2V + C_3V$ where $C$ is the total equivalent circuit capacitance,

i.e., $C = C_1 + C_2 + C_3$

It follows that for $n$ parallel-connected capacitors,

$$C = C_1 + C_2 + C_3 + \cdots + C_n$$

that is, the equivalent capacitance of a group of parallel-connected capacitors is the sum of the capacitances of the individual capacitors. (Note that this formula is similar to that used for *resistors* connected in *series*.)

### 4.9.2 Capacitors Connected in Series

Figure 4.7 shows three capacitors, $C_1$, $C_2$ and $C_3$, connected in series across a supply voltage $V$. Let the voltage across the individual capacitors be $V_1$, $V_2$, and $V_3$, respectively, as shown.

Let the charge on plate "a" of capacitor $C_1$ be $+Q$ coulombs. This induces an equal but opposite charge of $-Q$ coulombs on plate "b". The conductor between plates "b" and "c" is electrically isolated from the rest of the circuit so that an equal but opposite charge of $+Q$ coulombs must appear on plate "c", which, in turn, induces an equal and opposite charge of $-Q$ coulombs on plate "d", and so on.

When capacitors are connected in series the charge on each is the same.

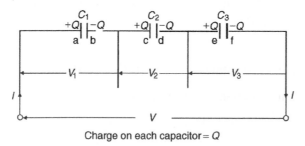

Charge on each capacitor $= Q$

**Figure 4.7: Three capacitors connected in series**

In a series circuit: $V = V_1 + V_2 + V_3$

Since $V = \dfrac{Q}{C}$ then $\dfrac{Q}{C} = \dfrac{Q}{C_1} + \dfrac{Q}{C_2} + \dfrac{Q}{C_3}$

where $C$ is the total equivalent circuit capacitance,

i.e., $\dfrac{1}{C} = \dfrac{1}{C_1} + \dfrac{1}{C_2} + \dfrac{1}{C_3}$

It follows that for $n$ series-connected capacitors:

$$\frac{1}{C} = \frac{1}{C_1} + \frac{1}{C_2} + \frac{1}{C_3} + \cdots + \frac{1}{C_n}$$

That is, for series-connected capacitors, the reciprocal of the equivalent capacitance is equal to the sum of the reciprocals of the individual capacitances. (Note that this formula is similar to that used for resistors connected in *parallel*.)

For the special case of two capacitors in series:

$$\frac{1}{C} = \frac{1}{C_1} + \frac{1}{C_2} = \frac{C_2 + C_1}{C_1 C_2}$$

Hence, $C = \dfrac{C_1 C_2}{C_1 + C_2}$ $\left( \text{i.e., } \dfrac{\text{product}}{\text{sum}} \right)$

### Example 4.10
Calculate the equivalent capacitance of two capacitors of $6\,\mu F$ and $4\,\mu F$ connected (a) in parallel and (b) in series.

### Solution
(a) In parallel, equivalent capacitance $C = C_1 + C_2 = 6\,\mu F + 4\,\mu F = \mathbf{10\,\mu F}$

(b) In series, equivalent capacitance $C$ is given by:

$$C = \frac{C_1 C_2}{C_1 + C_2}$$

This formula is used for the special case of *two* capacitors in series.

Thus, $C = \dfrac{6 \times 4}{6+4} = \dfrac{24}{10} = \mathbf{2.4\,\mu F}$

## Example 4.11

What capacitance must be connected in series with a $30\,\mu F$ capacitor for the equivalent capacitance to be $12\,\mu F$?

### Solution

Let $C = 12\,\mu F$ (the equivalent capacitance), $C_1 = 30\,\mu F$ and $C_2$ be the unknown capacitance.

For two capacitors in series $\dfrac{1}{C} = \dfrac{1}{C_1} + \dfrac{1}{C_2}$

Hence, $\dfrac{1}{C_2} = \dfrac{1}{C} - \dfrac{1}{C_1} = \dfrac{C_1 - C}{CC_1}$

$$\text{and} \quad C_2 = \frac{CC_1}{C_1 - C} = \frac{12 \times 30}{30 - 12}$$

$$= \frac{360}{18} = \mathbf{20\,\mu F}$$

## Example 4.12

Capacitances of $1\,\mu F$, $3\,\mu F$, $5\,\mu F$ and $6\,\mu F$ are connected in parallel to a direct voltage supply of $100\,V$. Determine (a) the equivalent circuit capacitance, (b) the total charge and (c) the charge on each capacitor.

### Solution

(a) The equivalent capacitance $C$ for four capacitors in parallel is given by:

$$C = C_1 + C_2 + C_3 + C_4$$

i.e., $C = 1 + 3 + 5 + 6 = \mathbf{15\,\mu F}$

(b) Total charge $Q_T = CV$ where $C$ is the equivalent circuit capacitance

i.e., $Q_T = 15 \times 10^{-6} \times 100 = 1.5 \times 10^{-3}\,C = \mathbf{1.5\,mC}$

(c) The charge on the $1\,\mu\text{F}$ capacitor

$$Q_1 = C_1 V = 1 \times 10^{-6} \times 100$$

$$= 0.1\,\text{mC}$$

The charge on the $3\,\mu\text{F}$ capacitor

$$Q_2 = C_2 V = 3 \times 10^{-6} \times 100$$

$$= 0.3\,\text{mC}$$

The charge on the $5\,\mu\text{F}$ capacitor

$$Q_3 = C_3 V = 5 \times 10^{-6} \times 100$$

$$= 0.5\,\text{mC}$$

The charge on the $6\,\mu\text{F}$ capacitor

$$Q_4 = C_4 V = 6 \times 10^{-6} \times 100$$

$$= 0.6\,\text{mC}$$

[Check: In a parallel circuit:

$$Q_T = Q_1 + Q_2 + Q_3 + Q_4$$

$$Q_1 + Q_2 + Q_3 + Q_4 = 0.1 + 0.3 + 0.5 + 0.6$$

$$= 1.5\,\text{mC} = Q_T]$$

### Example 4.13

Capacitances of $3\,\mu\text{F}$, $6\,\mu\text{F}$ and $12\,\mu\text{F}$ are connected in series across a $350\,\text{V}$ supply. Calculate (a) the equivalent circuit capacitance, (b) the charge on each capacitor and (c) the voltage across each capacitor.

### Solution

The circuit diagram is shown in Figure 4.8.

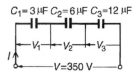

**Figure 4.8: Circuit diagram for Example 4.13**

(a) The equivalent circuit capacitance $C$ for three capacitors in series is given by:

$$\frac{1}{C} = \frac{1}{C_1} + \frac{1}{C_2} + \frac{1}{C_3}$$

i.e., $\qquad \frac{1}{C} = \frac{1}{3} + \frac{1}{6} + \frac{1}{12} = \frac{4+2+1}{12} =$

**So the equivalent circuit capacitance**

$$C = \frac{12}{7} = 1\frac{5}{7}\mu F$$

(b) Total charge $Q_T = CV$, hence,

$$Q_T = \frac{12}{7} \times 10^{-6} \times 350 = 600\ \mu C \text{ or } 0.6\ mC$$

**Since the capacitors are connected in series, 0.6 mC is the charge on each of them.**

(c) The voltage across the $3\,\mu F$ capacitor,

$$V_1 = \frac{Q}{C_1} = \frac{0.6 \times 10^{-3}}{3 \times 10^{-6}}$$

$$= \mathbf{200\,V}$$

The voltage across the $6\,\mu F$ capacitor,

$$V_2 = \frac{Q}{C_2} = \frac{0.6 \times 10^{-3}}{6 \times 10^{-6}}$$

$$= \mathbf{100\,V}$$

The voltage across the $12\,\mu F$ capacitor,

$$V_3 = \frac{Q}{C_3} = \frac{0.6 \times 10^{-3}}{12 \times 10^{-6}}$$

$$= \mathbf{50\,V}$$

[Check: In a series circuit

$$V = V_1 + V_2 + V_3$$

$$V_1 + V_2 + V_3 = 200 + 100 + 50 = 350\,V$$

$$= \text{supply voltage.}]$$

In practice, capacitors are rarely connected in series unless they are of the same capacitance. The reason for this can be seen from the above problem where the lowest valued capacitor (i.e., $3\,\mu F$) has the highest voltage across it (i.e., $200\,V$) which means that if all the capacitors have an identical construction they must all be rated at the highest voltage.

## 4.10  Dielectric Strength

The maximum amount of field strength that a dielectric can withstand is called the *dielectric strength of the material.*

Dielectric strength, $E_m = \dfrac{V_m}{d}$

### Example 4.14

A capacitor is to be constructed so that its capacitance is $0.2\,\mu F$ and to take a voltage of $1.25\,kV$ across its terminals. The dielectric is to be mica which, after allowing a safety factor of 2, has a dielectric strength of $50\,MV/m$. Find (a) the thickness of the mica needed, and (b) the area of a plate assuming a two-plate construction. (Assume $\varepsilon_r$ for mica to be 6.)

### Solution

(a)  Dielectric strength, $E = \dfrac{V}{d}$, i.e., $d = \dfrac{V}{E}$

$$= \frac{1.25 \times 10^3}{50 \times 10^6}\,m$$

$$= \mathbf{0.025\,mm}$$

(b) Capacitance, $C = \dfrac{\varepsilon_0 \varepsilon_r A}{d}$

hence, area $A = \dfrac{Cd}{\varepsilon_0 \varepsilon_r}$

$$= \frac{0.2 \times 10^{-6} \times 0.025 \times 10^{-3}}{8.85 \times 10^{-12} \times 6} \, m^2$$

$$= 0.09416 \, m^2 = \mathbf{941.6 \, cm^2}$$

## 4.11 Energy Stored

The energy, $W$, stored by a capacitor is given by:

$$W = \frac{1}{2} CV^2 \text{ joules}$$

### Example 4.15

(a) Determine the energy stored in a $3 \, \mu F$ capacitor when charged to $400 \, V$. (b) Find also the average power developed if this energy is dissipated in a time of $10 \, \mu s$.

### Solution

(a) **Energy stored** $W = \dfrac{1}{2} CV^2 \text{ joules}$

$$= \frac{1}{2} \times 3 \times 10^{-6} \times 400^2$$

$$= \frac{3}{2} \times 16 \times 10^{-2}$$

$$= \mathbf{0.24 \, J}$$

(b) **Power** $= \dfrac{\text{Energy}}{\text{time}} = \dfrac{0.24}{10 \times 10^{-6}} \, W = \mathbf{24 \, kW}$

### Example 4.16

A $12 \, \mu F$ capacitor is required to store $4 \, J$ of energy. Find the voltage to which the capacitor must be charged.

*Solution*

Energy stored $W = \dfrac{1}{2}CV^2$ hence, $V^2 = \dfrac{2W}{C}$

and $V = \sqrt{\left(\dfrac{2W}{C}\right)} = \sqrt{\left(\dfrac{2 \times 4}{12 \times 10^{-6}}\right)} = \sqrt{\left(\dfrac{2 \times 10^6}{3}\right)}$

$$= 816.5\,\text{V}$$

*Example 4.17*

A capacitor is charged with 10 mC. If the energy stored is 1.2 J find (a) the voltage and (b) the capacitance.

*Solution*

Energy stored $W = \dfrac{1}{2}CV^2$   and   $C = \dfrac{Q}{V}$

Hence,        $W = \dfrac{1}{2}\left(\dfrac{Q}{V}\right)V^2 = \dfrac{1}{2}QV$

from which    $V = \dfrac{2W}{Q}$

$Q = 10\,\text{mC} = 10 \times 10^{-3}\text{C}$   and   $W = 1.2\,\text{J}$

(a)  Voltage $V = \dfrac{2W}{Q} = \dfrac{2 \times 1.2}{10 \times 10^{-3}} = \mathbf{0.24\,kV}$ **or 240 V**

(b)  Capacitance $C = \dfrac{Q}{V} = \dfrac{10 \times 10^{-3}}{240}\,\text{F}$

$$= \dfrac{10 \times 10^6}{240 \times 10^3}\,\mu\text{F} = \mathbf{41.67\,\mu F}$$

## 4.12  Practical Types of Capacitors

Practical types of capacitors are characterized by the material used for their dielectric. The main types include: variable air, mica, paper, ceramic, plastic, titanium oxide, and electrolytic.

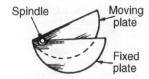

**Figure 4.9: End view of variable air capacitor**

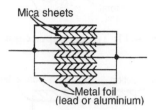

**Figure 4.10: Older construction mica capacitor**

1.  *Variable air capacitors.* These usually consist of two sets of metal plates (such as aluminum) one fixed, the other variable. The set of moving plates rotate on a spindle as shown by the end view of Figure 4.9.

    As the moving plates are rotated through half a revolution, the meshing, and therefore the capacitance, varies from a minimum to a maximum value. Variable air capacitors are used in radio and electronic circuits where very low losses are required, or where a variable capacitance is needed. The maximum value of such capacitors is between 500 pF and 1000 pF.

2.  *Mica capacitors.* A typical older type construction is shown in Figure 4.10.

    Usually the whole capacitor is impregnated with wax and placed in a bakelite case. Mica is easily obtained in thin sheets and is a good insulator. However, mica is expensive and is not used in capacitors above about 0.2 μF. A modified form of mica capacitor is the silvered mica type. The mica is coated on both sides with a thin layer of silver, which forms the plates. Capacitance is stable and less likely to change with age. Such capacitors have a constant capacitance with change of temperature, a high working voltage rating and a long service life and are used in high frequency circuits with fixed values of capacitance up to about 1000 pF.

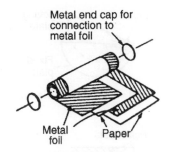

**Figure 4.11: Typical paper capacitor**

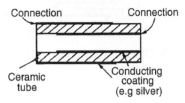

**Figure 4.12: Cross-section of tube of ceramic material**

3. *Paper capacitors.* A typical paper capacitor is shown in Figure 4.11 where the length of the roll corresponds to the capacitance required. The whole is usually impregnated with oil or wax to exclude moisture, and then placed in a plastic or aluminum container for protection. Paper capacitors are made in various working voltages up to about 150 kV and are used where loss is not very important. The maximum value of this type of capacitor is between 500 pF and 10 μF. Disadvantages of paper capacitors include variation in capacitance with temperature change and a shorter service life than most other types of capacitor.

4. *Ceramic capacitors.* These are made in various forms, each type of construction depending on the value of capacitance required. For high values, a tube of ceramic material is used as shown in the cross-section of Figure 4.12. For smaller values the cup construction is used as shown in Figure 4.13, and for still smaller values the disc construction shown in Figure 4.14 is used. Certain ceramic materials have a very high permittivity and this enables capacitors of high capacitance to be made which are of small physical size with a high working voltage rating. Ceramic capacitors are available in the range 1 pF to

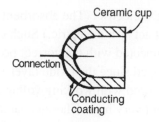

**Figure 4.13: Cup construction**

**Figure 4.14: Disc construction**

0.1 μF and may be used in high frequency electronic circuits subject to a wide range of temperatures.

5. *Plastic capacitors.* Some plastic materials such as polystyrene and Teflon can be used as dielectrics. Construction is similar to the paper capacitor but using a plastic film instead of paper. Plastic capacitors operate well under conditions of high temperature, provide a precise value of capacitance, a very long service life and high reliability.

6. *Titanium oxide capacitors* have a very high capacitance with a small physical size when used at a low temperature.

7. *Electrolytic capacitors.* Construction is similar to the paper capacitor with aluminum foil used for the plates and with a thick absorbent material, such as paper, impregnated with an electrolyte (ammonium borate), separating the plates. The finished capacitor is usually assembled in an aluminum container and hermetically sealed. Its operation depends on the formation of a thin aluminum oxide layer on the positive plate by electrolytic action when a suitable direct potential is maintained between the plates. This oxide layer is

very thin and forms the dielectric. (The absorbent paper between the plates is a conductor and does not act as a dielectric.) Such capacitors *must always be used on DC* and must be connected with the correct polarity; if this is not done the capacitor will be destroyed since the oxide layer will be destroyed. Electrolytic capacitors are manufactured with working voltage from 6 V to 600 V, although accuracy is generally not very high. These capacitors possess a much larger capacitance than other types of capacitors of similar dimensions due to the oxide film being only a few microns thick. The fact that they can be used only on DC supplies limit their usefulness.

## 4.13 Inductance

*Inductance* is the name given to the property of a circuit whereby there is an e.m.f. induced into the circuit by the change of flux linkages produced by a current change. When the e.m.f. is induced in the same circuit as that in which the current is changing, the property is called *self-inductance*, L. When the e.m.f. is induced in a circuit by a change of flux due to current changing in an adjacent circuit, the property is called *mutual inductance*, M. The unit of inductance is the *henry, H.*

*A circuit has an inductance of one henry when an e.m.f. of one volt is induced in it by a current changing at the rate of one ampere per second.*

Induced e.m.f. in a coil of N turns,

$$E = -N\frac{d\Phi}{dt} \text{ volts}$$

where $d\Phi$ is the change in flux in webers, and $dt$ is the time taken for the flux to change in seconds (i.e., $d\Phi/dt$ is the rate of change of flux).

Induced e.m.f. in a coil of inductance L henrys,

$$E = -L\frac{dI}{dt} \text{ volts}$$

where $dI$ is the change in current in amperes and $dt$ is the time taken for the current to change in seconds (i.e., $dI/dt$ is the rate of change of current). The minus signs in each of the above two equations remind us of its direction (given by Lenz's law).

*Example 4.18*

Determine the e.m.f. induced in a coil of 200 turns when there is a change of flux of 25 mWb linking with it in 50 ms.

**Solution**

$$\text{Induced e.m.f. } E = -N\frac{d\Phi}{dt} = -(200)\left(\frac{25\times10^{-3}}{50\times10^{-3}}\right)$$

$$= -100\,\text{volts}$$

*Example 4.19*

A flux of 400 μWb passing through a 150-turn coil is reversed in 40 ms. Find the average e.m.f. induced.

**Solution**

Since the flux reverses, the flux changes from +400 μWb to −400 μWb, a total change of flux of 800 μWb

$$\text{Induced e.m.f. } E = -N\frac{d\Phi}{dt} = -(150)\left(\frac{800\times10^{-6}}{40\times10^{-3}}\right)$$

$$= -\left(\frac{150\times800\times10^{3}}{40\times10^{6}}\right)$$

Hence the average e.m.f. induced $E = -3\,\text{V}$

*Example 4.20*

Calculate the e.m.f. induced in a coil of inductance 12 H by a current changing at the rate of 4 A/s.

**Solution**

$$\text{Induced e.m.f. } E = -L\frac{dI}{dt} = -(12)(4) = -48\,\text{volts}$$

*Example 4.21*

An e.m.f. of 1.5 kV is induced in a coil when a current of 4 A collapses uniformly to zero in 8 ms. Determine the inductance of the coil.

*Solution*

Change in current, $dI = (4 - 0) = 4\,A$; $dt = 8\,ms = 8 \times 10^{-3}\,s$;

$$\frac{dI}{dt} = \frac{4}{8 \times 10^{-3}} = \frac{4000}{8} = 500\,A/s;$$

$$E = 1.5\,kV = 1500\,V$$

Since $|E| = L\left(\dfrac{dI}{dt}\right)$

inductance, $L = \dfrac{|E|}{(dI/dt)} = \dfrac{1500}{500} = \mathbf{3\,H}$

(Note that $|E|$ means the "magnitude of $E$," which disregards the minus sign.)

## 4.14 Inductors

A component called an inductor is used when the property of inductance is required in a circuit. The basic form of an inductor is simply a coil of wire.

Factors which affect the inductance of an inductor include:

  (i)   the number of turns of wire—the more turns, the higher the inductance.

  (ii)  the cross-sectional area of the coil of wire—the greater the cross-sectional area the higher the inductance.

  (iii) the presence of a magnetic core—when the coil is wound on an iron core, the same current sets up a more concentrated magnetic field and the inductance is increased.

  (iv)  the way the turns are arranged—a short thick coil of wire has a higher inductance than a long thin one.

Two examples of practical inductors are shown in Figure 4.15, and the standard electrical circuit diagram symbols for air-cored and iron-cored inductors are shown in Figure 4.16.

An iron-cored inductor is often called a *choke* since, when used in AC circuits, it has a choking effect, limiting the current flowing through it. Inductance is often undesirable in a circuit. To reduce inductance to a minimum, the wire may be bent back on itself, as shown in Figure 4.17, so that the magnetizing effect of one conductor is neutralized by

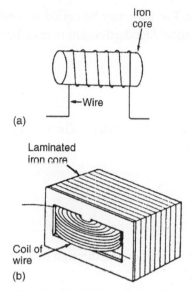

(a)

(b)

**Figure 4.15: Two examples of practical inductors**

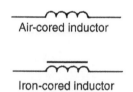

**Figure 4.16: Standard electrical symbols for air-cored and iron-cored inductors**

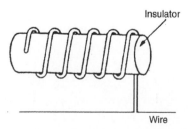

**Figure 4.17: Wire coiled around an insulator to form an inductor**

that of the adjacent conductor. The wire may be coiled around an insulator, as shown, without increasing the inductance. Standard resistors may be non-inductively wound in this manner.

## 4.15 Energy Stored

An inductor possesses an ability to store energy. The energy stored, $W$, in the magnetic field of an inductor is given by:

$$W = \frac{1}{2} LI^2 \text{ joules}$$

### Example 4.22
An 8 H inductor has a current of 3 A flowing through it. How much energy is stored in the magnetic field of the inductor?

### Solution
Energy stored, $W = \frac{1}{2} LI^2 = \frac{1}{2}(8)(3)^2 =$ **36 joules**

# CHAPTER 5

# DC Circuit Theory

John Bird

## 5.1 Introduction

The laws that determine the currents and voltage drops in DC networks are: (a) Ohm's law, (b) the laws for resistors in series and in parallel, and (c) Kirchhoff's laws (see Section 5.2). In addition, there are a number of circuit theorems that have been developed for solving problems in electrical networks. These include:

(i) the superposition theorem (see Section 5.3),

(ii) Thévenin's theorem (see Section 5.5),

(iii) Norton's theorem (see Section 5.7), and

(iv) the maximum power transfer theorem (see Section 5.9).

## 5.2 Kirchhoff's Laws

*Kirchhoff's laws* state:

(a) **Current Law**. *At any junction in an electric circuit the total current flowing towards that junction is equal to the total current flowing away from the junction,* i.e., $\Sigma I = 0$.

Thus, referring to Figure 5.1:

$I_1 + I_2 = I_3 + I_4 + I_5$ or,

$I_1 + I_2 - I_3 - I_4 - I_5 = 0$

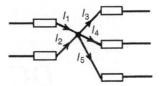

Figure 5.1: Junction showing Kirchhoff's current law

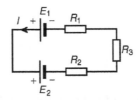

Figure 5.2: Loop showing Kirchhoff's voltage law

(b) **Voltage Law**. *In any closed loop in a network, the algebraic sum of the voltage drops (i.e., products of current and resistance) taken around the loop is equal to the resultant e.m.f. acting in that loop.*

Thus, referring to Figure 5.2:

$$E_1 - E_2 = IR_1 + IR_2 + IR_3$$

(Note that if current flows away from the positive terminal of a source, that source is considered by convention to be positive. Thus, moving anticlockwise around the loop of Figure 5.2, $E_1$ is positive and $E_2$ is negative.)

### Example 5.1
(a) Find the unknown currents marked in Figure 5.3(a). (b) Determine the value of e.m.f. $E$ in Figure 5.3(b).

### Solution
(a) Applying Kirchhoff's current law:

For junction B:   $50 = 20 + I_1 . \mathbf{I_1 = 30\ A}$

For junction C:   $20 + 15 = I_2 . \mathbf{I_2 = 35\ A}$

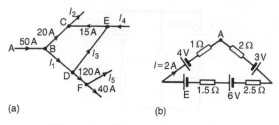

(a)  (b)

**Figure 5.3: Figures for Example 5.1**

For junction D:   $I_1 = I_3 + 120$

i.e.,            $30 = I_3 + 120. I_3 = -90\,\text{A}$

(i.e., in the opposite direction to that shown in Figure 5.3(a))

For junction E:   $I_4 + I_3 = 15$

i.e.,            $I_4 = 15 - (-90).$

$I_4 = 105\,\text{A}$

For junction F: $120 = I_5 + 40. I_5 = 80\,\text{A}$

(b)  Applying Kirchhoff's voltage law and moving clockwise around the loop of Figure 5.3 (b) starting at point A:

$$3 + 6 + E - 4 = (I)(2) + (I)(2.5) + (I)(1.5) + (I)(1)$$
$$= I(2 + 2.5 + 1.5 + 1)$$

i.e.,    $5 + E = 2(7)$, since $I = 2\,\text{A}$

$$E = 14 - 5 = 9\,\text{V}$$

### Example 5.2

Use Kirchhoff's laws to determine the currents flowing in each branch of the network shown in Figure 5.4.

### Solution
### Procedure

1.  Use Kirchhoff's current law and label current directions on the original circuit diagram. The directions chosen are arbitrary, but it is usual, as a starting point, to

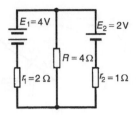

**Figure 5.4: Network for Example 5.2**

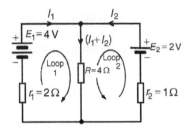

**Figure 5.5: Labeling current directions**

assume that current flows from the positive terminals of the batteries. This is shown in Figure 5.5 where the three branch currents are expressed in terms of $I_1$ and $I_2$ only, since the current through $R$ is $I_1 + I_2$.

2. Divide the circuit into two loops and apply Kirchhoff's voltage law to each. From loop one of Figure 5.5, and moving in a clockwise direction as indicated (the direction chosen does not matter), gives:

$$E_1 = I_1 r_1 + (I_1 + I_2) R, \text{ i.e., } 4 = 2I_1 + 4(I_1 + I_2),$$
$$\text{i.e., } 6I_1 + 4I_2 = 4 \tag{5.1}$$

From loop 2 of Figure 5.5, and moving in an anticlockwise direction as indicated (once again, the choice of direction does not matter; it does not have to be in the same direction as that chosen for the first loop), gives:

$$E_2 = I_2 r_2 + (I_1 + I_2) R, \text{ i.e., } 2 = I_2 + 4(I_1 + I_2),$$
$$\text{i.e., } 4I_1 + 5I_2 = 2 \tag{5.2}$$

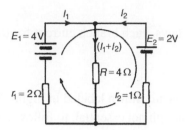

**Figure 5.6: Possible third loop**

3. Solve equations (1) and (2) for $I_1$ and $I_2$.

$2 \times$ (1) gives:   $12I_1 + 8I_2 = 8$                                          (5.3)

$3 \times$ (2) gives:   $12I_1 + 15I_2 = 6$                                         (5.4)

(3) $-$ (4) gives: $-7I_2 = 2$ hence, $I_2 = -\dfrac{2}{7} = \mathbf{-0.286\,A}$

(i.e., $I_2$ is flowing in the opposite direction to that shown in Figure 5.5.)

From (1) $6I_1 + 4\,(-0.286) = 4$

$\qquad\quad 6I_1 = 4 + 1.144$

Hence,    $I_1 = \dfrac{5.144}{6} = \mathbf{0.857\,A}$

Current flowing through resistance $R$ is,

$I_1 + I_2 = 0.857 + (-0.286) = \mathbf{0.571\,A}$

Note that a third loop is possible, as shown in Figure 5.6, giving a third equation that can be used as a check:

$E_1 - E_2 = I_1 r_1 - I_2 r_2$

$\quad 4 - 2 = 2I_1 - I_2$

$\qquad 2 = 2I_1 - I_2$

[Check: $2I_1 - I_2 = 2(0.857) - (-0.286) = 2$]

### *Example 5.3*

Determine, using Kirchhoff's laws, each branch current for the network shown in Figure 5.7.

### *Solution*

1.  Currents and their directions are shown labeled in Figure 5.8 following Kirchhoff's current law. It is usual, although not essential, to follow conventional current flow with current flowing from the positive terminal of the source.

2.  The network is divided into two loops as shown in Figure 5.8. Applying Kirchhoff's voltage law gives:

**For loop one:**

$$E_1 + E_2 = I_1 R_1 + I_2 R_2$$
$$\text{i.e., } 16 = 0.5I_1 + 2I_2 \tag{5.5}$$

**For loop two:**

$$E_2 = I_2 R_2 - (I_1 - I_2) R_3$$

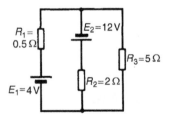

**Figure 5.7: Network for Example 5.3**

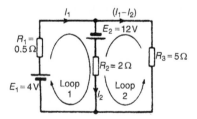

**Figure 5.8: Labeling current directions**

Note that since loop two is in the opposite direction to current $(I_1 - I_2)$, the voltage drop across $R_3$ (i.e., $(I_1 - I_2) (R_3)$) is by convention negative.

Thus, $\quad 12 = 2I_2 - 5(I_1 + I_2)$

i.e., $\quad 12 = -5I_1 + 7I_2$ $\hfill (5.6)$

3.  Solving equations (5.1) and (5.2) to find $I_1$ and $I_2$:

    $10 \times (1)$ gives $160 = 5I_1 + 20I_2$

    $(5.6) + (5.7)$ gives $172 = 27I_2$ hence, $I_2 = \dfrac{172}{27} = \mathbf{6.37\,A}$ $\hfill (5.7)$

    From (1): $\quad 16 = 0.5I_1 + 2(6.37)$

    $$I_1 = \frac{16 - 2(6.37)}{0.5} = \mathbf{6.52\,A}$$

    Current flowing in $R_3 = I_1 - I_2 = 6.52 - 6.3$
    $$= \mathbf{0.15\,A}$$

### Example 5.4

For the bridge network shown in Figure 5.9 determine the currents in each of the resistors.

### Solution

Let the current in the $2\,\Omega$ resistor be $I_1$; then by Kirchhoff's current law, the current in the $14\,\Omega$ resistor is $(I - I_1)$. Let the current in the $32\,\Omega$ resistor be $I_2$ as shown in Figure 5.10. Then the current in the $11\,\Omega$ resistor is $(I - I_2)$ and that in the $3\,\Omega$ resistor is $(I - I_1 + I_2)$.

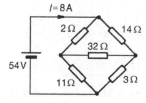

**Figure 5.9: Bridge network for Example 5.4**

Applying Kirchhoff's voltage law to loop one and moving in a clockwise direction as shown in Figure 5.10 gives:

$$54 = 2I_1 + 11\,(I_1 - I_2)$$
$$\text{i.e.,}\ \ 13I_1 - 11I_2 = 54 \tag{5.8}$$

Applying Kirchhoff's voltage law to loop two and moving in an anticlockwise direction as shown in Figure 5.10 gives:

$$0 = 2I_1 + 32I_2 - 14(I - I_1)$$

However, $I = 8\,\text{A}$

Hence,          $0 = 2I_1 + 32I_2 - 14(8 - I_1)$
$$\text{i.e.,}\ \ 16I_1 + 32I_2 = 112 \tag{5.9}$$

Equations (5.8) and (5.9) are simultaneous equations with two unknowns, $I_1$ and $I_2$.

$$16 \times (1)\ \text{gives:}\ \ 208I_1 - 176I_2 = 864 \tag{5.10}$$

$$13 \times (2)\ \text{gives:}\ \ 208I_1 + 416I_2 = 1456 \tag{5.11}$$

$$(4) - (3)\ \text{gives:}\qquad\qquad 592I_2 = 592$$
$$I_2 = 1\,\text{A}$$

Substituting for $I_2$ in (1) gives:

$$13I_2 - 11 = 54$$
$$I_1 = \frac{65}{13} = 5\,\text{A}$$

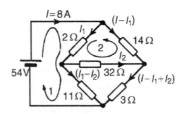

**Figure 5.10: Labeling directions**

the current flowing in the $2\,\Omega$ resistor $= I_1 = \mathbf{5\,A}$

the current flowing in the $14\,\Omega$ resistor $= I - I_1$

$$= 8 - 5 = \mathbf{3\,A}$$

the current flowing in the $32\,\Omega$ resistor $= I_2 = \mathbf{1\,A}$

the current flowing in the $11\,\Omega$ resistor $= I_1 - I_2 = 5 - 1$

$$= \mathbf{4\,A}\text{ and}$$

the current flowing in the $3\,\Omega$ resistor $= I - I_1 + I_2$

$$= 8 - 5 + 1$$

$$= \mathbf{4\,A}$$

## 5.3 The superposition Theorem

The *superposition theorem* states:

*"In any network made up of linear resistances and containing more than one source of e.m.f., the resultant current flowing in any branch is the algebraic sum of the currents that would flow in that branch if each source was considered separately, all other sources being replaced at that time by their respective internal resistances."*

### Example 5.5

Figure 5.11 shows a circuit containing two sources of e.m.f., each with their internal resistance. Determine the current in each branch of the network by using the superposition theorem.

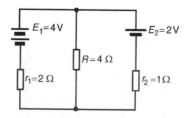

**Figure 5.11: Circuit for Example 5.5**

*Solution*

**Procedure:**

1. Redraw the original circuit with source $E_2$ removed, being replaced by $r_2$ only, as shown in Figure 5.12(a).

2. Label the currents in each branch and their directions as shown in Figure 5.12(a) and determine their values. (Note that the choice of current directions depends on the battery polarity, which, by convention is taken as flowing from the positive battery terminal as shown.) $R$ in parallel with $r_2$ gives an equivalent resistance of:

$$\frac{4 \times 1}{4 + 1} = 0.8 \,\Omega$$

From the equivalent circuit of Figure 5.12(b),

$$I_1 = \frac{E_1}{r_1 + 0.8} = \frac{4}{2 + 0.8} = 1.429 \text{ A}$$

From Figure 5.12(a),

$$I_2 = \left(\frac{1}{4 + 1}\right) I_1 = \frac{1}{5}(1.429) = 0.286 \text{ A}$$

And,

$$I_3 = \left(\frac{4}{4 + 1}\right) I_1 = \frac{4}{5}(1.429)$$

$$= 1.143 \text{A by current division}$$

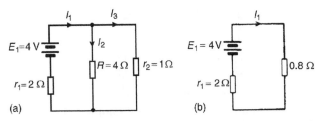

**Figure 5.12: (a) Redrawn circuit; (b) Equivalent circuit**

3. Redraw the original circuit with source $E_1$ removed, being replaced by $r_1$ only, as shown in Figure 5.13(a).

4. Label the currents in each branch and their directions as shown in Figure 5.13(a) and determine their values. $r_1$ in parallel with $R$ gives an equivalent resistance of:

$$\frac{2 \times 4}{2 + 4} = \frac{8}{6} = 1.333\,\Omega$$

From the equivalent circuit of Figure 5.13(b)

$$I_4 = \frac{E_2}{1.333 + r_2} = \frac{2}{1.333 + 1} = 0.857\,\text{A}$$

From Figure 5.13(a)

$$I_5 = \left(\frac{2}{2 + 4}\right)I_4 = \frac{2}{6}(0.857) = 0.286\,\text{A}$$

$$I_6 = \left(\frac{4}{2 + 4}\right)I_4 = \frac{4}{6}(0.857) = 0.571\,\text{A}$$

5. Superimpose Figure 5.13(a) on to Figure 5.12(a) as shown in Figure 5.14.

6. Determine the algebraic sum of the currents flowing in each branch.
   Resultant current flowing through source 1, i.e.,

   $$I_1 - I_6 = 1.429 - 0.571$$
   $$= \mathbf{0.858\,A\ (discharging)}$$

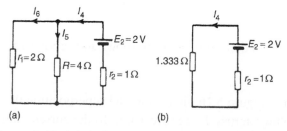

**Figure 5.13: (a) Redrawn circuit; (b) Equivalent circuit**

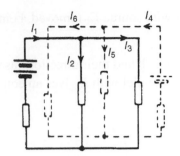

**Figure 5.14: Superimposed circuits**

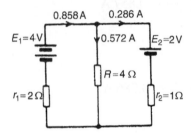

**Figure 5.15: Resultant currents and their directions**

Resultant current flowing through source 2, i.e.,

$$I_4 - I_3 = 0.857 - 1.143$$
$$= -0.286\,A \text{ (charging)}$$

Resultant current flowing through resistor $R$, i.e.,

$$I_2 + I_5 = 0.286 + 0.286$$
$$= 0.572\,A$$

The resultant currents with their directions are shown in Figure 5.15.

### Example 5.6

For the circuit shown in Figure 5.16, find, using the superposition theorem, (a) the current flowing in and the voltage across the $18\,\Omega$ resistor, (b) the current in the 8 V battery and (c) the current in the 3 V battery.

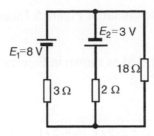

**Figure 5.16: Circuit for Example 5.6**

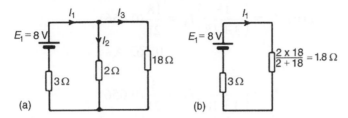

**Figure 5.17: (a) Redrawn circuit; (b) Equivalent circuit**

*Solution*

1. Removing source $E_2$ gives the circuit of Figure 5.17(a).

2. The current directions are labeled as shown in Figure 5.17(a), $I_1$ flowing from the positive terminal of $E_1$.

From Figure 5.17(b), $I_1 = \dfrac{E_1}{3+1.8} = \dfrac{8}{4.8} = 1.667$ A

From Figure 5.17(a), $I_2 = \left(\dfrac{18}{2+18}\right)I_1$

$$= \dfrac{18}{20}(1.667) = 1.500 \text{ A}$$

and $\qquad I_3 = \left(\dfrac{2}{2+18}\right)I_1$

$$= \dfrac{2}{20}(1.667) = 0.167 \text{ A}$$

3. Removing source $E_1$ gives the circuit of Figure 5.18(a), (which is the same as Figure 5.18(b)).

4. The current directions are labeled as shown in Figures 5.18(a) and 5.18(b), $I_4$ flowing from the positive terminal of $E_2$:

From Figure 5.18(c), $I_4 = \dfrac{E_2}{2 + 2.571} = \dfrac{3}{4.571}$

$$= 0.656 \text{ A}$$

From Figure 5.18(b), $I_5 = \left(\dfrac{18}{3 + 18}\right) I_4 = \dfrac{18}{21}(0.656)$

$$= 0.562 \text{ A}$$

$$I_6 = \left(\dfrac{3}{3 + 18}\right) I_4 = \dfrac{3}{21}(0.656)$$

$$= 0.094 \text{ A}$$

5. Superimposing Figure 5.18(a) on to Figure 5.17(a) gives the circuit in Figure 5.19.

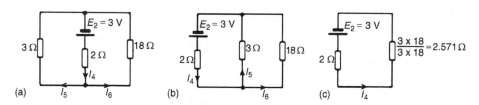

Figure 5.18: (a) Step 1; (b) Step 2; (c) Step 3

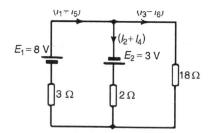

Figure 5.19: Result of superimposing

6.  (a) Resultant current in the $18\,\Omega$ resistor:

$$= I_3 - I_6$$
$$= 0.167 - 0.094$$
$$= \textbf{0.073A}$$

voltage across the $18\,\Omega$ resistor

$$= 0.073 \times 18 = \textbf{1.314V}$$

(b) Resultant current in the $8\,V$ battery:

$$= I_1 + I_5 = 1.667 + 0.562$$
$$= \textbf{2.229A (discharging)}$$

(c) Resultant current in the $3\,V$ battery:

$$= I_2 + I_4 = 1.500 + 0.656$$
$$= \textbf{2.156A (discharging)}$$

## 5.4 General DC Circuit Theory

The following points involving DC circuit analysis need to be appreciated before proceeding with problems using Thévenin's and Norton's theorems:

(i) The open-circuit voltage, $E$, across terminals AB in Figure 5.20 is equal to $10\,V$, since no current flows through the $2\,\Omega$ resistor; therefore, no voltage drop occurs.

(ii) The open-circuit voltage, $E$, across terminals AB in Figure 5.21(a) is the same as the voltage across the $6\,\Omega$ resistor. The circuit may be redrawn as shown in Figure 5.21(b).

$$E = \left( \frac{6}{6+4} \right)(50)$$

by voltage division in a series circuit, i.e., $E = \textbf{30V}$

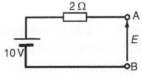

**Figure 5.20: Example circuit**

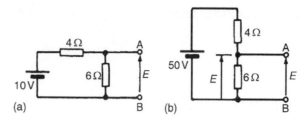

**Figure 5.21: (a) Example circuit; (b) Redrawn circuit**

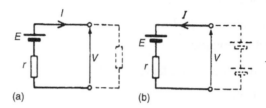

**Figure 5.22: (a) Example circuit; (b) Second example circuit**

(iii) For the circuit shown in Figure 5.22(a) representing a practical source supplying energy, $V = E - Ir$, where $E$ is the battery e.m.f., $V$ is the battery terminal voltage and $r$ is the internal resistance of the battery. For the circuit shown in Figure 5.22(b), $V = E - (-I)r$, i.e., $V = E + Ir$.

(iv) The resistance "looking-in" at terminals AB in Figure 5.23(a) is obtained by reducing the circuit in stages as shown in Figures 5.23(b) to (d). The equivalent resistance across AB is $7\,\Omega$.

(v) For the circuit shown in Figure 5.24(a), the $3\,\Omega$ resistor carries no current and the voltage across the $20\,\Omega$ resistor is $10\,V$. Redrawing the circuit gives Figure 5.24(b), from which,

$$E = \left(\frac{4}{4+6}\right) \times 10 = \mathbf{4\,V}$$

(vi) If the $10\,V$ battery in Figure 5.24(a) is removed and replaced by a short-circuit, as shown in Figure 5.24(c), then the $20\,\Omega$ resistor may be removed. The reason for this is that a short-circuit has zero resistance, and $20\,\Omega$ in parallel with zero ohms gives an equivalent resistance of: $(20 \times 0/20 + 0)$, i.e., $0\,\Omega$. The circuit

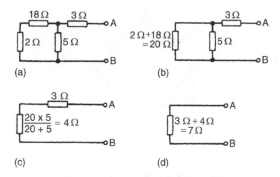

Figure 5.23: (a) Stage 1; (b) Stage 2; (c) Stage 3; (d) Stage 4, solution

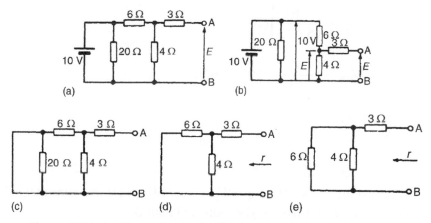

Figure 5.24: (a) Example circuit; (b) Step 1; (c) Step 2; (d) Step 3;
(e) Step 4, equivalent resistance

is then as shown in Figure 5.24(d), which is redrawn in Figure 5.24(e). From Figure 5.24(e), the equivalent resistance across AB,

$$r = \frac{6 \times 4}{6 \times 4} + 3 = 2.4 + 3 = \mathbf{5.4\,\Omega}$$

(vii) To find the voltage across AB in Figure 5.25: Since the 20 V supply is across the 5 Ω and 15 Ω resistors in series then, by voltage division, the voltage drop across AC,

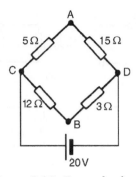

**Figure 5.25: Example circuit**

$$V_{AC} = \left(\frac{5}{5+15}\right)(20) = 5\,\text{V}$$

Similarly, $V_{CB} = \left(\frac{12}{12+3}\right)(20) = 16\,\text{V}.$

$V_C$ is at a potential of $+20\,\text{V}$.

$V_A = V_C - V_{AC} = +20 - 5 = 15\,\text{V}$ and

$V_B = V_C - V_{BC} = +20 - 16 = 4\,\text{V}.$

The voltage between AB is $V_A - V_B = 15 - 4 = 11\,\text{V}$ and current would flow from A to B since A has a higher potential than B.

(viii) In Figure 5.26(a), to find the equivalent resistance across AB the circuit may be redrawn as in Figures 5.26(b) and (c). From Figure 5.26(c), the equivalent resistance across AB,

$$= \frac{5 \times 15}{5 + 15} + \frac{12 \times 3}{12 + 3} = 3.75 + 2.4 = \mathbf{6.15\,\Omega}$$

(ix) In the worked problems in Sections 5.5 and 5.7 following, it may be considered that Thévenin's and Norton's theorems have no obvious advantages compared with, say, Kirchhoff's laws. However, these theorems can be used to analyze part of a circuit and in much more complicated networks the principle of

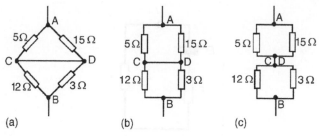

**Figure 5.26: (a) Example circuit; (b) Redrawn circuit; (c) Redrawn circuit**

replacing the supply by a constant voltage source in series with a resistance (or impedance) is very useful.

## 5.5 Thévenin's Theorem

*Thévenin's theorem* states:

*The current in any branch of a network is that which would result if an e.m.f. equal to the voltage across a break made in the branch, were introduced into the branch, all other e.m.f.'s being removed and represented by the internal resistances of the sources.*

The procedure adopted when using Thévenin's theorem is summarized below. To determine the current in any branch of an active network (i.e., one containing a source of e.m.f.):

(i)   remove the resistance $R$ from that branch,

(ii)  determine the open-circuit voltage, $E$, across the break,

(iii) remove each source of e.m.f. and replace them by their internal resistances and then determine the resistance, $r$, "looking-in" at the break,

(iv)  determine the value of the current from the equivalent

circuit shown in Figure 5.27, i.e., $I = \dfrac{E}{R + r}$

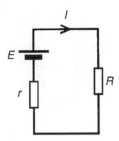

**Figure 5.27: Equivalent circuit**

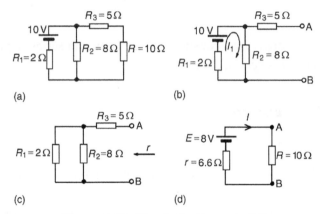

**Figure 5.28: Circuit for Example 5.7**

### Example 5.7

Use Thévenin's theorem to find the current flowing in the $10\,\Omega$ resistor for the circuit shown in Figure 5.28(a).

### Solution

Following the above procedure:

 (i) The $10\,\Omega$ resistance is removed from the circuit as shown in Figure 5.28(b)

(ii) There is no current flowing in the $5\,\Omega$ resistor and current $I_1$ is given by:

$$I_1 = \frac{10}{R_1 + R_2} = \frac{10}{2 + 8} = 1\,\text{A}$$

Voltage across $R_2 = I_1 R_2 = 1 \times 8 = 8\,\text{V}$

Voltage across AB, i.e., the open-circuit voltage across the break, $E = 8\,\text{V}$

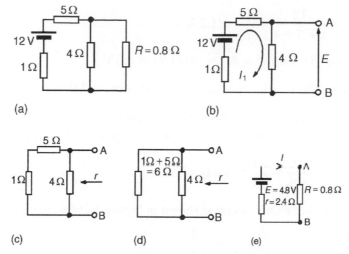

**Figure 5.29: Network for Example 5.8**

(iii) Removing the source of e.m.f. gives the circuit of Figure 5.28(c).

$$\text{Resistance, } r = R_3 + \frac{R_1 R_2}{R_1 + R_2} = 5 + \frac{2 \times 8}{2 + 8}$$
$$= 5 + 1.6 = 6.6\,\Omega$$

(iv) The equivalent Thévenin's circuit is shown in Figure 5.28(d).

$$\text{Current } I = \frac{E}{R + r} = \frac{8}{10 + 6.6} = \frac{8}{16.6} = 0.482\,\text{A}$$

The current flowing in the $10\,\Omega$ resistor of Figure 5.28(a) is **0.482 A**

### Example 5.8

For the network shown in Figure 5.29(a) determine the current in the $0.8\,\Omega$ resistor using Thévenin's theorem.

### Solution

Following the procedure:

(i) The $0.8\,\Omega$ resistor is removed from the circuit as shown in Figure 5.29(b).

(ii) Current $I_1 = \dfrac{12}{1+5+4} = \dfrac{12}{10} = 1.2\,\text{A}$

Voltage across $4\,\Omega$ resistor $= 4I_1 = (4)\,(1.2) = 4.8\,\text{V}$.

Voltage across AB, i.e., the open-circuit voltage across AB, $E = 4.8\,\text{V}$.

(iii) Removing the source of e.m.f. gives the circuit shown in Figure 5.29(c). The equivalent circuit of Figure 5.29(c) is shown in Figure 5.29(d), from which,

resistance $r = \dfrac{4\times6}{4+6} = \dfrac{24}{10} = 2.4\,\Omega$

(iv) The equivalent Thévenin's circuit is shown in Figure 5.29(e), from which,

current $I = \dfrac{E}{r+R} = \dfrac{4.8}{2.4+0.8} = \dfrac{4.8}{3.2}$

$I = 1.5\text{A} = $ **current in the $0.8\,\Omega$ resistor**

### Example 5.9
Use Thévenin's theorem to determine the current $I$ flowing in the $4\,\Omega$ resistor shown in Figure 5.30(a). Find also the power dissipated in the $4\,\Omega$ resistor.

### Solution
Following the procedure:

(i) The $4\,\Omega$ resistor is removed from the circuit as shown in Figure 5.30(b).

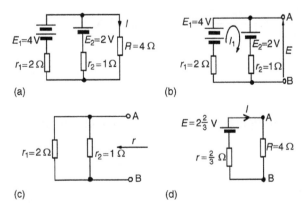

**Figure 5.30: Circuit for Example 5.9 showing steps for solution**

(ii) Current $I_1 = \dfrac{E_1 - E_2}{r_1 + r_2} = \dfrac{4 - 2}{2 + 1} = \dfrac{2}{3}$ A

Voltage across AB, $E = E_1 - I_1 r_1 = 4 - \left(\dfrac{2}{3}\right)(2) = 2\dfrac{2}{3}$ V

Alternatively, voltage across AB,

$E = E_2 - I_1 r_2$

$\quad = 2 - \left(\dfrac{2}{3}\right)(1) = 2\dfrac{2}{3}$ V

(iii) Removing the sources of e.m.f. gives the circuit shown in Figure 5.30(c), from which resistance,

$r = \dfrac{2 \times 1}{2 + 1} = \dfrac{2}{3}$ Ω

(iv) The equivalent Thévenin's circuit is shown in Figure 5.30(d), from which,

current, $I = \dfrac{E}{r + R} = \dfrac{2\frac{2}{3}}{\frac{2}{3} + 4} = \dfrac{8/3}{14/3}$

$\qquad = \dfrac{8}{14} = \mathbf{0.571\ A}$

$\qquad = $ **current in the 4 Ω resistor**

Power dissipated in 4 Ω resistor,

$P = I^2 R = (0.571)^2\,(4) = \mathbf{1.304\,W}$

### Example 5.10

Use Thévenin's theorem to determine the current flowing in the 3 Ω resistance of the network shown in Figure 5.31(a). The voltage source has negligible internal resistance.

### Solution

(Note the symbol for an ideal voltage source in Figure 5.31(a), which may be used as an alternative to the battery symbol.)

Following the procedure:

(i) The $3\,\Omega$ resistance is removed from the circuit as shown in Figure 5.31(b).

(ii) The $1\frac{2}{3}\,\Omega$ resistance now carries no current.

Voltage across $10\,\Omega$ resistor $= \left(\dfrac{10}{10+5}\right)(24)$

$= 16\ \text{V}$

Voltage across AB, $E = 16$ V.

(iii) Removing the source of e.m.f. and replacing it by its internal resistance means that the $20\,\Omega$ resistance is short-circuited as shown in Figure 5.31(c) since its internal resistance is zero. The $20\,\Omega$ resistance may thus be removed as shown in Figure 5.31(d).

From Figure 5.31(d), resistance,

$r = 1\dfrac{2}{3} + \dfrac{10 \times 5}{10+5}$

$= 1\dfrac{2}{3} + \dfrac{50}{15} = 5\,\Omega$

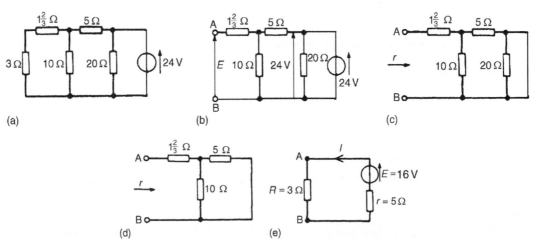

**Figure 5.31: Circuit for Example 5.10 showing steps for solution**

(iv)  The equivalent Thévenin's circuit is shown in Figure 5.31(e), from which

$$\text{current, } I = \frac{E}{r + R} = \frac{16}{3 + 5} = \frac{16}{8} = \textbf{2A}$$

$$= \textbf{current in the 3}\,\boldsymbol{\Omega}\textbf{ resistance.}$$

### Example 5.11

A Wheatstone Bridge network is shown in Figure 5.32(a). Calculate the current flowing in the 32 Ω resistor, and its direction, using Thévenin's theorem. Assume the source of e.m.f. to have negligible resistance.

### Solution

Following the procedure:

  (i)  The 32 Ω resistor is removed from the circuit as shown in Figure 5.32(b).

 (ii)  The voltage between A and C,

$$V_{AC} = \left(\frac{R_1}{R_1 + R_4}\right)(E) = \left(\frac{2}{2 + 11}\right)(54) = 8.31\,\text{V}$$

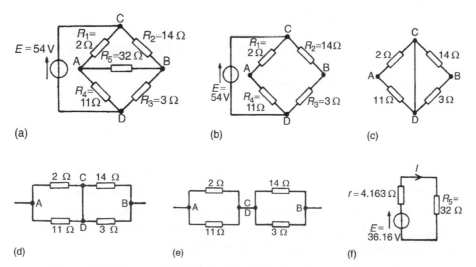

(a)          (b)          (c)

(d)          (e)          (f)

**Figure 5.32: Network for Example 5.11 showing steps for solution**

The voltage between B and C,

$$V_{BC} = \left(\frac{R_2}{R_2 + R_3}\right)(E) = \left(\frac{14}{14 + 3}\right)(54) = 44.47 \text{ V}$$

The voltage between A and B = 44.47 − 8.31 = **36.16 V**

Point C is at a potential of +54 V. Between C and A is a voltage drop of 8.31 V. The voltage at point A is 54 − 8.31 = 45.69 V. Between C and B is a voltage drop of 44.47 V. The voltage at point B is 54 − 44.47 = 9.53 V. Since the voltage at A is greater than at B, current must flow in the direction A to B.

(iii)  Replacing the source of e.m.f. with a short-circuit (i.e., zero internal resistance) gives the circuit shown in Figure 5.32(c). The circuit is redrawn and simplified as shown in Figure 5.32(d) and (e), from which the resistance between terminals A and B,

$$r = \frac{2 \times 11}{2 + 11} + \frac{14 \times 3}{14 + 3} = \frac{22}{13} + \frac{42}{17}$$
$$= 1.692 + 2.471 = \mathbf{4.163\,\Omega}$$

(iv)  The equivalent Thévenin's circuit is shown in Figure 5.32(f), from which,

$$\text{current, } I = \frac{E}{r + R_5} = \frac{36.16}{4.163 + 32} = 1 \text{ A}$$

The current in the 32 Ω resistor of Figure 5.32(a) is 1 A, flowing from A to B.

## 5.6  Constant-Current Source

A source of electrical energy can be represented by a source of e.m.f. in series with a resistance. In Section 5.5, the Thévenin constant-voltage source consisted of a constant e.m.f. $E$ in series with an internal resistance $r$. However, this is not the only form of representation. A source of electrical energy can also be represented by a constant-current source in parallel with a resistance. It may be shown that the two forms are equivalent. An *ideal constant-voltage generator* is one with zero internal resistance so that it supplies the same voltage to all loads. An *ideal constant-current generator* is one with infinite internal resistance so that it supplies the same current to all loads.

Note the symbol for an ideal current source (BS 3939, 1985), shown in Figure 5.33.

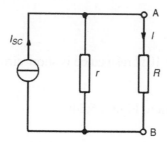

**Figure 5.33: Symbol for ideal current source**

## 5.7 Norton's Theorem

*Norton's theorem* states:

*The current that flows in any branch of a network is the same as that which would flow in the branch if it were connected across a source of electrical energy, the short-circuit current of which is equal to the current that would flow in a short-circuit across the branch, and the internal resistance of which is equal to the resistance which appears across the open-circuited branch terminals.*

The procedure adopted when using Norton's theorem is summarized below.

To determine the current flowing in a resistance $R$ of a branch AB of an active network:

 (i)   short-circuit branch AB,

 (ii)  determine the short-circuit current $I_{SC}$ flowing in the branch,

 (iii) remove all sources of e.m.f. and replace them by their internal resistance (or, if a current source exists, replace with an open-circuit), then determine the resistance $r$, "looking-in" at a break made between A and B,

 (iv)  determine the current $I$ flowing in resistance $R$ from the Norton equivalent network shown in Figure 5.33, i.e.,

$$I = \left(\frac{r}{r+R}\right)I_{SC}$$

### Example 5.12
Use Norton's theorem to determine the current flowing in the $10\,\Omega$ resistance for the circuit shown in Figure 5.34(a).

**Solution**

Following the above procedure:

(i) The branch containing the $10\,\Omega$ resistance is short-circuited as shown in Figure 5.34(b).

(ii) Figure 5.34(c) is equivalent to Figure 5.34(b).

$$I_{SC} = \frac{10}{2} = 5\text{A}$$

(iii) If the $10\,\text{V}$ source of e.m.f. is removed from Figure 5.34(b), the resistance "looking-in" at a break made between A and B is given by:

$$r = \frac{2 \times 8}{2 + 8} = 1.6\,\Omega$$

(iv) From the Norton equivalent network shown in Figure 5.34(d), the current in the $10\,\Omega$ resistance, by current division, is given by:

$$I = \left(\frac{1.6}{1.6 + 5 + 10}\right)(5) = \textbf{0.482 A}$$

as obtained previously in Example 5.7 using Thévenin's theorem.

**Example 5.13**

Use Norton's theorem to determine the current $I$ flowing in the $4\,\Omega$ resistance shown in Figure 5.35(a).

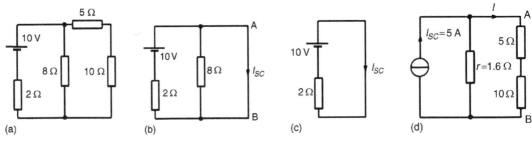

**Figure 5.34: Circuit for Example 5.12 showing steps**

**Solution**

Following the procedure:

(i)   The $4\,\Omega$ branch is short-circuited, as shown in Figure 5.35(b).

(ii)  From Figure 13.45(b), $I_{SC} = I_1 + I_2 = \dfrac{4}{2} + \dfrac{2}{1} = 4\,\text{A}$

(iii) If the sources of e.m.f. are removed the resistance "looking-in" at a break made between A and B is given by:

$$r = \frac{2 \times 1}{2 + 1} = \frac{2}{3}\,\Omega$$

(iv)  From the Norton equivalent network shown in Figure 5.35(c)the current in the $4\,\Omega$ resistance is given by:

$$I = \left(\frac{2/3}{(2/3) + 4}\right)(4) = \mathbf{0.571\,A}$$

as obtained previously in problems 2, 5 and 9 using Kirchhoff's laws and the theorems of superposition and Thévenin.

**Example 5.14**

Use Norton's theorem to determine the current flowing in the $3\,\Omega$ resistance of the network shown in Figure 5.36(a). The voltage source has negligible internal resistance.

**Solution**

Following the procedure:

(i)   The branch containing the $3\,\Omega$ resistance is short-circuited, as shown in Figure 5.36(b).

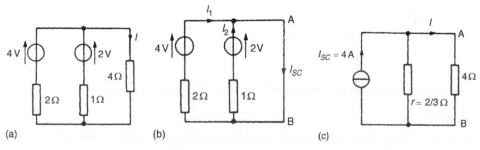

(a)                    (b)                    (c)

**Figure 5.35: Circuits for Example 5.13**

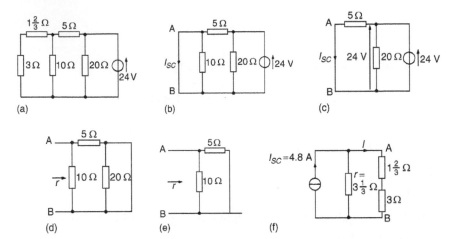

**Figure 5.36: Circuits for Example 5.14**

(ii) From the equivalent circuit shown in Figure 5.36(c),

$$I_{SC} = \frac{24}{5} = 4.8 \, \text{A}$$

(iii) If the 24 V source of e.m.f. is removed the resistance "looking-in" at a break made between A and B is obtained from Figure 5.36(d) and its equivalent circuit shown in Figure 5.36(e) and is given by:

$$r = \frac{10 \times 5}{10 + 5} = \frac{50}{15} = 3\frac{1}{3} \, \Omega$$

(iv) From the Norton equivalent network shown in Figure 5.36(f) the current in the 3 Ω resistance is given by:

$$I = \left( \frac{3\frac{1}{3}}{3\frac{1}{3} + 1\frac{2}{3} + 3} \right)(4.8) = \mathbf{2\,A},$$

as obtained previously in Example 5.10 using Thévenin's theorem.

**Example 5.15**
Determine the current flowing in the 2 Ω resistance in the network shown in Figure 5.37(a).

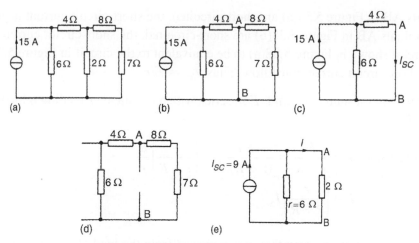

**Figure 5.37: Circuits for Example 5.15**

*Solution*

Following the procedure:

(i) The $2\,\Omega$ resistance branch is short-circuited as shown in Figure 5.37(b).

(ii) Figure 5.37(c) is equivalent to Figure 5.37(b).

(iii) If the 15 A current source is replaced by an open circuit then from Figure 5.37(d) the resistance "looking-in" at a break made between A and B is given by $(6 + 4)\,\Omega$ in parallel with $(8 + 7)\,\Omega$, i.e.,

$$r = \frac{(10)(15)}{10 + 15} = \frac{150}{25} = 6\,\Omega$$

(iv) From the Norton equivalent network shown in Figure 5.37(e) the current in the $2\,\Omega$ resistance is given by:

$$I = \left(\frac{6}{6 + 2}\right)(9) = \mathbf{6.75\,A}$$

## 5.8 Thévenin and Norton Equivalent Networks

The Thévenin and Norton networks shown in Figure 5.38 are equivalent to each other. The resistance "looking-in" at terminals AB is the same in each of the networks, i.e., $r$.

If terminals AB in Figure 5.38(a) are short-circuited, the short-circuit current is given by $E/r$. If terminals AB in Figure 5.38(b) are short-circuited, the short-circuit current is $I_{SC}$. For the circuit shown in Figure 5.38(a) to be equivalent to the circuit in Figure 5.38(b) the same short-circuit current must flow. Thus, $I_{SC} = E/r$.

Figure 5.39 shows a source of e.m.f. $E$ in series with a resistance $r$ feeding a load resistance $R$.

From Figure 13.50, $I = \dfrac{E}{r+R} = \dfrac{E/r}{(r+R)/r} = \left(\dfrac{r}{r+R}\right)\dfrac{E}{r}$

i.e.,  $I = \left(\dfrac{r}{r+R}\right)I_{SC}$

From Figure 5.40, it can be seen that, when viewed from the load, the source appears as a source of current $I_{SC}$, which is divided between $r$ and $R$ connected in parallel.

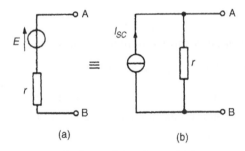

(a)                    (b)

**Figure 5.38: Equivalent Thévenin and Norton networks**

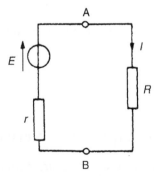

**Figure 5.39: Source $E$ in series with resistance $r$ feeding load resistance $R$**

Thus the two representations shown in Figure 5.38 are equivalent.

### Example 5.16
Convert the circuit shown in Figure 5.41 to an equivalent Norton network.

### Solution
If terminals AB in Figure 5.41 are short-circuited, the short-circuit current

$$I_{SC} = \frac{10}{2} = 5\,A$$

The resistance looking-in at terminals AB is $2\,\Omega$. The equivalent Norton network is shown in Figure 5.42.

### Example 5.17
Convert the network shown in Figure 5.43 to an equivalent Thévenin circuit.

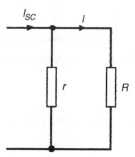

**Figure 5.40: Source when viewed from load**

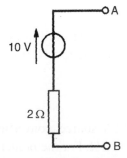

**Figure 5.41: Circuit for Example 5.16**

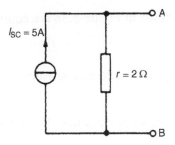

**Figure 5.42: Equivalent Norton network**

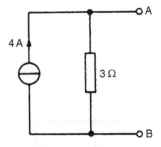

**Figure 5.43: Network for Example 5.17**

*Solution*

The open-circuit voltage $E$ across terminals AB in Figure 5.43 is given by:

$E = (I_{SC})\,(r) = (4)\,(3) = 12\,\text{V}.$

The resistance looking-in at terminals AB is $3\,\Omega$. The equivalent Thévenin circuit is as shown in Figure 5.44.

*Example 5.18*

(a) Convert the circuit to the left of terminals AB in Figure 5.45(a) to an equivalent Thévenin circuit by initially converting to a Norton equivalent circuit. (b) Determine the current flowing in the $1.8\,\Omega$ resistor.

*Solution*

(a) For the branch containing the 12 V source, converting to a Norton equivalent circuit gives $I_{SC} = 12/3 = 4\,\text{A}$ and $r_1 = 3\,\Omega$. For the branch containing the 24 V source, converting to a Norton equivalent circuit gives $I_{SC2} = 24/2 = 12\,\text{A}$ and $r_2 = 2\,\Omega$.

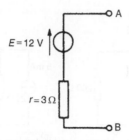

Figure 5.44: Equivalent Thévenin circuit

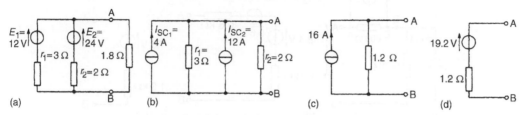

Figure 5.45: Circuits for Example 5.18

Thus Figure 5.45(b) shows a network equivalent to Figure 5.45(a).

From Figure 5.45(b) the total short-circuit current is $4 + 12 = 16\,\text{A}$, and the total resistance is given by: $\dfrac{3 \times 2}{3 + 2} = 1.2\,\Omega$

Thus Figure 5.45(b) simplifies to Figure 5.45(c). The open-circuit voltage across AB of Figure 5.45(c), $E = (16)(1.2) = 19.2\,\text{V}$, and the resistance "looking-in" at AB is $1.2\,\Omega$. The Thévenin equivalent circuit is as shown in Figure 5.45(d).

(b) When the $1.8\,\Omega$ resistance is connected between terminals A and B of Figure 5.45(d) the current $I$ flowing is given by:

$$I = \frac{19.2}{1.2 + 1.8} = \mathbf{6.4\,A}$$

### Example 5.19

Determine by successive conversions between Thévenin and Norton equivalent networks a Thévenin equivalent circuit for terminals AB of Figure 5.46(a). Determine the current flowing in the $200\,\Omega$ resistance.

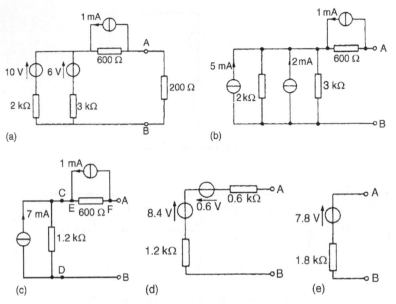

**Figure 5.46: Circuits for Example 5.19**

**Solution**

For the branch containing the 10 V source, converting to a Norton equivalent network gives:

$$I_{SC} = \frac{10}{2000} = 5 \text{ mA and } r_1 = 2 \text{ k}\Omega.$$

For the branch containing the 6 V source, converting to a Norton equivalent network gives:

$$I_{SC} = \frac{6}{3000} = 2 \text{ mA and } r_2 = 3 \text{ k}\Omega.$$

Thus, the network of Figure 5.46(a) converts to Figure 5.46(b).

Combining the 5 mA and 2 mA current sources gives the equivalent network of Figure 5.46(c) where the short-circuit current for the original two branches considered is 7 mA and the resistance is:

$$\frac{2 \times 3}{2 + 3} = 1.2 \text{ k}\Omega.$$

Both of the Norton equivalent networks shown in Figure 5.46(c) may be converted to Thévenin equivalent circuits. The open-circuit voltage across CD is: $(7 \times 10^{-3})$ $(1.2 \times 10^3) = 8.4\,\text{V}$ and the resistance looking-in at CD is 1.2 kΩ.

The open-circuit voltage across *EF* is $(1 \times 10^{-3})\,(600) = 0.6\,\text{V}$ and the resistance "looking-in" at *EF* is 0.6 kΩ. Thus, Figure 5.46(c) converts to Figure 5.46(d). Combining the two Thévenin circuits gives:

$E = 8.4 - 0.6 = \textbf{7.8 V}$, and the resistance,

$r = (1.2 + 0.6)\,\text{k}\Omega = \textbf{1.8 k}\boldsymbol{\Omega}.$

Thus, the Thévenin equivalent circuit for terminals AB of Figure 5.46(a) is as shown in Figure 5.46(e).

Therefore, the current *I* flowing in a 200 Ω resistance connected between A and B is given by:

$$I = \frac{7.8}{1800 + 200} = \frac{7.8}{2000} = \textbf{3.9 mA}$$

## 5.9 Maximum Power Transfer Theorem

The *maximum power transfer theorem* states:

*The power transferred from a supply source to a load is at its maximum when the resistance of the load is equal to the internal resistance of the source.*

In Figure 5.47, when $R = r$ the power transferred from the source to the load is a maximum.

Typical practical applications of the maximum power transfer theorem are found in stereo amplifier design, seeking to maximize power delivered to speakers, and in electric vehicle design, seeking to maximize power delivered to drive a motor.

### Example 5.20

The circuit diagram of Figure 5.48 shows dry cells of source e.m.f. 6 V, and internal resistance 2.5 Ω. If the load resistance *RL* is varied from 0 to 5 Ω in 0.5 Ω steps, calculate the power dissipated by the load in each case. Plot a graph of *RL* (horizontally) against power (vertically) and determine the maximum power dissipated.

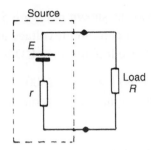

**Figure 5.47: When *r* = *R*, power transfer is maximum**

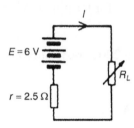

**Figure 5.48: Circuit for Example 5.20**

*Solution*

When $R_L = 0$, current $I = \dfrac{E}{r + R_L} = \dfrac{6}{2.5} = 2.4$ A, and

power dissipated in $R_L$, $P = I^2RL$,

i.e., $$P = (2.4)^2\,(0) = 0\,\text{W}$$

When $RL = 0.5\,\Omega$, current $I = \dfrac{E}{r + R_L} = \dfrac{6}{2.5 + 0.5} = 2\,\text{A}$

and $P = I^2R_L = (2)^2\,(0.5) = 2\,\text{W}$

When $RL = 1.0\,\Omega$, current $I = \dfrac{6}{2.5 + 1.0} = 1.714\,\text{A}$

and $P = (1.714)^2\,(1.0) = 2.94\,\text{W}$

With similar calculations the following table is produced:

| $R_L(\Omega)$ | $I = \dfrac{E}{r + R_L}$ | $P = I^2 R_L (W)$ |
|---|---|---|
| 0 | 2.4 | 0 |
| 0.5 | 2.0 | 2.00 |
| 1.0 | 1.714 | 2.94 |
| 1.5 | 1.5 | 3.38 |
| 2.0 | 1.333 | 3.56 |
| 2.5 | 1.2 | 3.60 |
| 3.0 | 1.091 | 3.57 |
| 3.5 | 1.0 | 3.50 |
| 4.0 | 0.923 | 3.41 |
| 4.5 | 0.857 | 3.31 |
| 5.0 | 0.8 | 3.20 |

A graph of $R_L$ against $P$ is shown in Figure 5.49. *The maximum value of power is 3.60 W*, which occurs when $R_L$ is 2.5 Ω, i.e., *maximum power occurs when $R_L = r$*, which is what the maximum power transfer theorem states.

### Example 5.21

A DC source has an open-circuit voltage of 30 V and an internal resistance of 1.5 Ω. State the value of load resistance that gives maximum power dissipation and determine the value of this power.

### Solution

The circuit diagram is shown in Figure 5.50. From the maximum power transfer theorem, for maximum power dissipation,

$$R_L = r = \textbf{1.5 } \boldsymbol{\Omega}$$

From Figure 5.50, current $I = \dfrac{E}{r + R_L} = \dfrac{30}{1.5 + 1.5} = 10 \text{ A}$

Power $P = I^2 R_L = (10)^2(1.5) = \textbf{150 W} = \textbf{maximum power dissipated}$

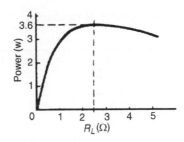

Figure 5.49: Graph of $R_L$ vs. *P*

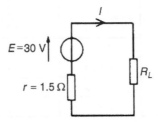

Figure 5.50: Circuit diagram for Example 5.21

### Example 5.22

Find the value of the load resistor $R_L$ shown in Figure 5.51(a) that gives maximum power dissipation and determine the value of this power.

### Solution

Using the procedure for Thévenin's theorem:

 (i) Resistance $R_L$ is removed from the circuit as shown in Figure 5.51(b).

 (ii) The voltage across AB is the same as the voltage across the 1 $\Omega$ resistor:

Hence, $E = \left(\dfrac{12}{12+3}\right)(15) = 12\,\text{V}$

(iii) Removing the source of e.m.f. gives the circuit of Figure 5.51(c),

from which resistance, $r = \dfrac{12 \times 3}{12+3} = \dfrac{36}{15} = 2.4\,\Omega$

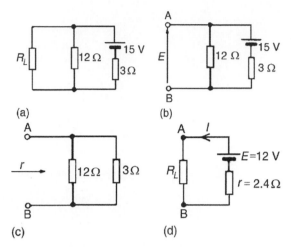

**Figure 5.51: Circuits for Example 5.22**

(iv) The equivalent Thévenin's circuit supplying terminals AB is shown in Figure 5.51(d), from which current $I = E/(r + R_L)$.

For maximum power, $R_L = r = \mathbf{2.4\,\Omega}$.

Thus, current, $I = \dfrac{12}{2.4 + 2.4} = 2.5\,\text{A}$.

Power, $P$, dissipated in load $R_L$,

$$P = I^2 R_L = (2.5)^2\,(2.4) = \mathbf{15\,W}$$

# CHAPTER 6

# Alternating Voltages and Currents

John Bird

## 6.1 The AC Generator

Let a single turn coil be free to rotate at constant angular velocity symmetrically between the poles of a magnet system as shown in Figure 6.1.

An e.m.f. is generated in the coil (from Faraday's laws) which varies in magnitude and reverses its direction at regular intervals. The reason for this is shown in Figure 6.2. In positions (a), (e) and (i) the conductors of the loop are effectively moving along the magnetic field, no flux is cut and hence, no e.m.f is induced. In position (c) maximum flux is cut and maximum e.m.f is induced. In position (g), maximum flux is cut and maximum e.m.f is again induced. However, using Fleming's right-hand rule, the induced e.m.f is in the opposite direction to that in position (c) and is shown as $-E$. In positions (b), (d), (f) and (h) some flux is cut and some e.m.f is induced. If all such positions

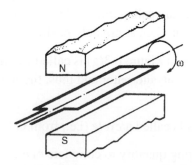

**Figure 6.1: Coil rotates at constant angular velocity**

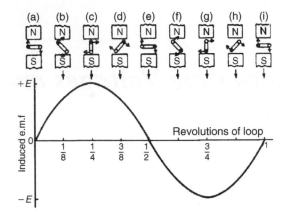

**Figure 6.2: One cycle of alternating e.m.f produced**

of the coil are considered, in one revolution of the coil, one cycle of alternating e.m.f is produced as shown. This is the principle of operation of the AC generator (i.e., the alternator).

## 6.2 Waveforms

If values of quantities that vary with time $t$ are plotted to a base of time, the resulting graph is called a *waveform*. Some typical waveforms are shown in Figure 6.3. Waveforms (a) and (b) are *unidirectional waveforms*, for, although they vary considerably with time, they flow in one direction only (i.e., they do not cross the time axis and become negative). Waveforms (c) to (g) are called *alternating waveforms* since their quantities are continually changing in direction (i.e., alternately positive and negative).

A waveform of the type shown in Figure 6.3(g) is called a *sine wave*. It is the shape of the waveform of e.m.f produced by an alternator and thus, the mains electricity supply is of "sinusoidal" form.

One complete series of values is called a *cycle* (i.e., from O to P in Figure 6.3(g)).

The time taken for an alternating quantity to complete one cycle is called the *period* or the *periodic time, T*, of the waveform.

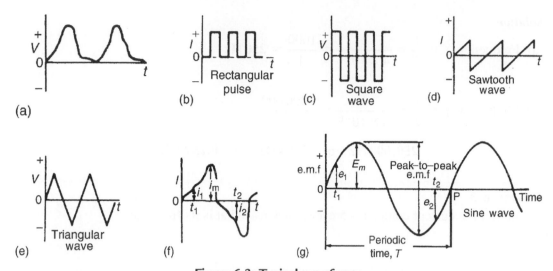

**Figure 6.3: Typical waveforms**

The number of cycles completed in one second is called the *frequency*, $f$, of the supply and is measured in *hertz*, Hz. (The standard frequency of the electricity supply in the U.S. is 60 Hz and in Great Britain is 50 Hz.)

$$T = \frac{1}{f} \quad \text{or} \quad f = \frac{1}{T}$$

*Example 6.1*

Determine the periodic time for frequencies of (a) 50 Hz and (b) 20 kHz.

*Solution*

(a)  Periodic time $T = \dfrac{1}{f} = \dfrac{1}{50} =$ **0.02 s or 20 msv**

(b)  Periodic time $T = \dfrac{1}{f} = \dfrac{1}{20\,000} =$ **0.000 05 s or 50 μs**

*Example 6.2*

Determine the frequencies for periodic times of (a) 4 ms, (b) 4 μs.

*Solution*

(a) Frequency $f = \dfrac{1}{T} = \dfrac{1}{4 \times 10^{-3}} = \dfrac{1000}{4} = \mathbf{250\,Hz}$

(b) Frequency $f = \dfrac{1}{T} = \dfrac{1}{4 \times 10^{-6}} = \dfrac{1\,000\,000}{4}$

$$= \mathbf{250,000\,Hz} \quad \text{or} \quad \mathbf{250\,kHz} \quad \text{or} \quad \mathbf{0.25\,MHz}$$

*Example 6.3*

An alternating current completes 5 cycles in 8 ms. What is its frequency?

*Solution*

Time for 1 cycle $= \dfrac{8}{5}$ ms $= 1.6$ ms $=$ periodic time $T$

Frequency $f = \dfrac{1}{T} = \dfrac{1}{1.6 \times 10^{-3}} = \dfrac{1000}{1.6} = \dfrac{10\,000}{16}$

$$= \mathbf{625\,Hz}$$

## 6.3  AC Values

*Instantaneous values* are the values of the alternating quantities at any instant of time. They are represented by small letters, $i$, $v$, $e$, etc. (See Figures 6.3(f) and (g).)

The largest value reached in a half cycle is called the *peak value* or the *maximum value* or the *amplitude* of the waveform. Such values are represented by $V_m$, $I_m$ etc. (See Figures 6.3(f) and (g).) A *peak-to-peak* value of e.m.f is shown in Figure 6.3(g) and is the difference between the maximum and minimum values in a cycle.

The *average* or *mean value* of a symmetrical alternating quantity (such as a sine wave), is the average value measured over a half cycle (since over a complete cycle the average value is zero).

$$\text{Average or mean value} = \frac{\text{area under the curve}}{\text{length of base}}$$

The area under the curve is found by approximate methods such as the trapezoidal rule, the mid-ordinate rule or Simpson's rule. Average values are represented by $V_{AV}$, $I_{AV}$, etc.

For a sine wave,
average value = 0.637 × maximum value
(i.e., $2/\pi$ × maximum value)

The *effective value* of an alternating current is that current which will produce the same heating effect as an equivalent direct current. The effective value is called the *root mean square (rms) value* and whenever an alternating quantity is given, it is assumed to be the rms value. The symbols used for rms values are $I$, $V$, $E$, etc. For a nonsinusoidal waveform as shown in Figure 6.4 the rms value is given by:

$$I = \sqrt{\left(\frac{i_1^2 + i_2^2 + \cdots + i_n^2}{n}\right)}$$

where $n$ is the number of intervals used.

For a sine wave,
rms value = 0.707 × maximum value
(i.e., $1/\sqrt{2}$ × maximum value)

$$\text{Form factor} = \frac{\text{rms value}}{\text{average value}} \qquad \text{For a sine wave, form factor} = 1.11$$

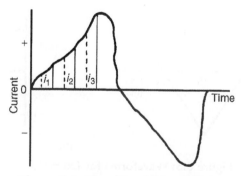

**Figure 6.4: Nonsinusoidal waveform**

$$\text{Peak factor} = \frac{\text{maximum value}}{\text{rms value}} \qquad \text{For a sine wave, peak factor} = 1.41$$

The values of form and peak factors give an indication of the shape of waveforms.

**Example 6.4**

For the periodic waveforms shown in Figure 6.5 determine for each: (i) frequency, (ii) average value over half a cycle, (iii) rms value, (iv) form factor, and (v) peak factor.

**Solution**

**(a) Triangular waveform (Figure 6.5(a))**

(i) Time for 1 complete cycle $= 20\,\text{ms} = $ periodic time, $T$.

Hence, frequency $f = \dfrac{1}{T} = \dfrac{1}{20 \times 10^{-3}} = \dfrac{1000}{20}$

$$= 50\,\text{Hz}$$

(ii) Area under the triangular waveform for a half cycle

$$= \tfrac{1}{2} \times \text{base} \times \text{height} = \tfrac{1}{2} \times (10 \times 10^{-3}) \times 200$$

$$= 1 \text{ volt second}$$

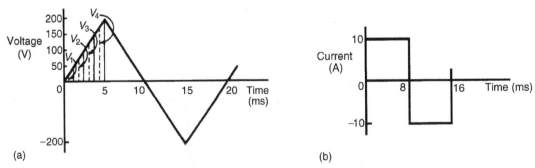

**Figure 6.5: Waveforms for Example 6.4**

Average value of waveform

$$= \frac{\text{area under curve}}{\text{length of base}} = \frac{1 \text{ volt second}}{10 \times 10^{-3} \text{ second}}$$

$$= \frac{1000}{10} = \mathbf{100\ V}$$

(iii) In Figure 6.5(a), the first 1/4 cycle is divided into 4 intervals.

$$\text{Thus, rms value} = \sqrt{\left(\frac{v_1^2 + v_2^2 + v_3^2 + v_4^2}{4}\right)}$$

$$= \sqrt{\left(\frac{25^2 + 75^2 + 125^2 + 175^2}{4}\right)}$$

$$= \mathbf{114.6\ V}$$

(Note that the greater the number of intervals chosen, the greater the accuracy of the result. For example, if twice the number of ordinates as that chosen above are used, the rms value is found to be 115.6 V)

(iv) Form factor $= \dfrac{\text{rms value}}{\text{average value}} = \dfrac{114.6}{100} = \mathbf{1.15}$

(v) Peak factor $= \dfrac{\text{maximum value}}{\text{rms value}} = \dfrac{200}{114.6} = \mathbf{1.75}$

## (b) Rectangular waveform (Figure 6.5(b))

(i) Time for 1 complete cycle $= 16\,\text{ms} = $ periodic time, $T$

$$\text{Hence, frequency}, f = \frac{1}{T} = \frac{1}{16 \times 10^{-3}} = \frac{1000}{16}$$

$$= \mathbf{62.5\ Hz}$$

(ii) Average value over half a cycle $= \dfrac{\text{area under curve}}{\text{length of base}}$

$$= \dfrac{10 \times (8 \times 10^{-3})}{8 \times 10^{-3}}$$

$$= \mathbf{10\,A}$$

(iii) The rms value $= \sqrt{\left( \dfrac{i_1^2 + i_2^2 + \cdots + i_n^2}{n} \right)} = \mathbf{10\,A}$

However, many intervals are chosen, since the waveform is rectangular.

(iv) Form factor $= \dfrac{\text{rms value}}{\text{average value}} = \dfrac{10}{10} = \mathbf{1}$

(v) Peak factor $= \dfrac{\text{maximum value}}{\text{rms value}} = \dfrac{10}{10} = \mathbf{1}$

### Example 6.5

The following table gives the corresponding values of current and time for a half cycle of alternating current.

| time $t$ (ms) | 0 | 0.5 | 1.0 | 1.5 | 2.0 | 2.5 | 3.0 | 3.5 | 4.0 | 4.5 | 5.0 |
|---|---|---|---|---|---|---|---|---|---|---|---|
| current $i$ (A) | 0 | 7 | 14 | 23 | 40 | 56 | 68 | 76 | 60 | 5 | 0 |

Assuming the negative half cycle is identical in shape to the positive half cycle, plot the waveform and find (a) the frequency of the supply, (b) the instantaneous values of current after 1.25 ms and 3.8 ms, (c) the peak or maximum value, (d) the mean or average value, and (e) the rms value of the waveform.

### Solution

The half cycle of alternating current is shown plotted in Figure 6.6.

(a) Time for a half cycle $= 5$ ms. The time for 1 cycle, i.e., the periodic time, $T = 10$ ms or 0.01 s.

Frequency, $f = \dfrac{1}{T} = \dfrac{1}{0.01} = \mathbf{100\,Hz}$

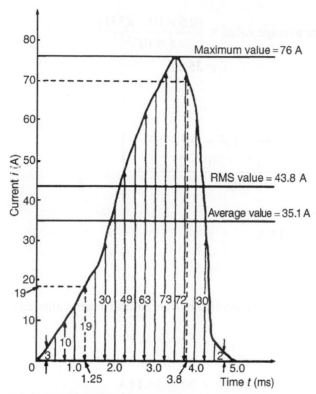

**Figure 6.6: Half cycle of alternating current for Example 6.5**

(b)  Instantaneous value of current after 1.25 ms is **19 A**, from Figure 6.6.

Instantaneous value of current after 3.8 ms is **70 A**, from Figure 6.6.

(c)  Peak or maximum value = **76 A**.

(d)  Mean or average value $= \dfrac{\text{area under curve}}{\text{length of base}}$

Using the mid-ordinate rule with 10 intervals, each of width 0.5 ms gives:

area under curve.

$= (0.5 \times 10^{-3})[3 + 10 + 19 + 30 + 49 + 63 +$
$73 + 72 + 30 + 2]$ (see Figure 6.6)
$= (0.5 \times 10^{-3})(351)$

Hence, mean or average value $= \dfrac{(0.5 \times 10^{-3})(351)}{5 \times 10^{-3}}$

$$= \mathbf{35.1\,A}$$

(e)  rms value

$$= \sqrt{\left(\dfrac{\begin{array}{c}3^2 + 10^2 + 19^2 + 30^2 + 49^2 + 63^2 +\\ 73^2 + 72^2 + 30^2 + 2^2\end{array}}{10}\right)}$$

$$= \sqrt{\left(\dfrac{19\,157}{10}\right)} = \mathbf{43.8\,A}$$

### Example 6.6
Calculate the rms value of a sinusoidal current of maximum value 20 A.

### Solution
For a sine wave, rms value $= 0.707 \times$ maximum value

$$= 0.707 \times 20 = \mathbf{14.14\,A}$$

### Example 6.7
Determine the peak and mean values for a 240 V mains supply.

### Solution
For a sine wave, rms value of voltage $V = 0.707 \times V_m$

A 240 V mains supply means that 240 V is the rms value,

$$V_m = \dfrac{V}{0.707} = \dfrac{240}{0.707} = \mathbf{339.5\,V} = \textbf{peak value}$$

Mean value $V_{AV} = 0.637\,V_m = 0.637 \times 339.5 = \mathbf{216.3\,V}$

### Example 6.8
A supply voltage has a mean value of 150 V. Determine its maximum value and its rms value.

### Solution

For a sine wave, mean value $= 0.637 \times$ maximum value

Hence, maximum value $= \dfrac{\text{mean value}}{0.637} = \dfrac{150}{0.637} = \mathbf{235.5\,V}$

rms value $= 0.707 \times$ maximum value $= 0.707 \times 235.5$

$$= \mathbf{166.5\,V}$$

## 6.4 The Equation of a Sinusoidal Waveform

In Figure 6.7, OA represents a vector that is free to rotate anticlockwise about 0 at an angular velocity of $\omega$ rad/s. A rotating vector is known as a *phasor*.

After time $t$ seconds the vector OA has turned through an angle $\omega t$. If the line BC is constructed perpendicular to OA as shown, then,

$$\sin \omega t = \dfrac{\text{BC}}{\text{OB}} \quad \text{i.e., BC} = \text{OB} \sin \omega t$$

If all such vertical components are projected onto a graph of $y$ against angle $\omega t$ (in radians), a sine curve results of maximum value OA. Any quantity that varies sinusoidally can be represented as a phasor.

A sine curve may not always start at 0°. To show this, a periodic function is represented by $y = \sin(\omega t + \phi)$, where $\phi$ is the phase (or angle) difference compared with $y = \sin \omega t$. In Figure 6.8(a), $y_2 = \sin(\omega t + \phi)$ starts $\phi$ radians earlier than $y_1 = \sin \omega t$ and is said to *lead* $y_1$ by $\phi$ radians. Phasors $y_1$ and $y_2$ are shown in Figure 6.8(b) at the time when $t = 0$.

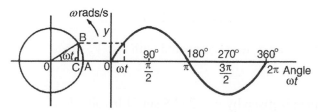

**Figure 6.7: Rotating vector OA and plot of rotation showing resulting sine curve**

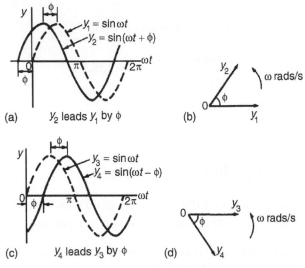

**Figure 6.8: Phase angle, leading and lagging**

In Figure 6.8(c), $y_4 = \sin(\omega t - \phi)$ starts $\phi$ radians later than $y_3 = \sin \omega t$ and is said to *log* $y_3$ by $\phi$ radians. Phasors $y_3$ and $y_4$ are shown in Figure 6.8(d) at the time when $t = 0$.

Given the general sinusoidal voltage, $\upsilon = V_m \sin(\omega t \pm \phi)$, then

    (i)  Amplitude or maximum value $= V_m$

    (ii)  Peak-to-peak value $= 2 V_m$

    (iii)  Angular velocity $= \omega$ rad/s

    (iv)  Periodic time, $T = 2\pi/\omega$ seconds

    (v)  Frequency, $f = \omega/2\pi$ Hz (since $= 2\pi f$)

    (vi)  $\phi =$ angle of lag or lead (compared with $v = V_m \sin \omega t$

### Example 6.9
An alternating voltage is given by $\upsilon = 282.8 \sin 314t$ volts. Find (a) the rms voltage, (b) the frequency and (c) the instantaneous value of voltage when $t = 4\,\mathrm{ms}$.

## Solution

(a) The general expression for an alternating voltage is
$v = V_m \sin(\omega t \pm \phi)$.

Comparing $v = 282.8 \sin 314t$ with this general expression gives the peak voltage as
282.8 V

The rms voltage = 0.707 × maximum value

$$= 0.707 \times 282.8 = \mathbf{200\,V}$$

(b) Angular velocity, $\omega = 314\,\text{rad/s}$, i.e. $2\pi f = 314$

Frequency, $f = \dfrac{314}{2\pi} = \mathbf{50\,Hz}$

(c) When $t = 4\,\text{ms}$, $v = 282.8 \sin(314 \times 4 \times 10^{-3})$

$$= 228.2 \sin(1.256) = \mathbf{268.9\,V}$$

Note that 1.256 radians $= \left(1.256 \times \dfrac{180}{\pi}\right)^{\circ}$

$$= 71.96^{\circ}$$

Hence, $v = 282.8 \sin 71.\,96^{\circ} = \mathbf{268.9\,V}$

## Example 6.10

An alternating voltage is given by

$v = 75 \sin(200\pi t - 0.25)$ volts.

Find (a) the amplitude, (b) the peak-to-peak value, (c) the rms value, (d) the periodic time, (e) the frequency, and (f) the phase angle (in degrees and minutes) relative to $75 \sin 200\pi t$.

## Solution

Comparing $v = 75 \sin(200\pi t - 0.25)$ with the general expression $v = V_m \sin(\omega t \pm \phi)$ gives:

(a) Amplitude, or peak value = **75 V**

(b) Peak-to-peak value = $2 \times 75 = \mathbf{150\,V}$

(c) The rms value $= 0.707 \times$ maximum value
$$= 0.707 \times 75 = \textbf{53 V}$$

(d) Angular velocity, $\omega = 200\pi$ rad/s

Hence, periodic time, $T = \dfrac{2\pi}{\omega} = \dfrac{2\pi}{200\pi} = \dfrac{1}{100}$
$$= \textbf{0.01 s or 10 ms}$$

(e) Frequency, $f = \dfrac{1}{T} = \dfrac{1}{0.01} = \textbf{100 Hz}$

(f) Phase angle, $\phi = 0.25$ radians lagging $75 \sin 200\pi t$

$$0.25 \text{ rads} = \left(0.25 \times \frac{180}{\pi}\right)^{\circ} = 14.32^{\circ} = 14^{\circ}19'$$

Hence, phase angle $= \textbf{14}^{\circ}\textbf{19}'$ **lagging**

## Example 6.11
An alternating voltage, $v$, has a periodic time of 0.01 s and a peak value of 40 V. When time $t$ is zero, $v = -20$ V. Express the instantaneous voltage in the form $v = V_m \sin(\omega t \pm \phi)$.

## Solution
Amplitude, $V_m = 40$ V

Periodic time $T = \dfrac{2\pi}{\omega}$ hence, angular velocity,

$$\omega = \frac{2\pi}{T} = \frac{2\pi}{0.01} = 200\pi \text{ rad/s}$$

$v = V_m \sin(\omega t + \phi)$ becomes $v = 40 \sin(200\pi t + \phi)$ V

When time $t = 0$, $v = -20$ V

i.e., $-20 = 40 \sin \phi$

so that $\sin \phi = \dfrac{-20}{40} = -0.5$

Hence, $\phi = \sin^{-1}(-0.5) = -30° = \left(-30 \times \dfrac{\pi}{180}\right)$ rads

$$= -\dfrac{\pi}{6} \text{ rads}$$

**Thus, $v = 40 \sin \left(200\pi t - \dfrac{\pi}{6}\right)$ V**

### Example 6.12

The current in an AC circuit at any time $t$ seconds is given by:

$i = 120 \sin(100\pi t + 0.36)$ amperes. Find:

(a) the peak value, the periodic time, the frequency and phase angle relative to $120 \sin 100\pi t$,

(b) the value of the current when $t = 0$,

(c) the value of the current when $t = 8\,\text{ms}$,

(d) the time when the current first reaches $60\,\text{A}$, and

(e) the time when the current is first a maximum.

### Solution

(a) Peak Value = **120 A**

Periodic time $T = \dfrac{2\pi}{\omega} = \dfrac{2\pi}{100\pi}$ (since $\omega = 100\pi$)

$$= \dfrac{1}{50} = \textbf{0.02 s or 20 ms}$$

Frequency, $f = \dfrac{1}{T} = \dfrac{1}{0.02} = \textbf{50 Hz}$

$$\text{Phase angle} = 0.36 \text{ rads} = \left(0.36 \times \frac{180}{\pi}\right)^{\circ}$$

$$= \mathbf{20°38' \text{ leading}}$$

(b) When $t = 0$, $i = 120 \sin(0 + 0.36) = 120 \sin 20°38'$

$$= \mathbf{49.3\,A}$$

(c) When $t = 8$, $i = 120 \sin \left[100\pi\left(\frac{8}{10^3}\right) + 0.36\right]$

$$= 120 \sin 2.8733 (= 120 \sin 164°38') = \mathbf{31.8\,A}$$

(d) When $i = 60$ A, $60 = 120 \sin(100\pi t + 0.36)$

thus, $\dfrac{60}{120} = \sin(100\pi t + 0.36)$

so that $(100\pi t + 0.36) = \sin^{-1} 0.5 = 30° = \dfrac{\pi}{6}$ rads

$$= 0.5236 \text{ rads}$$

Hence, time $t = \dfrac{0.5236 - 0.36}{100\pi} = \mathbf{0.521\,ms}$

(e) When the current is a maximum, $i = 120$ A

Thus, $120 = 120 \sin(100\pi t + 0.36)$

$$= \sin(100\pi t + 0.36)$$

$$(100\pi t + 0.36) = \sin^{-1} 1 = 90° = \frac{\pi}{2} \text{ rads}$$

$$= 1.5708 \text{ rads}$$

Hence, time $t = \dfrac{1.5708 - 0.36}{100\pi} = \mathbf{3.85\,ms}$

## 6.5  Combination of Waveforms

The resultant of the addition (or subtraction) of two sinusoidal quantities may be determined either:

(a) by plotting the periodic functions graphically (see worked Examples 6.13 and 6.16), or

(b) by resolution of phasors by drawing or calculation (see worked Examples 6.14 and 6.15).

### Example 6.13

The instantaneous values of two alternating currents are given by $i_1 = 20 \sin \omega t$ amperes and $i_2 = 10 \sin(\omega t + \pi/3)$ amperes. By plotting $i_1$ and $i_2$ on the same axes, using the same scale, over one cycle, and adding ordinates at intervals, obtain a sinusoidal expression for $i_1 + i_2$.

### Solution

$i_1 = 20 \sin \omega t$ and $i_2 = 10 \sin \left( \omega t + \dfrac{\pi}{3} \right)$ are shown plotted in Figure 6.9.

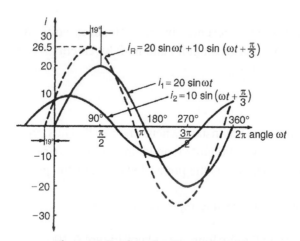

**Figure 6.9: Plots for Example 6.13**

Ordinates of $i_1$ and $i_2$ are added at, say, 15° intervals (a pair of dividers are useful for this).

For example,

at 30°, $i_1 + i_2 = 10 + 10 = 20$ A

at 60°, $i_1 + i_2 = 8.7 + 17.3 = 26$ A

at 150°, $i_1 + i_2 = 10 + (-5) = 5$ A, and so on.

The resultant waveform for $i_1 + i_2$ is shown by the broken line in Figure 6.9. It has the same period, and frequency, as $i_1$ and $i_2$. The amplitude or peak value is 26.5 A.

The resultant waveform leads the curve $i_1 = 20 \sin \omega t$ by 19°.

i.e. $\left(19 \times \dfrac{\pi}{180}\right)$ rads $= 0.332$ rads

The sinusoidal expression for the resultant $i_1 + i_2$ is given by:

$i_R = i_1 + i_2 = 26.5 \sin(\omega t + 0.332)$ A

### Example 6.14

Two alternating voltages are represented by $\upsilon_1 = 50 \sin \omega t$ volts and $\upsilon_2 = 100 \sin(\omega t - \pi/6)$ V. Draw the phasor diagram and find, by calculation, a sinusoidal expression to represent $\upsilon_1 + \upsilon_2$.

### Solution

Phasors are usually drawn at the instant when time $t = 0$. Thus, $\upsilon_1$ is drawn horizontally 50 units long and $\upsilon_2$ is drawn 100 units long lagging $\upsilon_1$ by $\pi/6$ rads, i.e., 30°. This is shown in Figure 6.10(a) where 0 is the point of rotation of the phasors.

Procedure to draw phasor diagram to represent $\upsilon_1 + \upsilon_2$:

(i)   Draw $\upsilon_1$ horizontal 50 units long, i.e., Oa of Figure 6.10(b).

(ii)  Join $\upsilon_2$ to the end of $\upsilon_1$ at the appropriate angle, i.e., ab of Figure 6.10(b).

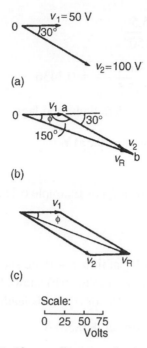

**Figure 6.10: Phasor diagrams for Example 6.14**

(iii)  The resultant $v_R = v_1 + v_2$ is given by the length Ob and its phase angle $\phi$ may be measured with respect to $v_1$.

Alternatively, when two phasors are being added the resultant is always the diagonal of the parallelogram, as shown in Figure 6.10(c).

From the drawing, by measurement, $v_R = 145\,V$ and angle $\phi = 20°$ lagging $v_1$.

A more accurate solution is obtained by calculation, using the cosine and sine rules. Using the cosine rule on triangle Oab of Figure 6.10(b) gives:

$$v_R^2 = v_1^2 + v_2^2 - 2v_1 v_2 \cos 150°$$

$$= 50^2 + 100^2 - 2(50)(100)\cos 150°$$

$$= 2500 + 10\,000 - (-8660)$$

$$v_R = \sqrt{(21160)} = 145.5\,V$$

Using the sine rule, $\dfrac{100}{\sin\phi} = \dfrac{145.5}{\sin 150°}$

from which $\qquad \sin\phi = \dfrac{100 \sin 150°}{145.5} = 0.3436$

and $\phi = \sin^{-1} 0.3436 = 20°6' = 0.35$ radians, and lags $v_1$

Hence, $v_R = v_1 + v_2 = \mathbf{145.5\ sin(\omega t - 0.35)\,V}$

### Example 6.15
Find a sinusoidal expression for $(i_1 + i_2)$ of Example 6.13, (a) by drawing phasors, (b) by calculation.

### Solution

(a) The relative positions of $i_1$ and $i_2$ at time $t = 0$ are shown as phasors in Figure 6.11(a). The phasor diagram in Figure 6.11(b) shows the resultant $i_R$, and $i_R$ is measured as 26A and angle $\phi$ as 19° or 0.33 rads leading $i_1$.

   **Hence, by drawing, $i_R = \mathbf{26\ sin(\omega t + 0.33)\,A}$**

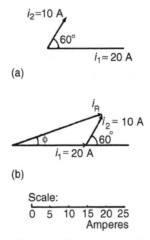

**Figure 6.11: Phasor diagrams for Example 6.15**

(b) From Figure 6.11(b), by the cosine rule:

$$i_R^2 = 20^2 + 10^2 - 2(20)(10)(\cos 120°)$$

from which $i_R$ = **26.46A**

By the sine rule: $\dfrac{10}{\sin \phi} = \dfrac{26.46}{\sin 120°}$

from which $\phi$ = 19.10° (i.e., 0.333 rads)

**By calculation $i_R$ = 26.46 sin($\omega t$ + 0.333) A**

An alternative method of calculation is to use *complex numbers*. (See Chapter 7.)

$$\text{Then } i_1 + i_2 = 20 \sin \omega t + 10 \sin\left(\omega t + \frac{\pi}{3}\right)$$

$$\equiv 20\angle 0 + 10\angle \frac{\pi}{3} \text{ rad}$$

$$\text{or } 20\angle 0° + 10\angle 60°$$

$$= (20 + j0) + (5 + j8.66)$$

$$= (25 + j8.66) = 26.46\angle 19.106° \text{ or } 26.46\angle 0.333 \text{ rad}$$

$$\equiv \mathbf{26.46 \ sin(\omega t + 0.333) \ A}$$

### Example 6.16

Two alternating voltages are given by $v_1 = 120 \sin \omega t$ volts and $v_2 = 200 \sin(\omega t - \pi/4)$ volts. Obtain sinusoidal expressions for $v_1 - v_2$ (a) by plotting waveforms, and (b) by resolution of phasors.

### Solution

(a) $v_1 = 120 \sin \omega t$ and $v_2 = 200 \sin(\omega t - \pi/4)$ are shown plotted in Figure 6.12. Care must be taken when subtracting values of ordinates especially when at least one of the ordinates is negative. For example:

at 30°, $v_1 - v_2 = 60 - (-52) = 112 \text{ V}$

at 60°, $v_1 - v_2 = 104 - 52 = 52 \text{ V}$

at 150°, $v_1 - v_2 = 60 - 193 = -133 \text{ V}$, and so on.

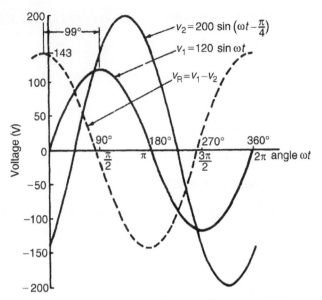

**Figure 6.12: Voltage plots for Example 6.16**

The resultant waveform, $\upsilon_R = \upsilon_1 - \upsilon_2$, is shown by the broken line in Figure 6.12. The maximum value of $\upsilon_R$ is 143 V and the waveform is seen to lead $\upsilon_1$ by 99° (i.e., 1.73 radians).

**By drawing,** $\upsilon_R = \upsilon_1 - \upsilon_2$

$$= \textbf{143 sin}(\omega t + \textbf{1.73}) \textbf{ volts}$$

(b)  The relative positions of $\upsilon_1$ and $\upsilon_2$ are shown at time $t = 0$ as phasors in Figure 6.13(a). Since the resultant of $\upsilon_1 - \upsilon_2$ is required, $-\upsilon_2$ is drawn in the opposite direction to $+\upsilon_2$ and is shown by the broken line in Figure 6.13(a). The phasor diagram with the resultant is shown in Figure 6.13(b) where $-\upsilon_2$ is added phasorially to $\upsilon_1$.

By resolution:

Sum of horizontal components of $\upsilon_1$ and $\upsilon_2$

$$= 120 \cos 0° + 200 \cos 135° = -21.42$$

Sum of vertical components of $\upsilon_1$ and $\upsilon_2$

$$= 120 \sin 0° + 200 \sin 135° = 141.4$$

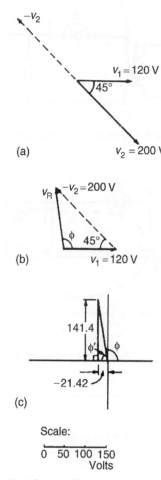

**Figure 6.13: Phasor diagrams for Example 6.16**

From Figure 6.13(c), resultant

$$\upsilon_R = \sqrt{[(-21.42)^2 + (141.4)^2]} = 143.0,$$

and $\tan \phi' = \dfrac{141.4}{21.42} = \tan 6.6013$, from which

$$\phi' = \tan^{-1} 6.6013 = 81°23' \text{ and}$$

$$\phi = 98°37' \text{ or } 1.721 \text{ radians}$$

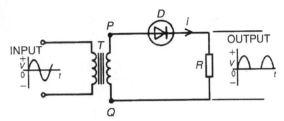

**Figure 6.14: Half-wave rectification**

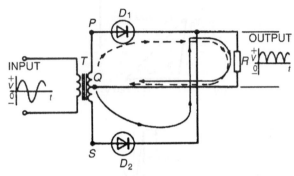

**Figure 6.15: Full-wave rectification**

**By resolution of phasors,**

$$v_R = v_1 - v_2 = 143.0 \sin(\omega t + 1.721) \text{ volts}$$

## 6.6 Rectification

The process of obtaining unidirectional currents and voltages from alternating currents and voltages is called *rectification*. Automatic switching in circuits is carried out by diodes.

Using a single diode, as shown in Figure 6.14, *half-wave rectification* is obtained. When *P* is sufficiently positive with respect to *Q*, diode *D* is switched on and current *i* flows. When *P* is negative with respect to *Q*, diode *D* is switched off. Transformer *T* isolates the equipment from direct connection with the mains supply and enables the mains voltage to be changed.

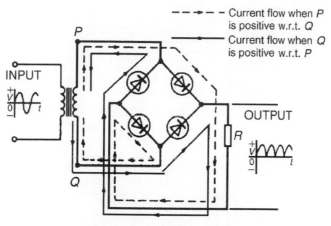

Figure 6.16: Bridge rectifier

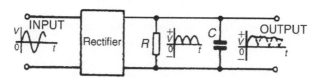

**Figure 6.17: Smoothing output using capacitors**

Two diodes may be used as shown in Figure 6.15 to obtain *full wave rectification*. A center-tapped transformer $T$ is used. When $P$ is sufficiently positive with respect to $Q$, diode $D_1$ conducts and current flows (shown by the broken line in Figure 6.15). When S is positive with respect to $Q$, diode $D_2$ conducts and current flows (shown by the continuous line in Figure 6.15). The current flowing in $R$ is in the same direction for both half cycles of the input. The output waveform is shown in Figure 6.15.

Four diodes may be used in a *bridge rectifier* circuit, as shown in Figure 6.16 to obtain *full wave rectification*. As for the rectifier shown in Figure 6.15, the current flowing in $R$ is in the same direction for both half cycles of the input giving the output waveform shown.

To smooth the output of the rectifiers described above, capacitors having a large capacitance may be connected across the load resistor $R$. The effect of this is shown on the output in Figure 6.17.

Figure 6.14 Bridge rectifier

Figure 6.15 Smoothing output using capacitors

Two diodes may be used as shown in Figure 6.13 to obtain full wave rectification. A centre-tapped transformer $T$ is used. When $P$ is sufficiently positive with respect to $Q$, diode $D_1$ conducts and current flows as shown by the broken line in Figure 6.15. When $Q$ is positive with respect to $P$, diode $D_2$ conducts and current flows as shown by the continuous line in Figure 6.14. The current flowing in $R_L$ is in the same direction for both halves of the cycle. The output waveform is shown in Figure 6.15.

Four diodes may be used in a bridge rectifier circuit, as shown in Figure 6.14 to give full wave rectification. As for the rectifier shown in Figure 6.14, the current flowing in $R_L$ is in the same direction for both half cycles of the input giving the output waveform shown.

To smooth the output of the rectifiers described above, capacitors having a large capacitance may be connected across the load resistor $R_L$. The effect of this is shown on the output in Figure 6.17.

# CHAPTER 7

# Complex Numbers

John Bird

## 7.1 Introduction

A *complex number* is of the form $(a + jb)$ where $a$ is a *real number* and $jb$ is an *imaginary number*. Therefore, $(1 + j2)$ and $(5 - j7)$ are examples of complex numbers.

By definition, $j = \sqrt{-1}$ and $j^2 = -1$

(Note: In electrical engineering, the letter j is used to represent $\sqrt{-1}$ instead of the letter $i$, as commonly used in pure mathematics, because $i$ is reserved for current.)

Complex numbers are widely used in the analysis of series, parallel and series-parallel electrical networks supplied by alternating voltages, in deriving balance equations with AC bridges, in analyzing AC circuits using Kirchhoff's laws, mesh and nodal analysis, the superposition theorem, with Thévenin's and Norton's theorems, and with delta-star and star-delta transforms, and in many other aspects of higher electrical engineering. The advantage of the use of complex numbers is that the manipulative processes become simply algebraic processes.

A complex number can be represented pictorially on an *Argand diagram*. In Figure 7.1, the line 0A represents the complex number $(2 + j3)$, 0B represents $(3 - j)$, 0C represents $(-2 + j2)$ and 0D represents $(-4 - j3)$.

A complex number of the form $a + jb$ is called a *Cartesian or rectangular complex number*. The significance of the $j$ operator is shown in Figure 7.2. In Figure 7.2(a) the number 4 (i.e., $4 + j0$) is shown drawn as a phasor horizontally to the right of the origin on the real axis. (Such a phasor could represent, for example, an alternating current, $i = 4 \sin \omega t$ amperes, when time $t$ is zero.)

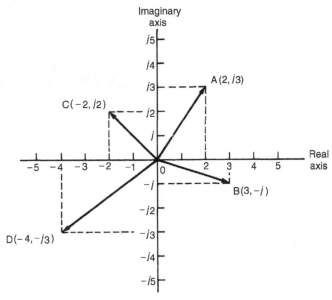

**Figure 7.1: The Argand diagram**

The number $j4$ (that is, $0 + j4$) is shown in Figure 7.2(b) drawn vertically upwards from the origin on the imaginary axis. Multiplying the number 4 by the operator $j$ results in an anticlockwise phase-shift of $90°$ without altering its magnitude.

Multiplying $j4$ by $j$ gives $j^24$, i.e., $-4$, and is shown in Figure 7.2(c) as a phasor four units long on the horizontal real axis to the left of the origin—an anticlockwise phase-shift of $90°$ compared with the position shown in Figure 7.2(b). Thus, multiplying by $j^2$ reverses the original direction of a phasor.

Multiplying $j^24$ by $j$ gives $j^34$, i.e., $-j4$, and is shown in Figure 7.2(d) as a phasor four units long on the vertical, imaginary axis downward from the origin—an anticlockwise phase-shift of $90°$ compared with the position shown in Figure 7.2(c).

Multiplying $j^34$ by $j$ gives $j^44$, i.e., 4, which is the original position of the phasor shown in Figure 7.2(a).

*Summarizing*, application of the operator $j$ to any number rotates it $90°$ anticlockwise on the Argand diagram, multiplying a number by $j^2$ rotates it $180°$ anticlockwise, multiplying a number by $j^3$ rotates it $270°$ anticlockwise and multiplication by $j^4$ rotates

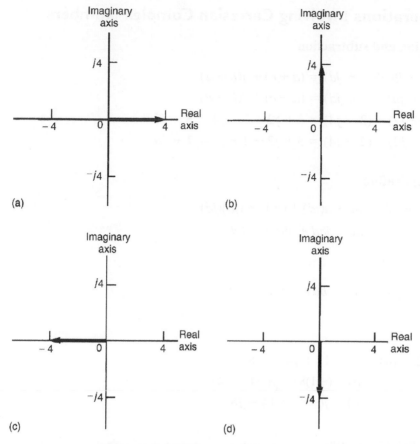

**Figure 7.2: Significance of the *j* operator**

it 360° anticlockwise, i.e., back to its original position. In each case, the phasor is unchanged in its magnitude.

By similar reasoning, if a phasor is operated on by $-j$ then a phase shift of $-90°$ (i.e., clockwise direction) occurs, again without change of magnitude.

In electrical circuits, 90° phase shifts occur between voltage and current with pure capacitors and inductors; this is the key as to why *j* notation is used so much in the analysis of electrical networks. This is explained later in this chapter.

## 7.2 Operations Involving Cartesian Complex Numbers

### (a) Addition and subtraction

$$(a + jb) + (c + jd) = (a + c) + j(b + d)$$

and $\quad (a + jb) - (c + jd) = (a - c) + j(b - d)$

Thus, $\quad (3 + j2) + (2 - j4) = 3 + j2 + 2 - j4 = \mathbf{5 - j2}$

and $\quad (3 + j2) - (2 - j4) = 3 + j2 - 2 + j4 = \mathbf{1 + j6}$

### (b) Multiplication

$$(a + jb)(c + jd) = ac + a(jd) + (jb)c + (jb)(jd)$$
$$= ac + jad + jbc + j^2bd$$

But $j^2 = -1$, thus,

$$(a + jb)(c + jd) = (ac - bd) + j(ad + bc)$$

For example,

$$(3 + j2)(2 - j4) = 6 - j12 + j4 - j^2 8$$
$$= (6 - (-1)8) + j(-12 + 4)$$
$$= 14 + j(-8) = \mathbf{14 - j8}$$

### (c) Complex conjugate

The *complex conjugate* of $(a + jb)$ is $(a - jb)$. For example, the conjugate of $(3 - j2)$ is $(3 + j2)$. The product of a complex number and its complex conjugate is always a real number, and this is an important property used when dividing complex numbers. Thus,

$$(a + jb)(a - jb) = a^2 - jab + jab - j^2b^2$$
$$= a^2 - (-b^2)$$
$$= a^2 + b^2 \text{ (i.e., a real number)}$$

For example, $(1 + j2)(1 - j2) = 1^2 + 2^2 = \mathbf{5}$

and $\qquad (3 - j4)(3 + j4) = 3^2 + 4^2 = \mathbf{25}$

## (d) Division

The expression of one complex number divided by another, in the form $a + jb$, is accomplished by multiplying the numerator and denominator by the complex conjugate of the denominator. This has the effect of making the denominator a real number. For example,

$$\frac{2 + j4}{3 - j4} = \frac{2 + j4}{3 - j4} \times \frac{3 + j4}{3 + j4} = \frac{6 + j8 + j12 + j^2 16}{3^2 + 4^2}$$

$$= \frac{6 + j8 + j12 - 16}{25}$$

$$= \frac{-10 + j20}{25}$$

$$= \frac{-10}{25} + j\frac{20}{25} \quad \text{or} \quad -0.4 + j0.8$$

The elimination of the imaginary part of the denominator by multiplying both the numerator and denominator by the conjugate of the denominator is often termed *rationalizing*.

### Example 7.1

In an electrical circuit the total impedance $Z_T$ is given by:

$$Z_T = \frac{Z_1 Z_2}{Z_1 + Z_2} + Z_3$$

Determine $Z_T$ in $(a + jb)$ form, correct to two decimal places, when $Z_1 = 5 - j3$, $Z_2 = 4 + j7$ and $Z_3 = 3.9 - j6.7$.

### Solution

$$Z_1 Z_2 = (5 - j3)(4 + j7) = 20 + j35 - j12 - j^2 21$$
$$= 20 + j35 - j12 + 21 = 41 + j23$$
$$Z_1 + Z_2 = (5 - j3) + (4 + j7) = 9 + j4$$

Hence, $\dfrac{Z_1 Z_2}{Z_1 + Z_2} = \dfrac{41 + j23}{9 + j4} = \dfrac{(41 + j23)(9 - j4)}{(9 + j4)(9 - j4)}$

$$= \frac{369 - j164 + j207 - j^2 92}{9^2 + 4^2}$$

$$= \frac{369 - j164 + j207 + 92}{97}$$

$$= \frac{461 + j43}{97} = 4.753 + j0.443$$

Thus, $\dfrac{Z_1 Z_2}{Z_1 + Z_2} + Z_3 = (4.753 + j0.443) + (3.9 - j6.7)$

$$= \mathbf{8.65 - j6.26}, \text{ correct to two decimal places.}$$

### Example 7.2

Given $Z_1 = 3 + j4$ and $Z_2 = 2 - j5$ determine in Cartesian form correct to three decimal places:

(a) $\dfrac{1}{Z_1}$   (b) $\dfrac{1}{Z_2}$   (c) $\dfrac{1}{Z_1} + \dfrac{1}{Z_2}$   (d) $\dfrac{1}{(1/Z_1) + (1/Z_2)}$

### Solution

(a) $\dfrac{1}{Z_1} = \dfrac{1}{3 + j4} = \dfrac{3 - j4}{(3 + j4)(3 - j4)} = \dfrac{3 - j4}{3^2 + 4^2}$

$$= \frac{3 - j4}{25} = \frac{3}{25} - j\frac{4}{25} = \mathbf{0.120 - j0.160}$$

(b) $\dfrac{1}{Z_2} = \dfrac{1}{2 - j5} = \dfrac{2 + j5}{(2 - j5)(2 + j5)} = \dfrac{2 + j5}{2^2 + 5^2} = \dfrac{2 + j5}{29}$

$$= \frac{2}{29} + j\frac{5}{29} = \mathbf{0.069 + j0.172}$$

(c) $\dfrac{1}{Z_1} + \dfrac{1}{Z_2} = (0.120 - j0.160) + (0.069 + j0.172)$

$$= \mathbf{0.189 + j0.012}$$

(d) $\dfrac{1}{(1/Z_1) + (1/Z_2)} = \dfrac{1}{0.189 + j0.012}$

$\qquad\qquad = \dfrac{0.189 - j0.012}{(0.189 + j0.012)(0.189 - j0.012)}$

$\qquad\qquad = \dfrac{0.189 - j0.012}{0.189^2 + 0.012^2}$

$\qquad\qquad = \dfrac{0.189 - j0.012}{0.03587}$

$\qquad\qquad = \dfrac{0.189}{0.03587} - \dfrac{j0.012}{0.03587}$

$\qquad\qquad = \mathbf{5.269 - j0.335}$

## 7.3 Complex Equations

If two complex numbers are equal, then their real parts are equal and their imaginary parts are equal. Hence, if $a + jb = c + jd$, then $a = c$ and $b = d$. This is a useful property, since equations having two unknown quantities can be solved from one equation. Complex equations are used when deriving balance equations with AC bridges.

### Example 7.3
Solve the following complex equations:

(a)  $3(a + jb) = 9 - j2$

(b)  $(2 + j)(-2 + j) = x + jy$

(c)  $(a - j2b) + (b - j3a) = 5 + j2$

### Solution
(a) $3(a + jb) = 9 - j2$. Thus, $3a + j3b = 9 - j2$

Equating real parts gives: $3a = 9$, i.e., $\boldsymbol{a = 3}$

Equating imaginary parts gives:
$3b = -2$, i.e., $\boldsymbol{b = -2/3}$

(b) $(2 + j)(-2 + j) = x + jy$

Thus,   $-4 + j2 - j2 + j2 = x + jy$
$$-5 + j0 = x + jy$$

Equating real and imaginary parts gives: $x = -5, y = 0$

(c) $(a - j2b) + (b - j3a) = 5 + j2$

Thus,   $(a + b) + j(-2b - 3a) = 5 + j2$

Hence,   $a + b = 5$                                                                                     (7.1)

and, $-2b - 3a = 2$                                                                                 (7.2)

We have two simultaneous equations to solve. Multiplying equation (7.1) by 2 gives:

$2a + 2b = 10$                                                                                          (7.3)

Adding equations (7.2) and (7.3) gives $-a = 12$, i.e., $a = -12$

From equation (7.1), $b = 17$

### Example 7.4

An equation derived from an AC bridge network is given by:

$$R_1 R_3 = (R_2 + j\omega L_2)\left[\frac{1}{(1/R_4) + (j\omega C)}\right]$$

$R_1$, $R_3$, $R_4$ and $C_4$ are known values. Determine expressions for $R_2$ and $L_2$ in terms of the known components.

### Solution

Multiplying both sides of the equation by $(1/R_4 + j\omega C_4)$ gives:

$$(R_1 R_3)(1/R_4 + j\omega C_4) = R_2 + j\omega L_2$$

i.e.,   $R_1 R_3/R_4 + jR_1 R_3 \omega C_4 = R_2 + j\omega L_2$

Equating the real parts gives: $R_2 = R_1 R_3/R_4$

Equating the imaginary parts gives:

$\omega L_2 = R_1 R_3 \omega C_4$, from which, $\boldsymbol{L_2 = R_1 R_3 C_4}$

## 7.4 The Polar Form of a Complex Number

In Figure 7.3(a), $Z = x + jy = r \cos \theta + jr \sin \theta$ from trigonometry,

$$= r(\cos \theta + j \sin \theta)$$

This latter form is usually abbreviated to $\boldsymbol{Z = r\angle\theta}$, and is called the polar form of a complex number.

$r$ is called the *modulus* (or magnitude of $Z$) and is written as mod $Z$ or $|Z|$. $r$ is determined from Pythagoras's theorem on triangle OAZ:

$$|Z| = r = \sqrt{(x^2 + y^2)}$$

The modulus is represented on the Argand diagram by the distance OZ. $\theta$ is called the argument (or amplitude) of $Z$ and is written as arg $Z$. $\theta$ is also deduced from triangle OAZ: arg $Z = \theta = \tan^{-1} y/x$.

For example, the cartesian complex number $(3 + j4)$ is equal to $r\angle\theta$ in polar form, where $r = \sqrt{(3^2 + 4^2)} = 5$ and,

$$\theta = \tan^{-1} \frac{4}{3} = \boldsymbol{53.13^\circ}$$

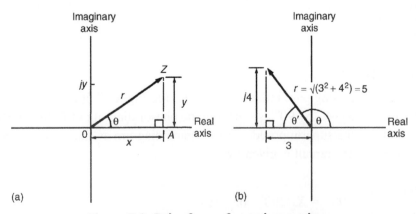

**Figure 7.3: Polar form of complex numbers**

Hence, $(3 + j4) = 5\angle 53.13°$

Similarly, $(-3 + j4)$ is shown in Figure 7.3(b),

where,    $r = \sqrt{(3^2 + 4^2)} = 5,$    $\theta' = \tan^{-1}\dfrac{4}{3} = 53.13°$

and,      $\theta = 180° - 53.13° = 126.87°$

Hence, $(-3 + j4) = 5\angle 126.87°$

## 7.5 Applying Complex Numbers to Series AC Circuits

Simple AC circuits may be analyzed by using phasor diagrams. However, when circuits become more complicated, analysis is considerably simplified by using complex numbers. It is essential that the basic operations used with complex numbers, as outlined in this chapter thus far, are thoroughly understood before proceeding with AC circuit analysis.

### 7.5.1 Series AC Circuits

#### 7.5.1.1 Pure Resistance

In an AC circuit containing resistance $R$ only (see Figure 7.4(a)), the current $I_R$ is *in phase* with the applied voltage $V_R$ as shown in the phasor diagram of Figure 7.4(b). The phasor diagram may be superimposed on the Argand diagram as shown in Figure 7.4(c). The impedance $\mathbf{Z}$ of the circuit is given by:

$$Z = \frac{V_R\angle 0°}{I_R\angle 0°} = R$$

#### 7.5.1.2 Pure Inductance

In an AC circuit containing pure inductance $L$ only (see Figure 7.5(a)), the current $I_L$ *lags* the applied voltage $V_L$ by 90° as shown in the phasor diagram of Figure 7.5(b). The phasor diagram may be superimposed on the Argand diagram as shown in Figure 7.5(c). The impedance $Z$ of the circuit is given by:

$$Z = \frac{V_L\angle 90°}{I_L\angle 0°} = \frac{V_L}{I_L}\angle 90° = X_L\angle 90°   \text{ or }   jX_L$$

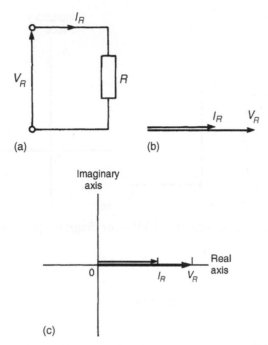

**Figure 7.4: (a) Circuit diagram; (b) Phasor diagram; (c) Argand diagram**

where $X_L$ is the *inductive reactance* given by:

$$X_L = \omega L = 2\pi f L \text{ ohms}$$

where $f$ is the frequency in hertz and $L$ is the inductance in henrys.

### 7.5.1.3 Pure Capacitance

In an AC circuit containing pure capacitance only (see Figure 7.5(a)), the current $I_C$ *leads* the applied voltage $V_C$ by 90° as shown in the phasor diagram of Figure 7.5(b). The phasor diagram may be superimposed on the Argand diagram as shown in Figure 7.5(c). The impedance $Z$ of the circuit is given by:

$$Z = \frac{V_C \angle -90°}{I_C \angle 0°} = \frac{V_C}{I_C} \angle -90° = X_C \angle -90° \quad \text{or} \quad -jX_C$$

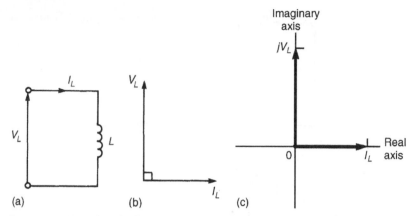

**Figure 7.5: (a) Circuit diagram; (b) Phasor diagram; (c) Argand diagram**

where $X_C$ is the *capacitive reactance* given by:

$$X_C = \frac{1}{\omega C} = \frac{1}{2\pi f C} \text{ ohms}$$

where $C$ is the capacitance in farads.

$$\left[ \text{Note: } -jX_C = \frac{-j}{\omega C} = \frac{-j(j)}{\omega C(j)} = \frac{-j^2}{j\omega C} = \frac{-(-1)}{j\omega C} = \frac{1}{j\omega C} \right]$$

### 7.5.1.4 R–L Series Circuit

In an AC circuit containing resistance $R$ and inductance $L$ in series (see Figure 7.7(a)), the applied voltage $V$ is the phasor sum of $V_R$ and $V_L$ as shown in the phasor diagram of Figure 7.7(b). The current $I$ lags the applied voltage $V$ by an angle lying between $0°$ and $90°$—the actual value depending on the values of $V_R$ and $V_L$, which depend on the values of $R$ and $L$. The circuit phase angle, that is, the angle between the current and the applied voltage, is shown as angle $\phi$ in the phasor diagram. In any series circuit the current is common to all components and is taken as the reference phasor in Figure 7.7(b). The phasor diagram may be superimposed on the Argand diagram as

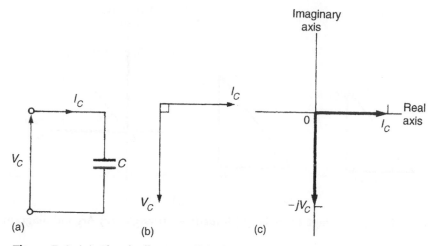

**Figure 7.6: (a) Circuit diagram; (b) Phasor diagram; (c) Argand diagram**

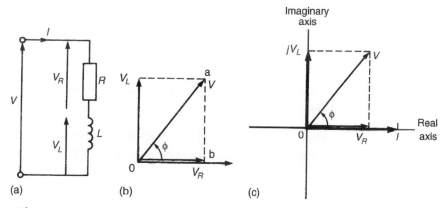

**Figure 7.7: (a) Circuit diagram; (b) Phasor diagram; (c) Argand diagram**

shown in Figure 7.7(c), where it may be seen that in complex form the supply voltage $V$ is given by:

$$V = V_R + jV_L$$

Figure 7.8(a) shows the voltage triangle that is derived from the phasor diagram of Figure 7.8(b) (triangle Oab). If each side of the voltage triangle is divided by current $I$,

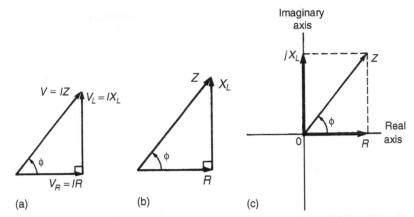

**Figure 7.8: (a) Voltage triangle; (b) Impedance triangle; (c) Argand diagram**

then the impedance triangle of Figure 7.8(b) is derived. The impedance triangle may be superimposed on the Argand diagram, as shown in Figure 7.8(c), where it may be seen that in complex form the impedance $Z$ is given by:

$$Z = R + jX_L$$

For example, an impedance expressed as $(3 + j4)\,\Omega$ means that the resistance is $3\,\Omega$ and the inductive reactance is $4\,\Omega$.

In polar form, $Z = |Z|\,\angle\phi$ where, from the impedance triangle, the modulus of impedance $|Z| = \sqrt{(R^2 + X_L^2)}$ and the circuit phase angle $\phi = \tan^{-1}(X_L/R)$ lagging.

### 7.5.1.5 R-C Series Circuit

In an AC circuit containing resistance $R$ and capacitance $C$ in series (see Figure 7.9(a)), the applied voltage $V$ is the phasor sum of $V_R$ and $V_C$ as shown in the phasor diagram of Figure 7.9(b). The current $I$ leads the applied voltage $V$ by an angle lying between $0°$ and $90°$—the actual value depending on the values of $V_R$ and $V_C$, which depend on the values of $R$ and $C$. The circuit phase angle is shown as angle $\phi$ in the phasor diagram. The phasor diagram may be superimposed on the Argand diagram as shown in Figure 7.9(c), where it may be seen that in complex form the supply voltage $V$ is given by:

$$V = V_R - jV_C$$

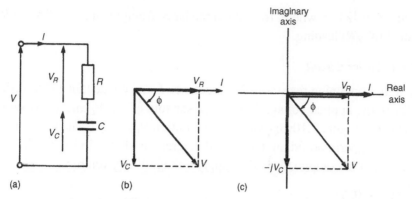

**Figure 7.9: (a) Circuit diagram; (b) Phasor diagram; (c) Argand diagram**

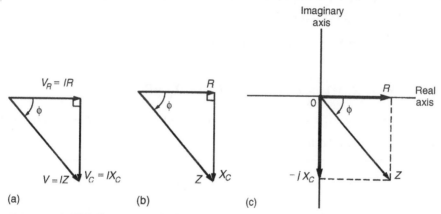

**Figure 7.10: (a) Voltage triangle; (b) Impedance triangle; (c) Argand diagram**

Figure 7.10(a) shows the voltage triangle that is derived from the phasor diagram of Figure 7.10(b). If each side of the voltage triangle is divided by current $I$, the impedance triangle is derived as shown in Figure 7.10(b). The impedance triangle may be superimposed on the Argand diagram as shown in Figure 7.10(c), where it may be seen that in complex form the impedance $Z$ is given by:

$$Z = R - jX_C$$

Thus, for example, an impedance expressed as $(9 - j14)\,\Omega$ means that the resistance is $9\,\Omega$ and the capacitive reactance $X_C$ is $14\,\Omega$.

In polar form, $Z = |Z|\angle\phi$ where, from the impedance triangle, angle, $|Z| = \sqrt{(R^2 + X_C^2)}$ and $\phi = \tan^{-1}(X_C/R)$ leading.

### 7.5.1.6 R-L-C Series Circuit

In an AC circuit containing resistance $R$, inductance $L$ and capacitance $C$ in series (see Figure 7.10(a)), the applied voltage $V$ is the phasor sum of $V_R$, $V_L$ and $V_C$ as shown in the phasor diagram of Figure 7.10(b) (where the condition $V_L > V_C$ is shown). The phasor diagram may be superimposed on the Argand diagram as shown in Figure 7.10(c), where it may be seen that in complex form the supply voltage $V$ is given by:

$$V = V_R + j(V_L - V_C)$$

From the voltage triangle the impedance triangle is derived and superimposing this on the Argand diagram gives, in complex form,

Impedance $Z = R + j(X_L - X_C)$ or $Z = |Z|\angle\phi$

where,
$|Z| = \sqrt{[R^2 + (X_L - X_C)^2]}$ and $\phi = \tan^{-1}(X_L - X_C)/R$

When $V_L = V_C$, $X_L = X_C$ and the applied voltage $V$ and the current $I$ are in phase. This effect is called *series resonance*.

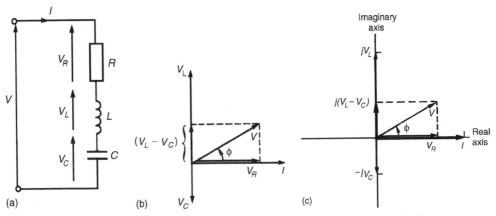

Figure 7.11: (a) Circuit diagram; (b) Phasor diagram; (c) Argand diagram

### 7.5.1.7 General Series Circuit

In an AC circuit containing several impedances connected in series, say, $Z_1, Z_2, Z_3, \ldots,$ $Z_n$, then the total equivalent impedance $Z_T$ is given by:

$$Z_T = Z_1 + Z_2 + Z_3 + \cdots + Z_n$$

### Example 7.5

Determine the values of the resistance and the series-connected inductance or capacitance for each of the following impedances: (a) $(12 + j5)\,\Omega$; (b) $-j40\,\Omega$; (c) $30\angle60°\,\Omega$; (d) $2.20 \times 10^6 \angle-30°\,\Omega$. Assume for each a frequency of $50\,\text{Hz}$.

### Solution

(a) From Section 24.2(d), for an $R$–$L$ series circuit, impedance $Z = R + jX_L$.

Thus, $Z = (12 + j5)\,\Omega$ represents a resistance of $12\,\Omega$ and an inductive reactance of $5\,\Omega$ in series.

Since inductive reactance $X_L = 2\pi fL$,

$$\text{Inductance } L = \frac{X_L}{2\pi f} = \frac{5}{2\pi(50)} = 0.0159\,\text{H}$$

So, the inductance is $15.9\,\text{mH}$.

**Thus, an impedance $(12 + j5)\,\Omega$ represents a resistance of $12\,\Omega$ in series with an inductance of $15.9\,\text{mH}$.**

(b) For a purely capacitive circuit, impedance $Z = -jX_C$.

Thus, $Z = -j40\,\Omega$ represents zero resistance and a capacitive reactance of $40\,\Omega$.

Since capacitive reactance $X_C = 1/(2\pi fC)$,

$$\begin{aligned}
\text{Capacitance } C &= \frac{1}{2\pi fX_C} = \frac{1}{2\pi(50)(40)}\,\text{F} \\
&= \frac{10^6}{2\pi(50)(40)}\,\mu\text{F} = 79.6\,\mu\text{F}
\end{aligned}$$

**Thus, an impedance $-j40\,\Omega$ represents a pure capacitor of capacitance $79.6\,\mu\text{F}$.**

(c) $30\angle 60° = 30(\cos 60° + j \sin 60°) = 15 + j25.98$

Thus, $Z = 30\angle 60° \,\Omega = (15 + j25.98)\,\Omega$ represents a resistance of $15\,\Omega$ and an inductive reactance of $25.98\,\Omega$ in series.

Since $X_L = 2\pi f L$,

$$\text{Inductance } L = \frac{X_L}{2\pi f} = \frac{25.98}{2\pi(50)}$$
$$= 0.0827 \, \text{H or } 82.7 \, \text{mH}$$

**Thus, an impedance $30\angle 60° \,\Omega$ represents a resistance of $15\,\Omega$ in series with an inductance of $82.7\,\text{mH}$.**

(d) $2.20 \times 10^6 \angle -30° = 2.20 \times 10^6[\cos(-30°) + j \sin(-30°)]$
$$= 1.905 \times 10^6 - j1.10 \times 10^6$$

Thus, $Z = 2.20 \times 10^6 \angle -30° \,\Omega$
$$= (1.905 \times 10^6 - j1.10 \times 10^6)\,\Omega$$

represents a resistance of $1.905 \times 10^6 \,\Omega$ (i.e., $1.905\,\text{M}\Omega$) and a capacitive reactance of $1.10 \times 10^6 \,\Omega$ in series.

Since capacitive reactance $X_C = 1/(2\pi f C)$,

$$\text{Capacitance } C = \frac{1}{2\pi f X_C} = \frac{1}{2\pi(50)(1.10 \times 10^6)}\,\text{F}$$
$$= 2.894 \times 10^{-9}\,\text{F} \quad \text{or} \quad 2.894\,\text{nF}$$

**Thus, an impedance $2.2 \times 10^6 \angle -30° \,\Omega$ represents a resistance of $1.905\,\text{M}\Omega$ in series with a $2.894\,\text{nF}$ capacitor.**

### Example 7.6

Determine, in polar and rectangular forms, the current flowing in an inductor of negligible resistance and inductance $159.2\,\text{mH}$ when it is connected to a $250\,\text{V}$, $50\,\text{Hz}$ supply.

### Solution

Inductive reactance

$$X_L = 2\pi f L = 2\pi(50)(159.2 \times 10^{-3}) = 50\,\Omega$$

Thus, circuit impedance $Z = (0 + j50)\,\Omega = 50\angle 90°\,\Omega$

Supply voltage, $V = 250\angle 0°$ V (or $(250 + j0)$V)

(Note that since the voltage is given as 250 V, this is assumed to mean $250\angle 0°$ V or $(250 + j0)$V.)

Hence, current $I = \dfrac{V}{Z} = \dfrac{250\angle 0°}{50\angle 90°} = \dfrac{250}{50}\angle(0° - 90°)$

$$= 5\angle -90°\text{A}$$

Alternatively, $I = \dfrac{V}{Z} = \dfrac{(250 + j0)}{(0 + j50)} = \dfrac{250(-j50)}{j50(-j50)}$

$$= \dfrac{-j(50)(250)}{50^2} = -j5\text{A}$$

which is the same as $5\angle -90°\text{A}$

### Example 7.7

A 3-μF capacitor is connected to a supply of frequency 1 kHz and a current of $2.83\angle 90°$A flows. Determine the value of the supply voltage.

### Solution

Capacitive reactance $X_C = \dfrac{1}{2\pi f C} = \dfrac{1}{2\pi(1000)(3 \times 10^{-6})}$

$$= 53.05\,\Omega$$

Hence, circuit impedance

$$Z = (0 - j53.05)\,\Omega = 53.05\angle -90°\,\Omega$$

Current $I = 2.83\angle 90°$ A (or $(0 + j2.83)$A)

Supply voltage, $V = IZ = (2.83\angle 90°)(53.05\angle -90°)$

i.e., **voltage** $= 150\angle 0°\,\text{V}$

Alternatively, $V = IZ = (0 + j2.83)(0 - j53.05)$

$$= -j^2(2.83)(53.05) = \textbf{150 V}$$

## Example 7.8

The impedance of an electrical circuit is $(30 - j50)$ ohms. Determine (a) the resistance, (b) the capacitance, (c) the modulus of the impedance, and (d) the current flowing and its phase angle, when the circuit is connected to a 240 V, 50 Hz supply.

### Solution

(a) Since impedance $Z = (30 - j50)\,\Omega$, *the resistance is 30 ohms* and the capacitive reactance is $50\,\Omega$.

(b) Since $X_C = 1/(2\pi fC)$, *capacitance*,

$$C = \frac{1}{2\pi f X_c} = \frac{1}{2\pi(50)(50)} = \textbf{63.66}\,\boldsymbol{\mu}\textbf{F}$$

(c) The modulus of impedance,

$$|Z| = \sqrt{(R^2 + X_C^2)} = \sqrt{(30^2 + 50^2)}$$
$$= \textbf{58.31}\,\boldsymbol{\Omega}$$

(d) Impedance $(30 - j50)\,\Omega = 58.31\angle\tan^{-1}\dfrac{X_C}{R}$

$$= \textbf{58.31}\angle\textbf{--59.04}°\,\boldsymbol{\Omega}$$

Hence, current $I = \dfrac{V}{Z} = \dfrac{240\angle 0°}{58.31\angle -59.04°}$

$$= \textbf{4.12}\angle\textbf{59.04}°\,\textbf{A}$$

## Example 7.9

A 200 V, 50 Hz supply is connected across a coil of negligible resistance and inductance 0.15 H connected in series with a $32\,\Omega$ resistor. Determine (a) the impedance of the circuit, (b) the current and circuit phase angle, (c) the voltage across the $32\,\Omega$ resistor, and (d) the voltage across the coil.

### Solution

(a) Inductive reactance $X_L = 2\pi fL = 2\pi(50)(0.15)$
$$= 47.1\,\Omega$$

Impedance $Z = R + jX_L$
$$= (32 + j47.1)\,\Omega \quad \text{or} \quad 57.0\angle 55.81°\,\Omega$$

The circuit diagram is shown in Figure 7.12.

(b) Current $I = \dfrac{V}{Z} = \dfrac{200\angle 0°}{57.0\angle 55.81°} = 3.51\angle -55.81°\,\text{A}$

i.e., the current is 3.51A lagging the voltage by 55.81°

(c) Voltage across the 32 $\Omega$ resistor,

$$V_R = IR = (3.51\angle -55.81°)(32\angle 0°)$$

i.e., $V_R = 112.3\angle -55.81°\,\text{V}$

(d) Voltage across the coil,

$$V_L = IX_L = (3.51\angle -55.81°)(47.1\angle 90°)$$

i.e., $V_L = 165.3\angle 34.19°\,\text{V}$

The phasor sum of $V_R$ and $V_L$ is the supply voltage $V$ as shown in the phasor diagram of Figure 7.13.

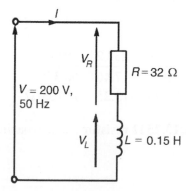

**Figure 7.12: Circuit diagram for Example 7.9**

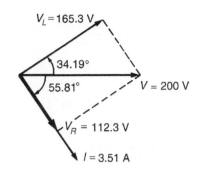

**Figure 7.13: Phasor diagram for Example 7.9**

$$V_R = 112.3\angle{-55.81°} = (63.11 - j92.89)\,\text{V}$$
$$V_L = 165.3\angle{34.19°}\,\text{V} = (136.73 + j92.89)\,\text{V}$$

Hence,

$$V = V_R + V_L = (63.11 - j92.89) + (136.73 + j92.89)$$

$$= (200 + j0)\,\text{V or } 200\angle{0°}\,\text{V, correct to three significant figures.}$$

### Example 7.10

Determine the value of impedance if a current of $(7 + j16)$A flows in a circuit when the supply voltage is $(120 + j200)$V. If the frequency of the supply is 5 MHz, determine the value of the components forming the series circuit.

### Solution

Impedance $Z = \dfrac{V}{I} = \dfrac{(120 + j200)}{(7 + j16)} = \dfrac{233.24\angle{59.04°}}{17.464\angle{66.37°}}$

$$= 13.36\angle{-7.33}\,\Omega \quad \text{or} \quad (13.25 - j1.705)\,\Omega$$

The series circuit consists of a **13.25 Ω resistor** and a capacitor of capacitive reactance **1.705 Ω.**

Since $X_C = \dfrac{1}{2\pi f C}$

Capacitance $C = \dfrac{1}{2\pi f X_C}$

$\qquad = \dfrac{1}{2\pi(5 \times 10^6)(1.705)}$

$\qquad = 1.867 \times 10^{-8}\,\text{F} = \mathbf{18.67\,nF}$

## 7.6 Applying Complex Numbers to Parallel AC Circuits

As with series circuits, parallel networks may be analyzed by using phasor diagrams. However, with parallel networks containing more than two branches, this can become very complicated. It is with parallel AC network analysis in particular that the full benefit of using complex numbers may be appreciated. The theory for parallel AC networks introduced previously is relevant; more advanced networks will be analyzed in this chapter using *j* notation. Before analyzing such networks admittance, conductance and susceptance are defined.

### 7.6.1 Admittance, Conductance and Susceptance

*Admittance* is defined as the current $I$ flowing in an AC circuit divided by the supply voltage $V$ (i.e., it is the reciprocal of impedance $Z$). The symbol for admittance is $Y$. Thus,

$$Y = \frac{I}{V} = \frac{1}{Z}$$

The unit of admittance is  the *siemen, S*.

An impedance may be resolved into a real part $R$ and an imaginary part $X$, giving $Z = R \pm jX$. Similarly, an admittance may be resolved into two parts—the real part being called the *conductance G*, and the imaginary part being called the *susceptance B*—and expressed in complex form. Thus, admittance,

$Y = G \pm jB$

When an AC circuit contains:

(a)  pure resistance, then,

$$Z = R \quad \text{and} \quad Y = \frac{1}{Z} = \frac{1}{R} = G$$

(b)  pure inductance, then,

$$Z = jX_L \quad \text{and} \quad Y = \frac{1}{Z} = \frac{1}{jX_L} = \frac{-j}{(jX_L)(-j)}$$

$$= \frac{-j}{X_L} = -jB_L$$

thus, a negative sign is associated with inductive susceptance, $B_L$.

(c)  pure capacitance, then,

$$Z = -jX_C \quad \text{and} \quad Y = \frac{1}{Z} = \frac{1}{-jX_C} = \frac{j}{(-jX_C)(j)}$$

$$= \frac{j}{X_C} = +jB_C$$

thus, a positive sign is associated with capacitive susceptance, $B_C$

(d)  resistance and inductance in series, then,

$$Z = R + jX_L \quad \text{and} \quad Y = \frac{1}{Z} = \frac{1}{R + jX_L}$$

$$= \frac{(R - jX_L)}{R^2 + X_L^2}$$

i.e., $Y = \dfrac{R}{R^2 + X_L^2} - j\dfrac{X_L}{R^2 + X_L^2}$  or  $Y = \dfrac{R}{|Z|^2} - j\dfrac{X_L}{|Z|^2}$

Thus, conductance, $G = R/|Z|^2$ and inductive susceptance, $B_L = -XL/|Z|^2$

(Note that in an inductive circuit, the imaginary term of the impedance, $X_L$, is positive, whereas the imaginary term of the admittance, $B_L$, is negative.)

(e)  resistance and capacitance in series, then,

$$Z = R - jX_C \quad \text{and} \quad Y = \frac{1}{Z} = \frac{1}{R - jX_C} = \frac{R + jX_C}{R^2 + X_C^2}$$

i.e., $Y = \dfrac{R}{R^2 + X_C^2} + j\dfrac{X_C}{R^2 + X_C^2}$ or

$Y = \dfrac{R}{|Z|^2} + j\dfrac{X_C}{|Z|^2}$

Thus, conductance, $G = R/|Z|^2$ and capacitive susceptance, $BC = XC/|Z|^2$

(Note that in a capacitive circuit, the imaginary term of the impedance, $X_C$, is negative, whereas the imaginary term of the admittance, $B_C$, is positive.)

(f)  resistance and inductance in parallel, then,

$$\frac{1}{Z} = \frac{1}{R} + \frac{1}{jX_L} = \frac{jX_L + R}{(R)(jX_L)}$$

from which, $Z = \dfrac{(R)(jX_L)}{R + jX_L}\left(\text{i.e., }\dfrac{\text{product}}{\text{sum}}\right)$

and, $Y = \dfrac{1}{Z} = \dfrac{R + jX_L}{jRX_L} = \dfrac{R}{jRX_L} + \dfrac{jX_L}{jRX_L}$

i.e., $Y = \dfrac{1}{jX_L} + \dfrac{1}{R} = \dfrac{(-j)}{(jX_L)(-j)} + \dfrac{1}{R}$

or, $Y = \dfrac{1}{R} - \dfrac{j}{X_L}$

Thus, conductance, $\mathbf{G = 1/R}$ and inductive susceptance, $\mathbf{B_L = -1/X_L}$.

(g)  resistance and capacitance in parallel, then,

$$Z = \frac{(R)(-jX_C)}{R - jX_C}\left(\text{i.e., }\frac{\text{product}}{\text{sum}}\right)$$

and $Y = \dfrac{1}{Z} = \dfrac{R - jX_C}{-jRX_C} = \dfrac{R}{-jRX_C} - \dfrac{jX_C}{-jRX_C}$

i.e., $Y = \dfrac{1}{-jX_C} + \dfrac{1}{R} = \dfrac{(j)}{(-jX_C)(j)} + \dfrac{1}{R}$

or,   $Y = \dfrac{1}{R} + \dfrac{j}{X_C}$    (7.1)

Thus, conductance, $G = 1/R$ and capacitive susceptance, $B_C = 1/X_C$

The conclusions that may be drawn from sections (d) to (g) above are:

(i) that a *series* circuit is more easily represented by an *impedance*,

(ii) that a *parallel* circuit is often more easily represented by an *admittance* especially when more than two parallel impedances are involved.

### Example 7.11

Determine the admittance, conductance and susceptance of the following impedances:
(a) $-j5\,\Omega$ (b) $(25 + j40)\,\Omega$ (c) $(3 - j2)\,\Omega$ (d) $50\angle40°\,\Omega$.

### Solution

(a) If impedance $Z = -j5\,\Omega$, then,

$\text{admittance } Y = \dfrac{1}{Z} = \dfrac{1}{-j5} = \dfrac{j}{(-j5)(j)} = \dfrac{j}{5}$

$\quad\quad = j\,0.2\,\text{S}$  or  $0.2\angle90°\,\text{S}$

Since there is no real part, *conductance*, $G = 0$, and *capacitive susceptance*, $B_C = 0.2\,\text{S}$.

(b) If impedance $Z = (25 + j40)\,\Omega$ then,

$\text{Admittance } Y = \dfrac{1}{Z} = \dfrac{1}{(25 + j40)} = \dfrac{25 - j40}{25^2 + 40^2}$

$\quad\quad = \dfrac{25}{2225} - \dfrac{j40}{2225} = (0.0112 - j0.0180)\,\text{S}$

Thus, *conductance*, $G = 0.0112$ S and *inductive susceptance*, $B_L = 0.0180\,\text{S}$.

(c) If impedance $Z = (3 - j2)\,\Omega$, then,

admittance $Y = \dfrac{1}{Z} = \dfrac{1}{(3 - j2)} = \dfrac{3 + j2}{3^2 + 2^2}$

$\qquad = \left(\dfrac{3}{13} + j\dfrac{2}{13}\right) \text{S or} \qquad \mathbf{(0.231 + j0.154)\,S}$

Thus, *conductance, $G = 0.231$ S* and *capacitive susceptance, $B_C = 0.154\,S$*

(d) If impedance $Z = 50\angle40°\,\Omega$, then,

admittance $Y = \dfrac{1}{Z} = \dfrac{1}{50\angle40°} = \dfrac{1\angle0°}{50\angle40°}$

$\qquad = \dfrac{1}{50}\angle{-40°} = \mathbf{0.02\angle{-40°}\,S} \quad \text{or}$

$$\mathbf{(0.0153 - j0.0129)\,S}$$

Thus, *conductance, $G = 0.0153\,S$* and *inductive susceptance, $B_L = 0.0129$ S.*

### Example 7.12

Determine expressions for the impedance of the following admittances: (a) $0.004\angle30°$ S (b) $(0.001 - j0.002)$ S (c) $(0.05 + j\,0.08)\,S$.

### Solution

(a) Since admittance $Y = 1/Z$, impedance $Z = 1/Y$.

Hence, impedance $Z = \dfrac{1}{0.004\angle30°} = \dfrac{1\angle0°}{0.004\angle30°}$

$\qquad = \mathbf{250\angle{-30°}\,\Omega} \quad \text{or} \quad \mathbf{(216.5 - j125)\,\Omega}$

(b) Impedance $Z = \dfrac{1}{(0.001 - j0.002)}$

$\qquad = \dfrac{0.001 + j0.002}{(0.001)^2 + (0.002)^2}$

$\qquad = \dfrac{0.001 + j0.002}{0.000\,005}$

$\qquad = \mathbf{(200 + j400)\,\Omega} \text{ or } \mathbf{447.2\,\angle63.43°\,\Omega}$

(c)  Admittance $Y = (0.05 + j0.08)$ S $= 0.094\angle 57.99°$ S

Hence, impedance  $Z = \dfrac{1}{0.0094\angle 57.99°}$

$$= \mathbf{10.64\angle{-}57.99°\,\Omega} \text{ or } \mathbf{(5.64 - j9.02)\,\Omega}$$

### Example 7.13

The admittance of a circuit is $(0.040 + j0.025)$ S. Determine the values of the resistance and the capacitive reactance of the circuit if they are connected (a) in parallel, (b) in series. Draw the phasor diagram for each of the circuits.

### Solution

(a)  Parallel connection

Admittance $Y = (0.040 + j0.025)$ S, therefore conductance, $G = 0.040$ S and capacitive susceptance, $B_C = 0.025$ S. From equation (7.1) when a circuit consists of resistance $R$ and capacitive reactance in parallel, then $Y = (1/R) + (j/X_C)$.

Hence, resistance  $R = \dfrac{1}{G} = \dfrac{1}{0.040} = \mathbf{25\,\Omega}$

and capacitive reactance  $X_C = \dfrac{1}{B_C} = \dfrac{1}{0.025} = \mathbf{40\,\Omega}$

The circuit and phasor diagrams are shown in Figure 7.14.

(b)  Series connection

Admittance $Y = (0.040 + j0.025)$ S, therefore,

Impedance  $Z = \dfrac{1}{Y} = \dfrac{1}{0.040 + j0.025}$

$$= \dfrac{0.040 - j0.025}{(0.040)^2 + (0.025)^2}$$

$$= (17.98 - j11.24)\,\Omega$$

Thus, the **resistance, $R = 17.98\,\Omega$** and **capacitive reactance, $X_C = 11.24\,\Omega$.**

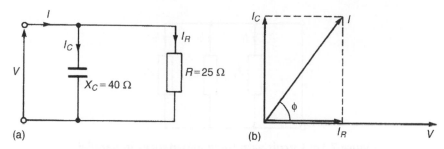

**Figure 7.14: (a) Circuit diagram; (b) Phasor diagram**

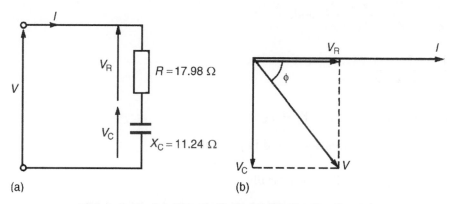

**Figure 7.15: (a) Circuit diagram; (b) Phasor diagram**

The circuit and phasor diagrams are shown in Figure 7.15.

The circuits shown in Figures 7.14(a) and 7.15(a) are equivalent in that they take the same supply current $I$ for a given supply voltage $V$; the phase angle $\phi$ between the current and voltage is the same in each of the phasor diagrams shown in Figures 7.14(b) and 7.15(b).

### 7.6.2 Parallel AC Networks

Figure 7.16 shows a circuit diagram containing three impedances, $Z_1$, $Z_2$ and $Z_3$ connected in parallel. The potential difference across each impedance is the same, i.e., the supply voltage $V$. Current $I_1 = V/Z_1$, $I_2 = V/Z_2$ and $I_3 = V/Z_3$. If $Z_T$ is the total equivalent impedance of the circuit then $I = V/Z_T$. The supply current, $I = I_1 + I_2 + I_3$ (phasorially).

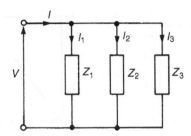

**Figure 7.16: Circuit with three impedances in parallel**

Thus, $\dfrac{V}{Z_T} = \dfrac{V}{Z_1} + \dfrac{V}{Z_2} + \dfrac{V}{Z_3}$ and,

$$\frac{1}{Z_T} = \frac{1}{Z_1} + \frac{1}{Z_2} + \frac{1}{Z_3}$$

or total admittance, $Y_T = Y_1 + Y_2 + Y_3$

In general, for $n$ impedances connected in parallel,

$$Y_T = Y_1 + Y_2 + Y_3 + \cdots + Y_n \quad \text{(phasorially)}$$

It is in parallel circuit analysis that the use of admittance has its greatest advantage.

### 7.6.2.1 Current Division in AC Circuits

For the special case of two impedances, $Z_1$ and $Z_2$, connected in parallel (see Figure 7.17),

$$\frac{1}{Z_T} = \frac{1}{Z_1} + \frac{1}{Z_2} = \frac{Z_2 + Z_1}{Z_1 Z_2}$$

The total impedance, $\mathbf{Z}_T = \mathbf{Z}_1\mathbf{Z}_2/(\mathbf{Z}_1 + \mathbf{Z}_2)$ (i.e., product/sum).

From Figure 7.17,

supply voltage, $V = IZ_T = I\left(\dfrac{Z_1 Z_2}{Z_1 + Z_2}\right)$

Also,           $V = I_1 Z_1$ (and $V = I_2 Z_2$)

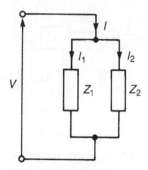

**Figure 7.17: Two impedances connected in parallel**

Thus, $\qquad I_1 Z_1 = I \left( \dfrac{Z_1 Z_2}{Z_1 + Z_2} \right)$

i.e., $\quad$ current $I_1 = I \left( \dfrac{Z_2}{Z_1 + Z_2} \right)$

Similarly, current $I_2 = I \left( \dfrac{Z_1}{Z_1 + Z_2} \right)$

Note that all of the above circuit symbols infer complex quantities either in Cartesian or polar form.

The following problems show how complex numbers are used to analyze parallel AC networks.

***Example 7.14***
Determine the values of currents $I$, $I_1$ and $I_2$ shown in the network of Figure 7.18.

***Solution***
Total circuit impedance,

$$Z_T = 5 + \frac{(8)(j6)}{8 + j6} = 5 + \frac{(j48)(8 - j6)}{8^2 + 6^2}$$

$$= 5 + \frac{j384 + 288}{100}$$

$$= (7.88 + j3.84)\,\Omega \quad \text{or} \quad 8.77\angle25.98^\circ\,\Omega$$

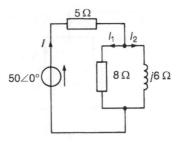

**Figure 7.18: Network for Example 7.14**

Current $I = \dfrac{V}{Z_T} = \dfrac{50\angle 0°}{8.77\angle 25.98°} = \mathbf{5.70\angle{-25.98°}A}$

Current $I_1 = I\left(\dfrac{j6}{8+j6}\right)$

$= (5.70\angle{-25.98°})\left(\dfrac{6\angle 90°}{10\angle 36.87°}\right)$

$= \mathbf{3.42\angle 27.15°\ A}$

Current $I_2 = I\left(\dfrac{8}{8+j6}\right)$

$= (5.70\angle{-25.98°})\left(\dfrac{8\angle 0°}{10\angle 36.87°}\right)$

$= \mathbf{4.56\angle{-62.85°}\ A}$

[Note: $I = I_1 + I_2 = 3.42\angle 27.15° + 4.56\angle{-62.85°}$

$= (3.043 + j1.561) + (2.081 - j4.058)$

$= (5.124 - j2.497)A$

$= 5.70\angle{-25.98°}\ A]$

### Example 7.15

For the parallel network shown in Figure 7.19, determine the value of supply current $I$ and its phase relative to the 40 V supply.

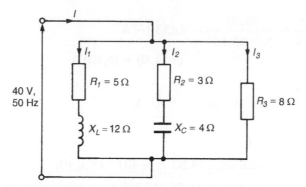

**Figure 7.19: Parallel network for Example 7.15**

*Solution*

Impedance $Z_1 = (5 + j12)\,\Omega$, $Z_2 = (3 - j4)\,\Omega$ and $Z_3 = 8\,\Omega$ Supply current

$I = \dfrac{V}{Z_T} = VY_T$ where $Z_T$ = total circuit impedance, and $Y_T$ = total circuit admittance.

$$Y_T = Y_1 + Y_2 + Y_3$$
$$= \frac{1}{Z_1} + \frac{1}{Z_2} + \frac{1}{Z_3} = \frac{1}{(5 + j12)} + \frac{1}{(3 - j4)} + \frac{1}{8}$$
$$= \frac{5 - j12}{5^2 + 12^2} + \frac{3 + j4}{3^2 + 4^2} + \frac{1}{8}$$
$$= (0.0296 - j0.0710) + (0.1200 + j0.1600) + (0.1250)$$

i.e., $Y_T = (0.2746 + j0.0890)\,\text{S}$ or $0.2887\angle17.96°\,\text{S}$

Current $I = VY_T = (40\angle0°)(0.2887\angle17.96°)$
$$= 11.55\angle17.96°\,\text{A}$$

**Hence, the current $I$ is 11.55A and is leading the 40 V supply by 17.96°.**

Alternatively, current $I = I_1 + I_2 + I_3$

Current $I_1 = \dfrac{40\angle0°}{5 + j12} = \dfrac{40\angle0°}{13\angle67.38°}$
$$= 3.077\angle{-67.38°}\,\text{A or } (1.183 - j2.840)\,\text{A}$$

Current $I_2 = \dfrac{40\angle 0°}{3 - j4} = \dfrac{40\angle 0°}{5\angle -53.13°} = 8\angle 53.13° \text{A}$

$$\text{or } (4.80 + j6.40)\,\text{A}$$

Current $I_3 = \dfrac{40\angle 0°}{8\angle 0°} = 5\angle 0° \text{A} \quad \text{or} \quad (5 + j0)\,\text{A}$

Thus, current $I = I_1 + I_2 + I_3$

$$= (1.183 - j2.840) + (4.80 + j6.40) + (5 + j0)$$

$$= 10.983 + j3.560 = \mathbf{11.55\angle 17.96°\,A}, \text{ as previously obtained.}$$

### Example 7.16

An AC network consists of a coil, of inductance 79.58 mH and resistance 18 Ω, in parallel with a capacitor of capacitance 64.96 μF. If the supply voltage is 250∠0°V at 50 Hz, determine (a) the total equivalent circuit impedance, (b) the supply current, (c) the circuit phase angle, (d) the current in the coil, and (e) the current in the capacitor.

### Solution

The circuit diagram is shown in Figure 7.20.

Inductive reactance, $X_L = 2\pi fL = 2\pi(50)(79.58 \times 10^{-3})$
$$= 25\,\Omega$$

Hence, the impedance of the coil,

$$Z_{\text{COIL}} = (R + jX_L) = (18 + j25)\,\Omega \text{ or } 30.81\angle 54.25°\,\Omega$$

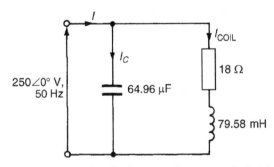

**Figure 7.20: Circuit diagram for Example 7.16**

Capacitive reactance, $X_C = \dfrac{1}{2\pi fC}$

$$= \dfrac{1}{2\pi(50)(64.96 \times 10^{-6})}$$

$$= 49\,\Omega$$

In complex form, the impedance presented by the capacitor $Z_C$ is $-jX_C$, i.e., $-j49\,\Omega$ or $49\angle-90°\,\Omega$.

(a) Total equivalent circuit impedance,

$$Z_T = \dfrac{Z_{\text{COIL}}\,X_C}{Z_{\text{COIL}} + Z_C} \left( \text{i.e., } \dfrac{\text{product}}{\text{sum}} \right)$$

$$= \dfrac{(30.81\angle 54.25°)(49\angle -90°)}{(18 + j25) + (-j49)}$$

$$= \dfrac{(30.81\angle 54.25°)(49\angle -90°)}{18 - j24}$$

$$= \dfrac{(30.81\angle 54.25°)(49\angle -90°)}{30\angle - 53.13°}$$

$$= 50.32\angle(54.25° - 90° - (-53.13°))$$

$$= \mathbf{50.32\angle 17.38°} \text{ or } \mathbf{(48.02 + j15.03)}\,\boldsymbol{\Omega}$$

(b) Supply current $I = \dfrac{V}{Z_T} = \dfrac{250\angle 0°}{50.32\angle 17.38°}$

$$= \mathbf{4.97\angle-17.38°\ A}$$

(c) Circuit phase angle $= 17.38°$ lagging, i.e., the current $I$ lags the voltage $V$ by $17.38°$.

(d) Current in the coil, $I_{\text{COIL}} = \dfrac{V}{Z_{\text{COIL}}} = \dfrac{250\angle 0°}{30.81\angle 54.25°}$

$$= \mathbf{8.11\angle-54.25°\ A}$$

(e) Current in the capacitor, $I_C = \dfrac{V}{Z_C} = \dfrac{250\angle 0°}{49\angle -90°}$

$$= \mathbf{5.10\angle 90°\ A}$$

# CHAPTER 8

# Transients and Laplace Transforms

**John Bird**

## 8.1 Introduction

A *transient state* will exist in a circuit containing one or more energy storage elements (i.e., capacitors and inductors) whenever the energy conditions in the circuit change, until the new *steady state* condition is reached. Transients are caused by changing the applied voltage or current, or by changing any of the circuit elements; such changes occur due to opening and closing switches. In this chapter, such equations are developed analytically by using both *differential equations* and *Laplace transforms* for different waveform supply voltages.

## 8.2 Response of *R-C* Series Circuit to a Step Input

### 8.2.1 Charging a Capacitor

A series *R-C* circuit is shown in Figure 8.1(a).

A step voltage of magnitude *V* is shown in Figure 8.1(b). The capacitor in Figure 8.1(a) is assumed to be initially uncharged.

From Kirchhoff's voltage law, supply voltage,

$$V = v_C + v_R \tag{8.1}$$

Voltage $v_R = iR$ and current $i = C\dfrac{dv_c}{dt}$, so, $v_R = CR\dfrac{dv_C}{dt}$

Therefore, from equation (8.1)

$$V = v_C + CR\frac{dv_C}{dt} \tag{8.2}$$

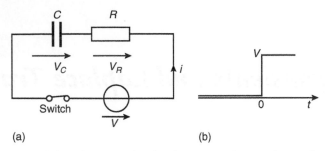

**Figure 8.1: (a) Series R-C circuit; (b) Step voltage of magnitude V**

This is a linear, constant coefficient, first order differential equation. Such a differential equation may be solved (find an expression for voltage $v_C$) by separating the variables. Rearranging equation (8.2) gives:

$$V - v_C = CR\frac{dv_C}{dt}$$

and

$$\frac{dv_C}{dt} = \frac{V - v_C}{CR}$$

from which,

$$\frac{dv_C}{V - v_c} = \frac{dt}{CR}$$

and integrating both sides gives $\int\frac{dv_C}{V - v_C} = \int\frac{dt}{CR}$

Hence, $-\ln(V - v_C) = \dfrac{t}{CR} + k$            (8.3)

where $k$ is the arbitrary constant of integration.

(To integrate $\int\dfrac{dv_C}{V - v_C}$ make an algebraic substitution, $u = V - v_C$—see *Engineering Mathematics* or *Higher Engineering Mathematics*, J.O. Bird, 2004, 4th edition, Elsevier.)

When time $t = 0$, $v_C = 0$, hence, $-\ln V = k$.

Thus, from equation (8.3), $-\ln(V - v_C) = \dfrac{t}{CR} - \ln V$

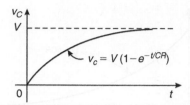

**Figure 8.2: Exponential growth curve of Equation 8.4**

Rearranging gives:

$$\ln V - \ln(V - v_C) = \frac{t}{CR}$$

$$\ln \frac{V}{V - v_C} = \frac{t}{CR} \text{ by the laws of logarithms}$$

i.e., $\qquad \dfrac{V}{V - v_C} = e^{t/CR}$

and $\qquad \dfrac{V - v_C}{V} = \dfrac{1}{e^{t/CR}} = e^{-t/CR}$

$$V - v_C = Ve^{-t/CR}$$

$$V - Ve^{-t/CR} = v_C$$

i.e., capacitor voltage, $v_c = V(1 - e^{-t/CR})$ \hfill (8.4)

This is an exponential growth curve, as shown in Figure 8.2.

From equation (8.1),

$$v_R = V - v_C$$

$$= V - [V(1 - e^{-t/CR})] \text{ from equation (8.4)}$$

$$= V - V + Ve^{-t/CR}$$

i.e., resistor voltage, $v_R = Ve^{-t/CR}$ \hfill (8.5)

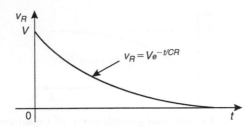

**Figure 8.3: Exponential decay curve of Equation 8.5**

This is an exponential decay curve, as shown in Figure 8.3.

In the circuit of Figure 8.1(a), current $i = C\dfrac{dv_C}{dt}$

Hence, $i = C\dfrac{d}{dt}[V(1 - e^{-t/CR})]$ from equation (8.4)

i.e., $i = C\dfrac{d}{dt}[V - Ve^{-t/CR}]$

$$= C\left[0 - (V)\left(\dfrac{-1}{CR}\right)e^{-t/CR}\right]$$

$$= C\left[\dfrac{V}{CR}e^{-t/CR}\right]$$

So, current, $i = \dfrac{V}{R}e^{-t/CR}$                                    (8.6)

where $\dfrac{V}{R}$ is the steady state current, $I$.

This is an exponential decay curve as shown in Figure 8.4.

After a period of time, it can be determined from equations (8.4) to (8.6) that the voltage across the capacitor, $v_C$, attains the value $V$, the supply voltage, while the resistor voltage, $v_R$, and current $i$ both decay to zero.

### Example 8.1
A 500 nF capacitor is connected in series with a 100 kΩ resistor and the circuit is connected to a 50 V, DC supply. Calculate (a) the initial value of current flowing,

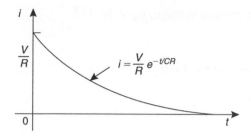

**Figure 8.4: Exponential decay curve of Equation 8.6**

(b) the value of current 150 ms after connection, (c) the value of capacitor voltage 80 ms after connection, and (d) the time after connection when the resistor voltage is 35 V.

***Solution***

(a) From equation (8.6), current, $i = \dfrac{V}{R} e^{-t/CR}$

**Initial current**, i.e., when $t = 0$,

$$i_0 = \frac{V}{R} e^0 = \frac{V}{R} = \frac{50}{100 \times 10^3} = \mathbf{0.5\,mA}$$

(b) Current, $i = \dfrac{V}{R} e^{-t/CR}$ so, when time $t = 150$ ms or 0.15 s,

$$i = \frac{50}{100 \times 10^3} e^{-0.5/(500 \times 10^{-9})(100 \times 10^3)}$$

$$= (0.5 \times 10^{-3}) e^{-3} = (0.5 \times 10^{-3})(0.049787)$$

$$= \mathbf{0.0249\ mA\ or\ 24.9\ \mu A}$$

(c) From equation (8.4), capacitor voltage, $v_C = V(1 - e^{-t/CR})$

When time t = 80 ms,

$$v_C = 50(1 - e^{-80 \times 10^{-3}/(500 \times 10^{-3} \times 100 \times 10^3)})$$

$$= 50(1 - e^{-1.6}) = 50(0.7981)$$

$$= \mathbf{39.91\ V}$$

(d) From equation (8.5), resistor voltage, $v_R = Ve^{-t/CR}$

When   $v_R = 35V$,

then    $35 = 50e^{-t/(500\times10^{-9}\times100\times10^3)}$

i.e.,   $\dfrac{35}{50} = e^{-t/0.05}$

and   $\ln\dfrac{35}{50} = \dfrac{-t}{0.05}$

from which, time $t = -0.05\ln 0.7$

$$= 0.0178s \text{ or } \mathbf{17.8\ ms}$$

### 8.2.2  Discharging a Capacitor

If after a period of time the step input voltage $V$ applied to the circuit of Figure 8.1 is suddenly removed, by opening the switch, then

from equation (8.1),          $v_R + v_C = 0$

or, from equation (8.2),   $CR\dfrac{dv_C}{dt} + v_C = 0$

Rearranging gives:          $\dfrac{dv_C}{dt} = \dfrac{-1}{CR}v_C$

and separating the variables gives: $\dfrac{dv_C}{v_C} = -\dfrac{dt}{CR}$

and integrating both sides gives: $\int\dfrac{dv_C}{v_C} = \int-\dfrac{dt}{CR}$

from which,      $\ln v_C = -\dfrac{t}{CR} + k$          (8.7)

where $k$ is a constant.

At time $t = 0$ (i.e., at the instant of opening the switch), $v_C = V$

Substituting $t = 0$ and $v_C = V$ in equation (8.7) gives:

$\ln V = 0 + k$

Substituting $k = \ln V$ into equation (8.7) gives:

$$\ln v_C = -\frac{t}{CR} + \ln V$$

and
$$\ln v_C - \ln V = -\frac{t}{CR}$$

$$\ln \frac{v_C}{V} = -\frac{t}{CR}$$

and
$$\frac{v_C}{V} = e^{-t/CR}$$

from which,     $v_C = Ve^{-t/CR}$     (8.8)

That is, the capacitor voltage, $v_C$, decays to zero after a period of time, the rate of decay depending on $CR$, which is the *time constant*, $\tau$. Since $v_R + v_C = 0$ then the magnitude of the resistor voltage, $v_R$, is given by:

$$v_R = Ve^{-t/CR}$$     (8.9)

and since $i = C\dfrac{dv_C}{dt} = C\dfrac{d}{dt}(Ve^{-t/CR})$

$$= (CV)\left(-\frac{1}{CR}\right)e^{-t/CR}$$

i.e., the magnitude of the current,

$$i = \frac{V}{R}e^{-t/CR}$$     (8.10)

### Example 8.2

A DC voltage supply of $200\,\mathrm{V}$ is connected across a $5\,\mu\mathrm{F}$ capacitor as shown in Figure 8.5. When the supply is suddenly cut by opening switch S, the capacitor is left isolated except for a parallel resistor of $2\,\mathrm{M}\Omega$. Calculate the voltage across the capacitor after $20\,\mathrm{s}$.

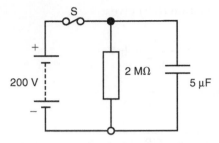

**Figure 8.5: Circuit for Example 8.2**

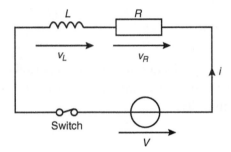

**Figure 8.6: Series R-L circuit**

*Solution*

From equation (8.8), $v_C = Ve^{-t/CR}$

After 20 s, $v_C = 200e^{-20/(5\times10^{-6}\times2\times10^{6})} = 200e^{-2}$

$$= 200(0.13534)$$

$$= \mathbf{27.07\,V}$$

## 8.3 Response of *R-L* Series Circuit to a Step Input

### 8.3.1 Current Growth

A series *R-L* circuit is shown in Figure 8.6. When the switch is closed and a step voltage *V* is applied, it is assumed that *L* carries no current.

From Kirchhoff's voltage law, $V = v_L + v_R$

Voltage $v_L = L\dfrac{di}{dt}$ and voltage $v_R = iR$

Hence, $V = L\dfrac{di}{dt} + iR$ (8.11)

This is a linear, constant coefficient, first order differential equation.

Again, such a differential equation may be solved by separating the variables.

Rearranging equation (8.11) gives: $\dfrac{di}{dt} = \dfrac{V - iR}{L}$

from which, $\dfrac{di}{V - iR} = \dfrac{dt}{L}$

and $\displaystyle\int \dfrac{di}{V - iR} = \int \dfrac{dt}{L}$

Hence, $-\dfrac{1}{R}\ln(V - iR) = \dfrac{t}{L} + k$ (8.12)

where $k$ is a constant.

(Use the algebraic substitution $u = V - iR$ to integrate $\displaystyle\int \dfrac{di}{V - iR}$ )

At time $t = 0$, $i = 0$, thus, $-\dfrac{1}{R}\ln V = 0 + k$

Substituting $k = -\dfrac{1}{R}\ln V$ in equation (8.12) gives:

$-\dfrac{1}{R}\ln(V - iR) = \dfrac{t}{L} - \dfrac{1}{R}\ln V$

Rearranging gives: $\dfrac{1}{R}[\ln V - \ln(V - iR)] = \dfrac{t}{L}$

and $\ln\left(\dfrac{V}{V - iR}\right) = \dfrac{Rt}{L}$

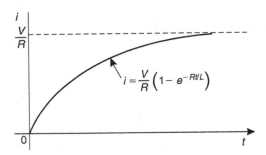

**Figure 8.7: Exponential growth curve of Equation 8.13**

Hence,    $\dfrac{V}{V - iR} = e^{Rt/L}$

and    $\dfrac{V - iR}{V} = \dfrac{1}{e^{Rt/L}} = e^{-Rt/L}$

$V - iR = Ve^{-Rt/L}$

$V - Ve^{-Rt/L} = iR$

and current,    $i = \dfrac{V}{R}(1 - e^{-Rt/L})$    (8.13)

This is an exponential growth curve as shown in Figure 8.7.

The voltage across the resistor in Figure 8.6, $v_R = iR$

Hence, $v_R = R\left[\dfrac{V}{R}(1 - e^{-Rt/L})\right]$ from equation (8.13)

i.e.,    $V_R = V(1 - e^{-Rt/L})$    (8.14)

which again represents an exponential growth curve.

The voltage across the inductor in Figure 8.6, $v_L = L\dfrac{di}{dt}$

i.e., $v_L = L\dfrac{d}{dt}\left[\dfrac{V}{R}(1 - e^{-Rt/L})\right] = \dfrac{LV}{R}\dfrac{d}{dt}[1 - e^{-Rt/L}]$

$= \dfrac{LV}{R}\left[0 - \left(-\dfrac{R}{L}\right)e^{-Rt/L}\right] = \dfrac{LV}{R}\left(\dfrac{R}{L}e^{-Rt/L}\right)$

i.e., $v_L = Ve^{-Rt/L}$ (8.15)

### Example 8.3

A coil of inductance 50 mH and resistance 5 $\Omega$ is connected to a 110 V, DC supply. Determine (a) the final value of current, (b) the value of current after 4 ms, (c) the value of the voltage across the resistor after 6 ms, (d) the value of the voltage across the inductance after 6 ms, and (e) the time when the current reaches 15 A.

### Solution

(a) From equation (8.13), when $t$ is large, the final, or steady state current $i$ is given by:

$$i = \frac{V}{R} = \frac{110}{5} = \mathbf{22\,A}$$

(b) From equation (8.13), current, $i = \dfrac{V}{R}(1 - e^{-Rt/L})$

When $t = 4$ ms, $i = \dfrac{110}{5}(1 - e^{(-(5)(4\times10^{-3})/50\times10^{-3})})$

$= 22(1 - e^{-0.40}) = 22(0.32968)$

$= \mathbf{7.25\,V}$

(c) From equation (8.14), the voltage across the resistor,
$v_R = V(1 - e^{-Rt/L})$

When $t = 6$ ms, $v_R = 110(1 - e^{(-(5)(6\times10^{-3})/50\times10^{-3})})$

$= 110(1 - e^{-0.60}) = 110(0.45119)$

$= \mathbf{49.63\,V}$

(d) From equation (8.15), the voltage across the inductance, $v_L = Ve^{-Rt/L}$

When $t = 6\,\text{ms}$,

$$v_L = 110e^{(-(5)(6\times10^{-3})/50\times10^{-3})} = 110e^{-0.60}$$

$$= \textbf{60.37 V}$$

(Note that at $t = 6\,\text{ms}$, $v_L + v_R = 60.37 + 49.63 = 110\text{V} = $ supply voltage, $V$.)

(e) When current $i$ reaches 15A,

$$15 = \frac{V}{R}(1 - e^{-Rt/L}) \quad \text{from equation (8.13)}$$

i.e.,    $$15 = \frac{110}{5}(1 - e^{-5t/(50\times10^{-3})})$$

$$15\left(\frac{5}{110}\right) = 1 - e^{-100t}$$

and    $$e^{-100t} = 1 - \frac{75}{110}$$

Hence, $$-100t = \ln\left(1 - \frac{75}{110}\right)$$

and    **time, $t$** $$= \frac{1}{-100}\ln\left(1 - \frac{75}{100}\right)$$

$$= \textbf{0.01145 s or 11.45 ms}$$

### 8.3.2  Current Decay

If after a period of time the step voltage $V$ applied to the circuit of Figure 8.6 is suddenly removed by opening the switch, then from equation (8.11),

$$0 = L\frac{di}{dt} + iR$$

Rearranging gives: $L\dfrac{di}{dt} = -iR$ or $\dfrac{di}{dt} = -\dfrac{iR}{L}$

Separating the variables gives: $\dfrac{di}{i} = -\dfrac{R}{L}\,dt$

and integrating both sides gives:

$$\int \dfrac{di}{i} = \int -\dfrac{R}{L}\,dt$$

$$\ln i = -\dfrac{R}{L}t + k \qquad\qquad (8.16)$$

At $t = 0$ (i.e., when the switch is opened),

$$i = I\left(= \dfrac{V}{R},\ \text{the steady state current}\right)$$

then $\ln I = 0 + k$

Substituting $k = \ln I$ into equation (8.16) gives:

$$\ln i = -\dfrac{R}{L}t + \ln I$$

Rearranging gives: $\ln i - \ln I = -\dfrac{R}{L}t$

$$\ln \dfrac{i}{I} = -\dfrac{R}{L}t$$

$$\dfrac{i}{I} = e^{-Rt/L}$$

and current, $\qquad i = Ie^{-Rt/L} \ \text{or} \ \dfrac{V}{R}e^{-Rt/L} \qquad\qquad (8.17)$

i.e., the current $i$ decays exponentially to zero.

From Figure 8.6, $v_R = iR = R\left(\dfrac{V}{R}e^{-Rt/L}\right)$ from equation (8.17)

So, $\qquad v_R = Ve^{-Rt/L} \qquad\qquad (8.18)$

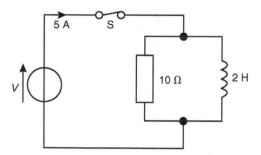

**Figure 8.8: Circuit for Example 8.4**

The voltage across the coil, $v_L = L\dfrac{di}{dt} = L\dfrac{d}{dt}\left(\dfrac{V}{R}e^{-Rt/L}\right)$ from equation (8.17)

$$= L\left(\frac{V}{R}\right)\left(-\frac{R}{L}\right)e^{-Rt/L}$$

The magnitude of $v_T$ is given by:

$$v_L = Ve^{-Rt/L} \tag{8.19}$$

Both $v_R$ and $v_L$ decay exponentially to zero.

### Example 8.4
In the circuit shown in Figure. 8.8, a current of 5 A flows from the supply source. Switch S is then opened. Determine (a) the time for the current in the 2 H inductor to fall to 200 mA and (b) the maximum voltage appearing across the resistor.

### Solution
(a) When the supply is cut off, the circuit consists of just the 10 Ω resistor and the 2 H coil in parallel. This is effectively the same circuit as Figure 8.6 with the supply voltage zero.

From equation (8.17), current $i = \dfrac{V}{R}e^{-Rt/L}$

In this case $\dfrac{V}{R} = 5\,\text{A}$, the initial value of current.

When $i = 200\,\text{mA}$ or $0.2\,\text{A}$,

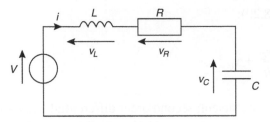

**Figure 8.9: L-R-C circuit**

$$0.2 = 5e^{-10t/2}$$

i.e., $\quad \dfrac{0.2}{5} = e^{-5t}$

thus, $\ln\dfrac{0.2}{5} = -5t$

and time, $t = -\dfrac{1}{5}\ln\dfrac{0.2}{5} = \mathbf{0.644}$ **s** or **644 ms**

(b) Since the current through the coil can only return through the $10\,\Omega$ resistance, the voltage across the resistor is a maximum at the moment of disconnection, i.e.,

$$v_{R_m} = IR = (5)(10) = \mathbf{50V}$$

## 8.4 *L-R-C* Series Circuit Response

*L-R-C* circuits are widely used in a variety of applications, such as in filters in communication systems, ignition systems in automobiles, and defibrillator circuits in biomedical applications (where an electric shock is used to stop the heart, in the hope that the heart will restart with rhythmic contractions).

For the circuit shown in Figure 8.9, from Kirchhoff's voltage law,

$$\mathbf{V} = v_L + v_R + v_C \tag{8.20}$$

$v_L = L\dfrac{di}{dt}$ and $i = C\dfrac{dv_C}{dt}$, hence,

$$v_L = L\frac{d}{dt}\left(C\frac{dv_C}{dt}\right) = LC\frac{d^2v_C}{dt^2}$$

$$v_R = iR = \left(C\frac{dv_C}{dt}\right)R = RC\frac{dv_C}{dt}$$

Hence, from equation (8.20):

$$V = LC\frac{d^2v_C}{dt^2} + RC\frac{dv_C}{dt} + v_C \tag{8.21}$$

This is a linear, constant coefficient, second order differential equation. (For the solution of second order differential equations, see *Higher Engineering Mathematics*).

To determine the transient response, the supply voltage, $V$, is made equal to zero,

i.e.,    $$LC\frac{d^2v_C}{dt^2} + RC\frac{dv_C}{dt} + v_C = 0 \tag{8.22}$$

A solution can be found by letting $v_C = Ae^{mt}$, from which,

$$\frac{dv_C}{dt} = Ame^{mt} \text{ and } \frac{dv_C}{dt^2} = Am^2e^{mt}$$

Substituting these expressions into equation (8.22) gives:

$$LC(Am^2\ e^{mt}) + RC(Ame^{mt}) + Ae^{mt} = 0$$

i.e.,    $$Ae^{mt}(m^2LC + mRC + 1) = 0$$

Thus, $v_C = Ae^{mt}$ is a solution of the given equation provided that

$$m^2LC + mRC + 1 = 0 \tag{8.23}$$

This is called the *auxiliary equation*.

Using the quadratic formula on equation (8.23) gives:

$$m = \frac{-RC \pm \sqrt{[(RC)^2 - 4(LC)(1)]}}{2LC}$$

$$= \frac{-RC \pm \sqrt{(R^2C^2 - 4LC)}}{2LC}$$

i.e., $$m = \frac{-RC}{2LC} \pm \sqrt{\frac{R^2C^2 - 4LC}{(2LC)^2}}$$

$$= -\frac{R}{2L} \pm \sqrt{\left( \frac{R^2C^2}{4L^2C^2} - \frac{4LC}{4L^2C^2} \right)}$$

$$= -\frac{R}{2L} \pm \sqrt{\left[ \left( \frac{R}{2L} \right)^2 - \frac{1}{LC} \right]} \qquad (8.24)$$

This equation may have either:

(i) *two different real roots*, when $(R/2L)^2 > (1/LC)$, when the circuit is said to be *overdamped* since the transient voltage decays very slowly with time, or,

(ii) *two real equal roots*, when $(R/2L)^2 = (1/LC)$, when the circuit is said to be *critically damped* since the transient voltage decays in the minimum amount of time without oscillations occurring, or,

(iii) *two complex roots*, when $(R/2L)^2 < (1/LC)$, when the circuit is said to be *underdamped* since the transient voltage oscillates about the final steady state value, the oscillations eventually dying away to give the steady state value, or,

(iv) if $R = 0$ in equation (8.24), the oscillations would continue indefinitely without any reduction in amplitude—this is the *undamped* condition.

Damping is discussed again in Section 8.8.

### Example 8.5

A series *L-R-C* circuit has inductance $L = 2\,\text{mH}$, resistance $R = 1\,\text{k}\Omega$ and capacitance, $C = 5\,\mu\text{F}$. (a) Determine whether the circuit is over, critical or underdamped. (b) If $C = 5\,\text{nF}$, determine the state of damping.

### Solution

(a) $$\left( \frac{R}{2L} \right)^2 = \left[ \frac{10^3}{2(2 \times 10^{-3})} \right]^2 = \frac{10^{12}}{16} = 6.25 \times 10^{10}$$

$$\frac{1}{LC} = \frac{1}{(2 \times 10^{-3})(5 \times 10^6)} = \frac{10^9}{10} = 10^8$$

Since, $\left(\dfrac{R}{2L}\right)^2 > \dfrac{1}{LC}$ the circuit is *overdamped*.

(b)  When $C = 5\,\text{nF}, \dfrac{1}{LC} = \dfrac{1}{(2 \times 10^{-3})(5 \times 10^{-9})} = 10^{11}$

Since, $\left(\dfrac{R}{2L}\right)^2 < \dfrac{1}{LC}$ the circuit is *underdamped*.

### Example 8.6
In the circuit of Example 8.5, what value of capacitance will give critical damping?

### Solution
For critical damping: $\left(\dfrac{R}{2L}\right)^2 = \dfrac{1}{LC}$

from which, **capacitance**,

$$C = \dfrac{1}{L\left(\dfrac{R}{2L}\right)^2} = \dfrac{1}{L\dfrac{R^2}{4L^2}} = \dfrac{4L^2}{LR^2} = \dfrac{4L}{R^2}$$

$$= \dfrac{4(2 \times 10^{-3})}{(10^3)^2} = 8 \times 10^{-9} \text{ F or } \mathbf{8\,nF}$$

### 8.4.1  Roots of the Auxiliary Equation

With reference to equation (8.24):

(i)  when the roots are real and different, say $m = \alpha$ and $m = \beta$, the general solution is:

$$v_C = Ae^{\alpha t} + Be^{\beta t} \qquad\qquad\qquad\qquad (8.25)$$

where, $\alpha = -\dfrac{R}{2L} + \sqrt{\left[\left(\dfrac{R}{2L}\right)^2 - \dfrac{1}{LC}\right]}$

$\beta = -\dfrac{R}{2L} - \sqrt{\left[\left(\dfrac{R}{2L}\right)^2 - \dfrac{1}{LC}\right]}$

(ii) when the roots are *real and equal*, say $m = \alpha$ twice, the general solution is

$$v_C = (At + B)e^{\alpha t} \tag{8.26}$$

where $\alpha = -\dfrac{R}{2L}$

(iii) when the roots are *complex*, say $m = \alpha \pm j\beta$, the general solution is

$$v_C = e^{\alpha t}\{A\cos\beta t + B\sin\beta t\} \tag{8.27}$$

where $\alpha = -\dfrac{R}{2L}$ and $\beta = \sqrt{\left[\dfrac{1}{LC} - \left(\dfrac{R}{2L}\right)^2\right]} \tag{8.28}$

To determine the actual expression for the voltage under any given initial condition, it is necessary to evaluate constants A and B in terms of $v_C$ and current $i$. The procedure is the same for each of the above three cases. Assuming in, say, case (iii) that at time $t = 0$, $v_C = v_0$ and $i(=C(dv_C/dt)) = i_0$ then substituting in equation (8.27):

$$v_0 = e^0\{A\cos 0 + B\sin 0\}$$

i.e., $\mathbf{v_0 = A}$ $\tag{8.29}$

Also, from equation (8.27),

$$\frac{dv_C}{dt} = e^{\alpha t}[-A\beta\sin\beta t + B\beta\cos\beta t] + [A\cos\beta t + B\sin\beta t](\alpha e^{\alpha t}) \tag{8.30}$$

by the product rule of differentiation.

When $t = 0$, $\dfrac{dv_C}{dt} = e^0[0 + B\beta] + [A](\alpha e^0) = B\beta + \alpha A$

Hence, at $t = 0$, $i_0 = C\dfrac{dv_C}{dt} = C(B\beta + \alpha A)$

From equation (8.29), $A = v_0$ hence $i_0 = C(B\beta + \alpha v_0)$
$$= CB\beta + C\alpha v_0$$

from which, $\mathbf{B = \dfrac{i_0 - C\alpha v_0}{C\beta}}$ $\tag{8.31}$

## Example 8.7

A coil has an equivalent circuit of inductance $1.5\,\text{H}$ in series with resistance $90\,\Omega$. It is connected across a charged $5\,\mu F$ capacitor at the moment when the capacitor voltage is $10\,\text{V}$. Determine the nature of the response and obtain an expression for the current in the coil.

### Solution

$$\left(\frac{R}{2L}\right)^2 = \left[\frac{90}{2(1.5)}\right]^2 = 900 \text{ and } \frac{1}{LC} = \frac{1}{(1.5)(5 \times 10^{-6})}$$

$$= 1.333 \times 10^5$$

Since $\left(\dfrac{R}{2L}\right)^2 < \dfrac{1}{LC}$ the circuit is *underdamped.*

From equation (8.28),

$$\alpha = -\frac{R}{2L} = -\frac{90}{2(1.5)} = -30$$

and $\beta = \sqrt{\left[\dfrac{1}{LC} - \left(\dfrac{R}{2L}\right)^2\right]}$

$$= \sqrt{[1.333 \times 10^5 - 900]} = 363.9$$

With $v_0 = 10\,\text{V}$ and $i_0 = 0$, from equation (8.29), $v_0 = A = 10$

and from equation (8.31),

$$B = \frac{i_0 - C\alpha v_0}{C\beta} = \frac{0 - (5 \times 10^{-6})(-30)(10)}{(5 \times 10^{-6})(363.9)}$$

$$= \frac{300}{363.9} = 0.8244$$

Current, $i = C\dfrac{dv_C}{dt}$, and from equation (8.30),

$$i = C\{e^{-30t}[-10(363.9)\sin \beta t + (0.8244)(363.9)\cos \beta t]$$

$$+ (10 \cos \beta t + 0.8244 \sin \beta t)(-30 e^{-30t})\}$$
$$= C\{e^{-30t}[-3639 \sin \beta t + 300 \cos \beta t - 300 \cos \beta t - 24.732 \sin \beta t]\}$$
$$= C e^{-30t}[-3663.732 \sin \beta t]$$
$$= -(5 \times 10^{-6})(3663.732) e^{-30t} \sin \beta t$$

i.e.,  **current, $i = -0.018\, e^{-30t} \sin 363.9t$ amperes**

## 8.5 Introduction to Laplace Transforms

The solution of most electrical problems can be reduced ultimately to the solution of differential equations and the use of *Laplace transforms* provides an alternative method to those used previously. Laplace transforms provide a convenient method for the calculation of the complete response of a circuit. In this section and in Section 8.6, the technique of Laplace transforms is developed and then used to solve differential equations. In Section 8.7, Laplace transforms are used to analyze transient responses directly from circuit diagrams.

### 8.5.1 Definition of a Laplace Transform

The Laplace transform of the function of time $f(t)$ is defined by the integral

$$\int_0^\infty e^{-st} f(t)\, dt \quad \text{where } s \text{ is a parameter}$$

There are various commonly used notations for the Laplace transform of $f(t)$ and these include $\mathcal{L}\{f(t)\}$ or $L\{f(t)\}$ or $\mathcal{L}(f)$ or Lf or $\overline{f}(s)$.

Also the letter $p$ is sometimes used instead of $s$ as the parameter. The notation used in this chapter will be $f(t)$ for the original function and $\mathcal{L}\{f(t)\}$ for its Laplace transform,

i.e., $\mathcal{L}\{f(t)\} = \displaystyle\int_0^\infty e^{-st} f(t)\, dt$ \hfill (8.32)

### 8.5.2 Laplace Transforms of Elementary Functions

Using equation (8.32):

(i)  when $f(t) = 1$, $\mathcal{L}\{1\} = \displaystyle\int_0^\infty e^{-st}(1)\, dt = \left[ \dfrac{e^{-st}}{-s} \right]_0^\infty$

$$= -\frac{1}{s}[e^{-s(\infty)} - e^0]$$

$$= -\frac{1}{s}[0 - 1]$$

$$= \frac{1}{s} \quad \text{(provided } s > 0\text{)}$$

(ii)  when $f(t) = \mathbf{k}, \mathcal{L}\{k\} = k\mathcal{L}\{1\} = k\left(\dfrac{1}{s}\right) = \dfrac{k}{s}$  from (i) above

(iii)  when $f(t) = \mathbf{e}^{at}, \mathcal{L}\{e^{at}\} = \int_0^\infty e^{-st}(e^{at})dt$

$$= \int e^{-(s-a)t}dt \text{ from the laws of indices}$$

$$= \left[\frac{e^{-(s-a)t}}{-(s-a)}\right]_0^\infty$$

$$= \frac{1}{-(s-a)}(0-1)$$

$$= \frac{1}{s-a} \quad \text{(provided } s > a\text{)}$$

(iv)  when $f(t) = t, \mathcal{L}\{t\} = \int_0^\infty e^{-st}t\,dt$

$$= \left[\frac{te^{-st}}{-s} - \int\frac{e^{-st}}{-s}dt\right]_0^\infty$$

$$= \left[\frac{te^{-st}}{-s} - \frac{e^{-st}}{-s^2}\right]_0^\infty \qquad \text{by integration by parts}$$

$$= \left[\frac{\infty e^{-s(\infty)}}{-s} - \frac{e^{-s(\infty)}}{s^2}\right] - \left[0 - \frac{e^0}{s^2}\right]$$

$$= (0 - 0) - \left(0 - \frac{1}{s^2}\right) \quad \text{since } (\infty \times 0) = 0$$

$$= \frac{1}{s^2} \quad \text{(provided } s > 0)$$

(v)  when $f(t) = \cos \omega t$,

$$\mathscr{L}\{\cos \omega t\} = \int_0^\infty e^{-st} \cos \omega t \, dt$$

$$= \left|\frac{e^{-st}}{s^2 + \omega^2} (\omega \sin \omega t - s \cos \omega t)\right|_0^\infty \quad \text{by integration by parts twice}$$

$$= \frac{s}{s^2 + \omega^2} \quad \text{(provided } s > 0)$$

A list of standard Laplace transforms is summarized in Table 8.1 below. It will not usually be necessary to derive the transforms as above—but merely to use them.

The following worked problems only require using the standard list of Table 8.1.

**Example 8.8**

Find the Laplace transforms of:

(a)  $1 + 2t - \frac{1}{3} t^4$

(b)  $5e^{2t} - 3e^{-t}$

**Solution**

(a)  $\mathscr{L}\left\{1 + 2t - \frac{1}{3} t^4\right\} = \mathscr{L}\{1\} + 2\mathscr{L}\{t\} - \frac{1}{3}\mathscr{L}\{t^4\}$

$$= \frac{1}{s} + 2\left(\frac{1}{s^2}\right) - \frac{1}{3}\left(\frac{4!}{s^{4+1}}\right) \quad \text{from 2, 7 and 9 of Table 8.1}$$

$$= \frac{1}{s} + \frac{2}{s^2} - \frac{1}{3}\left(\frac{4 \times 3 \times 2 \times 1}{s^5}\right)$$

$$= \frac{1}{s} + \frac{2}{s^2} - \frac{8}{s^5}$$

### Table 8.1: Standard Laplace transforms

| Time function $f(t)$ | Laplace transform $\mathcal{L}\{f(t)\} = \int_0^\infty e^{-st} f(t)\, dt$ |
|---|---|
| 1. | $\delta$ (unit impulse) | 1 |
| 2. | 1 (unit step function) | $\dfrac{1}{s}$ |
| 3. | $e^{at}$ (exponential function) | $\dfrac{1}{s-a}$ |
| 4. | unit step delayed by $T$ | $\dfrac{e^{-sT}}{s}$ |
| 5. | Sin $\omega t$ (sine wave) | $\dfrac{\omega}{s^2 + \omega^2}$ |
| 6. | cos $\omega t$ (cosine wave) | $\dfrac{s}{s^2 + \omega^2}$ |
| 7. | $t$ (unit ramp function) | $\dfrac{1}{s^2}$ |
| 8. | $t^2$ | $\dfrac{2!}{s^3}$ |
| 9. | $t^n$ ($n = 1, 2, 3...$) | $\dfrac{n!}{s^{n+1}}$ |
| 10. | cosh $\omega t$ | $\dfrac{s}{s^2 - \omega^2}$ |
| 11. | sinh $\omega t$ | $\dfrac{\omega}{s^2 - \omega^2}$ |
| 12. | $e^{at} t^n$ | $\dfrac{n!}{(s-a)^{n+1}}$ |
| 13. | $e^{-at} \sin \omega t$ (damped sine wave) | $\dfrac{\omega}{(s+a)^2 + \omega^2}$ |
| 14. | $e^{-at} \cos \omega t$ (damped cosine wave) | $\dfrac{s+a}{(s+a)^2 + \omega^2}$ |

**Table 8.1: (Continued)**

| Time function $f(t)$ | | Laplace transform $\mathcal{L}\{f(t)\} = \int_0^\infty e^{-st} f(t)\, dt$ |
|---|---|---|
| 15. | $e^{-at}\sinh \omega t$ | $\dfrac{\omega}{(s+a)^2 - \omega^2}$ |
| 16. | $e^{-at}\cosh \omega t$ | $\dfrac{s+a}{(s+a)^2 - \omega^2}$ |

(b)  $\mathcal{L}\{5e^{2t} - 3e^{-t}\} = 5\mathcal{L}\{e^{2t}\} - 3\mathcal{L}\{e^{-t}\}$

$$= 5\left(\frac{1}{s-2}\right) - 3\left(\frac{1}{s-(-1)}\right) \text{ from 3 of Table 8.1}$$

$$= \frac{5}{s-2} - \frac{3}{s+1}$$

$$= \frac{5(s+1) - 3(s-2)}{(s-2)(s+1)}$$

$$= \frac{2s+11}{s^2 - s - 2}$$

**Example 8.9**

Find the Laplace transform of $6\sin 3t - 4\cos 5t$.

**Solution**

$\mathcal{L}\{6\sin 3t - 4\cos 5t\} = 6\mathcal{L}\{\sin 3t\} - 4\mathcal{L}\{\cos 5t\}$

$$= 6\left(\frac{3}{s^2 + 3^2}\right) - 4\left(\frac{s}{s^2 + 5^2}\right) \text{ from 5 and 6 of Table 8.1}$$

$$= \frac{18}{s^2 + 9} - \frac{4s}{s^2 + 25}$$

**Example 8.10**

Use Table 8.1 to determine the Laplace transforms of the following waveforms:

(a)  a step voltage of 10 V which starts at time $t = 0$,

(b)  a step voltage of 10 V which starts at time $t = 5$ s,

(c) a ramp voltage which starts at zero and increases at 4 V/s,

(d) a ramp voltage which starts at time $t = 1$ s and increases at 4 V/s.

**Solution**

(a) From 2 of Table 8.1,

$$\mathscr{L}\{10\} = 10\mathscr{L}\{1\} = 10\left(\frac{1}{s}\right) = \frac{10}{s}$$

The waveform is shown in Figure 8.10(a).

(b) From 4 of Table 8.1, a step function of 10 V which is delayed by $t = 5$ s is given by:

$$10\left(\frac{e^{-sT}}{s}\right) = 10\left(\frac{e^{-5s}}{s}\right) = \frac{10}{s}e^{-5s}$$

This is, in fact, the function starting at $t = 0$ given in part (a), i.e., $(10/s)$ multiplied by $e^{-sT}$, where $T$ is the delay in seconds.

The waveform is shown in Figure 8.10(b).

(c) From 7 of Table 8.1, the Laplace transform of the unit ramp, $\mathscr{L}\{t\} = (1/s^2)$

Hence, the Laplace transform of a ramp voltage increasing at 4 V/s is given by:

$$4\mathscr{L}\{t\} = \frac{4}{s^2}$$

The waveform is shown in Figure 8.10(c).

(d) As with part (b), for a delayed function, the Laplace transform is the undelayed function, in this case $(4/s^2)$ from part (c), multiplied by $e^{-sT}$ where $T$ in this case is 1 s. The Laplace transform is given by:

$$\left(\frac{4}{s^2}\right)e^{-s}$$

The waveform is shown in Figure 8.10(d).

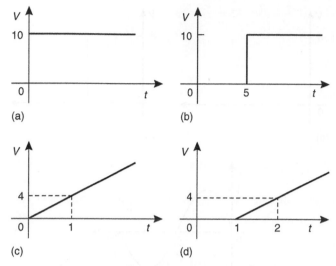

**Figure 8.10: Waveforms for Example 8.10**

*Example 8.11*

Determine the Laplace transforms of the following waveforms:

(a) an impulse voltage of 8 V, which starts at time $t = 0$,

(b) an impulse voltage of 8 V, which starts at time $t = 2\,$s,

(c) a sinusoidal current of 4 A and angular frequency 5 rad/s which starts at time $t = 0$.

*Solution*

(a) An *impulse* is an intense signal of very short duration. This function is often known as the *Dirac function*.

From 1 of Table 8.1, the Laplace transform of an impulse starting at time $t = 0$ is given by $\mathscr{L}\{\delta\} = 1$, hence, an impulse of 8 V is given by: $8\mathscr{L}\{\delta\} = \mathbf{8}$.

This is shown in Figure 8.11(a).

(b) From part (a) the Laplace transform of an impulse of 8 V is 8. Delaying the impulse by 2 s involves multiplying the undelayed function by $e^{-sT}$ where $T = 2\,$s.

Hence, the Laplace transform of the function is given by: $8e^{-2s}$

This is shown in Fig. 8.11(b).

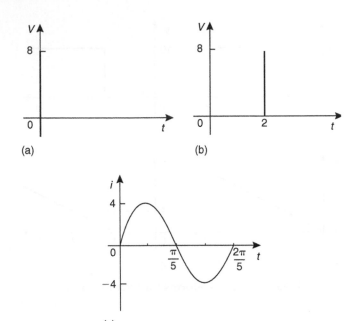

(c)

**Figure 8.11: Graphs for Example 8.11**

(c)  From 5 of Table 8.1, $\mathcal{L}\{\sin \omega t\} = \dfrac{\omega}{s^2 + \omega^2}$

When the amplitude is 4 A and $\omega = 5$, then

$$\mathcal{L}\{4 \sin \omega t\} = 4\left(\frac{5}{s^2 + 5^2}\right) = \frac{20}{s^2 + 25}$$

The waveform is shown in Figure 8.11(c).

**Example 8.12**

Find the Laplace transforms of:

(a)  $2t^4 e^{3t}$

(b)  $4e^{3t} \cos 5t$.

*Solution*

(a) From 12 of Table 8.1,

$$\mathcal{L}\{2t^4e^{3t}\} = 2\mathcal{L}\{t^4e^{3t}\}$$

$$= 2\left[\frac{4!}{(s-3)^{4+1}}\right]$$

$$= \frac{2(4 \times 3 \times 2 \times 1)}{(s-3)^5} = \frac{48}{(s-3)^5}$$

(b) From 14 of Table 8.1,

$$\mathcal{L}\{4e^{3t}\cos 5t\} = 4\mathcal{L}\{e^{3t}\cos 5t\}$$

$$= 4\left[\frac{s-3}{(s-3)^2 + 5^2}\right]$$

$$= \frac{4(s-3)}{s^2 - 6s + 9 + 25)} = \frac{4(s-3)}{s^2 - 6s + 34}$$

*Example 8.13*

Determine the Laplace transforms of:

(a) $2\cosh 3t$,

(b) $e^{-2t}\sin 3t$.

*Solution*

(b) From 10 of Table 8.1,

$$\mathcal{L}\{2\cosh 3t\} = 2\mathcal{L}\cosh 3t = 2\left[\frac{s}{s^2 - 3^2}\right] = \frac{2s}{s^2 - 9}$$

(c) From 13 of Table 8.1,

$$\mathcal{L}\{e^{-2t}\sin 3t\} = \frac{3}{(s+2)^2 + 3^2} = \frac{3}{s^2 + 4s + 4 + 9}$$

$$= \frac{3}{s^2 + 4s + 13}$$

### 8.5.3 Laplace Transforms of Derivatives

Using integration by parts, it may be shown that:

(a) for the first derivative:

$$\mathcal{L}\{f'(t)\} = s\mathcal{L}\{f(t)\} - f(0)$$

or
$$\mathcal{L}\left\{\frac{dy}{dx}\right\} = s\mathcal{L}\{y\} - y(0) \qquad (8.33)$$

where $y(0)$ is the value of $y$ at $x = 0$

(b) for the second derivative:

$$\mathcal{L}\{f''(t)\} = s^2 \mathcal{L}\{f(t)\} - sf(0) - f'(0)$$

or
$$\mathcal{L}\left\{\frac{d^2 y}{dx^2}\right\} = s^2\mathcal{L}\{y\} - sy(0) - y'(0) \qquad (8.34)$$

where $y'(0)$ is the value of $(dy/dx)$ at $x = 0$

Equations (8.33) and (8.34) are used in the solution of differential equations in Section 8.6.

### 8.5.4 The Initial and Final Value Theorems

The initial and final value theorems can often considerably reduce the work of solving electrical circuits.

(a) The *initial value* theorem states:

$$\lim_{t \to 0}[f(t)] = \lim_{s \to \infty}[s\mathcal{L}\{f(t)\}]$$

Thus, for example, if $\quad f(t) = v = Ve^{-t/CR}$ and if, say,
$$V = 10 \text{ and } CR = 0.5, \text{ then}$$
$$f(t) = v = 10e^{-2t}$$
$$\mathcal{L}\{f(t)\} = 10\left(\frac{1}{s+2}\right) \text{ from 3 of Table 8.1}$$
$$s\mathcal{L}\{f(t)\} = 10\left(\frac{s}{s+2}\right)$$

From the initial value theorem, the initial value of $f(t)$ is given by:

$$10\left[\frac{\infty}{\infty + 2}\right] = 10(1) = \mathbf{10}$$

(b) The *final value theorem* states:

$$\lim_{t \to \infty}[f(t)] = \lim_{s \to \infty}[s\mathcal{L}\{f(t)\}]$$

In the above example of $f(t) = 10e^{-2t}$, the final value is given by:

$$10\left[\frac{0}{0 + 2}\right] = \mathbf{0}$$

The initial and final value theorems are used in pulse circuit applications where the response of the circuit for small periods of time, or the behavior immediately the switch is closed, are of interest. The final value theorem is particularly useful in investigating the stability of systems (such as in automatic aircraft-landing systems) and is concerned with the steady state response for large values of time $t$, i.e., after all transient effects have died away.

## 8.6 Inverse Laplace Transforms and the Solution of Differential Equations

Since from 2 of Table 8.1, $\mathcal{L}\{1\} = \dfrac{1}{s}$   then,

$$\mathcal{L}^{-1} = \left\{\frac{1}{s}\right\} = 1$$

where $\mathcal{L}^{-1}$ means the inverse Laplace transform. Similarly, since from 5 of Table 8.1,

$$\mathcal{L}\{\sin \omega t\} = \frac{\omega}{s^2 + \omega^2} \text{ then } \mathcal{L}^{-1}\left\{\frac{\omega}{s^2 + \omega^2}\right\} = \sin \omega t$$

Finding an inverse transform involves locating the Laplace transform from the right-hand column of Table 8.1 and then reading the function from the left-hand column. The following worked problems demonstrate the method.

*Example 8.14*

Find the following inverse Laplace transforms:

(a) $\mathscr{L}^{-1}\left\{\dfrac{1}{s^2+9}\right\}$

(b) $\mathscr{L}^{-1}\left\{\dfrac{5}{3s-1}\right\}$

*Solution*

(a) $\mathscr{L}^{-1}\left\{\dfrac{1}{s^2+9}\right\} = \mathscr{L}^{-1}\left\{\dfrac{1}{s^2+3^2}\right\}$

$$= \frac{1}{3}\,\mathscr{L}^{-1}\left\{\frac{3}{s^2+3^2}\right\}$$

and from 5 of Table 8.1,

$$\frac{1}{3}\,\mathscr{L}^{-1}\left\{\frac{3}{s^2+3^2}\right\} = \frac{1}{3}\sin 3t$$

(b) $\mathscr{L}^{-1}\left\{\dfrac{5}{3s-1}\right\} = \mathscr{L}^{-1}\left\{\dfrac{5}{3\left(s-\dfrac{1}{3}\right)}\right\}$

$$= \frac{5}{3}\,\mathscr{L}^{-1}\left\{\frac{1}{s-\dfrac{1}{3}}\right\} = \frac{5}{3}e^{\frac{1}{3}t} \quad \text{from 3 of Table 8.1}$$

*Example 8.15*

Determine the following inverse Laplace transforms:

(a) $\mathscr{L}^{-1}\left\{\dfrac{6}{s^3}\right\}$

(b) $\mathcal{L}^{-1}\left\{\dfrac{3}{s^4}\right\}$

**Solution**

(a) From 8 of Table 8.1, $\mathcal{L}^{-1}\left\{\dfrac{2}{s^3}\right\} = t^2$

Hence, $\mathcal{L}^{-1}\left\{\dfrac{6}{s^3}\right\} = 3\mathcal{L}^{-1}\left\{\dfrac{2}{s^3}\right\} = \mathbf{3t^2}$

(b) From 9 of Table 8.1, if $s$ is to have a power of 4 then $n = 3$.

Thus, $\mathcal{L}^{-1}\left\{\dfrac{3!}{s^4}\right\} = t^3$, i.e., $\mathcal{L}^{-1}\left\{\dfrac{6}{s^4}\right\} = t^3$

Hence, $\mathcal{L}^{-1}\left\{\dfrac{3}{s^4}\right\} = \dfrac{1}{2}\mathcal{L}^{-1}\left\{\dfrac{6}{s^4}\right\} = \dfrac{\mathbf{1}}{\mathbf{2}}\mathbf{t^3}$

**Example 8.16**

Determine

(a) $\mathcal{L}^{-1}\left\{\dfrac{7s}{s^2 + 4}\right\}$

(b) $\mathcal{L}^{-1}\left\{\dfrac{4s}{s^2 - 16}\right\}$

**Solution**

(a) $\mathcal{L}^{-1}\left\{\dfrac{7s}{s^2 + 4}\right\} = 7\mathcal{L}^{-1}\left\{\dfrac{s}{s^2 + 2^2}\right\}$

$= \mathbf{7\cos 2t}$ from 6 of Table 8.1

(b) $\mathcal{L}^{-1}\left\{\dfrac{4s}{s^2 - 16}\right\} = 4\mathcal{L}^{-1}\left\{\dfrac{s}{s^2 - 4^2}\right\}$

$= \mathbf{4\cosh 4t}$ from 10 of Table 8.1

*Example 8.17*

Find $\mathcal{L}^{-1}\left\{\dfrac{2}{(s-3)^5}\right\}$

**Solution**

From 12 of Table 8.1, $\mathcal{L}^{-1}\left\{\dfrac{n!}{(s-a)^{n+1}}\right\} = e^{at}t^n$

Thus, $\mathcal{L}^{-1}\left\{\dfrac{1}{(s-a)^{n+1}}\right\} = \dfrac{1}{n!}e^{at}t^n$

and comparing with $\mathcal{L}^{-1}\left\{\dfrac{2}{(s-3)^5}\right\}$   shows that $n = 4$ and $a = 3$.

Hence, $\mathcal{L}^{-1}\left\{\dfrac{2}{(s-3)^5}\right\} = 2\mathcal{L}^{-1}\left\{\dfrac{1}{(s-3)^5}\right\}$

$$= 2\left[\dfrac{1}{4!}e^{3t}t^4\right]$$

$$= \dfrac{1}{12}e^{3t}t^4$$

*Example 8.18*

Determine

(a) $\mathcal{L}^{-1}\left\{\dfrac{3}{s^2 - 4s + 13}\right\}$

(b) $\mathcal{L}^{-1}\left\{\dfrac{2(s+1)}{s^2 + 2s + 10}\right\}$

**Solution**

(a) $\mathcal{L}^{-1}\left\{\dfrac{3}{s^2 - 4s + 13}\right\} = \mathcal{L}^{-1}\left\{\dfrac{3}{(s-2)^2 + 3^2}\right\}$

$$= e^{2t}\,\textbf{sin 3}t \text{ from 13 of Table 8.1}$$

(b)  $\mathcal{L}^{-1}\left\{\dfrac{2(s+1)}{s^2+2s+10}\right\} = \mathcal{L}^{-1}\left\{\dfrac{2(s+1)}{(s+1)^2+3^2}\right\}$

$= 2e^{-t}\cos 3t$ from 14 of Table 8.1

Note that in solving these examples, the denominator in each case has been made into a perfect square.

### 8.6.1  Use of Partial Fractions for Inverse Laplace Transforms

Sometimes the function whose inverse is required is not recognizable as a standard type, such as those listed in Table 8.1. In such cases it may be possible, by using *partial fractions*, to resolve the function into simpler fractions which may be inverted on sight. For example, the function $F(s) = \dfrac{2s-3}{s(s-3)}$ cannot be inverted on sight from Table 8.1. However, using partial fractions:

$$\frac{2s-3}{s(s-3)} \equiv \frac{A}{s} + \frac{B}{s-3} = \frac{A(s-3)+Bs}{s(s-3)}$$

from which $2s - 3 = A(s - 3) + Bs$

Letting $s = 0$ gives: $-3 = -3A$ from which $A = 1$

Letting $s = 3$ gives: $3 = 3B$ from which $B = 1$

Hence, $\dfrac{2s-3}{s(s-3)} \equiv \dfrac{1}{s} + \dfrac{1}{s-3}$

Thus, $\mathcal{L}^{-1}\left\{\dfrac{2s-3}{s(s-3)}\right\} = \mathcal{L}^{-1}\left\{\dfrac{1}{s} + \dfrac{1}{(s-3)}\right\}$

$= 1 + e^{3t}$ from 2 and 3 of Table 8.1

Partial fractions are explained in *Engineering Mathematics* and *Higher Engineering Mathematics*. The following worked problems demonstrate the method.

### Example 8.19

Determine $\mathcal{L}^{-1}\left\{\dfrac{4s-5}{s^2-s-2}\right\}$

*Solution*

$$\frac{4s-5}{s^2-s-2} \equiv \frac{4s-5}{(s-2)(s+1)} \equiv \frac{A}{(s-2)} + \frac{B}{(s+1)}$$

$$= \frac{A(s+1)+B(s-2)}{(s-2)(s+1)}$$

Hence, $4s - 5 = A(s + 1) + B(s - 2)$

When $s = 2, 3 = 3A$ from which, $A = 1$

When $s = -1, -9 = -3B$ from which, $B = 3$

Hence, $\mathcal{L}^{-1}\left\{\dfrac{4s-5}{s^2-s-2}\right\} \equiv \mathcal{L}^{-1}\left\{\dfrac{1}{s-2} + \dfrac{3}{s+1}\right\}$

$$= \mathcal{L}^{-1}\left\{\frac{1}{s-2}\right\} + \mathcal{L}^{-1}\left\{\frac{3}{s+1}\right\}$$

$$= e^{2t} + 3e^{-t} \text{ from 3 of Table 8.1}$$

**Example 8.20**

Find $\mathcal{L}^{-1}\left\{\dfrac{3s^3+s^2+12s+2}{(s-3)(s+1)^3}\right\}$

*Solution*

$$\frac{3s^3+s^2+12s+12}{(s-3)(s+1)^3} \equiv \frac{A}{s-3} + \frac{B}{s+1} + \frac{C}{(s+1)^2} + \frac{D}{(s+1)^3}$$

$$A(s+1)^3 + B(s-3)(s+1)^2 \quad = \frac{+C(s-3)(s+1) + D(s-3)}{(s-3)(s+1)^3}$$

Hence, $3s^3 + s^2 + 12s + 2 = A(s+1)^3 + B(s-3)(s+1)^2$
$$+ C(s-3)(s+1) + D(s-3)$$

When $s = 3, 128 = 64A$ from which $A = 2$

When $s = -1, -12 = -4D$ from which $D = 3$

Equating $s^3$ terms gives: $3 = A + B$ from which $B = 1$

Equating $s^2$ terms gives: $1 = 3A - B + C$ from which $C = -4$

Hence, $\mathcal{L}^{-1}\left\{\dfrac{3s^3 + s^2 + 12s + 2}{(s-3)(s+1)^3}\right\}$

$\equiv \mathcal{L}^{-1}\left\{\dfrac{2}{s-3} + \dfrac{1}{s+1} - \dfrac{4}{(s+1)^2} + \dfrac{3}{(s+1)^3}\right\}$

$= 2e^{3t} + e^{-t} - 4e^{-t}t + \dfrac{3}{2}e^{-t}t^2$ from 3 and 12 of Table 8.1.

### Example 8.21

Determine $\mathcal{L}^{-1}\left\{\dfrac{5s^2 + 8s - 1}{(s+3)(s^2+1)}\right\}$

### Solution

$\dfrac{5s^2 + 8s - 1}{(s+3)(s^2+1)} \equiv \dfrac{A}{s+3} + \dfrac{Bs + C}{s^2+1}$

$= \dfrac{A(s^2+1) + (Bs+C)(s+3)}{(s+3)(s^2+1)}$

Hence, $5s^2 + 8s - 1 = A(s^2 + 1) + (Bs + C)(s + 3)$

When $s = -3$, $20 = 10A$ from which $A = 2$

Equating $s^2$ terms gives: $5 = A + B$ from which $B = 3$

Equating $s$ terms gives: $8 = 3B + C$ from which $C = -1$

Hence,

$\mathcal{L}^{-1}\left\{\dfrac{5s^2 + 8s - 1}{(s+3)(s^2+1)}\right\} \equiv \mathcal{L}^{-1}\left\{\dfrac{2}{s+3} + \dfrac{3s-1}{s^2+1}\right\}$

$= \mathcal{L}^{-1}\left\{\dfrac{2}{s+3}\right\} + \mathcal{L}^{-1}\left\{\dfrac{3s}{s^2+1}\right\} - \mathcal{L}^{-1}\left\{\dfrac{1}{s^2+1}\right\}$

$= 2e^{-3t} + 3\cos t - \sin t$ from 3, 6 and 5 of Table 8.1.

## 8.6.2 Procedure to Solve Differential Equations by Using Laplace Transforms

(i) Take the Laplace transform of both sides of the differential equation by applying the formulae for the Laplace transforms of derivatives (i.e., equations (8.33) and (8.34) and, where necessary, using a list of standard Laplace transforms, such as Table 8.1.

(ii) Put in the given initial conditions, i.e., $y(0)$ and $y'(0)$.

(iii) Rearrange the equation to make $\mathcal{L}\{y\}$ the subject.

(iv) Determine $y$ by using, where necessary, partial fractions, and taking the inverse of each term by using Table 8.1.

This procedure is demonstrated in the following problems.

### Example 8.22

Use Laplace transforms to solve the differential equation:

$$2\frac{d^2y}{dx^2} + 5\frac{dy}{dx} - 3y = 0$$

given that when $x = 0$, $y = 4$ and $\dfrac{dy}{dx} = 7$

### Solution

(i) $\quad 2\mathcal{L}\left\{\dfrac{d^2y}{dx^2}\right\} + 5\mathcal{L}\left\{\dfrac{dy}{dx}\right\} - 3\mathcal{L}\{y\} = \mathcal{L}\{0\}$

$\quad\quad 2[s^2\mathcal{L}\{y\} - sy(0) - y'(0)] + 5[s\mathcal{L}\{y\} - y(0)] - 3\mathcal{L}\{y\} = 0$

From equation (8.33) and (8.34)

(ii) $\quad y(0) = 4$ and $y'(0) = 9$

Thus,

$\quad 2[s^2\mathcal{L}\{y\} - 4s - 9] + 5[s\mathcal{L}\{y\} - 4] - 3\mathcal{L}\{y\} = 0$

i.e.,

$\quad 2s^2\mathcal{L}\{y\} - 8s - 18 + 5s\mathcal{L}\{y\} - 20 - 3\mathcal{L}\{y\} = 0$

(iii) Rearranging gives: $(2s^2 + 5s - 3)\mathcal{L}\{y\} = 8s + 38$

$\quad$ i.e., $\mathcal{L}\{y\} = \dfrac{8s + 38}{2s^2 + 5s - 3}$

(iv) $\quad y = \mathcal{L}^{-1}\left\{\dfrac{8s + 38}{2s^2 + 5s - 3}\right\}$

Let $\dfrac{8s + 38}{2s^2 + 5s - 3} = \dfrac{8s + 38}{(2s - 1)(s + 3)}$

$\qquad\qquad = \dfrac{A}{2s - 1} + \dfrac{B}{s + 3}$

$\qquad\qquad = \dfrac{A(s + 3) + B(2s - 1)}{(2s - 1)(s + 3)}$

Hence, $8s + 38 = A(s + 3) + B(2s - 1)$

When $s = \frac{1}{2}$, $42 = 3\frac{1}{2}A$ from which, $A = 12$

When $s = -3$, $14 = -7B$ from which, $B = -2$

Hence, $y = \mathcal{L}^{-1}\left\{\dfrac{8s + 38}{2s^2 + 5s - 3}\right\}$

$\qquad\equiv \mathcal{L}^{-1}\left\{\dfrac{12}{2s - 1} - \dfrac{2}{s + 3}\right\}$

$\qquad = \mathcal{L}^{-1}\left\{\dfrac{12}{2(s - \frac{1}{2})}\right\} - \mathcal{L}^{-1}\left\{\dfrac{2}{s + 3}\right\}$

Hence, $\mathbf{y = 6e^{(1/2)x} - 2e^{-3x}}$ from 3 of Table 8.1.

## Example 8.23

Use Laplace transforms to solve the differential equation:

$$\frac{d^2y}{dx^2} + 6\frac{dy}{dx} + 13y = 0$$

given that when $x = 0$, $y = 3$ and $\dfrac{dy}{dx} = 7$

## Solution

Using the above procedure:

(i) $\mathcal{L}\left\{\dfrac{d^2y}{dx^2}\right\} + 6\mathcal{L}\left\{\dfrac{dy}{dx}\right\} + 13\mathcal{L}\{y\} = \mathcal{L}\{0\}$

Hence, $[s^2\mathcal{L}\{y\} - sy(0) - y'(0)] + 6[s\mathcal{L}\{y\} - y(0)] + 13\mathcal{L}\{y\} = 0$

from equations (8.33) and (8.34)

(ii)  $y(0) = 3$ and $y'(0) = 7$

Thus, $s^2\mathcal{L}\{y\} - 3s - 7 + 6s\mathcal{L}\{y\} - 18 + 13\mathcal{L}\{y\} = 0$

(iii)  Rearranging gives: $(s^2 + 6s + 13)\mathcal{L}\{y\} = 3s + 25$

i.e., $\mathcal{L}\{y\} = \dfrac{3s + 25}{s^2 + 6s + 13}$

(iv)  $y = \mathcal{L}^{-1}\left\{\dfrac{3s + 25}{s^2 + 6s + 13}\right\}$

$= \mathcal{L}^{-1}\left\{\dfrac{3s + 25}{(s + 3)^2 + 2^2}\right\} = \mathcal{L}^{-1}\left\{\dfrac{3(s + 3) + 16}{(s + 3)^2 + 2^2}\right\}$

$= \mathcal{L}^{-1}\left\{\dfrac{3(s + 3)}{(s + 3)^2 + 2^2}\right\} + \mathcal{L}^{-1}\left\{\dfrac{8(2)}{(s + 3)^2 + 2^2}\right\}$

$= 3e^{-3t}\cos 2t + 8e^{-3t}\sin 2t$ from 14 and 13 of Table 8.1.

Hence, $y = e^{-3t}(\textbf{3 cos } 2t + \textbf{8 sin } 2t)$

### Example 8.24

A step voltage is applied to a series *C-R* circuit. When the capacitor is fully charged the circuit is suddenly broken. Deduce, using Laplace transforms, an expression for the capacitor voltage during the transient period if the voltage when the supply is cut is $V$ volts.

### Solution

From Figure 8.1, $v_R + v_C = 0$ when the supply is cut,

i.e.,  $iR + v_c = 0$

i.e.,  $\left(c\dfrac{dv_c}{dt}\right)R + v_c = 0$

i.e.,  $CR\dfrac{dv_c}{dt} + v_c = 0$

Using the procedure:

(i)  $\mathcal{L}\left\{CR\dfrac{dv_c}{dt}\right\} + \mathcal{L}\{v_c\} = \mathcal{L}\{0\}$

   i.e., $CR[s\mathcal{L}\{v_c\} - v_0] + \mathcal{L}\{v_c\} = 0$

(ii)  $v_0 = V$, hence, $CR[s\mathcal{L}\{v_c\} - V] + \mathcal{L}\{v_c\} = 0$

(iii)  Rearranging gives: $CRs\mathcal{L}\{v_c\} - CRV + \mathcal{L}\{v_c\} = 0$

   i.e., $(CRs + 1)\mathcal{L}\{v_c\} = CRV$

   hence,     $\mathcal{L}\{v_c\} = \dfrac{CRV}{(CRs + 1)}$

(iv)  Capacitor voltage, $v_c = \mathcal{L}^{-1}\left\{\dfrac{CRV}{CRs + 1}\right\}$

$$= CRV\mathcal{L}^{-1}\left\{\dfrac{1}{CR\left(s + \dfrac{1}{CR}\right)}\right\}$$

$$= \dfrac{CRV}{CR}\mathcal{L}^{-1}\left\{\dfrac{1}{s + \dfrac{1}{CR}}\right\}$$

   i.e., $v_c = Ve^{(-t/CR)}$ as previously obtained in equation (8.8).

### Problem 8.25

A series $R$-$L$ circuit has a step input $V$ applied to it. Use Laplace transforms to determine an expression for the current $i$ flowing in the circuit given that when time $t = 0$, $i = 0$.

### Solution

From Figure 8.6 and equation (8.11),

$$v_R + v_L = V \text{ becomes } iR + L\dfrac{dt}{dt} = v$$

Using the procedure:

(i)  $\mathcal{L}\{iR\} + \mathcal{L}\left\{L\dfrac{di}{dt}\right\} = \mathcal{L}\{V\}$

i.e., $R\mathcal{L}\{i\} + L[s\mathcal{L}\{i\} - i(0)] = \dfrac{V}{s}$

(ii) $i(0) = 0$, hence, $R\mathcal{L}\{i\} + Ls\mathcal{L}\{i\} = \dfrac{V}{s}$

(iii) Rearranging gives: $(R + Ls)\mathcal{L}\{i\} = \dfrac{V}{s}$

i.e.,                     $\mathcal{L}\{i\} = \dfrac{V}{s(R + Ls)}$

(iv) $i = \mathcal{L}^{-1}\left\{\dfrac{V}{s(R + Ls)}\right\}$

Let $\dfrac{V}{s(R + Ls)} = \dfrac{A}{s} + \dfrac{B}{R + Ls} = \dfrac{A(R + Ls) + Bs}{s(R + Ls)}$

Hence, $V = A(R + Ls) + Bs$

When $s = 0, V = AR$ from which, $A = \dfrac{V}{R}$

When $s = -\dfrac{R}{L}, V = B\left(-\dfrac{R}{L}\right)$ from which,

$$B = -\dfrac{VL}{R}$$

Hence, $\mathcal{L}^{-1}\left\{\dfrac{V}{s(R + Ls)}\right\}$

$= \mathcal{L}^{-1}\left\{\dfrac{V/R}{s} + \dfrac{-VL/R}{R + Ls}\right\}$

$= \mathcal{L}^{-1}\left\{\dfrac{V}{Rs} - \dfrac{VL}{R(R + Ls)}\right\}$

$$= \mathcal{L}^{-1} \left\{ \left[ \frac{V}{R} \left( \frac{1}{s} \right) - \frac{V}{R} \left( \frac{1}{\frac{R}{L} + s} \right) \right] \right\}$$

$$= \frac{V}{R} \mathcal{L}^{-1} \left\{ \left[ \frac{1}{s} - \frac{1}{\left( s + \frac{R}{L} \right)} \right] \right\}$$

So current $i = \dfrac{V}{R}(1 - e^{-Rt/L})$   as previously obtained in equation (8.13).

### Example 8.26

If after a period of time, the switch in the *R-L* circuit of Example 25 is opened, use Laplace transforms to determine an expression to represent the current transient response. Assume that at the instant of opening the switch, the steady-state current flowing is *I*.

### Solution

From Figure 8.6, $v_L + v_R = 0$ when the switch is opened,

i.e., $L\dfrac{di}{dt} + iR = 0$

Using the procedure:

(i)  $\mathcal{L} \left\{ L\dfrac{di}{dt} \right\} + \mathcal{L}\{iR\} = \mathcal{L}\{0\}$

   i.e.,   $L[s\mathcal{L}\{i\} - i_0] + R\mathcal{L}\{i\} = 0$

(ii)  $i_0 = I$, hence, $L[s\mathcal{L}\{i\} - I] + R\mathcal{L}\{i\} = 0$

(iii)  Rearranging gives: $Ls\mathcal{L}\{i\} - LI + R\mathcal{L}\{i\} = 0$

   i.e., $(R + Ls)\,\mathcal{L}\{i\} = LI$

   and    $\mathcal{L}\{i\} = \dfrac{LI}{R + Ls}$

(iv)  Current, $i = \mathscr{L}^{-1}\left\{\dfrac{LI}{R+Ls}\right\}$

$$= LI\,\mathscr{L}^{-1}\left\{\dfrac{1}{L\left(\dfrac{R}{L}+s\right)}\right\}$$

$$= \dfrac{LI}{L}\,\mathscr{L}^{-1}\left\{\dfrac{1}{s+\dfrac{R}{L}}\right\}$$

i.e., $i = Ie^{(-Rt/L)}$ from 3 of Table 8.1.

Since $I = \dfrac{V}{R}$ then $i = \dfrac{V}{R}e^{-Rt/L}$   as previously derived in equation (8.17).

# CHAPTER 9

# Frequency Domain Circuit Analysis

Izzat Darwazeh
Luis Moura

## 9.1 Introduction

In this chapter, we present the main electrical analysis techniques for time-varying signals. We start by discussing sinusoidal alternating current (AC) signals and circuits. Phasor analysis is presented and it is shown that this greatly simplifies this analysis since it allows the introduction of the *generalized impedance*. The generalized impedance allows us to analyze AC circuits using all the circuit techniques and methods for DC circuits discussed previously. In section 9.3, we extend the phasor analysis technique to analyze circuits driven by nonsinusoidal signals. This is done by first discussing the Fourier series, which presents periodic signals as a sum of phasors. The Fourier series is a very important tool since it forms the basis of fundamental concepts in signal processing such as spectra and bandwidth. Finally, we present the Fourier transform, which allows the analysis of virtually any time-varying signal (periodic and nonperiodic) in the frequency domain.

## 9.2 Sinusoidal AC Electrical Analysis

AC sinusoidal electrical sources are time-varying voltages and currents described by functions of the form:

$$v_s(t) = V_s \sin(\omega t) \qquad (9.1)$$

$$i_s(t) = I_s \sin(\omega t) \qquad (9.2)$$

where $V_s$ and $I_s$ are the peak-amplitudes of the voltage and of the current waveforms, respectively, as illustrated in Figure 9.1. Here $\omega$ represents the *angular* frequency, in radians/second, equal to $2\pi/T$ where $T$ is the period of the waveform in seconds. The repetition rate

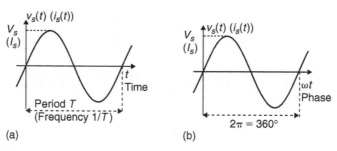

**Figure 9.1: (a) AC voltage (current) waveform versus time; (b) AC voltage (current) waveform versus phase**

of the waveform, that is the *linear* frequency, is equal to *1/T* in hertz. The quantity ($\omega t$) is an angle, in radians, usually called the instantaneous phase. Note that $\omega T$ corresponds to $2\pi$ rad. Here we interchangeably use the terms voltage/current sinusoidal signal or waveform, to designate the AC sinusoidal quantities.

By definition, all transient phenomena (such as those resulting, for example, from switching-on the circuit) have vanished in an AC circuit in its steady-state condition. Thus, the time origin in equations 9.1 and 9.2 can be "moved" so $v_s(t)$ and $i_s(t)$ are equally well described by cosine functions, that is:

$$v_s(t) = V_s \cos(\omega t) \tag{9.3}$$

$$i_s(t) = I_s \cos(\omega t) \tag{9.4}$$

Any signal varying with time is effectively an AC signal. We limit our definition of an AC signal here to a sinusoidal signal at specific frequency. This is particularly helpful to calculate impedances at specific frequencies as will be seen later in this chapter.

While the choice of the absolute time origin is of no relevance in AC analysis, the relative time difference between waveforms, which can also be quantified in terms of phase difference, is of vital importance. Figure 9.2(a) illustrates the constant phase difference between a voltage waveform and a current waveform at the same angular frequency $\omega$. If any two AC electrical waveforms have different angular frequencies, $\omega_1$ and $\omega_2$, then the phase difference between these two waveforms is a linear function of time; $(\omega_1 - \omega_2)t$. Assuming a time origin for the voltage waveform we can write the waveforms of Figure 9.2(a) as:

$$v_s(t) = V_s \sin(\omega t) \tag{9.5}$$

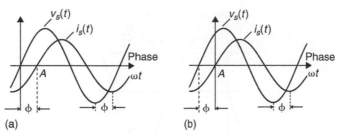

**Figure 9.2: Phase difference ($\phi = \pi/3$) between an AC voltage and an AC current**
**(a) The current lags the voltage (b) The voltage leads the current**

$$i_s(t) = I_s \sin(\omega t - \phi) \tag{9.6}$$

where $\phi = \pi/3$. In this situation, it is said that the current waveform *lags* the voltage waveform by $\phi$. In fact, the current waveform crosses the phase axis (point $A$) later than the voltage waveform. On the other hand, if we choose the time origin for the current waveform, as illustrated in Figure 9.2(b), we can write these waveforms as follows:

$$v_s(t) = V_s \sin(\omega t + \phi) \tag{9.7}$$

$$i_s(t) = I_s \sin(\omega t) \tag{9.8}$$

and it is said that the voltage waveform *leads* the current waveform.

### 9.2.1 Effective Electrical Values

By definition, the effective value of any voltage waveform is the DC voltage that, when applied to a resistance, would produce as much power dissipation (heat) as that caused by that voltage waveform. If we represent the AC voltage waveform by $V_s \sin(\omega t)$ and the effective voltage by $V_{eff}$, then we can write:

$$\frac{1}{T} \int_0^T \frac{V_{eff}^2}{R} \, dt = \frac{1}{T} \int_0^T \frac{v_s^2(t)}{R} \, dt$$

$$\Leftrightarrow \quad \frac{1}{T} \int_0^T \frac{V_{eff}^2}{R} \, dt = \frac{1}{T} \int_0^T \frac{V_s^2 \sin^2(\omega t)}{R} \, dt \tag{9.9}$$

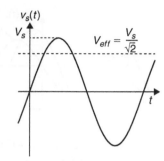

**Figure 9.3: Voltage AC waveform and its corresponding effective voltage**

The last equation can be written as follows:

$$\frac{1}{T}\frac{V_{eff}^2}{R}T = \frac{1}{T}\frac{V_s^2}{R}\int_0^T \frac{1-\cos(2\omega t)}{2}\,dt$$

$$\Leftrightarrow \qquad V_{eff}^2 = \frac{V_s^2}{2T}\left[t - \frac{1}{2\omega}\sin(2\omega t)\right]_0^T \tag{9.10}$$

Since $\omega = 2\pi/T$ the last equation can be written as:

$$V_{eff}^2 = \frac{V_s^2}{2} \tag{9.11}$$

or $V_{eff} = V_s/\sqrt{2} \approx 0.707\,V_s$. Figure 9.3 illustrates the effective voltage of an AC voltage waveform.

In a similar way it can be shown that the effective value of a sinusoidal current with peak-amplitude $I_s$ is $I_{eff} = I_s/\sqrt{2}$. The effective value of a sinusoidal voltage and/or current is also called the *root-mean-square* (RMS) value.

### Example 9.1

Show that the effective value of a triangular voltage waveform, like that shown in Figure 9.4, with peak amplitude $V_s$ is $V_{eff} = V_s/\sqrt{3}$.

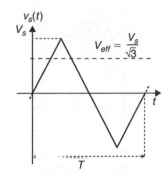

**Figure 9.4: Triangular voltage waveform and its corresponding effective voltage**

### Solution

Following the procedure described above we can write:

$$\frac{1}{T} \int_0^T \frac{V_{eff}^2}{R} \, dt = \frac{1}{T} \int_0^T \frac{v_s^2(t)}{R} \, dt$$

Looking at Figure 9.3, we see that the triangular waveform is symmetrical. Therefore, it is sufficient to consider the period of integration from $t = 0$ to $t = T/4$, giving

$$\frac{V_{eff}^2}{R} = \frac{4}{T} \int_0^{T/4} \frac{V_s^2 \, 4^2 \, t^2}{T^2 R} \, dt$$

$$\Leftrightarrow \quad \frac{V_{eff}^2}{R} = \frac{V_s^2 \, 4^3}{R T^3} \left[ \frac{t^3}{3} \right]_0^{T/4}$$

$$\Leftrightarrow \quad V_{eff}^2 = \frac{V_s^2}{3} \tag{9.12}$$

that is, $V_{eff} = V_s / \sqrt{3} \simeq 0.577 \, V_s$.

### 9.2.2 I-V Characteristics for Passive Elements

We now study the AC current-voltage (*I-V*) relationships for the main passive elements. We use cosine functions to represent AC currents and voltages waveforms. However, the same results would be obtained if sine functions were used instead.

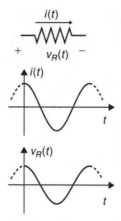

**Figure 9.5: Voltage and current in a resistance**

### 9.2.2.1 Resistance

Assuming a current, $i(t) = I_x \cos(\omega t)$ passing through a resistance $R$, the voltage developed across its terminals is, according to Ohm's law:

$$
\begin{aligned}
v_R(t) &= Ri(t) \\
&= RI_x \cos(\omega t) \\
&= V_r \cos(\omega t)
\end{aligned}
\tag{9.13}
$$

With,

$$
V_r = RI_x
\tag{9.14}
$$

Dividing both sides by $\sqrt{2}$ we obtain the RMS (or effective) value for the AC voltage as:

$$
\begin{aligned}
V_{r_{eff}} &= R \frac{I_x}{\sqrt{2}} \\
&= RI_{x_{eff}}
\end{aligned}
\tag{9.15}
$$

where $I_{x_{eff}}$ is the RMS (or effective) value for the AC current. From equation 9.13 and Figure 9.5 we observe that the voltage and the current are in phase, that is, the phase difference between the voltage and the current is zero.

### 9.2.2.2 Capacitance

If a current, $i(t) = I_x \cos(\omega t)$ passes through a capacitance $C$, the voltage developed across its terminals is:

$$v_C(t) = \frac{1}{C} \int_0^t i(t)\, dt + V_{co} \tag{9.16}$$

Note that since we are assuming steady-state conditions in the AC analysis we may set the initial condition $V_{co} = 0$, that is:

$$v_C(t) = \frac{1}{C} \int_0^t I_x \cos(\omega t)\, dt \tag{9.17}$$

Performing the integration we obtain:

$$
\begin{aligned}
v_C(t) &= \frac{I_x}{\omega C} \sin(\omega t) \\
&= \frac{I_x}{\omega C} \cos\left(\omega t - \frac{\pi}{2}\right) \\
&= V_c \cos\left(\omega t - \frac{\pi}{2}\right)
\end{aligned}
\tag{9.18}
$$

Where,

$$V_c = \frac{I_x}{\omega C} \tag{9.19}$$

In terms of RMS magnitudes we have:

$$
\begin{aligned}
V_{c_{eff}} &= \frac{I_{x_{eff}}}{\omega C} \\
&= X_C\, I_{x_{eff}}
\end{aligned}
\tag{9.20}
$$

where $I_{x_{eff}} = I_x/\sqrt{2}$. The quantity $X_C = (\omega C)^{-1}$ is called the *capacitive reactance* and is measured in ohms. It is important to note that the amplitude of $v_C(t)$ is inversely proportional to the capacitance and the angular frequency of the AC current. From

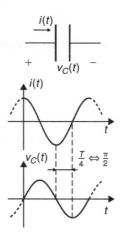

**Figure 9.6: Voltage and current in a capacitor**

equation 9.18 and Figure 9.6 we observe that the voltage waveform lags the current waveform by $\pi/2$ radians or 90 degrees.

### 9.2.2.3 Inductance

When a current, $i(t) = I_x \cos(\omega t)$ passes through an inductance $L$, the voltage developed across its terminals is given by:

$$
\begin{aligned}
v_L(t) &= L\frac{i(t)}{dt} \\
&= -L\omega I_x \sin(\omega t)\, dt \\
&= L\omega I_x \cos\left(\omega t + \frac{\pi}{2}\right) \\
&= V_l \cos\left(\omega t + \frac{\pi}{2}\right)
\end{aligned}
\tag{9.21}
$$

with $V_l = L\omega\, I_x$. In terms of RMS values we have:

$$
\begin{aligned}
V_{l_{eff}} &= I_{x_{eff}}\omega L \\
&= X_L I_{x_{eff}}
\end{aligned}
\tag{9.22}
$$

where $I_{x_{eff}} = I_x/\sqrt{2}$. The quantity $X_L = \omega L$ is called the *inductive reactance*, which is also measured in ohms. Note that now the amplitude of the voltage $v_L(t)$ is proportional

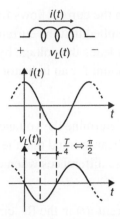

**Figure 9.7: Voltage and current in an inductor**

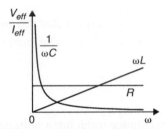

**Figure 9.8: $V_{eff}/I_{eff}$ versus $\omega$ for passive elements**

to the inductance and the angular frequency of the AC current. From equation 9.21 and Figure 9.7 we observe that the voltage waveform leads the current waveform by $\pi/2$ radians or 90 degrees.

Figure 9.8 illustrates the ratio $V_{eff}/I_{eff}$ versus the frequency, $\omega$, for the three passive elements discussed above. It is interesting to note that at DC ($\omega = 0$) the capacitor behaves as an open-circuit and the inductor behaves as a short-circuit. On the other hand, for very high frequencies ($\omega \to \infty$) the capacitor behaves as a short circuit and the inductor behaves as an open circuit.

### 9.2.2.4 A Note About Voltage Polarity and Current Direction in AC Circuits

Although voltages and currents in AC circuits continuously change polarity and direction it is important to set references for these two quantities. The convention we follow in this book is illustrated above. When the current flows from the positive to the negative terminal of a circuit element it is implied that the current and voltage are in phase for a resistor as in Figure 9.5; the current leads the voltage by 90 degrees for a capacitor as in Figure 9.6 and lags by the same amount for an inductor as in Figure 9.7.

### 9.2.2.5 Kirchhoff's Laws

Kirchhoff's laws can be applied to determine the voltage across or the current through any circuit element. However, we must bear in mind that the voltages and the currents in AC circuits will, in general, exhibit phase differences when capacitors or inductors are present.

### Example 9.2

Determine the amplitude of the current $i(t)$ in the RL circuit of Figure 9.9. Also, determine the phase difference between this current and the voltage source.

### Solution

Since the circuit contains an inductor, we expect that the current will exhibit a phase difference, $\phi$, with respect to the source voltage. The current $i(t)$ can be expressed as follows:

$$i(t) = I_s \cos(\omega t + \phi) \tag{9.23}$$

This current flows through the resistance inducing a voltage difference at its terminals that is in phase with $i(t)$:

$$\begin{aligned} v_R(t) &= Ri(t) \\ &= RI_s \cos(\omega t + \phi) \end{aligned} \tag{9.24}$$

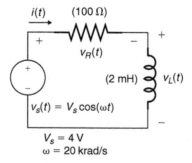

**Figure 9.9: RL circuit**

On the other hand, the flow of $i(t)$ through the inductor causes a voltage difference across its terminals that is in quadrature with $i(t)$, as expressed by equation 9.21:

$$v_L(t) = X_L I_s \cos\left(\omega t + \phi + \frac{\pi}{2}\right) \tag{9.25}$$

with $X_L = \omega L$. According to Kirchhoff's voltage law we can write:

$$
\begin{aligned}
v_s(t) &= v_R(t) + v_L(t) \\[4pt]
&= RI_s \cos(\omega t + \phi) + X_L I_s \cos\left(\omega t + \phi + \frac{\pi}{2}\right) \\[4pt]
&= RI_s \cos(\omega t + \phi) + X_L I_s \cos(\omega t + \phi) \cos\left(\frac{\pi}{2}\right) \\[4pt]
&\quad - X_L I_s \sin(\omega t + \phi) \sin\left(\frac{\pi}{2}\right) \\[4pt]
&= RI_s \cos(\omega t + \phi) - X_L I_s \sin(\omega t + \phi)
\end{aligned} \tag{9.26}
$$

The last equation can be written as follows:

$$V_s \cos(\omega t) = \sqrt{R^2 + X_L^2}\, I_s \cos(\omega t + \phi + \psi) \tag{9.27}$$

where,

$$\psi = \tan^{-1}\left(\frac{X_L}{R}\right) \tag{9.28}$$

In order for equation 9.27 to be an equality, the amplitude and the phase of the cosine functions on both sides of this equation must be equal. That is:

$$
\begin{cases}
V_s = \sqrt{R^2 + X_L^2}\, I_s \\
\omega t = \omega t + \phi + \psi
\end{cases} \tag{9.29}
$$

Solving the last set of equations in order to obtain $I_s$ and $\phi$ we have:

$$I_s = \frac{V_s}{\sqrt{R^2 + \omega^2 L^2}}$$
$$= 37\,\text{mA} \tag{9.30}$$

$$\phi = -\psi$$
$$= -0.38\,\text{rad}\,(-2.18°) \tag{9.31}$$

### 9.2.3 Phasor Analysis

In principle any AC circuit can be analyzed by applying Kirchhoff's laws with the trigonometric rules, as in the Example 9.2 above. However, the application of these trigonometric rules to analyze complex AC circuits can be a cumbersome task. Fortunately, the use of the complex exponential (the phasor) and complex algebra, provides a considerable simplification of AC circuit analysis.

From Euler's formula, a cosine alternating voltage waveform can be represented using the complex exponential function as follows:

$$V_s \cos(\omega t + \phi) = V_s \frac{e^{j(\omega t + \phi)} + e^{-j(\omega t + \phi)}}{2} \tag{9.32}$$

where we can see that the voltage expressed by equation 9.32 is the addition of two complex conjugated exponential functions (phasors). Note that either of these two complex exponential functions carries all the phase information, $\omega t$ and $\phi$, of the voltage waveform. In fact, the simplicity of analysis using phasors arises from each AC voltage and current being mathematically represented and manipulated as a *single* complex exponential function. However, in order to obtain the corresponding time domain waveform we must take the real part of the complex exponential waveform. Thus, the voltage waveform of equation 9.32 can be expressed as:

$$V_s \cos(\omega t + \phi) = \text{Real}\,[V_s e^{j(\omega t + \phi)}] \tag{9.33}$$

In order to illustrate that phasor analysis is similar to AC analysis using trigonometric rules, we reconsider the current-voltage relationships for the passive elements using the

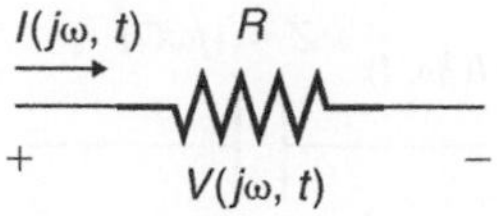

**Figure 9.10: Complex *V-I* relationship for a resistance**

complex exponential representation. We determine the voltage developed across each element when an AC current, $i(t)$, flows through them, $i(t)$ being expressed by its complex exponential representation, $I(j\omega,t)$, as follows:

$$i(t) = \text{Real} \left[ I(j\omega,t) \right] \tag{9.34}$$

$$I(j\omega,t) = I_x e^{j\omega t} \tag{9.35}$$

### 9.2.3.1 Resistance

The complex voltage (see also Figure 9.10) across the resistance terminals is determined by applying Ohm's law to the phasors representing the voltage across and the current flowing through the resistance, that is:

$$\begin{aligned} V_R(j\omega,t) &= RI(j\omega,t) \\ &= RI_x e^{j\omega t} \end{aligned} \tag{9.36}$$

Taking the real part of $V_R(j\omega,t)$ we obtain the corresponding voltage waveform;

$$v_R(t) = RI_x \cos(\omega t) \tag{9.37}$$

This equation is the same as equation 9.13.

### 9.2.3.2 Capacitance

Assuming a complex representation for the current flowing through a capacitor, $I(j\omega,t)$, the complex voltage across the capacitance is given by:

$$V_C(j\omega,t) = \frac{1}{C} \int_0^t I(j\omega,t)\, dt \tag{9.38}$$

$$= \frac{1}{j\omega C} I_x e^{j\omega t} \tag{9.39}$$

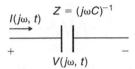

**Figure 9.11: Complex *V-I* relationship for a capacitance**

$$= \frac{1}{j\omega C} I(j\omega, t) \tag{9.40}$$

The quantity $(j\omega C)^{-1}$ is called the capacitive (complex) *impedance*. This impedance can be seen as $(-j)$ times the capacitive reactance $X_C = (\omega C)^{-1}$ discussed in section 9.2.2. Note that $(-j)$ accounts for the $-90°$ phase difference between the voltage and the current.

Taking the real part of $V_C(j\omega, t)$ we obtain the corresponding voltage waveform at the capacitor terminals:

$$v_C(t) = \text{Real}\left[\frac{1}{j\omega C} I_x e^{j\omega t}\, dt\right]$$
$$= \text{Real}\left[\frac{1}{\omega C} I_x e^{j(\omega t - \pi/2)}\, dt\right]$$

where we used the following equalities:

$$-j = e^{-j\pi/2} \tag{9.41}$$

Now $v_C(t)$ can be written as:

$$v_C(t) = \frac{I_x}{\omega C}\cos\left(\omega t - \frac{\pi}{2}\right) \tag{9.42}$$

Note that equation 9.42 is the same as equation 9.18.

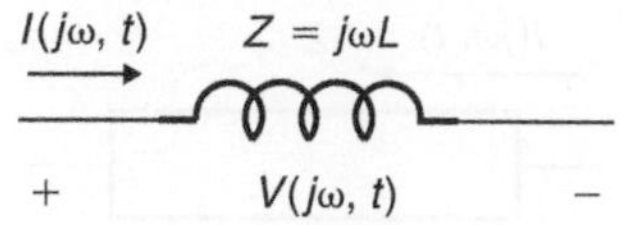

**Figure 9.12: Complex *V-I* relationship for an inductance**

### 9.2.3.3 Inductance

Assuming a complex representation for the current flowing through the inductor, $I(j\omega,t)$, the complex voltage across the inductance is given by:

$$V_L(j\omega,t) = L\frac{dI(j\omega,t)}{dt}$$
$$= j\omega L I_x e^{j\omega t} \tag{9.43}$$

$$= j\omega L I(j\omega,t) \tag{9.44}$$

The quantity $Z = j\omega L$ is called the inductive (complex) *impedance*. This impedance can be seen as $j$ times the inductive reactance $X_L = \omega L$ discussed in section 9.2.2. Note that now $j$ accounts for the 90° phase difference between the voltage and the current. Taking the real part of $V_L(j\omega,t)$ we obtain

$$v_C(t) = \text{Real}\,[j\omega L I_x e^{j\omega t}]$$
$$= \text{Real}\,[\omega L I_x e^{j(\omega t + \pi/2)}]$$
$$= I_x \omega L \cos\left(\omega t + \frac{\pi}{2}\right) \tag{9.45}$$

We note again that equation 9.45 is the same as equation 9.21.

### 9.2.4 The Generalized Impedance

The greatest advantage of using phasors in AC circuit analysis is that they allow for an Ohm's law type of relationship between the phasors describing the voltage and the current for each passive element:

$$\frac{V(j\omega,t)}{I(j\omega,t)} = Z \tag{9.46}$$

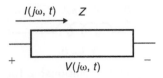

**Figure 9.13: Symbol of the general impedance**

where $Z$ is called the generalized impedance:

- $Z = R$ for a resistance
- $Z = (j\omega C)^{-1}$ for a capacitance
- $Z = j\omega L$ for an inductance

The generalized impedance concept is of great importance since it permits an extrapolation of the DC circuit analysis techniques discussed earlier to the analysis of AC circuits. For example, this means that we can apply the Nodal analysis technique to analyze AC circuits as illustrated by the next example. Figure 9.13 shows the symbol used to represent a general impedance.

### Example 9.3

Using the phasor analysis described above, determine the amplitude and phase of the current in the circuit of Figure 9.9 and show that the results are the same as those obtained in Example 9.2.

### Solution

The phasor describing the current can be written as follows:

$$I(j\omega, t) = I_s e^{j(\omega t + \phi)} \tag{9.47}$$

Applying Kirchhoff's voltage law, we can write:

$$V_s e^{j\omega t} = R I_s e^{(\omega t + \phi)} + j\omega L I_s e^{j(\omega t + \phi)} \tag{9.48}$$

or,

$$V_s e^{j\omega t} = (R + j\omega L) I_s e^{j(\omega t + \phi)} \tag{9.49}$$

The impedance $R + j\omega L$ can be expressed in the exponential form as follows:

$$R + j\omega L = \sqrt{R^2 + \omega^2 L^2}\, e^{j\tan^{-1}\left(\frac{\omega L}{R}\right)} \tag{9.50}$$

Hence, equation 9.49 can be written as:

$$V_s e^{j\omega t} = \sqrt{R^2 + \omega^2 L^2}\, I_s\, e^{j\left(\omega t + \phi + \tan^{-1}\left(\frac{\omega L}{R}\right)\right)} \tag{9.51}$$

In order for equation 9.51 to be an equality, the amplitude and the phase of the complex voltages on both sides of this equation must be equal. That is:

$$\begin{cases} V_s = \sqrt{R^2 + \omega^2 L^2}\, I_s \\ \omega t = \omega t + \phi + \psi \end{cases} \tag{9.52}$$

Solving, we have:

$$I_s = \frac{V_s}{\sqrt{R^2 + \omega^2 L^2}}$$

$$= 37\,\text{mA} \tag{9.53}$$

$$\phi = -\psi$$

$$= -0.38\,\text{rad}\,(-21.8°) \tag{9.54}$$

Note that these values are equal to those obtained in Example 9.2.

### 9.2.4.1 The Rotating and the Stationary Phasor

The concept of the rotating phasor arises from the time dependence of the complex exponential which characterizes AC voltages and currents. Let us consider the phasor representation for an AC voltage as shown below:

$$V(j\omega, t) = V_s\, e^{j(\omega t + \phi)} \tag{9.55}$$

This rotating phasor can be represented in the Argand diagram, as illustrated in Figure 9.14(a). Note that each instantaneous value for the rotating phasor (that is its position in

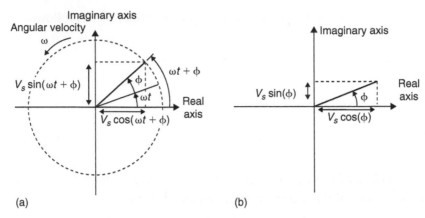

**Figure 9.14: The complex phasor represented in the Argand diagram (a) Instantaneous value of the rotating phasor; (b) The stationary phasor**

the Argand diagram) is located on a circle whose radius is given by the voltage amplitude, $V_s$, with an angle $\omega t + \phi$ at each instant of time. Each position in this circle is reached by the phasor every $2\pi/\omega$ seconds.

The rotating phasor described by equation 9.55 can be decomposed into the product of a stationary (or static) phasor with a rotating phasor as expressed by the equation below:

$$V(j\omega, t) = \underbrace{V_s\, e^{j\phi}}_{\text{Static phasor}} \times \underbrace{e^{j\omega t}}_{\text{Rotating phasor}} \tag{9.56}$$

$$= V_S \times e^{j\omega t} \tag{9.57}$$

where $V_S$ represents the static phasor. In the rest of this chapter, and unless stated otherwise, static phasors are represented by capital letters with capital subscripts.

In AC circuits where currents and voltages feature the same single tone or angular frequency, $\omega$, both sides of the equations describing the voltage and current relationships contain the complex exponential describing the rotating phasor, $exp(j\omega t)$, as illustrated by equations 9.48, 9.49, and 9.51 of Example 9.3. Thus, the phasor analysis of an AC circuit can be further simplified if we apply Ohm's law and the concept of the generalized impedance to only the static phasor to represent AC voltages and currents. Note that this mathematical manipulation is reasonable since, in AC circuits, what is important is to determine the *amplitude* and the *relative phase difference* between the AC quantities, both described by the *static phasor.* In the rest of this chapter, a phasor will mean a static phasor.

### Example 9.4

Determine the amplitude and phase of the current in the circuit of Figure 9.9 using the *static* phasor concept described above and show that the results are the same as those obtained in Example 9.2.

### Solution

The static phasor describing the current can be written as follows:

$$I_S = I_s e^{j\phi} \tag{9.58}$$

while the static phasor describing the source voltage can be written as:

$$V_S = V_s e^{j0}$$
$$= V_s \tag{9.59}$$

Applying Kirchhoff's voltage law we can write:

$$V_S = (R + j\omega L)I_S \tag{9.60}$$

$$= \sqrt{R^2 + \omega^2 L^2}\, e^{j \tan^{-1}\left(\frac{\omega L}{R}\right)} I_S \tag{9.61}$$

that is,

$$I_S = \frac{V_S}{\sqrt{R^2 + \omega^2 L^2}} e^{-j \tan^{-1}\left(\frac{\omega L}{R}\right)}$$

$$= \frac{V_s}{\sqrt{R^2 + \omega^2 L^2}} e^{-j \tan^{-1}\left(\frac{\omega L}{R}\right)}$$

$$= 37.0 \times 10^{-3} e^{-j\,0.38} \ \text{A}$$

Note that this result is equivalent to those obtained in Examples 9.2 and 9.3.

### 9.2.4.2 Series and parallel connection of complex impedances

As mentioned previously, the concept of the generalized impedance greatly simplifies the analysis of AC circuits. It is also important to note that the series of various impedances $Z_k$, $k = 1,2,... N$, can be characterized by an equivalent impedance, $Z_{eq}$, which is the sum of these impedances:

$$Z_{eq} = \sum_{k=1}^{N} Z_k \qquad (9.62)$$

For example, in the circuit of Figure 9.9 we observe that the impedance of the resistance is in a series connection with the impedance of the inductor. An equivalent impedance for this connection can be obtained by adding them:

$$Z_{eq} = R + j\omega L \qquad (9.63)$$

The real part of an impedance is called the resistance while the imaginary part of the impedance is called the reactance.

For a parallel connection of various electrical elements it is sometimes easier to work with the inverse of the complex impedance, the "admittance," $Y$:

$$Y = \frac{1}{Z} \qquad (9.64)$$

The parallel connection of admittances $Y_k$, $k = 1,2,..., N$, can be characterized by an equivalent admittance, $Y_{eq}$, which is equal to their sum:

$$Y_{eq} = \sum_{k=1}^{N} Y_k \qquad (9.65)$$

It follows that the parallel connection of two impedances $Z_1$ and $Z_2$ can be represented by an equivalent impedance $Z_{eq}$ given by:

$$Z_{eq} = \frac{Z_1 Z_2}{Z_1 + Z_2} \qquad (9.66)$$

### Example 9.5

Consider the AC circuit represented in Figure 9.15(a). Determine the amplitude and the phase of the voltage across the resistance $R_2$. Then, determine the average power dissipated in $R_2$.

### Solution

$v_{s1}(t)$ and $i_{s2}(t)$ can be expressed in their phasor representations as follows:

$$v_{s1}(t) = \text{Real} \left[ V_{S1} e^{j\omega t} \right]$$
$$V_{S1} = V_{s1} e^{j\frac{\pi}{4}}$$
$$i_{s2}(t) = \text{Real} \left[ I_{S2} e^{j\omega t} \right]$$
$$I_{S2} = I_{s2} e^{-j\frac{\pi}{2}}$$

where we have used the following equality: $\sin(\omega t) = \cos(\omega t - \pi/2)$. The impedances associated with the two inductances and two capacitances are calculated as follows:

$$Z_{L_1} = j\omega L_1 \big|_{\omega = 5 \times 10^3 \, \text{rad/s}}$$
$$= j\,150 \, \Omega$$
$$Z_{L_2} = j\omega L_2 \big|_{\omega = 5 \times 10^3 \, \text{rad/s}}$$
$$= j\,50 \, \Omega$$
$$Z_{C_1} = \frac{1}{j\omega C_1} \bigg|_{\omega = 5 \times 10^3 \, \text{rad/s}}$$
$$= -j\,66.7 \, \Omega$$
$$Z_{C_2} = \frac{1}{j\omega C_2} \bigg|_{\omega = 5 \times 10^3 \, \text{rad/s}}$$
$$= -j\,20 \, \Omega$$

From Figure 9.15(a) we observe that the impedance associated with the capacitance $C_2$ is in a parallel connection with the resistance $R_1$. We can determine an equivalent impedance for this parallel connection as follows (see equation 9.66):

$$Z_{C_2 R_1} = \frac{Z_{C_2} R_1}{Z_{C_2} + R_1}$$
$$= 3.2 - j\,19.5 \, \Omega$$

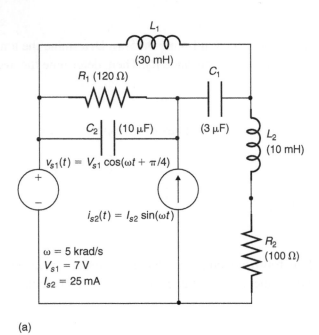

(a)

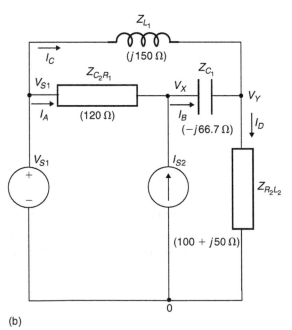

(b)

**Figure 9.15: (a) AC circuit; (b) Equivalent circuit represented as complex impedances**

Also, we can see that $R_2$ is in a series connection with the inductance $L_2$. The equivalent impedance for this connection can be calculated as shown below:

$$Z_{R_2 L_2} = R_2 + Z_{L_2}$$
$$= 100 + j50 \, \Omega$$

Figure 9.15(b) shows the reduced AC circuit with the various impedances associated with the inductances and capacitances as well as the phasor currents and phasor voltages at each node referenced to node 0. Applying Kirchhoff's current law, we can write:

$$I_A + I_{S2} = I_B$$
$$I_C + I_B = I_D$$

These can be rewritten after applying Ohm's law to the various impedances as shown below:

$$\frac{V_{S1} - V_X}{Z_{C_2 R_1}} + I_{S2} = \frac{V_X - V_Y}{Z_{C_1}}$$

$$\frac{V_{S1} - V_Y}{Z_{L_1}} + \frac{V_X - V_Y}{Z_{C_1}} = \frac{V_Y}{Z_{R_2 L_2}}$$

Solving in order to obtain $V_Y$, we have:

$$V_Y = Z_{R_2 L_2} \frac{V_{S1}(Z_{L_1} + Z_{C_1} + Z_{C_2 R_1}) + Z_{L_1} I_{S2} Z_{C_2 R_1}}{Z_{R_2 L_2}(Z_{L_1} + Z_{C_1} + Z_{C_2 R_1}) + Z_{L_1}(Z_{C_2 R_1} + Z_{C_1})}$$

Substituting complex values in the last equation we obtain:

$$V_Y = 3.5 e^{j2.3} \, \text{V}$$

The current that flows through $R_2$ is $I_D$ given by:

$$I_D = \frac{V_Y}{Z_{R_2 L_2}}$$
$$= 32 \times 10^{-3} e^{j1.80} \, \text{A}$$

and the voltage across the resistance $R_2$ is given by:

$$V_{R_2} = R_2 I_D$$

$$= 3.2e^{j1.80} \text{ V}$$

That is, the AC voltage across the resistance $R_2$ has a peak amplitude of 9.2 V. The phase of this voltage is 1.80 rad (103°).

The average power dissipated by $R_2$ can be calculated:

$$P_{AV_{R2}} = \frac{1}{R_2 T} \int_{t_o}^{t_o+T} v_{R_2}^2(t)\,dt \tag{9.67}$$

with $T = 2\pi/\omega = 1.3 \times 10^{-3} = 1.3\,\text{ms}$. $t_o$ is chosen to be zero. $v_{R_2}(t)$ can be obtained from its phasor value as follows:

$$v_{R_2}(t) = \text{Real}\,[V_{R_2} e^{j\omega t}]$$

$$= \text{Real}\,[3.2\,e^{j1.80} e^{j\omega t}]$$

$$= 3.2\cos(\omega t + 1.80)\,\text{V}$$

The average power dissipated by $R_2$ can be calculated as shown below:

$$P_{AV_{R2}} = \frac{3.2^2}{T\,R_2} \int_0^T \cos^2(\omega t + 1.80)\,dt$$

It is left to the reader to show that the $P_{AV_{R2}}$ is equal to:

$$P_{AV_{R2}} = \frac{3.2^2}{2} \frac{1}{R_2}$$

$$= 0.05 \text{ W}$$

It is important to note that the average power dissipated in the resistance can also be calculated directly from the phasor representation of the current flowing through and the voltage across $R_2$ as follows:

$$P_{AV_{R2}} = \frac{1}{2}\text{Real}\,[V_{R_2} I_{R_2}^*] = \frac{1}{2}\text{Real}\,[V_{R_2}^* I_{R_2}] \tag{9.68}$$

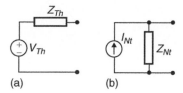

**Figure 9.16: (a) Thévenin equivalent AC circuit; (b) Norton equivalent AC circuit**

$$= \frac{1}{2} \frac{|V_{R_2}|^2}{R_2} \tag{9.69}$$

$$= \frac{1}{2} |I_{R_2}|^2 R_2$$

$$= 0.05 \, \text{W} \tag{9.70}$$

where the current flowing through $R_2$ is $I_{R_2} = I_D$.

### 9.2.4.3 Thévenin and Norton Theorems

Thévenin and Norton equivalent AC circuits can be obtained in a way similar to that described for DC resistive circuits. The main difference is that now the Thévenin equivalent AC circuit comprises an ideal AC voltage source in series with a complex impedance as shown in Figure 9.16(a). The Norton equivalent AC circuit is constituted by an ideal AC current source in parallel with a complex impedance as illustrated in Figure 9.16(b).

### Example 9.6
Consider the AC circuit represented in Figure 9.17(a). Determine the Thévenin equivalent AC circuit at the terminals $X$ and $Y$.

### Solution
Figure 9.17(b) shows the equivalent circuit for the calculation of the open-circuit voltage between terminals $X$ and $Y$. Firstly, the impedances for the capacitance and inductance are calculated for $\omega = 10^4$ rad/s, as shown below:

$$Z_L = j\omega L \big|_{\omega = 10^4 \, \text{rad/s}}$$
$$= j \, 400 \, \Omega$$

$$Z_C = \frac{1}{j\,\omega C}\bigg|_{\omega=10^4\,\text{rad/s}}$$
$$= -j\,1000\,\Omega$$

The phasor associated with the voltage $v_s(t)$ is $V_S = 3\,e^{-J\pi/5}$ V.

Note that the impedance associated with the capacitance is in a parallel connection with the resistance. We can replace these two impedances by an equivalent impedance given by:

$$Z_{RC} = \frac{Z_C R}{Z_C + R}$$
$$= 400 - j\,800\,\Omega \tag{9.71}$$

The voltage between terminals $X$ and $Y$ can be obtained from the voltage *impedance* divider formed by the impedances $Z_{RC}$ and $Z_L$ as follows:

$$V_{Th} = V_S \frac{Z_{RC}}{Z_{RC} + Z_L}$$
$$= 4.7\,e^{-j1.0}\,\text{V}$$

Figure 9.17(c) shows the equivalent circuit for the calculation of the Thévenin impedance, where the AC voltage source has been replaced by a short-circuit. From this figure, it is clear that the impedance $Z_L$ is in a parallel connection with $Z_{RC}$. Hence, $Z_{Th}$ can be calculated as follows:

$$Z_{Th} = \frac{Z_{RC} Z_L}{Z_{RC} + Z_L}$$
$$= 200 + j\,600\,\Omega$$

Figure 9.17(d) shows the Thévenin equivalent circuit for the circuit of 9.17(a). The Thévenin voltage $v_{Th}(t)$ can be determined from its phasor, $V_{Th}$, as follows:

$$v_{Th}(t) = \text{Real}\,[V_S\,e^{j\omega t}]_{\omega=10^4\,\text{rad/s}}$$
$$= 4.7\cos(10^4 t - 1.0)\,\text{V}$$

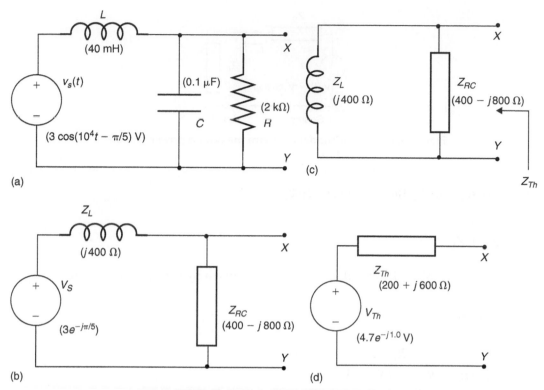

**Figure 9.17: (a) AC circuit; (b) Calculation of the Thévenin voltage; (c) Calculation of the Thévenin Impedance; (d) Equivalent Thévenin circuit**

## 9.2.5 Maximum Power Transfer

Whenever an AC signal is processed by an electrical network containing at least one resistance there is loss of power in the resistances. Since it is often important to ensure that this loss is minimal we consider the conditions which ensure maximum power transfer from two adjacent parts of a circuit. For this purpose we consider the circuit shown in Figure 9.18 where the section of the circuit providing the power is modeled as an AC voltage source with an output impedance $Z_S$ and the section where the power is transmitted is modeled as an impedance $Z_L$. We assume that the source impedance $Z_S$ has a resistive part given by $RS$ and a reactive part described by $j X_S$. Similarly, the load impedance has a resistive component, $R_L$ and a reactive component given by $j X_L$. The

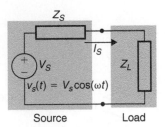

**Figure 9.18: Circuit model to derive maximum power transfer**

current $I_S$ supplied by the source is given by:

$$I_S = \frac{V_S}{Z_L + Z_S} \tag{9.72}$$

and the average power dissipated in the load, $P_L$, is given by (see equation 9.70):

$$
\begin{aligned}
P_L &= \frac{|I_S|^2}{2} R_L \\
&= \frac{V_s^2}{2} \frac{R_L}{(R_S + R_L)^2 + (X_S + X_L)^2}
\end{aligned}
\tag{9.73}
$$

From the last equation, we observe that the value of $X_L$ which maximizes the average power in the load is such that it minimizes the denominator, that is:

$$X_L = -X_S \tag{9.74}$$

Under this condition the average power in the load is given by:

$$P_L = \frac{V_s^2}{2} \frac{R_L}{(R_S + R_L)^2} \tag{9.75}$$

In order to find the value of $R_L$ that maximizes the power in the load we calculate $dP_L/dR_L$ and then we determine the value of $R_L$ for which $dP_L/dR_L$ is zero:

$$\frac{dP_L}{dR_L} = \frac{V_s^2}{2} \frac{(R_S + R_L) - 2R_L}{(R_S + R_L)^3} \tag{9.76}$$

Clearly, the value for $R_L$ which sets $dP_L/dR_L = 0$ is:

$$R_L = R_S \tag{9.77}$$

Hence, the maximum average power delivered to the load is:

$$P_{Lmax} = \frac{V_s^2}{8R_L} \tag{9.78}$$

It is clear that maximum power transfer occurs when $Z_L = Z_S^*$.

## 9.3 Generalized Frequency Domain Analysis

The analysis presented in the previous sections can be considered as a particular case of frequency domain analysis of single frequency signals. As discussed previously, those single frequency signals can be expressed in terms of phasors which, in turn, give rise to phasor analysis. It was seen that phasor analysis allows the application of Ohm's law to the generalized impedance associated with any passive element considerably simplifying electrical circuit analysis.

The analysis of circuits where the signal sources can assume other time-varying (that is non-sinusoidal) waveforms can be a cumbersome task since this gives rise to differential-integral equations. Therefore, it would be most convenient to be able to apply phasor analysis to such circuits. This analysis can indeed be employed using the *Fourier transform*, which allows us to express almost any time varying voltage and current waveform as a sum of phasors.

For reasons of simplicity, before we discuss the Fourier transform we present the Fourier series, which can be seen as a special case of the Fourier transform.

The term *signal* will be used to express either a voltage or a current waveform and we use the terms signal, waveform or function interchangeably to designate voltage or current quantities, which vary with time.

### 9.3.1 The Fourier Series

The Fourier series is used to express periodic signals in terms of sums of sine and cosine waveforms or in terms of sums of phasors. A periodic signal, with period $T$, is by definition a signal that repeats its shape and amplitude every $T$ seconds, that is:

$$x(t \pm kT) = x(t), \quad k = 1, 2, \ldots \tag{9.79}$$

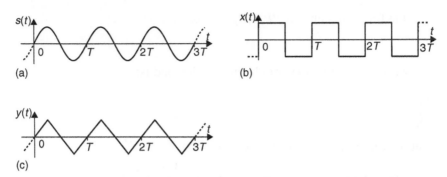

**Figure 9.19: Periodic waveforms (a) Sine; (b) Rectangular; (c) Triangular**

Examples of periodic waveforms are presented in Figure. 9.19 where we have drawn a sine wave, a periodic rectangular waveform, and a periodic triangular waveform. From this Figure it is clear that the waveforms repeat their shape and amplitude every $T$ seconds.

In order to show how the Fourier series provides representations of periodic waves as sums of sine or cosine waves we present, in Figure 9.20(a), the first two non-zero terms (sine waves) of the Fourier series for the periodic rectangular waveform of Figure 9.19(b). Figure 9.20(b) shows that the sum of these two sine waves starts to resemble the rectangular waveform. It will be shown that the addition of all the terms (harmonics) of a particular series converges to the periodic rectangular waveform. In a similar way, Figure 9.20(c) represents the first two non-zero terms of the Fourier series of the triangular waveform. Figure 9.20(d) shows that the sum of just these two sine waves produces a good approximation to the triangular waveform.

Since sine and cosine functions can be expressed as a sum of complex exponential functions (phasors), the Fourier series of a periodic waveform $x(t)$ with period $T$ can be expressed as a weighted sum, as shown below:

$$x(t) = \sum_{n=-\infty}^{\infty} C_n\, e^{j2\pi\frac{n}{T}t} \tag{9.80}$$

where the weights or Fourier coefficients, $C_n$, of the series can be determined as follows:

$$C_n = \frac{1}{T} \int_{t_o}^{t_o+T} x(t) e^{-j2\pi\frac{n}{T}t}\, dt \tag{9.81}$$

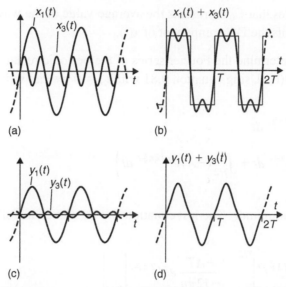

**Figure 9.20: (a) The first two non-zero terms of the Fourier series for the periodic rectangular waveform; (b) The sum of first two non-zero terms of the Fourier series as an approximation to the periodic rectangular waveform; (c) The first two non-zero terms of the Fourier series for the periodic triangular waveform; (d) The sum of first two non-zero terms of the Fourier series as an approximation to the periodic triangular waveform**

Here $t_o$ is a time instant that can be chosen to facilitate the calculation of these coefficients.

The existence of a convergent Fourier series of a periodic signal $x(t)$ requires only that the area of $x(t)$ per period to be finite and that $x(t)$ has a finite number of discontinuities and a finite number of maxima and minima per period. All periodic signals studied here and are to be found in any electrical system satisfy these requirements and, therefore, have a convergent Fourier series.

From equation 9.80 we observe that the phasors which compose the periodic signal $x(t)$ have an angular frequency $2\pi n/T$ which, for $|n| > 1$ is a multiple, or harmonic, of the fundamental angular frequency $\omega = 2\pi/T$. Note that, for $n = 0$ the coefficient $C_0$ is given by:

$$C_0 = \frac{1}{T} \int_{t_o}^{t_o+T} x(t)\, dt \tag{9.82}$$

This equation indicates that $C_0$ represents the average value of the waveform over its period $T$ and represents the DC component of $x(t)$.

As an example, we determine the Fourier series of the periodic rectangular waveform shown in Figure 9.19(b). Using equation 9.81 with $t_o = 0$ we can write:

$$C_n = \frac{1}{T} \int_0^T x(t)\, e^{-j2\pi\frac{n}{T}t}\, dt$$
$$= \frac{1}{T} \left( \int_0^{T/2} A e^{-j2\pi\frac{n}{T}t}\, dt + \int_{T/2}^T (-A)\, e^{-j2\pi\frac{n}{T}t}\, dt \right) \tag{9.83}$$

where $A$ is the peak amplitude. The last equation can be written as follows:

$$C_n = \frac{1}{T} \left( \frac{AT}{-j2\pi n} e^{-j2\pi\frac{n}{T}t} \Big|_0^{T/2} + \frac{-AT}{-j2\pi n} e^{-j2\pi\frac{n}{T}t} \Big|_{T/2}^T \right)$$
$$= \frac{A}{-j2\pi n} \left( e^{-j2\pi\frac{n}{T}\frac{T}{2}} - 1 \right) + \frac{-A}{-j2\pi n} \left( e^{-j2\pi\frac{n}{T}T} - e^{-j2\pi\frac{n}{T}\frac{T}{2}} \right)$$
$$= \frac{2A}{j2\pi n} (1 - e^{-j\pi n}) \tag{9.84}$$

where we have used the following equality:

$$e^{-j2\pi n} = 1, \quad n = 0, \pm 1, \pm 2, \pm 3, \ldots \tag{9.85}$$

However, we note that:

$$e^{-j\pi n} = \begin{cases} -1 & \text{if } n = \pm 1, \pm 3, \pm 5, \ldots \\ 1 & \text{if } n = 0, \pm 2, \pm 4, \pm 6, \ldots \end{cases} \tag{9.86}$$

and, therefore, the coefficients given by equation 9.84 can be written as follows:

$$C_n = \frac{A}{j\pi n} \times \begin{cases} 2 & \text{if } n = \pm 1, \pm 3, \pm 5, \ldots \\ 0 & \text{if } n = 0, \pm 2, \pm 4, \pm 6, \ldots \end{cases} \tag{9.87}$$

Note that for $n = 0$ the last equation cannot be determined as the result would be a non-defined number; 0/0. Hence, $C_0$ must be determined from equation 9.82:

$$C_0 = \frac{1}{T} \int_0^{T/2} A\, dt - \frac{1}{T} \int_{T/2}^T A\, dt = 0 \tag{9.88}$$

confirming that the average value of $x(t)$ is zero as is clear from Figure 9.19(b). From the above, equation 9.87, can be written as follows:

$$C_n = \begin{cases} \dfrac{2A}{j\pi n} & \text{if } |n| \text{ is odd} \\[2mm] 0 & \text{if } |n| \text{ is even} \end{cases} \tag{9.89}$$

It is clear that all even harmonics of the Fourier series are zero. Also, we observe that $C_n = C_{-n}^*$, a fact that applies to any real (non-complex) periodic signal. The coefficients $C_n$ can be written, in a general form, using the complex exponential form as follows:

$$C_n = |C_n| e^{j\, \angle(C_n)} \tag{9.90}$$

and equation 9.80 can be written as follows:

$$x(t) = \sum_{n=-\infty}^{\infty} |C_n| e^{j2\pi \frac{n}{T} t + j\, \angle(C_n)} \tag{9.91}$$

$$= C_0 + \sum_{n=1}^{\infty} 2|C_n| \frac{e^{j2\pi \frac{n}{T} t + j\, \angle(C_n)} + e^{-j2\pi \frac{n}{T} t - j\, \angle(C_n)}}{2}$$

$$= C_0 + \sum_{n=1}^{\infty} 2|C_n| \cos\left(2\pi \frac{n}{T} t + \angle(C_n)\right) \tag{9.92}$$

Expressing the coefficients $C_n$ of equation 9.89 in a complex exponential form (see also equation 9.90) we have:

$$C_n = \begin{cases} \dfrac{2A}{\pi n} e^{-j\frac{\pi}{2}} & \text{if } |n| \text{ is odd} \\[2mm] 0 & \text{if } |n| \text{ is even} \end{cases} \tag{9.93}$$

Hence, $x(t)$ can be written, using equation 9.92, as shown below:

$$x(t) = \sum_{\substack{n=1 \\ (n \text{ odd})}}^{\infty} \frac{4A}{\pi n} \cos\left(2\pi \frac{n}{T} t - \frac{\pi}{2}\right) \tag{9.94}$$

Figure 9.20(a) shows the first and the third harmonics of the rectangular signal as:

$$x_1(t) = \frac{4A}{\pi} \cos\left(2\pi \frac{1}{T} t - \frac{\pi}{2}\right) \tag{9.95}$$

$$x_3(t) = \frac{4A}{3\pi} \cos\left(2\pi \frac{3}{T} t - \frac{\pi}{2}\right) \tag{9.96}$$

from which Figure 9.20(b) was derived.

### Example 9.7

Determine the Fourier series of the periodic triangular waveform, $y(t)$, shown in Figure 9.19(c).

### Solution

From Figure 9.19(b) we observe that the average value of this waveform is zero. Hence, $C_0 = 0$. Using equation 9.81 with $t_o = 0$ we can write:

$$C_n = \frac{1}{T} \int_0^T y(t) e^{-j2\pi \frac{n}{T} t} \, dt$$

$$= \frac{1}{T} \left( \int_0^{T/4} \frac{4At}{T} e^{-j2\pi \frac{n}{T} t} \, dt \right.$$

$$+ \int_{T/4}^{3T/4} \left( 2A - \frac{4At}{T} \right) e^{-j2\pi \frac{n}{T} t} \, dt$$

$$\left. + \int_{3T/4}^{T} \left( \frac{4At}{T} - 4A \right) e^{-j2\pi \frac{n}{T} t} \, dt \right) \tag{9.97}$$

with $A$ representing the peak amplitude of the triangular waveform. Solving the integrals the coefficients can be written as follows:

$$C_n = \frac{A}{\pi^2 n^2}\left(2e^{-j\pi\frac{n}{2}} - 1 - 2e^{-j\pi\frac{3n}{2}} + e^{-j2\pi n}\right) \tag{9.98}$$

Using the result of equation 9.85 we express the coefficients $C_n$ as follows:

$$C_n = \frac{2A}{\pi^2 n^2}e^{-j\pi\frac{n}{2}}(1 - e^{-j\pi n}) \tag{9.99}$$

and using the result of equation 9.86 we can write these coefficients as:

$$C_n = \begin{cases} \frac{4A}{\pi^2 n^2}e^{-j\pi\frac{n}{2}} & \text{if } |n| \text{ is odd} \\ 0 & \text{if } |n| \text{ is even} \end{cases} \tag{9.100}$$

From equation 9.92 the Fourier series for the triangular periodic waveform can be written as:

$$y(t) = \sum_{\substack{n=1 \\ (n \text{ odd})}}^{\infty} \frac{8A}{\pi^2 n^2}\cos\left(2\pi\frac{n}{T}t - \frac{\pi n}{2}\right) \tag{9.101}$$

Figure 9.20(c) shows the first and the third harmonics given by:

$$y_1(t) = \frac{8A}{\pi^2}\cos\left(2\pi\frac{1}{T}t - \frac{\pi}{2}\right) \tag{9.102}$$

$$y_3(t) = \frac{8A}{3\pi^2}\cos\left(2\pi\frac{3}{T}t - \frac{3\pi}{2}\right) \tag{9.103}$$

Figure 9.20(d) clearly shows that the sum of these two harmonics, $y_1(t) + y_3(t)$, approximates the triangular periodic signal.

### 9.3.1.1 Normalized Power

The instantaneous power dissipated in a resistance $R$ with a voltage $v(t)$ applied to its terminals is $v^2(t)/R$, while the instantaneous power dissipated caused by a current $i(t)$ is

$i^2(t)R$. Since signals can be voltages or currents it is appropriate to define a normalized power by setting R = 1 $\Omega$. Then, the instantaneous power associated with a signal $x(t)$ is equal to:

$$p(t) = x^2(t) \tag{9.104}$$

Thus, if $x(t)$ represents a voltage, the instantaneous power dissipated in a resistance $R$ is obtained by dividing $p(t)$ by $R$ while if $x(t)$ represents a current, the instantaneous power dissipated in that resistance $R$ is obtained by multiplying $p(t)$ by $R$. It is also relevant to define a normalized average power (once again, $R = 1\,\Omega$) by integrating equation 9.104 as follows:

$$P_{AV} = \frac{1}{T} \int_{t_o}^{t_o+T} x^2(t)\, dt \tag{9.105}$$

### Example 9.8
Determine an expression for the average power associated with the periodic rectangular waveform shown in Figure 9.19(c).

### Solution
The average power associated with the periodic rectangular waveform is the normalized average power ($R = 1\,\Omega$), which can be determined according to equation 9.105, that is:

$$P_{AV} = \frac{1}{T} \left( \int_0^{T/2} A^2\, dt + \int_{T/2}^{T} (-A)^2\, dt \right)$$

$$= A^2\,(\text{Watts}) \tag{9.106}$$

where $A$ is the amplitude of the waveform.

### 9.3.1.2 Parseval's Power Theorem

Parseval's theorem relates the average power associated with a periodic signal, $x(t)$, with its Fourier coefficients, $C_n$:

$$\frac{1}{T} \int_{t_o}^{t_o+T} x^2(t)\, dt = C_0^2 + \sum_{n=1}^{\infty} 2|C_n|^2 \tag{9.107}$$

The proof of this theorem can be obtained as follows: The Fourier series indicates that $x(t)$ can be seen as a sum of a DC component with sinusoidal components as indicated by equation 9.92. The average power associated with $x(t)$ can be seen as the addition of the average power associated with the DC component with the average power associated with each of these components. It is known that the average power associated with a DC signal is the square of the amplitude of that DC signal. Also, it is known that the average power associated with a sinusoidal component is equal to half the square of its peak amplitude. Since the amplitude of each Fourier component of $x(t)$ is equal to $2|C_n|$ then the average power associated with each of these AC components is equal to:

$$
\begin{aligned}
P_{C_n} &= \frac{(2|C_n|)^2}{2}, \qquad n \geq 1 \\
&= (2|C_n|)^2, \qquad n \geq 1
\end{aligned}
\tag{9.108}
$$

and the total average power of $x(t)$ is:

$$
P_{AV_x} = C_0^2 + \sum_{n=1}^{\infty} 2|C_n|^2
\tag{9.109}
$$

### Example 9.9

Show that the fundamental and the third harmonic of the Fourier series of the periodic rectangular waveform, shown in Figure 9.19(c), contain approximately 90% of the power associated with this waveform.

### Solution

According to equation 9.106 the power associated with the rectangular periodic waveform with amplitude $\pm A$ is $A^2$W. From equation 9.108 the power associated with the fundamental component and the third harmonic of the Fourier series of the periodic rectangular waveform can be calculated as follows (see also equation 9.89):

$$
\begin{aligned}
P &= 2\left(\frac{2A}{\pi}\right)^2 + 2\left(\frac{2A}{3\pi}\right)^2 \\
&= 0.9A^2 \, (\text{W})
\end{aligned}
$$

### 9.3.1.3 Time Delay

If a periodic signal $x(t)$ has a Fourier series with coefficients $C_n$ we can obtain the Fourier series coefficients, $C'_n$, of a replica of $x(t)$ delayed by $\tau$ seconds,

i.e., $x(t - \tau)$ with $|\tau| < T/2$, as follows:

$$C'_n = \int_{t_0}^{t_0 + T} x(t - \tau) e^{-j2\pi \frac{n}{T}t} \, dt \tag{9.110}$$

using the change of variable $t' = t - \tau$, we can write:

$$dt = dt'$$
$$t = t_o \; ; \; t' = t_o - \tau$$
$$t = t_o + T \; ; \; t' = t_o - \tau + T$$

and equation 9.110 can be written as:

$$C'_n = \int_{t'_0}^{t'_0 + T} x(t') e^{-j2\pi \frac{n}{T}t'} \, dt' e^{-j2\pi \frac{n}{T}\tau}$$
$$= C_n \, e^{-j2\pi \frac{n}{T}\tau} \tag{9.111}$$

where $t'_o = t_o - \tau$. Note that the delay $\tau$ adds an extra linear phase to the Fourier series coefficients $C_n$.

### 9.3.2 Fourier Coefficients, Phasors, and Line Spectra

Each phasor that composes the Fourier series of a periodic signal can be seen as the product of a static phasor with a rotating phasor as indicated below:

$$\underbrace{|C_n| e^{j2\pi \frac{n}{T}t + j\angle(C_n)} = |C_n| e^{j\angle(C_n)}}_{\text{Static phasor}} \times \underbrace{e^{j2\pi \frac{n}{T}t}}_{\text{Rotating phasor}} \tag{9.112}$$

Comparing this equation with equation 9.56 we can identify each complex coefficient, $C_n$, as the static phasor corresponding to a rotating phasor with angular frequency $\omega = 2\pi n/T$. The phasor (static and rotating components), which is shown in Figure 9.21(a) can be represented in the *frequency domain* by associating its amplitude, $|C_n|$, and its phase, $\angle(C_n)$, with its angular frequency $\omega = 2\pi n/T$ (or with its linear frequency $f = n/T$). This gives rise to the so called line-spectrum, as illustrated in Figure 9.21(b).

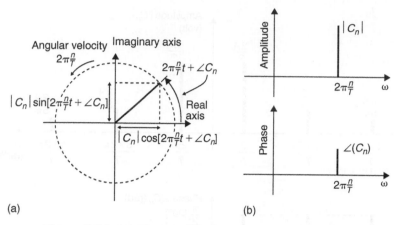

Figure 9.21: (a) Phasor; (b) Line spectrum of a phasor

This frequency representation consists of two plots; amplitude versus frequency and phase versus frequency.

Since the Fourier series expresses periodic signals as a sum of phasors we are now in a position to represent the line spectrum of any periodic signal. As an example, the line spectrum of the periodic square wave with period $T$ can be represented with $C_n$ given by equation 9.89. Figure 9.22 shows the line spectrum representing the fundamental component, the third and the fifth harmonics for this waveform. As mentioned previously, all the frequencies represented are integer multiples of the fundamental frequency $\omega = 2\pi n/T$. The spectral lines have a uniform spacing of $2\pi/T$. It is also important to note that the line spectrum of Figure 9.22 has positive and negative frequencies. Negative frequencies have no physical meaning and their appearance is a consequence of the mathematical representation of sine and of cosine functions by complex exponentials because these trigonometric functions (sine and cosine) are represented by the sum of a pair of complex conjugated phasors (see equation 9.92). We also note that the line spectrum has been plotted as a function of the angular frequency $\omega = 2\pi f$. However, we frequently plot line spectra versus the linear frequency $f = \omega/(2\pi)$.

### 9.3.3 Electrical Signal and Circuit Bandwidths

We discuss now the concepts of signal and electrical system bandwidths. In order to do so we consider the RC circuit of Figure 9.23 which is driven by a

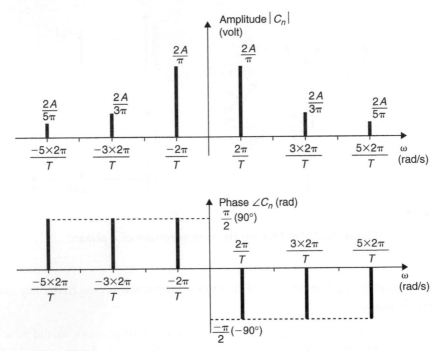

**Figure 9.22: Line spectrum of the rectangular waveform**

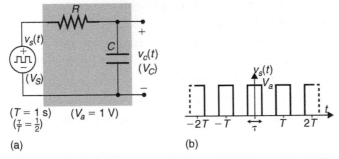

**Figure 9.23: (a) Periodic voltage applied to an RC circuit; (b) The periodic voltage $v(t)$**

square-wave voltage $v_s(t)$ as shown in Figure 9.23(b). This voltage waveform can be expressed as:

$$v_s(t) = \sum_{k=-\infty}^{\infty} V_a \, \text{rect}\left(\frac{t - kT}{\tau}\right) \tag{9.113}$$

where $V_a$ (V) is the amplitude and $T$ is the period. $\tau/T$ is called the *duty-cycle* of the waveform and is equal to 1/2 in this case. The function rect $(t/\tau)$ is defined as follows:

$$
\text{rect}\left(\frac{t}{\tau}\right) = \begin{cases} 1, & -\frac{1}{2} < \frac{t}{\tau} < \frac{1}{2} \\ 0, & \text{elsewhere} \end{cases}
\tag{9.114}
$$

The Fourier coefficients for $v_s(t)$, $V_{S_n}$, can be obtained from equation 9.81 where $t_o$ is chosen to be $-T/2$, that is:

$$
\begin{aligned}
V_{S_n} &= \frac{1}{T} \int_{-T/2}^{T/2} V_a \, \text{rect}\left(\frac{t}{\tau}\right) e^{-j2\pi\frac{n}{T}t} dt \\
&= \frac{1}{T} \int_{-T/2}^{T/2} V_a e^{-j2\pi\frac{n}{T}t} dt \\
&= \frac{TV_a}{-jT2\pi n} \left[ e^{-j2\pi\frac{n}{T}t} \right]_{-\tau/2}^{\tau/2} \\
&= \frac{V_a}{\pi n} \frac{e^{j\pi\frac{n}{T}\tau} - e^{-j\pi\frac{n}{T}\tau}}{2j} \\
&= \frac{V_a}{\pi n} \sin\left(\pi\frac{n}{T}\tau\right)
\end{aligned}
\tag{9.115}
$$

The last equation can be written as follows:

$$
V_{S_n} = \frac{V_a\tau}{T} \frac{\sin(\pi\frac{n}{T}\tau)}{\frac{\pi n\tau}{T}}
\tag{9.116}
$$

$$
= \frac{V_a\tau}{T} \text{sinc}\left(\frac{n\tau}{T}\right)
\tag{9.117}
$$

where the function sinc$(x)$ is defined as follows:

$$
\text{sinc}(x) \triangleq \frac{\sin(\pi x)}{\pi x}
\tag{9.118}
$$

Since $\tau/T = 1/2$, equation 9.117 can be further simplified to:

$$V_{S_n} = \frac{V_a}{2}\operatorname{sinc}\left(\frac{n}{2}\right) \tag{9.119}$$

It is left to the reader to show that the DC component of $v_s(t)$, $V_{S_0}$, is equal to $V_a\tau/T = V_a/2$.

The voltage signal $v_s(t)$ can be written as follows:

$$v_s(t) = \sum_{n=-\infty}^{\infty} \frac{V_a}{2}\operatorname{sinc}\left(\frac{n}{2}\right)e^{j2\pi\frac{n}{T}t} \tag{9.120}$$

Once again, it is left to the reader to show that the periodic square waveform of Figure 9.19(b) can be seen as a particular case of the rectangular waveform of Figure 9.23(b) when $\tau/T = 1/2$. Hint, assume that the average (or DC) component is zero and use a delay of $T/4$.

The *signal bandwidth* is a very important characteristic of any time-varying waveform since it indicates the spectral content and, of course, its minimum and maximum frequency components. From equation 9.120 we observe that the spectrum and therefore the bandwidth of the periodic square wave is infinite. However, it is clear that very high order harmonics have very small amplitudes and its impact on the series can be neglected. So a question arises; where do we truncate the Fourier series in order to determine the significant bandwidth of the signal? The criteria to perform such a truncation can vary depending on the application. One of these can be stated as the range of frequencies which contain a large percentage of the average power associated with this signal. For example, if this criterion defines this percentage as 95% of the total, then the bandwidth for the signal of Figure 9.23(b) is $3/T$. In fact $|V_{S_0}|^2 + 2|V_{S_1}|^2 + 2|V_{S_3}|^2 = 0.95 \times V_a^2/2$ where $V_a^2/2$ is the total average power associated with this signal. It is also important to realize that the signal bandwidth is a measure of how fast a signal varies in time. In order to illustrate this idea we consider Figure 9.24(a) where we see that the addition of higher order harmonics increases the "slope" of the reconstructed signal and that it varies more rapidly with time.

Now that we have determined the Fourier components of the input voltage signal, $v_s(t)$, of the circuit of Figure 9.23(a) we are in a position to determine the output voltage $v_c(t)$.

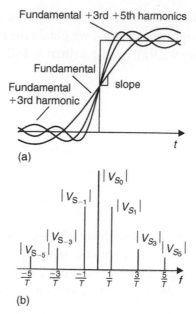

(a)

(b)

**Figure 9.24: Rectangular periodic waveform (a) Approximation by the various components; (b) Line spectrum of the approximation**

This voltage can be determined using the AC phasor analysis, discussed in section 9.2.3, and then applying the superposition theorem to all the voltage components (phasors) of the input signal $v_s(t)$.

The voltage phasor at the terminals of the capacitor, $V_C$, is determined using phasor analysis. This voltage can be obtained noting that the impedance associated with the capacitor and the resistor form an impedance voltage divider. Thus $V_C$ can be expressed as follows:

$$V_C = \frac{Z_c}{Z_c + R} V_S \tag{9.121}$$

where $Z_c = (j\omega C)^{-1}$ is the impedance associated with the capacitor. We can write:

$$V_C = \frac{1}{1 + j\omega RC} V_S \tag{9.122}$$

If we divide the phasor which represents the circuit output quantity, $V_C$, by the phasor which represents the circuit input quantity, $V_S$, we obtain the circuit *transfer function* which, for the circuit of Figure 9.23(a), can be written as follows:

$$H(\omega) = \frac{1}{1 + j\omega RC} \tag{9.123}$$

or

$$H(f) = \frac{1}{1 + j2\pi fRC} \tag{9.124}$$

The transfer function of a circuit is of particular relevance to electrical and electronic circuit analysis since it relates the output with the input by indicating how the amplitude and phase of the input phasors are modified. Figure 9.25 shows the magnitude (on a logarithmic scale) and phase of $H(f)$, given by equation 9.124, versus the frequency $f$,

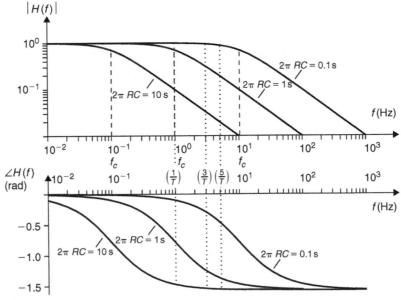

Figure 9.25: Magnitude and phase of the transfer function of the RC circuit of Figure 9.23

also on a logarithmic scale, for various values of the product *RC*. *RC* is called the *time constant* of the circuit. Close inspection of the transfer function $H(f)$ allows us to identify two distinct frequency ranges. The first is for $2\pi fRC \ll 1$, that is for $f \ll (2\pi RC)^{-1}$. Over this frequency range we can write:

$$H(f) \simeq 1 \quad \text{for } f \ll (2\pi RC)^{-1} \tag{9.125}$$

indicating that the circuit does not significantly change the amplitudes or phases of those components of the input signal with frequencies smaller than $(2\pi RC)^{-1}$.

The second frequency range is identified as $2\pi fRC \gg 1$. Now we can write:

$$H(f) \simeq \frac{1}{j2\pi f \ RC} \quad \text{for } f \gg (2\pi RC)^{-1} \tag{9.126}$$

indicating that the circuit significantly attenuates the amplitudes of those components of the input signal with frequencies larger than $(2\pi RC)^{-1}$. The attenuation of these high frequency components means that the circuit preferentially allows the passage of low-frequency components. Hence, this circuit is also called a *low-pass filter*. The frequency $f_c = (2\pi RC)^{-1}$ is called the cut-off frequency of the filter and it establishes its bandwidth. A more detailed discussion of the definition of circuit bandwidth is presented in section 9.3.5. Note that for frequencies $f \gg f_c$ this circuit introduces a phase shift of $-\pi/2$.

We are now in a position to apply the superposition theorem in order to obtain the output voltage. This can be effected by substituting the phasor $V_S$ in equation 9.124 by the sum of phasors (Fourier series) which represents the square wave and by evaluating the circuit transfer function at each frequency $f = n/T$. That is:

$$
\begin{aligned}
V_{C_n} &= [H(f)]_{f=\frac{n}{T}} \times V_{S_n} \\
&= \left[ \frac{1}{1 + j2\pi fRC} \right]_{f=\frac{n}{T}} \times V_{S_n} \\
&= \frac{1}{1 + j2\pi \frac{n}{T} RC} V_{S_n}
\end{aligned}
\tag{9.127}
$$

where the phasors $V_{C_n}$ are the coefficients of the Fourier series representing the voltage $v_c(t)$ and the phasors $V_{S_n}$ are the coefficients representing the periodic square voltage $v_s(t)$. The phasors $V_{C_n}$ can be written as:

$$V_{C_n} = \frac{1}{1 + j\,2\pi\frac{n}{T}RC} \frac{V_a}{2} \operatorname{sinc}\left(\frac{n}{2}\right) \quad \text{(V)} \tag{9.128}$$

which can also be written in the complex exponential form as:

$$V_{C_n} = \begin{cases} \dfrac{V_a e^{j\left(\frac{n\pi}{2} - \frac{\pi}{2}\right) - j\,\tan^{-1}\left(2\pi\frac{n}{T}RC\right)}}{\pi n \sqrt{1 + \left(\frac{2\pi n}{T}RC\right)^2}} & \text{for } |n| \text{ odd} \\[4mm] \dfrac{V_a}{2} & \text{for } n = 0 \\[2mm] 0 & \text{for } |n| \text{ even and } |n| > 1 \end{cases} \tag{9.129}$$

Figure 9.25 shows that if the low-pass filter features a time constant such that $2\pi RC = 10\,\text{s}$, corresponding to $f_c = 0.1\,\text{Hz}$, all frequency components of the input signal, with the exception of the DC component, are severely attenuated. Although for $2\pi RC = 1\,\text{s}$ ($f_c = 1\,\text{Hz}$) the fundamental frequency component is slightly attenuated, all higher order harmonics are considerably attenuated. This implies that for both situations described above the output voltage will be significantly different from the input voltage. On the other hand, for $2\pi RC = 0.1\,\text{s}$ ($f_c = 10\,\text{Hz}$) the fundamental, the third and the fifth order frequency components are hardly attenuated although higher-order harmonics suffer great attenuation. Note that, for this last situation ($f_c = 10\,\text{Hz}$), the significant bandwidth of the input voltage signal does not suffer significant attenuation. This means that the output voltage is very similar to the input voltage.

Since the Fourier coefficients of $v_c(t)$ are known, this voltage can be written using equation 9.92, that is:

$$v_c(t) = \frac{V_a}{2} + \sum_{\substack{n=1 \\ (n \text{ odd})}}^{\infty} \frac{2V_a}{\pi n \sqrt{1 + \left(\frac{2\pi n}{T}RC\right)^2}} \cos\left(2\pi\frac{n}{T}t + \phi_n\right) \tag{9.130}$$

with,

$$\phi_n = \begin{cases} \frac{n\pi}{2} - \frac{\pi}{2} - \tan^{-1}\left(2\pi\frac{n}{T}RC\right) & \text{for } n \text{ odd} \\ 0 & \text{for } n \text{ even} \end{cases} \qquad (9.131)$$

Figure 9.26 illustrates the output voltage $v_c(t)$ for the three time constants discussed above. As expected, for the two situations where $2\pi RC = 10$ s and $2\pi RC = 1$ s the output voltage $v_c(t)$ is very different from the input voltage due to the filtering effect of the input signal frequency components. However, for $2\pi RC = 0.1$ s the output voltage is very similar to the input signal since the main frequency components are not significantly attenuated.

It is also interesting to note that the effect of filtering all frequency components $(2\pi RC = 10$ s$)$ of the square voltage waveform results in a near-triangular periodic waveform, such as that of Figure 9.19(c), with an average value (DC component) equal to the DC value of the input square wave input voltage (see next example).

The waveforms of $v_c(t)$ illustrated in Figure 9.26 can be interpreted as the repetitive charging (towards $V_a$) and discharging (towards 0) of the capacitor. At the higher cut-off frequency $(2\pi RC = 0.1$ s$)$ the capacitor can charge and discharge in a rapid manner almost following the input signal. However, as the cut-off frequency (or bandwidth) of the filter is decreased the charging and discharging of the capacitor takes more time. It is as if the output voltage is suffering from an "electrical inertia" which opposes to the time-variations of that signal. In fact, the *bandwidth of a circuit* can actually be viewed as a qualitative measure of this "electrical inertia."

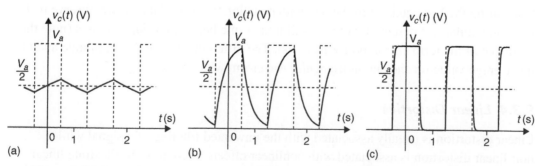

**Figure 9.26: Waveforms for $v_c(t)$ (a) $2\pi RC = 0.1$ s; (b) $2\pi RC = 1$ s; (c) $2\pi RC = 10$ s**

*Example 9.10*

Consider the circuit of Figure 9.23(a). Show that if the cut-off frequency is such that $f_c \ll T^{-1}$ then the resulting output voltage is a near-triangular waveform as shown in Figure 9.26(a).

*Solution*

If $(2\pi RC)^{-1} \ll T^{-1}$ this means that:

$$\frac{2\pi n RC}{T} \gg 1, \quad n \geq 1 \tag{9.132}$$

and we can write the Fourier coefficients of the output voltage, expressed by equation 9.128, as follows:

$$V_{C_n} \simeq \begin{cases} \dfrac{V_a}{j2\,\frac{2\pi n}{T}\,RC}\,\mathrm{sin\,c}\left(\frac{n}{2}\right) & \text{if } n \neq 0 \\[2mm] \dfrac{V_a}{2} & \text{if } n = 0 \end{cases} \tag{9.133}$$

This equation can be written in exponential form as follows:

$$V_{C_n} = \begin{cases} \dfrac{V_a T}{2RC}\,\dfrac{1}{\pi^2 n^2}\,e^{-jn\pi/2} & \text{if } |n| \text{ is odd} \\[2mm] \dfrac{V_a}{2} & \text{if } n = 0 \\[2mm] 0 & \text{if } |n| \text{ is even and } |n| > 0 \end{cases} \tag{9.134}$$

Comparing the last equation for $|n|$ odd with equation 9.100 for $|n|$ odd, we observe that they are similar in the sense that they exhibit the same behavior as $|n|$ increases (note the existence of the term $1/n^2$ in both equations). The difference lies in the amplitude and in the average value for the output triangular waveform which now is $V_a = 2$.

### 9.3.4 Linear Distortion

Linear distortion is usually associated with the unwanted filtering of a signal while non-linear distortion is associated with nonlinear effects in circuits. To illustrate linear distortion let us consider the transmission of a periodic signal $y(t)$ through an electrical channel with a transfer function $H(f)$. The output signal, $z(t)$, is said undistorted if it is a replica of $y(t)$, that is if $z(t)$ differs from $y(t)$ by a multiplying constant $A$, representing an amplification ($A > 1$) or attenuation ($A < 1$), and a time delay, $t_d$. Hence, $z(t)$ can be written as:

$$z(t) = Ay(t - t_d) \tag{9.135}$$

The relevant question is: What must $H(f)$ be in order to have such a distortionless transmission? To answer this we assume that $y(t)$ has a Fourier series given by:

$$y(t) = \sum_{n=-\infty}^{\infty} C_{Y_n} e^{j2\pi\frac{n}{T}t} dt \tag{9.136}$$

From equation 9.135 and from the time delay property of Fourier series (see equation 9.111) we can write the Fourier coefficients of $z(t)$ as follows:

$$C_{Z_n} = AC_{Y_n} e^{j2\pi\frac{n}{T}t_d} \tag{9.137}$$

From equation 9.127 we can determine $H(f)$ as follows:

$$[H(f)]_{f=\frac{n}{T}} = \frac{C_{Z_n}}{C_{Y_n}}$$
$$= Ae^{j2\pi\frac{n}{T}t_d} \tag{9.138}$$

that is,

$$H(f) = Ae^{j2\pi f t_d} \tag{9.139}$$

Figure 9.27 shows the magnitude and the phase of this transfer function. From this Figure we conclude that a distortionless system must provide the same amplification (or attenuation) to all frequency components of the input signal *and* must provide a linear phase shift to all these components.

The application of a sequence of rectangular pulses to an RC circuit illustrates what can be considered as linear distortion. Now, let us consider the transmission of those

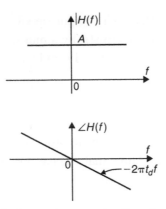

**Figure 9.27: Magnitude and phase of a transfer function of a distortionless system**

same pulses through an electrical channel that is modeled as the RC circuit of Figure 9.23(a). From the discussion above we saw that if the cut-off frequency of the RC circuit is smaller than the third harmonic frequency of the input signal, then the output signal is significantly different from the input signal. Severe linear distortion occurs since the various frequency components of the input signal are attenuated by different amounts *and* suffer different phase shifts. However, if the cut-off frequency of the RC circuit is larger than the third harmonic frequency then the output signal is approximately equal to the input signal, as illustrated by Figure 9.26(c). This is because the most significant frequency components of the input signal are affected by the same (unity) gain. Note that, in this situation, the phase shift is zero indicating that there is no delay between the input and output signals.

### 9.3.5 Bode Plots

In the previous section we saw that the complex nature of a transfer function, $H(f)$ (or $H(\omega)$), implies that the graphical representation of $H(f)$ requires two plots; the magnitude of $H(f)$, $|H(f)|$, and the phase of $H(f)$, $\angle H(f)$, versus frequency, as illustrated in Figure 9.25.

Often, it is advantageous to represent the transfer function, $|H(f)|$, on a logarithmic scale, given by:

$$|H_{dB}(f)| = 20 \log_{10}|H(f)| \quad \text{(dB)} \tag{9.140}$$

Here, $|H_{dB}(f)|$ and frequency are represented on logarithmic scales. The unit of the transfer function expressed in such a logarithmic scale is the decibel (dB).

The main advantage of this representation is that we can determine the asymptotes of the transfer function which, in turn facilitate its graphical representation. Note that the logarithmic operation also emphasises small differences in the transfer function which, if plotted in the linear scale, would not be so clearly visible. In order to illustrate this we again consider the transfer function of the RC circuit of Figure 9.23, given by:

$$H(f) = \frac{1}{1 + j2\pi fRC} \tag{9.141}$$

We can express this as:

$$|H_{dB}(f)| = 20 \log_{10} \left| \frac{1}{1 + j2\pi fRC} \right|$$

$$= 20 \log_{10} \frac{1}{\sqrt{1 + (2\pi fRC)^2}}$$

$$= 20 \log_{10}(1) - 20 \log_{10}(1 + (2\pi fRC)^2)^{\frac{1}{2}}$$

$$= -10 \log_{10}(1 + (2\pi fRC)^2) \tag{9.142}$$

We can now identify the two *asymptotes* of $|H_{dB}(f)|$, noting that:

$$1 + (2\pi fRC)^2 \simeq 1 \quad \text{if } 2\pi fRC \ll 1 \tag{9.143}$$

$$1 + (2\pi fRC)^2 \simeq (2\pi fRC)^2 \quad \text{if } 2\pi fRC \gg 1 \tag{9.144}$$

Hence, we can write:

$$|H_{dB}(f)| \simeq -10 \log_{10}(1)$$
$$\simeq 0 \text{ dB} \qquad \text{if } f \ll \frac{1}{2\pi RC} \tag{9.145}$$

$$|H_{dB}(f)| \simeq -10 \log_{10}(2\pi fRC)^2$$
$$\simeq -20 \log_{10}(2\pi fRC) \quad \text{if } f \gg \frac{1}{2\pi RC} \tag{9.146}$$

The phase of $H(f)$ is given by:

$$\angle H(f) = e^{-j \tan^{-1}(2\pi fRC)} \tag{9.147}$$

and it can also be approximated by asymptotes:

$$\angle H(f) \simeq \begin{cases} 0 & \text{if } f < \frac{1}{10} \times \frac{1}{2\pi RC} \\ -\frac{\pi}{4} \log_{10}(2\pi fRC) - \frac{\pi}{4} & \text{if } \frac{1}{10 \times 2\pi RC} < f < \frac{10}{2\pi RC} \\ -\frac{\pi}{2} & \text{if } f > \frac{10}{2\pi RC} \end{cases} \tag{9.148}$$

Figure 9.28(a) shows $|H_{dB}(f)|$ versus the frequency. In this figure we also show the corresponding values of $|H(f)|$. A gain of $-20$ dB (corresponding to an attenuation of 20 dB) is equivalent to a *linear* gain of 0.1 (or an attenuation of 10 times).

The two asymptotes given by equations 9.145 and 9.146 are represented in Figure 9.28(a), by dashed lines. Since the X-axis is also logarithmic the asymptote given by equation 9.146 is represented as a line whose slope is $-20$ dB/decade. A decade is a frequency range over which the ratio between the maximum and minimum frequency is 10. Note that this slope can be inferred by inspection of Figure 9.28(a) where we observe that for $f = (2\pi RC)^{-1}$ the asymptote given by equation 9.146 indicates 0 dB. From this figure we observe that these two asymptotes approximately describe the entire transfer function. The maximum error, $\Delta$, between $H(f)$ and the asymptotes occurs at the frequency $f = (2\pi RC)^{-1}$. It is given by:

$$\Delta = -20 \log_{10}(2\pi fRC)_{f=(2\pi RC)^{-1}} - |H_{dB}(f)|_{f=(2\pi RC)^{-1}}$$

$$= 0 + 10 \log_{10}(2)$$

$$\simeq 3 \text{ dB}$$

The *circuit or system bandwidth* is very often defined as the range of positive frequencies for which the magnitude of its transfer function is above the 3 dB attenuation value. This 3 dB value is equivalent to voltage or current output to input ratio of $1/\sqrt{2} \simeq 71\%$ (see Figure 9.28(a)) or, alternatively, output to input power ratio of 50%. Hence, the bandwidth for the RC circuit is from DC to $f = (2\pi RC)^{-1}$, the *cut-off* frequency.

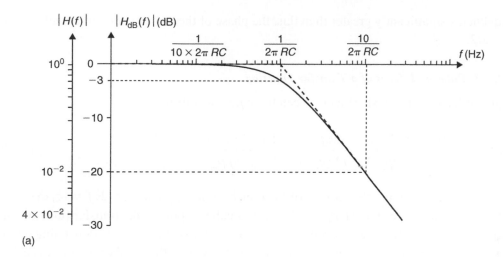

(a)

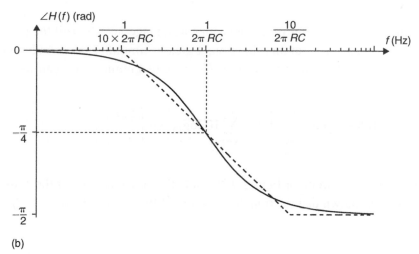

(b)

**Figure 9.28: Magnitude and phase of the transfer function of the RC circuit of Figure 9.23 (solid lines) and asymptotes (dashed lines)**

Figure 9.28(b) shows the angle of the transfer function, $\angle H(f)$, and also its asymptotes given by equation 9.148. From this figure we observe that for frequencies smaller than one tenth of the cut-off frequency the phase of the transfer function is close to zero. At the cut-off frequency $f = (2\pi RC)^{-1}$ the phase of the transfer function is $-\pi/4$ and for frequencies significantly greater than this, the phase of the transfer function tends to $-\pi/2$.

### 9.3.5.1 Poles and Zeros of a Transfer Function

In general, a circuit transfer function can be written as follows:

$$H(f) = A\frac{(1+ j2\pi f/z_1)(1+ j2\pi f/z_2)...(1+ j2\pi f/z_n)}{(1+ j2\pi f/p_1)(1+ j2\pi f/p_2)...(1+ j2\pi f/p_m)} \tag{9.149}$$

Each $z_i$, $I = 1, ..., n$, is called a zero of the transfer function, and, for $j2\pi f = -z_i$ the transfer function is zero. Each $p_i$, $i = 1,...,m$, is called a pole of the transfer function. At $j2\pi f = -p_i$ the transfer function is not defined since $H(jp_i/(2\pi)) \to \pm \infty$ depending on the sign of the DC gain, $A$. For a practical circuit $m \geq n$ and $m$, the number of poles, is called the order of the transfer function.

This representation of a transfer function is quite advantageous when all the poles and zeros are *real numbers* since, in this situation, it greatly simplifies the calculation of $|H_{dB}(f)|$. In fact, if all the poles and zeros of $H(f)$ are real numbers we can write:

$$|H_{dB}(f)| = \sum_{i=1}^{n}10 \log_{10}\left[1+\left(\frac{2\pi f}{z_i}\right)^2\right] - \sum_{k=1}^{m}10 \log_{10}\left[1+\left(\frac{2\pi f}{p_k}\right)^2\right] \tag{9.150}$$

Let us consider the CR circuit of Figure 9.29. Note the new positions of the resistor and capacitor. It can be shown that the transfer function of this circuit, $H_{CR}(f) = V_R/V_S$, can be written as:

$$H_{CR}(f) = \frac{j2\pi fRC}{1+ j2\pi fRC} \tag{9.151}$$

Relating this transfer function with equation 9.149 we observe that $H_{CR}(f)$ has one pole, equal to $(RC)^{-1}$, and a zero located at the origin. Since the pole and the zero are real numbers, we can use equation 9.150 to determine $|H_{CR_{dB}}(f)|$ as follows:

$$|H_{CR_{dB}}(f)| = 20 \log_{10}(2\pi fRC) - 10 \log_{10}(1+ (2\pi fRC)^2) \tag{9.152}$$

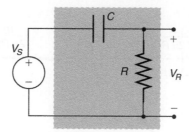

**Figure 9.29: CR circuit**

We can identify the two asymptotes of $|H_{CR_{dB}}(f)|$ (see also equations 9.143 and 9.144) which are given by:

$$|H_{CR_{dB}}(f)| \simeq 20 \log_{10}(2\pi f RC) \text{ dB} \quad \text{if } f \ll \frac{1}{2\pi RC} \tag{9.153}$$

$$|H_{CR_{dB}}(f)| \simeq 0 \text{ dB}, \quad \text{if } f \gg \frac{1}{2\pi RC} \tag{9.154}$$

The phase of $H_{CR}(f)$ is given by:

$$\angle H_{CR}(f) = e^{j\frac{\pi}{2} - j \tan^{-1}(2\pi f RC)} \tag{9.155}$$

and it can also be approximated by asymptotes:

$$\angle H_{CR}(f) \simeq \begin{cases} \frac{\pi}{2} & \text{if } f < \frac{1}{10} \times \frac{1}{2\pi RC} \\ \frac{\pi}{4} - \frac{\pi}{4} \log_{10}(2\pi f RC) & \frac{1}{10 \times 2\pi RC} < f < \frac{10}{2\pi RC} \\ 0 & f > \frac{10}{2\pi RC} \end{cases} \tag{9.156}$$

Figure 9.30(a) shows the magnitude, in dB, of this transfer function given by equation 9.152 and the asymptotes given by equations 9.153 and 9.154. We observe that this circuit attenuates frequencies smaller than the cut-off frequency, $f_c = (2\pi RC)^{-1}$, while it passes the frequency components higher than $f_c$. Hence, this circuit is called a *high-pass filter*. Note that, in theory, the bandwidth of this filter is infinity, although in practice unwanted circuit elements set a maximum operating frequency to this circuit.

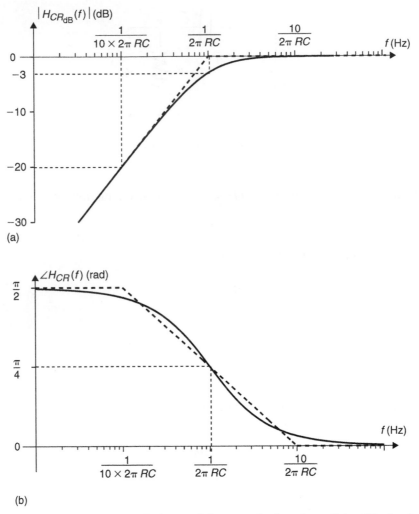

**Figure 9.30: Magnitude and phase of the transfer function of the CR circuit of Figure 9.29 (solid lines) and asymptotes (dashed lines)**

Figure 9.30(b) shows the phase of the transfer function. The three asymptotes for this phase given by equation 9.156 are also shown. At frequencies smaller than $f = (2\pi RC10)^{-1}$ the circuit imposes a phase of $\pi/2$ while at frequencies higher than $f = 10(2\pi RC)^{-1}$ the circuit does not change the phase.

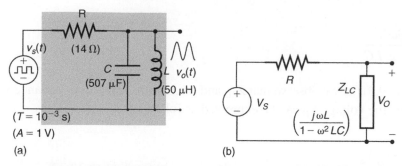

**Figure 9.31: (a) RLC circuit; (b) AC equivalent circuit**

### 9.3.5.2 Signal Filtering as Signal Shaping

Signal filtering can act as signal shaping as illustrated in Example 9.10 where a triangular waveform was obtained from the low-pass filtering of a square wave. This shaping is accomplished using at least one energy storage element in an electronic network, that is by using capacitors or inductors. Capacitive and inductive impedances are frequency dependent and different frequency components of a periodic signal suffer different amounts of attenuation (or amplification) and different amounts of phase shift giving rise to modified signals.

To further illustrate this idea, let us consider the circuit of Figure 9.31 where a square-wave voltage is applied (see Figure 9.19(b)). The purpose of this circuit is to reshape the input signal in order to obtain a sine wave voltage.

The output voltage, $v_o(t)$, is the voltage across the capacitor and inductor. Since the input voltage $v_s(t)$ can be decomposed as a sum of phasors the voltage $v_o(t)$ can be determined using AC phasor analysis together with the superposition theorem. We start by calculating the voltage at the output, $V_O$, using phasor analysis. Since the capacitor is in a parallel connection with the inductor we can determine an equivalent impedance,

$$Z_{LC} = \frac{Z_L Z_C}{Z_L + Z_C} \tag{9.157}$$

with,

$$Z_C = \frac{1}{j\omega C} \tag{9.158}$$

$$Z_L = j\omega L \tag{9.159}$$

that is:

$$Z_{LC} = \frac{j\omega L}{1 - \omega^2 LC} \qquad (9.160)$$

From Figure. 9.31(b) we observe that $Z_{LC}$ and the resistor form an impedance voltage divider. Thus the voltage $V_O$ can be expressed as follows:

$$
\begin{aligned}
V_O &= \frac{Z_{LC}}{Z_{LC} + R} V_S \\
&= \frac{\frac{j\omega L}{1 - \omega^2 LC}}{\frac{j\omega L}{1 - \omega^2 LC} + R} \\
&= \frac{j\omega L}{R(1 - \omega^2 LC) + j\omega L} V_S
\end{aligned}
\qquad (9.161)
$$

The transfer function is, therefore,

$$H_{RLC}(\omega) = \frac{j\omega L}{R(1 - \omega^2 LC) + j\omega L} \qquad (9.162)$$

Clearly, this can also be written as:

$$H_{RLC}(f) = \frac{j2\pi f L}{R(1 + (j2\pi f)^2 LC) + j2\pi f L} \qquad (9.163)$$

The two poles of $H_{RLC}(f)$ can be determined by setting the denominator of equation 9.163 to zero and solving this equation in order to obtain $j2\pi f$, that is:

$$R(1 + (j2\pi f)^2 LC) + j2\pi f L = 0 \qquad (9.164)$$

and since $L^2 - 4LCR^2 < 0$ we obtain:

$$j2\pi f_i = \frac{-L \pm j\sqrt{4LCR^2 - L^2}}{2RLC}, \qquad i = 1, 2 \qquad (9.165)$$

The two poles of the transfer function are obtained from the last equation (see also equation 9.149) as:

$$
\begin{aligned}
p_i &= -j\,2\,\pi f_i \\
&= \frac{+L \mp j\,\sqrt{4\,LCR^2 - L^2}}{2\,RLC}, \qquad i = 1,2
\end{aligned}
\tag{9.166}
$$

The two poles given by the last equation are complex conjugated. This means that we cannot apply equation 9.150 and we must determine $|H_{RLC_{dB}}(f)|$ using the standard procedure, that is:

$$
\begin{aligned}
|H_{RLC_{dB}}(f)| &= 20\,\log_{10}\left|\frac{j\,2\,\pi f\,L}{R(1 - (2\,\pi f)^2\,L\,C) + j\,2\,\pi f\,L}\right| \\
&= 20\,\log_{10}(2\pi f\,L) - 10\,\log_{10}\left[R^2(1 - (2\pi f)^2 L\,C)^2 + (2\pi f\,L)^2\right]
\end{aligned}
\tag{9.167}
$$

Figure 9.32 shows a plot of $|H_{RLC_{dB}}(f)|$. This figure indicates that the RLC circuit does not attenuate the component $f = (2\pi\sqrt{LC})^{-1} = 1\,\text{kHz}$ since $|H_{RLC_{dB}}((2\pi\sqrt{LC})^{-1})| = 0$ dB. However, it attenuates all frequency components around this frequency. Thus, this circuit is called a *band-pass* filter. The (3 dB) bandwidth of this circuit is 22 Hz centered in 1 kHz. For band-pass filters the Quality Factor, $Q$, is defined as the ratio of the central frequency, $f_o$, to its bandwidth, $BW$, that is

$$
Q = \frac{f_o}{BW}
\tag{9.168}
$$

The quality factor is a measure of the sharpness of the response of the circuit. A high quality factor indicates a high frequency selectivity of the band-pass filter. For this circuit the quality factor is $Q = 45$. Note that the third and the fifth harmonics suffer an attenuation greater than 40 dB resulting from the frequency selectivity of the circuit. This means that these frequency components have an amplitude (at least) 100 times smaller at the output of the circuit compared to its original amplitude at the input of the circuit.

We are now in a position to apply the superposition theorem to obtain $v_o(t)$. This can be effected by substituting the phasor $V_S$ in equation 9.163 by the Fourier series which

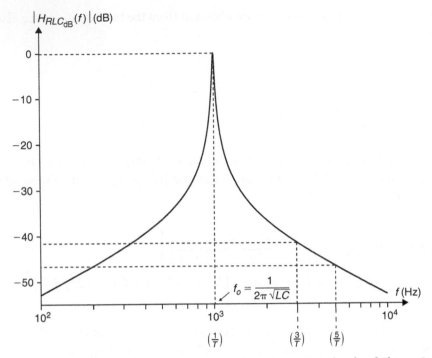

**Figure 9.32: Magnitude of the transfer function of the RLC circuit of Figure 9.31**

represents the periodic square wave and by evaluating the transfer function of the circuit, $H_{RLC}(f)$, at *each* frequency of these phasors, that is:

$$V_{O_n} = [H_{RLC}(f)]_{f=\frac{n}{T}} \times V_{S_n}$$

$$= \frac{j2\pi\frac{n}{T}L}{R(1 - 4\pi^2\frac{n^2}{T^2}LC) + j2\pi\frac{n}{T}L} V_{S_n} \tag{9.169}$$

where the phasors $V_{O_n}$ are the coefficients of the Fourier series representing $v_o(t)$ and the phasors $V_{S_n}$ are the coefficients of the Fourier series representing $v_s(t)$. Clearly, $V_{S_n}$

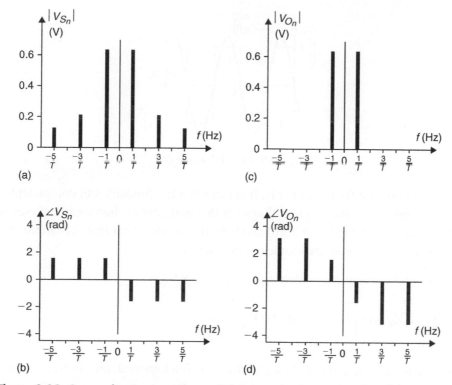

**Figure 9.33: Spectral representations of: (a) magnitude of $v_s(t)$; (b) phase of $v_s(t)$; (c) magnitude of $v_o(t)$; (d) phase of $v_o(t)$**

coincide with $C_n$ given by equation 9.87. However, the units for these coefficients are volts. The phasors $V_{O_n}$ can be written as:

$$
V_{O_n} = \begin{cases} \dfrac{j\,2\pi\dfrac{n}{T}L}{R(1 - 4\pi^2\dfrac{n^2}{T^2}LC) + j2\pi\dfrac{n}{T}L}\,\dfrac{2A}{j\pi n}\ \text{(V)} & \text{for } |n| \text{ odd} \\[6pt] 0 & \text{for } |n| \text{ even} \end{cases} \tag{9.170}
$$

Figures 9.33(a) and 9.33(b) show the magnitude and the phase of the spectral components of $v_s(t)$, respectively, while Figures 9.33(c) and 9.33(d) show the magnitude and the phase

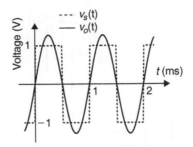

**Figure 9.34:** $v_s(t)$ and $v_o(t)$

of the components of $v_o(t)$, respectively. It is clear that the fundamental component (at $f = 1/T$) is present in the output voltage but that higher order harmonics are severely attenuated. Comparing Figures 9.33(b) and 9.33(d) it is also clear that the circuit changes the phase of the higher order harmonics of the input signal.

The voltage $v_o(t)$ can now be written using equation 9.92 as:

$$v_o(t) = \sum_{\substack{n=1 \\ (n \text{ odd})}}^{\infty} 2|V_{O_n}| \cos\left(2\pi \frac{n}{T} t + \text{angle}\,(V_{O_n})\right) \tag{9.171}$$

Since the harmonics, at frequencies higher than the fundamental, are strongly attenuated, we can write $v_o(t)$ as:

$$v_o(t) \simeq 2|V_{O_1}| \cos\left(2\pi \frac{1}{T} t + \text{angle}\,(V_{O_1})\right)$$

$$= \frac{4}{\pi} \cos\left(2\pi \frac{1}{T} t - \frac{\pi}{2}\right) \tag{9.172}$$

Finally Figure 9.34 shows $v_s(t)$ and $v_o(t)$ given by equation 9.171. From this figure it is clear that the output voltage is a sine wave corresponding to the fundamental component of the input periodic voltage signal $v_s(t)$.

### 9.3.6 The Fourier Transform

In the previous section we have seen that the Fourier series is a very powerful signal analysis tool since it allows us to decompose periodic signals into a sum of phasors. Such a decomposition, in turn, allows the analysis of electrical circuits using the AC

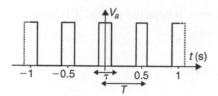

**Figure 9.35: Periodic voltage rectangular waveform**

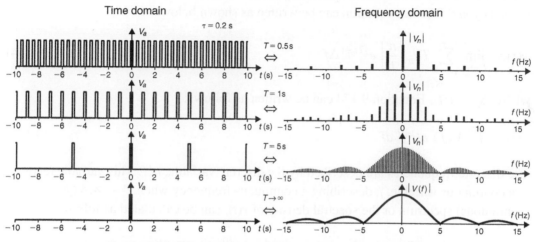

**Figure 9.36: The Fourier transform of a rectangular pulse**

phasor technique with the superposition theorem. While the Fourier series applies only to periodic waveforms, the Fourier transform is a far more powerful tool since, in addition to periodic signals, it can represent non-periodic signals as a "sum" of phasors. In order to illustrate the difference between the Fourier series and the Fourier transform we recall the Fourier series of a rectangular waveform like that depicted in Figure 9.35 with amplitude $V_a$ and duty-cycle $\tau/T$. Figure 9.36(a) shows the waveform and its correspondent line spectrum (magnitude). If we now increase the period $T$ (maintaining $\tau$ and the amplitude constant) we observe that the density of phasors increases (Figure 9.36(b) and 9.36(c)). Note that the amplitude of these phasors decreases since the power of the signal decreases. If we let the period tend to infinity this is equivalent to having a non-periodic signal, that is, we have a situation where the signal $v(t)$ is just a single rectangular pulse. In this situation, the signal spectrum is no longer discrete and no longer constituted by equally spaced discrete phasors. Instead the spectrum becomes *continuous*. In this situation, the spectrum is often referred to as having a continuous spectral density.

The procedure described above, where the period $T$ is increased, can be written, in mathematical terms, as follows:

$$v(t) = \lim_{T \to \infty} \sum_{n=-\infty}^{\infty} V_n\left(\frac{n}{T}\right) e^{j2\pi \frac{n}{T} t}$$

(9.173)

where we indicate the explicit dependency of the Fourier coefficients $V_n$ on the discrete frequency $n/T$. The last equation can be written as shown below:

$$v(t) = \lim_{T \to \infty} \sum_{n=-\infty}^{\infty} T V_n\left(\frac{n}{T}\right) e^{j2\pi \frac{n}{T} t} \Delta f$$

(9.174)

where $\Delta f = 1/T$. Equation 9.174 can be written as follows:

$$v(t) = \int_{-\infty}^{\infty} V(f)\, e^{j2\pi f t}\, df$$

(9.175)

The discrete frequencies are described by the discrete variable, $n/T$. This variable tends to a continuous variable, $f$, describing a continuous frequency when $T \to \infty$. $V(f)$, the (continuous) spectrum or the spectral density of $v(t)$, can be calculated as follows:

$$V(f) = \lim_{T \to \infty} T\, V_n\left(\frac{n}{T}\right)$$

(9.176)

$$= \lim_{T \to \infty} \int_{-T/2}^{T/2} v(t)\, e^{-j2\pi \frac{n}{T} t}\, dt$$

(9.177)

Where we chose $t_o = -T/2$. Finally, the last equation can be written as:

$$V(f) = \int_{-\infty}^{\infty} v(t)\, e^{-j2\pi f t}\, dt$$

(9.178)

A sufficient condition (but not strictly necessary) for the existence of the Fourier transform of a signal $x(t)$ is that the integral expressed by equation 9.178 has a finite value for every value of $f$.

### Example 9.11

Consider the single square voltage pulse shown in Figure 9.36. Show that the Fourier transform of this pulse is the same as that obtained from equation 9.176, which is derived from the Fourier series of a periodic sequence of rectangular pulses (see equation 9.117), when $T \rightarrow \infty$.

**Solution**

Using equation 9.178 we can write:

$$
\begin{aligned}
V(f) &= \int_{-\infty}^{\infty} V_a \, \mathrm{rect}\left(\frac{t}{\tau}\right) e^{-j2\pi ft} \, dt \\
&= \int_{-\tau/2}^{\tau/2} V_a \, e^{-j2\pi ft} \, dt \\
&= \frac{V_a}{-j2\pi f} \frac{e^{j\pi f\tau} - e^{-j\pi f\tau}}{2j} \\
&= V_a \, \tau \, \mathrm{sinc}(f\tau)
\end{aligned}
\tag{9.179}
$$

From equation 9.176 we can write:

$$
\begin{aligned}
V(f) &= \lim_{T \rightarrow \infty} T \, V_n\left(\frac{n}{T}\right) \\
&= \lim_{T \rightarrow \infty} T \, \frac{V_a \tau}{T} \mathrm{sinc}\left(\frac{n\tau}{T}\right) \\
&= V_a \, \tau \, \mathrm{sinc}(f\,\tau)
\end{aligned}
\tag{9.180}
$$

where $n/T \rightarrow f$ as $T \rightarrow \infty$.

From the above it should be clear that the Fourier transform, $V(f)$, represents a density of phasors which completely characterize $v(t)$ in the frequency domain. Such a representation is similar to the Fourier series coefficients in the context of periodic signals. However, it is important to note that while the unit of the voltage phasors (Fourier coefficients), $V_n$ is the volt, the unit of the spectral density, $V(f)$, is volt/hertz (or volt $\times$ second). $v(t)$ and $V(f)$, as given by equations 9.175 and 9.178 respectively, from the so-called Fourier transform pair:

$$
v(t) \overset{\mathfrak{F}}{\Leftrightarrow} V(f)
\tag{9.181}
$$

where $\mathfrak{F}$ denotes the Fourier integral operation.

#### 9.3.6.1 Linearity

The Fourier transform is a linear operator. Given two distinct signals $x_1(t)$ and $x_2(t)$ with Fourier transforms $X_1(f)$ and $X_2(f)$, respectively, then the Fourier transform of $y(t) = ax_1(t) + bx_2(t)$ is given by:

$$
\begin{aligned}
V(f) &= \int_{-\infty}^{\infty} [ax_1(t) + bx_2(t)] \, e^{-j2\pi ft} \, dt \\
&= aX_1(f) + bX_2(f)
\end{aligned}
\tag{9.182}
$$

#### 9.3.6.2 Duality

Another important property of Fourier transform pairs is the so-called duality. Let us consider a signal $x(t)$ with a Fourier transform represented by $X(f)$. If there is a signal $y(t) = X(t)$ then its Fourier transform is given by:

$$
\begin{aligned}
Y(f) &= \int_{-\infty}^{\infty} X(t) \, e^{-j2\pi ft} \, dt \\
&= \int_{-\infty}^{\infty} X(t) \, e^{j2\pi(-f)t} \, dt
\end{aligned}
\tag{9.183}
$$

and, according to equation 9.175 we have that:

$$
Y(f) = x(-f)
\tag{9.184}
$$

that is:

$$
X(t) \overset{\mathfrak{F}}{\Leftrightarrow} x(-f)
\tag{9.185}
$$

#### Example 9.12
Use the duality property of Fourier transform pairs to calculate the Fourier transform of $y(t) = A\,\mathrm{sinc}(t\eta)$.

#### Solution
From equation 9.179 and from equation 9.185 we can write:

$$
Y(f) = \frac{A}{\eta}\,\mathrm{rect}\left(\frac{f}{\eta}\right)
\tag{9.186}
$$

Note that the rectangular function is an even function, that is $\mathrm{rect}(-f) = \mathrm{rect}(f)$.

### 9.3.6.3 Time Delay

If a function $x(t)$ has a Fourier transform $X(f)$ then the Fourier transform of a delayed replica of $x(t)$ by a time $\tau$, $x(t - \tau)$, is given by:

$$\mathfrak{F}[x(t - \tau)] = \int_{-\infty}^{\infty} x(t - \tau)e^{-j2\pi ft}\, dt \tag{9.187}$$

using the change of variable $t' = t - \tau$ we can write:

$$dt = dt'$$
$$t \rightarrow -\infty;\, t' \rightarrow -\infty$$
$$t \rightarrow \infty;\, t' \rightarrow \infty$$

and equation 9.187 can be written as:

$$\mathfrak{F}[x(t - \tau)] = \int_{-\infty}^{\infty} x(t')\, e^{-j2\pi ft'}\, dt'\, e^{-j2\pi f\tau}$$
$$= X(f)\, e^{-j2\pi f\tau} \tag{9.188}$$

Note that the delay $\tau$ causes an addition of a linear phase to $X(f)$. If $\tau$ is negative this means that the signal is advanced in time and the linear phase added to the spectrum has a positive slope. It is worth noting the similarity between the delay property of the Fourier transform with the delay property of the Fourier series (see equation 9.111).

### 9.3.6.4 The Dirac Delta Function

The Dirac delta function, $\delta(t)$ can be visualized as an extremely narrow pulse located at $t = 0$. However, the area of this pulse is unity which implies that its amplitude tends to infinity. A common way of defining this function is to start with a rectangular waveform with unity area, such as that depicted in Figure 9.37(a), which can be expressed as follows:

$$z(t) = \frac{1}{\tau}\,\text{rect}\left(\frac{t}{\tau}\right) \tag{9.189}$$

with $\tau = 1$. If we now decrease the value of $\tau$, as shown in Figures 9.37(b) and (c), we observe that the width of the rectangle decreases while its amplitude increases in order to preserve unity area. When we let $\tau$ tend to zero we obtain the Dirac delta function:

$$\delta(t) = \lim_{\tau \to 0} \frac{1}{\tau}\,\text{rect}\left(\frac{t}{\tau}\right) \tag{9.190}$$

which is depicted in Figure 9.37(d). Note that:

$$\int_{-\infty}^{\infty} \delta(t)\, dt = 1 \tag{9.191}$$

The *area* is represented by the bold value next to the arrow representing the delta function. An important property of the Dirac delta function is called the *sampling property* which states that the multiplication of this function, centered at $t_o$, by a signal $v(t)$ results in a Dirac delta function centered in $t_o$ with an area given by the value of $v(t)$ at $t = t_o$, that is:

$$v(t) \times \delta(t - t_o) = v(t_o) \times \delta(t - t_o) \tag{9.192}$$

We emphasize that the area of $v(t) \times \delta(t - t_o)$ is equal to $v(t_o)$, that is:

$$\int_{-\infty}^{\infty} v(t) \times \delta(t - t_o) = v(t_o) \tag{9.193}$$

Figure 9.38 illustrates this last property expressed by equations 9.192 and 9.193.

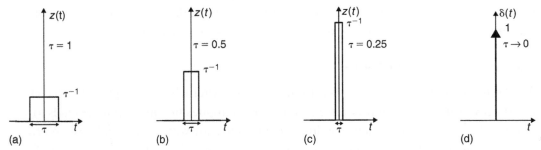

Figure 9.37: Rectangular function (a) $\tau = 1$; (b) $\tau = 0.5$; (c) $\tau = 0.25$; (d) $\tau \to 0$ (Dirac delta function)

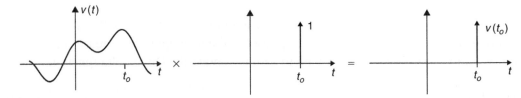

Figure 9.38: Illustration of the sampling property of the Dirac delta function

### 9.3.6.5 The Fourier Transform of a DC Signal

Let us calculate the Fourier transform of a DC signal, $w(t)$, with amplitude $A$. According to equation 9.178 this transform would be given by:

$$W(f) = \int_{-\infty}^{\infty} A\, e^{-j2\pi ft}\, dt \tag{9.194}$$

However, the definite integral cannot be determined because it does not converge for any value of $f$. The calculation of this Fourier transform requires the following mathematical manipulation. We express the DC value as follows:

$$w(t) = \lim_{\eta \to 0} A\, \mathrm{sinc}(t\eta) \tag{9.195}$$

Figure 9.39(a) illustrates equation 9.195 where we observe that as $\eta \to 0$, $w(t) \to A$. Taking the Fourier transform of $w(t)$, expressed by equation 9.195, we obtain:

$$W(f) = \int_{-\infty}^{\infty} \lim_{\eta \to 0} A\, \mathrm{sinc}(t\eta)\, e^{-j2\pi ft}\, dt \tag{9.196}$$

Since the integrand is a continuous function, we can change the order of the limit and the integral, that is:

$$W(f) = \lim_{\eta \to 0} \int_{-\infty}^{\infty} A\, \mathrm{sinc}(t\eta)\, e^{-j2\pi ft}\, dt \tag{9.197}$$

From equation 9.186 we can write $W(f)$ as follows:

$$W(f) = \lim_{\eta \to 0} \frac{A}{\eta}\, \mathrm{rect}\left(\frac{f}{\eta}\right) \tag{9.198}$$

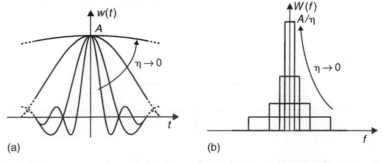

Figure 9.39: (a) Representation of the DC value $w(t) = A$; (b) Fourier transform of $w(t)$

This equation is, by definition (see equation 9.190), the Dirac delta function multiplied by *A* (see also Figure 9.39(b)), that is:

$$W(f) = A\delta(f) \tag{9.199}$$

This type of mathematical manipulation yields what is called the *generalized Fourier transform* and it allows for the calculation of Fourier transforms of a broad class of functions such as that illustrated in the next example.

### Example 9.13

Determine the Fourier transform of the unit-step function depicted in Figure 9.40.

### Solution

The unit-step function is defined as follows:

$$u(t) = \begin{cases} 1 & \text{if } t \geq 0 \\ 0 & \text{elsewhere} \end{cases} \tag{9.200}$$

This function can also be seen as the addition of a DC value of 1/2 with the signum function multiplied by a factor 1/2, as illustrated by Figure 9.40, and can be written as:

$$u(t) = \frac{1}{2} + \frac{1}{2}\operatorname{sign}(t) \tag{9.201}$$

where the signum function, $\operatorname{sign}(t)$, is defined as:

$$\operatorname{sign}(t) = \begin{cases} 1 & \text{if } t \geq 0 \\ -1 & \text{if } t < 0 \end{cases} \tag{9.202}$$

The Fourier transform of $u(t)$ is the addition of the Fourier transform of a DC value (discussed above in detail) with the Fourier transform of the signum function. We need a

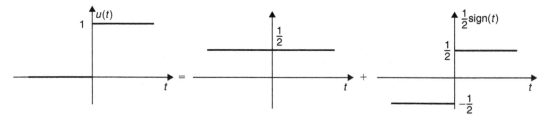

**Figure 9.40: Unit-step function as the addition of a constant value with the signum function**

mathematical manipulation so that the calculation of the transform of the signum function converges to its correct value. Figure 9.41 shows that sign($t$) can also be written as follows:

$$\text{sign}(t) = \lim_{\alpha \to 0} \begin{cases} \left(1 - e^{-\frac{t}{\alpha}}\right) e^{-\alpha t} & \text{if } t \geq 0 \\ \left(e^{\frac{t}{\alpha}} - 1\right) e^{\alpha t} & \text{if } t < 0 \end{cases} \tag{9.203}$$

with $\alpha > 0$. Figure 9.41 shows equation 9.203 for $\alpha = 0.5, 0.1$ and $0.02$. From this figure it is clear that as $\alpha$ tends to zero, equation 9.203 tends to equation 9.202.

The Fourier transform of the signum function can now be calculated as follows:

$$\text{Sign}(f) = \lim_{\alpha \to 0} \int_{-\infty}^{0} \left(e^{\frac{t}{\alpha}} - 1\right) e^{\alpha t} e^{-j2\pi ft} \, dt + \lim_{\alpha \to 0} \int_{0}^{\infty} \left(1 - e^{-\frac{t}{\alpha}}\right) e^{-\alpha t} e^{-j2\pi ft} \, dt$$

$$= \lim_{\alpha \to 0} \left[ \left( \frac{-1}{\alpha - 2j\pi f} + \frac{\alpha \, e^{\frac{t}{\alpha}}}{\alpha^2 - 2j\pi f\alpha + 1} \right) e^{\alpha t - j2\pi ft} \right]_{-\infty}^{0}$$

$$+ \lim_{\alpha \to 0} \left[ \left( \frac{\alpha \, e^{-\frac{t}{\alpha}}}{\alpha^2 + 2j\pi f\alpha + 1} - \frac{1}{\alpha + 2j\pi f)} \right) e^{-\alpha t - j2\pi ft} \right]_{0}^{\infty}$$

$$= \frac{1}{j\pi f} \tag{9.204}$$

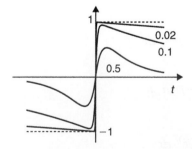

**Figure 9.41: The signum function obtained from equation 9.203.** $\alpha = 0.5, 0.1$ and $0.02$

where we have used the following equalities:

$$\lim_{\alpha \to 0} e^{\frac{t}{\alpha}} = 0 \qquad \text{for } t < 0, \, (\alpha > 0)$$

$$\lim_{\alpha \to 0} e^{\frac{-t}{\alpha}} = 0 \qquad \text{for } t > 0, \, (\alpha > 0)$$

Using equations 9.201, 9.199 and 9.204, we can write the Fourier transform of the unit step function as follows:

$$U(f) = \frac{1}{2} \delta(f) + \frac{1}{2} \text{Sign}(f)$$

$$= \frac{1}{2} \delta(f) + \frac{1}{j \, 2\pi f} \tag{9.205}$$

The generalized Fourier transform also allows us to perform the calculation of the Fourier transforms of periodic functions. Let us consider, for example, a periodic voltage signal, $v(t)$ with period $T$, which has a Fourier series such that:

$$v(t) = \sum_{n=-\infty}^{\infty} V_n \, e^{j 2\pi \frac{n}{T} t} \tag{9.206}$$

The Fourier transform of $v(t)$, $V(f)$, can be related to its Fourier series coefficients, $V_n$, as follows:

$$V(f) = \int_{-\infty}^{\infty} v(t) \, e^{-j 2\pi f t} \, dt$$

$$= \int_{-\infty}^{\infty} \sum_{n=-\infty}^{\infty} V_n \, e^{j 2\pi \frac{n}{T} t} \, e^{-j 2\pi f t} \, dt$$

$$= \sum_{n=-\infty}^{\infty} V_n \int_{-\infty}^{\infty} e^{j 2\pi \frac{n}{T} t} \, e^{-j 2\pi f t} \, dt$$

$$= \sum_{n=-\infty}^{\infty} V_n \int_{-\infty}^{\infty} e^{-j 2\pi (f - \frac{n}{T}) t} \, dt \tag{9.207}$$

This integral can be related to the Fourier transform of a DC quantity. According to equation 9.199 we have:

$$\int_{-\infty}^{\infty} 1 \times e^{-j2\pi f t}\, dt = \delta(f) \qquad (9.208)$$

and, therefore, the integral of equation 9.207 can be calculated as:

$$\int_{-\infty}^{\infty} e^{-j2\pi\left(f-\frac{n}{T}\right)t}\, dt = \delta\left(f - \frac{n}{T}\right) \qquad (9.209)$$

Finally, equation 9.207 which represents the spectrum of the periodic waveform $v(t)$ can be expressed as:

$$V(f) = \sum_{n=-\infty}^{\infty} V_n\, \delta\left(f - \frac{n}{T}\right) \qquad (9.210)$$

which is a discrete series of phasors as expected.

### Example 9.14

Determine the spectrum $V(f)$ of the periodic voltage waveform, $v(t)$ of Figure 9.35 with $\tau = T/3$.

### Solution

From equations 9.117 and 9.210 we can write $V(f)$ as follows:

$$
\begin{aligned}
V(f) &= \sum_{n=-\infty}^{\infty} \frac{V_A \tau}{T}\, \mathrm{sinc}\left(\frac{n\tau}{T}\right) \delta\left(f - \frac{n}{T}\right) \\
&= \sum_{n=-\infty}^{\infty} \frac{V_A}{3}\, \mathrm{sinc}\left(\frac{n}{3}\right) \delta\left(f - \frac{n}{T}\right)
\end{aligned}
\qquad (9.211)
$$

### 9.3.6.6 Rayleigh's Energy Theorem

This theorem states that the energy, $E_x$, of a signal $x(t)$ can be calculated from its spectrum $X(f)$ according to the following equation:

$$E_x = \int_{-\infty}^{\infty} x(t)^2\, dt = \int_{-\infty}^{\infty} |X(f)|^2\, dt \qquad (9.212)$$

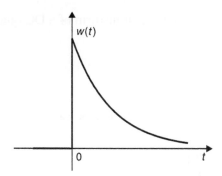

**Figure 9.42: Causal exponential**

### Example 9.15

Determine the energy of the causal exponential, $w(t)$ shown in Figure 9.42, using Rayleigh's energy theorem. (A causal signal $x(t)$ is any signal that is zero for $t < 0$.) Then, show that this result is the same as that obtained from the integration of $w^2(t)$.

### Solution

The causal exponential $w(t)$ of Figure 9.42 can be written as:

$$
w(t) = \begin{cases} e^{-\sigma t} & \text{for } t \geq 0 \\ 0 & \text{elsewhere} \end{cases}
\tag{9.213}
$$

where $\sigma > 0$. Hence, the spectrum of $w(t)$ can be calculated as:

$$
\begin{aligned}
W(f) &= \int_{-\infty}^{\infty} w(t) e^{-j2\pi ft} \, dt \\
&= \int_{0}^{\infty} e^{-\sigma t} e^{-j2\pi ft} \, dt \\
&= \left[ \frac{1}{-\sigma - j2\pi f} e^{-j2\pi ft} \right]_{0}^{\infty} \\
&= \frac{1}{\sigma + j2\pi f}
\end{aligned}
\tag{9.214}
$$

From equation 9.212 the energy of $w(t)$ can be calculated as follows,

$$
\begin{aligned}
E_w &= \int_{-\infty}^{\infty} \frac{1}{\sigma^2 + (2\pi f)^2}\, df \\
&= \left[ \frac{1}{\sigma 2\pi} \tan^{-1}\left( \frac{2\pi f}{\sigma} \right) \right]_{-\infty}^{\infty} \\
&= \frac{1}{\sigma 2\pi}\left( \frac{\pi}{2} + \frac{\pi}{2} \right) = \frac{1}{2\sigma}
\end{aligned}
\tag{9.215}
$$

The energy can also be calculated according to:

$$
\begin{aligned}
E_w &= \int_{-\infty}^{\infty} w^2(t)\, dt \\
&= \int_{0}^{\infty} e^{-2\sigma t}\, dt \\
&= \left[ \frac{1}{-2\sigma} e^{-2\sigma t} \right]_{0}^{\infty} \\
&= \frac{1}{2\sigma}
\end{aligned}
\tag{9.216}
$$

This result is the same as that given by Rayleigh's energy theorem.

### 9.3.7 Transfer Function and Impulse Response

The transfer function, $H(\omega)$ or $H(f)$, of a circuit has been introduced in section 9.3.3 where we saw that it can be obtained from phasor analysis, more specifically by evaluating the ratio of the phasor of the output signal with that of the input signal for all frequencies, $\omega$ or $f = \omega/(2\pi)$. There are four fundamental types of transfer functions:

- **Voltage transfer function:** In this situation both input and output phasors are voltages. The transfer function represents a voltage gain (or *voltage* attenuation if this gain is less than one) versus the frequency. This transfer function is dimensionless.

- **Current transfer function:** Both input and output phasors are currents. Hence, the transfer function represents a *current* gain (or current attenuation if this gain is less than one) versus the frequency. This transfer function is also dimensionless.
- **Impedance transfer function:** In this situation the input phasor is a current while the output phasor is a voltage. Note that now the gain versus the frequency has units of ohms. This transfer function is usually called *transimpedance gain.*
- **Admittance transfer function:** The input phasor is a voltage while the output phasor is a current. Now the gain versus the frequency, represented by this transfer function, has units of siemens. This transfer function is usually called *transconductance gain.*

From the discussion about the Fourier series we have concluded that knowledge of the transfer function of a circuit allows the calculation of the spectrum of the output signal for a given periodic input signal, using equation 9.127. In similar way, the spectrum of the output signal, $X_o(f)$, for a given input signal with $X_i(f)$ can be calculated as:

$$X_o(f) = H(f) \times X_i(f) \tag{9.217}$$

Taking the inverse Fourier transform of $X_o(f)$ and $X_i(f)$ we obtain the time domain representation for the output and input signals respectively. We can also take the inverse Fourier transform of the transfer function, $H(f)$, which is defined as the circuit *impulse response* represented by $h(t)$. The impulse response of a circuit is the circuit response when a Dirac delta function (with unit area) is applied to this circuit.

### Example 9.16

Determine the impulse response of the circuit of Figure 9.43.

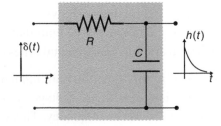

**Figure 9.43: Impulse response of an RC circuit**

*Solution*

The impulse response can be obtained calculating the inverse Fourier transform of the transfer function $H(f)$ which is given by equation 9.124:

$$H(f) = \frac{1}{1 + j\,2\pi fRC} \tag{9.218}$$

From Example 9.15 we know that:

$$\frac{1}{\sigma + j2\pi f} \underset{\Longleftrightarrow}{\mathfrak{F}} \begin{cases} e^{-\sigma t} & \text{for } t \geq 0 \\ 0 & \text{elsewhere} \end{cases} \tag{9.219}$$

Since $H(f)$ in equation 9.218 can be written as:

$$H(f) = \frac{1}{RC} \times \frac{1}{\frac{1}{RC} + j2\pi f} \tag{9.220}$$

$h(t)$ is given by:

$$h(t) = \begin{cases} \frac{1}{RC}\, e^{-\frac{t}{RC}} & \text{for } t \geq 0 \\ 0 & \text{elsewhere} \end{cases} \tag{9.221}$$

This equation can also be written as:

$$h(t) = \frac{1}{RC}\, e^{-\frac{t}{RC}}\, u(t) \tag{9.222}$$

where $u(t)$ represents the unit step function defined by equation 9.200.

From a theoretical point-of-view the impulse response $h(t)$ of a circuit is obtained applying a Dirac delta function, as illustrated in Figure 9.43. It should be clear to the reader that, in a practical situation, it is not possible to apply a Dirac delta pulse to a circuit to observe its impulse response; first because extremely narrow pulses with infinite amplitude are physically impossible to create and secondly because if this were possible the circuit would most certainly get damaged with the application of such a pulse! Hence, the application of a Dirac delta pulse should be understood as a mathematical model or abstraction which helps us to identify $h(t)$. However, as we show in Example 9.17, if we apply a narrow pulse whose bandwidth is much greater than that of the circuit then the output is a very good estimate of its impulse response, $h(t)$.

### Example 9.17
Show that if we apply a finite narrow pulse, whose bandwidth is much greater than the circuit bandwidth then the output produced by the circuit is a good estimate of its impulse response, $h(t)$.

### Solution
Let us consider a circuit with a transfer function $H(f)$ with maximum frequency $f_M$ as illustrated in Figure 9.44. If we apply to the circuit a narrow rectangular pulse, $x_i(t)$, such that:

$$x_i(t) = A \, \text{rect}\left(\frac{t}{\tau}\right) \tag{9.223}$$

with $\tau \ll f_M^{-1}$ then the spectrum of $x_i(t)$, that is $X_i(f) = A\tau \, \text{sinc}\,(f\tau)$, is nearly constant in the frequency range $-f_M < f < f_M$, as shown in Figure 9.44. The output spectrum $X_o(f)$ is:

$$\begin{aligned} X_o &= X_i(f) \, H(f) \\ &\simeq A\tau \, H(f) \end{aligned} \tag{9.224}$$

Taking the inverse Fourier transform the output signal is $x_o(t) \simeq A\tau h(t)$.

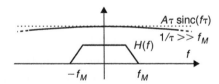

**Figure 9.44: A narrow pulse applied to a circuit. Frequency domain representation**

## 9.3.8 The Convolution Operation

The time domain waveform for $x_o(t)$, in equation 9.217, can be obtained by calculating the following inverse Fourier transform:

$$x_o(t) = \int_{-\infty}^{\infty} H(f)X_i(f) \, e^{j2\pi ft} \, df \tag{9.225}$$

Since the input signal, in the time domain, is represented by $x_i(t)$, this can be written as:

$$x_o(t) = \int_{-\infty}^{\infty} H(f) \underbrace{\int_{-\infty}^{\infty} x_i(\lambda)e^{j2\pi f\lambda}d\lambda}_{X_i(f)} \, e^{j2\pi ft} \, df \tag{9.226}$$

Changing the order of integration this equation can be written as follows:

$$x_o(t) = \int_{-\infty}^{\infty} x_i(\lambda) \underbrace{\int_{-\infty}^{\infty} H(f) e^{j2\pi f(t-\lambda)} df}_{h(t-\lambda)} \, d\lambda \tag{9.227}$$

(This change of the order of integration is possible whenever the functions are absolutely integrable. The variety of signals of interest and their corresponding spectra obey this requirement. For more details, see Oppenheim/Willsky, Signals and Systems, listed in the references at the end of this chapter.)

Because $H(f)$ has an inverse Fourier transform represented by $h(t)$, then $x_o(t)$ can be calculated as:

$$x_o(t) = \int_{-\infty}^{\infty} x_i(\lambda) h(t-\lambda) \, d\lambda \tag{9.228}$$

This represents the *convolution* operation between $x(t)$ and $h(t)$. This operation is also represented as follows:

$$x_o(t) = x_i(t) * h(t) \tag{9.229}$$

with $*$ indicating the convolution operation. It can be shown that:

$$x_i(t) * h(t) = h(t) * x_i(t) \tag{9.230}$$

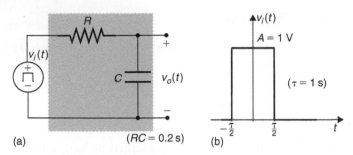

**Figure 9.45: Square voltage pulse, $v_i(t)$, is applied to an RC circuit**

In order to understand the convolution operation we consider the RC circuit of Figure 9.45 where now a single square voltage pulse, $v_i(t)$, is applied to its input. The output voltage $v_o(t)$ can be determined according to equation 9.228. However, we shall evaluate $v_o(t)$ by first approximating the input square pulse by a sum of $(N + 1)$ Dirac delta functions, as illustrated in Figure 9.46. Now $v_i(t)$ is approximated by the following expression:

$$v_i(t) \simeq \frac{A\tau}{N+1} \sum_{k=0}^{N} \delta\left(t + \frac{\tau}{2} - k\frac{\tau}{N}\right)$$

$$= \frac{1}{N+1} \sum_{k=0}^{N} \delta\left(t + \frac{N-2k}{2N}\right) \tag{9.231}$$

Note that the sum of the areas of the $(N + 1)$ delta functions is equal to the area of the rectangular pulse, $A\tau = 1$. Using equation 9.228 the voltage at the output of the RC circuit, $v_o(t)$ is given by:

$$v_o(t) = \int_{-\infty}^{\infty} v_i(\lambda) h(t - \lambda) \, d\lambda \tag{9.232}$$

$$= \frac{1}{N+1} \sum_{k=0}^{N} \int_{-\infty}^{\infty} \delta\left(\lambda + \frac{N-2k}{2N}\right) h(t - \lambda) \, d\lambda \tag{9.233}$$

where $h(t)$ is given by equation 9.222 with $RC = 0.2$ s. From equation 9.193 we can write this as,

$$v_o(t) = \frac{5}{N+1} \sum_{k=0}^{N} e^{-5\left(t + \frac{N-2k}{2N}\right)} u\left(t + \frac{N-2k}{2N}\right) \tag{9.234}$$

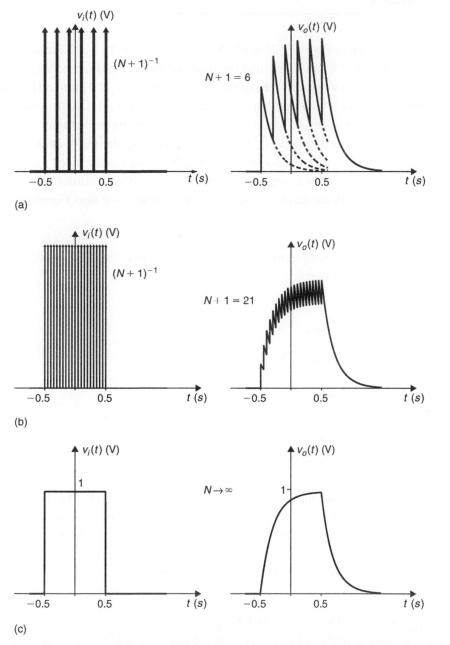

**Figure 9.46: Illustration of the convolution operation with the input voltage signal in the circuit of Figure 9.45 being approximated as a sum of ($N + 1$) Dirac delta functions (a) $N + 1 = 6$; (b) $N + 1 = 21$; (c) $N \to \infty$**

This equation is shown in Figure 9.46 with $(N + 1) = 6$, $(N + 1) = 21$ and $N \to \infty$. From Figure 9.46(a) $(N + 1 = 6)$ it is clear that the result of the convolution between $v_i(t)$ and $h(t)$ can be seen as a weighted sum of the impulse response $h(t)$ induced by each of the Dirac delta functions which approximates the input signal $v_i(t)$. By increasing $N$ we increase the number of delta functions and, of course, we increase their density in the time interval $\tau$. If $N \to \infty$ then $v_i(t)$ "becomes" the rectangular pulse as shown in Figure 9.45(c) and $v_o(t)$ is now a smooth waveform. Note the similarity of $v_o(t)$ obtained now, when the input voltage is a single rectangular pulse, with the output voltage when the input voltage is a periodic sequence of rectangular pulses (see also Figures 9.23 and 9.26).

Figure 9.47 illustrates the computation of $v_o(t)$ given by equation 9.232. According to the definition of $h(t)$ we can write:

$$h(t - \lambda) = \begin{cases} \dfrac{1}{RC} e^{-\frac{t-\lambda}{RC}} & \text{for } t - \lambda \geq 0 \\ 0 & \text{for } t - \lambda < 0 \end{cases} \tag{9.235}$$

and since $RC = 0.2$ we have:

$$h(t - \lambda) = \begin{cases} 5e^{-5(t-\lambda)} & \text{for } \lambda \leq \\ 0 & \text{for } \lambda > \end{cases} \tag{9.236}$$

Figure 9.47(a) illustrates the integrand of equation 9.232 for $t = -0.75$ s. Note the inversion of $h(-0.75 - \lambda)$ in the $\lambda$ axis. In this figure it is clear that the product of $h(-0.75 - \lambda)$ with $v_i(\lambda)$ is zero and, accordingly, $v_o(-0.75) = 0$. In fact, the output voltage is zero until $t > -0.5$ s as illustrated by Figure 9.47(b). For $-0.5 < t < 0.5$ the output voltage can be obtained from the following expression:

$$v_o(t) = 5 \int_{-0.5}^{t} e^{-5(t-\lambda)} \, d\lambda, \quad -0.5 \leq t < 0.5 \tag{9.237}$$

$$= 5 \left[ \frac{1}{5} e^{-5(t-\lambda)} \right]_{-0.5}^{t}, \quad -0.5 \leq t < 0.5$$

$$= 1 - e^{-5(t+0.5)}, \quad -0.5 \leq t < 0.5 \tag{9.238}$$

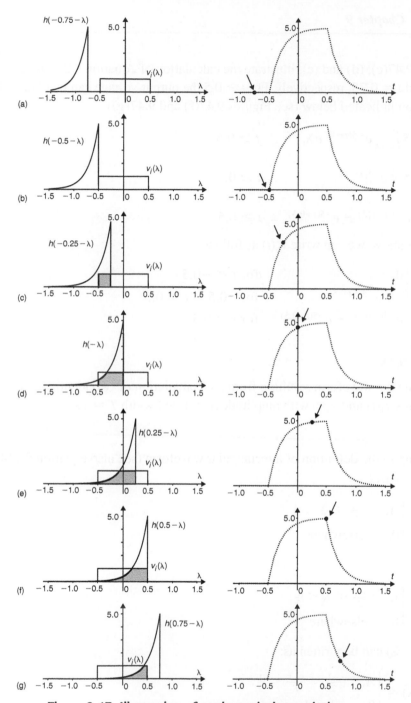

**Figure 9.47: Illustration of mathematical convolution**

Figures 9.47(c), (d) and (e), illustrate the calculation of equation 9.237 for $t = -0.25$, $t = 0$ and $t = 0.25$, respectively. For $t \geq 0.5$ the output voltage can be obtained from the expression indicated below (see Figures 9.47(f) and 9.47(g)):

$$v_o(t) = 5 \int_{-0.5}^{0.5} e^{-5(t-\lambda)} \, d\lambda, \qquad t \geq 0.5$$

$$= 5 \left[ \tfrac{1}{5} e^{-5(t-\lambda)} \right]_{-0.5}^{0.5}, \qquad t \geq 0.5$$

$$= e^{-5(t-0.5)} - e^{-5(t+0.5)}, \qquad t \geq 0.5 \tag{9.239}$$

From the above we can write $v_o(t)$ as follows:

$$v_o(t) = \begin{cases} 0 & \text{for } t < -0.5 \\ 1 - e^{-5(t+0.5)} & \text{for } -0.5 \leq t < 0.5 \\ e^{-5(t-0.5)} - e^{-5(t+0.5)} & \text{for } t \geq 0.5 \end{cases} \tag{9.240}$$

### Example 9.18

Determine the waveform resulting from the convolution of two identical rectangular waveforms $x_1(t)$ and $x_2(t)$ with amplitude $A = 1$ and width $T = 1$ s.

### Solution
According to the definition of a rectangular waveform (see also equation 9.114) we can write $x_1(\lambda)$ as:

$$x_1(\lambda) = \begin{cases} A, & \frac{-1}{2} < \frac{\lambda}{T} < \frac{1}{2} \\ 0, & \text{elsewhere} \end{cases} \tag{9.241}$$

that is,

$$x_1(\lambda) = \begin{cases} 1, & \frac{-1}{2} < \lambda < \frac{1}{2} \\ 0, & \text{elsewhere} \end{cases} \tag{9.242}$$

and $x_2(t - \lambda)$ can be written as:

$$x_2(t - \lambda) = \begin{cases} A, & \frac{-1}{2} < \frac{t-\lambda}{T} < \frac{1}{2} \\ 0, & \text{elsewhere} \end{cases} \tag{9.243}$$

that is,

$$x_2(t - \lambda) = \begin{cases} 1, & t - \frac{1}{2} < \lambda < t + \frac{1}{2} \\ 0, & \text{elsewhere} \end{cases} \tag{9.244}$$

The convolution of $x_1(t)$ and $x_2(t)$ is given by:

$$y(t) = \int_{-\infty}^{\infty} x_1(\lambda) x_2(t - \lambda)\, d\lambda \tag{9.245}$$

Figure 9.48(a) shows the functions whose product forms the integrand of equation 9.245 for $t = -1$ s, that is, this Figure shows $x_1(\lambda)$ and $x_2(-1 - \lambda)$. From this Figure it is clear that the product of these two functions is zero and so is the result of its integration. Note that for $t \leq -1$ the product of $x_1(\lambda)$ with $x_2(t - \lambda)$ is zero. Figures 9.48(b), (c) and (d) indicate that for the time interval $-1 < t \leq 1$ the two functions overlap. This overlap is maximum for $t = 0$ as shown by Figure 9.48(c). For the time interval, $-1 < t \leq 0$, we can write equation 9.245 as follows:

$$\begin{aligned} y(t) &= \int_{-0.5}^{t+0.5} d\lambda, & -1 < t \leq 0 \\ &= t + 1, & -1 < t \leq 0 \end{aligned} \tag{9.246}$$

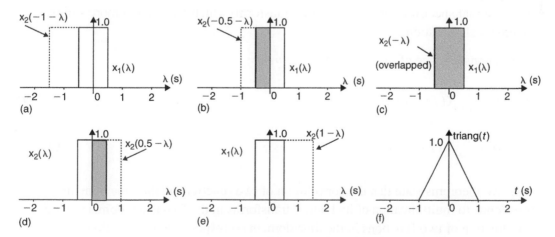

**Figure 9.48: Convolution of two identical rectangular waveforms**

For the time interval $0 < t < 1$ the overlap of the two functions decreases as illustrated by Figure 9.48(d) for $t = 0.5$. For this time interval we can write equation 9.245 as follows:

$$
\begin{aligned}
y(t) &= \int_{t-0.5}^{0.5} d\lambda, \quad 0 < t < 1 \\
&= 1 - t, \qquad 0 < t < 1
\end{aligned}
\tag{9.247}
$$

For $t \geq 1$ there is no overlap between $x_1(\lambda)$ and $x_2(-1 - \lambda)$ and $y(t)$ is again zero. From the above we can write $y(t)$ as:

$$
y(t) =
\begin{cases}
0 & \text{if } t \leq -1 \\
1 + t & \text{if } -1 < t \leq 0 \\
1 - t & \text{if } 0 < t < 1 \\
0 & \text{if } t \geq 1
\end{cases}
\tag{9.248}
$$

Figure 9.48(f) shows that $y(t)$ represents a triangle. In fact equation 9.248 defines the triangular function, triang($t$).

The discussion presented above reveals, once again, the advantage of analyzing circuits and signals in the frequency domain. While time domain analysis involves the calculation of convolution integrals using the circuit impulse response and the time domain signal, the frequency domain involves the multiplication of the circuit transfer functions with the signal spectrum (or signal Fourier transform) which is, by far, a more simple mathematical operation.

This is a consequence of the *convolution theorems*:

$$
x(t) * y(t) \overset{\Im}{\Longleftrightarrow} X(f) \times Y(f)
\tag{9.249}
$$

$$
x(t) \times y(t) \overset{\Im}{\Longleftrightarrow} X(f) * Y(f)
\tag{9.250}
$$

These two theorems state that the convolution of two functions in the time domain corresponds to multiplication of its Fourier transforms in the frequency domain while multiplication of two functions in the time domain corresponds to convolution of its Fourier transforms in the frequency domain.

# References

Carlson AB, Crilly PB and Rutledge JC. *Communication Systems: An Introduction to Signals and Noise in Electrical Communication*, 4th edition. McGraw-Hill Series in Electrical Engineering, 2001.

Chen C. *System and Signal Analysis*, 2nd edition. Saunders College Publishing, 1994.

Oppenheim, A. V. and A.S. Willsky, *Signals and Systems*, 1996 (Prentice Hall Signal Processing Series), 2nd edition.

Roberts, M. J., *Signals and Systems: Analysis using Transform Methods and Matlab®*, 2003, (McGraw-Hill International Editions).

Smith KCA and Alley RE. *Electrical Circuits, an Introduction*. Cambridge University Press, 1992.

# CHAPTER 10

# Digital Electronics

Clive Maxfield

## 10.1 Semiconductors

Most materials are conductors, insulators, or something in-between, but a special class of materials known as *semiconductors* can be persuaded to exhibit both conducting and insulating properties. The first semiconductor to undergo evaluation was the element germanium (chemical symbol Ge). However, for a variety of reasons, silicon (chemical symbol Si) replaced germanium as the semiconductor of choice. As silicon is the main constituent of sand and one of the most common elements on earth (silicon accounts for approximately 28% of the earth's crust), we aren't in any danger of running out of it in the foreseeable future.

Pure crystalline silicon acts as an insulator; however, scientists at Bell Laboratories in the United States found that, by inserting certain impurities into the crystal lattice, they could make silicon act as a conductor. The process of inserting the impurities is known as *doping*, and the most commonly used *dopants* are boron atoms with three electrons in their outermost electron shells and phosphorus atoms with five.

If a pure piece of silicon is surrounded by a gas containing boron or phosphorus and heated in a high-temperature oven, the boron or phosphorus atoms will permeate the crystal lattice and displace some silicon atoms without disturbing other atoms in the vicinity. This process is known as *diffusion*. Boron-doped silicon is called *P-type* silicon and phosphorus-doped silicon is called *N-type* (Figure 10.1).

Because boron atoms have only three electrons in their outermost electron shells, they can only make bonds with three of the silicon atoms surrounding them. Thus, the site (location) occupied by a boron atom in the silicon crystal will accept a free electron with relative ease and is therefore known as an *acceptor*. Similarly, because phosphorus atoms have five electrons in their outermost electron shells, the site of a phosphorus atom in the silicon crystal will donate an electron with relative ease and is therefore known as a *donor*.

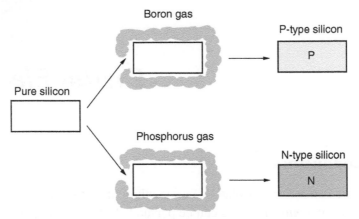

**Figure 10.1: Creating P-type and N-type silicon**

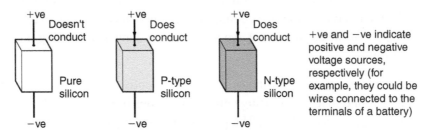

**Figure 10.2: Pure P-type and N-type silicon**

## 10.2  Semiconductor Diodes

As was noted above, pure crystalline silicon acts as an insulator. By comparison, both P-type and N-type silicon are reasonably good conductors (Figure 10.2).

When things start to become really interesting, however, is when a piece of silicon is doped such that part is P-type and part is N-type (Figure 10.3).

The silicon with both P-type and N-type conducts electricity in only one direction; in the other direction it behaves like an OPEN (OFF) switch. These structures, known as *semiconductor diodes*, come in many shapes and sizes; an example could be as shown in Figure 10.4. (Note that the "semiconductor" portion of *semiconductor diode* was initially

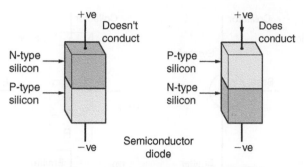

**Figure 10.3: Mixing P-type and N-type silicon**

**Figure 10.4: Diode: Component and symbol (a) Diode component; (b) Symbol**

used to distinguish these components from their vacuum tube-based cousins. As semiconductors took over, everyone started to just refer to them as *diodes*.)

If the triangular body of the symbol is pointing in the classical direction of current flow (more positive to more negative), the diode will conduct. An individually packaged diode consists of a piece of silicon with connections to external leads, all encapsulated in a protective package (the silicon is typically smaller than a grain of sand). The package protects the silicon from moisture and other impurities and, when the diode is operating, helps to conduct heat away from the silicon.

Due to the fact that diodes (and transistors as discussed below) are formed from solids— as opposed to vacuum tubes, which are largely formed from empty space—people started to refer to them as *solid-state* electronics.

## 10.3  Bipolar Junction Transistors

More complex components called *transistors* can be created by forming a sandwich out of three regions of doped silicon. One family of transistors is known as *bipolar junction*

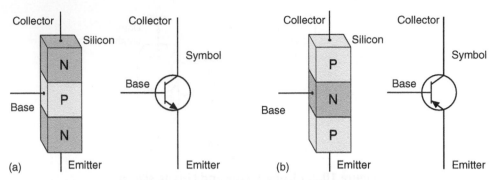

Figure 10.5: Bipolar junction transistors (BJTs); (a) NPN bipolar junction transistor;
(b) PNP bipolar junction transistor

*transistors (BJTs)* of which there are two basic types called *NPN* and *PNP*; these names relate to the way in which the silicon is doped (Figure 10.5).

In the analog world, a transistor can be used as a voltage amplifier, a current amplifier, or a switch; in the digital world, a transistor is primarily considered to be a switch. The structure of a transistor between the *collector* and *emitter* terminals is similar to that of two diodes connected back-to-back. Two diodes connected in this way would typically not conduct; however, when signals are applied to the *base* terminal, the transistor can be turned ON or OFF. If the transistor is turned ON, it acts like a CLOSED switch and allows current to flow between the collector and the emitter; if the transistor is turned OFF, it acts like an OPEN switch and no current flows. We may think of the collector and emitter as *data* terminals, and the base as the *control* terminal.

As for a diode, an individually packaged transistor consists of the silicon, with connections to external leads, all encapsulated in a protective package (the silicon is typically smaller than a grain of sand). The package protects the silicon from moisture and other impurities and helps to conduct heat away from the silicon when the transistor is operating. Transistors may be packaged in plastic or in little metal cans about a quarter of an inch in diameter with three leads sticking out of the bottom (Figure 10.6).

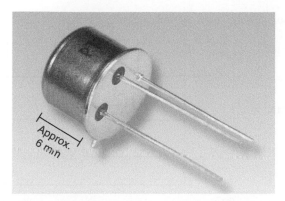

**Figure 10.6: Individually packaged transistor (photo courtesy of Alan Winstanley)**

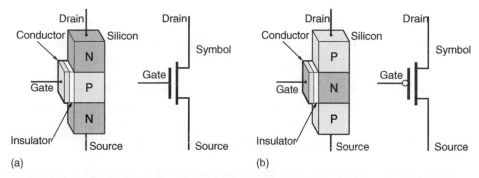

**Figure 10.7: Metal-oxide semiconductor field-effect transistors (MOSFETs)**
**(a) NMOS field-effect transistor; (b) PMOS field-effect transistor**

## 10.4 Metal-Oxide Semiconductor Field-Effect Transistors

Another family of transistors is known as *metal-oxide semiconductor field-effect transistors (MOSFETs)* of which there are two basic types called *n-channel* and *p-channel*; once again these names relate to the way in which the silicon is doped (Figure 10.7).

In the case of these devices, the *drain* and *source* form the *data* terminals and the *gate* acts as the *control* terminal. Unlike bipolar devices, the control terminal is connected to a conducting plate, which is insulated from the silicon by a layer of non-conducting oxide. In the original devices the conducting plate was metal—hence, the term *metal-oxide*.

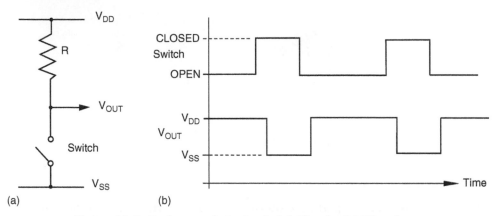

**Figure 10.8: Resistor-switch circuit (a) Circuit; (b) Waveform**

When a signal is applied to the gate terminal, the plate, insulated by the oxide, creates an electromagnetic field, which turns the transistor ON or OFF—hence, the term *field-effect*.

Now this is the bit that always confuses the unwary, because the term *channel* refers to the piece of silicon under the gate terminal, that is, the piece linking the drain and source regions. But the channel in the n-channel device is formed from P-type material, while the channel in the p-channel device is formed from N-type material.

At first glance, this would appear to be totally counterintuitive, but there is reason behind the madness. Let's consider the n-channel device. In order to turn this ON, a positive voltage is applied to the gate. This positive voltage attracts negative electrons in the P-type material and causes them to accumulate beneath the oxide layer where they form a negative channel—hence, the term *n-channel*. In fact, saying "n-channel" and "p-channel" is a bit of a mouthful, so instead we typically just refer to these as NMOS and PMOS transistors, respectively.

This chapter concentrates on MOSFETs, because their symbols, construction, and operation are easier to understand than those of bipolar junction transistors.

## 10.5 The Transistor as a Switch

To illustrate the application of a transistor as a switch, first consider a simple circuit comprising a resistor and a real switch (Figure 10.8).

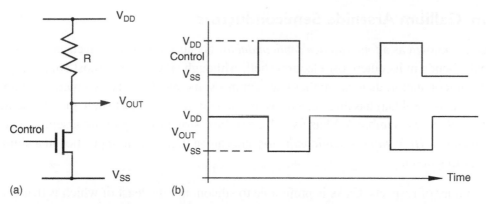

**Figure 10.9: Resistor-NMOS transistor circuit (a) Circuit; (b) Waveform**

The labels $V_{DD}$ and $V_{SS}$ are commonly used in circuits employing MOSFETs. At this point we have little interest in their actual values and, for the purpose of these examples, need only assume that $V_{DD}$ is more positive than $V_{SS}$.

When the switch is OPEN (OFF), $V_{OUT}$ is connected via the resistor to $V_{DD}$; when the switch is CLOSED (ON), $V_{OUT}$ is connected via the switch directly to $V_{SS}$. In this latter case, $V_{OUT}$ takes the value $V_{SS}$ because, like people, electricity takes the path of least resistance, and the resistance to $V_{SS}$ through the closed switch is far less than the resistance to $V_{DD}$ through the resistor. The waveforms in the illustration above show a delay between the switch operating and $V_{OUT}$ responding. Although this delay is extremely small, it is important to note that there will always be some element of delay in any physical system.

Now consider the case where the switch is replaced with an NMOS transistor whose control input can be switched between $V_{DD}$ and $V_{SS}$ (Figure 10.9).

When the control input to an NMOS transistor is connected to $V_{SS}$, the transistor is turned OFF and acts like an OPEN switch; when the control input is connected to $V_{DD}$, the transistor is turned ON and acts like a closed switch. Thus, the transistor functions in a similar manner to the switch. However, a switch is controlled by hand and can only be operated a few times a second, but a transistor's control input can be driven by other transistors, allowing it to be operated millions of times a second.

## 10.6  Gallium Arsenide Semiconductors

Silicon is known as a *four-valence semiconductor* because it has four electrons available to make bonds in its outermost electron shell. Although silicon is the most commonly used semiconductor, there is another that requires some mention. The element gallium (chemical symbol Ga) has three electrons available in its outermost shell and the element arsenic (chemical symbol As) has five. A crystalline structure of gallium arsenide (GaAs) is known as a *III-V valence semiconductor* and can be doped with impurities in a similar manner to silicon.

In a number of respects, GaAs is preferable to silicon, not the least of which is that GaAs transistors can switch approximately eight times faster than their silicon equivalents. However, GaAs is hard to work with, which results in GaAs transistors being more expensive than their silicon cousins.

## 10.7  Light-Emitting Diodes

On February 9, 1907, one of Marconi's engineers, Mr. H.J. Round of New York, NY, had a letter published in "Electrical World" magazine as follows:

*A Note on Carborundum*

*To the editors of Electrical World:*

*Sirs: During an investigation of the unsymmetrical passage of current through a contact of carborundum and other substances a curious phenomenon was noted. On applying a potential of 10 volts between two points on a crystal of carborundum, the crystal gave out a yellowish light.*

Mr. Round went on to note that some crystals gave out green, orange, or blue light. This is quite possibly the first documented reference to the effect upon which special components called *light-emitting diodes (LEDs)* are based.

Sad to relate, no one seemed particularly interested in Mr. Round's discovery, and nothing really happened until 1922, when the same phenomenon was observed by O.V. Losov in Leningrad. Losov took out four patents between 1927 and 1942, but he was killed during the Second World War and the details of his work were never discovered.

**Figure 10.10: Symbol for a LED**

In fact, it wasn't until 1951, following the discovery of the bipolar transistor, that researchers really started to investigate this effect in earnest. They found that by creating a semiconductor diode from a compound semiconductor formed from two or more elements—such as gallium arsenide (GaAs)—light is emitted from the PN junction, that is, the junction between the P-type and N-type doped materials.

As for a standard diode, a LED conducts electricity in only one direction (and it emits light only when it's conducting). Thus, the symbol for an LED is similar to that for a normal diode, but with two arrows to indicate light being emitted (Figure 10.10).

A LED formed from pure gallium arsenide emits infrared light, which is useful for sensors, but which is invisible to the human eye. It was discovered that adding aluminum to the semiconductor to give aluminum gallium arsenide (AlGaAs) resulted in red light humans could see. Thus, after much experimentation and refinement, the first red LEDs started to hit the streets in the late 1960s.

LEDs are interesting for a number of reasons, not the least of which is that they are extremely reliable, they have a very long life (typically 100,000 hours as compared to 1,000 hours for an incandescent light bulb), they generate very pure, saturated colors, and they are extremely energy efficient (LEDs use up to 90% less energy than an equivalent incandescent bulb).

Over time, more materials were discovered that could generate different colors. For example, gallium phosphide gives green light, and aluminum indium gallium phosphite can be used to generate yellow and orange light. For a long time, the only color missing was blue. This was important because blue light has the shortest wavelength of visible light, and engineers realized that if they could build a blue laser diode, they could quadruple the amount of data that could be stored on, and read from, a CD-ROM or DVD.

However, although semiconductor companies spent hundreds of millions of dollars desperately trying to create a blue LED, the little rapscallion remained elusive for more

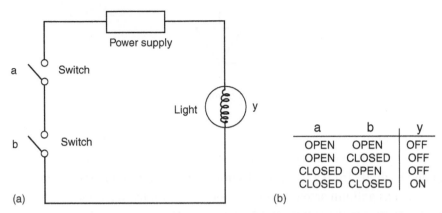

**Figure 10.11: Switch representation of a 2-input AND function (a) Circuit; (b) Truth table**

than three decades. In fact, it wasn't until 1996 that the Japanese electrical engineer Shuji Nakamura demonstrated a blue LED based on gallium nitride. Quite apart from its data storage applications, this discovery also makes it possible to combine the output from a blue LED with its red and green cousins to generate white light. Many observers believe that this may ultimately relegate the incandescent light bulb to the museum shelf.

### 10.7.1 Primitive Logic Functions

Consider an electrical circuit consisting of a power supply, a light, and two switches connected in series (one after the other). The switches are the inputs to the circuit and the light is the output. A truth table provides a convenient way to represent the operation of the circuit (Figure 10.11).

As the light is only ON when both the a *and* b switches are CLOSED (ON), this circuit could be said to perform a 2-input AND function. In fact, the results depend on the way in which the switches are connected; consider another circuit in which two switches are connected in *parallel* (side by side) (Figure 10.12).

In this case, as the light is ON when either a *or* b are CLOSED (ON), this circuit could be said to provide a 2-input OR function.[11] In a limited respect, we might consider that these circuits are making simple logical decisions; two switches offer four combinations of OPEN (OFF) and CLOSED (ON), but only certain combinations cause the light to be turned ON.

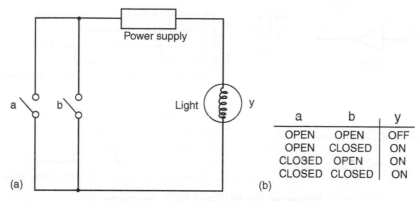

**Figure 10.12: Switch representation of a 2-input OR function (a) Circuit; (b) Truth table**

Logic functions such as AND and OR are generic concepts that can be implemented in a variety of ways, including switches as illustrated above, transistors for use in computers, and even pneumatic devices for use in hostile environments such as steel works or nuclear reactors. Thus, instead of drawing circuits using light switches, it is preferable to make use of more abstract forms of representation. This permits designers to specify the function of systems with minimal consideration as to their final physical realization. To facilitate this, special symbols are employed to represent logic functions, and truth table assignments are specified using the abstract terms FALSE and TRUE. This is because assignments such as OPEN, CLOSED, ON, and OFF may imply a particular implementation.

## 10.8 BUF and NOT Functions

The simplest of all the logic functions are known as *BUF* and *NOT* (Figure 10.13).

The F and T values in the truth tables are shorthand for FALSE and TRUE, respectively. The output of the BUF function has the same value as the input to the function; if the input is FALSE the output is FALSE, and if the input is TRUE the output is TRUE. By comparison, the small circle, or bobble, on the output of the NOT symbol indicates an inverting function; if the input is FALSE the output is TRUE, and if the input is TRUE the output is FALSE.

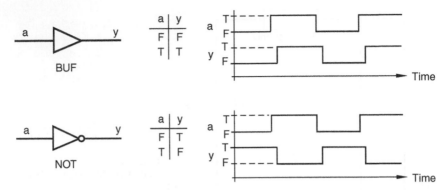

Figure 10.13: BUF and NOT functions

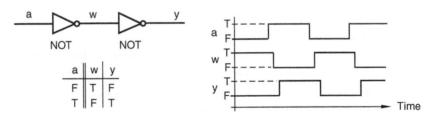

**Figure 10.14: Two NOT functions connected together in series**

As a reminder that these abstract functions will eventually have physical realizations, the waveforms show delays between transitions on the inputs and corresponding responses at the outputs. The actual values of these delays depend on the technology used to implement the functions, but it is important to note that in any physical implementation there will always be some element of delay.

Now consider the effect of connecting two NOT functions in *series* (one after the other) as shown in Figure 10.14.

The first NOT gate inverts the value from the input, and the second NOT gate inverts it back again. Thus, the two negations cancel each other out (sort of like *"two wrongs* do *make a right")*. The end result is equivalent to that of a BUF function, except that each NOT contributes an element of delay.

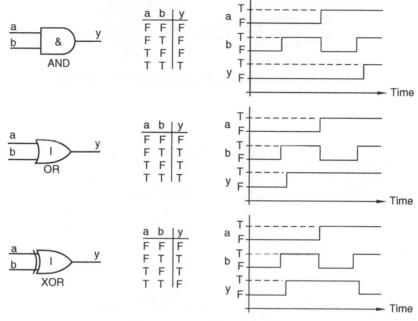

**Figure 10.15: AND, OR, and XOR functions**

## 10.9 AND, OR, and XOR Functions

Three slightly more complex functions are known as *AND, OR,* and *XOR* (Figure 10.15).

The AND and OR representations shown here are the abstract equivalents of our original switch examples. In the case of the AND, the output is only TRUE if *both* a *and* b are TRUE; in the case of the OR, the output is TRUE if *either* a *or* b are TRUE. In fact, the OR should more properly be called an *inclusive-OR*, because the TRUE output cases *include* the case when both inputs are TRUE. Contrast this with the *exclusive-OR*, or XOR, where the TRUE output cases *exclude* the case when both inputs are TRUE.

## 10.10 NAND, NOR, and XNOR Functions

Now consider the effect of appending a NOT function to the output of the AND function (Figure 10.16).

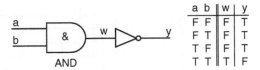

Figure 10.16: AND function followed by a NOT function

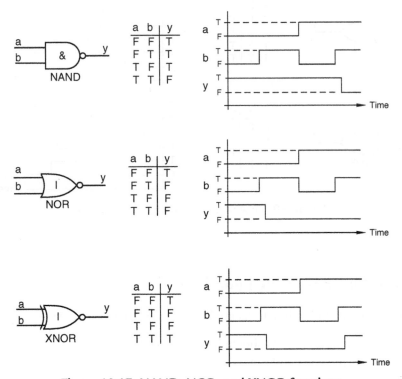

Figure 10.17: NAND, NOR, and XNOR functions

This combination of functions occurs frequently in designs. Similarly, the outputs of the OR and XOR functions are often inverted with NOT functions. This leads to three more primitive functions called *NAND (NOT-AND), NOR (NOT-OR)* and *NXOR (NOT-XOR)*. However, in practice the NXOR is almost always referred to as an XNOR (exclusive-NOR) (Figure 10.17).

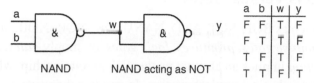

| a | b | y |
|---|---|---|
| F | F | T |
| T | T | F |

NAND acting as NOT

**Figure 10.18: Forming a NOT from a NAND**

| a | b | w | y |
|---|---|---|---|
| F | F | T | F |
| F | T | T | F |
| T | F | T | F |
| T | T | F | T |

NAND          NAND acting as NOT

**Figure 10.19: Forming an AND from two NANDs**

The bobbles on their outputs indicate that these are inverting functions. One way to visualize this is that the symbol for the NOT function has been forced back into the preceding symbol until only the bobble remains visible.

Of course, if we appended a NOT function to the output of a NAND, we'd end up back with our original AND function again. Similarly, appending a NOT to a NOR or an XNOR results in an OR and XOR, respectively.

## 10.11 Not a Lot

And that's about it. In reality there are only eight simple functions (BUF, NOT, AND, NAND, OR, NOR, XOR, and XNOR) from which everything else is constructed. In fact, some might argue that there are only seven core functions because you can construct a BUF out of two NOTs, as was discussed earlier.

Actually, if you want to go down this path, you can construct all of the above functions using one or more NAND gates (or one or more NOR gates). For example, if you connect the two inputs of a NAND gate together, you end up with a NOT as shown in Figure 10.18 (you can achieve the same effect by connecting the two inputs of a NOR gate together).

As the inputs a and b are connected together, they have to carry identical values, so we end up showing only two rows in the truth table. We also know that if we invert the output from a NAND, we end up with an AND. So we could append a NAND configured as a NOT to the output of another NAND to generate an AND (Figure 10.19).

Later on, we'll discover how to transform functions formed from ANDs into equivalent functions formed from ORs and vice versa. Coupled with what we've just seen here, this would allow us to build anything we wanted out of a bunch of 2-input NAND (or NOR) functions.

## 10.12 Functions Versus Gates

Simple functions such as BUF, NOT, AND, NAND, OR, NOR, XOR, and XNOR are often known as *primitive gates*, *primitives*, *logic gates*, or simply *gates*. Strictly speaking, the term *logic function* implies an abstract mathematical relationship, while *logic gate* implies an underlying physical implementation. In practice, however, these terms are often used interchangeably.

More complex functions can be constructed by combining primitive gates in different ways. A complete design—say a computer—employs a great many gates connected together to achieve the required result. When the time arrives to translate the abstract representation into a particular physical implementation, the logic symbols are converted into appropriate equivalents such as switches, transistors, or pneumatic valves. Similarly, the FALSE and TRUE logic values are mapped into appropriate equivalents such as switch positions, voltage levels, or air pressures. The majority of designs are translated into a single technology. However, one of the advantages of abstract representations is that they allow designers to implement different portions of a single design in dissimilar technologies with relative ease. Throughout the remainder of this book we will be concentrating on electronic implementations.

Finally, if some of the above seems to be a little esoteric, consider a real-world example from your home, such as two light switches mounted at opposite ends of a hallway controlling the same light. If both of the switches are UP or DOWN the light will be ON; for any other combination the light will be OFF. Constructing a truth table reveals a classic example of an XNOR function.

### 10.12.1  Using Transistors to Build Primitive Logic Functions

There are several different families of transistors available to designers and, although the actual implementations vary, each can be used to construct primitive logic gates. This chapter concentrates on the *metal-oxide semiconductor field-effect transistors (MOSFETs)*

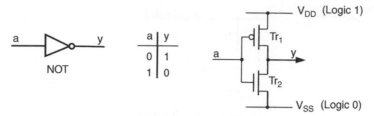

**Figure 10.20: CMOS implementation of a NOT gate**

introduced earlier, because their symbols, construction, and operation are easier to understand than are *bipolar junction transistors (BJTs)*.

Logic gates can be created using only NMOS or only PMOS transistors; however, a popular implementation called *complementary metal-oxide semiconductor (CMOS)* makes use of both NMOS and PMOS transistors connected in a complementary manner.

CMOS gates operate from two voltage levels, which are usually given the labels $V_{DD}$ and $V_{SS}$. To some extent the actual values of $V_{DD}$ and $V_{SS}$ are irrelevant as long as $V_{DD}$ is sufficiently more positive than $V_{SS}$. There are also two conventions known as *positive logic* and *negative logic*.[19] Under the positive logic convention used throughout this book, the more positive $V_{DD}$ is assigned the value of logic 1, and the more negative $V_{SS}$ is assigned the value of logic 0. Previously it was noted that truth table assignments can be specified using the abstract values FALSE and TRUE. However, electronic designers usually represent FALSE and TRUE as 0 and 1, respectively.

## 10.13 NOT and BUF Gates

The simplest logic function to implement in CMOS is a NOT gate (Figure 10.20). The small circle, or bobble, on the control input of transistor $Tr_1$ indicates a PMOS transistor. The bobble is used to indicate that this transistor has an active-low control, which means that a logic 0 applied to the control input turns the transistor ON and a logic 1 turns it OFF. The lack of a bobble on the control input of transistor $Tr_2$ indicates an NMOS transistor. The lack of a bobble says that this transistor has an active-high control, which means that a logic 1 applied to the control input turns the transistor ON and a logic 0 turns it OFF.

Thus, when a logic 0 is applied to input $a$, transistor $Tr_1$ is turned ON, transistor $Tr_2$ is turned OFF, and output $y$ is connected to logic 1 via $Tr_1$. Similarly, when a logic 1 is

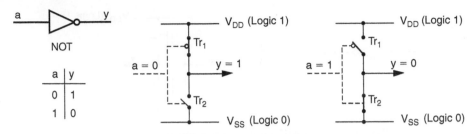

Figure 10.21: NOT gate's operation represented in terms of switches

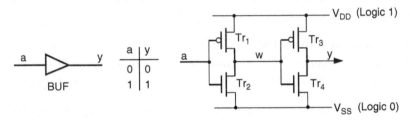

Figure 10.22: CMOS implementation of a BUF gate

applied to input $a$, transistor $Tr_1$ is turned OFF, transistor $Tr_2$ is turned ON, and output $y$ is connected to logic 0 via $Tr_2$.

Don't worry if all this seems a bit confusing at first. The main points to remember are that a logic 0 applied to its control input turns the PMOS transistor ON and the NMOS transistor OFF, while a logic 1 turns the PMOS transistor OFF and the NMOS transistor ON. It may help to visualize the NOT gate's operation in terms of switches rather than transistors (Figure 10.21).

Surprisingly, a non-inverting BUF gate is more complex than an inverting NOT gate. This is due to the fact that a BUF gate is constructed from two NOT gates connected in series (one after the other), which means that it requires four transistors (Figure 10.22).

The first NOT gate is formed from transistors $Tr_1$ and $Tr_2$, while the second is formed from transistors $Tr_3$ and $Tr_4$. A logic 0 applied to input $a$ is inverted to a logic 1 on $w$, and then inverted back again to a logic 0 on output $y$. Similarly, a logic 1 on $a$ is inverted to a logic 0 on $w$, and then inverted back again to a logic 1 on $y$.

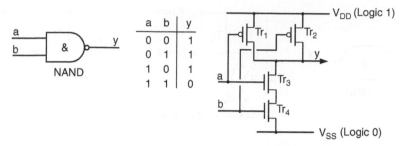

Figure 10.23: CMOS implementation of a 2-input NAND gate

Around this stage it is not unreasonable to question the need for BUF gates in the first place—after all, their logical function could be achieved using a simple piece of wire. But there's method to our madness, because BUF gates may actually be used for a number of reasons: for example, to isolate signals, to provide increased drive capability, or to add an element of delay.

## 10.14 NAND and AND Gates

The implementations of the NOT and BUF gates shown above illustrate an important point, which is that it is generally easier to implement an inverting function than its non-inverting equivalent. In the same way that a NOT is easier to implement than a BUF, a NAND is easier to implement than an AND, and a NOR is easier to implement than an OR. More significantly, inverting functions typically require fewer transistors and operate faster than their non-inverting counterparts. This can obviously be an important design consideration. Consider a 2-input NAND gate, which requires four transistors (Figure 10.23). (Note that a 3-input version could be constructed by adding an additional PMOS transistor in parallel with $Tr_1$ and $Tr_2$, and an additional NMOS transistor in series with $Tr_3$ and $Tr_4$.)

When both $a$ and $b$ are presented with logic 1s, transistors $Tr_1$ and $Tr_2$ are turned OFF, transistors $Tr_3$ and $Tr_4$ are turned ON, and output $y$ is connected to logic 0 via $Tr_3$ and $Tr_4$. Any other combination of inputs results in one or both of $Tr_3$ and $Tr_4$ being turned OFF, one or both of $Tr_1$ and $Tr_2$ being turned ON, and output $y$ being connected to logic 1 via $Tr_1$ and/or $Tr_2$. Once again, it may help to visualize the gate's operation in terms of switches rather than transistors (Figure 10.24).

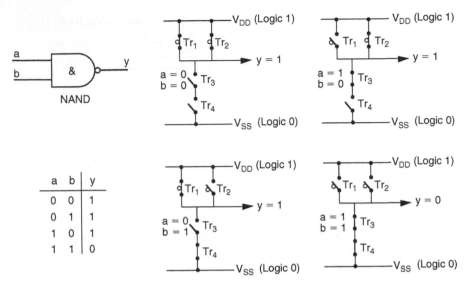

Figure 10.24: NAND gate's operation represented in terms of switches

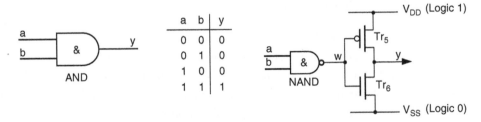

Figure 10.25: CMOS implementation of a 2-input AND gate

Now consider an AND gate. This is formed by inverting the output of a NAND with a NOT, which means that a 2-input AND requires six transistors (Figure 10.25).

## 10.15 NOR and OR Gates

A similar story occurs in the case of NOR gates and OR gates. First, consider a 2-input NOR, which requires four transistors (Figure 10.26). (A 3-input version could be constructed by adding an additional PMOS transistor in series with $Tr_1$ and $Tr_2$, and an additional NMOS transistor in parallel with $Tr_3$ and $Tr_4$.)

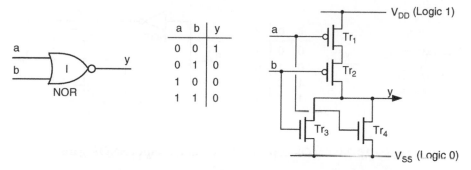

**Figure 10.26: CMOS implementation of a 2-input NOR gate**

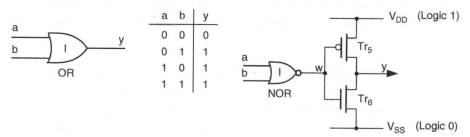

**Figure 10.27: CMOS implementation of a 2-input OR gate**

When both *a* and *b* are set to logic 0, transistors $Tr_3$ and $Tr_4$ are turned OFF, transistors $Tr_1$ and $Tr_2$ are turned ON, and output *y* is connected to logic 1 via $Tr_1$ and $Tr_2$. Any other combination of inputs results in one or both of $Tr_1$ and $Tr_2$ being turned OFF, one or both of $Tr_3$ and $Tr_4$ being turned ON, and output *y* being connected to logic 0 via $Tr_3$ and/or $Tr_4$.

Once again, an OR gate is formed by inverting the output of a NOR with a NOT, which means that a 2-input OR requires six transistors (Figure 10.27).

## 10.16 XNOR and XOR Gates

The concepts of NAND, AND, NOR, and OR are relatively easy to understand because they map onto the way we think in everyday life. For example, a textual equivalent of a NOR could be: *"If it's windy or if it's raining then I'm not going out."* By comparison, the concepts of XOR and XNOR can be a little harder to grasp because we don't usually consider things in these terms. A textual equivalent of an XOR could be: *"If it is*

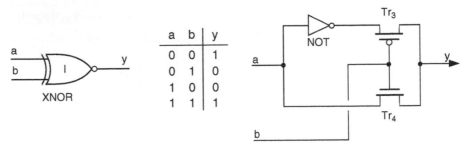

**Figure 10.28: CMOS implementation of a 2-input XNOR gate**

*windy and it's* not *raining, or if it's* not *windy and it is* raining, then I will go out.*"* Although this does make sense (in a strange sort of way), we don't often find ourselves making decisions in this manner.

For this reason, it is natural to assume that XNOR and XOR gates would be a little more difficult to construct. However, these gates are full of surprises, both in the way in which they work and the purposes for which they can be used. For example, a 2-input XNOR can be implemented using only four transistors (Figure 10.28). Unlike AND, NAND, OR, and NOR gates, there are no such beasts as XNOR or XOR primitives with more than two inputs. However, equivalent functions with more than two inputs can be formed by connecting a number of 2-input primitives together.

The NOT gate would be constructed in the standard way using two transistors as described above, but the XNOR differs from the previous gates in the way that transistors $Tr_3$ and $Tr_4$ are utilized. First, consider the case where input $b$ is presented with a logic 0: transistor $Tr_4$ is turned OFF, transistor $Tr_3$ is turned ON, and output $y$ is connected to the output of the NOT gate via $Tr_3$. Thus, when input $b$ is logic 0, output $y$ has the inverse of the value on input $a$. Now consider the case where input $b$ is presented with a logic 1: transistor $Tr_3$ is turned OFF, transistor $Tr_4$ is turned ON, and output $y$ is connected to input $a$ via $Tr_4$. Thus, when input $b$ is logic 1, output $y$ has the same value as input $a$. The end result of all these machinations is that wiring the transistors together in this way does result in a function that satisfies the requirements of the XNOR truth table.

Unlike the other complementary gates, it is not necessary to invert the output of the XNOR to form an XOR (although we could if we wanted to, of course). A little judicious

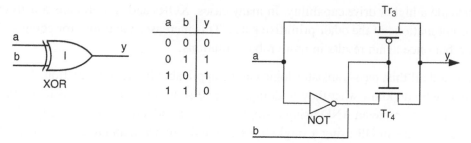

**Figure 10.29: CMOS implementation of a 2-input XOR gate**

rearranging of the components results in a 2-input XOR that also requires only four transistors (Figure 10.29).

First, consider the case where input $b$ is presented with a logic 0: transistor $Tr_4$ is turned OFF, transistor $Tr_3$ is turned ON, and output $y$ is connected to input $a$ via $Tr_3$. Thus, when input $b$ is logic 0, output $y$ has the same value as input $a$. Now consider the case where input $b$ is presented with a logic 1: transistor $Tr_3$ is turned OFF, transistor $Tr_4$ is turned ON, and output $y$ is connected to the output of the NOT gate via $Tr_4$. Thus, when input $b$ is logic 1, output $y$ has the inverse of the value on input $a$. Once again, this results in a junction that satisfies the requirements of the XOR truth table.

## 10.17 Pass-Transistor Logic

In the BUF, NOT, AND, NAND, OR, and NOR gates described earlier, the input signals and internal data signals are only used to drive control terminals on the transistors. By comparison, transistors $Tr_3$ and $Tr_4$ in the XOR and XNOR gates shown above are connected so that input and internal data signals pass between their data terminals. This technique is known as *pass-transistor logic*. It can be attractive in that it minimizes the number of transistors required to implement a function, but it's not necessarily the best approach. Strange and unexpected effects can ensue if you're not careful and you don't know what you're doing.

An alternative solution for an XOR is to invert the output of the XNOR shown above with a NOT. Similarly, an XNOR can be constructed by inverting the output of the XOR shown above with a NOT. Although these new implementations each now require six transistors rather than four, they are more robust because the NOT gates buffer the outputs and provide a higher drive capability. In many cases, XORs and XNORs are constructed from combinations of the other primitive gates. This increases the transistor count still further, but once again results in more robust solutions.

Having said all this, pass-transistor logic can be applicable in certain situations for designers who do know what they're doing. In the discussions above, it was noted that it is possible to create an AND using a single transistor and a resistor. Similarly, it's possible to create an OR using a single transistor and a resistor, and to create an XOR or an XNOR using only two transistors and a resistor. If you're feeling brave, try to work out how to achieve these minimal implementations for yourself.

### 10.17.1 Boolean Algebra

One of the most significant mathematical tools available to electronics designers was actually invented for quite a different purpose. Around the 1850s, a British mathematician, George Boole (1815–1864), developed a new form of mathematics that is now known as *Boolean algebra*. Boole's intention was to use mathematical techniques to represent and rigorously test logical and philosophical arguments. His work was based on the following: a *statement* is a sentence that asserts or denies an attribute about an object or group of objects:

> **Statement:** Your face resembles a cabbage.
>
> Depending on how carefully you choose your friends, they may either agree or disagree with the sentiment expressed; therefore, this statement cannot be proved to be either true or false.

By comparison, a *proposition* is a statement that is either true or false with no ambiguity:

> **Proposition:** I just tipped a bucket of burning oil into your lap.
>
> This proposition may be true or it may be false, but it is definitely one or the other and there is no ambiguity about it.

Propositions can be combined together in several ways; a proposition combined with an AND operator is known as a *conjunction*:

**Conjunction:** You have a parrot on your head AND you have a fish in your ear.

The result of a conjunction is true if all of the propositions comprising that conjunction are true.

A proposition combined with an OR operator is known as a *disjunction*:

**Disjunction:** You have a parrot on your head OR you have a fish in your ear.

The result of a disjunction is true if at least one of the propositions comprising that disjunction is true.

From these humble beginnings, Boole established a new mathematical field known as *symbolic* logic, in which the logical relationship between propositions can be represented symbolically by such means as equations or truth tables. Sadly, this work found little application outside the school of symbolic logic for almost one hundred years.

In fact, the significance of Boole's work was not fully appreciated until the late 1930s, when a graduate student at MIT, Claude Shannon, submitted a master's thesis that revolutionized electronics. In this thesis, Shannon showed that Boolean algebra offered an ideal technique for representing the logical operation of digital systems. Shannon had realized that the Boolean concepts of FALSE and TRUE could be mapped onto the binary digits 0 and 1, and that both could be easily implemented by means of electronic circuits.

Logical functions can be represented using graphical symbols, equations, or truth tables, and these views can be used interchangeably (Figure 10.30).

There are a variety of ways to represent Boolean equations. In this chapter, the symbols &, |, and ^ are used to represent AND, OR, and XOR respectively; a negation, or NOT, is represented by a horizontal line, or bar, over the portion of the equation to be negated.

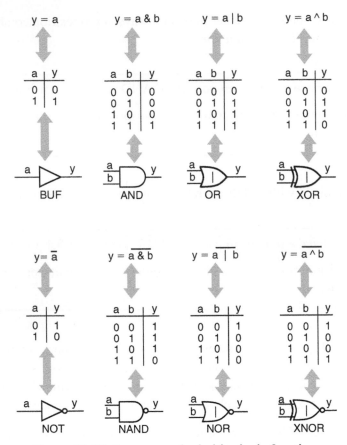

**Figure 10.30: Summary of primitive logic functions**

## 10.18 Combining a Single Variable With Logic 0 or Logic 1

A set of simple but highly useful rules can be derived from the combination of a single variable with a logic 0 or logic 1 (Figure 10.31).

## 10.19 The Idempotent Rules

The rules derived from the combination of a single variable with itself are known as the *idempotent rules* (Figure 10.32).

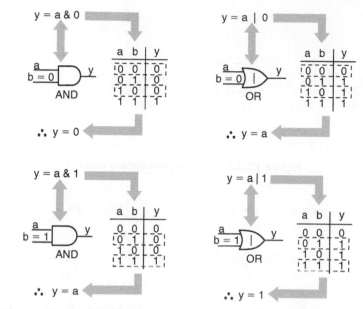

**Figure 10.31: Combining a single variable with a logic 0 or logic 1**

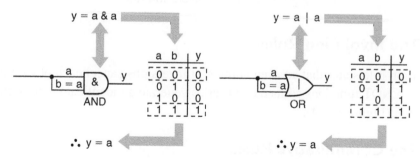

**Figure 10.32: The idempotent rules**

## 10.20 The Complementary Rules

The rules derived from the combination of a single variable with the inverse of itself are
known as the *complementary rules* (Figure 10.33).

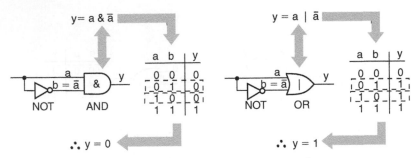

Figure 10.33: The complementary rules

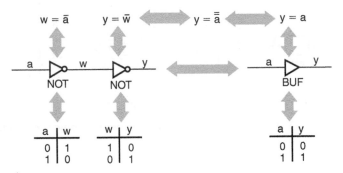

Figure 10.34: The involution rules

## 10.21 The Involution Rules

The involution rule states that an even number of inversions cancel each other out; for example, two NOT functions connected in series generate an identical result to that of a BUF function (Figure 10.34).

## 10.22 The Commutative Rules

The commutative rules state that the order in which variables are specified will not affect the result of an AND or OR operation (Figure 10.35).

## 10.23 The Associative Rules

The associative rules state that the order in which pairs of variables are associated together will not affect the result of multiple AND or OR operations (Figure 10.36).

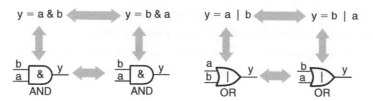

**Figure 10.35: The commutative rules**

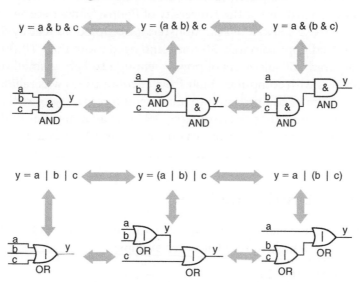

**Figure 10.36: The associative rules**

## 10.24 Precedence of Operators

In standard arithmetic, the multiplication operator is said to have a higher *precedence* than the addition operator. This means that, if an equation contains both multiplication and addition operators without parentheses, then the multiplication is performed before the addition; for example:

$$6 + 2 \times 4 \equiv 6 + (2 + 4)$$

Similarly, in Boolean algebra, the & (AND) operator has a higher precedence than the | (OR) operator:

$$a \mid b \, \& \, c \equiv a \mid (b \, \& \, c)$$

Due to the similarities between these arithmetic and logical operators, the & (AND) operator is known as a *logical multiplication or product*, while the | (OR) operator is known as a *logical addition or sum*. To avoid any confusion as to the order in which logical operations will be performed, this book will always make use of parentheses.

The first true electronic computer, ENIAC (Electronic Numerical Integrator and Calculator), was constructed at the University of Pennsylvania between 1943 and 1946. In many ways ENIAC was a monster; it occupied 30 feet by 50 feet of floor space, weighed approximately 30 tons, and used more than 18,000 vacuum tubes which required 150 kilowatts of power—enough to light a small town. One of the big problems with computers built from vacuum tubes was reliability; 90% of ENIAC's down-time was attributed to locating and replacing burnt-out tubes. Records from 1952 show that approximately 19,000 vacuum tubes had to be replaced in that year alone; that averages out to about 50 tubes a day!

## 10.25  The First Distributive Rule

In standard arithmetic, the multiplication operator will distribute over the addition operator because it has a higher precedence; for example:

$$6 \times (5 + 2) \equiv (6 \times 5) + (6 \times 2)$$

Similarly, in Boolean algebra, the & (AND) operator will distribute over an | (OR) operator because it has a higher precedence; this is known as *the first distributive rule* (Figure 10.37).

## 10.26  The Second Distributive Rule

In standard arithmetic, the addition operator will not distribute over the multiplication operator because it has a lower precedence:

$$6 + (5 \times 2) \neq (6 + 5) \times (6 + 2)$$

However, Boolean algebra is special in this case. Even though the | (OR) operator has lower precedence than the & (AND) operator, it will still distribute over the & operator; this is known as *the second distributive rule* (Figure 10.38).

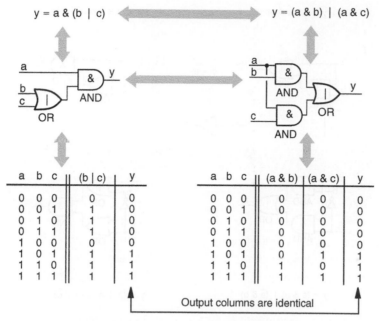

**Figure 10.37: The first distributive rule**

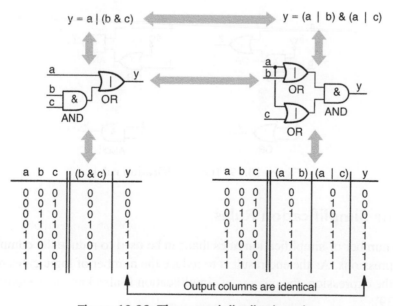

**Figure 10.38: The second distributive rule**

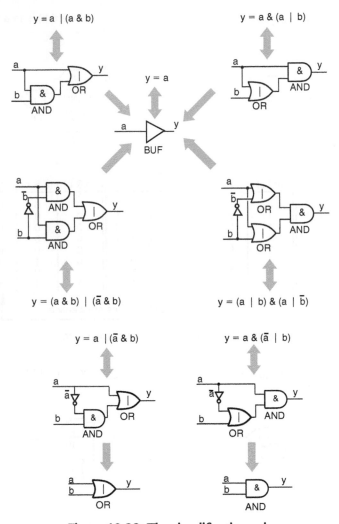

**Figure 10.39: The simplification rules**

## 10.27 The Simplification Rules

There are a number of simplification rules that can be used to reduce the complexity of Boolean expressions. As the end result is to reduce the number of logic gates required to implement the expression, the process of simplification is also known as *minimization* (Figure 10.39).

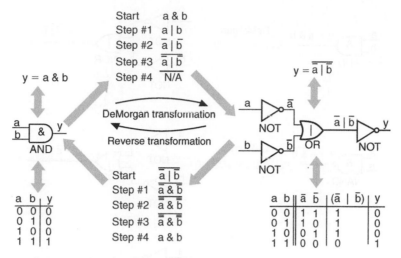

**Figure 10.40: DeMorgan Transformation of an AND function**

## 10.28 DeMorgan Transformations

A contemporary of Boole's, Augustus DeMorgan (1806–1871), also made significant contributions to the field of symbolic logic, most notably a set of rules which facilitate the conversion of Boolean expressions into alternate and often more convenient forms. A *DeMorgan Transformation* comprises four steps:

1. Exchange all of the & operators for | operators and vice versa.

2. Invert all the variables; also exchange 0s for 1s and vice versa.

3. Invert the entire function.

4. Reduce any multiple inversions.

Consider the DeMorgan Transformation of a 2-input AND function (Figure 10.40). Note that the NOT gate on the output of the new function can be combined with the OR to form a NOR.

Similar transformations can be performed on the other primitive functions (Figure 10.41).

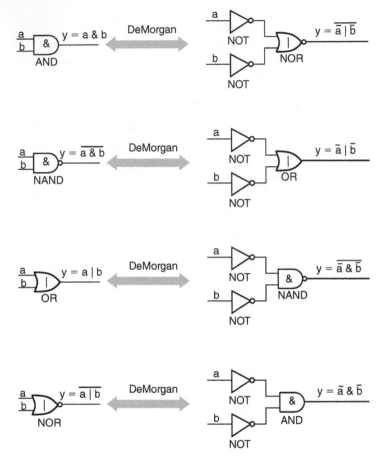

**Figure 10.41: DeMorgan Transformations of AND, NAND, OR, and NOR functions**

| a | b | c | Minterms | Maxterms |
|---|---|---|----------|----------|
| 0 | 0 | 0 | $(\overline{a} \,\&\, \overline{b} \,\&\, \overline{c})$ | $(a \mid b \mid c)$ |
| 0 | 0 | 1 | $(\overline{a} \,\&\, \overline{b} \,\&\, c)$ | $(a \mid b \mid \overline{c})$ |
| 0 | 1 | 0 | $(\overline{a} \,\&\, b \,\&\, \overline{c})$ | $(a \mid \overline{b} \mid c)$ |
| 0 | 1 | 1 | $(\overline{a} \,\&\, b \,\&\, c)$ | $(a \mid \overline{b} \mid \overline{c})$ |
| 1 | 0 | 0 | $(a \,\&\, \overline{b} \,\&\, \overline{c})$ | $(\overline{a} \mid b \mid c)$ |
| 1 | 0 | 1 | $(a \,\&\, \overline{b} \,\&\, c)$ | $(\overline{a} \mid b \mid \overline{c})$ |
| 1 | 1 | 0 | $(a \,\&\, b \,\&\, \overline{c})$ | $(\overline{a} \mid \overline{b} \mid c)$ |
| 1 | 1 | 1 | $(a \,\&\, b \,\&\, c)$ | $(\overline{a} \mid \overline{b} \mid \overline{c})$ |

**Figure 10.42: Minterms and maxterms**

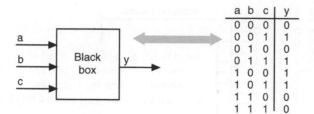

| a | b | c | y |
|---|---|---|---|
| 0 | 0 | 0 | 0 |
| 0 | 0 | 1 | 1 |
| 0 | 1 | 0 | 0 |
| 0 | 1 | 1 | 1 |
| 1 | 0 | 0 | 1 |
| 1 | 0 | 1 | 1 |
| 1 | 1 | 0 | 0 |
| 1 | 1 | 1 | 0 |

**Figure 10.43: Black box with associated truth table**

## 10.29 Minterms and Maxterms

For each combination of inputs to a logical function, there is an associated *minterm* and an associated *maxterm*. Consider a truth table with three inputs: *a*, *b*, and *c* (Figure 10.42).

The minterm associated with each input combination is the & (AND), or product, of the input variables, while the maxterm is the | (OR), or sum, of the inverted input variables. Minterms and maxterms are useful for deriving Boolean equations from truth tables as discussed below.

## 10.30 Sum-of-Products and Product-of-Sums

A designer will often specify portions of a design using truth tables, and determine how to implement these functions as logic gates later. The designer may start by representing a function as a "black box" with an associated truth table (Figure 10.43). Note that the values assigned to the output *y* in the truth table shown in Figure 10.43 were selected randomly, and have no significance beyond the purposes of this example.

There are two commonly used techniques for deriving Boolean equations from a truth table. In the first technique, the minterms corresponding to each line in the truth table for which the output is a logic 1 are extracted and combined using | (OR) operators; this method results in an equation said to be in *sum-of-products* form. In the second technique, the maxterms corresponding to each line in the truth table for which the output is a logic 0 are combined using & (AND) operators; this method results in an equation said to be in *product-of-sums form* (Figure 10.44).

For a function whose output is logic 1 fewer times than it is logic 0, it is generally easier to extract a sum-of-products equation. Similarly, if the output is logic 0 fewer times than it is logic 1, it is generally easier to extract a product-of-sums equation.

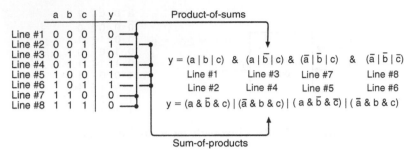

**Figure 10.44: Sum-of-products versus product-of-sums equations**

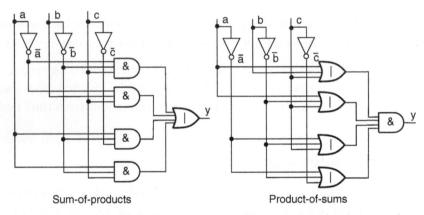

Sum-of-products　　　　　　　　　　　　Product-of-sums

**Figure 10.45: Sum-of-products versus product-of-sums implementations**

The sum-of-products and product-of-sums forms complement each other and return identical results. An equation in either form can be transformed into its alternative form by means of the appropriate DeMorgan Transformation.

Once an equation has been obtained in the required form, the designer would typically make use of the appropriate simplification rules to minimize the number of logic gates required to implement the function. However, neglecting any potential minimization, the equations above could be translated directly into their logic gate equivalents (Figure 10.45).

## 10.31  Canonical Forms

In a mathematical context, the term *canonical form* is taken to mean a generic or basic representation. Canonical forms provide the means to compare two expressions without

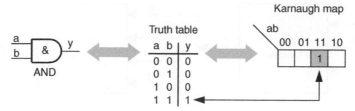

**Figure 10.46: Karnaugh map for a 2-input AND function**

falling into the trap of trying to compare "apples" with "oranges." The sum-of-products and product-of-sums representations are different canonical forms. Thus, to compare two Boolean equations, both must first be coerced into the same canonical form; either sum-of-products or product-of-sums.

## 10.32  Karnaugh Maps

In 1953, Maurice Karnaugh (pronounced "car-no") invented a form of logic diagram called a *Karnaugh map*, which provides an alternative technique for representing Boolean functions; for example, consider the Karnaugh map for a 2-input AND function (Figure 10.46).

The Karnaugh map comprises a box for every line in the truth table. The binary values above the boxes are those associated with the $a$ and $b$ inputs. Unlike a truth table, in which the input values typically follow a binary sequence, the Karnaugh map's input values must be ordered such that the values for adjacent columns vary by only a single bit: for example, $00_2$, $01_2$, $11_2$, and $10_2$. This ordering is known as a *Gray code* and it is a key factor in the way in which Karnaugh maps work.

The $y$ column in the truth table shows all the 0 and 1 values associated with the gate's output. Similarly, all of the output values could be entered into the Karnaugh map's boxes. However, for reasons of clarity, it is common for only a single set of values to be used (typically the 1s).

Similar maps can be constructed for 3-input and 4-input functions. In the case of a 4-input map, the values associated with the $c$ and $d$ inputs must also be ordered as a Gray code: that is, they must be ordered in such a way that the values for adjacent rows vary by only a single bit (Figure 10.47).

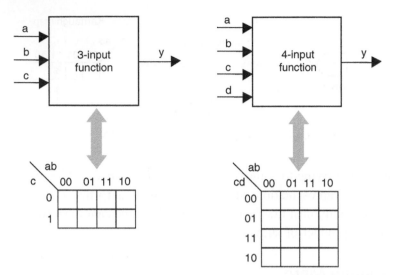

**Figure 10.47: Karnaugh maps for 3-input and 4-input functions**

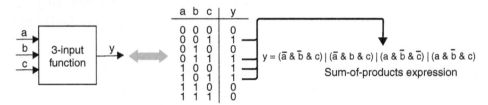

$y = (\bar{a} \,\&\, \bar{b} \,\&\, c) \,|\, (\bar{a} \,\&\, b \,\&\, c) \,|\, (a \,\&\, \bar{b} \,\&\, \bar{c}) \,|\, (a \,\&\, \bar{b} \,\&\, c)$

Sum-of-products expression

**Figure 10.48: Example 3-input function**

## 10.33 Minimization Using Karnaugh Maps

Karnaugh maps often prove useful in the simplification and minimization of Boolean functions. Consider an example 3-input function represented as a black box with an associated truth table (Figure 10.48). (The values assigned to output *y* in the truth table were selected randomly and have no significance beyond the purposes of this example.)

The equation extracted from the truth table in sum-of-products form contains four minterms, one for each of the 1s assigned to the output. Algebraic simplification techniques could be employed to minimize this equation, but this would necessitate every minterm being compared to each of the others, which can be somewhat time-consuming.

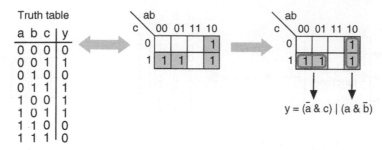

**Figure 10.49: Karnaugh map minimization of example 3-input function**

This is where Karnaugh maps enter the game. The 1s assigned to the map's boxes represent the same minterms as the 1s in the truth table's output column; however, as the input values associated with each row and column in the map differ by only one bit, any pair of horizontally or vertically adjacent boxes corresponds to minterms that differ by only a single variable. Such pairs of minterms can be grouped together and the variable that differs can be discarded (Figure 10.49).

In the case of the horizontal group, input $a$ is 0 for both boxes, input $c$ is 1 for both boxes, and input $b$ is 0 for one box and 1 for the other. Thus, for this group, changing the value on $b$ does not affect the value of the output. This means that $b$ is redundant and can be discarded from the equation representing this group. Similarly, in the case of the vertical group, input $a$ is 1 for both boxes, input $b$ is 0 for both boxes, and input $c$ is 0 for one box and 1 for the other. Thus, input $c$ is redundant for *this* group and can be discarded.

## 10.34  Grouping Minterms

In the case of a 3-input Karnaugh map, any two horizontally or vertically adjacent minterms, each composed of three variables, can be combined to form a new product term composed of only two variables. Similarly, in the case of a 4-input map, any two adjacent minterms, each composed of four variables, can be combined to form a new product term composed of only three variables. Additionally, the 1s associated with the minterms can be used to form multiple groups. For example, consider the 3-input function shown in Figure 10.50, in which the minterm corresponding to $a = 1$, $b = 1$, and $c = 0$ is common to three groups.

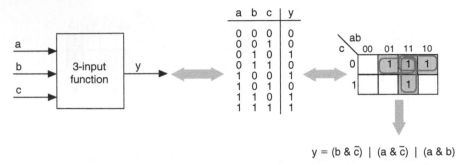

$$y = (b \, \& \, \bar{c}) \mid (a \, \& \, \bar{c}) \mid (a \, \& \, b)$$

Figure 10.50: Karnaugh map minterms used to form multiple groups

Groupings can also be formed from four adjacent minterms, in which case two redundant variables can be discarded; consider some 4-input Karnaugh map examples (Figure 10.51).

In fact, any group of $2^n$ adjacent minterms can be gathered together where $n$ is a positive integer. For example, $2^1 = $ two minterms, $2^2 = 2 \times 2 = $ four minterms, $2^3 = 2 \times 2 \times 2 = $ eight minterms, etc.

As was noted earlier, Karnaugh map input values are ordered so that the values associated with adjacent rows and columns differ by only a single bit. One result of this ordering is that the top and bottom rows are also separated by only a single bit (it may help to visualize the map rolled into a horizontal cylinder such that the top and bottom edges are touching). Similarly, the left and right columns are separated by only a single bit (in this case it may help to visualize the map rolled into a vertical cylinder such that the left and right edges are touching). This leads to some additional groupings, a few of which are shown in Figure 10.7.

Note especially the last example. Diagonally adjacent minterms generally cannot be used to form a group: however, remembering that the left-right columns and the top-bottom rows are logically adjacent, this means that the four corner minterms are also logically adjacent, which in turn means that they can be used to form a single group.

## 10.35 Incompletely Specified Functions

In certain cases a function may be incompletely specified: that is, the output may be undefined for some of the input combinations. For example, if the designer knows that

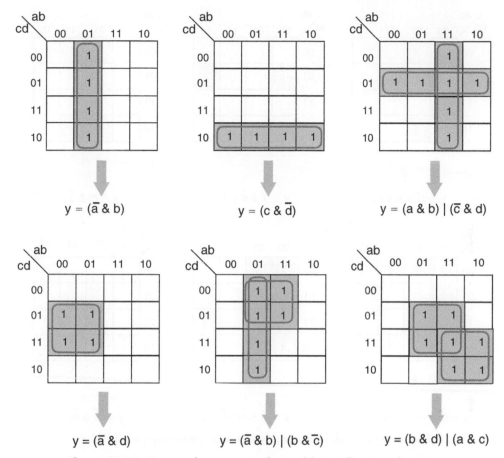

**Figure 10.51: Karnaugh map groupings of four adjacent minterms**

certain input combinations will never occur, then the value assigned to the output for these combinations is irrelevant. Alternatively, for some input combinations the designer may simply not care about the value on the output. In both cases, the designer can represent the output values associated with the relevant input combinations as question marks in the map (Figure 10.53).

The ? characters indicate *don't care* states, which can be considered to represent either 0 or 1 values at the designer's discretion. In the example shown in Figure 10.8, we have no interest in the ? character at $a = 0$, $b = 0$, $c = 1$, $d = 0$ or the ? character at $a = 0$, $b = 1$,

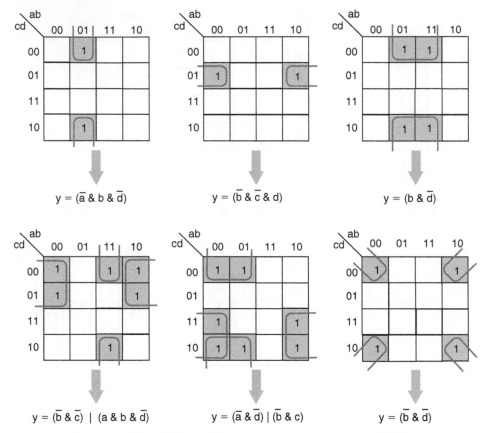

**Figure 10.52: Additional Karnaugh map grouping possibilities**

$c = 1$, $d = 1$, because neither of these can be used to form a larger group. However, if we decide that the other three ? characters are going to represent 1 values, then they can be used to form larger groups, which allows us to minimize the function to a greater degree than would otherwise be possible.

It should be noted that many electronics references use X characters to represent don't care states. Unfortunately, this may lead to confusion as design tools such as logic simulators use X characters to represent *don't know* states. Unless otherwise indicated, this chapter will use ? and X to represent *don't care* and *don't know* states, respectively.

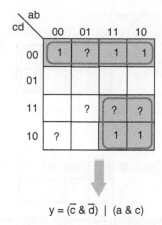

$$y = (\overline{c} \, \& \, \overline{d}) \mid (a \, \& \, c)$$

**Figure 10.53: Karnaugh map for an incompletely specified function**

## 10.36 Populating Maps Using 0s Versus 1s

When we were extracting Boolean equations from truth tables, we noted that in the case of a function whose output is logic 1 fewer times than it is logic 0, it is generally easier to extract a *sum-of-products* equation. Similarly, if the output is logic 0 fewer times than it is logic 1, it is generally easier to extract a *product-of-sums* equation.

The same thing applies to a Karnaugh map. If the output is logic 1 fewer times than it is logic 0, then it's probably going to be a lot easier to populate the map using logic 1s. Alternatively, if the output is logic 0 fewer times than it is logic 1, then populating the map using logic 0s may not be a bad idea.

When a Karnaugh map is populated using the 1s assigned to the truth table's output, the resulting Boolean expression is extracted from the map in sum-of-products form. By comparison, if the Karnaugh map is populated using the 0s assigned to the truth table's output, then the groupings of 0s are used to generate expressions in product-of-sums form (Figure 10.54).

Although the sum-of-products and product-of-sums expressions appear to be somewhat different, they do produce identical results. The expressions can be shown to be equivalent using algebraic means, or by constructing truth tables for each expression and comparing the outputs.

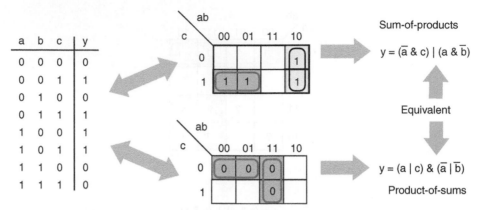

**Figure 10.54: Populating Karnaugh maps with 0s versus 1s**

Karnaugh maps are most often used to represent 3-input and 4-input functions. It is possible to create similar maps for 5-input and 6-input functions, but these maps can quickly become unwieldy and difficult to use. Thus, the Karnaugh technique is generally not considered to have any application for functions with more than six inputs.

### 10.36.1 Using Primitive Logic Functions to Build More Complex Functions

The primitive functions NOT, AND, OR, NAND, NOR, XOR, and XNOR can be connected together to build more complex functions which may, in turn, be used as building blocks in yet more sophisticated systems. The examples introduced in this chapter were selected because they occur commonly in designs, are relatively simple to understand, and will prove useful in later discussions.

## 10.37  Scalar Versus Vector Notation

A single signal carrying one bit of binary data is known as a *scalar* entity. A set of signals carrying similar data can be gathered together into a group known as a *vector*.

Consider the circuit fragments shown in Figure 10.55. Each of these fragments represents four 2-input AND gates. In the case of the scalar notation, each signal is assigned a unique name: for example, $a3$, $a2$, $a1$, and $a0$. By comparison, when using

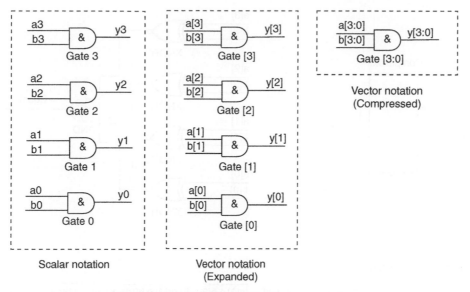

**Figure 10.55: Scalar versus vector notation**

vector notation, a single name is applied to a group of signals, and individual signals within the group are referenced by means of an index: for example, $a[3]$, $a[2]$, $a[1]$, and $a[0]$. This means that if we were to see a schematic (circuit) diagram containing two signals called $a3$ and $a[3]$, we would understand this to represent two completely different signals (the former being a scalar named "$a3$" and the latter being the third element of a vector named "$a$").

A key advantage of vector notation is that it allows all of the signals comprising the vector to be easily referenced in a single statement: for example, $a[3:0]$, $b[3:0]$, and $y[3:0]$. Thus, vector notation can be used to reduce the size and complexity of a circuit diagram while at the same time increasing its clarity.

## 10.38 Equality Comparators

In some designs it may be necessary to compare two sets of binary values to see if they contain the same data. Consider a function used to compare two 4-bit vectors: $a[3:0]$ and

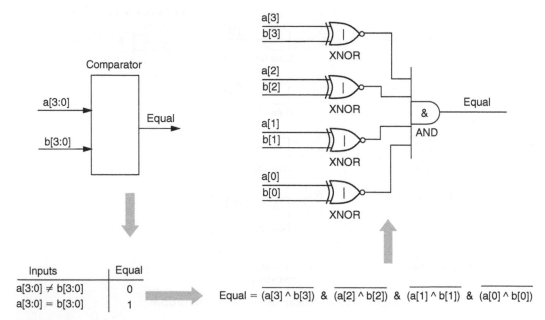

**Figure 10.56: Equality comparator**

$b[3:0]$. A scalar output called equal is to be set to logic 1 if each bit in $a[3:0]$ is equal to its corresponding bit in $b[3:0]$: that is, the vectors are equal if $a[3] = b[3]$, $a[2] = b[2]$, $a[1] = b[1]$, and $a[0] = b[0]$ (Figure 10.56).

The values on $a[3]$ and $b[3]$ are compared using a 2-input XNOR gate. If the values on its inputs are the same (both 0s or both 1s), the output of an XNOR will be 1, but if the values on its inputs are different, the output will be 0. Similar comparisons are performed between the other inputs: $a[2]$ with $b[2]$, $a[1]$ with $b[1]$, and $a[0]$ with $b[0]$. The final AND gate is used to gather the results of the individual comparisons. If all the inputs to the AND gate are 1, the two vectors are the same, and the output of the AND gate will be 1. Correspondingly, if any of the inputs to the AND gate are 0, the two vectors are different, and the output of the AND gate will be 0.

Note that a similar result could have been obtained by replacing the XNORs with XORs and the AND with a NOR, and that either of these implementations could be easily extended to accommodate input vectors of greater width.

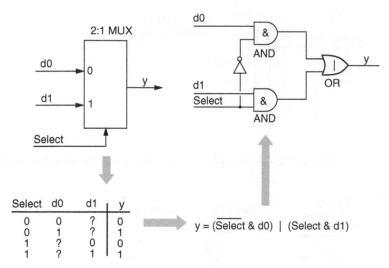

**Figure 10.57: A 2:1 multiplexer**

## 10.39 Multiplexers

A multiplexer uses a binary value, or *address*, to select between a number of inputs and to convey the data from the selected input to the output. For example, consider a 2:1 ("two-to-one") multiplexer (Figure 10.57).

The 0 and 1 annotations on the multiplexer symbol represent the possible values of the select input and are used to indicate which data input will be selected.

The ? characters in the truth table indicate *don't care* states. When the select input is presented with $a0$, the output from the function depends only on the value of the $d0$ data input, and we don't care about the value on $d1$. Similarly, when select is presented with $a1$, the output from the function depends only on the value of the $d1$ data input, and we don't care about the value on $d0$. The use of don't care states reduces the size of the truth table, better represents the operation of this particular function, and simplifies the extraction of the sum-of-products expression because the don't cares are ignored.

An identical result could have been achieved using a full truth table combined with a Karnaugh map minimization (Figure 10.58).

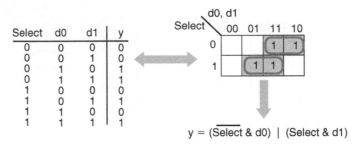

**Figure 10.58: Deriving the 2:1 multiplexer equation by means of a Karnaugh Map**

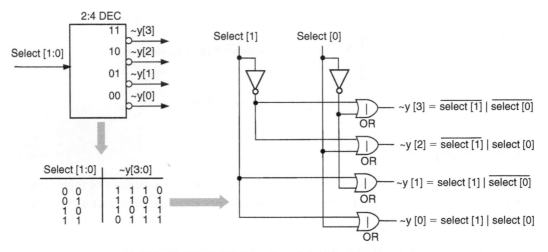

**Figure 10.59: A 2:4 decoder with active-low outputs**

Larger multiplexers are also common in designs: for example, 4:1 multiplexers with four data inputs feeding one output and 8:1 multiplexers with eight data inputs feeding one output. In the case of a 4:1 multiplexer, we will require two select inputs to choose between the four data inputs (using binary patterns of 00, 01, 10, and 11). Similarly, in the case of an 8:1 multiplexer, we will require three select inputs to choose between the eight data inputs (using binary patterns of 000, 001, 010, 011, 100, 101, 110, and 111).

## 10.40 Decoders

A decoder uses a binary value, or *address*, to select between a number of outputs and to assert the selected output by placing it in its active state. For example, consider a 2:4 ("two-to-four") decoder (Figure 10.59).

The 00, 01, 10 and 11 annotations on the decoder symbol represent the possible values that can be applied to the select[1:0] inputs and are used to indicate which output will be asserted.

The truth table shows that when a particular output is selected, it is asserted to $a0$, and when that output is not selected, it returns to $a1$. Because the outputs are asserted to 0s, this device is said to have *active-low* outputs. An *active-low* signal is one whose active state is considered to be logic 0. (Similar functions can be created with *active-high* outputs, which means that when an output is selected it is asserted to a logic 1.) The active-low nature of this particular function is also indicated by the bobbles (small circles) associated with the symbol's outputs and by the tilde ("~") characters in the names of the output signals. Additionally, we know that as each output is 0 for only one input combination, it is simpler to extract the equations in product-of-sums form.

Larger decoders are also commonly used in designs: for example, 3:8 decoders with three select inputs and eight outputs, 4:16 decoders with four select inputs and sixteen outputs, etc.

## 10.41 Tri-State Functions

There is a special category of gates called *tri-state functions* whose outputs can adopt three states: 0, 1, and Z. Lets first consider a simple tri-state buffer (Figure 10.60).

The tri-state buffer's symbol is based on a standard buffer with an additional control input known as the *enable*. The active-low nature of this particular function's enable is indicated by the bobble associated with this input on the symbol and by the tilde character in its name, ~enable. (Similar functions with active-high enables are also commonly used in designs.)

The Z character in the truth table represents a state known as *high-impedance*, in which the gate is not driving either of the standard 0 or 1 values. In fact, in the high-impedance state the gate is effectively disconnected from its output.

Although Boolean algebra is not well equipped to represent the Z state, the implementation of the tri-state buffer is relatively easy to understand. When the ~enable input is presented with a 1 (its inactive state), the output of the OR gate is forced to 1 and the output of the NOR gate is forced to 0, thereby turning both the $Tr_1$ and $Tr_2$ transistors OFF, respectively. With both transistors turned OFF, the output $y$ is disconnected from $V_{DD}$ and $V_{SS}$, and is therefore in the high-impedance state.

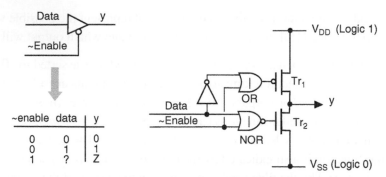

Figure 10.60: Tri-state buffer with active-low enable

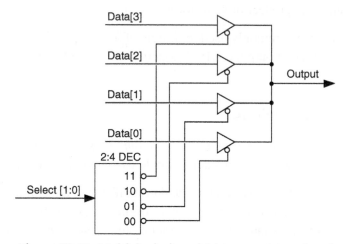

Figure 10.61: Multiple devices driving a common signal

When the ~enable input is presented with a 0 (its active state), the outputs of the OR and NOR gates are determined by the value on the data input. The circuit is arranged so that only one of the $Tr_1$ and $Tr_2$ transistors can be ON at any particular time. If the data input is presented with a 1, transistor $Tr_1$ is turned ON, thereby connecting output $y$ to $V_{DD}$ (which equates to logic 1). By comparison, if the data input is presented with a 0, transistor $Tr_2$ is turned ON, thereby connecting output $y$ to $V_{SS}$ (which equates to logic 0).

Tri-state buffers can be used in conjunction with additional control logic to allow the outputs of multiple devices to drive a common signal. For example, consider the simple circuit shown in Figure 10.61.

The use of a 2:4 decoder with active-low outputs ensures that only one of the tri-state buffers is enabled at any time. The enabled buffer will propagate the data on its input to the common output, while the remaining buffers will be forced to their tri-state condition.

With hindsight it now becomes obvious that the standard primitive gates (AND, OR, NAND, NOR, etc.) depend on internal $Z$ states to function (when any transistor is turned OFF, its output effectively goes to a $Z$ state). However, the standard primitive gates are constructed in such a way that at least one of the transistors connected to the output is turned ON, which means that the output of a standard gate is always driving either 0 or 1.

## 10.42 Combinational Versus Sequential Functions

Logic functions are categorized as being either *combinational* (sometimes referred to as *combinatorial*) or sequential. In the case of a *combinational* function, the logic values on that function's outputs are directly related to the current *combination* of values on its inputs. All of the previous example functions have been of this type.

In the case of a sequential function, the logic values on that function's outputs depend not only on its current input values, but also on previous input values. That is, the output values depend on a *sequence* of input values. Because sequential functions remember previous input values, they are also referred to as *memory elements*.

## 10.43 RS Latches

One of the simpler sequential functions is that of an RS latch, which can be implemented using two NOR gates connected in a back-to-back configuration (Figure 11.8). In this NOR implementation, both reset and set inputs are active-high as indicated by the lack of bobbles associated with these inputs on the symbol. The names of these inputs reflect the effect they have on the $q$ output; when reset is active $q$ is reset to 0, and when set is active $q$ is set to 1.

The $q$ and $\sim q$ outputs are known as the *true* and *complementary* outputs, respectively. In the latch's normal mode of operation, the value on $\sim q$ is the inverse, or complement, of the value on $q$. This is also indicated by the bobble associated with the $\sim q$ output on the symbol. The only time $\sim q$ is not the inverse of $q$ occurs when both reset and set are active at the same time (this unstable state is discussed in more detail below).

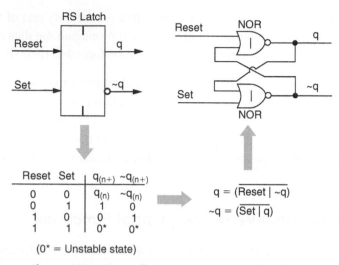

$$q = \overline{(\text{Reset} \mid \sim q)}$$

$$\sim q = \overline{(\text{Set} \mid q)}$$

(0* = Unstable state)

**Figure 10.62: NOR implementation of an RS latch**

The truth table column labels $q_{(n+)}$ and $\sim q_{(n+)}$ indicate that these columns refer to the future values on the outputs. The $n+$ subscripts represent some future time, or "now-plus." By comparison, the labels $q_{(n)}$ and $\sim q_{(n)}$ used in the body of the truth table indicate the current values on the outputs. In this case the $n$ subscripts represent the current time, or "now." Thus, the first row in the truth table indicates that when both reset and set are in their inactive states (logic 0s), the future values on the outputs will be the same as their current values.

The secret of the RS latch's ability to remember previous input values is based on a technique known as *feedback*. This refers to the feeding back of the outputs as additional inputs into the function. In order to see how this works, let's assume that both the reset and set inputs are initially in their inactive states, but that some previous input sequence placed the latch in its *set condition*; that is, $q$ is 1 and $\sim q$ is 0. Now consider what occurs when the reset input is placed in its active state and then returns to its inactive state (Figure 10.63).

As a reminder, if *any* input to a NOR is 1, its output will be forced to 0, and it's only if *both* inputs to a NOR are 0 that the output will be 1. Thus, when reset is placed in its active (logic 1) state ⬜1⬜, the $q$ output from the first gate is forced to 0 ⬜2⬜. This 0 on

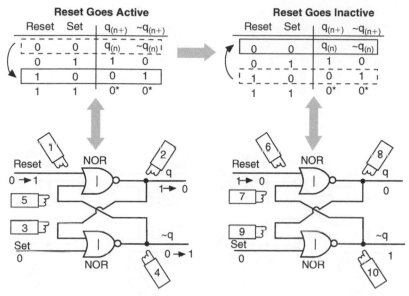

**Figure 10.63: RS latch: reset input goes active then inactive**

$q$ is fed back into the second gate $\boxed{3}$ and, as *both* inputs to this gate are now 0, the $\sim q$ output is forced to 1 $\boxed{4}$. The key point to note is that the 1 on $\sim q$ is now fed back into the first gate $\boxed{5}$.

When the reset input returns to its inactive (logic 0) state $\boxed{6}$, the 1 from the $\sim q$ output continues feeding back into the first gate $\boxed{7}$, which means that the $q$ output continues to be forced to 0 $\boxed{8}$. Similarly, the 0 on $q$ continues feeding back into the second gate $\boxed{9}$, and as both of this gate's inputs are now at 0, the $\sim q$ output continues to be forced to 1 $\boxed{10}$. The end result is that the 1 from $\boxed{7}$ causes the 0 at $\boxed{8}$ which is fed back to $\boxed{9}$, and the 0 on the set input combined with the 0 from $\boxed{9}$ causes the 1 at $\boxed{10}$ which is fed back to $\boxed{7}$.

Thus, the latch has now been placed in its *reset condition*, and a self-sustaining loop has been established. Even though both the reset and set inputs are now inactive, the $q$ output remains at 0, indicating that reset was the last input to be in its active state. Once the function has been placed in its *reset condition*, any subsequent activity on the reset input will have no effect on the outputs, which means that the only way to affect the function is by means of its set input.

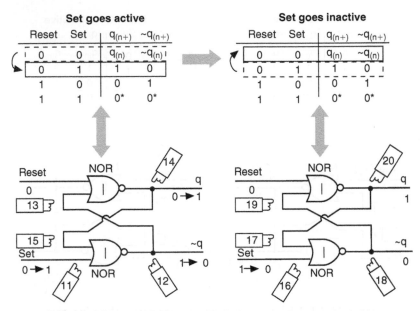

**Figure 10.64: RS latch: set input goes active then inactive**

Now consider what occurs when the set input is placed in its active state and then returns to its inactive state (Figure 10.64).

When set is placed in its active (logic 1) state ⌊11⌋☞, the ~q output from the second gate is forced to 0 ⌊12⌋☞. This 0 on ~q is fed back into the first gate ⌊13⌋☞ and, as both inputs to this gate are now 0, the q output is forced to 1 ⌊14⌋☞. The key point to note is that the 1 on q is now fed back into the second gate ⌊15⌋☞.

When the set input returns to its inactive (logic 0) state ⌊16⌋☞, the 1 from the q output continues feeding back to the second gate ⌊17⌋☞ and the ~q output continues to be forced to 0 ⌊18⌋☞. Similarly, the 0 on the ~q output continues feeding back into the first gate ⌊19⌋☞, and the q output continues to be forced to 1 ⌊20⌋☞. The end result is that the 1 at ⌊17⌋☞ causes the 0 at ⌊18⌋☞ which is fed back to ⌊19⌋☞, and the 0 on the reset input combined with the 0 at ⌊19⌋☞ causes the 1 at ⌊20⌋☞ which is fed back to ⌊17⌋☞.

Thus, the latch has been returned to its *set condition* and, once again, a self-sustaining loop has been established. Even though both the reset and set inputs are now inactive, the q output remains at 1, indicating that set was the last input to be in its active state. Once

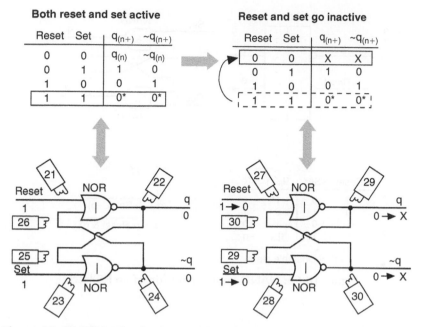

**Figure 10.65: RS latch: the reset and set inputs go inactive simultaneously**

the function has been placed in its *set condition*, any subsequent activity on the set input will have no effect on the outputs, which means that the only way to affect the function is by means of its reset input.

The unstable condition indicated by the fourth row of the RS latch's truth table occurs when both the reset and set inputs are active at the same time. Problems occur when both reset and set return to their inactive states simultaneously or too closely together (Figure 10.65).

When both reset and set are active at the same time, the 1 on reset [21] forces the $q$ output to 0 [22] and the 1 on set [23] forces the $\sim q$ output to 0 [24]. The 0 on $q$ is fed back to the second gate [25], and the 0 on $\sim q$ is fed back to the first gate [26].

Now consider what occurs when reset and set go inactive simultaneously ([27] and [28], respectively). When the new 0 values on reset and set are combined with the 0 values fed back from $q$ [29] and $\sim q$ [30], each gate initially sees both of its inputs at 0

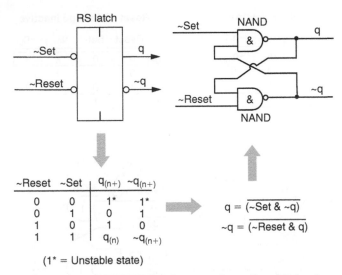

(1* = Unstable state)

**Figure 10.66: NAND implementation of an RS latch**

and therefore both gates attempt to drive their outputs to 1. After any delays associated with the gates have been satisfied, both of the outputs will indeed go to 1.

When the output of the first gate goes to 1, this value is fed back to the input of the second gate. While this is happening, the output of the second gate goes to 1, and this value is fed back to the input of the first gate. Each gate now has its fed-back input at 1, and both gates therefore attempt to drive their outputs to 0. As we see, the circuit has entered a *metastable* condition in which the outputs oscillate between 0 and 1 values.

If both halves of the function were exactly the same, these metastable oscillations would continue indefinitely. But there will always be some differences (no matter how small) between the gates and their delays, and the function will eventually collapse into either its *reset condition* or its *set condition*. As there is no way to predict the final values on the $q$ and $\sim q$ outputs, they are indicated as being in X, or *don't know*, states ( 29  and  30 ). These X states will persist until a valid input sequence occurs on either the reset or set inputs.

An alternative implementation for an RS latch can be realized using two NAND gates connected in a back-to-back configuration (Figure 10.66).

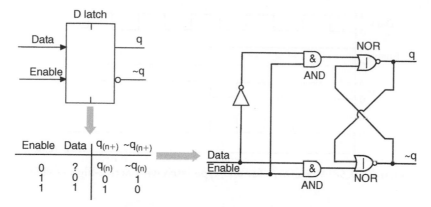

**Figure 10.67: D-type latch with active-high enable**

In a NAND implementation, both the ~reset and ~set inputs are active low, as is indicated by the bobbles associated with these inputs on the symbol and by the tilde characters in their names. As a reminder, if *any* input to a NAND is 0, the output is 1, and it's only if *both* inputs to a NAND are 1 that the output will be 0. Working out how this version of the latch works is left as an exercise to the reader.

## 10.44 D-Type Latches

A more sophisticated function called a *D-type ("data-type") latch* can be constructed by attaching two ANDs and a NOT to the front of an RS latch (Figure 10.67).

The enable input is active high for this configuration, as is indicated by the lack of a bobble on the symbol. When enable is placed in its active (logic 1) state, the true and inverted versions of the data input are allowed to propagate through the AND gates and are presented to the back-to-back NOR gates. If the data input changes while enable is still active, the outputs will respond to reflect the new value.

When enable returns to its inactive (logic 0) state, it forces the outputs of both ANDs to 0, and any further changes on the data input have no effect. Thus, the back-to-back NOR gates remember the last value they saw from the data input prior to the enable input going inactive.

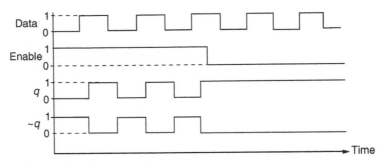

**Figure 10.68: Waveform for a D-type latch with active-high enable**

Consider an example waveform (Figure 10.68). While the enable input is in its active state, whatever value is presented to the data input appears on the $q$ output and an inverted version appears on the $\sim q$ output. As usual, there will always be some element of delay between changes on the inputs and corresponding responses on the outputs. When enable goes inactive, the outputs remember their previous values and no longer respond to any changes on the data input. As the operation of the device depends on the logic value, or level, on enable, this input is said to be *level-sensitive*.

## 10.45 D-Type Flip-Flops

In the case of a D-type flip-flop (which may also be referred to as a *register*), the data *appears* to be loaded when a transition, or edge, occurs on the clock input, which is therefore said to be *edge-sensitive* (the reason we say *"appears to be loaded when an edge occurs"* is discussed in the sidebar on the next page). A transition from 0 to 1 is known as a *rising-edge* or a *positive-edge*, while a transition from 1 to 0 is known as a *falling-edge* or a *negative-edge*. A D-type flip-flop's clock input may be positive-edge or negative-edge triggered (Figure 10.69).

The chevrons (arrows ">") associated with the clock inputs on the symbols indicate that these are edge-sensitive inputs. A chevron without an associated bobble indicates a positive-edge clock, and a chevron with a bobble indicates a negative-edge clock. The last rows in the truth tables show that an inactive edge on the clock

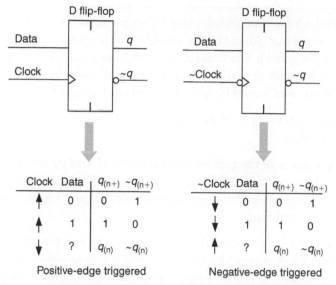

**Figure 10.69: Positive-edge and negative-edge D-type flip-flops**

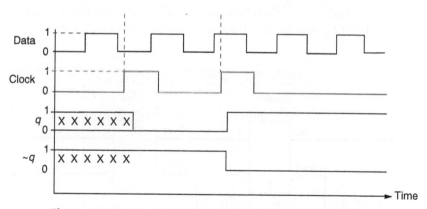

**Figure 10.70: Waveform for positive-edge D-type flip-flop**

leaves the contents of the flip-flops unchanged (these cases are often omitted from the truth tables).

Consider an example waveform for a positive-edge triggered D-type flip-flop (Figure 10.70). As the observer initially has no knowledge as to the contents of the flop-flop, the $q$ and $\sim q$ outputs are initially shown as having X, or *don't know*, values.

There are a number of ways to implement a D-type flip-flop. The most understandable from our point of view would be to use two D-type latches in series (one after the other). The first latch could have an active-low enable and the second could have an active-high enable. Both of these enables would be connected together, and would be known as the *clock input* to the outside world.

This is known as a *master-slave* relationship, where the first latch is the "master" and the second is the "slave."

When the clock input is 0, the master latch is enabled and passes whatever value is presented to its data input through to its outputs (only its *q* output is actually used in this example). Meanwhile, the slave latch is disabled and continues to store (and output) its existing contents.

When the clock input is subsequently driven to a 1, the master latch is disabled and continues to store (and output) its existing contents. Meanwhile the slave latch is now enabled and passes whatever value is presented to *its* data input (the value from the output of the master latch) through to *its* outputs.

Thus, everything is really controlled by voltage *levels*, but from the outside world it *appears* that the flip-flop was loaded by a *rising-edge* on the clock input.

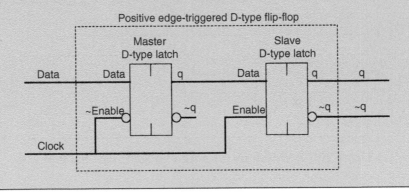

The first rising edge of the clock loads the 0 on the data input into the flip-flop, which (after a small delay) causes *q* to change to 0 and ~*q* to change to 1. The second rising edge of the clock loads the 1 on the data input into the flip-flop; *q* goes to 1 and ~*q* goes to 0.

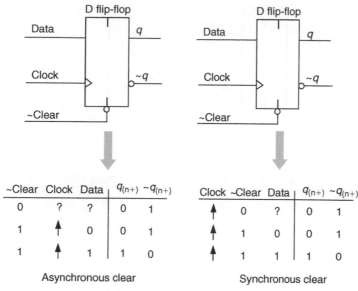

**Figure 10.71: D-type flip-flops with asynchronous and synchronous clear inputs**

Some flip-flops have an additional input called ~clear or ~reset which forces $q$ to 0 and ~$q$ to 1, irrespective of the value on the data input (Figure 10.71). Similarly, some flip-flops have a ~preset or ~set input, which forces $q$ to 1 and ~$q$ to 0, and some have both ~clear and ~preset inputs.

The examples shown in Figure 10.71 reflect active-low ~clear inputs, but active-high equivalents are also available. Furthermore, as is illustrated in Figure 10.71, these inputs may be either *asynchronous* or *synchronous*. In the more common asynchronous case, the effect of ~clear going active is immediate and overrides both the clock and data inputs (the "asynchronous" qualifier reflects the fact that the effect of this input is *not* synchronized to the clock). By comparison, in the synchronous case the effect of ~clear *is* synchronized to the active edge of the clock.

## 10.46 JK and T Flip-Flops

The majority of examples in this book are based on D-type flip-flops. However, for the sake of completeness, it should be noted that there are several other flavors of flip-flops available. Two common types are the JK and T (for Toggle) flip-flops (Figure 11.18).

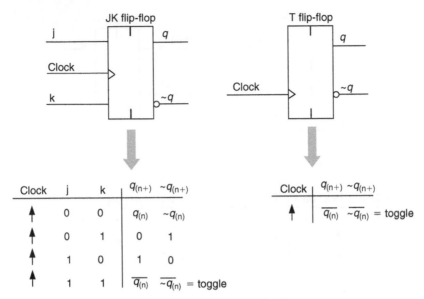

**Figure 10.72: JK and T flip-flops**

The first row of the JK flip-flop's truth table shows that when both the $j$ and $k$ (data) inputs are 0, an active edge on the clock input leaves the contents of the flip-flop unchanged. The two middle rows of the truth table show that if the $j$ and $k$ inputs have opposite values, an active edge on the clock input will effectively load the flip-flop (the $q$ output) with the value on $j$ (the $\sim q$ output will take the complementary value). The last line of the truth table shows that when both the $j$ and $k$ inputs are 1, an active edge on the clock causes the outputs to toggle to the inverse of their previous values. By comparison, the T flip-flop doesn't have any data inputs; the outputs simply toggle to the inverse of their previous values on each active edge of the clock input.

## 10.47  Shift Registers

As was previously noted, another term for a flip-flop is *register*. Functions known as *shift registers*—which facilitate the shifting of binary data one bit at a time—are commonly used in digital systems. Consider a simple 4-bit shift register constructed using D-type flip-flops (Figure 10.73).

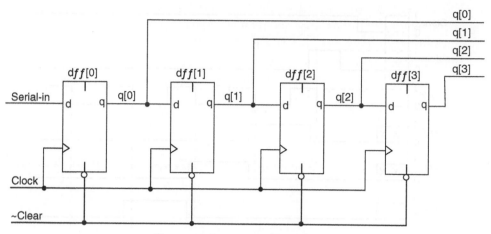

**Figure 10.73: SIPO shift register**

This particular example is based on positive-edge triggered D-type flip-flops with active-low ~*clear* inputs (in this case we're only using each register's *q* output). Also, this example is classed as a *serial-in-parallel-out (SIPO)* shift register, because data is loaded in serially (one after the other) and read out in parallel (side by side).

When the ~*clear* input is set to 1 (its inactive state), a positive-edge on the clock input loads the value on the *serial_in* input into the first flip-flop, dff[0]. At the same time, the value that used to be in dff[0] is loaded into dff[1], the value that used to be in dff[1] is loaded into dff[2], and the value that used to be in dff[2] is loaded into dff[3].

This may seem a bit weird and wonderful the first time you see it, but the way in which this works is actually quite simple (and of course capriciously cunning). Each flip-flop exhibits a delay between seeing an active edge on its clock input and the ensuing response on its *q* output. These delays provide sufficient time for the next flip-flop in the chain to load the value from the previous stage before that value changes. Consider an example waveform where a single logic 1 value is migrated through the shift register (Figure 10.74).

Initially all of the flip-flops contain don't know X values. When the ~*clear* input goes to its active state (logic 0), all of the flip-flops are cleared to 0. When the first active edge occurs on the clock input, the *serial_in* input is 1, so this is the value that's loaded into

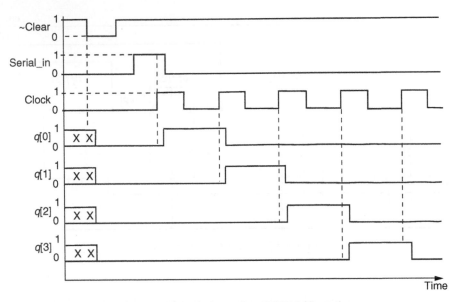

**Figure 10.74: Waveform for SIPO shift register**

the first flip-flop. At the same time, the original 0 value from the first flip-flop is loaded into the second, the original 0 value from the second flip-flop is loaded into the third, and the original 0 value from the third flip-flop is loaded into the fourth.

When the next active edge occurs on the clock input, the *serial_in* input is 0, so this is the value that's loaded into the first flip-flop. At the same time, the original 1 value from the first flip-flop is loaded into the second, the 0 value from the second flip-flop is loaded into the third, and the 0 value from the third flip-flop is loaded into the fourth.

Similarly, when the next active edge occurs on the clock input, the *serial_in* input is still 0, so this is the value that's loaded into the first flip-flop. At the same time, the 0 value from the first flip-flop is loaded into the second, the 1 value from the second flip-flop is loaded into the third, and the 0 value from the third flip-flop is loaded into the fourth. And so it goes . . .

Other common shift register variants are the *parallel-in-serial-out (PISO)*, and the *serial-in-serial-out (SISO)*; for example, consider a 4-bit SISO shift register (Figure 10.75).

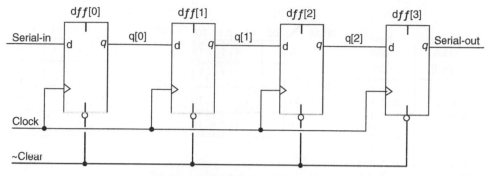

**Figure 10.75: SISO shift register**

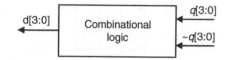

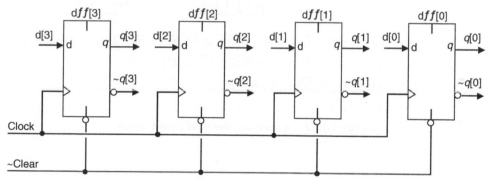

**Figure 10.76: Modulo-16 binary counter**

## 10.48 Counters

Counter functions are also commonly used in digital systems. The number of states that the counter will sequence through before returning to its original value is called the *modulus* of the counter. For example, a function that counts from 0000₂ to 1111₂ in binary (or 0 to 15 in decimal) has a modulus of sixteen and would be called a *modulo-16*, or *mod-16*, counter. Consider a modulo-16 counter implemented using D-type flip-flops (Figure 10.76).

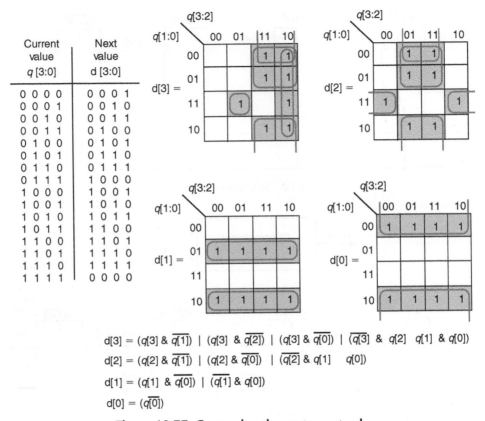

| Current value q [3:0] | Next value d [3:0] |
|---|---|
| 0 0 0 0 | 0 0 0 1 |
| 0 0 0 1 | 0 0 1 0 |
| 0 0 1 0 | 0 0 1 1 |
| 0 0 1 1 | 0 1 0 0 |
| 0 1 0 0 | 0 1 0 1 |
| 0 1 0 1 | 0 1 1 0 |
| 0 1 1 0 | 0 1 1 1 |
| 0 1 1 1 | 1 0 0 0 |
| 1 0 0 0 | 1 0 0 1 |
| 1 0 0 1 | 1 0 1 0 |
| 1 0 1 0 | 1 0 1 1 |
| 1 0 1 1 | 1 1 0 0 |
| 1 1 0 0 | 1 1 0 1 |
| 1 1 0 1 | 1 1 1 0 |
| 1 1 1 0 | 1 1 1 1 |
| 1 1 1 1 | 0 0 0 0 |

$$d[3] = (q[3] \& \overline{q[1]}) \mid (q[3] \& \overline{q[2]}) \mid (q[3] \& \overline{q[0]}) \mid (\overline{q[3]} \& q[2] \; q[1] \& q[0])$$

$$d[2] = (q[2] \& \overline{q[1]}) \mid (q[2] \& \overline{q[0]}) \mid (\overline{q[2]} \& q[1] \quad q[0])$$

$$d[1] = (q[1] \& \overline{q[0]}) \mid (\overline{q[1]} \& q[0])$$

$$d[0] = (q\overline{[0]})$$

**Figure 10.77: Generating the next count value**

This particular example is based on positive-edge triggered D-type flip-flops with active-low ~*clear* inputs. The four flip-flops are used to store the current count value which is displayed on the q[3:0] outputs. When the ~*clear* input is set to 1 (its inactive state), a positive-edge on the clock input causes the counter to load the next value in the count sequence.

A block of combinational logic is used to generate the next value, d[3:0], which is based on the current value q[3:0] (Figure 10.77). Note that there is no need to create the inverted versions of q[3:0], because these signals are already available from the flip-flops as ~q[3:0].

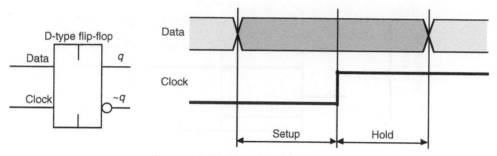

**Figure 10.78: Setup and hold times**

## 10.49 Setup and Hold Times

One point we've glossed over thus far is the fact that there are certain timing requirements associated with flip-flops. In particular, there are two parameters called the *setup* and *hold times*, which describe the relationship between the flip-flop's data and clock inputs (Figure 10.78).

The waveform shown here is a little different to those we've seen before. What we're trying to indicate is that when we start (on the left-hand side), the value presented to the data input may be a 0 or a 1, and it can change back and forth as often as it pleases. However, it must settle one way or the other before the *setup* time; otherwise when the active edge occurs on the clock we can't guarantee what will happen. Similarly, the value presented to the data input must remain stable for the *hold* time following the clock, or once again we can't guarantee what will happen. In our illustration, the period for which the value on the data input must remain stable is shown as being the darker gray.

The setup and hold times shown above are reasonably understandable. However things can sometimes become a little confusing, especially in the case of today's deep submicron (DSM) integrated circuit technologies. The problem is that we may sometimes see so-called *negative* setup and hold times (Figure 10.79).

Once again, the periods for which the value on the data input must remain stable are shown as being the darker gray. These effects, which may seem a little strange at first, are caused by internal delay paths inside the flip-flop.

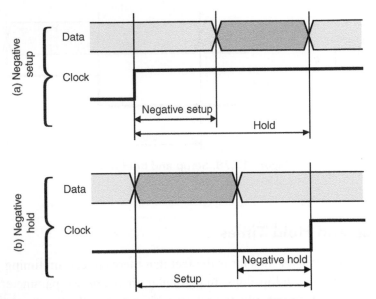

**Figure 10.79: Negative setup and hold times**

Last but not least, we should note that there will also be setup and hold times between the clear (or reset) and preset (or set) inputs and the clock input. Also, there will be corresponding setup and hold times between the data, clear (or reset) and preset (or set) inputs and the enable input on D-type latches (phew!).

## 10.50 Brick by Brick

Let us pause here for a brief philosophical moment. Consider, if you will, a brick formed from clay. Now, there's not a lot you can do with a single brick, but when you combine thousands and thousands of bricks together you can create the most tremendous structures. At the end of the day, the Great Wall of China is no more than a pile of bricks molded by man's imagination.

In the world of the electronics engineer, transistors are the clay, primitive logic gates are the bricks, and the functions described above are simply building blocks. Any digital system, even one as complex as a supercomputer, is constructed from building blocks like comparators, multiplexers, shift registers, and counters. Once you understand the building blocks, there are no ends to the things you can achieve!

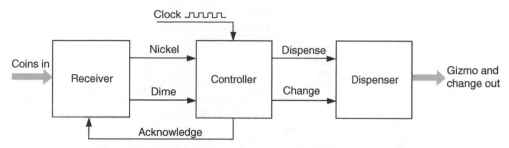

**Figure 10.80: Block diagram of a coin-operated machine**

### 10.50.1 *State Diagrams, State Tables, and State Machines*

Consider a coin-operated machine that accepts nickels and dimes and, for the princely sum of fifteen cents, dispenses some useful article called a "gizmo" that the well-dressed man-about-town could not possibly be without. We may consider such a machine to comprise three main blocks: a *receiver* that accepts money, a *dispenser* that dispenses the "gismo" and any change, and a *controller* that oversees everything and makes sure things function as planned (Figure 10.80).

The connections marked *nickel, dime, dispense, change,* and *acknowledge* represent digital signals carrying logic 0 and 1 values. The user can deposit nickels and dimes into the receiver in any order, but may only deposit one coin at a time. When a coin is deposited, the receiver determines its type and sets the corresponding signal (nickel or dime) to a logic 1.

The operation of the controller is synchronized by the clock signal. On a rising edge of the clock, the controller examines the nickel and dime inputs to see if any coins have been deposited. The controller keeps track of the amount of money deposited and determines if any actions are to be performed.

Every time the controller inspects the nickel and dime signals, it sends an *acknowledge* signal back to the receiver. The *acknowledge* signal informs the receiver that the coin has been accounted for, and the receiver responds by resetting the nickel and dime signals to 0 and awaiting the next coin. The *acknowledge* signal can be generated in a variety of ways which are not particularly relevant here.

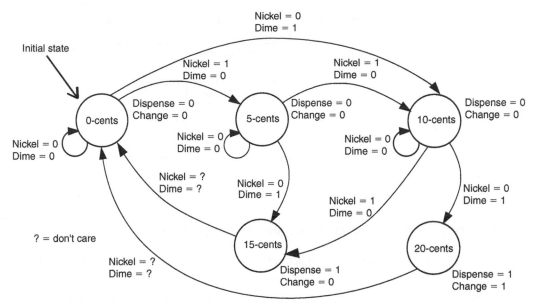

**Figure 10.81: State diagram for the controller**

When the controller decides that sufficient funds have been deposited, it instructs the dispenser to dispense a "gizmo" and any change (if necessary) by setting the *dispense* and *change* signals to 1, respectively.

## 10.51 State Diagrams

A useful level of abstraction for a function such as the controller is to consider it as consisting of a set of *states* through which it sequences. The *current state* depends on the *previous state* combined with the *previous* values on the nickel and dime inputs. Similarly, the *next state* depends on the *current state* combined with the *current* values on the nickel and dime inputs. The operation of the controller may be represented by means of a *state diagram*, which offers a way to view the problem and to describe a solution (Figure 10.81).

The states are represented by the circles labeled 0-cents, 5-cents, 10-cents, 15-cents, and 20-cents, and the values on the dispense and change outputs are associated with these states. The arcs connecting the states are called *state transitions* and the values of the nickel and dime inputs associated with the state transitions are called *guard conditions*. The controller will only sequence between two states if the values on the nickel and dime inputs match the guard conditions.

Let's assume that the controller is in its initial state of *0-cents*. The values of the *nickel* and *dime* inputs are tested on every rising edge on the clock. (The controller is known to sequence between states only on the rising edge of the clock, so displaying this signal on every state transition would be redundant.) As long as no coins are deposited, the *nickel* and *dime* inputs remain at 0 and the controller remains in the *0-cents* state. Once a coin is deposited, the next rising edge on the clock will cause the controller to sequence to the *5-cents* or the *10-cents* states depending on the coin's type. It is at this point that the controller sends an *acknowledge* signal back to the receiver instructing it to reset the *nickel* and *dime* signals back to 0 and to await the next coin.

Note that the *0-cents*, *5-cents*, and *10-cents* states have state transitions that loop back into them (the ones with associated *nickel* = 0 and *dime* = 0 guard conditions). These indicate that the controller will stay in whichever state it is currently in until a new coin is deposited. So at this stage of our discussions, the controller is either in the *5-cents* or the *10-cents* state depending on whether the first coin was a nickel or dime, respectively. What happens when the next coin is deposited? Well this depends on the state we're in and the type of the new coin. If the controller is in the *5-cents* state, then a nickel or dime will move it to the *10-cents* or *15-cents* states, respectively. Alternatively, if the controller is in the *10-cents* state, then a nickel or dime will move it to the *15-cents* or *20-cents* states, respectively.

When the controller reaches either the *15-cents* or *20-cents* states, the next clock will cause it to dispense a "gizmo" and return to its initial *0-cents* state (in the case of the *20-cents* state, the controller will also dispense a nickel in change).

## 10.52 State Tables

Another form of representation is that of a *state table*. This is similar to a truth table (inputs on the left and corresponding outputs on the right), but it also includes the *current state* as an input and the *next state* as an output (Figure 10.82).

| Current state | Clock | Nickel | Dime | Dispense | Change | Next state |
|---|---|---|---|---|---|---|
| 0-cents | ↑ | 0 | 0 | 0 | 0 | 0-cents |
| 0-cents | ↑ | 1 | 0 | 0 | 0 | 5-cents |
| 0-cents | ↑ | 0 | 1 | 0 | 0 | 10-cents |
| 5-cents | ↑ | 0 | 0 | 0 | 0 | 5-cents |
| 5-cents | ↑ | 1 | 0 | 0 | 0 | 10-cents |
| 5-cents | ↑ | 0 | 1 | 0 | 0 | 15-cents |
| 10-cents | ↑ | 0 | 0 | 0 | 0 | 10-cents |
| 10-cents | ↑ | 1 | 0 | 0 | 0 | 15-cents |
| 10-cents | ↑ | 0 | 1 | 0 | 0 | 20-cents |
| 15-cents | ↑ | ? | ? | 1 | 0 | 0-cents |
| 20-cents | ↑ | ? | ? | 1 | 1 | 0-cents |

**Figure 10.82: State table for the controller**

In this instance the clock signal has been included for purposes of clarity (it's only when there's a rising edge on the clock that the outputs are set to the values shown in that row of the table). However, as for the state diagram, displaying this signal is somewhat redundant and it is often omitted.

## 10.53 State Machines

The actual implementation of a function such as the controller is called a *state machine*. In fact, when the number of states is constrained and finite, this is more usually called a *finite state machine (FSM)*. The heart of a state machine consists of a set of registers known as the *state variables*. Each state, *0-cents*, *5-cents*, *10-cents*, . . . is assigned a unique binary pattern of 0s and 1s, and the pattern representing the *current state* is stored in the state variables.

The two most common forms of synchronous, or clocked, state machines are known as *Moore* and *Mealy* machines after the men who formalized them. A Moore machine is distinguished by the fact that the outputs are derived only from the values in the state variables (Figure 10.83). The controller function featured in this discussion is a classic example of a Moore machine.

By comparison, the outputs from a Mealy machine may be derived from a combination of the values in the state variables and one or more of the inputs (Figure 10.84).

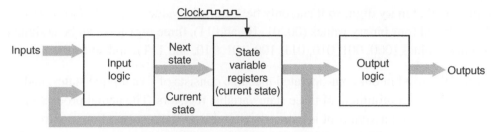

**Figure 10.83: Block diagram of a Moore machine**

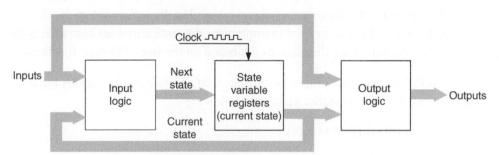

**Figure 10.84: Block diagram of a Mealy machine**

In both of the Moore and Mealy forms, the input logic consists of primitive gates such as AND, NAND, OR, and NOR. These combine the values on the inputs with the *current state* (which is fed back from the state variables) to generate the pattern of 0s and 1s representing the *next state*. This new pattern of 0s and 1s is presented to the inputs of the state variables and will be loaded into them on the next rising edge of the clock.

The output logic also consists of standard primitive logic gates that generate the appropriate values on the outputs from the *current state* stored in the state variables.

## 10.54 State Assignment

A key consideration in the design of a state machine is that of *state assignment*, which refers to the process by which the states are assigned to the binary patterns of 0s and 1s that are to be stored in the state variables. A common form of state assignment requiring the minimum number of registers is known as *binary encoding*. Each register can only contain a single binary digit, so it can only be assigned a value of 0 or 1. Two registers can be assigned four binary values (00, 01, 10, and 11), three registers can be assigned eight binary values (000, 001, 010, 011, 100, 101, 110, and 111), and so forth.

The controller used in our coin-operated machine consists of five unique states, and therefore requires a minimum of three state variable registers. The actual process of binary encoded state assignment is a nontrivial problem. In the case of our controller function, there are 6,720 possible combinations by which five states can be assigned to the eight binary values provided by three registers. Each of these solutions may require a different arrangement of primitive gates to construct the input and output logic, which in turn affects the maximum frequency that can be used to drive the system clock. Additionally, the type of registers used to implement the state variables also affects the supporting logic; the following discussions are based on the use of D-type flip-flops.

Assuming that full use is made of *don't care* states, an analysis of the various binary encoded solutions for our controller yields the following . . .

138 solutions requiring 7 product terms

852 solutions requiring 8 product terms

1,876 solutions requiring 9 product terms

3,094 solutions requiring 10 product terms

570 solutions requiring 11 product terms

190 solutions requiring 12 product terms

. . . where a *product term* is a group of *literals* linked by & (AND) operators—for example, (*a* & *b* & *c*)—and a *literal* is any true or inverted variable. Thus, the product term (*a* & *b* & *c*) contains three literals (*a*, *b*, and *c*).

But wait, there's more! A further analysis of the 138 solutions requiring seven product terms yields the following:

66 solutions requiring 17 literals

24 solutions requiring 18 literals

48 solutions requiring 19 literals

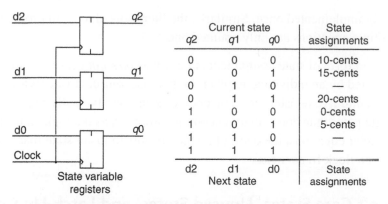

| | Current state | | | State |
|---|---|---|---|---|
| | q2 | q1 | q0 | assignments |
| | 0 | 0 | 0 | 10-cents |
| | 0 | 0 | 1 | 15-cents |
| | 0 | 1 | 0 | — |
| | 0 | 1 | 1 | 20-cents |
| | 1 | 0 | 0 | 0-cents |
| | 1 | 0 | 1 | 5-cents |
| | 1 | 1 | 0 | — |
| | 1 | 1 | 1 | — |
| | d2 | d1 | d0 | State |
| | | Next state | | assignments |

**Figure 10.85: Example binary encoded state assignment**

Thus, the chances of a random assignment resulting in an optimal solution is relatively slight. Fortunately, there are computer programs available to aid designers in this task. One solution resulting in the minimum number of product terms and literals is shown in Figure 10.85.

A truth table for the controller function can now be derived from the state table shown in Figure 10.82 by replacing the assignments in the *current state* column with the corresponding binary patterns for the state variable outputs ($q2$, $q1$, and $q0$), and replacing the assignments in the *next state* column with the corresponding binary patterns for the state variable inputs ($d2$, $d1$, and $d0$). The resulting equations can then be derived from the truth table by means of standard algebraic or Karnaugh map techniques. As an alternative, a computer program can be used to obtain the same results in less time with far fewer opportunities for error. Whichever technique is employed, the state assignments above lead to the following minimized Boolean equations:

$$d0 = (\overline{q0}\ \&\ \overline{q2}\ \&\ \textbf{dime})\ |\ (q0\ \&\ q2\ \&\ \overline{\text{nickel}})\ |\ (\overline{q0}\ \&\ \text{nickel})$$
$$d1 = (\overline{q0}\ \&\ \overline{q2}\ \&\ \textbf{dime})$$
$$d2 = (\overline{q0}\ \&\ \overline{q2})\ |\ (q2\ \&\ \overline{\text{nickel}}\ \&\ \overline{\text{dime}})\ |\ (\overline{q0}\ \&\ q2\ \&\ \overline{\text{dime}})$$
$$\text{Dispense} = (q0\ \&\ \overline{q2})$$
$$\text{Change} = (q1)$$

The product terms shown in bold appear in multiple equations. However, regardless of the number of times a product term appears, it is only counted once because it only has to be physically implemented once. Similarly, the literals used to form product terms that appear in multiple equations are only counted once.

Another common form of state assignment is known as *one-hot encoding*, in which each state is represented by an individual register. In this case, our controller with its five states would require five register bits. The one-hot technique typically requires a greater number of logic gates than does binary encoding. However, as the logic gates are used to implement simpler equations, the one-hot method results in faster state machines that can operate at higher clock frequencies.

## 10.55 Don't Care States, Unused States, and Latch-Up Conditions

It was previously noted that the analysis of the binary encoded state assignment made full use of *don't care* states. This allows us to generate a solution that uses the least number of logic gates, but there are additional considerations that must now be discussed in more detail.

The original definition of our coin-operated machine stated that it is only possible for a single coin to be deposited at a time. Assuming this to be true, then the nickel and dime signals will never be assigned 1 values simultaneously. The designer (or a computer program) can use this information to assign *don't care* states to the outputs for any combination of inputs that includes a 1 on both nickel and dime signals.

Additionally, the three binary encoded state variable registers provide eight possible binary patterns, of which only five were used. The analysis above was based on the assumption that *don't care* states can be assigned to the outputs for any combination of inputs that includes one of the unused patterns on the state variables. This assumption also requires further justification.

When the coin-operated machine is first powered-up, each state variable register can potentially initialize with a random logic 0 or 1 value. The controller could therefore power-up with its state variables containing any of the eight possible patterns of 0s and 1s. For some state machines this would not be an important consideration, but this is not true in the case of our coin-operated machine. For example, the controller could power-up in the *20-cents* state, in which case it would immediately dispense a "gizmo" and five cents change. The owner of such a machine may well be of the opinion that this was a less than ideal feature.

Alternatively, the controller could power-up with its state variables in one of the unused combinations. Subsequently, the controller could sequence directly—or via one or more of the other unused combinations—to any of the defined states. In a worst-case scenario, the controller could remain in the unused combination indefinitely or sequence endlessly between unused combinations; these worst-case scenarios are known as *latch-up conditions*.

One method of avoiding latch-up conditions is to assign additional, dummy states to each of the unused combinations and to define state transitions from each of these dummy states to the controller's initialization state of *0-cents*. Unfortunately, in the case of our coin-operated machine, this technique would not affect the fact that the controller could wake up in a valid state other than 0-cents. An alternative is to provide some additional circuitry to generate a *power-on reset* signal—for example, a single pulse that occurs only when the power is first applied to the machine. The power-on reset can be used to force the state variable registers into the pattern associated with the *0-cents* state. The analysis above assumed the use of such a power-on-reset.

# CHAPTER 11

# Analog Electronics

Mike Tooley
Tim Williams

The operational amplifier is the basic building block for analog circuits, and progress in op-amp performance is the "litmus test" for analog IC electronics technology in much the same way as progress in memory devices is for digital electronics technology. This chapter will be devoted to op-amps and comparators, with a tailpiece on voltage references.

## 11.1 Operational Amplifiers Defined

Operational amplifiers (Figure 11.1) are analog integrated circuits designed for linear amplification that offer near-ideal characteristics (virtually infinite voltage gain and input resistance coupled with low output resistance and wide bandwidth).

Operational amplifiers can be thought of as universal "gain blocks" to which external components are added in order to define their function within a circuit. By adding two resistors, we can produce an amplifier having a precisely defined gain. Alternatively, with two resistors and two capacitors we can produce a simple band-pass filter. From this you might begin to suspect that operational amplifiers are really easy to use. The good news is that they are!

## 11.2 Symbols and Connections

The symbol for an operational amplifier is shown in Figure 11.2. There are a few things to note about this. The device has two inputs and one output and no common connection. Furthermore, we often don't show the supply connections—it is often clearer to leave them out of the circuit altogether!

In Figure 11.2, one of the inputs is marked "−" and the other is marked "+". These polarity markings have nothing to do with the supply connections—they indicate the

Figure 11.1: A typical operational amplifier. This device is supplied in an 8-pin dual-in-line (DIL) package. It has a JFET input stage and produces a typical open-loop voltage gain of 200,000

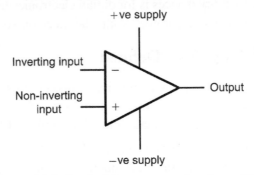

Figure 11.2: Symbol for an operational amplifier

overall phase shift between each input and the output. The "+" sign indicates zero phase shift while the "−" sign indicates 180° phase shift. Since 180° phase shift produces an inverted waveform, the "−" input is often referred to as the *inverting input*. Similarly, the "+" input is known as the *non-inverting* input.

Most (but not all) operational amplifiers require a symmetrical supply (of typically ±6 V to ±15 V) which allows the output voltage to swing both positive (above 0 V) and negative (below 0 V). Figure 11.3 shows how the supply connections would appear if we decided to include them. Note that we usually have two separate supplies; a positive supply and an equal, but opposite, negative supply. The common connection to these two supplies (i.e., the 0 V supply connection) acts as the *common rail* in our circuit. The input and output voltages are usually measured relative to this rail.

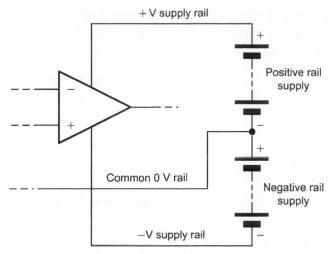

**Figure 11.3: Supply connections for an operational amplifier**

## 11.3 Operational Amplifier Parameters

Before we take a look at some of the characteristics of "ideal" and "real" operational amplifiers it is important to define some of the terms and parameters that we apply to these devices.

### 11.3.1 Open-Loop Voltage Gain

The open-loop voltage gain of an operational amplifier is defined as the ratio of output voltage to input voltage measured with no feedback applied. In practice, this value is exceptionally high (typically greater than 100,000) but is liable to considerable variation from one device to another.

Open-loop voltage gain may thus be thought of as the "internal" voltage gain of the device, thus:

$$A_{V(OL)} = \frac{V_{OUT}}{V_{IN}}$$

where $A_{V(OL)}$ is the open-loop voltage gain, $V_{OUT}$ and $V_{IN}$ are the output and input voltages respectively under open-loop conditions.

In linear voltage amplifying applications, a large amount of negative feedback will normally be applied and the open-loop voltage gain can be thought of as the internal voltage gain provided by the device.

The open-loop voltage gain is often expressed in *decibels (dB)* rather than as a ratio. In this case:

$$A_{V(OL)} = 20 \log_{10} \frac{V_{OUT}}{V_{IN}}$$

Most operational amplifiers have open-loop voltage gains of 90 dB, or more.

### 11.3.2 Closed-Loop Voltage Gain

The closed-loop voltage gain of an operational amplifier is defined as the ratio of output voltage to input voltage measured with a small proportion of the output fed back to the input (i.e., with feedback applied). The effect of providing negative feedback is to reduce the loop voltage gain to a value that is both predictable and manageable. Practical closed-loop voltage gains range from one to several thousand but note that high values of voltage gain may make unacceptable restrictions on bandwidth.

Closed-loop voltage gain is once again the ratio of output voltage to input voltage but with negative feedback is applied, hence:

$$A_{V(CL)} = \frac{V_{OUT}}{V_{IN}}$$

where $A_{V(CL)}$ is the open-loop voltage gain, $V_{OUT}$ and $V_{IN}$ are the output and input voltages respectively under closed-loop conditions. The closed-loop voltage gain is normally very much less than the open-loop voltage gain.

### Example 11.1
An operational amplifier operating with negative feedback produces an output voltage of 2 V when supplied with an input of 400 μV. Determine the value of closed-loop voltage gain.

### Solution

Now:

$$A_{V(CL)} = \frac{V_{out}}{V_{IN}}$$

Thus:

$$A_{V(CL)} = \frac{2}{400 \times 10^{-6}} = \frac{2 \times 10^6}{400} = 5{,}000$$

Expressed in decibels (rather than as a ratio) this is:

$$A_{V(CL)} = 20 \log_{10}(5{,}000) = 20 \times 3.7 = 74\,dB$$

### 11.3.3 Input Resistance

The input resistance of an operational amplifier is defined as the ratio of input voltage to input current expressed in ohms. It is often expedient to assume that the input of an operational amplifier is purely resistive though this is not the case at high frequencies where shunt capacitive reactance may become significant. The input resistance of operational amplifiers is very much dependent on the semiconductor technology employed. In practice values range from about $2\,M\Omega$ for common bipolar types to over $10^{12}\,\Omega$ for FET and CMOS devices.

Input resistance is the ratio of input voltage to input current:

$$R_{IN} = \frac{V_{IN}}{I_{IN}}$$

where $R_{IN}$ is the input resistance (in ohms), $V_{IN}$ is the input voltage (in volts) and $I_{IN}$ is the input current (in amps). Note that we usually assume that the input of an operational amplifier is purely resistive though this may not be the case at high frequencies where shunt capacitive reactance may become significant.

The input resistance of operational amplifiers is very much dependent on the semiconductor technology employed. In practice, values range from about $2\,M\Omega$ for bipolar operational amplifiers to over $10^{12}\,\Omega$ for CMOS devices.

*Example 11.2*

An operational amplifier has an input resistance of $2\,M\Omega$. Determine the input current when an input voltage of $5\,mV$ is present.

*Solution*

Now:

$$R_{IN} = \frac{V_{IN}}{I_{IN}}$$

thus,

$$I_{IN} = \frac{V_{IN}}{R_{IN}} = \frac{5 \times 10^{-3}}{2 \times 10^{6}} = 2.5 \times 10^{-9}\,A = 2.5\,nA$$

### 11.3.4  Output Resistance

The output resistance of an operational amplifier is defined as the ratio of open-circuit output voltage to short-circuit output current expressed in ohms. Typical values of output resistance range from less than $10\,\Omega$ to around $100\,\Omega$ depending upon the configuration and amount of feedback employed.

Output resistance is the ratio of open-circuit output voltage to short-circuit output current, hence:

$$R_{OUT} = \frac{V_{OUT(OC)}}{I_{OUT(SC)}}$$

where $R_{OUT}$ is the output resistance (in ohms), $V_{OUT(OC)}$ is the open-circuit output voltage (in volts) and $I_{OUT(SC)}$ is the short-circuit output current (in amps).

#### 11.3.4.1  Input Offset Voltage

An ideal operational amplifier would provide zero output voltage when $0\,V$ difference is applied to its inputs. In practice, due to imperfect internal balance, there may be some small voltage present at the output. The voltage that must be applied differentially to the operational amplifier input in order to make the output voltage exactly zero is known as the input offset voltage.

Input offset voltage may be minimized by applying relatively large amounts of negative feedback or by using the offset null facility provided by a number of operational amplifier devices. Typical values of input offset voltage range from 1 mV to 15 mV. Where AC rather than DC coupling is employed, offset voltage is not normally a problem and can be happily ignored.

### 11.3.5 Full-Power Bandwidth

The full-power bandwidth for an operational amplifier is equivalent to the frequency at which the maximum undistorted peak output voltage swing falls to 0.707 of its low frequency (DC) value (the sinusoidal input voltage remaining constant). Typical full-power bandwidths range from 10 kHz to over 1 MHz for some high-speed devices.

### 11.3.6 Slew Rate

Slew rate is the rate of change of output voltage with time, when a rectangular step input voltage is applied (as shown in Figure 11.4). The slew rate of an operational amplifier is the rate of change of output voltage with time in response to a perfect step-function input.

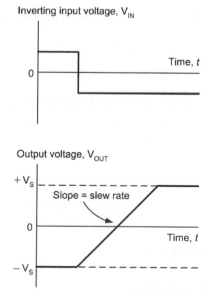

**Figure 11.4: Slew rate for an operational amplifier**

Hence:

$$\text{Slew rate} = \frac{\Delta V_{OUT}}{\Delta t}$$

where $\Delta V_{OUT}$ is the change in output voltage (in volts) and $\Delta t$ is the corresponding interval of time in s).

Slew rate is measured in V/s (or V/μs) and typical values range from 0.2 V/μs to over 20 V/μs. Slew rate imposes a limitation on circuits in which large amplitude pulses rather than small amplitude sinusoidal signals are likely to be encountered.

## 11.4 Operational Amplifier Characteristics

Having now defined the parameters that we use to describe operational amplifiers, we shall now consider the desirable characteristics for an "ideal" operational amplifier. These are:

(a)  The open-loop voltage gain should be very high (ideally infinite).

(b)  The input resistance should be very high (ideally infinite).

(c)  The output resistance should be very low (ideally zero).

(d)  Full-power bandwidth should be as wide as possible.

(e)  Slew rate should be as large as possible.

(f)  Input offset should be as small as possible.

The characteristics of most modern integrated circuit operational amplifiers (i.e., "real" operational amplifiers) come very close to those of an "ideal" operational amplifier, as witnessed by the data shown in Table 11.1.

### Example 11.3
A perfect rectangular pulse is applied to the input of an operational amplifier. If it takes 4 μs for the output voltage to change from −5 V to +5 V, determine the slew rate of the device.

### Solution
The slew rate can be determined from:

$$\text{Slew rate} = \frac{\Delta V_{OUT}}{\Delta t} = \frac{10\,\text{V}}{4\,\mu\text{s}} = 2.5\,\text{V}/\mu\text{s}$$

**Table 11.1: Comparison of operational amplifier
parameters for "ideal" and "real" devices**

| Parameter | Ideal | Real |
|---|---|---|
| Voltage gain | Infinite | 100,000 |
| Input resistance | Infinite | 100 MΩ |
| Output resistance | Zero | 20 Ω |
| Bandwidth | Infinite | 2 MHz |
| Slew-rate | Infinite | 10 V/μs |
| Input offset | Zero | Less than 5 mV |

### Example 11.4

A wideband operational amplifier has a slew rate of 15 V/μs. If the amplifier is used in a circuit with a voltage gain of 20 and a perfect step input of 100 mV is applied to its input, determine the time taken for the output to change level.

### Solution

The output voltage change will be $20 \times 100 = 2{,}000\,\text{mV}$ (or 2 V). Rearranging the formula for slew rate gives:

$$\Delta t = \frac{\Delta V_{\text{OUT}}}{\text{Slew rate}} = \frac{2\,\text{V}}{15\,\text{V/μs}} = 0.133\,\text{μs}$$

## 11.5 Operational Amplifier Applications

Table 11.2 shows abbreviated data for some common types of integrated circuit operational amplifier together with some typical applications.

### Example 11.5

Which of the operational amplifiers in the table would be most suitable for each of the following applications?

(a) amplifying the low-level output from a piezoelectric vibration sensor,

(b) a high-gain amplifier that can be used to faithfully amplify very small signals,

(c) a low-frequency amplifier for audio signals.

Table 11.2: Some common examples of integrated circuit operation

| Device | Type | Open-loop voltage gain (dB) | Input bias current | Slew rate (V/µs) | Application |
|---|---|---|---|---|---|
| AD548 | Bipolar | 100 min. | 0.01 nA | 1.8 | Instrumentation amplifier |
| AD711 | FET | 100 | 25 pA | 20 | Wideband amplifier |
| CA3140 | CMOS | 100 | 5 pA | 9 | Low-noise wideband amplifier |
| LF347 | FET | 110 | 50 pA | 13 | Wideband amplifier |
| LM301 | Bipolar | 88 | 70 nA | 0.4 | General-purpose operational amplifier |
| LM348 | Bipolar | 96 | 30 nA | 0.6 | General-purpose operational amplifier |
| TL071 | FET | 106 | 30 pA | 13 | Wideband amplifier |
| 741 | Bipolar | 106 | 80 nA | 0.5 | General-purpose operational amplifier |

*Solution*

(a) AD548 (this operational amplifier is designed for use in instrumentation applications and it offers a very low input offset current which is important when the input is derived from a piezoelectric transducer).

(b) CA3140 (this is a low-noise operational amplifier that also offers high gain and fast slew rate).

(c) LM348 or LM741 (both are general-purpose operational amplifiers and are ideal for non-critical applications such as audio amplifiers).

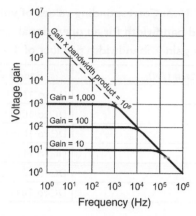

**Figure 11.5: Frequency response curves for an operational amplifier**

## 11.6  Gain and Bandwidth

It is important to note that, since the product of gain and bandwidth is a constant for any particular operational amplifier. Hence, an increase in gain can only be achieved at the expense of bandwidth, and vice versa.

Figure 11.5 shows the relationship between voltage gain and bandwidth for a typical operational amplifier (note that the axes use logarithmic, rather than linear scales). The open-loop voltage gain (i.e., that obtained with no feedback applied) is 100,000 (or 100 dB) and the bandwidth obtained in this condition is a mere 10 Hz. The effect of applying increasing amounts of negative feedback (and consequently reducing the gain to a more manageable amount) is that the bandwidth increases in direct proportion.

The frequency response curves in Figure 11.5 show the effect on the bandwidth of making the closed-loop gains equal to 10,000, 1,000, 100, and 10. Table 11.3 summarizes these results. You should also note that the (gain $\times$ bandwidth) product for this amplifier is $1 \times 10^6$ Hz (i.e., 1 MHz).

We can determine the bandwidth of the amplifier when the closed-loop voltage gain is set to 46 dB by constructing a line and noting the intercept point on the response curve. This shows that the bandwidth will be 10 kHz. Note that, for this operational amplifier, the (gain $\times$ bandwidth) product is $2 \times 10^6$ Hz (or 2 MHz).

**Table 11.3: Corresponding values of voltage
gain and bandwidth for an operational amplifier
with a gain $\times$ bandwidth product of $1 \times 10^6$**

| Voltage gain (Av) | Bandwidth |
|---|---|
| 1 | DC to 1 MHz |
| 10 | DC to 100 kHz |
| 100 | DC to 10 kHz |
| 1,000 | DC to 1 kHz |
| 10,000 | DC to 100 Hz |
| 100,000 | DC to 10 Hz |

## 11.7  Inverting Amplifier With Feedback

Figure 11.6 shows the circuit of an inverting amplifier with negative feedback applied.
For the sake of our explanation we will assume that the operational amplifier is "ideal."
Now consider what happens when a small positive input voltage is applied. This voltage
($V_{IN}$) produces a current ($I_{IN}$) flowing in the input resistor $R1$.

Since the operational amplifier is "ideal" we will assume that:

(a)  the input resistance (i.e., the resistance that appears between the inverting and
non-inverting input terminals, $R_{IC}$) is infinite, and

(b)  the open-loop voltage gain (i.e., the ratio of $V_{OUT}$ to $V_{IN}$ with no feedback
applied) is infinite.

As a consequence of (a) and (b):

(i)  the voltage appearing between the inverting and non-inverting inputs ($V_{IC}$) will
be zero, and

(ii)  the current flowing into the chip ($I_{IC}$) will be zero (recall that $I_{IC} = V_{IC}/R_{IC}$ and
$R_{IC}$ is infinite).

Applying Kirchhoff's Current Law at node A gives:

$$I_{IN} = I_{IC} + I_F \text{ but } I_{IC} = 0 \text{ thus } I_{IN} = I_F \qquad (11.1)$$

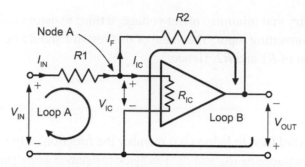

**Figure 11.6: Operational amplifier with negative feedback applied**

(This shows that the current in the feedback resistor, $R2$, is the same as the input current, $I_{IN}$).

Applying Kirchhoff's Voltage Law to loop A gives:

$$V_{IN} = (I_{IN} \times R1) + V_{IC}$$

but $V_{IC} = 0$ thus $V_{IN} = I_{IN} \times R1$                 (11.2)

Using Kirchhoff's Voltage Law in loop B gives:

$$V_{OUT} = -V_{IC} + (I_F \times R2)$$

but $V_{IC} = 0$ thus $V_{OUT} = I_F \times R2$            (11.3)

Combining (11.1) and (11.3) gives:

$$V_{OUT} = I_{IN} \times R2 \qquad (11.4)$$

The voltage gain of the stage is given by:

$$A_v = \frac{V_{OUT}}{V_{IN}} \qquad (11.5)$$

Combining (11.4) and (11.2) with (11.5) gives:

$$A_v = \frac{I_{IN} \times R2}{I_{IN} \times R1} = \frac{R2}{R1}$$

To preserve symmetry and minimize offset voltage, a third resistor is often included in series with the non-inverting input. The value of this resistor should be equivalent to the parallel combination of $R1$ and $R2$. Hence:

$$R3 = \frac{R1 \times R2}{R1 + R2}$$

From this point onward (and to help you remember the function of the resistors), we shall refer to the input resistance as $R_{IN}$ and the feedback resistance as $R_F$ (instead of the more general and less meaningful $R1$ and $R2$, respectively).

## 11.8 Operational amplifier configurations

The three basic configurations for operational voltage amplifiers, together with the expressions for their voltage gain, are shown in Figure 11.7. Supply rails have been omitted from these diagrams for clarity but are assumed to be symmetrical about 0 V.

All of the amplifier circuits described previously have used direct coupling and thus have frequency response characteristics that extend to DC. This, of course, is undesirable for many applications, particularly where a wanted AC signal may be superimposed on an unwanted DC voltage level or when the bandwidth of the amplifier greatly exceeds that of the signal that it is required to amplify. In such cases, capacitors of appropriate value may be inserted in series with the input resistor, $R_{IN}$, and in parallel with the feedback resistor, $R_F$, as shown in Figure 11.8.

The value of the input and feedback capacitors, $C_{IN}$ and $C_F$, respectively, are chosen so as to roll-off the frequency response of the amplifier at the desired lower and upper cut-off frequencies, respectively. The effect of these two capacitors on an operational amplifier's frequency response is shown in Figure 11.9.

By selecting appropriate values of capacitor, the frequency response of an inverting operational voltage amplifier may be very easily tailored to suit a particular set of requirements.

The lower cut-off frequency is determined by the value of the input capacitance, $C_{IN}$, and input resistance, $R_{IN}$. The lower cut-off frequency is given by:

$$f_1 = \frac{1}{2\pi C_{IN} R_{IN}} = \frac{0.159}{C_{IN} R_{IN}}$$

where $f_1$ is the lower cut-off frequency in Hz, $C_{IN}$ is in farads and $R_{IN}$ is in ohms.

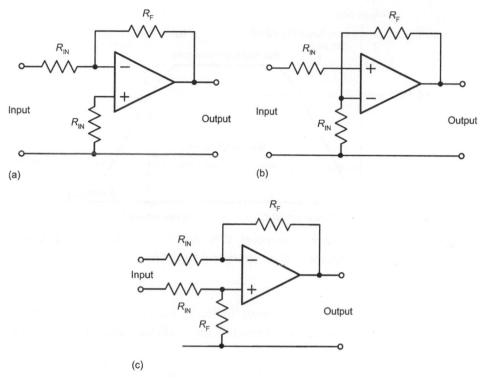

Figure 11.7: The three basic configurations for operational voltage amplifiers. (a) Inverting amplifier; (b) Non-inverting amplifier; (c) Differential amplifier.

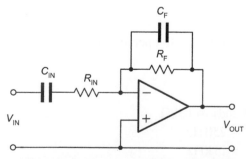

Figure 11.8: Adding capacitors to modify the frequency response of an inverting operational amplifier

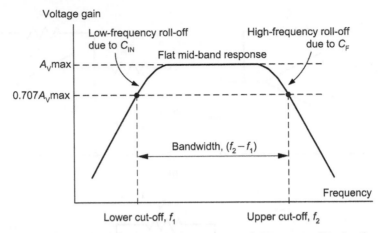

**Figure 11.9: Effect of adding capacitors, CIN and CF, to modify the frequency response of an operational amplifier**

Provided the upper frequency response is not limited by the gain × bandwidth product, the upper cut-off frequency will be determined by the feedback capacitance, $C_F$, and feedback resistance, $R_F$, such that:

$$f_2 = \frac{1}{2\pi C_F R_F} = \frac{0.159}{C_F R_F}$$

where $f_2$ is the upper cut-off frequency in Hz, $C_F$ is in farads and $R_2$ is in ohms.

### Example 11.6
An inverting operational amplifier is to operate according to the following specification:

Voltage gain = 100
Input resistance (at mid-band) = $10\,k\Omega$
Lower cut-off frequency = $250\,Hz$
Upper cut-off frequency = $15\,kHz$

Devise a circuit to satisfy the above specification using an operational amplifier.

### Solution

To make things a little easier, we can break the problem down into manageable parts. We shall base our circuit on a single operational amplifier configured as an inverting amplifier with capacitors to define the upper and lower cut-off frequencies, as shown in the previous figure.

The nominal input resistance is the same as the value for $R_{IN}$. Thus:

$$R_{IN} = 10\,k\Omega$$

To determine the value of $R_F$ we can make use of the formula for mid-band voltage gain:

$$A_v = \frac{R2}{R1}$$

Thus, $R2 = A_v \times R1 = 100 \times 10\,k\Omega = 100\,k\Omega$

To determine the value of $C_{IN}$ we will use the formula for the low-frequency cut-off:

$$f_1 = \frac{0.159}{C_{IN}R_{IN}}$$

from which:

$$C_{IN} = \frac{0.159}{f_1 R_{IN}} = \frac{0.159}{250 \times 10 \times 10^3}$$

hence:

$$C_{IN} = \frac{0.159}{2.5 \times 10^6} = 63 \times 10^{-9}\,F = 63\,nF$$

Finally, to determine the value of $C_F$ we will use the formula for high-frequency cut-off:

$$f_2 = \frac{0.159}{C_F R_F}$$

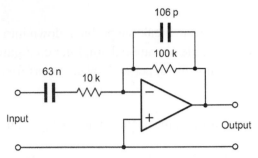

**Figure 11.10: See Example 11.6. This operational amplifier has a mid-band voltage gain of 10 over the frequency range 250 Hz to 15 kHz**

from which:

$$C_F = \frac{0.159}{f_2 R_{IN}} = \frac{0.159}{15 \times 10^3 \times 100 \times 10^3}$$

hence:

$$C_F = \frac{0.159}{1.5 \times 10^9} = 0.106 \times 10^{-9}\,F = 106\ \text{pF}$$

For most applications the nearest preferred values (68 nF for $C_{IN}$ and 100 pF for $C_F$) would be perfectly adequate. The complete circuit of the operational amplifier stage is shown in Figure 11.10.

## 11.9  Operational Amplifier Circuits

As well as their application as a general-purpose amplifying device, operational amplifiers have a number of other uses, including voltage followers, differentiators, integrators, comparators, and summing amplifiers. We shall conclude this section by taking a brief look at each of these applications.

### 11.9.1  Voltage Followers

A voltage follower using an operational amplifier is shown in Figure 11.11. This circuit is essentially an inverting amplifier in which 100% of the output is fed back

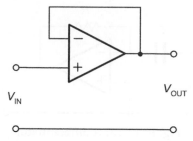

**Figure 11.11: A voltage follower**

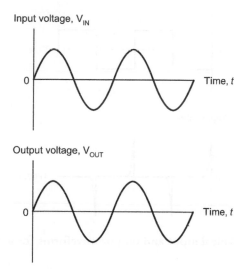

**Figure 11.12: Typical input and output waveforms for a voltage follower**

to the input. The result is an amplifier that has a voltage gain of 1 (i.e., unity), a very high input resistance and a very high output resistance. This stage is often referred to as a buffer and is used for matching a high-impedance circuit to a low-impedance circuit.

Typical input and output waveforms for a voltage follower are shown in Figure 11.12. Notice how the input and output waveforms are both in-phase (they rise and fall together) and that they are identical in amplitude.

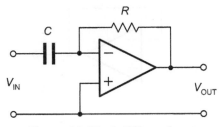

**Figure 11.13: A differentiator**

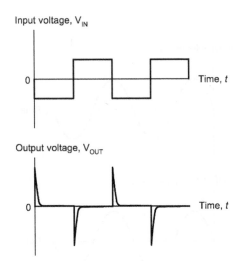

**Figure 11.14: Typical input and output waveforms for a differentiator**

### 11.9.2 Differentiators

A differentiator using an operational amplifier is shown in Figure 11.13. A differentiator produces an output voltage that is equivalent to the rate of change of its input. This may sound a little complex but it simply means that, if the input voltage remains constant (i.e., if it isn't changing) the output also remains constant. The faster the input voltage changes the greater will the output be. In mathematics this is equivalent to the differential function.

Typical input and output waveforms for a differentiator are shown in Figure 11.14. Notice how the square wave input is converted to a train of short duration pulses at the output.

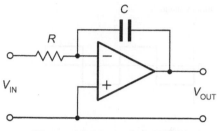

**Figure 11.15: An integrator**

Note also that the output waveform is inverted because the signal has been applied to the inverting input of the operational amplifier.

### 11.9.3 Integrators

An integrator using an operational amplifier is shown in Figure 11.15. This circuit provides the opposite function to that of a differentiator (see earlier) in that its output is equivalent to the area under the graph of the input function rather than its rate of change. If the input voltage remains constant (and is other than 0 V) the output voltage will ramp up or down according to the polarity of the input. The longer the input voltage remains at a particular value the larger the value of output voltage (of either polarity) will be produced.

Typical input and output waveforms for an integrator are shown in Figure 11.16. Notice how the square wave input is converted to a wave that has a triangular shape. Once again, note that the output waveform is inverted.

### 11.9.4 Comparators

A comparator using an operational amplifier is shown in Figure 11.17. Since no negative feedback has been applied, this circuit uses the maximum gain of the operational amplifier. The output voltage produced by the operational amplifier will thus rise to the maximum possible value (equal to the positive supply rail voltage) whenever the voltage present at the non-inverting input exceeds that present at the inverting input. Conversely, the output voltage produced by the operational amplifier will fall to the minimum possible value (equal to the negative supply rail voltage) whenever the voltage present at the inverting input exceeds that present at the non-inverting input.

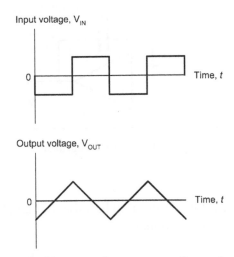

Figure 11.16: Typical input and output waveforms for an integrator

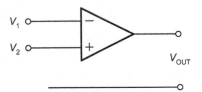

Figure 11.17: A comparator

Typical input and output waveforms for a comparator are shown in Figure 11.18. Notice how the output is either $+15\,V$ or $-15\,V$ depending on the relative polarity of the two input. A typical application for a comparator is that of comparing a signal voltage with a reference voltage. The output will go high (or low) in order to signal the result of the comparison.

### 11.9.5 Summing Amplifiers

A summing amplifier using an operational amplifier is shown in Figure 11.19. This circuit produces an output that is the sum of its two input voltages. However, since the operational amplifier is connected in inverting mode, the output voltage is given by:

$$V_{OUT} = -(V_1 + V_2)$$

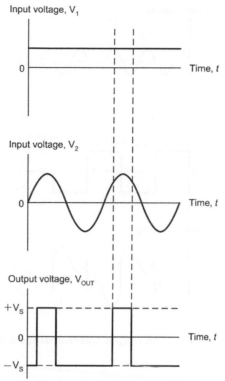

Figure 11.18: Typical input and output waveforms for a comparator

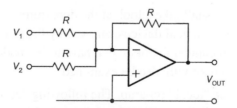

Figure 11.19: A summing amplifier

where $V_1$ and $V_2$ are the input voltages (note that all of the resistors used in the circuit have the same value). Typical input and output waveforms for a summing amplifier are shown in Figure 11.20. A typical application is that of "mixing" two input signals to produce an output voltage that is the sum of the two.

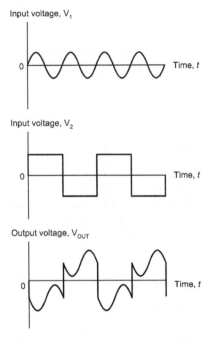

**Figure 11.20: Typical input and output waveforms for a summing amplifier**

## 11.10 The Ideal Op-Amp

In the following sections, we shall take a look at the departures from the ideal op-amp parameters that are found in practical devices, and survey the trade-offs—including cost and availability, as well as technical factors—that have to be made in real designs. Some instances of anomalous behavior will also be examined.

But first we will examine the "ideal" op-amp. The following set of characteristics (in no particular order, since they are all unattainable) defines the ideal voltage gain block:

- infinite input impedance, no bias current

- zero output impedance

- arbitrarily large input and output voltage range

- arbitrarily small supply current and/or voltage

- infinite operating bandwidth

- infinite open-loop gain

- zero input offset voltage and current

- zero noise contribution

- absolute insensitivity to temperature, power rail and common mode input fluctuations

- zero cost

- off-the-shelf availability in any package

- compatibility between different manufacturers

- perfect reliability

Since none of these features is achievable, you have to select a practical op-amp from the multitude of imperfect types on the market to suit a given application. Some basic examples of trade-offs are:

- a high-frequency AC amplifier will need maximum gain-bandwidth product but won't be interested in bias current or offset voltage,

- a battery-powered circuit will want the best of all the parameters but at minimum supply current and voltage,

- a consumer design will need to minimize the cost at the expense of technical performance,

- a precision instrumentation amplifier will need minimum input offsets and noise but can sacrifice speed and cheapness.

Device data sheets contain some but not all of the necessary information to make these trade-offs (most crucially, they say nothing about cost and availability, which you must get from the distributor). The functional characteristics often need some interpretation and critical parameters can be hidden or even absent. In general, if a particular parameter you are interested in is not given in the data sheet, it is safest to assume a pessimistic figure. It means that the manufacturer is not prepared to test his devices for that parameter or to certify a minimum or maximum value.

**Table 11.4: Parameters for applications categories**

| Category | GBW MHz | Slew rate V/μs | $V_{OS}$ mV | ICC mA | $V_{OS}$ drift μV/°C | Noise nV/√Hz | Gain/ phase error% |
|---|---|---|---|---|---|---|---|
| General purpose | 1−30 | 0.5−40 | 0.5−20 | | | | |
| Low power | 0.05−5 | 0.03−3 | 0.5−20 | 0.015−1 | | | |
| Precision | | 0.3−10 | 0.06−0.5 | | 0.5−4 | 3−30 | |
| High speed & video | 30−1000 | 100−5000 | 1−25 | 3−15 | | | 0.01−0.3 |

### 11.10.1 Applications Categories

In fact, although there is a bewildering variety of devices available, op-amps are divided into a few broad categories based on their application, in which the above trade-offs are altered in different directions. Table 11.4 suggests a reasonable range over which you might expect to find a spread of certain critical parameters for op-amps in each category.

## 11.11 The Practical Op-Amp

### 11.11.1 Offset Voltage

Input offset voltage $V_{OS}$ can be defined as that differential DC voltage required between the inverting and non-inverting inputs of an amplifier to drive its output to zero. In the perfect amplifier, zero volts in will give zero volts out; practical devices will show offsets ranging from tens of millivolts down to a few microvolts. The offset appears as an error voltage in series with the actual input voltage. Definitions vary, but a "precision" op-amp is usually considered to be one that has a $V_{OS}$ of less than 200 μV and a $V_{OS}$ temperature coefficient (see later) of less than 2 μV/°C. Bipolar input opamps are the best for very-low-offset voltage applications unless you are prepared to limit the bandwidth to a few tens of Hz, in which case the CMOS chopper-stabilized types come into their own. The chopper technique achieves very low values of $V_{OS}$ and drift by repeatedly nulling the

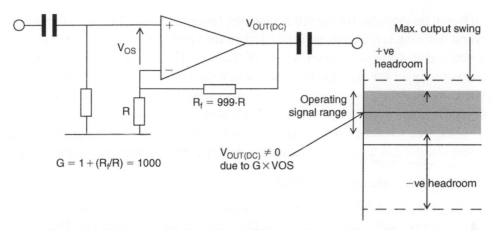

**Figure 11.21: Non-inverting AC amplifier and the problem of headroom**

amplifier's actual $V_{OS}$ several hundred times a second with the aid of charge storage capacitors.

Offsets are always quoted referenced to the input. The output offset voltage is the input offset times the closed-loop gain. This can have embarrassing consequences particularly in high-gain AC amplifiers where the designer has neglected offset errors because, for performance purposes, they are unimportant. Consider a non-inverting accoupled amplifier with a gain of 1000 as depicted in Figure 11.21.

Let's say the circuit is for audio applications and the op-amp is one half of a TL072 selected for low noise and wide bandwidth, running on supply voltages of $\pm 12\,$V. The TL072 has a maximum quoted $V_{OS}$ of 10 mV. In the circuit shown, this will be amplified by the closed loop gain to give a DC offset at the output of 10 V—which is far too close to the supply rail to leave any headroom to cope with overloads. In fact, the TL072 is likely to saturate at $9-10\,$V anyway with $\pm 12\,$V power rails.

### 11.11.1.1 Output Saturation Due to Amplified Offset

The designer may be wanting 2 mV pk-pk AC signals at the input to be amplified up to 2 V pk-pk signals at the output. If the DC conditions are taken for granted then you might expect at least 20 dB of headroom: $\pm 1\,$V output swing with $\pm 10\,$V available. But, with a worst-case $V_{OS}$ device virtually no headroom will be available for one polarity of input and 20 V will be available for the other. Unipolar (asymmetrical) clipping will result. The worst outcome is if the design is checked on the bench with a device which has a much-better-than-worst-case offset, say 1 mV. Then the DC output voltage will only be 1 V and virtually all the expected headroom will be available. If this design is let through to production then the scene is set for unexpected customer complaints of distortion! An additional problem presents itself if the output coupling capacitor is polarized: the DC output voltage can assume either polarity depending on the polarity of the offset. If this isn't recognized it can lead to early failure of the capacitor in some production units.

### 11.11.1.2  Reducing the Effect of Offset

The solutions are plentiful. The easiest is to change the feedback to AC-coupling which gives a DC gain of unity so that the output DC voltage offset is the same as the input offset (Figure 11.22). The inverting configuration is simpler in this respect. The difficulty with this solution is that the time constant $R_f \cdot C$ can be inordinately long, leading to power-on delays of several seconds.

The second solution is to reduce the gain to a sensible value and cascade gain blocks. For instance, two AC-coupled gain blocks with a gain of 33 each, cascaded, would have the same performance but the offsets would be easily manageable. The bandwidth would also be improved, along with the out-of-band roll-off, if this were necessary. Unfortunately, this solution adds components and therefore cost.

A third solution is to use an amplifier with a better $V_{OS}$ specification. This will either involve a trade-off in gain-bandwidth, power consumption or other parameters, or cost. For instance, in the above example AD's OP-227G with a maximum offset of $180\,\mu V$ might be a suitable candidate, though it is noticeably more expensive. The overall cost

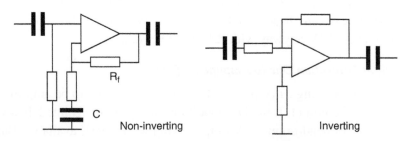

**Figure 11.22: AC coupling to reduce offset**

might work out the same though, given the reduction in components over the second solution.

### 11.11.1.3 Offset Drift

Offset voltage drift is closely related to initial offset voltage and is a measure of how $V_{OS}$ changes with temperature and time. Most manufacturers will specify drift with temperature, but only those offering precision devices will specify drift over time. Present technology for standard devices allows temperature coefficients of between 5 and $40\,\mu V/^\circ C$, with $10\,\mu V/^\circ C$ being typical. For bipolar inputs, the magnitude of drift is directly related to the initial offset at room temperature. A rule of thumb is $3.3\,\mu V/^\circ C$ for each millivolt of initial offset. This drift has to be added to the worst case offset voltage when calculating offset effects and can be significant when operating over a wide temperature range.

Early MOS-input op-amps suffered from poor offset voltage performance due to gate threshold voltage shifts with time, temperature and applied gate voltage. New processes, particularly developments in silicon gate technology, have overcome these problems and CMOS op-amps (Texas Instruments' LinCMOS™ range for instance) can achieve bipolar-level $V_{OS}$ figures with extremely good drift, $1-2\,\mu V/^\circ C$ being quoted.

### 11.11.1.4 Circuit Techniques to Remove the Effect of Drift

Microprocessor control has allowed new analog techniques to be developed and one of these is the nulling of input amplifier offsets, as in Figure 11.23. With this technique the initial circuit offsets can be calibrated out of the system by applying a zero input, storing the resultant input value (which is the sum of the offsets) in non-volatile memory and subsequently subtracting this from real-time input values. With this technique, only offset drifts, not absolute offset values, are important. Alternatively, for the cost of a few extra components—analog switches and interfacing—the nulling can be done repetitively in real time and even the drift can be subtracted out. (This is the microprocessor equivalent of the chopper op-amps discussed earlier.)

### 11.11.2 Bias and Offset Currents

Input bias current is the average DC current required by the inputs of the amplifier to establish correct bias conditions in the first stage. Input offset current is the difference in the bias

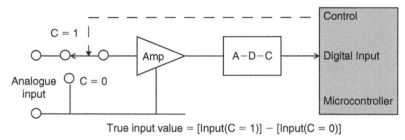

True input value = [Input(C = 1)] − [Input(C = 0)]

**Figure 11.23: Offset nulling with a microcontroller**

current requirements of the two input terminals. A bipolar input stage requires a bias current which is directly related to the current flowing in the collector circuit, divided by the transistor gain. FET-input (or BiFET) op-amps on the other hand do not require a bias current as such, and their input currents are determined only by leakage and the need for input protection.

### 11.11.2.1 Bias Current Levels

Input bias currents of bipolar devices range from a few microamps down to a few nanoamps, with most industry-standard devices offering better than $0.5\,\mu A$. There is a well-established trade-off between bias current and speed; high speeds require higher first-stage collector currents to charge the internal node capacitance faster, which in turn requires higher bias currents. Precision bipolar op-amps achieve less than $20\,nA$ while some devices using current nulling techniques can boast picoamp levels. JFET and CMOS devices routinely achieve input currents of a few picoamps or tens of picoamps at 25°C, but because this is almost entirely reverse-bias junction leakage it increases exponentially with temperature. Industry standard JFET op-amps are therefore no better than bipolar ones at high temperatures, though precision JFET and CMOS still show nanoamp levels at the 125°C extreme. Note that even the 25°C figure for JFETs can be misleading, because it is quoted at 25°C *junction* temperature: many JFET op-amps take a fairly high supply current and warm up significantly in operation, so that the junction temperature is actually several degrees or tens of degrees higher than ambient.

The significance of input bias and offset currents is twofold: they determine the steady-state input impedance of the amplifier and they result in added voltage offsets. Input impedance is rarely quoted as a parameter on op-amp data sheets since bias currents are a better measure of actual effects. It is irrelevant for the closed-loop inverting configuration,

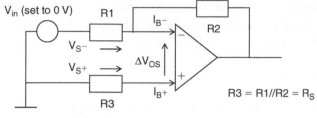

Ideal situation: $I_B$, $R_S$ equal at $+$ and $-$ terminals, so $V_S+ = V_S- = I_B \cdot R_S$ and $\Delta V_{OS} = 0$

Bad design: $R_S$ not equal at $+$ and $-$ terminals so, neglecting $I_{OS}$,
$V_S+ = I_B \cdot R3$, $V_S- = I_B \cdot R1//R2$ and $\Delta V_{OS} = I_B \cdot (R1//R2 - R3)$

Practical op-amp: $I_B-$ differs from $I_B+$ by $I_{OS}$, $R_S$ equal at both terminals, so
$V_S+ = I_B+ \cdot R_S$, $V_S- = (I_B + I_{OS}) \cdot R_S$ and $\Delta V_{OS} = I_{OS} \cdot R_S$

**Figure 11.24: Bias and offset currents**

since the actual impedance seen at the op-amp input terminals is reduced to near zero by feedback. The input impedance of the non-inverting configuration is determined by the change in input voltage divided by the change in bias current due to it.

### 11.11.2.2 Output Offsets Due to Bias and Offset Currents

Of more importance is the bias current's contribution to offsets. The bias current flowing in the source resistance $R_S$ at each terminal generates a voltage in series with the input; if the bias currents and source resistances were equal the voltages would cancel out and no extra offset would be added. (See Figure 11.24.)

As it is, the offset current generates an effective offset voltage given by $I_{OS} \cdot R_S$ (with a temperature coefficient determined by both) which adds to, or subtracts from, the inherent offset voltage $V_{OS}$ of the op-amp. Clearly, whichever dominates the output depends on the magnitude of $R_S$. Higher values demand an op-amp with lower bias and offset currents. For instance, the current and voltage offsets generated by a 741's input circuit are equal when $R_S = 33\,\mathrm{K}\Omega$ (typical $V_{OS} = 1\,\mathrm{mV}$, $I_{OS} = 30\,\mathrm{nA}$). The same value for the TL081 JFET op-amp is $1000\,\mathrm{M}\Omega$ ($V_{OS} = 5\,\mathrm{mV}$, $I_{OS} = 5\,\mathrm{pA}$).

$I_B$ itself does not contribute to offset *provided* that the source resistances are equal at each terminal. If they are not then the offset contribution is $I_B \cdot \Delta R_S$. Since IB can be an order of magnitude higher than $I_{OS}$ for bipolar op-amps, it pays to equalize $R_S$: this is the function of R3 in the circuit above. R3 can be omitted or changed in value if current offset is not calculated to be a problem. Apart from the disadvantage of an extra component, R3 is also an extra source of noise (generated by the noise component of $I_B$) which can weigh heavily against it in low-noise circuits.

### 11.11.3  Common Mode Effects

Two factors, which because they don't appear in op-amp circuit theory can be overlooked until late in the design, are common mode rejection ratio (CMRR) and power supply rejection ratio (PSRR). Figure 11.25 shows these schematically. Related to these is common mode input voltage range.

#### 11.11.3.1  CMRR

An ideal op-amp will not produce an output when both inputs, ignoring offsets, are at the same (common mode) potential throughout the input range. In practice, gain differences between the two inputs, and variations in offset with common mode voltage, combine to produce an error at the output as the common mode voltage varies. This error is referred to the input (that is, divided by the gain) to produce an equivalent input common mode error voltage. The ratio of this voltage to the actual common mode input voltage is the common mode rejection ratio (CMRR), usually expressed in dB. For example, a CMRR of 80 dB would give an equivalent input voltage error of $100\,\mu V$ for every 1 V change at both $+$ and $-$ inputs together. The inverting amplifier configuration is inherently immune to common mode errors since the inputs stay at a constant level, whereas the non-inverting and differential circuits are susceptible.

CMRR is not necessarily a constant. It will vary with common mode input level and temperature, and always worsens with increasing frequency. Individual manufacturers may specify an average or a worst-case value, and will always specify it at DC.

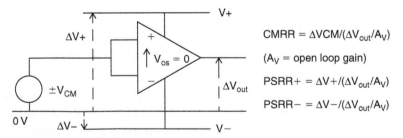

**Figure 11.25: Common mode and power supply rejection ratio**

### 11.11.3.2 PSRR

Power supply rejection is similar to CMRR but relates to error voltages referred to the input as a result of changes in the power rail voltages. As before, a PSRR of 80 dB with a rail voltage change of 1 V would result in an equivalent input error of 100 μV. Again, PSRR worsens with increasing frequency and may be only 20–30 dB in the tens-to hundreds of kilo Hertz range, so that high-frequency noise on the power rails is easily reflected on the output. There may also be a difference of several tens of dB between the PSRRs of the positive and negative supply rails, due to the difference in internal biasing arrangements. For this reason it is unwise to expect equal but anti-phase power rail signals, such as mains frequency ripple, to cancel each other out.

## 11.11.4 Input Voltage Range

Common mode input voltage range is usually defined as the range of input voltages over which the quoted CMRR is met. Errors quickly increase as it is exceeded. The input range may or may not include the negative supply rail, depending on the type of input. The popular LM324 range and its derivatives have a pnp emitter coupled pair at the input, which allows operation down to slightly below the negative rail. The CMOS-input devices from Texas, National, STM and Intersil also allow operation down to the negative rail. Some of these op-amps stop a few volts short of the positive rail, as they are optimized for operation from a single positive supply, but there are also some devices available which include both rails within their input range, known unsurprisingly as "rail-to-rail" input op-amps. Conventional bipolar devices of the 741 type, designed for ±15 V rails, cannot swing to within less than 2 V of each rail, and BiFET types are even more restricted.

### 11.11.4.1 Absolute Maximum Input

The common mode operating input voltage is normally different from the absolute maximum input voltage range, which is usually equal to the supply voltage. If you exceed the maximum input voltage without current limiting then you are likely to destroy the device; this can quite easily happen inadvertently, apart from circuits connected to external inputs, if for instance a large value capacitor is discharged directly through the input. Even if current is limited to a safe value, overvoltages on the input can lead to unpredictable behavior. Latch-up, where the IC locks itself into a quasi-stable state and may draw large currents from the power supply, leading to burnout, is one possibility.

Another is that the sign of the inputs may change, so that the inverting input suddenly becomes non-inverting. (This was a well known fault on early devices such as the 709.) These problems most frequently arise with capacitive coupling direct to one or other input, or when power rails to different parts of the circuit are turned on or off at different times. The safe way to guard against them is to include a reasonable amount of resistance at each input, directly in series with the input pin.

### 11.11.5 Output Parameters

Two factors constrain the output voltage available from an op-amp: the power rail voltage, and the load impedance.

#### 11.11.5.1 Power Rail Voltage

It should be obvious that the output cannot swing to a greater value than either power rail. Unfortunately it is often easy to overlook this fact, particularly as the power connections are frequently omitted from circuit diagrams, and with different quad op-amp packages being supplied from different rails it is hard to keep track of which device is powered from what voltage. More seriously, with unregulated supplies the actual voltage may be noticeably less than the nominal. The required output must be calculated for the worst-case supply voltage.

Historically, most op-amps could not swing their output right up to either supply rail. The profusion of CMOS-output devices have dealt with this limitation, as have many of the types intended for single-supply operation which have a current sink at the output and can reach within a few tens of millivolts of the negative (or ground) supply terminal. Other conventional bipolar and biFET parts cannot swing to within less than 2 V of either rail. The classic output stage (Figure 11.26) is a complementary emitter follower pair that gives low output impedance, but the output available in either direction is limited by $(V_{DR(min)} + V_{BE})$. Depending on the detailed design of the output, the swing may or may not be symmetrical in either polarity. This fact is disguised in some data sheets where the maximum peak-to-peak output voltage swing is quoted, rather than the maximum output voltage relative to the supply terminals.

#### 11.11.5.2 Load Impedance

Output also depends on the circuit load impedance. This may again seem obvious, but there is an erroneous belief that because feedback reduces the output impedance of an

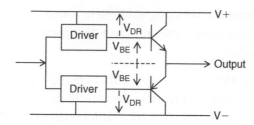

**Figure 11.26: Output voltage swing restrictions**

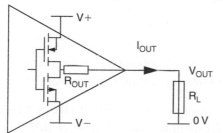

$V_{OUT}$ can only swing to within $R_L/(R_L + R_{OUT})$ of either rail, due to voltage drop $I_{OUT} \cdot R_{OUT}$

$R_{OUT}$ is the equivalent output resistance of the CMOS output transistors, usually dependent on supply voltage (reduces with increasing $V_{supply}$)

**Figure 11.27: Limits on rail-to-rail swing with CMOS outputs**

op-amp in proportion to the ratio of open- to closed-loop gains, it should be capable of driving very low load resistors. Well of course to an extent it is, but Ohm's Law is not so easily flouted and a low output resistance can only be driven to a low output voltage swing, depending entirely on the current drive capability of the output stage. The maximum output current that can be obtained from most devices is limited by package dissipation considerations to about $\pm 10$ mA. In some cases, the output current spec is given as a particular output voltage swing when driving a stated value of load, typically $2-10$ k$\Omega$. The "rail-to-rail" op-amps with CMOS output will in fact only give a full rail-to-rail swing if they are driven into an open circuit; any output load, including, of course, the feedback resistor, reduces the total available swing in proportion to the ratio of output resistance to load resistance (Figure 11.27).

If you want more output current, it is quite in order to buffer the output with an external complementary emitter follower or something similar, provided that feedback is taken from the final output. Take care with short-circuit protection when doing this (or else don't be surprised if you have to keep replacing transistors) and also bear in mind that you have changed the high frequency response of the combination and the closed-loop circuit may now be unstable.

Some single-supply op amps are not designed both to source and sink current and, when used with split supplies, may have some crossover distortion as the output signal passes through the midsupply value.

Output current protection is universally provided in op-amps to prevent damage when driving a short circuit. This does not work in the reverse direction, that is when the output voltage is forced outside either supply rail by a fault condition. In this case there will be one or two forward-biased diode junctions to the power rail and current will flow through these limited only by the fault source impedance.

### 11.11.6 AC Parameters

The performance of an op-amp at high frequency is described by a motley collection of parameters, each of which refers to slightly different operating conditions. They are:

- Large-signal bandwidth, or full-power response: the maximum frequency, at unity closed-loop gain, for which a sinusoidal input signal will produce full output at rated load without exceeding a given distortion level. This bandwidth figure is normally determined by the slew-rate performance.

- Small-signal or unity-gain bandwidth, or gain-bandwidth product: the frequency at which the open-loop gain falls to unity (0 dB). The "small-signal" label means that the output voltage swing is small enough that slew-rate limitations do not apply.

- Slew rate: the maximum rate of change of output voltage for a large input step change, quoted in volts per microsecond.

- Settling time: elapsed time from the application of a step input change to the point at which the output has entered and remained within a specified error band about the final steady-state value.

These parameters are illustrated in Figure 11.28.

### 11.11.7 Slew Rate and Large Signal Bandwidth

These two specifications are intimately related. All conventional voltage feedback op-amps can be modeled by a transconductance gain block driving a transimpedance amplifier with capacitive feedback (Figure 11.29).

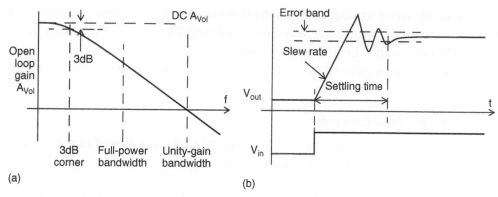

(a)

(b)

**Figure 11.28: AC op-amp specifications. (a) Frequency domain specifications; (b) Time domain specifications**

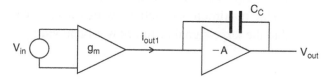

**Figure 11.29: Op-amp slewing model**

The compensation capacitor $C_C$ is the dominant factor setting the op-amp's frequency response. It is necessary because a feedback circuit would be unstable if the gain block's high frequency response was not limited. Digital designers avoid capacitors in the signal path because they slow the response time, but this is the price for freedom from unwanted oscillations when working with linear circuits.

### 11.11.7.1 Slew Rate

The exact value of the price is measured by the slew rate. From the above circuit, you can see that the rate of change of $V_{out}$ is determined entirely by $i_{out1}$ and $C_C$ (remember $dV/dt = I/C$). As an example, the 741's input section current source can supply $20\,\mu A$ and its compensation capacitor is $30\,pF$, so its maximum slew rate is $0.67\,V/\mu s$. Op-amp designers have the freedom to set both these parameters within certain limits, and this is what distinguishes a fast, high-supply-current device from a slow, low-supply current one. "Programmable" devices such as the LM4250 or LM346 make the trade-off more obvious by putting it in the circuit designer's hands.

If $i_{out1}$ can be increased without affecting the transconductance *gm*, then slew rate can be improved without a corresponding reduction in stability. This is one of the major virtues of the biFET range of op-amps. The JFET input stage can be run at high currents for a low gm relative to the bipolar and so can provide an order of magnitude or more increase in slew rate.

### 11.11.7.2 Large-Signal Bandwidth

Slew-rate limitations on $dV_{out}/dt$ can be equated to the maximum rate-of-change of a sinewave output. The time derivative of a sinewave is

$$d/dt\,[V_p \sin \omega t] = \omega \cdot V_p \cos \omega t \quad \text{where } \omega = 2\pi f$$

This has a maximum value of $2nf \cdot V_p$, which relates frequency directly to peak output voltage. If $V_p$ is equated to the maximum DC output swing then $f_{max}$ can be inferred from the slew rate and is equal to the large signal or full power bandwidth,

$$2\pi \cdot f_{max} = \text{slew rate}/V_p$$

### 11.11.7.3 Slewing Distortion

Operating an op-amp above the slew-rate limit will cause slewing distortion on the output. In the limit the output will be a triangle wave (Figure 11.30) as it alternately switches between positive and negative slewing, which will decrease in amplitude as the frequency is raised further. If the positive and negative slew rates differ there will be asymmetrical distortion on the output. This can generate an unexpected effect equivalent to a DC offset voltage, due to rectification of the asymmetrical feedback waveform or overloading of the input stage by large distortion signals at the summing junction. Also, slewing is not always linear from start to finish but may exhibit a fast rise for the first part of the change followed by a reversion to the expected rate for the latter part.

### 11.11.8 Small-Signal Bandwidth

The op-amp frequency response shown in Figure 11.28(a) exhibits the same characteristic as a simple low-pass RC filter. The 3 dB frequency or corner frequency is that point at which the open-loop gain has dropped by 3 dB from its DC value. It is set by the compensation capacitor $C_C$ and is in the low Hertz or tens of Hertz range for most

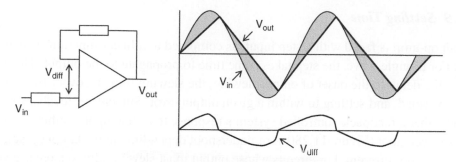

**Figure 11.30: Slewing distortion**

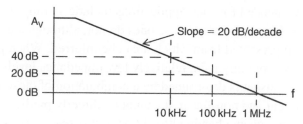

**Figure 11.31: Gain-bandwidth roll-off**

devices. The gain then "rolls off" at a constant rate of 20 dB per decade (a ten-times increase in frequency produces a tenfold gain reduction) until at some higher frequency the gain has dropped to 1. This frequency therefore represents the unity-gain bandwidth of the part, also called the small-signal bandwidth.

The fact of a constant roll-off means that it is possible to speak of a constant "gainband width product" (GBW) for a device. The LM324's op-amps for instance have a typical unity-gain bandwidth of 1 MHz, so if you wanted to use them at this frequency you could only use them as voltage followers—and small-signal ones at that, since large output swings would be slew-rate limited. A gain of 10 would be achievable up to 100 kHz, a gain of 100 up to 10 kHz and so on (but see the comments on open-loop gain later). This gain-bandwidth trade-off is illustrated in Figure 11.31. On the other hand, many more recent devices have unity-gain bandwidths of 5–30 MHz and can therefore offer reasonable gains up to the MHz region. Anything with a GBW of more than 30 MHz is justifiably offered as a "high-speed" device.

### 11.11.9  Settling Time

When an op-amp is faced with a step input, as compared to a linear function such as a sinusoid or triangle wave, the step takes some time to propagate to the output. This time includes the delay to the onset of output slewing, the slewing time, recovery from slew limited overload, and settling to within a given output error. Students of feedback theory will know that a feedback-controlled system's response to a step input exhibits some degree of overshoot (Figure 11.28(b)) or undershoot depending on its damping factor. Op-amps are no different. For circuits whose output must slew rapidly to a precise value, particularly analog-to-digital converters and sample-and-hold buffers, the settling time is an important parameter.

Op-amps specifically intended for such applications include settling time parameters in their specifications. Most general-purpose ones do not, although a graph of output pulse response is often presented from which it can be inferred. When present, settling time is usually specified for unity gain, relatively low impedance levels, and low or no capacitive loading. Because it is determined by a combination of closed-loop amplifier characteristics both linear and non-linear, it cannot be directly predicted from the open-loop specs of slew rate and bandwidth, although it is reasonable to assume that an amplifier which performs well in these respects will also have a fast settling time.

### 11.11.10  The Oscillating Amplifier

Just about every analog designer has been bugged by the problem of the feedback amplifier that oscillates (and its converse, the oscillator that doesn't) at some time or other. There are really only a few fundamental causes of unwanted oscillations, they are all curable, and they can be listed as follows:

- feedback-loop instability,
- incorrect grounding,
- power supply coupling,
- output stage instability, and
- parasitic coupling.

The most important clue in tracking down instability is the frequency of oscillation. If this is near the unity-gain bandwidth of the device then you are most probably suffering feedback-induced instability. This can be checked by temporarily increasing the closed-loop gain. If feedback is the problem, then the oscillation should stop or at least decrease in frequency. If it doesn't, look elsewhere.

Feedback-loop instability is caused by too much feedback at or near the unity-gain frequency, where the op-amp's phase margin is approaching a critical value. (Many books on feedback circuit theory deal with the question of stability, gain and phase margin, using tools such as the Bode plot and the Nyquist diagram, so this isn't covered here.)

### 11.11.10.1 Ground Coupling

Ground loops or other types of incorrect grounding cause coupling from output back to input of the circuit via a common impedance in its grounded segment. The circuit topology is illustrated in Figure 11.32. If the resulting feedback sense gives an output component in-phase with the input then positive feedback occurs, and if this overrides the intended negative feedback you will have oscillation. The frequency will depend on the phase contribution of the common impedance, which will normally be inductive, and can vary over a wide range.

### 11.11.10.2 Power Supply Coupling

Power supplies should be properly bypassed to avoid similar coupling through the common mode power supply impedance. Power supply rejection ratio falls with frequency, and typical $0.01 - 0.1 \, \mu F$ decoupling capacitors may resonate with the parasitic inductance of long power leads in the MHz region, so these problems usually show up in the $1 - 10 \, MHz$ range. Using $1 - 10 \mu F$ tantalum capacitors for power rail bypassing will drop the resonant frequency and stray circuit Q to the level at which problems are unlikely (compare Figure 3.19 for capacitor resonances).

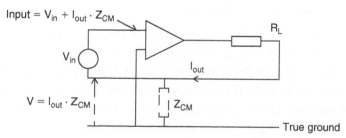

**Figure 11.32: Common-impedance ground coupling**

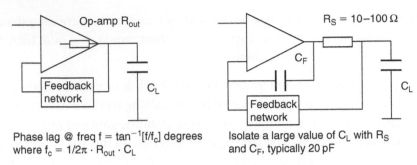

Phase lag @ freq f = $\tan^{-1}[f/f_c]$ degrees
where $f_c = 1/2\pi \cdot R_{out} \cdot C_L$

Isolate a large value of $C_L$ with $R_S$
and $C_F$, typically 20 pF

**Figure 11.33: Output capacitive loading**

### 11.11.10.3 Output Stage Instability

Localized output-stage instability is most common when the device is driving a capacitive load. This can create output oscillations in the high-MHz range which are generally cured by good power-rail decoupling close to the power supply pins, with the decoupling ground point close to the return point of the load impedance, or by including a low-value series resistor in the output within the feedback loop.

Capacitive loads also cause a phase lag in the output voltage by acting in combination with the op-amp's open-loop output resistance (Figure 11.33). This increased phase shift reduces the phase margin of a feedback circuit. A typical capacitive load, often invisible to the designer because it is not treated as a component, is a length of coaxial cable. Until the length starts to approach a quarter-wavelength at the frequency of interest, coax looks like a capacitor: for instance, 10 meters of the popular RG58C/U 50Ω type will be about 1000 pF. The capacitance can be decoupled from the output with a low-value series resistor, and high-frequency feedback provided by a small direct feedback capacitor $C_F$ compensates for the phase lag caused by $C_L$.

### 11.11.10.4 Stray Capacitance at the Input

A further phase lag is introduced by the stray capacitance $C_S$ at the op-amp's inverting input. With normal layout practice this is of the order of $3-5$ pF which becomes significant when high-value feedback resistors are used, as is common with MOS- and JFET-input amplifiers. The roll-off frequency due to this capacitance is determined by the feedback network impedance as seen from the inverting input. The small-value direct feedback capacitance $C_F$ of Figure 11.34 can be added to combat this roll-off, by roughly equating time constants in the feedback loop and across the input. In fact this technique

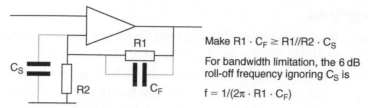

**Figure 11.34: Adding feedback capacitance**

is recommended for all low-frequency circuits as with it you can restrict loop bandwidth to the minimum necessary, thereby cutting down on noise, interference susceptibility and response instability.

### 11.11.10.5 Parasitic Feedback

Finally in the catalogue of instability sources, remember to watch out for parasitic coupling mechanisms, especially from the output to the non-inverting input. Any coupling here creates unwanted positive feedback. Layout is the most important factor: keep all feedback and input components close to the amplifier, separate input and output components, keep all pc tracks short and direct, and use a ground plane and/or shield tracks for sensitive circuits.

### 11.11.11 Open-Loop Gain

One of the major features of the classical feedback equation which is used in almost all op-amp design,

$$A_{CL} = A_{OL}/(1 + A_{OL} \cdot \beta)$$

where $\beta$ is the feedback factor,
$A_{OL}$ is the open-loop gain,
$A_{CL}$ is the closed-loop gain

is that if you assume a very high $A_{OL}$ then the closed-loop gain is almost entirely determined by $\beta$, the feedback factor. This is set by external (passive) components and can therefore be very tightly defined. Op-amps always offer a very high DC open-loop gain (80 dB as a minimum, usually $100-120$ dB) and this can easily tempt the designer into ignoring the effect of $A_{OL}$ entirely.

### 11.11.11.1 Sagging $A_{OL}$

$A_{OL}$ does, in fact, change quite markedly with both frequency and temperature. We have already seen (Figure 11.31) that the AC $A_{OL}$ rolls off at a constant rate, usually $20\,dB/$ decade, and this determines the gain that can be achieved for any given bandwidth. In fact when the frequency starts approaching the maximum bandwidth the excess gain available becomes progressively lower and this affects the validity of the high-$A_{OL}$ approximation. If your circuit has a requirement for precise gain then you need to evaluate the actual gain that will be achieved.

As an example, take $\beta = 0.01$ (for a gain of 100) and $A_{OL} = 10^5$ (100 dB) at DC. The actual gain, from the feedback equation, is

$$A_{CL} = 10^5/(1 + 10^5 \cdot 0.01) = 99.9$$

Now raise the frequency to the point at which it is a decade below the maximum expected bandwidth at this gain. This will have reduced $A_{OL}$ to ten times the closed-loop gain or 1000. The actual gain is now:

$$A_{CL} = 1000/(1 + 1000 \cdot 0.01) = 90.9$$

which shows a 10% gain reduction at one-tenth the desired bandwidth!

$A_{OL}$ also changes with temperature. The data sheet will not always tell you how much, but it is common for it to halve when going from the low temperature extreme to the high extreme. If your circuit is sensitive to changes in closed-loop gain, it would be wise to check whether the likely changes it will experience in $A_{OL}$ are acceptable and if not, either reduce the closed-loop gain to give more gain margin, or find an op-amp with a higher value for $A_{OL}$.

### 11.11.12 Noise

A perfect amplifier with perfect components would be capable of amplifying an infinitely small signal to, say, 10 V p-p with perfect resolution. The imperfection which prevents it from doing so is called noise. The noise contribution of the amplifier circuit places a lower limit on the resolution of the desired signal, and you will need to account for it when working with low-level (sub-millivolt) signals or when the signal-to-noise ratio needs to be high, as in precision amplifiers and audio or video circuits.

There are three noise sources which you need to consider:

- amplifier-generated noise,
- thermal noise, and
- electromagnetic interference.

The third of these is either electromagnetically coupled into the circuit conductors at RF, or by common mode mechanisms at lower frequencies. It can be minimized by good layout and shielding and by keeping the operating bandwidth low, and is mentioned here only to warn you to keep it in mind when thinking about noise. Chapter 25 discusses EMC in more depth.

### 11.11.12.1  Thermal Noise

The other two sources, like the DC offset and bias error components discussed earlier, are conventionally referred to the op-amp input. Thermal, or "Johnson" noise is generated in the resistive component of any circuit impedance by thermal agitation of the electrons. All resistors around the input circuit contribute this. It is given by:

$$e_n = \sqrt{(4kTBR)}$$

where $e_n$ = rms value of noise voltage

$k$ = Boltzmann's constant, $1.38 \times 10^{-23}$ joules/°K

$T$ = absolute temperature

$B$ = bandwidth in which the noise is measured

$R$ = circuit resistance

As a rule of thumb, it is easier to remember that the noise contribution of a $1\,k\Omega$ resistor at room temperature (298°K) in a 1 Hz bandwidth is 4 nV rms. The noise is proportional to the square root of bandwidth and resistance, so a $100\,k\Omega$ resistor in 1 Hz, or a $1\,k\Omega$ resistor in 100 Hz, will generate 40 nV. Noise is a statistical process. To convert the rms noise to peak-to-peak, multiply by 6.6 for a probability of less than 0.1% that a peak will exceed the calculated limit, or 5 for a probability of less than 1%.

### 11.11.12.2 Amplifier noise

Amplifier noise is what you will find specified in the data sheet (sometimes; where it is not specified it can be $2-4$ times worse than an equivalent low-noise part). It is characterized as a voltage source in series with one input, and a current source in parallel with each input, with the amplifier itself being considered noiseless. The values are specified at unity bandwidth, as rms nanovolts or nanoamps per root-Hertz; alternatively they may be specified over a given bandwidth. Because you need to add together all noise contributions, it is usually easiest to calculate them at unity bandwidth and then multiply the overall result by the square root of the bandwidth. This assumes a constant noise spectral density over the bandwidth of interest, which is true for resistors but may not be for the op-amp (see later). Noise, being statistical, is added on a root-mean-square basis. So the general noise model for an op-amp circuit is as shown in Figure 11.35.

When the noise is added in rms fashion, if any noise source is less than a third of another it can be neglected with an error of less than 5%. This is a useful feature to remember with complex circuits where it is difficult to account accurately for all generator resistances.

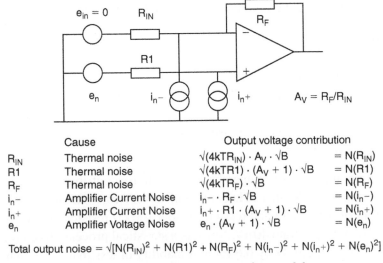

| | Cause | Output voltage contribution | |
|---|---|---|---|
| $R_{IN}$ | Thermal noise | $\sqrt{(4kTR_{IN})} \cdot A_V \cdot \sqrt{B}$ | $= N(R_{IN})$ |
| R1 | Thermal noise | $\sqrt{(4kTR1)} \cdot (A_V + 1) \cdot \sqrt{B}$ | $= N(R1)$ |
| $R_F$ | Thermal noise | $\sqrt{(4kTR_F)} \cdot \sqrt{B}$ | $= N(R_F)$ |
| $i_n-$ | Amplifier Current Noise | $i_n- \cdot R_F \cdot \sqrt{B}$ | $= N(i_n-)$ |
| $i_n+$ | Amplifier Current Noise | $i_n+ \cdot R1 \cdot (A_V + 1) \cdot \sqrt{B}$ | $= N(i_n+)$ |
| $e_n$ | Amplifier Voltage Noise | $e_n \cdot (A_V + 1) \cdot \sqrt{B}$ | $= N(e_n)$ |

Total output noise $= \sqrt{[N(R_{IN})^2 + N(R1)^2 + N(R_F)^2 + N(i_n-)^2 + N(i_n+)^2 + N(e_n)^2]}$

**Figure 11.35: The op-amp noise model**

As an example of how to apply the noise model, let us examine the trade-offs between a high-impedance and a low-impedance circuit for different op-amps. The circuit is the standard inverting configuration with R1 sized according to the principle laid out earlier for minimization of bias current errors (R3 in Figure 11.24). $R_{IN}$ is the sum of generator output impedance and amplifier input resistor. The op-amps chosen have the following noise characteristics (at 1 kHz):

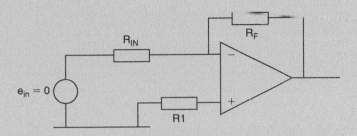

| OP27: | $e_n = 3\,nV/\sqrt{Hz}$ | $i_n = 0.4\,pA/\sqrt{Hz}$ | (low noise precision bipolar) |
|---|---|---|---|
| TL071: | $e_n = 18\,nV/\sqrt{Hz}$ | $i_n = 0.01\,pA/\sqrt{Hz}$ | (low noise biFET) |
| LMV324: | $e_n = 39\,nV/\sqrt{Hz}$ | $i_n = 0.21\,pA/\sqrt{Hz}$ | (industry standard low voltage bipolar) |

Working from the noise model of Figure 11.34, the contributions (in nV/$\sqrt{Hz}$) are tabulated for a low-impedance circuit and a high-impedance circuit, with the major contributor in each case shown emphasized and the negligible contributors shown in brackets:

Low impedance, $R_{IN} = 200\,\Omega$, $R1 = 180\,\Omega$, $R_F = 2\,K\Omega$

| Noise contributor | OP27 | TL071 | LMV324 |
|---|---|---|---|
| N($R_{IN}$) | 17.9 | 17.9 | 17.9 |
| N(R1) | 18.7 | 18.7 | 18.7 |
| N($R_F$) | (5.6) | (5.6) | (5.6) |
| N(in⁻) | (0.8) | (0.02) | (0.42) |
| N(in⁺) | (0.79) | (0.02) | (0.46) |
| N($e_n$) | **33** | **198** | **429** |
| Total noise voltage | 41.9 | 200 | 430 |

High impedance, $R_{IN} = 200\,K\Omega$, R1 $= 180\,K\Omega$, $R_F = 2\,M\Omega$

| Noise contributor | OP27 | TL071 | LMV324 |
|---|---|---|---|
| $N(R_{IN})$ | 565 | 565 | 565 |
| N(R1) | 590 | **590** | **590** |
| $N(R_F)$ | 178 | 178 | 178 |
| $N(in^-)$ | **800** | (20) | 420 |
| $N(in^+)$ | 792 | (19.8) | 460 |
| $N(e_n)$ | (33) | 198 | 429 |
| Total noise voltage | 1402 | 836 | 1127 |

Some further rules of thumb follow from this example:

- high impedance circuits are noisy,

- in low impedance circuits, op-amp voltage noise will be the dominant factor,

- in high-impedance circuits, one or other of resistor noise or op-amp current noise will dominate: use a biFET or CMOS device and delete R1, and

- don't expect a low-noise-voltage op-amp to give you any advantage in a high impedance circuit.

### 11.11.12.3 Noise Bandwidth

Deciding the actual noise bandwidth is not always simple. The bandwidth used in the noise calculations is a notional "brick-wall" value, which assumes infinite attenuation above the cut-off frequency. This of course is not achievable in practice, and the circuit bandwidth has to be adjusted to reflect this fact. For a single-pole response with a cutoff frequency $f_c$ and a roll-off of 6 dB/octave, the noise bandwidth is $1.57f_c$. For a cascade of single-pole filters the ratio of the noise bandwidth to cut-off frequency decreases.

For more complex circuits it is usually enough to make some approximation to the actual bandwidth. If the low-frequency cut-off is more than a decade below the high frequency one then it can be neglected with little error, and the noise bandwidth can be taken as from DC to the high-frequency cut-off. The exception to this is in very-low frequency and DC applications (below a few tens of Hz), because at some point the op-amp noise contribution starts to rise with decreasing frequency. This region is known as 1/f or "flicker" noise. All op-amps show this characteristic, but the point at which the noise starts to rise (the 1/f noise corner) can be reduced from a few hundred Hz to below 10 Hz by careful design of the device.

### 11.11.13 *Supply Current and Voltage*

Circuit diagrams often leave out the supply connections to op-amp packages, for the very good reason that they create extra clutter, and the purpose of a circuit diagram is to communicate information as clearly as it can. When a single supply or a dual-rail supply is used throughout a circuit then confusion is unlikely, but with several different voltage levels in use it becomes difficult to work out exactly which op-amp is supplied by what voltage, and it is then better practice to show supplies to each package.

#### 11.11.13.1 *Supply Voltage*

By far the largest number of recent op-amp introductions are aimed at low-power, single-supply applications where the circuit is battery-operated. The lithium battery voltage of 3 V is a major driving force in this trend. Although "low power" and "single supply" are independent parameters, they usually coexist as circuit specifications. A few years ago, op-amps were associated with $\pm 15$ V supplies, which shrank to $\pm 5$ V then to just $+5$ V; now, nominal $+3$ V supplies are common, with surprisingly little sacrifice in device performance. But low-power, low-voltage devices are not as forgiving of system design shortcomings because they have less input range to accommodate poorly behaved input signals, less head room to deal with dynamic range requirements, and less output drive capacity. System design decisions should still favour higher voltage rails where these are possible.

#### 11.11.13.2 *Supply Current*

One of the dis-benefits of not showing supply connections is that it is easy to forget about supply current ($I_S$). Data sheets will normally give typical and maximum figures for $I_S$ at a specified voltage, and no load. If the supply voltages are the same in the circuit as on the data sheet, and if none of the outputs are required to deliver any significant current, then it is reasonable just to add the maximum figures for all the devices in circuit to arrive at a worst-case power consumption. At other supply voltages you will have to make some estimate of the true supply current, and some data sheets include a graph of typical $I_S$

versus supply voltage to aid in this. Also, note that $I_S$ varies with temperature, usually increasing with cold.

When an op-amp output drives a load, be it resistive, capacitive or inductive, the current needed to do so is drawn from one supply rail or the other, depending on the polarity of the output. In the worst case of a short circuit load, $I_S$ is limited by the device's output current limiting. It is quite possible for the load currents to dominate power supply drain. With typical quiescent $I_S$ figures of a milliamp or so, you only need an output load resistance of $10\,k\Omega$ being driven with a $\pm 10\,V$ swing to double the actual current consumption of the circuit. When calculating worst-case load currents in these circumstances, you need to know not only the maximum output swing into resistive loads, but also the current that may be needed to drive capacitive loads.

### 11.11.13.3 $I_S$ vs. Speed and Dissipation

Op-amp supply current is usually a trade-off against speed. You can find devices which are spec'd at $10\,\mu A$ $I_S$, but such a part can only offer a slew rate of $0.03\,V/\mu s$. Conversely, fast devices require more current, often up to $10\,mA$. At these levels, package dissipation rears its head. An op-amp run at $\pm 15\,V$ with $10\,mA$ $I_S$ is dissipating $300\,mW$. With a thermal resistance of $100-150°C/W$ (the data sheet will give you the exact value) its junction temperature will be $30-45°C$ above ambient, and this is before it drives any load! This could well prevent the use of the part at high ambient temperatures, and will also affect other parameters which are temperature sensitive. With such a device, make sure you know what its operating temperature will be before getting deeply involved in performance calculations.

### 11.11.14 Temperature Ratings

And so we come naturally to the question of over what temperature range can you use a particular device. Analog ICs historically have been marketed for three distinct sectors, with three specified temperature ranges:

- Commercial: 0 to $+70°C$

- Industrial: $-40$ to $+85°C$ (occasionally $-25$ to $+85°C$)

- Military: $-55$ to $+125°C$

The picture is nowadays slightly blurred with the introduction of parts for the automotive market, which may be spec'd over −40 to +125°C, and with some Japanese suppliers (predominantly in the digital rather than analog area) offering non-standard ratings such as −20 to +75°C.

If you are designing equipment for the typical commercial environment of 0 to 50°C then you are not going to worry much about device temperature ratings: just about every IC ever made will operate within this range. At the other extreme, if you are designing for military use then you will be buying military qualified components, paying the earth for them and this book will be of little use to you. But the question quite frequently arises, what parts should I use when my ambient temperature range goes a few degrees below zero or above 70°C?

In theory, you should use industrial-temperature-rated devices. Unhappily, there are three good reasons why you might not:

- the part you want to use may not be available in the industrial range;

- if it is available, it may be too expensive;

- even if it is listed as available, it may actually prove to be on a long lead-time or otherwise hard to get.

So the question resolves itself into: Can I use commercial parts outside their specified temperature range? And the answer is: Maybe. No IC manufacturer will give you a guarantee that the part will operate outside the temperature range that he specifies. But the fact is, most parts will, and there are two main factors which limit such use, namely specifications and reliability.

### 11.11.14.1 Specification Validity

The manufacturer will specify temperature-sensitive parameters (which is most of them) either at a nominal temperature (25°C) or over the temperature range. These specifications have bite, in that if the part fails to meet them the customer is entitled to return it and ask for a replacement. So the manufacturer will test the parts at the specification limits. However, he is not responsible for what happens outside the temperature range, and it is more than likely that some parameters will drift out of their specification when the temperature limits are over-stepped. Very often these parameters are unimportant in the application, such as offset voltage in an AC amplifier. Therefore you can with care design a circuit with wider tolerances than would be needed for the published figures and trust that these will be sufficient for wide temperature range abuse.

It is of course a risky approach, and two extra risks are that some parameters may change much more outside the specified temperature range than they do within it, and that you may successfully test a sample of manufacturer A's product, but manufacturer B's nominally identical parts behave quite differently. We shall comment on this again in section 11.2.15.

### 11.11.14.2  Package Reliability

The second factor is reliability. The reliability of any semiconductor device worsens with increasing temperature; a temperature rise of 10°C halves the expected lifetime. So operating ICs at high temperatures is to be avoided wherever possible, but there is no magic cut-off at 70°C or 85°C. The maximum junction temperature should always be observed, but this is usually in the region of 100–150°C.

At low temperatures the problem is included moisture. Molded plastic packages allow some moisture to creep along the lead-to-plastic interface (this is worse at high temperatures and humidities) and this can accumulate over the surface of the chip, where it is a long-term corrosive influence. When the operating temperature dips below 0°C the moisture freezes, and the resulting change in conductivity and volume can give sudden changes in parameters which are well outside the drift specifications. The effect is very much less with "hermetic" packages using a glass-ceramic-metal seal, and in fact progress in plastic packages has advanced to the point where included moisture is not as serious a problem as it used to be. Other board-related problems arise when equipment is used below 0°C due to condensation of airborne moisture on the cold pcb surface, as ambient temperature rises.

### 11.11.15  Cost and Availability

The subtitle to this section could be, why use industry standards? Basically, the application of op-amps (along with virtually all other components) follows the 80/20 rule beloved of management consultants: 80% of applications can be met with 20% of the available types. These devices, because of their popularity, become "industry standards" and are sourced by several manufacturers. Their costs are low and their availability is high. The majority of other parts are too specialised to fulfil more than a handful of applications and they are only produced by one or perhaps two manufacturers. Because they are only made in small quantities their cost is high, and they can sometimes be out of stock for months.

### 11.11.15.1  When to Use Industry Standards

The virtue of selecting industry standards is that the parts are well established, unlikely in the extreme to run into sourcing problems or be withdrawn (the humble 741 has been around for over 30 years!) and, because of the competition between manufacturers, they will remain cheap. If they will do the job, use them in preference to a sole sourced device. For companies with many different designs of product, keeping the variety of component parts low and re-using them in new designs has the benefit of increasing the total purchase of any given part. This potentially reduces its price further.

Another hidden advantage of older, more established devices is that their quirks and idiosyncrasies are well known to the suppliers' applications support engineers, and you are less likely to run into unusual effects that are peculiar to your usage and that take days of design time to resolve.

But nothing comes for free: the negative aspect of multisourcing is that many parameters go unspecified for cheap devices, and this leaves open the possibility that different manufacturers' nominally identical parts can differ substantially in those parameters that are omitted from the common spec. If you've designed and tested a circuit with manufacturer A's devices, and they happen to be quite fast, you will be heading for production problems when your purchasing manager buys a few thousand of manufacturer B's devices which are slower. For instance, TI's data sheet for the LM324 gives a typical slew rate of $0.5\,\mathrm{V/\mu s}$ at 5 V supply; but National, who could fairly be said to have invented the part, do not mention slew rate at all in theirs.

To deal with this, design the circuit from the outset to be insensitive to those parameters which are badly specified, un-specified or (worse) specified differently in the data sheets of each manufacturer. Or, look for a more tightly specified part.

### 11.11.15.2  When Not to Use Industry Standards

Within the last few years there has been a countertrend to the imperative for multisourcing and the use of industry standards. Hundreds of new types have been introduced, and many of them are much better than their predecessors. They not only minimize the trade-offs between speed, power, precision and cost, but are also more fully specified. You can select a part by function and application—for instance, a DAC buffer or 75-ohm cable driver—rather than by comparing technologies, or by looking at a particular specification such as gain bandwidth product. Following the manufacturer's selection guides on the basis of application will often lead quickly to the most suitable part.

Selecting more application-specific ICs in this way steers the design process away from industry standards. But there are a number of reasons why alternate sourcing has become less of a necessity, despite its advantages given above. The average product life cycle—sometimes months rather than years—is much shorter than the lifetime of a good op-amp. In addition, qualifying multiple sources is a task that many designers don't have time or expertise to do fully. Finally, for highly competitive products, you'll have to choose parts that give your design the edge (even if they are proprietary) and for which there may be nothing comparable in performance, cost, or functionality.

### 11.11.15.3 Quad or Dual Packages

Comparing prices, the LM324 does offer, in fact, the lowest cost-per-op-amp (5p). This points up another factor to bear in mind when selecting devices: choose a quad or dual package in preference to a single device, when your circuit uses several gain stages. This reduces both unit cost and production cost. Such parts often have quiescent supply currents only slightly greater than a single-channel device, but with better offset, temperature tracking of drift, matching, and other specs. The disadvantages are inflexibility in supply voltage and pc layout, and possible thermal, power rail or RF interaction between gain blocks on a single substrate.

Some parts are available only in dual or quad configurations because single-channel versions would not have enough applications. Conversely, highest speed op amps, with bandwidths above several hundred megahertz, are often available in singles only, because of internal crosstalk. However, pinouts in multichannel configurations are less standardised than the basic single-channel unit, so substitutes are harder to find.

### 11.11.16 Current Feedback Op-Amps

There are also op-amps that use current feedback topology instead of the more familiar voltage feedback (Figure 11.36). Voltage feedback is the classic, well-understood

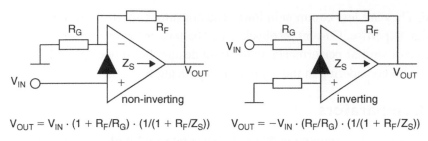

$$V_{OUT} = V_{IN} \cdot (1 + R_F/R_G) \cdot (1/(1 + R_F/Z_S)) \qquad V_{OUT} = -V_{IN} \cdot (R_F/R_G) \cdot (1/(1 + R_F/Z_S))$$

**Figure 11.36: The current feedback circuit**

mechanism which we have been discussing all the way through this section so far. In current feedback, the error signal is a current flowing into the inverting input; the input buffer's low impedance, in contrast to a voltage amplifier's high input impedance, allows large currents to flow into it with negligible voltage offset. This current is the slewing current, and slew rate is a function of the feedback resistor and change in output voltage. Therefore, the current-feedback amplifier has nearly constant output transition times, regardless of amplitude.

A very small change in current at the inverting input will cause a large change in output voltage. Instead of open-loop voltage gain, the current feedback op-amp is characterized by current gain or "transimpedance" $Z_S$. As long as $Z_S \gg R_F$, the feedback resistor, the steady-state (non-slewing) current at the inverting input is small and it is still possible to use the usual op-amp assumptions as initial approximations for circuit analysis, i.e., the differential voltage between the inputs is negligible, as is the differential current.

In performance, current feedback generally offers higher slew rate for a given power consumption than voltage feedback, and voltage feedback offers you flexibility in selecting a feedback resistor, two high-impedance inputs, and better DC specifications. With a current-feedback op-amp, you first set the desired *bandwidth* via the feedback resistor, and then the gain is set according to the usual resistive ratios. This means that the wider the bandwidth, the lower will be the operating impedances. If $R_F$ is doubled, the bandwidth will be halved. The circuit becomes *less* stable when capacitance is added across the feedback resistor.

Current feedback devices tend to be used only at higher frequencies, for applications such as professional video and high-performance wideband instrumentation. The same part can be used in several applications for quantity cost savings, using only as much bandwidth as needed. They are less common in lower end consumer applications because they need more design expertise. Current feedback is no "better" or "worse" than voltage, which is also capable of similar performance in the right design, but it does provide an alternative which is worth considering in the appropriate application.

## 11.12 Comparators

A comparator is just an op-amp with a faster slew rate, and with its output optimized for switching. It is intended to be used open-loop, so that feedback stability considerations don't apply. The device exploits the very large open-loop gain of the op-amp circuit so that the output swings between "fully-on" and "fully-off," depending on the polarity of the differential input voltage, and there should be no stable state in between. Input-referred and open-loop parameters—offsets, bias currents, temperature drift, noise, common mode and power supply rejection ratio, supply current and open-loop gain—are all specified in the same way as op-amps. Output and AC parameters are specified differently.

### 11.12.1 Output Parameters

The most frequent use of a comparator is to interface with logic circuitry, so the output circuit is designed to facilitate this. Two configurations are common: the open collector, and the totem pole (Figure 11.37). The open-collector type requires a pull-up resistor externally, while the totem-pole does not. Both types interface readily to the classical LSTTL logic input, which requires a higher pull-down current than is needed to pull it up. The CMOS input, which only takes a small current at the transition due to its input capacitance, is even easier. The output is specified either in terms of its saturation voltage, sink current, leakage current and maximum collector voltage for the open-collector type or in terms of high- and low-level output voltages at specified load currents for the totem-pole type.

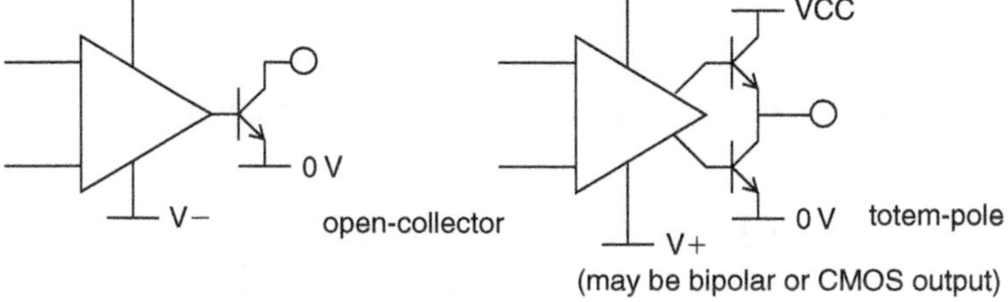

**Figure 11.37: Comparator outputs**

Because the totem-pole type is invariably aimed at logic applications, it is always specified for 3.3 or 5 V output levels. The open-collector type, which includes the highly popular LM339/393 and is derivatives, is more flexible since any output voltage can be obtained simply by pulling up to the required rail, which can be separate from the analog supply rails.

### 11.12.2 AC Parameters

Because the comparator is used as a switch, the only AC parameter which is specified is the response time. This is the time between an input step function and the point at which the output crosses a defined threshold. It includes the propagation delay through the IC and the slewing rate of the output. Outside of the device itself, two factors have a large effect on the response time:

- the input overdrive, and

- the output load impedance.

### 11.12.2.1 Overdrive

For the specifications, an input step function is applied which forces the differential input voltage from one polarity to the other. The overdrive, as in Figure 11.38, is the final steady-state differential voltage. Usually, the step amplitude is held constant and its offset is varied to give different overdrive values. The greater the overdrive, the more current is available from the differential input stage to propagate the change of state through to the output, although beyond a certain point there is no gain to be had from increasing it. Small overdrives can lead to suprisingly long response times and you should check the data sheet carefully to see if the device is being specified in similar fashion to how your circuit will drive it.

The specification test assumes that the step function has a much shorter rise time than the response to be measured. Response time specs are virtually meaningless when the

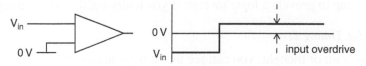

**Figure 11.38: Comparator overdrive**

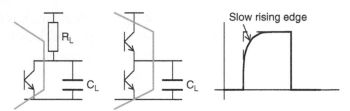

**Figure 11.39: Output slewing vs. load capacitance**

comparator is driven by slow rise time analog signals. We shall discuss this more fully under the heading of hysteresis.

### 11.12.2.2 Load Impedance

The output load resistance $R_L$ (for open-collector types) and capacitance $C_L$ have a major influence on the output slewing rate. The capacitance includes the device output capacitance, circuit strays and the input capacitance of the driven circuit (this last is usually the most significant). The slewing rate is determined by the current that is available to charge and discharge the capacitance, following the rule $dV/dt = I/C$. For the negative-going transition this current is supplied by the output sink transistor and is in the region of 10–50 mA, assuring a fast edge, but the current available to charge the positive transition is supplied by the pull-up device or resistor and may be an order of magnitude lower. The choice of output resistor directly affects the positive-going rise time (Figure 11.39) and the power dissipation of the circuit.

### 11.12.2.3 The Advantages of the Active Low

On this latter point, it is worth remembering that if you expect low-duty-cycle pulses at the output, want low power drain and a fast leading edge and have a choice of logic polarity, that the preferable configuration is to use an active-low output as in Figure 11.40(a). The signal is normally off so that power drain is low, and the leading edge transition depends on the output transistor rather than the pull-up. If a fast trailing edge is also needed, the pull-up can be reduced in value without significantly affecting power drain if the duty cycle is low. It is easy and cheap to provide a logic inverter if you really need positive going pulses.

### 11.12.2.4 Pulse Timing Error

Continuing this train of thought, you can see that it is quite easy for the pulse timing to be affected by the output rise- and fall-times. This is quite often the source of unexpected

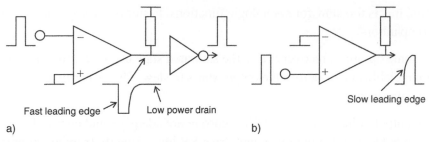

**Figure 11.40: Comparator output configurations. (a) Preferred configuration; (b) Poor configuration**

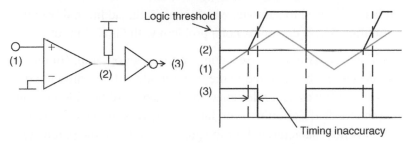

**Figure 11.41: Timing error through pull-up delay**

errors in circuits which convert analog levels into pulse widths for timing measurement. Because the pulse rising edge is slowed to a greater extent than the falling edge, the point at which it crosses the following logic gate's switching threshold is different, so that rising and falling analog inputs result in different switching points. This effect is demonstrated in Figure 11.41. The problem is generally more visible with CMOS-input-level gates than it is with TTL-input-level ones, as TTL's switching threshold is closer to 0 V whereas the CMOS threshold is ill-defined, being anywhere between 0.3 and 0.7 times its supply rail. The difference can amount to a microsecond or more in low power circuits.

### 11.12.3 Op-Amps as Comparators (and Vice Versa)

You may often be faced with a circuit full of multiple op-amp packages and the need for a single comparator. Rather than invest in an extra package for the comparator function, it is quite in order to use a spare op-amp as a comparator with the following provisos:

- the response time and output slew-rate are adequate. Typical cheap op-amp slew rates of 0.5 V/μs will traverse the logic "gray area" from 0.8 to 2 V in about

$3\,\mu s$; this is too slow for some logic functions. Faster op-amps make better comparators.

- in some op-amps, recovery from the saturated state can take some time, causing appreciable delays before the output starts to slew. This is hardly ever specified on data sheets.

- the output voltage swing and drive current are adequate and correct for the intended load. Clearly you cannot drive 5 V logic directly from an op-amp output that swings to within 2 V of $\pm 15\,V$ supply rails. Some form of interface clamping is needed; this could take the form of a feedback zener arrangement so that the output is not allowed to saturate, which confers the additional benefit of reduced response time. Drive current is not a problem with CMOS inputs.

It is also possible, if you have to, to use a comparator as an op-amp. (In most cases: some totem-pole outputs cannot be operated in the linear mode without drawing destructively large supply currents.) It was never designed for this, and will be hideously unstable unless you slug the feedback circuitry with large capacitors, in which case it will be slow. Also, of course, it is not characterized for the purpose, so for some parameters you are dealing with an unknown quantity. Unless the application is completely noncritical it is best to design op-amp circuits with op-amps.

### 11.12.4 Hysteresis and Oscillations

When the analog input signal is changing relatively slowly, the comparator may spend appreciable time in the linear mode while the output swings from one saturation point to the other. This is dangerous. As the input crosses the linear-gain region the device suddenly becomes a very-high-gain open-loop amplifier. Only a small fraction of stray positive feedback is needed for the open-loop amplifier to become a high frequency oscillator (Figure 11.42).

The frequency of oscillation is determined by the phase shift introduced by the stray feedback and is generally of the same order as the equivalent unity gain bandwidth. This is not specified for comparators, but for typical industry-standard devices is several megahertz. The term "relatively slowly" as used above means relative to the period of the oscillation, so that any traverse of the linear region which takes longer than a few hundred nanoseconds must be regarded as slow: this of course applies to a very large proportion of analog input signals!

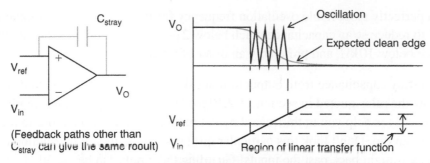

**Figure 11.42: Oscillation during output transitions**

### 11.12.4.1 The Subtle Effects of Edge Oscillation

This oscillation can be particularly troublesome if you are interfacing to fast logic circuits, especially when connecting to a clock input. It can be hard to spot on the scope, as you will probably have the timebase set low for the analog signal frequency, but the oscillations appear to the digital input as multiple edges and are treated as such: so for instance a clock counter might advance several counts when it appears to have had only one edge, or a positive-going clock input might erroneously trigger on a negative-going edge.

Even when you don't have to contend with high-speed logic circuits, the oscillation generated by the comparator can be an unexpected and unwelcome source of RF interference.

### Minimize Stray Feedback

The preferred solution to this problem is to reduce the stray feedback path to a minimum so that the comparator remains stable even when crossing the linear region. This is achieved by following three golden rules:

- keep the input drive impedance low;

- minimize stray feedback capacitance by careful layout;

- avoid introducing other spurious feedback paths, again by careful layout and grounding.

The lower the input impedance, the more feedback capacitance is needed to generate enough phase shift for instability. For instance, $2\,\mathrm{pF}$ and $10\,\mathrm{k\Omega}$ gives a pole frequency of

8 MHz, a perfectly respectable oscillation frequency for many high-speed comparators. It is hard to reduce stray capacitance much below 2 pF, so the moral is, keep the drive impedance below 10 kΩ, and preferably an order of magnitude lower.

Minimum stray capacitance from output to input should always be the layout designer's aim; follow the rules quoted in section 11.2.10 for high-frequency op-amp stability. Most IC packages help you in this regard by not putting the output pin close to the non-inverting input pin. Don't look this particular gift horse in the mouth by running the output track straight back past the inputs! Guarding the inputs can be useful. And, again as with op-amp circuits, do not introduce ground-loop or common mode feedback paths by incorrect layout.

### 11.12.4.2 Hysteresis

Another approach to the problem of unwanted oscillation is to kill it with hysteresis. This approach is used when the above methods fail or cannot be applied, and you can also use it as a legitimate circuit technique in its own right, as in the well-known Schmitt trigger. Hysteresis is the application of deliberate positive feedback in order to propel the output speedily and predictably through the linear region. The principle of hysteresis is shown in Figure 11.43.

Note that although this looks superficially like the classic inverting op-amp configuration, feedback is applied to the *non*-inverting input and is therefore operating in the positive sense. Note also that the application of hysteresis modifies the switching threshold in both directions, and that it is modified differently in either direction by the presence of R3. This resistor is shown in the circuit of Figure 11.44 to emphasize that it must be included in calculating hysteresis; we have assumed that the comparator is the open-collector type. If the output is the totem-pole type, then R3 is omitted but the output levels and impedance must be taken into consideration. These values directly affect the switching threshold and can cause surprisingly large inaccuracies.

Because hysteresis deliberately alters the switching threshold, it cannot be indiscriminately applied to all comparator circuits to clean up their oscillatory tendencies, nor should it. The techniques outlined previously should be the first priority. But it is not always possible to keep drive impedances low and where high impedance is necessary, hysteresis is a valuable tool. If the minimum input *dV/dt* is predictable, you can also apply a judicious amount of AC hysteresis (by substituting a capacitor for R2), which will

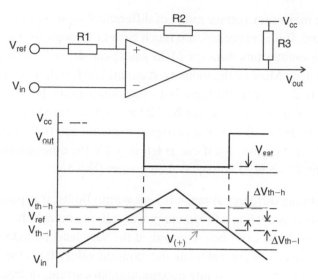

Ignoring input and output leakage currents,

$V_{out}(H) = V_{cc} - (V_{cc} - V_{ref})(R3/[R1 + R2 + R3])$

$V_{out}(L) = V_{sat}$

$V_{th-h} = \alpha \cdot V_{cc} + (1 + \alpha) \cdot V_{ref}$   where $\alpha = R1/(R1 + R2 + R3)$

$\Delta V_{th-h} = \alpha \cdot (V_{cc} + V_{ref})$

$V_{th-l} = \beta \cdot V_{sat} + (1 - \beta) \cdot V_{ref}$   where $\beta = R1/(R1 + R2)$

$\Delta V_{th-l} = \beta \cdot (V_{sat} - V_{ref})$

A common simplification is that $R3 \ll R1 + R2$ so that $\alpha = \beta$, and that $V_{ref}$ is half of $V_{cc}$ and $V_{sat} = 0$, in which case $\Delta V_{th}$ (the total hysteresis band) reduces to $\beta \cdot V_{cc}$.

**Figure 11.43: Hysteresis**

prevent oscillation without affecting the DC threshold: but beware slow-moving inputs or you will simply end up with a longer time-constant oscillator!

## 11.12.5  Input Voltage Limits

When an op-amp is operating closed-loop, the differential voltage at its inputs is theoretically zero. If it isn't then the feedback loop is open, either by design or because of one or another form of overloading. Comparators on the other hand are intended for open-loop operation and their differential input voltage is never expected to be zero.

Data sheets specify the maximum voltage range of differential input signals and this should not be overlooked. If it is exceeded, too much current through the breakdown of the input transistor base-emitter junctions (or MOS gates) can degrade the input offset and bias current parameters. Most of the industry-standard LM339 derivatives have a differential limit equal to the supply rail limit, but some comparators have quite restricted differential input ranges. For instance the fast NE529 has a differential input restriction of $\pm 5$ V, with a common mode of $\pm 6$ V. These two quantities interact: both inputs at $+4$ V will satisfy the common mode limit, but if one is left at $+4$ V the other cannot be taken below $-1$ V because the differential voltage is then greater than 5 V.

Even if the normal operating differential range is kept within limits, it is possible for abnormal conditions (such as cycling of separate power rails) to breach the limit. If this is at all likely, and if the condition can't be prevented, at the very least include some input current protection resistance. You can calculate the required values from the expected or possible overvoltage divided by the absolute maximum input current, or from the power dissipation, which is always quoted on device data sheets.

### 11.12.5.1 Comparator Parameters vs. Input Voltage

Also, while considering large differential input voltages, remember that unexpected things can happen to the comparator even when the limits are not exceeded. Response time is usually specified for a common mode voltage of zero and may degrade when the common mode limits are approached; this applies equally to bias currents. Some data sheets show curves of input bias current which have step changes (Figure 11.44) at certain differential input voltages, due to internal DC feedback. Notice these and make sure your circuit can cope!

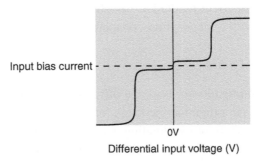

Figure 11.44: Input bias current steps

In multichannel packages, some comparators may remain unused. Never leave unused inputs open, as that device could oscillate on its own, which would then be coupled into the other devices in the same package. If both inputs are grounded, the unpredictable offset voltage will mean that the output voltage, and hence unit supply current, will vary. The safest course is to ground one input and supply the other from another fixed voltage within its differential and common mode limit (which might include the supply rail), so that the device is always saturated.

### 11.12.6 Comparator Sourcing

Exactly the same comments about sourcing apply to comparators as were made earlier about op-amps (see section 11.2.15). Like the LM324 op-amp, the most popular and cheapest part per comparator is the quad LM339, with its dual counterpart the LM393 not far behind.

## 11.13 Voltage References

The need for a stable reference voltage is found in power supplies, measurement instrumentation, DAC/ADC systems and calibration standards. Two techniques exist to provide such references, one based on the precision zener diode and the other on the band-gap voltage of silicon.

### 11.13.1 Zener References

We have already discussed the operation of the basic zener diode. To produce a reference from a zener, it must be temperature compensated, fed from a constant current and buffered. Temperature compensation is achieved by selecting a low-tempco zener voltage in the range $5.5-7$ V and mating it with a silicon diode so that the voltage tempcos cancel. The combination is driven from a constant current generator and buffered to give a constant output voltage regardless of load.

Since surface breakdown increases noise and degrades stability, a precision zener is usually fabricated below the surface of the IC which contains its support circuitry, but this gives a greater spread of tempco and absolute voltage. The overall reference must therefore allow adjustment of these parameters, normally by laser wafer trimming. Such references can offer long-term stability of 50 ppm/year and absolute accuracy of 0.1% with $\pm 10$ ppm/°C tempco. Better performance is obtained if the reference can be stabilized with an on-chip heater, as in the LM399 for example. This takes a comparatively large power drain and has a warm-up time measured in seconds but offers sub-ppm tempcos.

### 11.13.2 Band-Gap References

A significant disadvantage of the zener reference is that its output voltage is set at around 6.9 V and it therefore needs a comparatively high supply voltage. A competing type of reference overcomes this and other problems, notably cost and supply current, and has become extremely widespread since its invention by Robert Widlar in 1971. The fundamental circuit is shown in Figure 11.45. In this circuit I1 and I2 differ by a fixed ratio and $V_{ref}$ is given (neglecting base currents) by

$$V_{ref} = V_{BE3} + I2 \cdot R2 = V_{BE3} + (V_{BE1} - V_{BE2}) \cdot R2/R1$$

The temperature coefficient of the second term can be arranged by suitable selection of I1, R1 and R2 to cancel that of the $V_{BE3}$ term. This turns out to occur when $V_{ref}$ is in the neighborhood of 1.2 V, which is equivalent to the "band-gap" voltage of a silicon junction at 0°K.

Such a band-gap reference, relying only on matched transistors, is easily integrated along with biasing, buffer and amplifier circuitry to give a complete reference in a single package. It is capable of a lower minimum operating current and a sharper knee than any zener. As well as the unprocessed band-gap voltage of 1.2 V (actual voltage depends on detailed internal design and process variations and varies between 1.205 and 1.26 V)

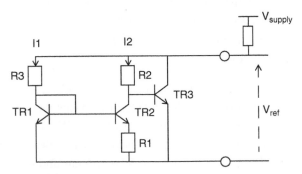

**Figure 11.45: The band-gap reference**

devices are available with trimmed outputs of 2.5, 5 and 10 V, principally for use in digital-to-analog/analog-to-digital conversion circuits. Other voltages are available, and there are several adjustable parts offered as well.

### 11.13.2.1 Costs and Interchangeability

There is an obvious trade-off between initial voltage tolerance and tempco on the one hand, and cost and availability on the other, since the manufacturer has to accept a lower yield and longer test and trim time for the closer tolerances. Initial voltage can be trimmed exactly with a potentiometer, but this method adds both parts and production cost which will offset the higher cost of a tighter-tolerance part. Trimming the reference voltage can also worsen the reference temperature coefficient in some configurations, and there is the extra tempco of the trimming components to include. Table 11.5 shows a sample of typical two-terminal 1.2 V references, including their tolerance, tempco, minimum operating current and cost. Most of these are available in different grades, corresponding to tighter or looser tolerances and tempcos.

Although it would appear from this table that there is a wide choice of types offering much the same performance, not all of these are directly interchangeable. The minor differences in regulation voltage may catch you out if you have designed a circuit for a given voltage tolerance and subsequently want to change to a different type. The preferable solution is to allow as wide a tolerance as possible in the first place.

Table 11.5: Some voltage references

| Type | Output voltage | Tolerance | Tempco | Min. current | Cost £, 25+ |
|------|----------------|-----------|--------|--------------|-------------|
| MAX6520EUR-T | 1.2 V | ±1% | 20 ppm/°C typ | 50 µA | 1.29 |
| LM4041B-1.2 | 1.225 V | ±0.2% | 100 ppm/°C | 45 µA | 0.97 |
| ICL8069DCZR | 1.23 V | ±1.6% | 100 ppm/°C | 50 µA | 0.78 |
| ICL8069CCZR | 1.23 V | ±1.6% | 50 ppm/°C | 50 µA | 1.27 |
| LM385Z-1.2 | 1.235 V | ±2% | 20 ppm/°C avg | 10 µA | 0.30 |
| LM385Z-1.2 | 1.235 V | ±1% | 20 ppm/°C avg | 10 µA | 0.55 |
| LT1004CZ-1.2 | 1.235 V | ±4 mV | 20 ppm/°C | 10 µA | 1.68 |
| ZRA124A01 | 1.24 V | ±1% | 30 ppm/°C | 50 µA | 0.67 |
| ZRA125F02 | 1.25 V | ±2% | 30 ppm/°C | 50 µA | 0.55 |

Also, there are variations in the allowable or required capacitive loading. Some parts require a decoupling capacitor of $0.1-1\,\mu F$ across them, others require that such a capacitor is *not* included. The parts are mostly supplied in the TO-92 package or the small outline SOT23, but not all pin-outs are the same. Again, check before specifying alternatives.

### 11.13.3  Reference Specifications

#### 11.13.3.1  Line and Load Regulation

Line regulation is the change in output voltage due to a specified change in input voltage, normally quoted in microvolts per volt. Load regulation is a similar change due to a change in load current, expressed either in percent for a given current change or as a dynamic resistance in ohms. It should, but doesn't always, include self-heating effects due to dissipation change.

#### 11.13.3.2  Output Voltage Tolerance

This is the deviation from nominal output voltage. It is quoted at a given temperature and input voltage or current, and the nominal voltage will differ under other conditions. Generally it is expressed as a percentage figure, but the asymmetry of device yields may persuade a manufacturer to quote upper and lower bounds and the nominal figure may not be in the middle of them. In your circuit design, it is best to ignore the nominal voltage and work everything out for upper and lower limits.

#### 11.13.3.3  Output Voltage Temperature Coefficient

This is the output voltage change due to an ambient temperature difference, usually from 25°C. Because neither band-gap nor zener references exhibit a straight line voltage-temperature curve (see Figure 11.46), manufacturers choose different ways to express their tempcos, sometimes as an average value across the range in ppm/°C, sometimes as different values at a series of spot temperatures, and sometimes as a worst-case error band in mV. To evaluate different manufacturers' references properly you need to correct for these differences in specification.

#### 11.13.3.4  Long-Term Stability

Usually expressed in ppm/1000hr or in microvolts change from the nominal voltage, this is a difficult specification to verify and so is often quoted as a typical figure based on characterisation of a sample. It is rarely specified on the cheaper components. Zeners

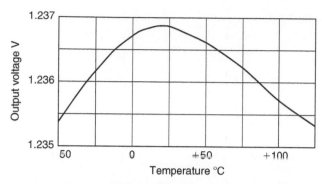

**Figure 11.46: Typical band-gap reference temperature characteristics**

tend to stabilise after a couple of years, so for ultra-precision applications the practice of burning in zener references at high temperatures to speed up the settling process is sometimes followed.

### 11.13.3.5 Settling Time

This is the time taken for the output to settle within a specified error band after application of power. It is typically in the tens to hundreds of microseconds region, and is normally only of interest if you are concerned about the dynamic performance of the reference circuit—for instance, if the application has to wake up rapidly from "sleep" mode. It does not include any long-term effects due to thermal shifts, but of course these do occur, more noticeably at higher operating currents.

### 11.13.3.6 Minimum Supply Current

The regulation of a two-terminal device is not maintained below a certain minimum current. Typical values for band-gap references are $50-100\,\mu A$, with $10\,\mu A$ being available although some earlier devices are much higher than this. The very low useable operating currents combined with low dynamic resistance at these currents make band-gap devices very much preferable to zener types for low-power circuitry. The maximum operating current is usually based on the point at which the device goes outside its regulation specification, but may also be determined by allowable power dissipation.

# CHAPTER 12

# Circuit Simulation

Mike Tooley

Computer simulation provides you with a powerful and cost-effective tool for designing, simulating, and analyzing a wide variety of electronic circuits. In recent years, the computer software packages designed for this task have not only become increasingly sophisticated but also have become increasingly easy to use. Furthermore, several of the most powerful and popular packages are now available at low cost either in evaluation, "lite" or student versions. In addition, there are several excellent freeware and shareware packages.

Whereas early electronic simulation software required that circuits be entered using a complex *netlist* that described all of the components and connections present in a circuit, most modern packages use an on-screen graphical representation of the circuit on test. This, in turn, generates a netlist (or its equivalent) for submission to the computational engine that actually performs the circuit analysis using mathematical models and *algorithms*. In order to describe the characteristics and behavior of components such as diodes and transistors, manufacturers often provide models in the form of a standard list of parameters.

Most programs that simulate electronic circuits use a set of algorithms that describe the behavior of electronic components. The most commonly used algorithm was developed at the Berkeley Institute in the United States and it is known as *SPICE* (Simulation Program with Integrated Circuit Emphasis).

Results of circuit analysis can be displayed in various ways, including displays that simulate those of real test instruments (these are sometimes referred to as *virtual instruments*). A further benefit of using electronic circuit simulation software is that, when a circuit design has been finalized, it is usually possible to export a file from the

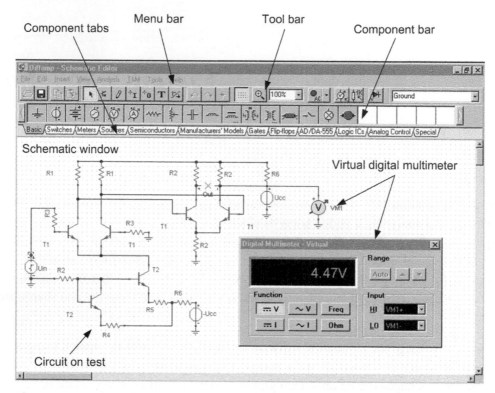

**Figure 12.1: Using Tina Pro to construct and test a circuit prior to detailed analysis**

design/simulation software to a PCB layout package. It may also be possible to export files for use in screen printing or CNC drilling. This greatly reduces the time that it takes to produce a finished electronic circuit.

## 12.1 Types of Analysis

Various types of analysis are available within modern SPICE-based circuit simulation packages. These are discussed in the following sections.

### 12.1.1 DC Analysis

DC analysis determines the DC operating point of the circuit under investigation. In this mode any wound components (e.g., inductors and transformers) are short-circuited and any capacitors that may be present are left open-circuit. In order to determine the initial conditions, a DC analysis is usually automatically performed prior to a transient analysis. It is also usually performed prior to an AC small-signal analysis in order to obtain the linearized, small-signal models for nonlinear devices. Furthermore, if specified, the DC small-signal value of a transfer function (ratio of output variable to input source), input resistance, and output resistance is also computed as a part of the DC solution. The DC analysis can also be used to generate DC transfer curves in which a specified independent voltage or current source is stepped over a user-specified range and the DC output variables are stored for each sequential source value.

### 12.1.2 AC Small-Signal Analysis

The AC small-signal analysis feature of SPICE software computes the AC output variables as a function of frequency. The program first computes the DC operating point of the circuit and determines linearized, small-signal models for all of the non-linear devices in the circuit (e.g., diodes and transistors). The resultant linear circuit is then analyzed over a user-specified range of frequencies. The desired output of an AC small-signal analysis is usually a transfer function (voltage gain, transimpedance, etc.). If the

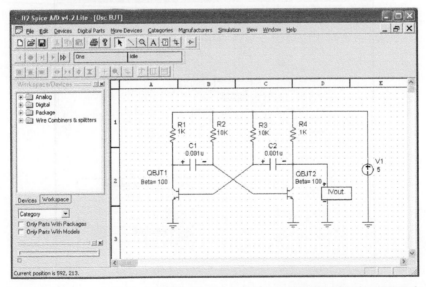

**Figure 12.2: An astable multivibrator circuit being simulated using B2 Spice**

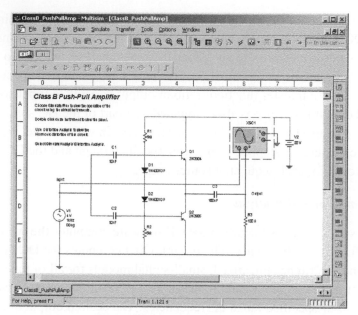

Figure 12.3: A Class B push-pull amplifier circuit being simulated by Multisim

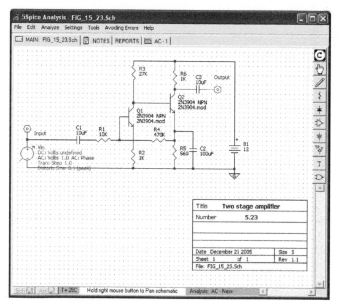

Figure 12.4: High-gain amplifier being analyzed using the 5Spice Analysis package

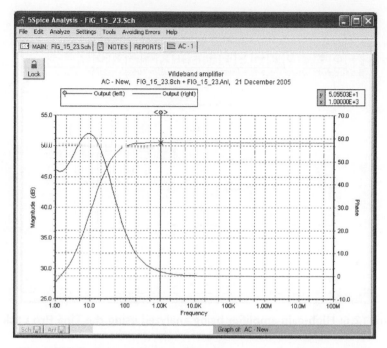

**Figure 12.5: Gain and phase plotted as a result of small-signal AC analysis of the circuit in Fig. 12.4**

circuit has only one AC input, it is convenient to set that input to unity and zero phase, so that output variables have the same value as the transfer function of the output variable with respect to the input.

### 12.1.3  Transient Analysis

The transient analysis feature of a SPICE package computes the transient output variables as a function of time over a user-specified time interval. The initial conditions are automatically determined by a DC analysis. All sources that are not time dependent (for example, power supplies) are set to their DC value.

### 12.1.4  Pole-zero Analysis

The pole-zero analysis facility computes the poles and/or zeros in the small-signal AC transfer function. The program first computes the DC operating point and then determines

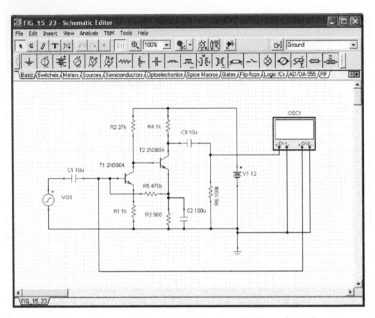

Figure 12.6: High-gain amplifier being analyzed using the Tina Pro package

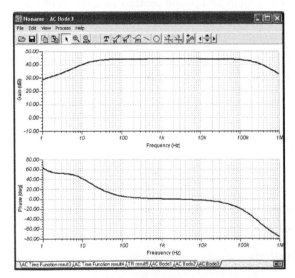

Figure 12.7: Gain and phase plotted as a result of small-signal
AC analysis of the circuit in Fig. 12.6

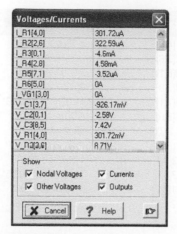

**Figure 12.8: Results of DC analysis of circuit shown in Fig. 12.6**

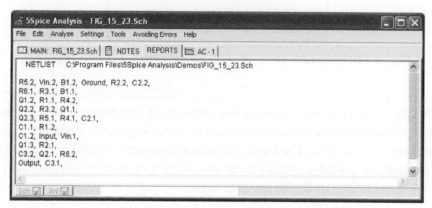

**Figure 12.9: Computer generated netlist for the circuit shown in Fig. 12.4**

the linearized, small-signal models for all the nonlinear devices in the circuit. This circuit is then used to find the poles and zeros of the transfer function.

Two types of transfer functions are usually supported. One of these determines the voltage transfer function (i.e., output voltage divided by input voltage) and the other usually computes the output *transimpedance* (i.e., output voltage divided by input current) or *transconductance* (i.e., output current divided by input voltage). These two

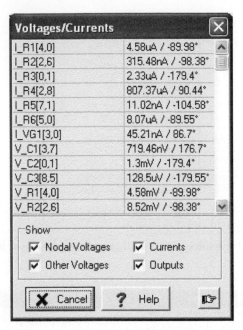

**Figure 12.10: Results of AC analysis of the circuit shown in Fig. 12.6**

transfer functions cover all the cases and one can make it possible to determine the poles/zeros of functions like impedance ratio (i.e., input impedance divided by output impedance) and voltage gain. The input and output ports are specified as two pairs of nodes. Note that, for complex circuits it can take some time to carry out this analysis and the analysis may fail if there is an excessive number of poles or zeros.

### 12.1.5 Small-Signal Distortion Analysis

The distortion analysis facility provided by SPICE-driven software packages computes steady-state harmonic and inter-modulation products for small input signal magnitudes. If signals of a single frequency are specified as the input to the circuit, the complex values of the second and third harmonics are determined at every point in the circuit. If there are signals of two frequencies input to the circuit, the analysis finds out the complex values of the circuit variables at the sum and difference of the input frequencies, and at the difference of the smaller frequency from the second harmonic of the larger frequency.

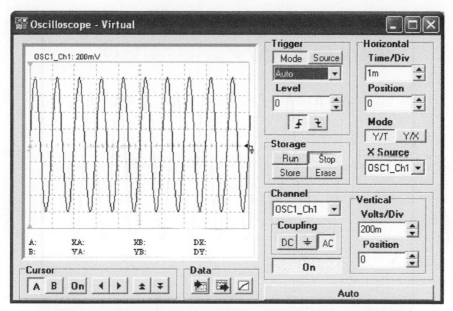

**Figure 12.11: Using the virtual oscilloscope in Tina Pro to display an
output voltage waveform for the circuit shown in Fig. 12.6**

## 12.1.6 Sensitivity Analysis

Sensitivity analysis allows you to determine either the DC operating-point sensitivity or
the AC small-signal sensitivity of an output variable with respect to all circuit variables,
including model parameters. The software calculates the difference in an output variable
(either a node voltage or a bran ch current) by perturbing each parameter of each
device independently. Since the method is a numerical approximation, the results may
demonstrate second order affects in highly sensitive parameters, or may fail to show very
low but nonzero sensitivity. Further, since each variable is perturbed by a small fraction of
its value, zero-valued parameters are not analyzed (this has the benefit of reducing what is
usually a very large amount of data).

## 12.1.7 Noise Analysis

The noise analysis feature determines the amount of noise generated by the components
and devices (e.g., transistors) present in the circuit that is being analyzed. When provided

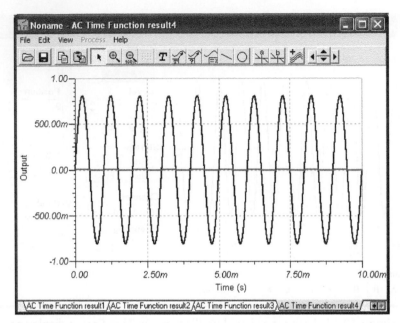

**Figure 12.12: Alternative waveform plotting facility provided in Tina Pro**

with an input source and an output port, the analysis calculates the noise contributions of each device (and each noise generator within the device) to the output port voltage. It also calculates the input noise to the circuit, equivalent to the output noise referred to the specified input source. This is done for every frequency point in a specified range. After calculating the spectral densities, noise analysis integrates these values over the specified frequency range to arrive at the total noise voltage/current (over this frequency range).

### 12.1.8 Thermal Analysis

Many SPICE packages will allow you to determine the effects of temperature on the performance of a circuit. Most analyses are performed at normal ambient temperatures (e.g., 27°C) but it can be advantageous to look at the effects of reduced or increased temperatures, particularly where the circuit is to be used in an environment in which there is a considerable variation in temperature.

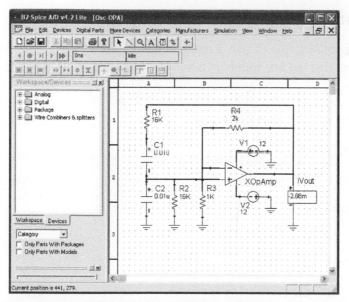

**Figure 12.13: Analysis of a Wien Bridge oscillator using B2 Spice**

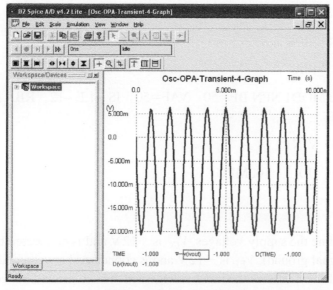

**Figure 12.14: Transient analysis of the circuit in Fig. 12.13 produced the output waveform plot**

## 12.2 Netlists and Component Models

The following is an example of how a netlist for a simple *differential amplifier* is constructed (note that the line numbers have been included solely for explanatory purposes):

1. SIMPLE DIFFERENTIAL PAIR

2. VCC  7  0    12

3. VEE  8  0    −12

4. VIN  1  0    AC 1

5. RS1  1  2    1K

6. RS2  6  0    1K

7. Q1  3  2  4    MOD1

8. Q2  5  6  4    MOD1

9. RC1  7  3    10K

10. RC2  7  5    10K

11. RE  4  8    10K

12. MODEL MOD1 NPN BF=50    VAF=50    IS=1.E−12    RB=100 CJC=.5PF TF=.6NS

13. .TF  V(5)  VIN

14. .AC  DEC  10  1    100MEG

15. .END

Lines 2 and 3 define the supply voltages. $V_{CC}$ is $+12\,V$ and is connected between node 7 and node 0 (signal ground). $V_{EE}$ is $-12\,V$ and is connected between node 8 and node 0 (signal ground). Line 4 defines the input voltage which is connected between node 1 and node 0 (ground) while lines 5 and 6 define 1 k$\Omega$ resistors (RS1 and RS2) connected between 1 and 2, and 6 and 0.

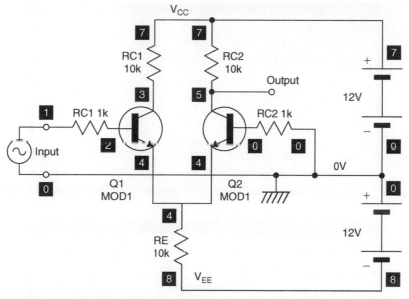

Figure 12.15: Differential amplifier with the nodes marked for generating a netlist

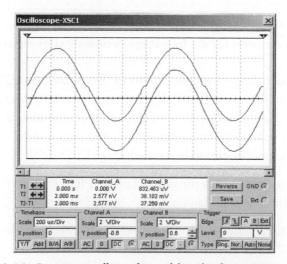

Figure 12.16: Cross-over distortion evident in the output waveform
from the Class B amplifier shown in Fig. 12.3

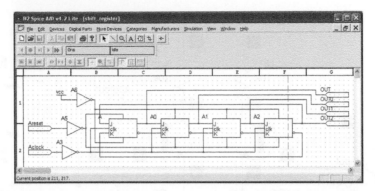

**Figure 12.17: Four-stage circulating shift register simulated using B2 Spice**

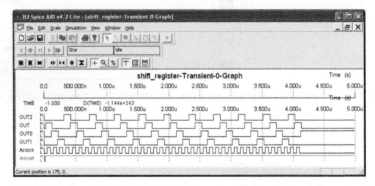

**Figure 12.18: Waveforms for the four-stage circulating shift register in Fig. 12.17**

Lines 7 and 8 are used to define the connections of two transistors (Q1 and Q2). The characteristics of these transistors (both identical) are defined by MOD1 (see line 12). Lines 9, 10 and 11 define the connections of three further resistors (RC1, RC2 and RE, respectively). Line 12 defines the transistor model. The device is NPN and has a current gain of 50. The corresponding circuit is shown in Fig. 12.15.

Most semiconductor manufacturers provide detailed SPICE models for the devices that they produce. The following is a manufacturer's SPICE model for a 2N3904 transistor:

NPN (Is=6.734f   Xti=3    Eg=1.11    Vaf=74.03    Bf=416.4    Ne=1.259
Ise=6.734    Ikf=66.78m    Xtb=1.5 Br=.7371    Nc=2Isc=0    Ikr=0    Rc=1
Cjc=3.638p    Mjc=.3085    Vjc=.75    Fc=.5    Cje=4.493p    Mje=.2593    Vje=.75
Tr=239.5n    Tf=301.2pItf=.4    Vtf=4    Xtf=2    Rb=10)

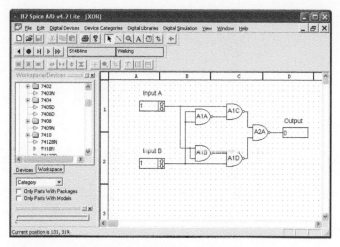

**Figure 12.19: Using B2 Spice to check the function of a simple combinational logic circuit**

## 12.3 Logic Simulation

As well as an ability to carry out small-signal AC and transient analysis of linear circuits (see Figs 12.3 and 12.16), modern SPICE software packages usually incorporate facilities that can be used to analyze logic and also "mixed-mode" (i.e., analog and digital) circuits. Several examples of digital logic analysis are shown in Figs 12.17, 12.18 and 12.19.

Figure 12.17 shows a four-stage shift register based on J-K bistables. The result of carrying out an analysis of this circuit is shown in Figure 12.18.

Finally, Figure 12.19 shows how a simple combinational logic circuit can be rapidly "assembled" and tested and its logical function checked. This circuit arrangement shows how the exclusive-OR function can be realized using only two-input NAND gates.

# CHAPTER 13

# Interfacing

Tim Williams

## 13.1 Mixing Analog and Digital

The two main problems which face designers who have to integrate analog and digital circuits on the same printed circuit board (PCB) are:

- preventing digital switching noise from contaminating the analog signal, and
- interfacing the wide range of analog input voltages to the digital circuit.

Generating analog outputs from digital signals is not usually a problem. Generating digital inputs from analog signals is.

### 13.1.1 Ground Noise

The high-frequency switching noise discussed earlier must be kept out of analog circuits at all costs. An analog-to-digital interface quantizes a variable analog signal into a digital word, and the number of bits in the word determines the resolution that can be achieved of the signal. Assuming a full-scale voltage range of 0 to 10 V, which is typical of many analog-digital converters (ADCs), Table 13.1 shows the voltage levels that correspond to one bit change in the digital word.

You can see that the more resolution is demanded of the interface, the smaller the voltage change that will cause one bit change. 8 bits is regarded as commonplace in ADC circuits, 12 bits as reasonably high resolution (0.025%) and 16 bits as precision.

The significance of these diminishing voltage levels is that any noise that is coupled into the analog input will cause unwanted fluctuation of the digital value. For a 12-bit converter, a 1-bit uncertainty will be given by noise of 2.4 mV at the converter input; for a 16-bit,

Table 13.1: ADC resolution voltage for different
word lengths, 10 V full-scale

| Word length | Resolution voltage |
|-------------|--------------------|
| 8 bit | 39 mV |
| 10 bit | 10 mV |
| 12 bit | 2.4 mV |
| 14 bit | 0.6 mV |
| 16 bit | 0.15 mV |

this reduces to 150 microvolts. By contrast, the switching noise on the digital ground line is normally tens of millivolts and frequently hundreds of millivolts peak amplitude. If this noise were coupled into the converter input—and it is hard to keep ground noise out of the input—you would be unable to use a converter of greater precision than 8–10 bits.

### 13.1.2 Filtering

One partial solution to this problem is to filter the bandwidth of the analog signal to well below that of the noise so that the effective noise signal is reduced. For slowly-varying analog signals this works reasonably well, especially if the noise injection occurs at the input of the signal-processing amplifier so that bandwidth limitation has maximum effect. Filtering is in any case good practice to minimize susceptibility to external noise.

Filtering the input amplifier is no use if the noise is injected into the ADC itself. For fast ADCs and wide-bandwidth analog signals you cannot take this approach anyway and the only available solution is to prevent the injection of digital noise at its source.

### 13.1.3 Segregation

The basic rule to follow when designing an analog-to-digital interface is to segregate the circuits, including grounds, completely. This means that:

- separate analog and digital grounds should be established, connected only at one point;
- the analog and digital sections of the circuit should be physically separated, with no digital tracks traversing the analog section or vice versa. This will minimize crosstalk between the circuits.

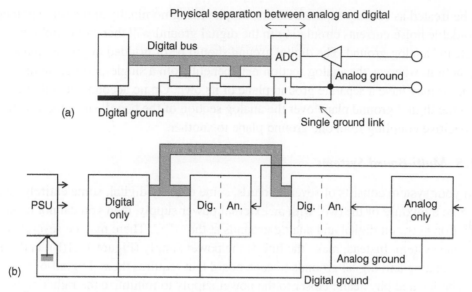

**Figure 13.1: Layout for separate analog and digital grounds (a) Single-board; (b) Multi-board**

It should be appreciated that no grounding scheme which establishes a multiplicity of different grounds can ever be optimum, because there will always be circuits which need to communicate signals across different ground areas. These signals are then particularly exposed to the nuances of both internal and external interference, or indeed may be the source of it. You should always strive to make such circuits low-risk in terms of their bandwidth and sensitivity, or else keep a single ground system for all circuits (both digital and analog) and take extreme care in its layout so that ground noise from one noisy part of the system does not circulate in another sensitive part.

### 13.1.4  Single-Board Systems

The appropriate grounding schemes for single-board and multi-board systems are shown in Figure 13.1. If your system has a single analog-to-digital converter, perhaps with a multiplexer to select from several analog inputs, then the connection between analog and digital grounds can be made at this ADC as in Figure 13.1(a). This scheme requires that the analog and digital power supply returns are not linked together anywhere else, so that two separate power supply circuits are needed. The analog and digital grounds must be treated as entirely separate tracks, despite being nominally at the same potential; unavoidable noise currents circulating in the digital ground will then not couple into the "clean" analog ground. The digital ground should be of gridded or ground plane construction, whereas the analog section may benefit from a single-point grounding system, or may have a separate ground plane of its own. On no account should you extend the digital ground plane over the analog section of the board, since there will then be capacitive coupling from one ground plane to another.

### 13.1.5 Multi-Board Systems

When your system consists of several boards, some entirely digital, some entirely analog and some a mixture of the two, with an external power supply, then you cannot make the connection between digital and analog grounds at the ADC. There may be several ADCs in the one system. Instead, make the link at the power supply (Figure 13.1(b)) and run separate analog and digital grounds to each board that requires them. Digital-only boards should be located physically closer to the power supply to minimize the radiating loop area or length.

## 13.2 Generating Digital Levels From Analog Inputs

The first rule when you want to use a varying analog voltage to generate an on/off digital signal—as distinct from an analog-to-digital conversion—is: always use either a comparator or a Schmitt-trigger gate. Never feed an analog signal straight into an ordinary TTL or CMOS gate input.

The reason is that ordinary gates do not have well-defined input voltage switching thresholds. Not only that, but they are also very critical of slow rise-time inputs. Very few analog input signals have the slew rate, typically faster than $5\,V/\mu s$, required to produce a clean output from an ordinary logic gate. The result of applying a slow analog voltage to a logic gate is shown in Figure 13.2.

A Schmitt trigger gate, or a comparator with hysteresis, will get over the slow rise time problem. A Schmitt trigger gate has the same output characteristics as an ordinary gate but it includes input hysteresis to ensure a fast transition. The threshold levels of typical Schmitt devices, such as the 74HC14, are specified within wide tolerances and so do not overcome the variability of the actual switching point. When the analog levels corresponding to high and low states can be kept above $V_{IH}$ and below $V_{IL}$, respectively,

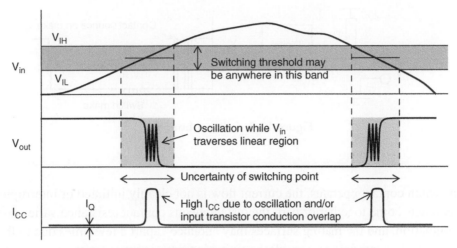

**Figure 13.2: The effect of a slow input to a logic gate**

a Schmitt is adequate. For more precision you will need to use a comparator with an accurately specified reference voltage.

Secondly, if the analog supply rail range is greater than the logic supply, interfacing the analog signal straight to the logic input will threaten the gate with damage. This is possible even if the normal signal range is within the logic supply range; abnormal conditions such as turn-on or turn-off may exceed the rails. This, of course, is also a problem with Schmitt trigger gates. Normally, the inputs are protected by clamp diodes to the supply and ground rails, but the current through these must be limited to a safe level so a resistor in series with the input is essential. More positive steps to limit the input voltage, such as running the analog section from the same supply voltage as the logic (heeding the earlier advice about separate digital and analog ground rails), are to be preferred.

### 13.2.1 Debouncing Switch Inputs

On the face of it, switch inputs to digital circuitry must be the easiest of interfaces. All you should need are an input port or gate, a pull-up resistor and a single pole switch (Figure 13.3). This circuit, though it undoubtedly works, is prone to a serious problem because of the electromechanical nature of the switch and the speed of logic devices.

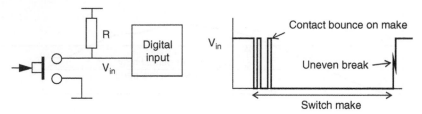

**Figure 13.3: Contact bounce**

When a switch contact operates, the current flow is not cleanly initiated or interrupted. As the contacts come together or part, the instantaneous contact resistance varies due to contamination, and the mating surfaces may "bounce" apart a few times due to the springiness of the material. As a result the switching edge is irregular and may easily consist of several discrete edges, extending over a period of typically 1 ms. You can verify this behavior simply by observing the input waveform of Figure 13.3 on a storage scope.

Of course, the digital input responds very fast to each crossing of the switching threshold, and consequently the port or gate sees several transitions each time the switch is operated, before it settles to a steady-state 1 or 0. This may not be a problem for level-sensitive inputs, but it undoubtedly is for edge-sensitive ones such as counter or latch clock inputs. Mistriggering of counter circuits that are fed from a switch input is commonly caused by this phenomenon.

The simple solution to contact bounce is to filter the logic input with an RC network (Figure 13.4(a)). The RC time constant must be significantly longer than the bounce period to effectively attenuate the contact noise. This has the extra advantage of protecting against induced impulsive or RF interference, but it requires additional discrete components and demands that the logic input must be a Schmitt-trigger type, since the input rise time has been deliberately slowed.

If the switch input may change state quickly, an RC time constant which is sufficiently long to cure the bounce will slow the response to the switch unacceptably. This can be overcome in two ways: the R-S latch, Figure 13.4(b), which requires a changeover rather than single-throw switch, or a software- or hardware-implemented delay. Figure 13.4(c) shows the hardware delay, which uses a continuously-clocked shift register and OR gate

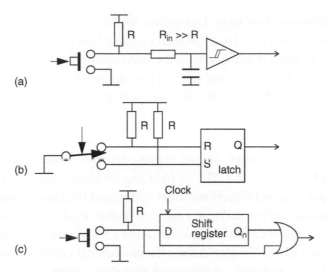

**Figure 13.4: Switch de bouncing circuits**

to effectively "window out" the bounce. The delay can be adjusted to suit the bounce period. These two solutions are most suited to realization with semi-custom logic arrays or ASICs, where the overhead of the extra logic is low.

## 13.3  Classic Data Interface Standards

When you want to connect logic signals from one piece of equipment to another, it is not sufficient to use standard logic devices and make direct gate-to-gate connections, even if they are isolated from the main system. Standard logic is not suited to driving long lines; line terminations are unspecified and noise immunity is low, so that reflections and interference would give unacceptably high data corruption. External logic interfaces must be specially designed for the purpose.

At the same time, it is essential that there is some commonality of interface between different manufacturers' equipment. This allows the user to connect, say, a computer from manufacturer A to a printer from manufacturer B without worrying about electrical compatibility. There is therefore a need for a standard definition for electrical interface signals.

This need has been recognized for many years, and there are a wide variety of data interchange standards available. The logic of the marketplace has dictated that only a small number of these are dominant. This section will consider the two main commercial ones: EIA-232F and EIA-422. EIA-232F is an update of the popular RS-232C standard published in 1969, to bring it into line with the international CCITT V.24 and V.28 and ISO IS2110 standards. EIA-422 is the same as the earlier RS-422 standard. The prefix changes are cosmetic, purely to identify the source of the standards as the EIA.

### 13.3.1 EIA-232F

The boom in data communications has led to many products which make interface conformity claims by quoting "RS-232" in their specifications. Some of these claims are in fact quite spurious, and discerning users will regard interface conformity as an indicator of product quality, and test it early on in their evaluation. The major characteristics of the specification are given in Table 13.2. As well as specifying the electrical parameters, EIA-232F also defines the mechanical connections and pin configuration, and the functional description of each data circuit.

By modern standards the performance of EIA-232F is primitive. It was originally designed to link data terminal equipment (DTE) to modems, known as data communications equipment (DCE). It was also used for data terminal-to-mainframe interfaces. These early applications were relatively low speed, less than 20 kbaud, and used cables shorter than 50 feet. Applications which call for such limited capability are now abundant, hence the standard's great popularity. Its new revision recognizes this by replacing the phrase "data communication equipment" with "data circuit-terminating equipment," also abbreviated to DCE. It does not clarify exactly what is a DTE and what is a DCE, and since many applications are simple DTE (computer) to DTE (terminal or printer) connections, it is often open to debate as to what is at which end of the interface. Although a point-to-point connection provides the correct pin terminations for DTE-to-DCE, a useful extra gadget is a cable known as a "null modem" (Figure 13.5) which creates a DTE-to-DTE connection. The common sight of an installation technician crouched over a 9-way connector swapping pins 2 and 3, to make one end's receiver listen to the other end's driver, has yet to disappear.

EIA-232's transmission distance is limited by its unbalanced design and restricted drive current. The unbalanced design is very susceptible to external noise pick-up and to ground shifts between the driver and receiver. The limited drive current means that the slew rate must be kept slow enough to prevent the cable becoming a transmission line, and this puts a limit on the fastest data rate that can be accommodated. Maximum cable length,

Table 13.2: Major electrical characteristics of EIA-232F, EIA-422 and EIA-485

| Interface | EIA-232F | EIA-422 | EIA-485 |
|---|---|---|---|
| Line type | Unbalanced, point-to-point | Balanced, differential, multi-drop (one driver per bus) | Balanced, differential, multiple drivers per bus (half duplex) |
| Line impedance | Not applicable | $100\,\Omega$ | $120\,\Omega$ |
| Max. line length | Load dependent, typically 15 m depending on capacitance | $L \approx 10^5/B$ meters $B$ = bit rate, kB/s | Max. recommended 1200 m, depending on attenuation |
| Max data rate | 20 kB/s | 10 MB/s | 10 MB/s |
| **Driver** | | | |
| Output voltage | $\pm 5$ to $\pm 15$V loaded with 3–7 k$\Omega$ +V = logic 0, −V = logic 1 | $\pm 10$V max differential unloaded, $\pm 2$V min loaded with 100 $\Omega$ | $\pm 6$V max differential Unloaded, $\pm 1.5$V min loaded with 54 |
| Short circuit current | 500 mA max | 150 mA max | 150 mA to gnd, 250 mA to −7 +12V |
| Rise time | 4% of unit interval (1 ms max) 30 V/µs max. slew rate | 10% of unit interval (min 20 ns) | 30% of unit interval |
| Output with power off | $>300\,\Omega$ output resistance | $\pm 100$ µA max leakage | $>12$ k$\Omega$ output resistance |
| **Receiver** | | | |
| Sensitivity | $\pm 3$V max thresholds | $\pm 200$ mV | $\pm 200$ mV |
| Input impedance | 3 k$\Omega$ −7 k$\Omega$, <2500 pF | 4 k$\Omega$ min | 12 k$\Omega$ |
| Common mode range | Not applicable | $\pm 7$V | +12 to −7V |

originally fixed at 50 feet, is now restricted by a requirement for maximum load capacitance (including receiver input) for each circuit of 2500 pF. As the line length increases so does its capacitance, requiring more current to maintain the same transition time. The graph of Figure 13.6 shows the drive current versus load capacitance required to maintain the 4% transition time relationship at different data rates. In practice, the line length is limited to 3 meters or less for data rates more than 20 kb/s. Most drivers can handle the higher transmission rates over such a short length without drawing excessive supply current.

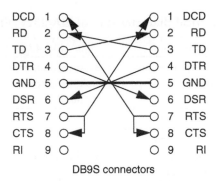

DB9S connectors

**Figure 13.5: The null modem**

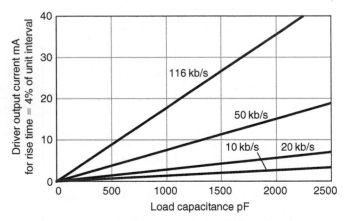

**Figure 13.6: EIA-232F transmit driver output current versus C$_L$**

Note that there are several common "enhancements" that are not permitted by strict adherence to the standard. EIA-232F makes no provision for tri-stating the driver output, so multiple driver access to one line is not possible. Similarly, paralleling receivers is not allowed unless the combined input impedance is held between $3\,\text{k}\Omega$ and $7\,\text{k}\Omega$. It does not consider electrically isolated interfaces: no specification is offered for isolation requirements, despite their desirability. It does not specify the communication data format. The usual "one start bit, eight data bits, two stop bits" format is not part of the standard, just its most common application. It is not directly compatible with another common single-ended standard, EIA-423, although such connections will usually work. Also, you cannot legitimately run EIA-232F off a $\pm 5\,\text{V}$ supply rail—the minimum driver output voltage is specified as $\pm 5\,\text{V}$, loaded with $3–7\,\text{k}\Omega$ and with an output impedance of $300\,\Omega$.

The standard calls for slew-rate limiting to 30 V/μs maximum. Although you can do this with an output capacitor, which operates in conjunction with the output transistor's current limit while it is slewing, this will increase the dissipation, and reduces the maximum possible cable length. It is preferable to use a driver which has on-chip slew rate limiting, requiring no external capacitors and making the slew rate independent of cable length.

## 13.3.2 EIA-422

Many data communications applications now require data rates in the megabaud region, for which EIA-232F is inadequate. This need is fulfilled by the EIA-422 standard, which is an electrical specification for drivers and receivers for use in a balanced or differential, point-to-point or multi-drop high speed interface using twisted pair cable. Table 13.2 summarizes the EIA-422 specification in comparison with EIA-232F. One driver and up to ten receivers are allowed. The maximum data rate is specified as 10 Mbaud, with a trade-off against cable length; maximum cable length at 100 kbaud is 4000 feet. Note that unlike EIA-232F, EIA-422 does not specify functional or mechanical parameters of the interface. These are included in other standards which incorporate it, notably EIA-449 and EIA-530.

EIA-422 achieves its high-speed and long-distance capabilities by specifying a balanced and terminated design. The balanced design reduces sensitivity to external common mode noise and allows a ground differential of up to a few volts to exist between the driver and one or more of the receivers without affecting the receiver's thresholds. A cable termination, together with increased driver current, allows fast slew rates which in turn allows high data rates. If the cable is not terminated, serious ringing on the edges occurs which may cause spurious switching in the receiver. The specified termination of 100 Ω is closely matched to the characteristic impedance of typical twisted pair cables. Only one termination is used, at the receiver at the far end of the cable.

## 13.3.3 Interface Design

By far the easiest way to realize either EIA-232F or EIA-422 interfaces is to use one of the many specially tailored driver and receiver chip sets that are available. The more common ones, such as the 1488 driver/1489 receiver for EIA-232F or the 26LS31 driver/26LS32 receiver for EIA-422, are available competitively from many sources and in low-power CMOS versions. You can also obtain combined driver/receiver parts so that a small interface can be handled with one IC. Because the 9-pin implementation of EIA-232F is so common, a single package 3-transmitter plus 5-receiver part is also widely sourced. The high-voltage requirement of EIA-232F, typically $\pm12\,\text{V}$ supplies, is addressed by some suppliers who offer on-chip DC-to-DC converters from the $+5\,\text{V}$ rail.

Figure 13.7 suggests typical interface circuits for the two standards. Note the inclusion of power supply isolating diodes, to protect the rest of the circuit against the inevitable over voltages that will come its way. You can also construct an interface, particularly the simpler EIA-232F, using standard components such as op-amps, comparators, CMOS buffer devices or discrete components if you are prepared to spend some time characterizing the circuit against the requirements of the standard and against expected overload conditions. This may turn out to be marginally cheaper in component cost, but its overall worth is somewhat questionable.

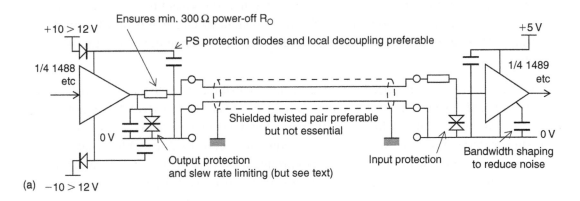

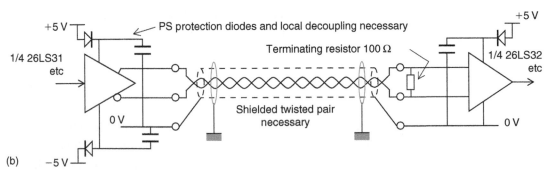

**Figure 13.7: Typical EIA-232F and EIA-422 interface circuits (a) EIA-232F; (b) EIA-422**

## 13.4 High Performance Data Interface Standards

This section briefly reviews some of the newer data interface standards that have grown up for high-speed purposes around particular applications and have subsequently become more widely entrenched.

### 13.4.1 EIA-485

EIA-485 shares many similarities with EIA-422, and is widely used as the basis for in-house and industrial datacom systems. For instance, one variant of the SCSI interface (HVD-SCSI: high voltage differential—small computer systems interface) uses 485 as the basis for its electrical specification. 485-compliant devices can be used in 422 systems, though the reverse is not necessarily true. The principal difference is that 485 allows multiple transmitters on the same line, driving up to 32 unit loads, with half-duplex (bidirectional) communication. One Unit Load is defined as a steady-state load allowing 1 mA of current under a maximum common mode voltage of 12 V or 0.8 mA at –7 V. ULs may consist of drivers or receivers and failsafe resistors (see below), but do not include the termination resistors. The bidirectional communication means that 485 drivers must allow for line contention and for driving a line that is terminated at each end with 120 Ω. The two specifications are compared in Table 13.2.

One further problem that arises in a half-duplex system is that there will be periods when no transmitters are driving the line, so that it becomes high impedance, and it is desirable for the receivers to remain in a fixed state in this situation. This means that a differential voltage of more than 200 mV should be provided by a suitable passive circuit that complies with both the termination impedance requirements and the unit load constraints. A network designed to do this is called a "failsafe" network.

### 13.4.2 CAN

The Controller Area Network standard was originally developed within the automotive industry to replace the complex electrical wiring harness with a two-wire data bus. It has since been standardized in ISO 11898. The specification allows signaling rates up to 1 MB/s, high immunity from electrical interference, and an ability to self-diagnose and repair errors. It is now widespread in many sectors, including factory automation, medical, marine, aerospace and of course automotive. It is particularly suited to applications requiring many short messages in a short period of time with high reliability in noisy operating environments.

The ISO 11898 architecture defines the lowest two layers of the OSI/ISO seven layer model, that is, the data-link layer and the physical layer. The communication protocol is carrier-sense multiple access, with collision detection and arbitration on message priority (CSMA/CD+AMP). The first version of CAN was defined in ISO 11519 and allowed applications up to 125 kB/s with an 11-bit message identifier. The 1 MB/s ISO 11898:1993 version is standard CAN 2.0 A, also with an 11-bit identifier, while Extended CAN 2.0B is provided in a 1995 amendment to the standard and provides a 29-bit identifier.

The physical CAN bus is a single twisted pair, shielded or unshielded, terminated at each end with 120 $\Omega$. Balanced differential signaling is used. Nodes may be added or removed at any time, even while the network is operating. Unpowered nodes should not disturb the bus, so transceivers should be configured so that their pins are in a high impedance state with the power off. The standard specification allows a maximum cable length of 40 m with up to 30 nodes, and a maximum stub length (from the bus to the node) of 0.3 m. Longer stub and line lengths can be implemented, with a tradeoff in signaling rates. The recessive (quiescent) state is for both bus lines to be biased equally to approximately 2.5 V relative to ground; in the dominant state, one line (CANH) is taken positive by 1 V while the other (CANL) is taken negative by the same amount, giving a 2 V differential signal. The required common mode voltage range is from −2 V to +7 V, i.e., ±4.5 V about the quiescent state.

### 13.4.3 USB

The Universal Serial Bus is a cable bus that supports data exchange between a host computer and a wide range of simultaneously accessible peripherals. The attached peripherals share USB bandwidth through a host scheduled, token-based protocol. The bus allows peripherals to be attached, configured, used, and detached while the host and other peripherals are in operation. There is only one host in any USB system. The USB interface to the host computer system is referred to as the Host Controller, which may be implemented in a combination of hardware, firmware, or software.

USB devices are either hubs, which act as wiring concentrators and provide additional attachment points to the bus, or system functions such as mice, storage devices or data sources or outputs. A root hub is integrated within the host system to provide one or more attachment points.

The USB transfers signal and power over a four-wire point-to-point cable. A differential input receiver must be used to accept the USB data signal. The receiver has an input sensitivity of at least 200 mV when both differential data inputs are within the common mode range of 0.8 V to 2.5 V. A differential output driver drives the USB data signal with a static output swing in its low state of <0.3 V with a 1.5 kΩ load to 3.6 V and in its high state of >2.8 V with a 15 kΩ load to ground. A full-speed USB connection is made through a shielded, twisted pair cable with a characteristic impedance ($Z_0$) of 90 Ω 15% and a maximum one way delay of 26 ns. The impedance of each of the drivers must be between 28 and 44 Ω. The detailed specification controls the rise and fall times of the output drivers for a range of load capacitances.

In version 1.1, there are two data rates:

- the full-speed signaling bit rate is 12 Mb/s;
- a limited capability low-speed signaling mode is also defined at 1.5 Mb/s.

Both modes can be supported in the same USB bus by automatic dynamic mode switching between transfers. The low-speed mode is defined to support a limited number of low-bandwidth devices, such as mice. In order to provide guaranteed input voltage levels and proper termination impedance, biased terminations are used at each end of the cable. The terminations also allow detection of attachment at each port and differentiate between full-speed and low-speed devices. The USB 2.0 specification adds a high-speed data rate of 480 MB/s between compliant devices using the same cable as 1.1, with both source and load terminations of 45 Ω.

The cable also carries supply wires, nominally +5 V, on each segment to deliver power to devices. Cable segments of variable lengths, up to several meters, are possible. The specification defines connectors, and the cable has four conductors: a twisted signal pair of standard gauge and a power pair in a range of permitted gauges.

The clock is transmitted, encoded along with the differential data. The clock encoding scheme is non-return-to-zero with bit stuffing to ensure adequate transitions. A SYNC field precedes each packet to allow the receiver(s) to synchronize their bit recovery clocks.

### 13.4.4 Ethernet

Ethernet is a well established specification for serial data transmission. It was first published in 1980 by a multivendor consortium that created the DEC-Intel-Xerox (DIX) standard. In 1985 Ethernet was standardized in IEEE 802.3, since when it has been extended a number of times. "Classic" Ethernet operates at a data transmission rate of 10 Mbit/s. Since the 1990s, Ethernet has developed in the following areas:

- Transmission media

- Data transmission rates
    - Fast Ethernet at 100 Mbit/s (1995)
    - Gigabit Ethernet at 1 Gbit/s (1999)
- Network topologies.

Nowadays Ethernet is the most widespread networking technology in the world in commercial information technology systems, and is also gaining importance in industrial automation. All network users have the same rights under Ethernet. Any user can exchange data of any size with another user at any time, and any network device that is transmitting is heard by all other users. Each Ethernet user filters the data packets that are intended for it out from the stream, ignoring all the others.

In the standard Ethernet, all the network users share one collision domain. Network access is controlled by the CSMA/CD procedure (Carrier Sense Multiple Access with Collision Detection). Before transmitting data, a network user first checks whether the network is free (carrier sense). If so, it starts to transmit data. At the same time it checks whether other users have also begun to transmit (collision detection). If that is the case, a collision occurs. All the network users concerned now stop their transmission, wait for a period of time determined according to a randomizing principle, and then start transmission again. The result of this is that the time required to transmit data packets depends heavily on the network loading, and cannot be determined in advance. The more collisions occur, the slower the entire network becomes.

This lack of determinism can be overcome by a variant of the basic approach known as *switched* Ethernet. This refers to a network in which each Ethernet user is assigned a port in a switch, which analyses all the data packets as they arrive, directing them on to the appropriate port. Switches separate former collision domains into individual point-to-point connections between the network components and the relevant user equipment. Preventing collisions makes the full network bandwidth available to each point-to-point connection. The second pair of conductors in the four-wire Ethernet cable, which otherwise is needed for collision detection, can now be used for transmission, so providing a significant increase in data transfer rate.

The Ethernet interface at each user is defined according to Figure 13.8. It is usual to find structured twisted pair local area network wiring already integrated within a building, and the cabling characteristics are given in IEC 11801 and related standards; hence, the 10Base-T and 100Base-T variants are the most popular of the Ethernet implementations, and the appropriate MAU/MDI using the RJ45 connector are included in most types of computer. The maximum lengths are set by signal timing limitations in the Fast Ethernet implementation, and an Ethernet system implementation relies on correct integration of cable lengths, types and terminations.

In contrast to the coaxial versions of Ethernet, which may be connected in multidrop, each segment of twisted pair or fiber route is a point-to-point connection between hosts; this means that a network system that is more than simply two hosts requires a number of hubs or switches, which integrate the connections to each user. A hub will simply pass through the Ethernet traffic between its ports without controlling it in any way, but a switch does control the traffic, separating packets to their destination ports.

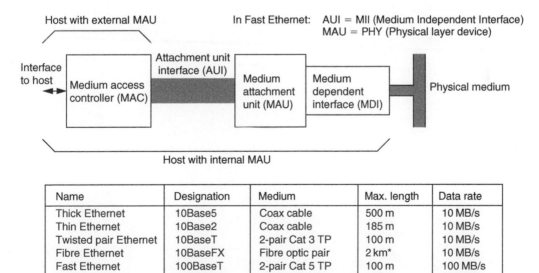

| Name | Designation | Medium | Max. length | Data rate |
|------|-------------|--------|-------------|-----------|
| Thick Ethernet | 10Base5 | Coax cable | 500 m | 10 MB/s |
| Thin Ethernet | 10Base2 | Coax cable | 185 m | 10 MB/s |
| Twisted pair Ethernet | 10BaseT | 2-pair Cat 3 TP | 100 m | 10 MB/s |
| Fibre Ethernet | 10BaseFX | Fibre optic pair | 2 km* | 10 MB/s |
| Fast Ethernet | 100BaseT | 2-pair Cat 5 TP | 100 m | 100 MB/s |
| Fast fibre Ethernet | 100BaseFX | Fibre optic pair | 412 m | 100 MB/s |

\* depends on fibre type

**Figure 13.8: Ethernet interface and media**

The 100Base-T electrical characteristics are a peak differential output signal of 1V into a 100 Ω characteristic impedance twisted pair; the 10Base-T level is 2.5 V. The rise and fall time and amplitude symmetries are also defined to achieve a high level of balance and hence common mode performance. It is normal to use a transformer and common mode choke to isolate the network connection from the driver electronics.

# CHAPTER 14

# Microcontrollers and Microprocessors

Mike Tooley

Many of today's complex electronic systems are based on the use of a microprocessor or microcontroller. Such systems comprise hardware that is controlled by software. If it is necessary to change the way that the system behaves it is the software (rather than the hardware) that is changed.

In this chapter we provide an introduction to microprocessors and explain, in simple terms, both how they operate and how they are used. We shall start by explaining some of the terminology that is used to describe different types of system that involve the use of a microprocessor or a similar device.

## 14.1 Microprocessor Systems

Microprocessor systems are usually assembled on a single PCB comprising a microprocessor CPU together with a number of specialized support chips. These very large scale integrated (VLSI) devices provide input and output to the system, control and timing as well as storage for programs and data.

Typical applications for microprocessor systems include the control of complex industrial processes. Typical examples are based on families of chips such as the Z80CPU plus Z80PIO, Z80CTC, and Z80SIO. Figure 11.1 shows a block diagram of a microprocessor system and the photograph in Figure 11.2 shows an actual Z80 microprocessor chip.

## 14.2 Single-Chip Microcomputers

A single-chip microcomputer is a complete computer system (comprising CPU, RAM and ROM, etc.) in a single VLSI package. A single-chip microcomputer requires very little external circuitry in order to provide all of the functions associated with a complete computer system (but usually with limited input and output capability).

Single-chip microcomputers may be programmed using in-built programmable memories or via external memory chips. Typical applications of single-chip microcomputers include computer printers, instrument controllers, and displays. A typical example is the Z84C.

## 14.3  Microcontrollers

A microcontroller is a single-chip microcomputer that is designed specifically for control rather than general-purpose applications. They are often used to satisfy a particular control requirement, such as controlling a motor drive. Single-chip microcomputers, on the other hand, usually perform a variety of different functions and may control several processes at the same time.

Typical applications include control of peripheral devices such as motors, drives, printers, and minor subsystem components. Typical examples are the Z86E, 8051, 68705 and 89C51.

## 14.4  PIC Microcontrollers

A PIC microcontroller is a general-purpose microcontroller device that is normally used in a stand-alone application to perform simple logic, timing and input/output control. PIC devices provide a flexible low-cost solution that very effectively bridges the gap between single-chip computers and the use of discrete logic and timer chips.

A number of PIC and microcontroller devices have been produced that incorporate a high-level language interpreter. The resident interpreter allows developers to develop their programs languages such as BASIC rather than having to resort to more complex assembly language. This feature makes PIC microcontrollers very easy to use. PIC microcontrollers are used in "self-contained" applications involving logic, timing and simple analog to digital and digital to analog conversion. Typical examples are the PIC12C508 and PIC16C620.

## 14.5  Programmed Logic Devices

While not an example of a microprocessor device, a programmed logic device (PLD) is a programmable chip that can carry out complex logical operations. For completeness, we have included a reference to such devices here. PLDs are capable of replacing a large number of conventional logic gates, thus minimizing chip-count and reducing printed circuit board sizes. Programming is relatively straightforward and simply requires the derivation of complex logic functions using Boolean algebra or truth tables. Typical examples are the 16L8 and 22V10.

## 14.6 Programmable Logic Controllers

Programmable logic controllers (PLC) are microprocessor based systems that are used for controlling a wide variety of automatic processes, from operating an airport baggage handling system to brewing a pint of your favorite lager. PLCs are rugged and modular and they are designed specifically for operation in the process control environment.

The control program for a PLC is usually stored in one or more semiconductor memory devices. The program can be entered (or modified) by means of a simple hand-held programmer, a laptop controller, or downloaded over a local area network (LAN). PLC manufacturers include Allen Bradley, Siemens and Mitsubishi.

## 14.7 Microprocessor Systems

The basic components of any microprocessor system (see Figure 14.1) are:

(a) a central processing unit (CPU);

(b) a memory, comprising both "read/write" (RAM) and "read only" (ROM) devices; and

(c) a means of providing input and output (I/O), such as a keypad for input and a display for output.

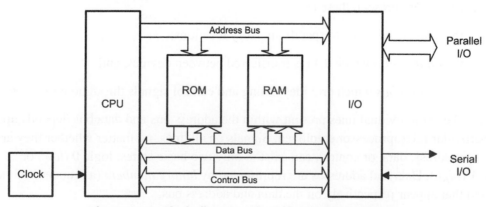

**Figure 14.1: Block diagram of a microprocessor system**

**Figure 14.2: A Z80 microprocessor**

In a microprocessor system, the functions of the CPU are provided by a single very large scale integrated (VLSI) microprocessor chip (see Figure 14.2). This chip is equivalent to many thousands of individual transistors. Semiconductor devices are also used to provide the read/write and read-only memory. Strictly speaking, both types of memory permit "random access" since any item of data can be retrieved with equal ease regardless of its actual location within the memory. Despite this, the term RAM has become synonymous with semiconductor read/write memory.

The basic components of the system (CPU, RAM, ROM, and I/O) are linked together using a multiple-wire connecting system known as a *bus* (see Figure 14.1). Three different buses are present, these are:

(a) the *address bus* used to specify memory locations;

(b) the *data bus* on which data is transferred between devices; and

(c) the *control bus* which provides timing and control signals throughout the system.

The number of individual lines present within the address bus and data bus depends upon the particular microprocessor employed. Signals on all lines, no matter whether they are used for address, data, or control, can exist in only two basic states: logic 0 (*low*) or logic 1 (*high*). Data and addresses are represented by *binary numbers* (a sequence of 1s and 0s) that appear respectively on the data and address bus.

Many microprocessors designed for control and instrumentation applications make use of an 8-bit data bus and a 16-bit address bus. Others have data and address buses that can operate with as many as 128-bits at a time.

The largest binary number that can appear on an 8-bit data bus corresponds to the condition when all eight lines are at logic 1. Therefore, the largest value of data that can be present on the bus at any instant of time is equivalent to the binary number 11111111 (or 255). Similarly, most the highest address that can appear on a 16-bit address bus is 1111111111111111 (or 65,535). The full range of data values and addresses for a simple microprocessor of this type is thus:

| | | |
|---|---|---|
| *Data* | from | 00000000 |
| | to | 11111111 |
| *Addresses* | from | 0000000000000000 |
| | to | 1111111111111111 |

## 14.8 Data Representation

Binary numbers—particularly large ones—are not very convenient. To make numbers easier to handle we often convert binary numbers to *hexadecimal* (base 16). This format is easier for mere humans to comprehend and offers the advantage over denary (base 10) in that it can be converted to and from binary with ease. The first sixteen numbers in binary, denary, and hexadecimal are shown in the table below. A single hexadecimal character (in the range zero to F) is used to represent a group of four binary digits (bits). This group of four bits (or single hex. character) is sometimes called a *nibble*.

A *byte* of data comprises a group of eight bits. Thus a byte can be represented by just two hexadecimal (hex) characters. A group of sixteen bits (a word) can be represented by four hex characters, thirty-two bits (a double word by eight hex. characters, and so on).

The value of a byte expressed in binary can be easily converted to hex by arranging the bits in groups of four and converting each nibble into hexadecimal using Table 14.1.

Note that, to avoid confusion about whether a number is hexadecimal or decimal, we often place a $ symbol before a hexadecimal number or add an H to the end of the

**Table 14.1: Binary, denary, and hexadecimal**

| Binary (base 2) | Denary (base 10) | Hexadecimal (base 16) |
|---|---|---|
| 0000 | 0 | 0 |
| 0001 | 1 | 1 |
| 0010 | 2 | 2 |
| 0011 | 3 | 3 |
| 0100 | 4 | 4 |
| 0101 | 5 | 5 |
| 0110 | 6 | 6 |
| 0111 | 7 | 7 |
| 1000 | 8 | 8 |
| 1001 | 9 | 9 |
| 1010 | 10 | A |
| 1011 | 11 | B |
| 1100 | 12 | C |
| 1101 | 13 | D |
| 1110 | 14 | E |
| 1111 | 15 | F |

number. For example, 64 means decimal "sixty-four" whereas $64 means hexadecimal "six-four," which is equivalent to decimal 100. Similarly, 7FH means hexadecimal "seven-F," which is equivalent to decimal 127.

### Example 14.1
Convert hexadecimal A3 into binary.

### Solution
From Table 14.1, A = 1010 and 3 = 0101. Thus, A3 in hexadecimal is equivalent to 10100101 in binary.

### Example 14.2
Convert binary 11101000 binary to hexadecimal.

**Table 14.2: Data types**

| Data type | Bits | Range of values |
|-----------|------|-----------------|
| Unsigned byte | 8 | 0 to 255 |
| Signed byte | 8 | −128 to +127 |
| Unsigned word | 16 | 0 to 65,535 |
| Signed word | 16 | −32,768 to +32,767 |

*Solution*

From Table 14.1, 1110 = E and 1000 = 8. Thus, 11101000 in binary is equivalent to E8 in hexadecimal.

## 14.9  Data Types

A byte of data can be stored at each address within the total memory space of a microprocessor system. Hence, one byte can be stored at each of the 65,536 memory locations within a microprocessor system having a 16-bit address bus.

Individual bits within a byte are numbered from 0 (least significant bit) to 7 (most significant bit). In the case of 16-bit words, the bits are numbered from 0 (least significant bit) to 15 (most significant bit).

Negative (or signed) numbers can be represented using *two's complement* notation where the leading (most significant) bit indicates the sign of the number (1 = negative, 0 = positive). For example, the signed 8-bit number 10000001 represents the denary number −1.

The range of integer data values that can be represented as bytes, words, and long words are shown in Table 14.2.

## 14.10  Data Storage

The semiconductor ROM within a microprocessor system provides storage for the program code as well as any permanent data that requires storage. All of this data is referred to as non-volatile because it remains intact when the power supply is disconnected.

The semiconductor RAM within a microprocessor system provides storage for the transient data and variables that are used by programs. Part of the RAM is also be used by the microprocessor as a temporary store for data whilst carrying out its normal processing tasks.

It is important to note that any program or data stored in RAM will be lost when the power supply is switched off or disconnected. The only exception to this is CMOS RAM that is kept alive by means of a small battery. This *battery-backed memory* is used to retain important data, such as the time and date.

When expressing the amount of storage provided by a memory device we usually use Kilobytes (Kbyte). It is important to note that a Kilobyte of memory is actually 1,024 bytes (not 1,000 bytes). The reason for choosing the Kbyte rather than the kbyte (1,000 bytes) is that 1,024 happens to be the nearest power of 2 (note that $2^{10} = 1,024$).

The capacity of a semiconductor ROM is usually specified in terms of an address range and the number of bits stored at each address. For example, $2\,K \times 8$ bits (capacity 2 Kbytes), $4\,K \times 8$ bits (capacity 4 Kbytes), and so on. Note that it is not always necessary (or desirable) for the entire memory space of a microprocessor to be populated by memory devices.

## 14.11 The Microprocessor

The microprocessor *central processing unit (CPU)* forms the heart of any microprocessor or microcomputer system computer and, consequently, its operation is crucial to the entire system.

The primary function of the microprocessor is that of fetching, decoding, and executing instructions resident in memory. As such, it must be able to transfer data from external memory into its own internal registers and vice versa. Furthermore, it must operate predictably, distinguishing, for example, between an operation contained within an instruction and any accompanying addresses of read/write memory locations. In addition, various system housekeeping tasks need to be performed including being able to suspend normal processing in order to respond to an external device that needs attention.

The main parts of a microprocessor CPU are:

(a) *registers* for temporary storage of addresses and data;

(b) an *arithmetic logic unit (ALU)* that performs arithmetic and logic operations;

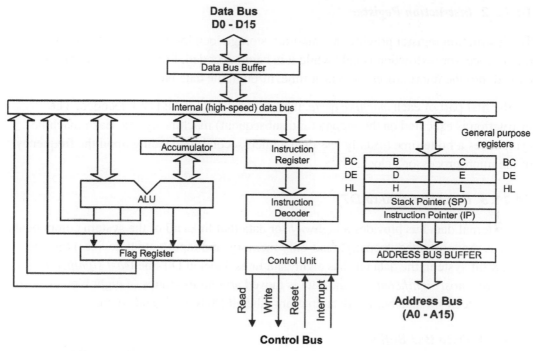

**Figure 14.3: Internal architecture of a typical 8-bit microprocessor CPU**

(c)  a unit that receives and decodes instructions; and

(d)  a means of controlling and timing operations within the system.

Figure 14.3 shows the principal internal features of a typical 8-bit microprocessor. We will briefly explain each of these features in turn.

## 14.11.1  Accumulator

The accumulator functions as a source and destination register for many of the basic microprocessor operations. As a *source register* it contains the data that will be used in a particular operation while as a *destination register* it will be used to hold the result of a particular operation. The accumulator (or *A-register*) features in a very large number of microprocessor operations; consequently, more reference is made to this register than any others.

### 14.11.2 Instruction Register

The instruction register provides a temporary storage location in which the current microprocessor instruction is held while it is being decoded. Program instructions are passed into the microprocessor, one at time, through the data bus.

On the first part of each *machine cycle*, the instruction is fetched and decoded. The instruction is executed on the second (and subsequent) machine cycles. Each machine cycle takes a finite time (usually less than a microsecond) depending upon the frequency of the microprocessor's clock.

### 14.11.3 Data Bus (D0 to D7)

The external data bus provides a highway for data that links all of the system components (such as random access memory, read-only memory, and input/output devices) together. In an 8-bit system, the data bus has eight data lines, labeled D0 (the *least significant bit*) to D7 (*the most significant bit*) and data is moved around in groups of eight bits, or *bytes*. With a sixteen-bit data bus the data lines are labeled D0 to D15, and so on.

### 14.11.4 Data Bus Buffer

The data bus buffer is a temporary register through which bytes of data pass on their way into, and out of, the microprocessor. The buffer is thus referred to as *bidirectional* with data passing out of the microprocessor on a *write operation* and into the processor during a *read operation*. The direction of data transfer is determined by the *control unit* as it responds to each individual program instruction.

### 14.11.5 Internal Data Bus

The internal data bus is a high-speed data highway that links all of the microprocessor's internal elements together. Data is constantly flowing backwards and forwards along the internal data bus lines.

### 14.11.6 General-Purpose Registers

Many microprocessor operations (for example, adding two 8-bit numbers together) require the use of more than one register. There is also a requirement for temporarily storing the partial result of an operation whilst other operations take place. Both of these needs can be met by providing a number of general-purpose registers. The use to which these registers are put is left mainly up to the programmer.

### 14.11.7 Stack Pointer

When the time comes to suspend a particular task in order to briefly attend to something else, most microprocessors make use of a region of external random access memory (RAM) known as a *stack*. When the main program is interrupted, the microprocessor temporarily places in the stack the contents of its internal registers together with the address of the next instruction in the main program. When the interrupt has been attended to, the microprocessor recovers the data that has been stored temporarily in the stack together with the address of the next instruction within the main program. Therefore, it is able to return to the main program exactly where it left off and with all the data preserved in its registers. The stack pointer is simply a register that contains the address of the last used stack location.

### 14.11.8 Program Counter

Programs consist of a sequence of instructions that are executed by the microprocessor. These instructions are stored in external random access memory (RAM) or read-only memory (ROM). Instructions must be fetched and executed by the microprocessor in a strict sequence. By storing the address of the next instruction to be executed, the program counter allows the microprocessor to keep track of where it is within the program. The program counter is automatically incremented when each instruction is executed.

### 14.11.9 Address Bus Buffer

The address bus buffer is a temporary register through which addresses (in this case comprising 16-bits) pass on their way out of the microprocessor. In a simple microprocessor, the address buffer is unidirectional with addresses placed on the address bus during both read and write operations. The address bus lines are labeled A0 to A15, where A0 is the least-significant address bus line and A16 is the most significant address bus line. Note that a 16-bit address bus can be used to communicate with 65,536 individual memory locations. At each location a single byte of data is stored.

### 14.11.10 Control Bus

The control bus is a collection of signal lines that are both used to control the transfer of data around the system and also to interact with external devices. The control signals used by microprocessors tend to differ with different types, however the following are commonly found:

READ    An output signal from the microprocessor that indicates that the current operation is a read operation.

WRITE   An output signal from the microprocessor that indicates that the current operation is a write operation.

RESET   A signal that resets the internal registers and initializes the program counter so that the program can be re-started from the beginning.

IRQ     Interrupt request from an external device attempting to gain the attention of the microprocessor (the request may be obeyed or ignored according to the state of the microprocessor at the time that the interrupt request is received).

NMI     Nonmaskable interrupt (i.e., an interrupt signal that cannot be ignored by the microprocessor).

### 14.11.11  Address Bus (A0 to A15)

The address bus provides a highway for addresses that links with all of the system components (such as random access memory, read-only memory, and input/output devices). In a system with a 16-bit address bus, there are sixteen address lines, labeled A0 (the least significant bit) to A15 (the most significant bit). In a system with a 32-bit address bus there are 32 address lines, labeled A0 to A31, and so on.

### 14.11.12  Instruction Decoder

The instruction decoder is nothing more than an arrangement of logic gates that acts on the bits stored in the instruction register and determines which instruction is currently being referenced. The instruction decoder provides output signals for the microprocessor's control unit.

### 14.11.13  Control Unit

The control unit is responsible for organizing the orderly flow of data within the microprocessor as well as generating, and responding to, signals on the control bus. The control unit is also responsible for the timing of all data transfers. This process is synchronized using an internal or external clock signal (not shown in Figure 14.3).

### 14.11.14 Arithmetic Logic Unit (ALU)

As its name suggests, the ALU performs arithmetic and logic operations. The ALU has two inputs (in this case these are both 8-bits wide). One of these inputs is derived from the Accumulator while the other is taken from the internal data bus via a temporary register (not shown in Figure 14.3). The operations provided by the ALU usually include addition, subtraction, logical AND, logical OR, logical exclusive-OR, shift left, shift right, etc. The result of most ALU operations appears in the accumulator.

### 14.11.15 Flag Register (or Status Register)

The result of an ALU operation is sometimes important in determining what subsequent action takes place. The flag register contains a number of individual bits that are set or reset according to the outcome of an ALU operation. These bits are referred to as flags. The following flags are available in most microprocessors:

ZERO  The zero flag is set when the result of an ALU operation is zero (i.e., a byte value of 00000000).

CARRY  The carry flag is set whenever the result of an ALU operation (such as addition) generates a carry bit (in other words, when the result cannot be contained within an 8-bit register).

INTERRUPT The interrupt flag indicates whether external interrupts are currently enabled or disabled.

### 14.11.16 Clocks

The clock used in a microprocessor system is simply an accurate and stable square wave generator. In most cases the frequency of the square wave generator is determined by a quarts crystal. A simple 4-MHz square wave clock oscillator (together with the clock waveform that is produces) is shown in Figure 14.4. Note that one complete clock cycle is sometimes referred to as a T-state.

Microprocessors sometimes have an internal clock circuit in which case the quartz crystal (or other resonant device) is connected directly to pins on the microprocessor chip. In Figure 14.5(a) an external clock is shown connected to a microprocessor while in Figure 14.5(b) an internal clock oscillator is used.

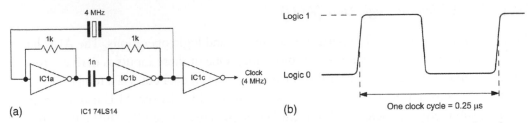

(a)                IC1 74LS14                        (b)

**Figure 14.4: (a) A typical microprocessor clock circuit; and (b) waveform produced by the clock circuit**

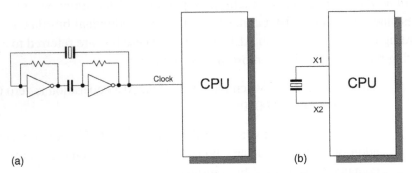

(a)                                                (b)

**Figure 14.5: (a) An external CPU clock; and (b) an internal CPU clock**

## 14.12 Microprocessor Operation

The majority of operations performed by a microprocessor involve the movement of data. Indeed, the program code (a set of instructions stored in ROM or RAM) must itself be fetched from memory prior to execution. The microprocessor thus performs a continuous sequence of instruction fetch and execute cycles. The act of fetching an instruction code (or operand or data value) from memory involves a read operation while the act of moving data from the microprocessor to a memory location involves a write operation (see Figure 14.6).

Each cycle of CPU operation is known as a *machine cycle*. Program instructions may require several machine cycles (typically between two and five). The first machine

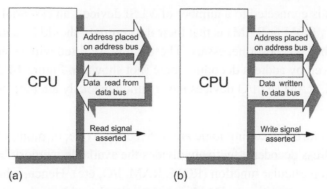

Figure 14.6: (a) Read; and (b) write operations

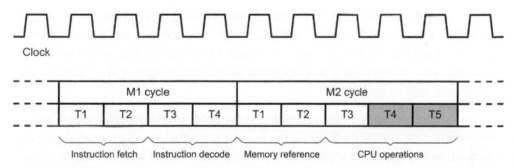

Figure 14.7: A typical timing diagram for a microprocessor's fetch-execute cycle

cycle in any cycle consists of an instruction fetch (the instruction code is read from the memory) and it is known as the *M1 cycle*. Subsequent cycles M2, M3, and so on, depend on the type of instruction that is being executed. This fetch-execute sequence is shown in Figure 14.7.

Microprocessors determine the source of data (when it is being read) and the destination of data (when it is being written) by placing a unique address on the address bus. The address at which the data is to be placed (during a write operation), or from which it is to be fetched (during a read operation), can either constitute part of the memory of the system (in which case it may be within ROM or RAM), or it can be considered to be associated with input/output (I/O).

Since the data bus is connected to a number of VLSI devices, an essential requirement of such chips (e.g., ROM or RAM) is that their data outputs should be capable of being isolated from the bus whenever necessary. These chips are fitted with select or enable inputs that are driven by address decoding logic (not shown in Figure 14.2). This logic ensures that ROM, RAM and I/O devices never simultaneously attempt to place data on the bus!

The inputs of the address decoding logic are derived from one, or more, of the address bus lines. The address decoder effectively divides the available memory into blocks corresponding to a particular function (ROM, RAM, I/O, etc). Hence, where the processor is reading and writing to RAM, for example, the address decoding logic will ensure that only the RAM is selected whilst the ROM and I/O remain isolated from the data bus.

Within the CPU, data is stored in several registers. Registers themselves can be thought of as a simple pigeon-hole arrangement that can store as many bits as there are holes available. Generally, these devices can store groups of sixteen or thirty-two bits. Additionally, some registers may be configured as either one register of sixteen bits or two registers of thirty-two bits.

Some microprocessor registers are accessible to the programmer whereas others are used by the microprocessor itself. Registers may be classified as either general purpose or dedicated. In the latter case a particular function is associated with the register, such as holding the result of an operation or signalling the result of a comparison. A typical microprocessor and its register model is shown in Figure 14.8.

### 14.12.1  The Arithmetic Logic Unit

The ALU can perform arithmetic operations (addition and subtraction) and logic (complementation, logical AND, logical OR, etc). The ALU operates on two inputs (sixteen or thirty-two bits in length depending upon the CPU type) and it provides one output (again of sixteen or thirty-two bits). In addition, the ALU status is preserved in the *flag register* so that, for example, an overflow, zero or negative result can be detected.

The control unit is responsible for the movement of data within the CPU and the management of control signals, both internal and external. The control unit asserts the requisite signals to read or write data as appropriate to the current instruction.

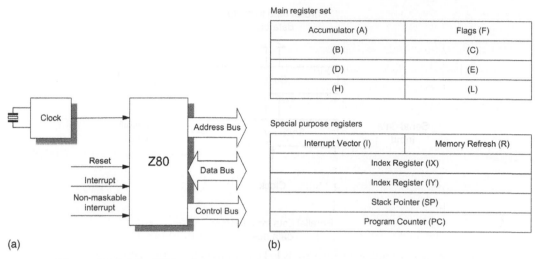

**Figure 14.8: The Z80 microprocessor (showing some of its more important control signals) together with its register model**

## 14.12.2 Input and Output

The transfer of data within a microprocessor system involves moving groups of 8, 16 or 32 bits using the bus architecture described earlier. Consequently it is a relatively simple matter to transfer data into and out of the system in parallel form. This process is further simplified by using a *Programmable Parallel I/O* device (a Z80PIO, 8255, or equivalent VLSI chip). This device provides registers for the temporary storage of data that not only buffer the data but also provide a degree of electrical isolation from the system data bus.

Parallel data transfer is primarily suited to high-speed operation over relatively short distances, a typical example being the linking of a microcomputer to an adjacent dot matrix printer. There are, however, some applications in which parallel data transfer is inappropriate, the most common example being data communication by means of telephone lines. In such cases data must be sent serially (one bit after another) rather than in parallel form.

To transmit data in serial form, the parallel data from the microprocessor must be reorganized into a stream of bits. This task is greatly simplified by using an LSI interface

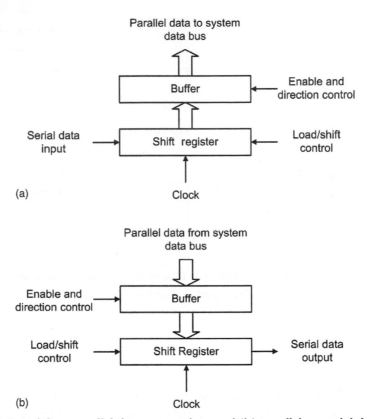

**Figure 14.9: (a) Serial-to-parallel data conversion; and (b) parallel-to-serial data conversion**

device that contains a shift register that is loaded with parallel data from the data bus. This data is then read out as a serial bit stream by successive shifting. The reverse process, serial-to-parallel conversion, also uses a shift register. Here data is loaded in serial form, each bit shifting further into the register until it becomes full. Data is then placed simultaneously on the parallel output lines. The basic principles of parallel-to-serial and serial-to-parallel data conversion are illustrated in Figure 14.9.

### 14.12.3 An Example Program

The following example program (see Table 14.3) is written in assembly code. The program transfers 8-bit data from an input port (Port A), complements (i.e., inverts) the

**Table 14.3: A simple example program**

| Address | Data | Assembly code | Comment |
|---------|------|---------------|---------|
| 2002 | DB FF | IN A, (FFH) | Get a byte from Port A |
| 2002 | 2F | CPL | Invert the byte |
| 2003 | D3 FE | OUT (FEH), A | Output the byte to Port B |
| 2005 | C3 00 20 | JP 2000 | Go round again |

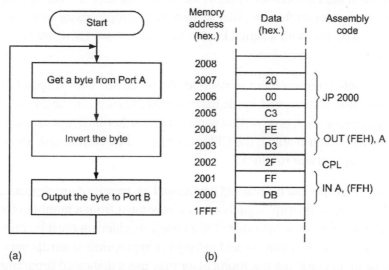

(a)                                                    (b)

**Figure 14.10: (a) Flowchart for the example program; and (b) the eight bytes of program code stored in memory**

data (by changing 0's to 1's and 1's to 0's in every bit position) and then outputs the result to an output port (Port B). The program repeats indefinitely.

Just three microprocessor instructions are required to carry out this task together with a fourth (jump) instruction that causes the three instructions to be repeated over and over again. A program of this sort is most easily written in assembly code, which consists of a series of easy to remember mnemonics. The flowchart for the program is shown in Figure 14.10(a).

The program occupies a total of eight bytes of memory, starting at a hexadecimal address of 2000 as shown in Figure 14.10(b). You should also note that the two ports, A and B, each have unique addresses; Port A is at hexadecimal address FF, while Port B is at hexadecimal address FE.

### 14.12.4 Interrupts

A program that simply executes a loop indefinitely has a rather limited practical application. In most microprocessor systems we want to be able to interrupt the normal sequence of program flow in order to alert the microprocessor to the need to do something. We can do this with a signal known as an *interrupt*. There are two types of interrupt: maskable and nonmaskable.

When a *nonmaskable interrupt* input is asserted, the processor must suspend execution of the current instruction and respond immediately to the interrupt. In the case of a *maskable interrupt*, the processor's response will depend upon whether interrupts are currently enabled or disabled (when enabled, the CPU will suspend its current task and carry out the requisite interrupt service routine).

The response to interrupts can be enabled or disabled by means of appropriate program instructions. In practice, interrupt signals may be generated from a number of sources and since each will require its own customized response a mechanism must be provided for identifying the source of the interrupt and calling the appropriate interrupt service routine. In order to assist in this task, the microprocessor may use a dedicated programmable interrupt controller chip.

## 14.13  A Microcontroller System

Figure 14.11 shows the arrangement of a typical microcontroller system. The sensed quantities (temperature, position, etc.) are converted to corresponding electrical signals by means of a number of sensors. The outputs from the sensors (in either digital or analog form) are passed as input signals to the microcontroller. The microcontroller also accepts inputs from the user. These user set options typically include target values for variables (such as desired room temperature), limit values (such as maximum shaft speed), or time constraints (such as "on" time and "off" time, delay time, etc).

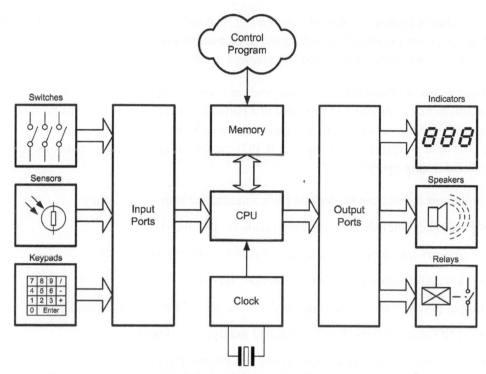

**Figure 14.11: A microcontroller system with typical inputs and outputs**

The operation of the microcontroller is controlled by a sequence of software instructions known as a *control program*. The control program operates continuously, examining inputs from sensors, user settings, and time data before making changes to the output signals sent to one or more controlled devices.

The controlled quantities are produced by the controlled devices in response to output signals from the microcontroller. The controlled device generally converts energy from one form into energy in another form. For example, the controlled device might be an electrical heater that converts electrical energy from the AC mains supply into heat energy, thus producing a given temperature (the controlled quantity).

In most real-world systems there is a requirement for the system to be automatic or self-regulating. Once set, such systems will continue to operate without continuous operator intervention. The output of a self-regulating system is fed back to its input in order to produce what is known as a *closed-loop system*. A good example of a closed-loop system is a heating control system that is designed to maintain a constant room temperature and humidity within a building regardless of changes in the outside conditions.

In simple terms, a microcontroller must produce a specific state on each of the lines connected to its output ports in response to a particular combination of states present on each of the lines connected to its input ports (see Figure 14.11). Microcontrollers must also have a central processing unit (CPU) capable of performing simple arithmetic, logical and timing operations.

The input port signals can be derived from a number of sources including:

- switches (including momentary action pushbuttons),
- sensors (producing logic-level compatible outputs), and
- keypads (both encoded and unencoded types).

The output port signals can be connected to a number of devices including:

- LED indicators (both individual and multiple bar types),
- LED seven segment displays (via a suitable interface),
- motors and actuators (both linear and rotary types) via a suitable buffer/driver or a dedicated interface),
- relays (both conventional electromagnetic types and optically couple solid-state types), and
- transistor drivers and other solid-state switching devices.

### 14.13.1 Input Devices

Input devices supply information to the computer system from the outside world. In an ordinary personal computer, the most obvious input device is the keyboard. Other input devices available on a PC are the mouse (pointing device), scanner, and modem. Microcontrollers use much simpler input devices. These need be nothing more than individual switches or contacts that make and break but many other types of device are also used including many types of sensor that provide logic level outputs (such as float switches, proximity detectors, light sensors, etc).

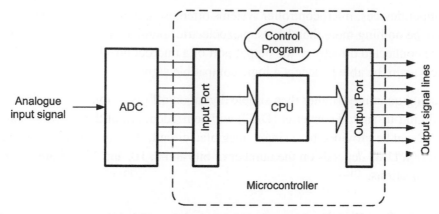

**Figure 14.12: An analog input signal can be connected to a microcontroller input port via an analog-to-digital converter (ADC)**

It is important to note that, in order to be connected directly to the input port of a microcontroller, an input device must provide a logic compatible signal. This is because microcontroller inputs can only accept digital input signals with the same voltage levels as the logic power source. The 0 V ground level (often referred to as $V_{SS}$ in the case of a CMOS microcontroller) and the positive supply $V_{DD}$ in the case of a CMOS microcontroller) is invariably 5 V $\pm$ 5%. A level of approximately 0 V indicates a logic 0 signal and a voltage approximately equal to the positive power supply indicates a logic 1 signal.

Other input devices may sense analog quantities (such as velocity) but use a digital code to represent their value as an input to the microcontroller system. Some microcontrollers provide an internal analog-to-digital converter (ADC) in order to greatly simplify the connection of analog sensors as input devices, but where this facility isn't available it will be necessary to use an external ADC, which usually takes the form of a single integrated circuit. The resolution of the ADC will depend upon the number of bits used and 8, 10, and 12-bit devices are common in control applications.

### 14.13.2  Output Devices

Output devices are used to communicate information or actions from a computer system to the outside world. In a personal computer system, the most common output device is the CRT (cathode ray tube) display. Other output devices include printers and modems.

As with input devices, microcontroller systems often use much simpler output devices. These may be nothing more than LEDs, piezoelectric sounders, relays and motors. In order to be connected directly to the output port of a microcontroller, an output device must, once again, be able to accept a logic compatible signal.

Where analog quantities (rather than simple digital on/off operation) are required at the output a digital-to-analog converter (DAC) will be needed. All of the functions associated with a DAC can be provided by a single integrated circuit. As with an ADC, the output resolution of a DAC depends on the number of bits and 8, 10, and 12 bits are common in control applications.

### 14.13.3 Interface Circuits

Finally, where input and output signals are not logic compatible (i.e., when they are outside the range of signals that can be connected directly to the microcontroller) some additional interface circuitry may be required in order to shift the voltage levels or to provide additional current drive. Additional circuitry may also be required when a load (such as a relay or motor) requires more current than is available from a standard logic device or output port. For example, a common range of interface circuits (solid-state

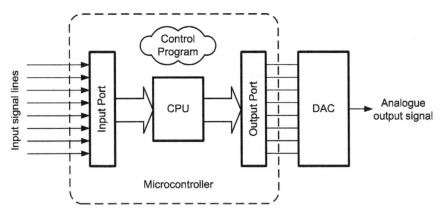

**Figure 14.13: An analog output signal can be produced by connecting a digital-to-analog converter (DAC) to a microcontroller output power**

relays) is available that will allow a microcontroller to be easily interfaced to an AC mains-connected load. It then becomes possible for a small microcontroller (operating from only a 5 V DC supply) to control a central heating system operating from 240 V AC mains.

## 14.14 Symbols Introduced in this Chapter

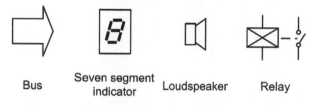

Bus          Seven segment       Loudspeaker        Relay
             indicator

**Figure 14.14: Symbols introduced in this chapter**

# CHAPTER 15

# Power Electronics

Keith H. Sueker
Tim Williams

Much of the design work in power electronics involves specification of ancillary apparatus in a system. It is essential to a successful design that the engineer knows the general characteristics of these components well enough to permit selection of a suitable device for the intended application. The components in this chapter are usually described in detail in vendor catalog information, but the designer must know the significance of the ratings and how they apply to the job at hand. Competent vendors can be valuable partners in the design process.

Commonly used symbols in power electronics diagrams are shown in Figure 15.1. The utility breaker symbol is generally used in single line drawings of power sources, whereas the industrial symbol is used on schematics. There are no hard and fast rules; however, there are a number of variations on this symbol set.

## 15.1 Switchgear

The equipments intended to connect and disconnect power circuits are known collectively as *switchgear* (please—not *switchgears* and not *switch-gear*). Switchgear units range from the small, molded-case circuit breakers in a household panelboard to the huge, air break switches on 750-kV transmission lines. They are generally divided into the four groups of disconnect or isolator switches, load break switches, *circuit breakers*, and *contactors*.

Disconnect or isolator switches are used to connect or disconnect circuits at no load or very light loads. They have minimum arc-quenching capability and are intended to interrupt only transmission line charging currents or transformer exciting currents

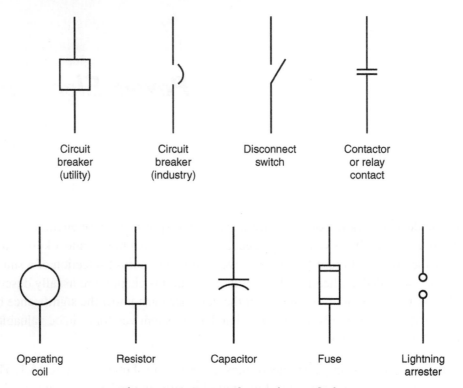

**Figure 15.1: Power electronics symbols**

at most. They are usually the least expensive type of switch. Mechanically, they are designed to provide sufficient contact pressure to remain closed through fault currents despite the high mechanical forces these currents may cause. Simple knife switches rely on multiple leaves for contact and frictional forces to maintain contact. Others types have over-center latches, while still others have clamping locks that toggle toward the end of the closing cycle. All operate in air and have visible contacts as a safety provision, although low-voltage safety switches rely on handle position. All have provisions for lockout.

Medium- and high-voltage disconnect switches are available as indoor designs that are typically mounted in metal switchgear enclosures or as outdoor switches incorporated into elevated structures. Both horizontally and vertically operating switches are available in outdoor designs, and most are available with motor operators. Some have optional pneumatic operators.

Load break switches generally follow the basic design arrangements of disconnect switches except that they are equipped with arc chutes that enable them to interrupt the current they are designed to carry. They are not designed to interrupt fault currents; they must remain closed through faults. Again, motor operators are available in most designs. Motor-operated load break switches can be a lower-cost alternative to circuit breakers in some applications where remote control is required.

Circuit breakers are the heavy-duty members of the switchgear family. They are rated thermally for a given continuous load current, as well as a maximum fault current that they can interrupt. The arcing contacts are in air with small breakers, but the larger types have contacts in a vacuum or in oil. High-voltage utility breakers may utilize sulfur hexafluoride ($SF_6$) gas. Most breakers have a stored energy operating mechanism in which a heavy spring is wound up by a motor and maintained in a charged state. The spring energy then swiftly parts the contacts on a trip operation. Typically, the circuit is cleared in 3 to 5 cycles, since rapid interruption is essential to minimize arc heating and contact erosion. Indoor breakers are usually in metal cabinets as part of a switchgear lineup, whereas outdoor breakers may be stand-alone units.

Some caution should be used when specifying vacuum circuit breakers. When these breakers interrupt an arc, the voltage across the contacts is initially quite low. As the current drops to a low value, however, it is suddenly extinguished with a very high *di/dt*. This current is termed the *chop current*, and it can be as high as 3 to 5 A. If the breaker is ahead of a transformer, the high *di/dt* level can generate a high voltage through the exciting inductance of the transformer, and this can be passed on to secondary circuits. The required voltage control can be obtained with arresters on the primary or metal oxide varistors (MOVs) on the secondary of the transformer. The MOV should be rated to dissipate the transformed chop current at the clamping voltage rating of the MOV. It also must be rated for repeated operations while dissipating the $1/2\ LI^2$ energy of the primary inductance where I is the chop current.

Molded case breakers are equipped with thermal and magnetic overload elements that are self-contained. They are rated by maximum load current and interrupt capacity. Thermal types employ selectable heaters to match the load current for overload protection. Larger breakers are operated from external protective relays that can provide both overload and short circuit protection through time overcurrent elements and instantaneous elements. Nearly all relays are operated from current transformers and most are now solid-state.

Because of their heavy operating mechanisms, circuit breakers are not rated for frequent operation. Most carry a maximum number of recommended operations before being inspected and repaired if necessary. Also, after clearing a fault, breakers should be inspected for arc damage or any mechanical problems.

The real workhorses of switchgear are the contactors. These are electromagnetically operated switches that can be used for motor starting and general-purpose control. They are rated for many thousands of operations. Contactors can employ air breaks at low voltages or vacuum contacts at medium voltages. Most have continuously energized operating coils and open when control power is removed. Motor starters can handle overloads of five times rated or more, and lighting contactors also have overload ratings for incandescent lamps. The operating coils often have a magnetic circuit with a large air gap when open and a very small gap when closed. The operating coils may have a high inrush current when energized, and the control power source must be able to supply this current without excessive voltage drop. Some types have optional DC coils that use a contact to insert a current reducing resistor into the control circuit as the contactor closes.

Any piece of electrically operated switchgear, whether breaker or contactor, has inductive control circuits that can develop high voltages in control circuits when interrupted. Good design practice calls for R/C transient suppressors on operating coils or motors. MOVs will limit the developed voltage on opening but will be of no help in limiting the *di/dt* that may interfere with other circuits. Contactors may be mounted within equipment cabinets or as standalone items.

## 15.2  Surge Suppression

Transient overvoltages can arise from a number of sources. Power disturbances result from lightning strokes or switching operations on transmission and distribution lines. Switching of power factor correction capacitors for voltage control is a major cause of switching transients. All utility lines are designed for a certain *basic insulation level (BIL)* that defines the maximum surge voltage that will not damage the utility equipment, but which may be passed on to the customer. Some consideration should be given to the supply system BIL in highpower electronics with direct exposure to medium-voltage utility lines. Such information is generally available from the utility representative. The standard test waveform for establishing BIL capability is a voltage that rises to the instantaneous BIL value in $1.2\,\mu s$ and decays to half that value in another $50\,\mu s$.

Other sources of transient overvoltages may lie within power electronics equipment itself. Interrupting contactor coils has already been mentioned. Diode and SCR reverse recovery current transients can also propagate within equipment. Arcing loads may require shielding of control circuits. In general, a solid grounding system will minimize problems.

Apparatus for surge protection covers the range from the little discs in 120 V power strips for computers to the giant lightning arresters on 765-kV transmission lines. Many types now utilize the nonlinear characteristics of MOVs. These ZnO ceramic elements have a low leakage current as the applied voltage is increased until a threshold is reached at which the current will increase rapidly for higher voltages. The operating voltage is controlled by the thickness of the ceramic disk and the processing. MOVs may be stacked in series for higher voltages and in parallel for higher currents.

Lightning arresters are classified by their current rating at a given clamping voltage. Station-class arresters can handle the highest currents and are the type used by utilities on transmission and subtransmission lines. Intermediate-class arresters have a lesser clamping ability and are used on substations and some power electronics that are directly connected to a substation. The lowest clamping currents are in distribution-class arresters that are used on distribution feeders and the smaller power electronics equipment. The cost, of course, is related to the clamping current. Arresters are rated for their clamping voltage by class and for their *maximum continuous operating voltage, MCOV*. They are typically connected line-to-ground. Lightning arresters are often used to protect dry-type transformers in power electronic equipment, because such transformers may have a lower BIL rating than the supply switchgear. In 15-kV-class equipment, for example, the switchgear may be rated for 95 or 110 kV BIL, whereas the transformer may be rated for only 60 kV.

As a design rule, MOVs used for the protection of power electronics will limit peak voltage transients to 2-1/2 times their maximum continuous rated rms voltage. They may be connected either line-to-line or line-to-ground in three-phase circuits. Line-to-line connections limit switching voltage transients best but do not protect against common-mode (all three lines to ground) transients. On the other hand, the line-to-ground connection that protects against common-mode transients does not do as good a job on applied line transients. For optimum protection in equipments with exposure to severe lightning or switching transients, both may be appropriate. The volt-ampere curves for a

MOV should be checked to be sure the device can sink sufficient current at the maximum tolerable circuit voltage to handle the expected transient energies. This current will be a function of the MOV size, and a wide range of diameters is available to handle nearly any design need. Small units are supplied with wire leads, whereas the larger units are packaged in molded cases with mounting feet and screw terminals for connections.

Another device in the protection arsenal is the surge capacitor. Transient voltages with fast rise times, high *dv/dt*, may not distribute the voltage evenly among the turns on a transformer or motor winding. This effect arises because of the turn-to-turn and turn-to-ground capacitance distributions in the winding. Surge capacitors can be used to slow the *dv/dt* and minimize the overvoltages on the winding ends. These are generally in the range of 0.5 to 1.0 kF for medium-voltage service. Some care should be exercised when these are used with SCR circuits because of the possibility of serious overvoltages from ringing. Damping resistors may be required.

## 15.3 Conductors

Current-carrying conductors range from the small wires of home circuits to massive bus bar sets that may carry several hundred kiloamperes. Copper is the primary conductor, with aluminum often used for bus bars and transformer windings. Conductor cross-sectional areas are designated by American Wire Gauge (AWG) number in the smaller sizes, with a decrease of three numbers representing a doubling of the cross-sectional area. Numbered sizes go up to #0000, 4Ø (four aught). For larger conductors, the cross sections are expressed directly in circular mils, $D^2$, where D is the conductor diameter in thousandths of an inch. For example, a conductor 1/2 inch in diameter would be 250,000 circular mils. This would usually be expressed as 250 kcm, although older tables may use 250 mcm. For noncircular conductors, the area in circular mils is the area in square inches times $(4/\pi) \times 10^6$.

High-current conductors are usually divided into a number of spaced parallel bus bars to facilitate cooling. A rough guide to current capacity for usual conditions is 1000 A/in$^2$ of cross section. Connections between bus bar sections should be designed to avoid problems from differential expansion between the conductors and the bolts that fasten them, as both heat up from current or ambient temperature. Silicon bronze bolts are a good match for the temperature coefficient of expansion of copper, and they have sufficient strength for good connections. However, highly reliable connections can be made between copper or aluminum bus sections with steel bolts and heavy Belleville washers on top of larger-diameter steel flat washers. The joint should be tightened until the Belleville washer is just flat. Ordinary split washers are not recommended. If the bus is subjected to high magnetic fields, stainless steel hardware should be used, but the field from the bus itself does not usually require this. Environmental conditions, however, may favor stainless.

All joints in buswork must be clean and free of grease. Joints can be cleaned with fine steel wool and coated with a commercial joint compound before bolting. Aluminum bus must be cleaned free of all oxide and then immediately protected with an aluminum-rated joint compound to prevent oxide formation.

Most control wiring is made with bare copper stranded conductors having 300- or 600-V insulation, much of which is polyvinyl chloride (PVC). These conductors are generally listed by Underwriter's Laboratories, the Canadian Standards Association, or both. Most equipment standards require labeled wire that carries a UL or CSA printed listing number along with AWG gauge and insulation temperature rating (see Figure 15.2). The National Electric Code should be followed for the required current rating of the conductors. Power wiring is similar to control wiring except, of course, for being much larger. Cabinet wiring is often limited to about 250 kcm because of the necessary tight bending radii, although there are no hard rules on this.

In sequence, these identify the vendor, appliance wire, wire size, voltage rating, fire retardant class, insulation temperature, Underwriter's Laboratories as a listing agency, appliance wire listing number, CSA as a listing agency, alternate use as control circuit wire, maximum operating temperature, and listing identification.

Stranded conductors should be terminated in pressure-swaged crimp connectors that then can be bolted to bus work or terminal blocks. Circuit breakers and other power devices often have provisions for fastening stranded conductors with clamp plates or pressure bolts with rounded ends. Swaged connectors should not be used on these terminals. Fine-strand, extra-flexible welding cable should never be used with clamp plates. Pressure-crimped connectors are imperative.

ROME AWM 20 AWG 600 V FR-1 105°C (UL) AWM E-11755 CSA TEW 105°C ZZ 15213

**Figure 15.2: Typical wire labeling**

Medium-voltage conductors rated to 7.5 kV are available either shielded or unshielded, but higher-voltage cables must be shielded unless air spaced from other conductors and ground. Spacings must follow standards. Shielded conductors have a center current-carrying conductor, a layer of insulation, and then a conductive shield covered by an insulated protective layer. The shield is grounded. This arrangement assures that the radial electrostatic field is uniform along the length and that there are no voids in the insulation to cause corona deterioration. Terminations are made with stress cones, devices of several types that gradually increase the insulation radius to an extended shield while maintaining void-free conditions. When the radius is sufficient to reduce the voltage stress to allowable levels, the shield can be ended and conventional terminal lugs attached to the extended insulated conductor. Some stress cones have shrink-fit tubing and others a silicone grease to eliminate voids. Figure 15.3 shows a typical arrangement.

The forces between current-carrying conductors vary as the square of the current, so bracing for fault currents becomes a serious issue in high-power equipment. Electronic systems such as motor starters that are connected directly to a power line may face especially high fault currents. Circuit breakers require several cycles to trip and are of no use in limiting initial fault currents. Ordinary fuses also have relatively long melting times and do not help. On the other hand, semiconductor-type fuses will melt subcycle and limit fault current, the magnitude of which is a function of the prospective fault current

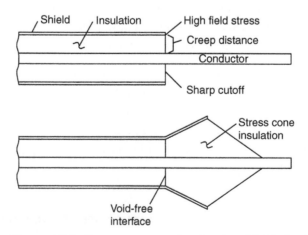

**Figure 15.3: Stress cone termination for shielded cable**

without a fuse. The force in pounds per linear foot developed between two parallel round conductors with spacing $d$ in inches is:

$$F = 5.41 \, I^2 \times 10^{-7} \, l d$$

where $I$ is the rms fault current in each. The force is dependent on the conductor geometry. Forces are attractive for currents in the same direction and repulsive for opposite polarities.

When equipment is supplied from an internal transformer rated for the load current, the steady-state fault current will seldom exceed twenty times rated current ($1/Xpu$). However, an inductive source causes an asymmetric fault current that theoretically may reach a maximum of twice the steady-state peak value. Although L/R current decay makes a peak of around 1.5 times steady-state peak more likely, this still allows more than twice the steady-state peak force, since the force is proportional to current squared. Circuit breakers are rated for a maximum peak current that will allow them to close and latch the mechanism.

High-current conductors are sometimes made with liquid cooling, one form utilizing copper tubing soldered or brazed into grooves that are milled into the edge of the bus. An advantage of liquid cooling in general is that most of the heat generated in the equipment can be transferred to the water, thus minimizing heating of the air in cabinets with power electronics. Liquid cooling also saves on copper.

Buswork carrying high levels of AC currents, especially with a high harmonic content, may cause parasitic heating of adjacent steel cabinet parts due to induced eddy currents. One solution to the problem is to replace the cabinet sections with stainless steel, aluminum, or fiberglass sheet and structural members. Another solution is to interpose a copper plate between the bus and the offending cabinet member. The plate will have high eddy currents, but the low resistance of the copper will minimize losses. Eddy currents in the copper will generate a flux in opposition to the incident flux to shield the cabinet steel.

## 15.4 Capacitors

The three major dielectric types of capacitors are those with various types of film dielectrics used mostly for power factor correction and R/C snubbers, electrolytic types used for filters, and ceramic types in the smaller ratings. The electrolytics have a much higher energy storage for a given volume, but they are not available in voltages above about 500 V and are generally rated for DC service only. They further have leakage currents and limited ratings for ripple current. Still, their high energy density makes them popular for filters on DC power supplies. Even when operated at rated conditions, electrolytic capacitors have a definite lifetime, because the electrolyte will evaporate over time, especially if the capacitors are operated at high ripple currents or in high ambient temperatures. Design consideration should be given to adequate ventilation or heat sinking.

Film dielectric power factor correction capacitors have replaced most of the earlier types made with paper dielectric. These capacitors are rated by kilovar (kvar) at rated voltage and are available both as single units and three-phase assemblies in one can. Power factor correction capacitors are always fused, either with standard medium-voltage fuses or with expulsion fuses in outdoor installations. The latter discharge a plume of water vapor when ablative material in the fuse tube is evaporated as the fuse clears a fault.

Capacitors applied to a power system can create problems in the presence of harmonics generated by nonlinear loads. The capacitor bank will form a parallel resonance with the source inductance of the utility supply, and if this resonance falls on a harmonic of the line frequency at which harmonic currents are present, the result can be serious overvoltages and/or overcurrents. Good engineering practice is to make a harmonic voltage survey before installing power factor correction capacitors.

Power factor capacitor ratings are described in IEEE 18-2002, *IEEE Standard for Shunt Power Capacitors*. In summary, they may be operated at maximum conditions of 110% rated rms voltage, 120% of rated peak voltage, 135% of rated kvar, and 180% of rated rms current. Each of these ratings must include any harmonic voltages or currents. When a capacitor is used with a series inductor to form a series resonant harmonic current trap, the increase in voltage at power frequency caused by the inductor must be considered. Most third-harmonic filters and some fifth-harmonic filters may require capacitors rated above the nominal circuit voltage.

Energizing a section of a capacitor bank when the remainder of the bank is on line can result in damaging transient currents. When a single capacitor is connected to a power line, the surge current is limited by the impedance of the source. Within a capacitor bank, however, the only impedance limiting switching current is the small inductance and resistance of the buswork between sections. The charged capacitors will discharge into the incoming capacitor with little current limiting. Each switched section within a capacitor bank should be protected with a current-limiting reactor. Surge currents should be kept within the instantaneous ratings of the capacitors and switchgear.

Some capacitors designed for DC operation are made with a very long sandwich of conductive and dielectric strips rolled into a cylinder. Connections are made at one end of the two conductive strips, a "tab foil" design. Other types are made from a dielectric strip with a foil or deposited film of metal on one side. The film type can evaporate a small area of the metal on an internal failure without damage, and they are advertised as being self-healing. Capacitors designed for R/C snubber circuits, however, are often required to carry high rms currents and must be so rated. These capacitors are also formed from a sandwich of aluminum foil strips and film dielectric rolled into a cylinder, but the foil layers are offset axially so that the connections to the two foil windings can be made all along the two edges of the winding. This arrangement, known as *extended foil*, lowers the inductance of the capacitor, and the resistive losses are much lower because the current does not have to flow in from one end of the winding. The two constructions are shown in Figure 15.4. In general, DC-rated capacitors should not be used for AC service or R/C snubbers unless they also have an acceptable AC voltage and current rating. Note that snubber capacitors are subjected to repetitive charge and discharge that results in much higher rms currents than would be expected from their capacitance and applied voltage.

All capacitors can be connected in series or parallel for higher voltages or capacitances. They may be freely paralleled, but series connections may require the use of a voltage-sharing resistor connected in parallel with each capacitor. Film types operated on AC

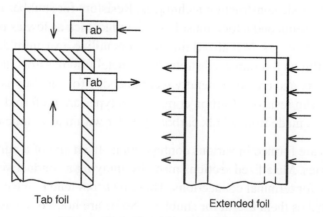

Tab foil              Extended foil

**Figure 15.4: Capacitor construction**

circuits may not require sharing resistors for series operation, but resistors are required if DC voltage components are present. Without sharing resistors, the DC voltage will distribute in proportion to the highly variable leakage resistances. Sharing resistors must have a resistance low enough to swamp out the leakage resistance variations to a sufficient degree of voltage uniformity. Design guidance is available from vendors.

Yet another version of capacitors is the ceramic type. Made from ceramic material with a high dielectric constant, ceramic capacitors generally have smaller capacitances but are available in high voltage ratings. Such capacitors have a very low self-inductance and may be desirable for some types of snubbers.

## 15.5 Resistors

Power electronic systems employ a large variety of resistor types and ratings. At the low-power end, they are used in R/C snubber circuits, in voltage dividers, and as damping elements for various resonant circuits. The two general resistor classes in the lower power ranges are wirewound and metallized film. Wirewound resistors are wound from a resistance alloy wire, usually on a cylindrical ceramic body. Terminal connections are welded at each end of a solenoidal winding. Noninductive wirewound resistors are made with two paralleled windings wound in opposite directions around the body so that their magnetic fields tend to cancel. Another construction technique is to wind the resistor from an elongated hairpin with the loop anchored to one end of the body and the leads brought out at the other end, the two wires being insulated from each other. There are many variations on these basic construction techniques. Resistors for snubber use, especially with fast switching semiconductors, must have an inductance as low as possible to minimize transient voltages. Metallized film resistors utilize a vacuum deposited resistance metal film on a ceramic substrate. Such metal film resistors have little transient heat storage capacity and are not generally recommended for snubber use. The same is true for carbon film resistors. Carbon composition types are preferred for low-power snubbers. These are made from a bulk carbon cylinder within a ceramic tube.

Ceramic resistors are formed in various configurations from any of a number of conductive ceramics. Metallized sections made by spraying a conductive metal onto the ceramic allow for terminal connections. These resistors tend to have a low inherent inductance that makes them useful for snubbers. Some are housed in cast metal bodies that provide an insulated heat sink for power dissipation.

High-power resistors take on several forms, all of which are designed to permit efficient cooling (see Figure 15.5). Some in the power ranges up to a few kilowatts are made with rectangular conductors of resistance alloy wound into an air core cylinder with appropriate insulators and supports. Resistors with still higher power ratings are made from stamped sheet metal resistance alloys, sometimes stainless steel, assembled into

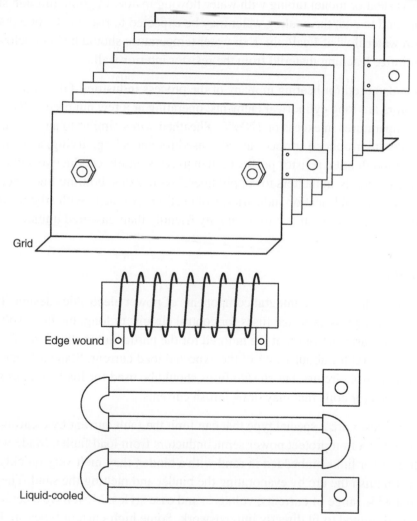

Grid

Edge wound

Liquid-cooled

**Figure 15.5: Power resistor types**

stacks with series, parallel, or series/parallel connections for the desired resistance. The general description is *grid resistor*. Iron grid castings preceded this type of construction, and such resistors were often used for starting DC motors on trolley cars.

Water-cooled resistors are useful in equipment with water-cooled semiconductors or for the manufacture of compact testing loads for power electronic systems. Many are made from stainless steel or monel tubing with water flowing inside. In going through such a resistor from end to end, the cooling water may be expected to rise 3.8°C for a dissipation of 1 kW with water flow at 1 gal/min. Exit water temperature should be kept below about 70°C to minimize leaching material from the resistor interior wall.

Resistors are also used for heating in many of the process industries. Globar® silicon carbide resistors are long cylindrical elements, operating at a few hundred volts, that can create temperatures in excess of 1200°C. Sheathed wires similar to an electric stove element with grounded surfaces are also used for annealing, drying, and similar processes. Although not a resistor per se, molten glass is highly conductive and is held at temperature electrically in melters to supply fiberglass nozzles, bottling lines, float glass, and many other glass fabrication industries. Connections are made with silicon carbide rods. Electric melters are more environmentally friendly than gas-fired units.

## 15.6 Fuses

These protective elements are integral components of power electronics design. They range from the tiny glass cartridge fuses for control circuits to long, medium-voltage types. Each has characteristics that are tailored for the particular applications. Control fuses should be rated for about 125% of the expected load current. Standard types can be used for most control circuits, but *slo-blo* fuses should be used for loads such as small motors and contactor coils that may draw inrush currents.

Semiconductor fuses are a special type that can limit the fault current by clearing subcycle, and they often protect power semiconductors from load faults. Made with multiple thin, silver links embedded in sand with a binder, they melt very quickly on faults and extinguish the arc by evaporating the binder and melting the sand. They are available in a wide range of currents, voltages, and case styles. Most have a ceramic case, and many are designed to fit directly into buswork. Some high-current types are built as matched units, paralleled by the vendor. In pulsed applications, they should not be loaded with an rms pulse current more than 60 to 70% of the melting current for the pulse duration. Steady-state current should not exceed 80% of rated.

If protection of semiconductors is a design objective, the fuse $I^2t$ rating should be well under the $I^2t$ rating of the semiconductor. Better coordination can be obtained in SCR converters if each SCR path is fused rather than the supply lines. This arrangement also offers protection from internal bus-to-bus faults when the load can source power.

Medium-voltage fuses are available as "E" rated for transformers and general-purpose applications, and "R" rated for use with applications such as motors with high starting currents. Most mount in clip assemblies. These fuses may be matched in resistance and paralleled by the vendor for higher currents.

All high-current fuses should be bolted into sanded buswork with joint compound and sufficient pressure to ensure a minimum resistance. Fuses are rated under the assumption that the buswork to which they are mounted will sink heat from the fuse and not source heat into it.

## 15.7 Supply Voltages

The primary operating voltage for most power electronics is divided into two general classes: low-voltage, service voltages of 600 V or less, and medium-voltage, service voltages of 601 V through 34.5 kV. The vast majority of power electronics will wind up on either 600 V-, 5-kV, or 15-kV-class supplies in the U.S., but there are applications at 2400 V and 6900 V, especially in older plants. Overseas, many other voltages may be encountered, with 400 V, 3300 V, and 11 kV being popular, all at 50 Hz.

## 15.8 Enclosures

Equipment enclosures are described in NEMA standard ICS 1–110. Briefly, the designer may be expected to encounter Type 1, Type 4, and Type 12 enclosures most often. Type 1 is a general-purpose indoor, ventilated enclosure that protects personnel from accidental exposure to high voltages and protects equipment from dripping water. Type 4 is a watertight, dusttight, nonventilated indoor or outdoor enclosure. Type 12 is a dusttight, driptight indoor enclosure. Type 12 may have nonventilated sections that are dusttight and ventilated sections that are not.

Most enclosures are made with 10 to 12 ga steel, although smaller wall mount cabinets may be 14 ga. Corners and seams are welded, and free-standing enclosures are equipped with three-point door latches. The rear wall of a cabinet has welded studs that support a removable panel so that component assembly can be done outside the cabinet. All doors should be connected to the enclosure frame with flexible grounding straps for safety. The industry standard for free standing enclosures is 90 inches in height.

## 15.9 Hipot, Corona, and BIL

Any insulation system must be able to tolerate a continually applied voltage, a transient overvoltage, and a surge voltage. Furthermore, it must be free of partial discharge (corona) under the worst-case operating conditions. The hipot test is typically a 1-min application of a 50- or 60-Hz voltage between all conductors and ground, during which the system must not fail shorted or show a fluctuating leakage current. There may, of course, be displacement currents from the capacitance to ground.

Absent a specific high-potential test specification, a rule of thumb is a 1-min, 60-Hz applied sinusoidal voltage of twice rated rms voltage plus 1000 V for equipment rated 600 V or less and 2.25 times rated voltage plus 2000 V for ratings of 601 V and above.

The ability to withstand surge voltages is defined by a test wave with a 1.2 μs rise time to peak and a 50 μs fall to half voltage. This test approximately defines a basic insulation level (BIL) for the system. The test is a single application of this wave, and the requirement to pass is simply freedom from breakdown.

Yet another test is the voltage at which a certain level of corona begins. This is detected by the appearance of impulse discharge currents on an oscilloscope as the applied voltage is slowly raised. The voltage at which these currents appear is the onset or inception level, and the cessation of the impulses as the voltage is reduced is the offset or extinction voltage. Standardized metering circuits in commercial corona testers allow these impulse currents to be quantified in micro-coulombs of current-time integral.

A simple corona tester can be made that is sufficient for most purposes with only a hipot tester, a filter, and a coupling circuit as shown in Figure 15.6. The noise filter can be made with a high-voltage resistor and capacitor, and the current demand should be kept below the maximum rating of the hipot tester. The RF choke (RFC) can be any small inductor of from 1 to 100 mH inductance, and the high-pass R/C filter can be used to eliminate the

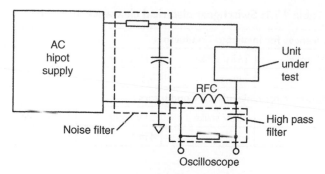

**Figure 15.6: Simple corona tester**

fundamental current from the oscilloscope. Some tinkering of these components can be expected. In operation, corona will be indicated by the appearance of noise spikes as the voltage is raised. The unit can be tested with some twisted hookup wire.

## 15.10 Spacings

Even the lowest-voltage systems require some consideration for the electrical clearances between conductors of different voltage. Standards have been developed by the Canadian Standards Association (CSA), Institute of Electrical and Electronics Engineers (IEEE), National Electrical Manufacturers Association (NEMA), and Underwriter's Laboratories (UL). These standards cover everything from PC boards to high-voltage switchgear.

Spacings are generally considered in two classes: *strike*, the clearance through air paths, and *creep*, the clearance along insulating surfaces. Whereas the strike capability of an air path between spherical conductors may be much larger then the standards allow, the standards recognize the imperfect world of sharp-edged conductors, conductor movement on faults, voltage transients, and safety margins. Similarly, the creep standards recognize that insulating surfaces may become contaminated by conductive dust or moisture.

Understanding these standards is especially important in applying medium-voltage transformers that are directly connected to customer switchgear. The switchgear is the first line of defense and must cope with lightning and switching transient voltages, but it will pass along these transients to connected equipment. Unless equipment connected to customer switchgear is protected by auxiliary arresters and/or surge capacitors, it

Table 15.1: Switchgear electrical clearance standards

| Clearances for Insulated Conductors | | | |
|---|---|---|---|
| 5 kV strike | 15 kV strike | 5 kV creep | 15 kV creep |
| 2 in | 3 in | 3.5 in | 5.5 in |
| Clearances for Uninsulated Conductors | | | |
| 5 kV strike | 15 kV strike | 5 kV creep | 15 kV creep |
| 3 in | 6 in | 4 in | 6.5 in |

must meet the same standards as the switchgear itself. Table 15.1 is taken from the Westinghouse document, "Electrical Clearances for Switchgear," and, although some years old, it is typical of the several extant standards.

The insulated conductors include extruded insulations, insulating boots, and high-voltage taping. The standards recognize that these insulating materials may degrade with continued exposure to high voltages.

## 15.11 Metal Oxide Varistors

Metal oxide varistors (MOVs) are components that have a nonlinear V/I characteristic. In the case of varistors used for voltage protection, the voltage varies but little over a very wide range of current. The types used for power electronics are made by pressing and sintering wafers of zinc oxide ceramic with the characteristics determined by the process, the diameter, and the thickness. These devices are available in sizes from those suitable for surface mounting on PC boards to those for large station-type lightning arresters. The range spans sizes from a few millimeters to 90 mm in diameter.

The V/I curve for a typical 60-mm dia., 480 V rated MOV is shown in Figure 15.7. Note that the current is only 1 A at 1000 V peak and virtually zero at the 680 V peak in a 480 V circuit. However, it will limit the peak voltage to about 1200 V at 1000 A. This means it will protect a 1200 V SCR or other semiconductor from peak transient currents as high as 1000 A. MOVs are generally applied at their nominal rms voltage rating and are expected to clamp transients to a peak voltage of 2.5 times their rms rating.

MOVs have little power dissipation capability, and they can be easily destroyed by repetitive transients such as produced by SCR commutation. MOV catalogs show the

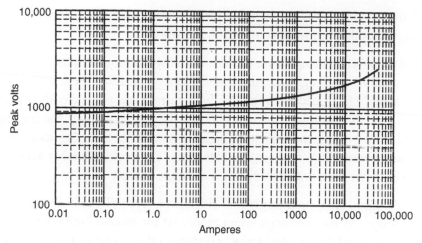

**Figure 15.7: 480 V, 60-mm MOV characteristic**

lifetime characteristics as a function of the current magnitude and duration. When used for suppressing breaker chop, for example, the maximum lifetime exposure should be calculated so that a suitably sized MOV can be specified.

## 15.12 Protective Relays

Utilities and large industrial plants use a variety of relay types to protect the system and its components against fault currents. The most basic types are overcurrent relays, which are available in a number of styles. All will trip a breaker on overcurrent, but the timing is widely variable among the several types. Relays are available from inverse to extremely inverse according to the design. All trip with a delay on low-current faults but trip more quickly as the fault current rises. Many are available with an auxiliary instantaneous element that will trip subcycle. Nearly all types are now electronic, with power derived from the protected circuit itself. They are usually cascaded with decreasing trip current settings as the system branches out from source to load through a succession of buses and circuit breakers. This allows an overcurrent to be cleared as close to the fault as possible so as to avoid disturbing other loads.

Another useful type is the differential relay. This relay has two sets of current coils and will trip on current imbalance between the two sets. When equipped with suitable current transformer ratios on the two sets, it can protect a transformer or generator from internal faults and distinguish between them and external faults. Most differential relays have delay elements to allow for inrush currents in transformers.

Electric utilities often use impedance or distance relays to protect transmission and distribution circuits. Although computers now take over many of these tasks, the principle remains the same. The impedance relay has both current and voltage coils, with the voltage coils used as restraint elements. If the voltage is high enough, the current coils are inhibited from tripping the associated breaker. In a sense, this relay measures the impedance and hence the distance to the fault, and it can decide whether a downstream breaker can clear the fault with less disturbance to the system.

Relays are identified on system single-line diagrams as type 50 for instantaneous overcurrent relays, 51 for time overcurrent relays, 64 for ground fault relays, 87 for differential relays, and 21 for impedance or distance relays. The relay designations are usually shown adjacent to the circuit breaker they trip, with instantaneous and time overcurrent relays shown as 50/51. Undervoltage, phase balance, phase sequence, directional power, and frequency relays are but a few of the many other types available.

This essay has been a bit cavalier in equating, by implication, impedance to reactance. In most power systems work, the resistive losses are small enough to have little effect on fault currents or regulation, so impedance is often considered as reactance in calculations. The same is true of commutation in converters where resistance does play a small role.

### 15.12.1  Analytical Tools

Several specialized analytical tools have been developed to aid in the solution of power and power electronics circuits. Learning these tools can make the design job easier, especially when studying the interaction between a power electronics system and the supplying utility system. Also, it is necessary to understand these analytical tools and their nomenclature to converse with utility and vendor engineers associated with the power electronics field.

## 15.13  Symmetrical Components

Analysis of a three-phase AC circuit with unbalanced currents or voltages gets into some rather messy complex numbers. In 1918, Dr. C. L. Fortesque delivered a paper before the AIEE, predecessor organization to the IEEE, which laid the groundwork for symmetrical components, a method of representing unbalanced voltage or current phasors by symmetrical sets of phasors. These symmetrical components are positive- and negative-sequence three-phase components as well as a zero-sequence single-phase component. This latter phasor is involved with four-wire systems, usually involving ground circuits. The network can be solved in the usual fashion with each of the symmetrical components, and then the individual solutions combined to represent the unbalanced system. Symmetrical components are universally used by power company engineers for system parameters.

Symmetrical component analysis uses a complex operator, a, where $a = -0.5 + j\,0.866$, a unit phasor at 120°. Then, $a^2 = -0.5 - j\,0.866$, and $a^3 = 1.0$. If a set of asymmetric phasors are given as $x$, $y$, and $z$, then:

$$Ex0 = \frac{(x + y + z)}{3}$$
$$Ex1 = \frac{(x + ay + a^2 z)}{3}$$
$$Ex2 = \frac{(x + a^2 y + az)}{3}$$

where all quantities are phasors. $Ex0$, $Ex1$, and $Ex2$ are referred to as the zero-sequence, positive-sequence, and negative-sequence components of $x$, respectively. Then, $Ex0 = Ey0 = Ez0$, $Ey1 = a^2\,Ex1$, $Ez1 = a\,Ex1$, $Ey2 = a\,Ex2$ and $Ez2 = a^2\,Ex2$.

This process is shown in Figure 15.8 where a (very) unbalanced set of phasors are $x = 6.0$, $y = -j2.0$ and $z = -0.707 + j0.707$. The sequence networks are shown at the right. In this case,

$$Ex0 = 1.764 - j0.431$$
$$Ex1 = 2.899 + j0.419$$
$$Ex2 = 1.337 + j0.011$$

The original asymmetric phasors may then be reconstituted as:

$$x = Ex0 + Ex1 + Ex2$$
$$y = Ey0 + Ey1 + Ey2 = Ex0 + a^2\,Ex1 + a\,Ex2$$
$$z = Ez0 + Ez1 + Ez2 = Ex0 + a\,Ex1 + a^2\,Ex2$$

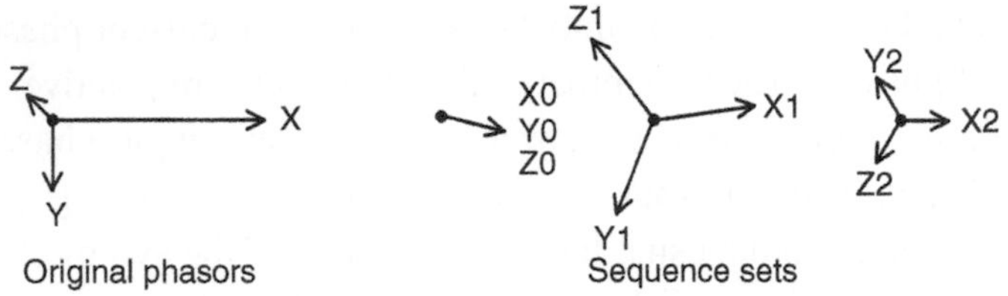

**Figure 15.8: Symmetrical components**

If the set of phasors just resolved were to represent load impedances, the line currents could be determined by impressing the balanced line voltages onto the three sequence networks separately and adding the three components of each line current.

Symmetrical components are often used to describe the characteristics of overhead transmission lines. For example, the familiar set of three conductors in a horizontal row has equal couplings from the two outer lines to the center line, but they have a different coupling to each other. Hence, the mutual inductances and capacitances of the set are different. The use of symmetrical components of these impedances allows the line to be analyzed as two balanced positive- and negative sequence networks. The resultant currents can then be combined. Absent a grounded circuit, the zero-sequence network is not present.

The many circuit simulation software packages now available can reduce the need for using symmetrical components for circuit solutions, but they are still valuable for defining the unbalanced loading and fault performances of synchronous machines.

## 15.14 Per Unit Constants

Per unit quantities greatly simplify comparisons between items of power apparatus and aid in solving fault calculations. Per unit is a method of normalizing the characteristics of elements in a power electronics system so they can be represented independent of the particular voltage at that point in the system. Their characteristics are translated relative to a common base so that extended calculations can be made easily.

In its simplest form, a per unit quantity is merely the percent quantity divided by 100. It spares one the nonsense of 50% voltage times 50% current equals 2500% power. In per unit notation, 0.5 pu voltage times 0.5 pu current equals 0.25 pu power as it should be. A transformer with 6% impedance would have a per unit impedance of 0.06 pu. Although not described as such, this impedance is based on the rated voltage and current of the transformer. It accommodates the differences in primary and secondary voltages by describing the percent rated voltage in either winding required to produce rated current in that winding with the other winding shorted. The regulation characteristics of the transformer are completely described by this figure. When other elements are added to a system, however, there will be a whole set of different ratings of the various elements. A 500-kVA transformer at 4160 V with 6% reactance may serve a 50-kVA transformer at 480 V with 4% reactance that, in turn, serves a 5-kVA lighting transformer at 120 V with 3% reactance. It is a real nuisance to chase the various voltages and currents back through the system to find, for example, the short circuit current at the final transformer. Per unit quantities make it easy.

First, one must choose a particular power level as a base quantity. The selection is completely arbitrary but is usually related to the rating of one of the component items. In this case, the 50-kVA transformer will be used as the base, and its leakage impedance will be 4% on that base, 0.04 pu. To relate the 5-kVA lighting transformer to this quantity, one simply multiplies the 5-kVA impedance of 0.03 pu on its own base by the power ratio of 50 kVA/5 kVA = 10. With the two in cascade, the total impedance is now $0.04 + 0.03 \times 10 = 0.34$ pu. The 500-kVA transformer by the same procedure becomes 0.06 50/500 = 0.006 pu on the 50-kVA base. The series string impedance is then $0.006 + 0.04 + 0.30 = 0.346$ pu on the 50 kVA base. This total series impedance is 0.0346 pu on a 5-kVA base, and a fault on the secondary of the 5-kVA transformer will result in 1/0.0346 = 28.9 times rated current, 28.9 pu on the 5-kVA base. At the 50-kVA transformer, this fault will result in 1/0.346 pu = 2.89 pu current on a 50 kVA base, and at the 500-kVA transformer the fault is 1/3.46 = 0.289 pu on a 500-kVA base. At any point in the system, one can define a base impedance as $Zbase = V_{LL}^2/VA$ or $Zbase = V_{LL}^2/(1000 \times kVA)$ where *VA* or *kVA* is a three-phase rating. Then, in ohms, $Zohms = Zbase\ Zpu$ at that base. The base impedance is the impedance that, when connected to each line of a three-phase system at rated voltage, will draw rated load current and develop rated voltamperes.

It is worth the effort to develop a familiarity with the per unit system, because it greatly eases conversations with utility engineers, motor designers, transformer designers, and others associated with power electronics. It is universally used.

## 15.15 Circuit Simulation

Many power electronics circuits can be simulated and studied with relatively simple computer programs. While many engineers prefer to use commercial circuit simulation

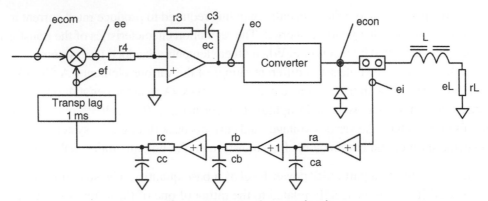

**Figure 15.9: Arc heater circuit**

software packages, there is some merit in being able to write simple code to study circuit operation under transient conditions. The example that follows is written in BASIC, but it can be easily translated to C or any other preferred programming language. It is the concepts of handling the circuit that count.

The schematic of Figure 15.9 shows a circuit the author designed some years ago. The circuit is an arc heater with a current regulator, and the concern was to define the current overshoot when the arc ignited from applied voltage. The converter was a 24-pulse system that permitted a relatively fast current loop of 2000 rad/s. The system had a nonlinear output inductor, a feedback filter, a transport lag from the SCRs and serial optical links, a negative slew rate limit, and an arc strike voltage. The BASIC program follows. It is heavily annotated to illustrate the approach. Figure 15.10 shows the output waveform with a starting current transient of some 270% of initial setpoint, entirely acceptable in this case. Note that the current shows a slight undershoot and then overshoot when falling to the command level once the arc is ignited.

### 15.15.1  Circuit Simulation Notes

Initial voltage:            0 – current integrator enabled at $t = 0$
Initial current:            0.01 A (to get a finite inductance)
Ignition voltage:            600 V
Equivalent arc resistance:  0.25 $\Omega$

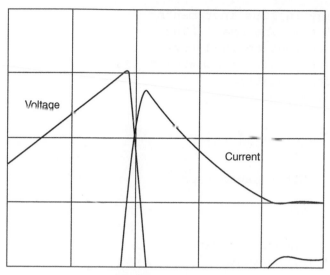

Horizontal 20 ms/div, V = 200 V/div, I = 100 A/div
ARC heater starting characteristic

**Figure 15.10: Circuit voltage and current waveforms**

| | |
|---|---|
| Inductor: | Inductance inversely proportional to current to 1.1 power, bounded by 10 mH maximum and 1.1 mH minimum |
| Feedback: | Three cascaded filter poles at 2000 rad/s. Filter will handle both feedback and anti-aliasing in the digital system. |
| Transport lag: | 1 ms delay in SCRs and digital system, simulated as an actual delay |
| Regulator: | Lead at 250 rad/s to match 4-ms load time constant |
| Negative slew rate: | −10%/ms to approximate 50-Hz sine wave |
| Commanded current: | 100 A, a low initial current setting to minimize overshoot |

This program will run in BASIC 4.5 or higher as well as QuickBASIC.

```
'Arc heater starting program
SCREEN 12 ' set 640 x 480 screen
PALETTE 0,4144959 ' set reverse palette colors
PALETTE 15,0 'set background white
DIM ed(10000)
```

```
td = 100 'delay in 10µs increments
dt = .00001 ' time increment 10µs
icom = 100 ' current command level
ecom = icom/300 ' current command = .33v for 100A
rL = .25 ' load resistance
r3 = 40000 ' lead resistor
r4 = 10000 ' input resistor
ra = 500 ' filter resistors
rb = 500
rc = 500
ca = .000001 ' filter capacitors
cb = .000001
cc = .000001
cd = .000001
c3 = .000001 ' integrator capacitor
i = .01 ' initial current
again:
IF i <= 0 THEN i = 1 ' clip undershoot
L = .01 * (250/i ) ^ 1.1 ' inductor characteristic
IF L > .01 THEN L = .01 ' maximum inductance 10mH
IF L < .0011 THEN L = .0011 ' minimum saturated inductance 1.1mH
ee = ecom - ef ' error signal of command minus feedback
ia = (ei - ea)/ra ' filter capacitor current for Euler integration
ib = (ea - eb)/rb ' same, next stage
ic = (eb -ec)/rc ' same, final stage
p = n - td ' digital system transport lag
IF p <1 THEN p = 1 ' initialize
ed(n) = ec ' last stage filter voltage
ef = ed(p) ' delay of td/100ms
ec = ec + ic * dt/cc ' three cascaded poles of filter with poles
  at 2000rad/s
eb = eb + ib * dt/cb ' sections are isolated
ea = ea + ia * dt/ca ' Euler integration
ei = .0033 * i ' 1500A = 5V feedback from shunt
econ = 240 * eo ' converter gain, 5v = 1200V
IF econ < 0 THEN econ = 0 ' commutating diode prevents negative
  voltage
IF econ > 1200 THEN econ = 1200 ' voltage ceiling
IF (i < icom) AND (econ > 650) THEN econ = 650 'starting voltage
  limit
```

```
IF econx-econ > 1 THEN econ = econx-1 ' negative slew rate limit
   10%/ms
i = i + (econ - eL) * dt/L ' load current
IF econ > 600 THEN k = 1 ' flag to detect first current above
   isetpoint
IF k = 0 THEN i = 0 ' no current until econ> 600V arc ignition
   voltage
eL = i * rL ' load voltage
eo = ee * ( r3/r4 ) + ecap ' output voltage of opamp
IF eo > 10 THEN eo = 10 ' opamp limit
ecap = ecap + (eo - ecap )/r3 * dt/c3 ' voltage on integrator cap
PSET ( n/20 + 50, 400 - i ) ' plot current
PSET ( n/20 + 50, 400 - .5 * econ ) ' plot voltage
n = n + 1
IF n > 10000 GOTO quit: ' end of display
econx = econ ' set econx for prior voltage to set negative slew
   rate maximum
GOTO again:
quit:
FOR n = 0 to 400 STEP 100
LINE ( 50, n ) - ( 550, n ) ' ordinate scale
NEXT
FOR n = 50 to 550 STEP 100
LINE (n, 0 ) - ( n, 400 ) ' abscissa scale
NEXT
LOCATE 27, 15: PRINT "Horizontal 20ms/div, V=200V/ div,
   I=100A/div"
LOCATE 28, 20: PRINT "ARC HEATER STARTING CHARACTERISTIC"
LOCATE 10,10: PRINT "Voltage"
LOCATE 15,50: PRINT "Current"
```

## 15.16 Simulation Software

A number of software packages are now available to simulate the operation of nearly any power electronic circuit. Component characteristics are included, and the programs are set up so that representation of a circuit is relatively easy. All are described on the Internet in some detail, and most have student versions, limited-capability versions, limited-time versions, or introductory packages. The comments that follow must be taken at a point in time, since the software evolves rapidly.

MATLAB—An interactive program for numerical computation and data visualization that is used by control engineers for analysis and design. Numerous "toolboxes" such as SIMULINK, a differential equation solver, are available for simulation of dynamic systems. It provides an interactive graphical environment and a customizable set of block libraries that allow for the design, simulation, and implementation of control, signal processing, communications, and other time-varying systems.

MATHCAD—An equation-based program that allows one to document, perform, and share calculation and design work. It can integrate mathematical notation, text, and graphics in a single worksheet. It allows capture of the critical methods and values of engineering projects.

SPICE—One of the early simulation programs, SPICE allows a circuit to be built directly on the display screen in schematic form. Libraries are available for the various circuit elements. Both steady-state and transient behavior can be analyzed. Many related programs are also available—PSpice, Saber, and Micro-Cap to name just a few. Some are directly compatible with SPICE.

ElectroMagnetic Transients Program (EMTP)—Devoted primarily to the solution of transient effects in electric power systems, variants are available for circuit work. It is developed and maintained by a consortium of international power companies and associated organizations. The core program is in the public domain.

The above is only a sampling of the more popular software available for circuit analysis. Most packages can be purchased on the Internet and some have student versions that can be downloaded at no cost.

## 15.17  Feedback Control Systems

Nearly all systems in power electronics rely on feedback control systems for their operation. This chapter presents the basic analog analysis of such systems because, in this author's opinion, it offers a more intuitive understanding of their behavior than can be obtained from modern control theory with digital techniques.

### 15.17.1  Basics

Figure 15.11 shows the simplest feedback control system. A command signal is received by a summing junction and compared to a feedback signal of opposite polarity.

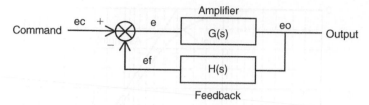

**Figure 15.11: Basic feedback system**

The difference signal is sent to an amplifier that produces the system output with a feedback signal derived from the amplifier output. Both the amplifier characteristic and the feedback characteristic are functions of frequency and are shown as G(s) and H(s), respectively.

The performance of such system can be derived from an equation that relates output to input. The equation is developed as follows:

1. $e = ec - ef$

2. $eo = e \times G(s)$

3. $ef = eo \times H(s)$

4. $eo = G(s) \times [ec - eo \times H(s)]$

5. $eo/ec = G(s)/[1 + G(s) \times H(s)] = A$

where *eo/ec* is the closed-loop system gain as a function of frequency.

If the feedback is disconnected from the summing junction, then $A = G(s) \times H(s)$, the open-loop gain. Simple systems such as the one shown in Figure 15.11 can be analyzed for stability and performance by an examination of the open-loop gain characteristic as the frequency is varied. For most purposes, the asymptotic response will suffice.

### 15.17.2 Amplitude Responses

Figure 15.12 shows the actual and asymptotic responses of a simple R/C circuit consisting of a series 1 MΩ resistor and a 3.3 μF shunt capacitor. Such frequency response characteristics for systems are referred to as *Bode (bodey) plots*.

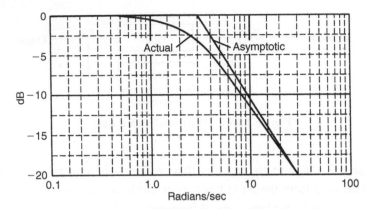

**Figure 15.12: R/C frequency response**

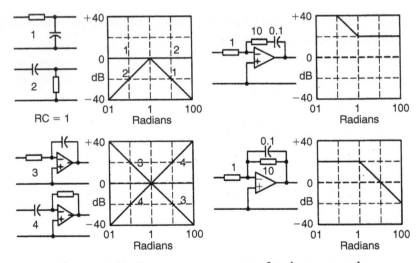

**Figure 15.13: Frequency responses of various networks**

The plot is in decibels (dB) equal to $20 \times \log 10$ (vo/vi) where vo and vi are the output and input voltages, respectively. These may just as easily be currents or currents translated to voltages through shunts or CTs. The asymptotic response is useful, because it can be quickly drawn and has a maximum error of only 3 dB at the break point. The break point in radians per second is simply the reciprocal of the time constant in seconds. Figure 15.13 shows a number of circuit elements and their asymptotic frequency responses.

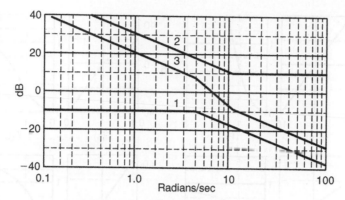

**Figure 15.14: Composite response**

The asymptotic response of cascaded circuit elements can be determined by simply adding their individual responses. Figure 15.14 shows the process for two elements with different asymptotic responses, 1 and 2, and the response, 3, when they are cascaded.

The time response of a closed-loop feedback system can be deduced from the open-loop frequency response. The primary factor affecting the time response is the slope of the frequency response as it crosses the zero-dB line, the line of unity gain, and its response in the frequency decade before and after the crossover. Several normalized frequency response characteristics are shown in Figure 15.15 along with their corresponding time responses. Frequency plots are in radians/sec and time plots in seconds.

At upper left, the gain crosses the zero-dB axis with a slope of −2, −40 dB per decade. The time response is dramatically unstable, and the system takes off for the moon. At upper right, the gain curve approaches the zero-dB axis with a slope of −1 then goes to −2. The system is stable but with an overshoot. The lack of high-frequency gain results in a poor rise time.

The curve at lower right shows a similar behavior with overshoot. Now, however, the high-frequency gain is better, and the system has a good rise time. Finally, at lower left, the system crosses with a slope of −1, −20 dB per decade and is critically damped with a good rise time and no overshoot. These response characteristics can yield some insight into the behavior of more complex systems.

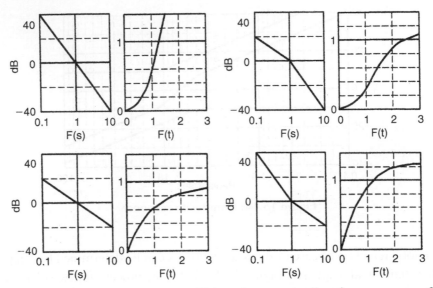

**Figure 15.15: Frequency responses, F(s), and corresponding time responses, f(t)**

### 15.17.3 Phase Responses

The amplitude response with frequency is only part of the story, the remainder being the phase response. The curves at lower left in Figure 15.15 result from a pure integrator, and the phase shift is a constant 90° lag independent of frequency. The characteristic at upper left is equivalent to two integrators in cascade and has a phase shift of 180°. The reason it is unstable is that the feedback voltage now adds directly to the command voltage. Instead of being negative feedback, it is positive, and it makes the system regenerative. The output rises until something saturates, and then the process repeats. The result is an oscillator.

A low-pass filter such as shown at the lower right of Figure 15.13 has a phase lag of 45° at the break frequency, and the lag approaches 90° for higher frequencies. Filtering signals will always result in a lagging phase characteristic. The actual and asymptotic phase responses of an R/C low-pass filter normalized to one radian per second are shown in Figure 15.16.

Filters are not the only sources of phase lag. Any sort of a time delay, termed a *transport lag*, also contributes to phase lag. In an SCR bridge converter, for example, an SCR

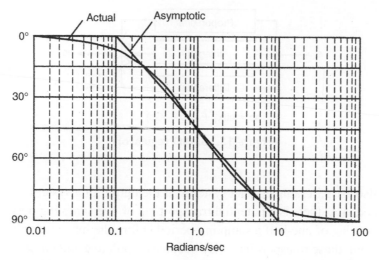

**Figure 15.16: Phase responses of an R/C low-pass filter**

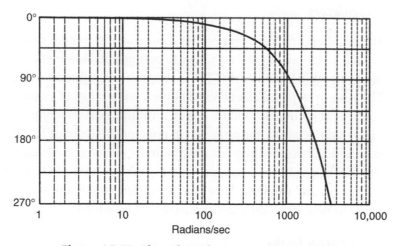

**Figure 15.17: Phase lag of a 1.4-ms transport lag**

cannot respond immediately to a command change unless it has a positive anode voltage. If the command is a sudden large decrease, the time delay may approach 240°, 11 ms in a 60-Hz system, if the previous SCR has just been fired. The average delay will be 30° for a small change in command, 1.4 ms. Phase shifts for such a 1.4-ms time delay are shown in Figure 15.17.

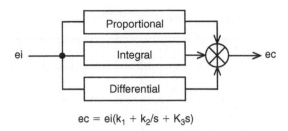

$$ec = ei(k_1 + k_2/s + K_3s)$$

**Figure 15.18: PID regulator**

Time delays also arise in sampled data systems. If the output of a system is periodically sampled for feedback, there is a potential transport lag of one sampling period and an average transport lag of one-half a sampling period before the information is available. In a complex system, these transport lags may become cumulative and constitute a source of instability.

### 15.17.4  PID Regulators

Many industrial controllers employ a proportional, integral, differential regulator arrangement that can be tailored by the customer to optimize a particular control system. The basic layout is shown in Figure 15.18. Three channels are summed with a variable gain on each. The system response can be varied over a wide range of characteristics. A system with only a proportional response has an error that is inversely proportional to that gain. If an integrator is added, the error can in principle be reduced to zero. The "in principle" must be added, because there are always limits on accuracy in any system. The differentiator can be used to compensate for lags in the system and to improve the high-frequency response and rise time. However, the differentiator amplifies noise, and there will be a limit to how much differential control can be added.

### 15.17.5  Nested Control Loops

Many systems require nested control loops to control several variables. One example is a DC motor drive that must have a very fast current control loop to limit the armature current but also requires a voltage loop for speed control. The voltage control cannot override the current loop, but it will set the required current so long as it is within the limits set by the current loop. In short, the voltage or speed loop commands the current that is required to satisfy the voltage, but the current loop sets the current limit.

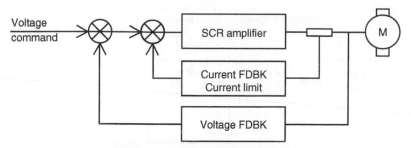

**Figure 15.19: Nested control loops**

Both loops must be unconditionally stable. Figure 15.19 shows a typical system. The armature current is regulated by feedback from a current shunt and isolator amplifier. The frequency of such a current regulator using SCRs can have a crossover as high as 1000 radians/sec, but 500 radians/sec is easier to handle and less critical on feedback. If the current loop is set up for 500 radians/sec, the voltage loop must, generally, crossover at a decade lower in frequency, 50 radians/sec, for stability on a 50- or 60-Hz system.

## 15.18 Power Supplies

The power supply is a vital but often neglected part of any electronic product. It is the interface between the noisy, variable and ill-defined power source from the outside world and the hopefully clear-cut requirements of the internal circuitry. For the purposes of this discussion it is assumed that power is taken from the conventional AC mains supply. Other supply options are possible, for instance a low-voltage DC bus, or the standard aircraft supply of 400 Hz 48 V. Batteries we shall discuss separately at the end of this chapter.

### 15.18.1 General

A conceptual block diagram for the two common types of power supply linear and switch-mode is given in Figure 15.20.

#### 15.18.1.1 The Linear Supply

The component blocks of a linear supply are common to all variants, and can be described as follows:

- input circuit: conditions the input power and protects the unit, typically voltage selector, fuse, on-off switching, filter and transient suppressor;

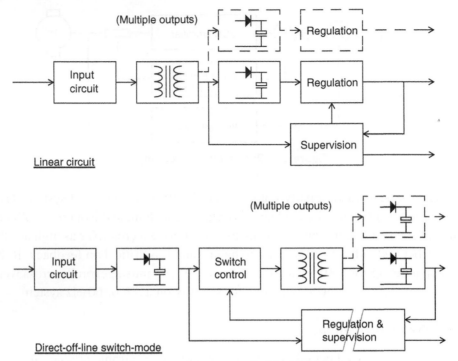

**Figure 15.20: Power supply block diagrams**

- transformer: isolates the output circuitry from the AC input, and steps down (or up) the voltage to the required operating level;

- rectifier and reservoir: converts the AC transformer voltage to DC, reduces the AC ripple component of the DC and determines the output hold-up time when the input is interrupted;

- regulation: stabilizes the output voltage against input and load fluctuations;

- supervision: protects against over-voltage and over-current on the output and signals the state of the power supply to other circuitry; often omitted on simpler circuits.

### 15.18.1.2 The Switch-Mode Supply

The advantage of the direct-off-line switch-mode supply is that it eliminates the 50 Hz mains transformer and replaces it with one operating at a much higher frequency, typically 30–300 kHz. This greatly reduces its weight and volume. The component blocks are somewhat different from a linear supply. The input circuit performs a similar function but requires more stringent filtering. This is followed immediately by a rectifier and reservoir that must work at the full line voltage, and feeds the switch element that chops the high-voltage DC at the chosen switching frequency.

The transformer performs the same function as in a linear supply but now operates with a high-frequency squarewave instead of a low-frequency sinewave. The secondary output needs only a small-value reservoir capacitor because of the high frequency. Regulation can now be achieved by controlling the switch duty cycle against feedback from the output; the feedback path must be isolated so that the separation of the output circuit from the mains input is not compromised. The supervision function, where it is needed, can be combined with the regulation circuitry.

### 15.18.1.3 Specifications

The technical and commercial considerations that apply to a power supply can add up to a formidable list. Such a list might run as follows:

- input parameters: minimum and maximum voltage maximum allowable input current, surge and continuous frequency range, for AC supplies permissible waveform distortion and interference generation;

- efficiency: output power divided by input power, over the entire range of load and line conditions;

- output parameters: minimum and maximum voltage(s) minimum and maximum load current(s) maximum allowable ripple and noise load and line regulation transient response;

- abnormal conditions: performance under output overload performance under transient input conditions such as spikes, surges, dips and interruptions performance on turn-on and turn-off: soft start, power-down interrupts;

- mechanical parameters: size and weight thermal and environmental requirements input and output connectors screening;

- safety approval requirements;

- cost and availability requirements.

### 15.18.1.4  Off the Shelf Versus Roll Your Own

The first rule of power supply design is: do not design one yourself if you can buy it off the shelf. There are many specialist power supply manufacturers who will be only too pleased to sell you one of their standard units or, if this doesn't fit the bill, to offer you a custom version.

The advantages of using a standard unit are that it saves a considerable amount of design and testing time, the resources for which may not be available in a small company with short timescales. This advantage extends into production—you are buying a completed and tested unit. Also, your supplier should be able to offer a unit that is already known to meet safety and EMC regulations, which can be a very substantial hidden bonus.

### Costs

The major disadvantage will be unit cost, which will probably though not necessarily be more than the cost of an in-house designed and built power supply. The supplier must, after all, be able to make a profit. The exact economics depend very much on the eventual quantity of products that will be built; for lower volumes of a standard unit it will be cheaper to buy off the shelf, for high volumes or a custom-designed unit it may be cheaper to design your own. It may also be that a standard unit won't fulfill your requirements, though it is often worth bending the requirements by judicious circuit redesign until they match. For instance, the vast majority of standard units offer voltages of 3.3 V or 5 V (for logic) and $\pm 12$ V or 15 V (for analog and interface). Life is much easier if you can design your circuit around these voltages.

A graph of unit costs versus power rating for a selection of readily-available single output standard units is shown in Figure 15.21. Typically, you can budget for £1 per watt in the 50 to 200 W range. There is little cost difference between linear and switch-mode types. On the assumption that this has convinced you to roll your own, the next section will examine the specification parameters from the standpoint of design.

### 15.18.2  Input and Output Parameters

### 15.18.2.1  Voltage

Typically you will be designing for 230 V AC in the UK and continental Europe and 115 V in the U.S. Other countries have frustratingly minor differences. The usual supply voltage variability is $\pm 10\%$, or sometimes $+10\%/-15\%$. In the UK the

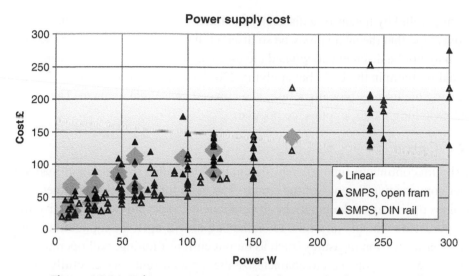

**Figure 15.21: Price versus power rating for standard power supplies**

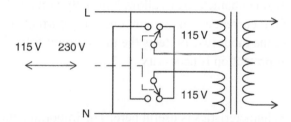

**Figure 15.22: Split-primary transformer wiring**

supply authorities are obliged to maintain their voltage at the point of connection to the customer's premises within ±6%, to which is added an allowance for local loading effects. If the voltage tolerance is applied to the UK/Europe nominal then the input voltage range becomes 207–253 V or 195–253V. This range must be handled transparently by the power supply circuitry.

To cope simultaneously with both the American supply voltage, which may drop below 100 V, and the European voltages is difficult for a linear supply although it is possible to design "universal" switch-mode circuits that can accept such a wide range (see the comment at the end of section 15.18.2.5). Historically, this problem was handled by using a mains transformer with a split primary (Figure 15.22) which can be connected in series or parallel by means of a discreetly mounted voltage selector switch. This has the disadvantage that the switch may be so discreet that the user doesn't know about it, or else it may not be discreet enough and the user may be tempted to fiddle with it. This is not a real problem in the U.S., but applying 230 V to a unit that is set for 115 V will at least annoy the user by blowing a fuse, and at worst cause real damage. Therefore, universal switching mode supplies are popular.

### 15.18.2.2 Current

The maximum continuous input current should be determined by the output load and the power conversion efficiency of the circuit. The main interest in this parameter is that it determines the rating of the input circuit components, especially the protective fuse. You have to decide whether an overload on the output will open the input circuit fuse or whether other protection measures, such as output current limiting, will operate. If the input fuse must blow, you need to characterize the input current very carefully over the entire range of input voltages. It is quite possible that the difference between maximum continuous current at full load, and minimum overload current at which the fuse should blow, is less than the fusing characteristics allow. Normally you need at least a 2:1 ratio between prospective fault current and maximum operating current. This may not be possible, in which case the input fuse protects the input circuit from faults only and some extra secondary circuit protection is necessary.

### 15.18.2.3 Fuses

A brief survey of fuse characteristics is useful here. The important characteristics that are specified by fuse manufacturers are the following:

- *Rated current $I_N$*: that value by which the fuse is characterized for its application and which is marked on the fuse. For fuses to IEC 60127, this is the maximum value that the fuse can carry continuously without opening and without reaching too high a temperature, and is typically 60% of its minimum fusing current. For fuses to the American UL-198-G standard the rated current is 85–90% of its minimum fusing current, so that it runs hotter when carrying its rated current. The minimum fusing current is that at which the fusing element just reaches its melting temperature.

- *Time-current characteristic*: the pre-arcing time is the interval between the application of a current greater than the minimum fusing current and the instant at which an arc is initiated. This depends on the over-current to which the fuse is subjected and manufacturers will normally provide curves of the time-current characteristic, in which the fuse current is normalized to its rated current as shown in Figure 15.22. Several varieties of this characteristic are available:

FF: very fast acting
F:   fast acting
M:  medium time lag
T:   time lag (or anti-surge, slow-blow)
TT: long time lag

Most applications can be satisfied with either type F or type T and it is best to specify these if at all possible, since replacements are easily obtainable. Type FF is mainly used for protecting semiconductor circuits.

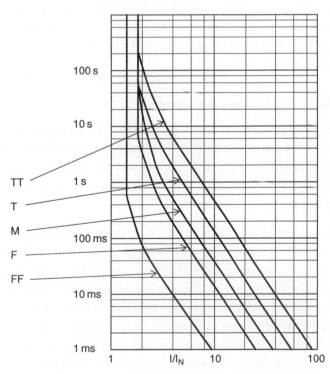

**Figure 15.23: Typical fuse time-current curves**

The total operating time of the fuse is the sum of the pre-arcing time and the time for which the arc is maintained. Normally the latter must be taken into account only when interrupting high currents, typically more than ten times the rated current.

The energy in a short-duration surge required to open the fuse depends on $I^2t$, and for pulse or surge applications you should consult the fuse's published $I^2t$ rating. Current pulses that are not to open the fuse should have an $I^2t$ value less than 50–80% of the $I^2t$ value of the fuse.

- *Breaking capacity*: breaking capacity is the maximum current the fuse can interrupt at its rated voltage. The rated voltage of the fuse should exceed the maximum system voltage. To select the proper breaking capacity you need to know the maximum prospective fault current in the circuit to be protected — which is usually determined in mains-powered electronic products by the characteristics of the next fuse upstream in the supply. Cartridge fuses fall into one of two categories, high breaking capacity (HBC) that are sandfilled to quench the arc and have breaking capacities in the 1000s of amps, and low breaking capacity (LBC), which are unquenched and have breaking capacities of a few tens of amps or less.

### 15.18.2.4  Switch-On Surge, or Inrush Current

Continuous maximum input current is usually less than the input current experienced at switch-on. An unfortunate characteristic of mains power transformers is their low impedance when power is first applied. At the instant that voltage is applied to the primary, the current through it is limited only by the source resistance, primary winding resistance and the leakage inductance.

The effect is most noticeable on toroidal mains transformers when the mains voltage is applied at its peak halfway through the cycle, as in Figure 15.24. The typical mains supply has an extremely low source impedance, so that the only current limiting is provided by the transformer primary resistance and by the fuse. Toroidals are particularly efficient and can be wound with relatively few turns, so that their series resistance and leakage inductance is low; the surge current can be more than ten times the operating current of the transformer. (The effect happens with all transformers, but is more of a problem with toroidals.) In these circumstances, the fuse usually loses out. The actual value of surge depends on where in the cycle the switch is closed, which is random; if it

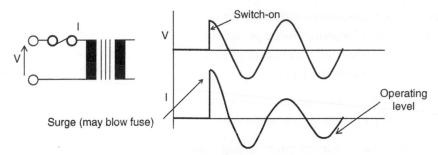

**Figure 15.24: Switch-on surge**

is near the zero crossing the surge is small or nonexistent, so it is possible for the problem to pass unnoticed if it is not thoroughly tested.

A separate component of this current is the abnormal secondary load due to the low impedance of the uncharged power supply reservoir capacitor. For the same reason, inrush current is also a problem in direct-off-line switch-mode supplies, where the reservoir capacitor is charged directly through the mains rectifier, and comparatively complex "soft-start" circuits may be needed in order to protect the input components.

Several simpler solutions are possible. One is to specify an anti-surge or time-lag (type T or TT) fuse. This will rupture at around twice its rated current if sustained for tens or hundreds of seconds, but will carry a short overload of ten or twenty times rated current for a few milliseconds. Even so, it is not always easy to size the fuse so that it provides adequate protection without eventually failing in normal use, particularly with the high ratios of surge to operating current that can occur. A resettable thermal circuit breaker is sometimes more attractive than a fuse, especially as it is inherently insensitive to switch-on surges.

### Current Limiting
A more elegant solution is to use a negative-temperature-coefficient (NTC) thermistor in series with the transformer primary and fuse. The device has a high initial resistance that limits the inrush current but in so doing dissipates power, which heats it up. As it heats, its resistance drops to a point at which the power dissipated is just sufficient to maintain the low resistance and most of the applied voltage is developed across the transformer. The heating takes one or two seconds during which the primary current increases gradually rather than instantaneously.

NTC thermistors characterised especially for use as inrush current limiters are available, and can be used also for switch-mode power supply inputs, motor soft-start and filament lamp applications. Although the concept of an automatic current-limiter is attractive, there are three major disadvantages:

- because the devices operate on temperature rise they are difficult to apply over a wide ambient temperature range;

- they run at a high temperature during normal operation, so require ventilation and must be kept away from other heat-sensitive components;

- they have a long cool-down period of several tens of seconds and so do not provide good protection against a short supply interruption.

### PTC Thermistor Limiting

Another solution to the inrush current problem is to use instead a positive–temperature-coefficient thermistor in place of the fuse. These are characterised such that provided the current remains below a given value self-heating is negligible, and the resistance of the device is low. When the current exceeds this value under fault conditions the thermistor starts to self-heat significantly and its resistance increases until the current drops to a low value. Such a device does not protect against electric shock and so cannot replace a fuse in all applications, but because of its inherent insensitivity to surges it can be useful in local protection of a transformer winding.

A further more complex solution is to switch the AC input voltage only at the instant of zero crossing, using a triac. This results in a predictable switch-on characteristic, and may be attractive if electronic switching is required for other reasons such as standby control. Similarly, DC input supplies can use a power MOSFET to provide a controlled resistance at turn-on, as well as other circuitry such as reverse polarity protection and standby switching.

### 15.18.2.5 Waveform Distortion and Interference

#### Interference

Electrical interference generated within equipment and conducted out through the mains supply port was already subject to regulation for some product sectors in some countries, and with the adoption of the European EMC Directive it is mandatory for all electrical or electronic products to comply with interference limits. The usual method of reducing such interference is to use a radio frequency filter at the mains supply inlet, but good design practice also plays a substantial part. Switch-mode power supplies are normally the worst offenders, because they generate large interference currents at harmonics of the switching frequency well into the HF region. The size and weight advantages of switch-mode supplies are balanced by the need to fit larger filters to meet the interference limits.

### Peak Current Summation

An increasing problem for electricity supply systems is the proportion of semiconductor-based equipment in the supply load. This is because the load current that such equipment takes is pulsed rather than sinusoidal. Current is only drawn at the peak of the input voltage, in order to charge the reservoir capacitors in the power supply. The normal RMS-to-average ratio of 1.11 for a sinusoidal current is considerably higher for this type of waveform (Figure 15.25).

The ratio of the peak load current $I_{pk}$ to $I_{rms}$ is called the *crest factor* and here it depends on the input impedance of the reservoir circuit. The lower the impedance, the faster the reservoir capacitor(s) will charge, which results in lower output voltage ripple but higher peak current.

The significance of crest factor is that it affects the power handling capability of the supply network. A network of a given sinusoidal RMS current rating will show considerable extra losses when faced with loads of a high crest factor. The supply mains does not have zero impedance, and the result of the extra network voltage drop at each crest is a waveform distortion in which the sinusoidal peak is flattened. This is a form of harmonic distortion and its seriousness depends on the susceptibility of other loads and components in the network.

Large systems installations, in which there are many electronic power supplies of fairly high rating fed from the same supply, are the main threat. In domestic premises, the

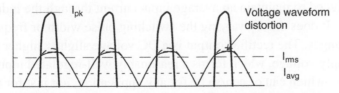

**Figure 15.25: Peak input current in a rectifier/reservoir power supply**

switch-mode supplies of TV sets are the main offenders; in commercial buildings, the problem is worst with switch-mode supplies of PCs and their monitors, and fluorescent lamps with electronic ballasts. The current peaks always occur together and so reinforce each other. A network that is dominated by resistive loads such as heating and filament-lamp lighting can tolerate a proportion of high crest factor loads more easily.

### Power Factor Correction

The "peakiness" of the input current waveform is best described in terms of its harmonic content and legislation now exists in Europe, under the EMC Directive, to control this. The European standard EN 61000-3-2:2000 places limits on the amplitude of each of the harmonic components of the mains input current up to the 40th (2 kHz at 50 Hz mains frequency), and it applies to virtually all electrical and electronic apparatus up to an input current of 16 A, although products other than lighting equipment with a rated power of less than 75 W are exempt. The limits, although not particularly stiff, are pretty much impossible for a switched-mode power supply to meet without some treatment of the input current. This treatment is generically known as *power factor correction* (PFC).

In this context, *power factor* (PF) is the ratio between the real power, as transferred through the power supply to its load with associated losses, and the apparent power drawn from the mains: RMS line voltage times RMS line current. A purely resistive load will have a PF of unity, but since peaks increase the RMS current, one drawing a peaky waveform will have a PF of 0.5–0.75. "Correcting" the PF toward unity requires that the input current waveform is made nearly sinusoidal, so that its harmonic content is much reduced. This is done by a second switching "pre-regulator" operating directly at the mains input. The usual topology is a boost regulator, as shown in Figure 15.26.

The input rectifier supplies a full-wave-rectified half-sine voltage across $C_{IN}$. This capacitor is too low a value to affect the 50-Hz input current significantly, but high enough to act as an effective reservoir at the switching frequency (typically 50–100 kHz). One sense input of the switching controller comes from this input voltage, and the controller is designed to maintain an average input current through the inductor in phase with this voltage. It does this by varying the switching pulse width or frequency as the input voltage changes. The rectified output is a DC voltage slightly higher than the highest peak supply voltage, which forms a reasonably well-regulated input to the main SMPS converter—which can of course be for any application, not just for an electronic power supply.

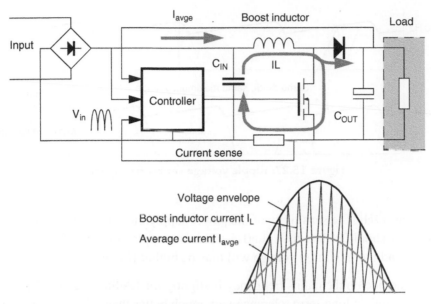

**Figure 15.26: Power factor correction**

Naturally the addition of a second switching converter increases the cost of the total power supply, and contributes to more interference, which must be filtered out at the mains input. Neither of these disadvantages are excessive, and commercial PFC power supply modules are now widely available. If you are designing your own, several IC manufacturers offer controllers specifically for the purpose, such as the L4981A/B, L6561, UC3853-5, and MC33626/33368. An extra advantage of the PFC pre-regulator is that almost by definition it will work over a wide input voltage range; so that a byproduct of including it is that a single power supply will cover all worldwide markets (section 15.18.2.1), and will also have a uniform and predictable response to dips and interruptions (section 15.18.3.2).

### 15.18.2.6 Frequency

The UK and European mains frequency is held to 50 Hz $\pm$ 1%. The American supply standard is 60 Hz. The difference in frequencies does not generally cause any problem for equipment that has to operate off either supply (provided that it's designed in Europe!), since mains transformers and reservoir circuits that perform correctly at 50 Hz will have

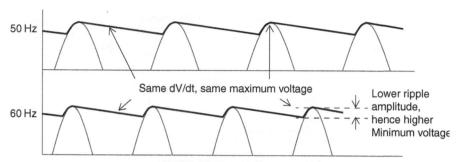

**Figure 15.27: Ripple voltage versus frequency**

no difficulty at 60 Hz. The sensitivity of the power supply circuits to supply voltage droops at 60 Hz should be less than at 50 Hz since the ripple amplitude is only 83% of the 50 Hz figure, and the minimum voltage will thus be higher (Figure 15.27).

The ±1% tolerance on the mains frequency is slightly misleading because the supply authorities maintain a long-term tolerance very much better than this. Diurnal variations are arranged to cancel out, and this allows the mains to be used as a timing source for clocks and other purposes. If you are planning to use the mains frequency for internal timing then you will need to incorporate some kind of switching arrangement if the equipment will be used on both U.S. and European systems.

### 15.18.2.7 Efficiency

The efficiency of a power supply module is its output power divided by its input power. The difference between the two quantities is accounted for by power losses in the various components in the power supply.

$$\text{Efficiency } \eta = P_{out}/P_{in} = P_{out}/(P_{out} + P_{loss})$$

The efficiency normally worsens as the load is reduced, because the various losses and quiescent operating currents assume a greater proportion of the input power. Therefore, if you are concerned about efficiency, do not use a power supply that is heavily overrated for its purpose. Linear supply efficiency also varies considerably with its input voltage, being worst at high voltages, because the excess must be lost across the regulator. Switch-mode supplies do not have this problem.

Normally efficiency is not of prime concern for mains power supplies, since it is not essential to make optimum use of the available power, although at higher powers the heat generated by an inefficient unit can be troublesome. It is far more important that a power converter for a portable instrument should be efficient because this directly affects useable battery life.

Linear power supplies are rarely more than 50% efficient unless they can be matched to a narrow input voltage range, whereas switch-mode supplies can easily exceed 70% and with careful design can reach 90%. This makes switch-mode supplies more popular, despite their greater complexity, at the higher power levels and for battery-powered units.

### Sources of Power Loss

The components in a power supply that make the major contribution to losses are:

- the transformer: core losses, determined by the operating level and core material, and copper losses, determined by $I^2R$ where R is the winding resistance;

- the rectifiers: diode forward voltage drop, $V_F$, multiplied by operating current; more significant at low output voltages;

- linear regulator: the voltage dropped across the series pass element multiplied by the operating current; greatest at high input voltages;

- switching regulator: power dissipated in the switching element due to saturation voltage, plus switching losses in this and in snubber and suppressor components, proportional to switching frequency.

If you sum the approximate contribution of each of these factors you can generally make a reasonable forecast of the efficiency of a given power supply design. The actual figure can be confirmed by measurement and if it is wildly astray then you should be looking for the cause.

### 15.18.2.8 Deriving the Input Voltage from the Output

In a linear supply with a series pass regulator element, the design must proceed from the minimum acceptable output voltage at maximum load current and minimum input voltage. These are the worst-case conditions and determine the input voltage step-down required. The minimum DC input voltage is given by the minimum output voltage plus all the tolerances and voltage drops in series:

$$V_{in,dc} = V_{out(min)} + V_{tol,reg} + V_{series,reg} + V_{series, CS} \dots$$

Where, $V_{out(min)}$ is the minimum acceptable output voltage

$V_{tol,reg}$ is the regulator voltage tolerance, assuming it is not adjustable

$V_{series,reg}$ is the voltage drop across the regulator series pass element

$V_{series,CS}$ is the voltage drop across the current sense element if fitted

All the above parameters are specified at full load current. This value for $V_{in,dc}$ is then the minimum input voltage allowed for a DC input supply, or it is the voltage at the minimum of the ripple trough for a rectified and smoothed AC input supply. This is related to the transformer secondary voltage as follows:

$$V_{tx} = (V_{in,dc} + V_{ripple} + V_D)/0.92 \cdot (V_{ac(nom)} / V_{ac(min)}) \cdot 1 / \sqrt{2}$$

Where, $V_{ripple}$ is the peak ripple voltage across the reservoir capacitor

$V_D$ is the voltage drop across the rectifier diode(s)

$V_{tx}$ is the RMS transformer secondary voltage

$V_{ac(nom)}$ is the specified transformer input voltage

$V_{ac(min)}$ is the minimum line input voltage

All parameters at full load current.

The figure of 0.92 is an approximate allowance for full-wave rectifier efficiency with a single-capacitor reservoir. It can be more accurately derived using curves published by Schade[1]. Complications set in because the current drawn through the secondary is not sinusoidal, but occurs at the crest of the waveform (see section 15.18.2.5). The extra peak current reduces the peak secondary voltage from its quoted value, if this value is specified for a resistive load. You can get around this either by knowing the transformer's losses in advance and allowing for the extra IR drop, or by specifying the transformer for a given circuit and letting the transformer supplier do the work for you, if you're buying a custom

---

[1] O.H. Schade, *Analysis of Rectifier Operation*, Proc. IRE, vol 31, 1943, pp 341–361.

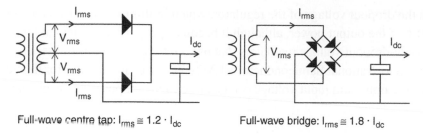

Full-wave centre tap: $I_{rms} \cong 1.2 \cdot I_{dc}$     Full-wave bridge: $I_{rms} \cong 1.8 \cdot I_{dc}$

**Figure 15.28: Rectifier configuration**

component. The transformer secondary RMS current rating is determined by the rectifier configuration (Figure 15.28).

### Example 15.1

Take as an example a typical linear regulator circuit supplying $5\,V \pm 5\%$ at $1\,A$.

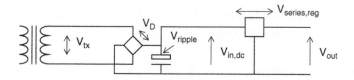

### Solution

Here, $V_{out(min)}$ is allowed to be $5\,V - 5\% = 4.75\,V$. The regulator we shall use is a standard 7805 type with $\pm 4\%$ tolerance and so $V_{tol,reg}$ is $5\,V \cdot 0.04 = 0.2\,V$. Its specified minimum series voltage drop (or dropout voltage) at $1\,A$ and a junction temperature of $25°C$ (note the temperature restriction) is $2.5\,V$ maximum. The required minimum input voltage is:

$$V_{in,dc} = 4.75 + 0.2 + 2.5 = 7.45V$$

If the peak ripple voltage is $2\,V$ and each diode forward drop in the bridge is $1\,V$, then the transformer voltage with a $240\,V$ nominal spec but a minimum line voltage of $195\,V$ will need to be:

$$V_{tx} = [7.45 + 2 + (2 \times 1)]/0.92 \cdot 240/195 \cdot 1/\sqrt{2} = 10.83\,Vrms$$

From this example you can see that the secondary-side input voltage needed to assure a given output voltage is very much higher than the actual output voltage. One of the major culprits is the dropout voltage of the regulator, which in this example accounts for at least 50% of the output power, although it becomes proportionally less at higher output voltages. Low-dropout voltage regulators that use a PNP transistor as the series pass element, such as National Semiconductor's LM2930 range, are popular for this reason and also where the minimum input voltage can be close to the output level, as in automotive applications.

### Power Losses at High Input Voltage

You can also see more clearly in the above example where the power losses are which contribute to reduced efficiency. When the input voltage is increased to its maximum value the dissipation in the series-pass element is worst. In the above example with the mains input at 264 V, the average value of $V_{in,dc}$ rises to 12.5 V, and 7.45 V of this must be lost across the regulator, which because it is passing the full load current amounts to one-and-a-half times the load power! The advantage of the switch-mode supply is that it adjusts to varying input voltages by modifying its switching duty cycle, so that an increased input voltage automatically reduces the input current and the overall power taken by the unit remains roughly constant.

### 15.18.2.9 Low-Load Condition

When the output load is removed or substantially reduced then the dissipation in the power supply will drop. This is good news for almost all parts of the circuit, except for the voltage rating of the components around $V_{in,dc}$. When there is a combination of low load and maximum supply input voltage, the peak value of $V_{in,dc}$ is highest. A crucial factor here is the transformer regulation. This is the ratio:

$$\text{Regulation} = (V_{sec,unloaded} - V_{sec,loaded})/V_{sec,loaded}$$

and a small or poorly designed transformer can have a regulation exceeding 20%. If this figure is used for the transformer in the above example then the peak $V_{tx}$ off-load at maximum input voltage will rise to 20.2 V. At the same time the diode forward voltage drops at low current will be much less, say 0.6 V each, so the possible maximum voltage at the reservoir capacitor could be around 19 V. Thus, even the common 16 V rated electrolytic will not be adequate for this circuit. For higher voltage outputs, the maximum input voltage can even exceed the voltage rating of the regulator itself, and you have to invest in a pre-regulator to hold the maximum to a manageable level. Note that this condition is not the worst-case for regulator power dissipation, because the regulator is not passing significant load current.

### Maximum Regulator Dissipation

In fact maximum series-pass dissipation does not necessarily occur at full load current, because as the current rises the voltage across the series-pass element falls. The maximum dissipation will occur at less than full output if the voltage dropped across the DC supply's equivalent series resistance is greater than half the difference between the no-load input voltage and the output voltage. Figure 15.29 shows this graphically.

### Minimum Load Requirement

A further problem, particularly with switch-mode supplies, is that the stability of the regulator cannot always be assured down to zero load. For this reason some rails have to be run with a minimum load, such as a bleed resistor, to remain within specification, and this represents an unnecessary additional power drain. Many circuits, of course, always take a minimum current and so the minimum load is not then a problem.

### 15.18.2.10 Rectifier and Capacitor Selection

The specification of the rectifiers and capacitors is dominated by surge and ripple current concerns.

### Reservoir Capacitor

The minimum capacitor value is easily decided from the required ripple voltage:

$$C = I_L/V_{ripple} \cdot t$$

where, $I_L$ is the DC load current,

$V_{ripple}$ is the acceptable ripple voltage

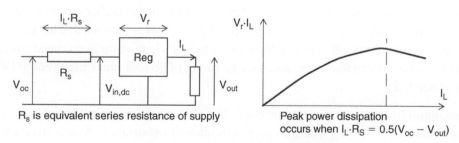

**Figure 15.29: Peak power dissipation**

For mains inputs, $t$ is about 2 ms less than the AC input period, 8 ms for 50 Hz or 6 ms for 60 Hz full-wave.

A more accurate value can be derived from Schade's curves (see previous footnote), which have been reprinted in numerous textbooks, but remember that the tolerance on reservoir capacitors is wide (typically $\pm 20\%$) and accuracy is rarely needed.

For load currents exceeding 1 A, ripple current rating tends to determine capacitor selection rather than ripple voltage. As is made clear throughout this chapter, the peak current flow through the rectifier/capacitor circuit is many times higher than the DC current, due to the short time in each cycle for which the capacitor is charging. The RMS ripple current is 2–3 times higher than the DC load. Ripple current rating is directly related to temperature and you may need to derate the component further if you need high ambient temperature and/or high reliability operation.

As an example, a load current of 2 A and a permissible ripple voltage of 3 V at 100 Hz suggests a 5300 μF capacitor. Typical capacitors of the next value up from this, 6800 μF, have 85°C ripple current ratings from 2 to 4 A. The higher ratings are larger and more expensive. But actual ripple current requirements will be 4–6 A. To meet this you will need to use either a much larger capacitor (typically 22,000 μF), or two smaller capacitors in parallel, or derate the operating temperature and use a slightly larger capacitor. If you don't do this, your design will become yet another statistic to prove that electrolytic capacitors are the prime cause of power supply failure.

### Rectifiers

Although in the full-wave arrangements (Figure 15.28) the diodes only conduct on alternate half cycles, because the RMS current is 2–3 times higher than the DC load current a rating of *at least* the full load current, and preferably twice it, is necessary. Surge current on turn-on may be much higher, especially in the higher power supplies where the ratio of reservoir capacitance to operating current is increased. This is of even greater concern in direct-off-line switch-mode supplies where there is no transformer series resistance to limit the surge, and a diode rating of up to 5 times the average DC current is needed.

The maximum instantaneous surge current is $V_{max}/R_s$ and the capacitor charges with a time constant of $\tau = C \cdot R_s$, where $R_s$ is the circuit series resistance. As a conservative guide, the surge won't damage the diode if ô is less than a half-cycle at mains frequency and $V_{max}/R_s$ is less than the diode's rated $I_{FSM}$. All diode manufacturers publish $I_{FSM}$ ratings for a given time constant; for example, the typical 1N5400 series with 3 A average rating have an $I_{FSM}$ of 200 A. You may discover that you have to incorporate a small extra series resistance to limit the surge current, or use a larger diode, or apply the techniques discussed in section 15.18.2.4.

The rectifier's peak-inverse-voltage (PIV) rating needs to be at least equivalent to the peak AC voltage for the full-wave bridge circuit, or twice this for the full-wave centre tap. But you should increase this considerably (by 50 to 100%) to allow for line transients. This is easy for low-voltage circuits, since 200 V diodes cost hardly any more than 50 V ones, and does not normally make much cost difference in mains circuits. For 240 V, a minimum of 600 V PIV and preferably 800 V PIV should be specified, even if you are using a transient suppressor at the input.

### 15.18.2.11 Load and Line Regulation

Load regulation refers to the permissible shift in output voltage when the load is varied, usually from none to full. Line (or input) regulation similarly refers to the permissible shift in output voltage when the input is varied, usually from maximum to minimum. Provided that the design of the input circuit has been properly considered as described above, so that the input voltage never goes outside the regulator's operational range, these parameters should be wholly a function of the regulator circuit itself. The regulator is essentially a feedback circuit that compares its output voltage against a reference voltage, so the regulation depends on two parameters: the stability of the reference, and the gain of the feedback error amplifier. If you use a monolithic regulator IC, then these factors are taken into account by the manufacturer who will specify regulation as a data sheet parameter.

### Thermal Regulation

A monolithic regulator IC includes the voltage reference on-chip, along with other circuitry and the series pass element. This means that the reference is subject to a thermal shift when the power dissipation of the series pass element changes. This gives rise to a separate longer term component of regulation, called *thermal regulation*, defined as the change in output voltage caused by a change in dissipated power for a specified time. Provided the chip has been well-designed, thermal regulation is not a significant factor for most purposes, but it is rarely specified in data sheets and for some precision applications may render monolithic regulators unsuitable.

### Load Sensing

No three-terminal regulator can maintain a constant voltage at anywhere other than its output terminals. It is common in larger systems for the load to be located at some distance from the power supply module, so that load-dependent voltage drops occur in the wiring connecting the load to the power supply output. This directly impacts the achievable load regulation.

The accepted way to overcome this problem is to split the regulator feedback path, and incorporate two extra "sensing" terminals that are connected so as to sense the output voltage at the load itself (Figure 15.30). The voltage drop across this extra pair of wires is negligible because they only carry the signal current. The voltage at the regulator output is adjusted so as to regulate the voltage at the sensing terminals.

The minimum voltage at the regulator input must be increased to allow for the extra output voltage drop. It is wise to connect coupling resistors (shown shaded in Figure 15.29) from the output to sense terminals, so as to ensure correct operation when the

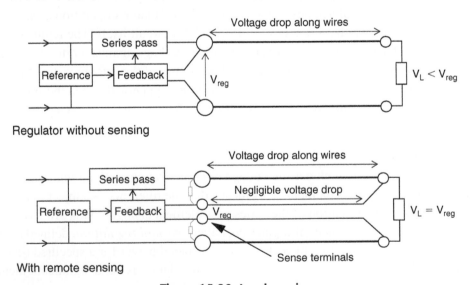

**Figure 15.30: Load sensing**

sense terminals are accidentally or deliberately disconnected. Sensing can only offer remote load regulation at one point and so is not really suited when one power supply module feeds several loads at different points.

### 15.18.2.12 Ripple and Noise

Ripple is the component of the AC supply frequency (or more often its second harmonic) that is present on the output voltage; noise is all other AC contamination on the output. In a linear power supply, ripple is the predominant factor and is given by the AC across the reservoir capacitor reduced by the ripple rejection (typically 70–80 dB) of the regulator circuit. A figure of less than 1 mV RMS should be easy to obtain. HF noise is filtered by the reservoir and output capacitors and there are no significant internal noise sources, provided that the regulator isn't allowed to oscillate, so that apart from supply-frequency ripple linear power supplies are very "quiet" units.

### Switching Noise

The same cannot be said for switch-mode power supplies. Here the noise is mainly due to output voltage spikes at the switching frequency, caused by fast-rise-time edges and HF ringing at these edges feeding through, or past, filtering components to the output. The ESR and ESL of typical output filter capacitors limits their ability to attenuate these spikes, while the self-inductance of ground wiring limits the high frequency effectiveness of ground decoupling anyway. Switch-mode output ripple and noise is typically 1% of the rail voltage, or 100–200 mV. In fact comparing ripple and noise specifications is the easiest way to distinguish a linear from a switch-mode unit, if there is no other obvious indication. The bandwidth over which the specification applies is important, since there is significant energy in the high-order harmonics of the switching noise, and at least 10 MHz is needed to get a true picture. Because of stray coupling over this extended bandwidth the noise frequently appears in common mode, on both supply and 0 V simultaneously, and is then very difficult to control. Differential mode noise spikes can be reduced dramatically by including a ferrite bead in series, and a small ceramic capacitor in parallel with the output capacitor.

The presence of switching noise is not a problem for digital circuits, but it creates difficulties for sensitive analog circuits if their bandwidth exceeds the switching frequency. It can cause interference on video signals, misclocking in pulse circuits and voltage shifts in DC amplifiers. These effects have to be treated as EMC phenomena and can be cured by suitable layout, filtering and shielding, but if you have the option in the

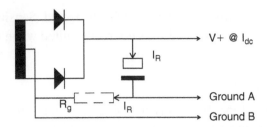

**Figure 15.31: Incorrect reservoir connection**

early stages to choose a linear supply instead, take it—you will save yourself a lot of trouble.

### Layout to Avoid Ripple

Power supply output ripple is aggravated by incorrect layout of the wiring around the reservoir capacitor. This is a specific instance of the common-impedance interference coupling discussed earlier.

At first sight grounds A and B in Figure 15.31 look equivalent. But there will be a potential between them of $I_R \cdot R_g$, where $I_R$ is the capacitor ripple current and $R_g$ is the track or wiring resistance common to the two grounds through which the ripple current flows. (The ripple current path is through the transformer, the two diodes and the capacitor.) This current is only drawn on peaks of the AC input waveform to charge the reservoir capacitor, and its magnitude is only limited by the combined series resistance of the transformer winding, the diodes, capacitor and track or wiring. If the steady-state DC current supplied is 1 A then the peak ripple current may be of the order of 5 A; thus 10 mΩ of $R_g$ will give a peak difference of 50 mV between grounds A and B. If some parts of the circuit are grounded to A and some to B, then tens of millivolts of hum injection are included in the design at no additional cost, and increasing the reservoir value to try and reduce it will actually make matters worse as the peak ripple current is increased. You can check the problem easily, by observing the output ripple on a scope; if it has a pulse shape then wiring is the problem, if it looks more like a sawtooth then you need more smoothing.

### Correct Reservoir Connection

The solution to this problem, and the correct design approach, is to ground all parts of the supplied circuit on the supply side of the reservoir capacitor, so that the ripple current ground path is not common to any other part of the circuit (Figure 15.32). The same

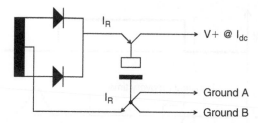

**Figure 15.32: Correct reservoir connection**

applies to the V+ supply itself. The common impedance path is now reduced to the capacitor's own ESR, which is the best you can do.

### 15.18.2.13 Transient Response

The transient response of a power supply is a measure of how fast it reacts to a sudden change in load current. This is primarily a function of the bandwidth of the regulator's feedback loop. The regulator has to maintain a constant output in the face of load changes, and the speed at which it can do this is set by its frequency response as with any conventional operational amplifier. The trade-off that the power supply designer has to worry about is against the stability of the regulator under all load conditions; a regulator with a very fast response is likely to be unstable under some conditions of load, and so its bandwidth is "slugged" by a compensation capacitor within the regulator circuit. Too much of this and the transient response suffers. The same effect can be had by siting a large capacitor at the regulator output, but this is a brute-force and inefficient approach because its effect is heavily load-dependent. Note that the 78XX series of three-terminal regulators should have a small, typically $0.1\,\mu F$ capacitor at the output for good transient response and HF noise decoupling. This is separate from the required $0.33\text{-}1\,\mu F$ capacitor at the input to ensure stability.

### Switch-Mode vs. Linear

The transient response of a switch-mode power supply is noticeably worse than that of a linear because the bandwidth of the feedback loop has to be considerably less than the switching frequency. Typically, switch-mode transient recovery time is measured in milliseconds while linear is in the tens of microseconds.

If your circuit only presents slowly-varying loads then the power supply's transient response will not interest you. It becomes important when a large proportion of the load

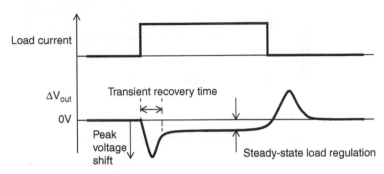

**Figure 15.33: Load transient response**

can be instantaneously switched—a relay coil or bank of LEDs for example—and the rest of the load is susceptible to short-duration over- or under-voltages.

Although load transient response is usually the most significant, a regulator also exhibits a delayed response to line transients, and this may become important when you are feeding it from a DC input, which can change quickly. The line transient response is normally of the same order as the load response.

### 15.18.3  Abnormal Conditions

#### 15.18.3.1  Output Overload

At some point in its life, a power supply is almost sure to be faced with an overload on its output. This can take the form of a direct short circuit across its output due to the slip of a technician's screwdriver, or a reduced load resistance due to component failure in the load circuit, or incorrect connection of too many loads. It may also be mistaken connection to the output of another power supply. The overload can be transient or sustained, and at the very least any power supply should be designed to withstand a continuous short circuit at its output(s) without damage. This is almost universally achieved with one of two techniques: constant current limiting or foldback current limiting.

*Constant Current Limiting*
Output overloads threaten mainly the series pass element in a linear supply, or the switching element in a switch-mode supply. In either case, an output over-current will subject the device to the maximum current that the input can supply while it is sustaining

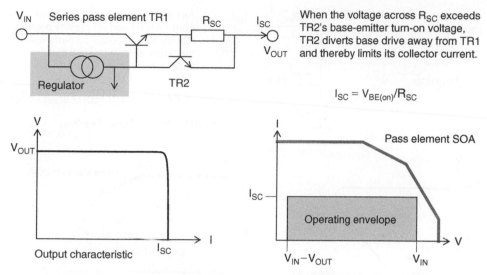

**Figure 15.34: Constant current limiting**

the full input-output differential voltage, and this will put its dissipation well outside its safe operating area (SOA) boundary (see Figure 15.21). Swift destruction will ensue.

Constant current limiting operates by ensuring that the output current available from the power supply limits at a maximum that is only marginally above the full load rating of the unit. Figure 15.34 shows this operation for a linear supply. This simple circuit works quite well but the actual value of $I_{SC}$ is very dependent on TR2's $V_{BE}$, and hence, on temperature, so that either you must allow a large margin over full load current or use a more complex circuit.

Switch-mode current limiting is more complex yet because you need to limit on a cycle-by-cycle basis to protect the switching element properly; current sensing on the output line is insufficient. Several techniques have been evolved to achieve this; consult switching regulator design manuals for details.

### Foldback Current Limiting

A disadvantage of constant current limiting is that to obtain sufficient SOA the pass element must have a much higher collector current capability than is needed for normal operation. "Foldback" current limiting reduces the short circuit current while still

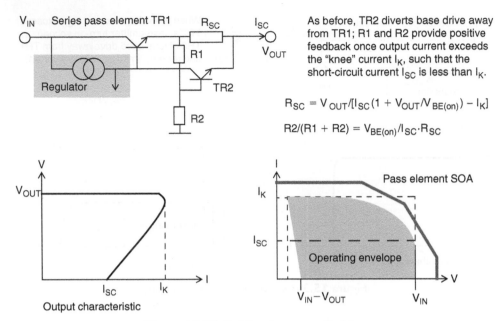

As before, TR2 diverts base drive away from TR1; R1 and R2 provide positive feedback once output current exceeds the "knee" current $I_K$, such that the short-circuit current $I_{SC}$ is less than $I_K$.

$$R_{SC} = V_{OUT}/[I_{SC}(1 + V_{OUT}/V_{BE(on)}) - I_K]$$

$$R2/(R1 + R2) = V_{BE(on)}/I_{SC}\cdot R_{SC}$$

**Figure 15.35: Foldback current limiting**

allowing full output current during normal regulator operation, thereby giving more efficient use of the pass element's SOA.

The development of the constant-current circuit to give foldback operation is shown in Figure 15.35. Although foldback allows the use of a smaller series pass element, it has its limitations. As the foldback ratio, $I_K/I_{SC}$, is increased, the required value of $R_{SC}$ increases and this calls for a greater input voltage at high foldback ratios. There is an absolute limit to the foldback ratio when $R_{SC}$ is infinite of:

$$[I_K/I_{SC}]_{max} = 1 + (V_{OUT}/V_{BE(on)})$$

and so foldback ratios of greater than 2 or 3 are impractical for low voltage regulators.

### 15.18.3.2 Input Transients

Under this heading we need to consider spikes, surges and interruptions on the input supply.

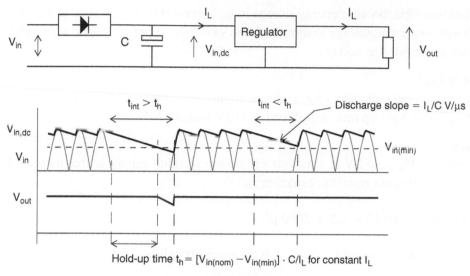

Hold-up time $t_h = [V_{in(nom)} - V_{in(min)}] \cdot C/I_L$ for constant $I_L$

**Figure 15.36: Hold-up time**

### Interruptions

On the mains supply, dips ("brownouts") and outages of up to 500 ms are fairly common, due to surge currents and fault clearing in the supply network. Other sources of supply may also experience such interruptions. The occurrence of longer supply breaks depends very much on location. In the UK, the average consumer loses power for 90 minutes in the year, but a rural consumer on the end of a long overhead line may experience much longer interruptions, while an urban consumer with several redundant supply routes may see none at all.

A power supply should be able to cope with short interruptions and brownouts transparently, so that the load is unaffected by them. The "hold-up time" (Figure 15.36) specifies for how long the output remains stable after loss of input, and it can be anywhere from a few to several hundred milliseconds. It is almost entirely determined by the size of the main reservoir capacitor, since this provides the only source of power when the input is removed. A linear regulator can be considered as a constant-current sink discharging this capacitor and therefore it is easy to calculate the hold-up time for a given load and input voltage. A switch-mode regulator draws more current as its input voltage drops, so accurately determining hold-up time for this type requires the solution of a current-time integral. The higher the operating voltage, the easier it is to obtain a long hold-up time, because energy storage in the reservoir is proportional to $0.5 \cdot C \cdot V2$. This gives another advantage to direct-off-line switching supplies, whose main reservoir operates at the full line voltage.

### Example 15.2

Taking the quoted parameters for the linear supply in section 15.18.2.8, what values does this give for its hold-up time at full load at 240 V and 204 V?

### Solution

The ripple on $V_{in,dc}$, of 2 V at 1 A, with a full-wave rectified supply so that its period is 10 ms, means that the reservoir capacitor is:

$$C = I \cdot t/V = 1 \cdot 10.10 - 3/2 = 5000 \ \mu F$$

At 240 V the minimum value for $V_{in,dc}$ at the ripple trough is:

$$V_{in,dc(min)} = 14.05 - 2_{(ripple)} - 2_{(diode)} = 10.05 \text{ V}$$

so the hold-up time at this voltage given a minimum requirement of 7.45V at the regulator is:

$$t_h = (10.05 - 7.45) \cdot 5000.10^{-6}/1 = 13 \text{ ms}$$

At 204 V (240 V – 15%) the minimum value for $V_{in,dc}$ is 7.94 V, so the hold-up time is now:

$$t_h = (7.94 - 7.45) \cdot 5000.10^{-6}/1 = 2.5 \text{ ms}$$

It is clear that hold-up time specified at nominal input voltage may be considerably less when the power supply is running at its minimum input voltage. In fact, the minimum input voltage as calculated in section 15.18.2.8 is that for which the hold-up time is zero. All this is assuming the worst-case condition, that the supply is interrupted at the minimum of the ripple trough. If hold-up time is important for your circuit, you must decide at what input voltage it is to be specified.

### Spikes and Surges

Earlier we discussed the occurrence of transient overvoltages on mains and automotive supplies. Some precautions need to be taken to prevent these as far as possible from propagating through the power supply and impacting the load circuit. Short, low-energy but fast rise-time transients can only be dealt with by good circuit layout, minimizing ground inductance and stray coupling, and by input filtering. Slower but higher-energy transients call for the use of transient suppressor devices at various points in the power supply, and for overvoltage protection.

### 15.18.3.3 Transient Suppressors

Figure 15.37 shows three positions for transient suppressors in a linear supply. The advantages and disadvantages of each position can be summarized as follows:

- Z1: protects all components in the unit from differential-mode surges but is subject to the lowest source impedance. This means that it must have a high energy rating to withstand the maximum expected surge without destruction, and it will have a fairly high ratio of clamped voltage to normal running voltage. In effect, voltage surges up to about twice the peak operating voltage will be let through.

- Z2: this is a more effective position as it still protects the vulnerable rectifiers, but is itself protected by the additional source impedance of the transformer. It can therefore be a smaller component but still have a good ratio of clamped to peak operating voltage. It has no effect on spikes that may have been converted from differential to common mode by the interwinding capacitance of the transformer.

- Z3: this protects the regulator and subsequent circuitry but not the rectifiers. It is something of a "belt-and-braces" position, but it does suppress input common mode spikes that the previous positions would have let through. It should be sized so that its peak clamping voltage is just less than the absolute maximum input voltage of the regulator. Smaller surges then rely on the transient response of the regulator to contain them.

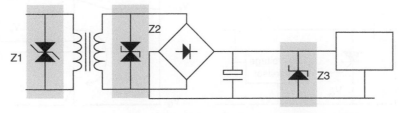

**Figure 15.37: Transient suppression in a linear supply**

### 15.18.3.4  Overvoltage Protection

If the circuit that your power supply is driving is very expensive and susceptible to overvoltages—for instance it may include an expensive microprocessor that must not be subject to more than 7 V—then it is worth including extra circuitry at the power supply output for overvoltage protection. The first time that it operates, it will have saved the extra expense of designing it in.

This might be as simple as a 6.2 or 6.8 V zener diode across the output of a 5 V supply. This does not offer foolproof protection, because if the overvoltage is sustained and derives from a low source impedance—perhaps the series-pass element has failed—then the zener is likely to fail itself, and may fail open-circuit, in which case it has been wasted. Something more drastic is called for, and the conventional solution is a crowbar.

This gets its name from the time-honored method of ensuring that no voltage is present between two live terminals, by the simple expedient of putting a crowbar—which is assumed to be able to carry any likely short-circuit current-across them. In its more sophisticated version in electronic power supplies, the crowbar takes the form of a triggered thyristor. The thyristor is permanently in place across the output, or in some designs across the reservoir, but it is only triggered when a supervisory circuit detects the presence of an overvoltage. It then stays triggered, holding the output voltage to VH, until the current through it is interrupted by external circumstances such as a power supply reset. Although this current may be high, the voltage across it is not, so its dissipation is fairly low. Obviously the power supply itself must be protected against a sustained output short circuit, either by current limiting or a fuse or preferably both. Figure 15.38 shows the operating principle.

### Crowbar Circuit Requirements
The thyristor must be capable of dumping, virtually instantaneously, both the continuous short-circuit current of the supply and the energy stored in the reservoir capacitors. It must

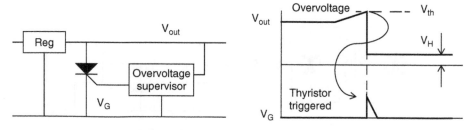

**Figure 15.38: Overvoltage protection**

therefore have a high single-pulse $I^2t$ and *di/dt* rating. Some manufacturers characterize devices especially for this purpose, and the *di/dt* performance is helped by making sure the trigger pulse has a fast edge and is well in excess of the minimum gate current requirement.

Both the supervisory circuit and the thyristor itself must be immune from false triggering due to short transients, as the nuisance value of an unnecessary shutdown may exceed that of a real overvoltage in some instances. Some degree of delay in the trigger pulse is essential, and characterising the overall system (power supply plus crowbar protection plus load) for the acceptable and necessary delay and overvoltage threshold is the most critical part of overvoltage protection design.

### 15.18.3.5 Turn-On and Turn-Off

Sometimes, the behavior of the power rails when the input power is applied or removed is important to the load circuit. The power rail never instantaneously reaches its operating level as soon as the input is applied. It will ramp up to the rail voltage as the reservoir and other capacitors charge, and it may overshoot its nominal voltage briefly if the regulator frequency compensation has not been optimized—this is a particular danger with some switch-mode circuits. It may suffer from noise or oscillation due to the switch-on process as it ramps up. Particularly if the load circuit includes a microprocessor, it will not be safe to start the circuit operation until the rail voltage has settled. You may require the power supply to have a flag output that signals to the load circuit that all is well. This output is often connected to the micro's RESET input.

Similarly, when the power is switched off, the microprocessor needs to be able to power down in an orderly fashion. This is best achieved by generating a power-fail interrupt as soon as a power failure is detected, followed by an undervoltage warning when the power rail starts to droop. The time delay between the two will be roughly equivalent to the hold-up time as discussed earlier, and this delay must be long enough to enable the micro to perform its power-down housekeeping functions. Required outputs are shown in Figure 15.39.

### PSU Supervisor Circuits

All the functions of undervoltage and power fail monitoring, and overvoltage protection, can be gathered up into a single power supply supervisory circuit, and several ICs are on the market for this purpose. Examples are the MC3423, ICL7665 and 7673, TL7705 and MAX690 series. These chips are basically a collection of comparators and delay circuits, integrated into one package for ease of use. Unfortunately, there are a multiplicity of

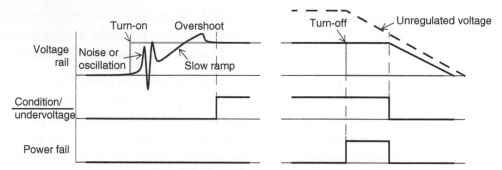

**Figure 15.39: Power rail supervision waveforms**

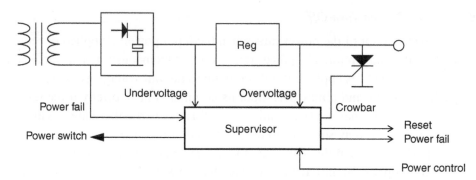

**Figure 15.40: Configuration of power supply supervisor**

types and few second sources, and the parts cost may be greater than you would suffer when using standard comparators such as the LM339. In many cases you may still prefer to design your own supervisory circuit from such standard components.

A typical application will require the supervisory circuit to have inputs from the DC output rail for overvoltage protection, the reservoir capacitor for undervoltage warning, and the low-voltage AC input for power fail detection. Outputs will go to the crowbar device and the load circuit. Bear in mind that the supervisor needs to operate reliably down to very low supply voltages.

### 15.18.4 Mechanical Requirements

#### 15.18.4.1 Case Size and Construction

If you are designing a supply as an integral part of the rest of the equipment then generally you won't need to consider its mechanical characteristics separately from the equipment design. If you are buying in a standard unit, or designing yourself a modular unit, which will be used for different products, then case construction becomes important. Standard products tend to fall into one of four categories:

- open frame, chassis mounting;

- enclosed, chassis mounting;

- encapsulated, PCB or chassis mounting;

- rack module.

Both linear and switch-mode types are available in all these variants, but power rating, connections and the need for screening play an important role in the final selection.

### Open Frame

This is normally the cheapest option, since all that is supplied is a PCB mounted on a simple metal chassis, which serves as a rudimentary heatsink. Connections are made by wiring to terminals or spade lugs mounted on the board. No environmental protection or screening is offered and the power unit must be enclosed completely within the equipment it is supplying. Open frame units are most popular in the 10 to 100 W range, with models available up to 250 W.

### Enclosed

Cased power supplies are more common for power ratings above 100 W. They offer reasonably effective screening that is important for switch-mode supplies, and can incorporate a fan for efficient convective cooling, which is not possible with open or encapsulated types but is necessary at high powers. The greater electronics cost tends to mask the cost of the extra mechanical components. Connections are made to screw terminals on the outside of the case, and the internal circuitry is guarded from wandering fingers and other foreign bodies.

### Encapsulated

Fully encapsulated units are available up to 40 W, and can be either PCB-mounted via pins or chassis-mounted using screw terminals for the connections. Their great advantage is that they can be treated as just another component during equipment production, and do not need any further environmental protection for their internal circuit. EMI screening can be incorporated as part of the encapsulating box. Higher power ratings than 40 W require a heatsink outside the encapsulation. The encapsulation tends to provoke reliability problems when much heat has to be dissipated, and if you are going for a higher-power unit it would be wise to seek concrete reliability data. Encapsulation is particularly popular for low-power DC-to-DC converters, which can be incorporated within a system at board level, to generate different and/or regulated supplies from a common DC bus.

### Rack Mounting Modules or Cassettes

With the increasing popularity of rack-mounted modular processor equipment, usually based on the Eurocard rack and DIN-41612 connector standard, there is a corresponding need for power supply modules that can share the same rack. These are available from 25 up to 500 W. All but the smallest are switch-mode types, since space and thermal capacity are strictly limited, and applications are mainly digital. Connection is by mating plug and socket, mounted on the card frame, and it is vital to ensure that the connector used is capable of carrying the load current without loss, and is rated for mains voltages. The DIN-41612 H15 connector is widely used, with a leading earth pin to maintain safety when withdrawing or inserting the unit.

### 15.18.4.2 Heatsinking

A necessary requirement for the continued health of any semiconductor device, be it monolithic IC regulator, rectifier diode or power transistor, is that its junction temperature should stay within safe limits. Junction temperature is directly related to power dissipation, thermal resistance and ambient temperature, and the function of a heatsink is to provide the lowest possible thermal resistance between the junction and its environment—assuming the environment is always cooler.

Suffice it to say that the power supply often represents the most concentrated source of heat in an item of equipment. As soon as its efficiency is roughly known, you should calculate the heat output and take steps to ensure that the mechanical arrangement will allow an efficient heat flow. At the minimum this will involve ensuring that all components that will need heatsinking are positioned to allow this, and that the power unit's positioning within the overall equipment gives adequate thermal conductivity to the environment. Too many designs end up with a fan tacked onto the case as an afterthought!

### 15.18.4.3 Safety Approvals

Major safety risks for power supplies are the threat of electric shock due to contact with "live parts," and the threat of overheating and fire due to a fault. One of the important but forgotten functions of a power supply is to ensure a safe segregation of the low-voltage circuitry, which may be accessible to the user, from the high-voltage input, which must be inaccessible. Segregation is normally assured in a power supply by maintaining a minimum distance around all parts that are connected to the mains, including spacing between the primary and secondary of the transformer. This, of course, adds extra space to the design requirements. Insulation of at least a minimum thickness may be substituted for empty space.

There are many national and international authorities concerned with setting safety requirements. Foremost among these are UL in the United States, CSA in Canada, and the CENELEC safety standards, implementing the Low Voltage Directive in Europe. As designer, you can either choose to apply a particular set of requirements for your company's market, or if you plan to export worldwide, you can discover the most stringent requirements and apply these across the board. A common specification is EN 60950-1 (IEC 60950-1), which is the safety standard for information technology equipment and which is quoted by default by most off-the-shelf power supplies. If no safety specification is quoted, beware.

Most of the time it is legally necessary to have your product approved to safety regulations, often it is also commercially desirable. Using a bought-in supply that already has the right safety approval goes a long way to helping your own equipment achieve it. Note that there is a difference, on data sheets, between the words "designed to meet..." and "certified to..." The former means that, when you go for your own safety approval, the approvals agency will still want to satisfy themselves, at your expense, that the power supply does indeed meet their requirements. The latter means that this part of the approvals procedure can be bypassed. It therefore puts the unit cost of the power supply up, but saves you some part of your own approval expenses.

## 15.18.5 Batteries

Battery power is mainly used for portability or stand-by (float) purposes. All batteries operate on one or another variant of the principle of electro-chemical reaction, in which anode (negative) and cathode (positive) terminals are separated by an electrolyte, which is the vehicle for the reaction. This basic arrangement forms a "cell," and a battery consists of one or more cells. The chemistry of the materials involved is such that a potential is developed between the electrodes, which is capable of sustaining a discharge current.

The voltage output of a particular cell type is a complex function of time, temperature, discharge history and state of charge.

The basic distinction is between primary (nonrechargeable) and secondary (rechargeable) cells. This section will survey the various types of each shortly, but first we shall make a few general observations on designing with batteries.

### 15.18.5.1 Initial Considerations

When you know you are going to use a battery, select the cell type as early as possible in the circuit and mechanical design. This allows you to take the battery's properties into account and increases the likelihood of a cost-effective result, as otherwise you will probably need a larger, or more expensive, battery or will suffer a reduced equipment specification. Having made the selection, you can then design the circuit so that it works over the widest possible part of the battery's available voltage range. Some of the cheaper types deliver useful power over quite a wide range, with an endpoint voltage of 60–70% of nominal, and some of this energy will be lost if the design cannot cope with it. Also, check that the battery can deliver the circuit's load current requirements over the working temperature. This capability varies considerably for different chemical systems. Rechargeable batteries can often be recharged only over a narrower temperature range than they can be discharged.

Always aim to use standard types if your specification calls for the user to be able to replace the battery. Not only are they cheaper and better documented, but they are widely distributed and are likely to remain so for many years. You should only need to use special batteries if your environmental conditions or energy density requirements are extreme, in which case you have to make special provisions for replacement or else consider the equipment as a throwaway item.

### Voltage and Capacity Ratings

Different types of battery have different nominal open-circuit voltages, and the actual terminal voltage falls as the stored energy is used. Manufacturers provide a discharge characteristic curve for each type, which indicates the behavior of voltage against time for given discharge conditions. Note that the open-circuit voltage can exceed the voltage under load by up to 15%, and the operating voltage may be significantly less than the nominal battery voltage for some of the duration.

The capacity of a battery is expressed in ampere-hours (Ah) or milliampere-hours (mAh). It may also be expressed in normalized form as the "C" figure, which is the nominal capacity at a given discharge rate. This is more frequently applied to rechargeable types. Capacity will be less than the C rating if the battery is discharged at a faster rate; for instance, a 15 Ah lead-acid type discharged at 15 amps (1C) will only last for about 20 minutes (Figure 15.42).

Three typical modes under which a battery can be discharged are constant resistance, constant current and constant power. For batteries with a sloping discharge characteristic, such as alkaline manganese, the constant power mode is the most efficient user of the battery's energy but also needs the most complex voltage regulating system to power the actual circuit.

### Series and Parallel Connection

Cells can be connected in series to boost voltage output, but doing so decreases the reliability of the overall battery and there is a risk of the weakest cell being driven into reverse voltage at the end of its life. This increases the likelihood of leakage or rupture, and is the reason why manufacturers recommend that all cells should be replaced at the same time. Good design practice minimizes the number of series-connected cells. There are now several ICs that can be used to multiply the voltage output of even a single cell with high efficiency. It is not difficult to design a switching converter that simultaneously boosts and regulates the battery voltage.

Parallel connection can be used for some types to increase the capacity or discharge capability, or the reliability of the battery. Increased reliability requires a series diode in each parallel path to isolate failed cells. Recharging parallel cells is rarely recommended because of the uncertainty of charge distribution between the cells. It is therefore best to restrict parallel connection to specially-assembled units.

On the same subject, reverse insertion of the whole battery will threaten your circuit, and if it is possible, the user will do it. Either incorporate assured polarity into the battery compartment or provide reverse polarity protection, such as a fuse, series diode or purpose-designed circuit, at the equipment power input.

### Mechanical Design

Choose the battery contact material with care to avoid corrosion in the presence of moisture. The recommended materials for primary cells are nickel-plated steel, austenitic stainless steel or inconel, but definitely not copper or its alloys. The contacts should be springy in order to take up the dimensional tolerances between cells. Singlepoint contacts are adequate for low current loads, but you should consider multiple contacts for higher current loads. The simplest solution is to use ready-made battery compartments or holders, provided that they are properly matched to the types of cell you are using. PCB-mounting batteries have to be hand soldered in place after the rest of the board has been built, and you need to liaise well with the production department if you are going to specify these types.

Rechargeable batteries when under charge, and all types when under overload, have a tendency to out-gas. Always allow for safe venting of any gas, and since some gases will be flammable, don't position a battery near to any sparking or hot components. In any case, heat and batteries are incompatible: service life and efficiency will be greatly improved if the battery is kept cool. If severe vibration or shock is part of the environment, remember that batteries are heavy and will probably need extra anchorage and shock absorption material. Organic solvents and adhesives may affect the case material and should be kept away.

Dimensions of popular sizes of battery are shown in Table 15.2.

### Storage, Shelf Life, and Disposal

Maximum shelf life is obtained if batteries are stored within a fairly restricted temperature and humidity range. Self discharge rate invariably increases with temperature. Different chemical systems have varying requirements, but extreme temperature cycling should be avoided, and you should arrange for tight stock control with proper rotation of incoming and outgoing units, to ensure that an excessively aged battery is not used. Rechargeable types should be given a regular top-up charge.

In the early 90s, legislation appeared in many countries banning the use of some substances in batteries, particularly mercury, for environmental reasons. Thus mercuric oxide button cells were effectively outlawed and are not now obtainable. In Europe, this was achieved through the Batteries and Accumulators Directive (91/157/EEC).

This Directive also encourages the collection of spent NiCad batteries with a view to recovery or disposal, and their gradual reduction in household waste. In fact, what it has achieved is rather the development of alternative rechargeable technologies to NiCad, particularly NiMH and lithium. NiCads, though, are still widely used, despite the

### Table 15.2: Sizes of popular primary batteries

| Designation | | | | Dimensions mm | |
|---|---|---|---|---|---|
| IEC | ANSI | Size | Voltage | Dia (or L x W) | Height |
| Alkaline manganese dioxide | | | | | |
| LR03 | 24A | AAA | 1.5 | 10.5 | 44.5 |
| LR6 | 15A | AA | 1.5 | 14.5 | 50.5 |
| LR14 | 14A | C | 1.5 | 26.2 | 50 |
| LR20 | 13A | D | 1.5 | 34.2 | 61.5 |
| 6LR61 | 1604A | PP3 | 9 | 26.5 × 17.5 | 48.5 |
| 4LR25X | 908A | Lamp | 6 | 67 × 67 | 115 |
| 4LR25-2 | 918A | Lamp | 6 | 136 × 73 | 127 |
| Lithium manganese dioxide—cylindrical cell | | | | | |
| CR17345 | 5018LC | 2/3A | 3 | 17 | 34.5 |
| CR11108 | 5008LC | 1/3N | 3 | 11.6 | 10.8 |
| 2CR11108 | 1406LC | 2 × 1/3N | 6 | 25.2 | 13 |
| 2CR5 | 5032LC | 2 × 2/3A | 6 | 17 × 34 | 45 |
| CR-P2 | 5024LC | 2 × 2/3A | 6 | 19.5 × 35 | 36 |
| Lithium manganese dioxide—coin cell | | | | | |
| CR2016 | 5000LC | | 3 | 20 | 1.6 |
| CR2025 | 5003LC | | 3 | 20 | 2.5 |
| CR2032 | 5004LC | | 3 | 20 | 3.2 |
| CR2430 | 5011LC | | 3 | 24.5 | 3 |
| CR2450 | 5029LC | | 3 | 24.5 | 5 |
| Silver oxide button cells mAh | | | | | |
| SR41 | 1135S0 | 42 | 1.55 | 7.87 | 3.6 |
| SR43 | 1133S0 | 120 | 1.55 | 11.56 | 4.19 |
| SR44 | 1131S0 | 165 | 1.55 | 11.56 | 5.58 |
| SR48 | 1137S0 | 70 | 1.55 | 7.87 | 5.38 |
| SR54 | 1138S0 | 70 | 1.55 | 11.56 | 3.05 |
| SR55 | 1160S0 | 40 | 1.55 | 11.56 | 2.21 |

(*Continued*)

**Table 15.2: Continued**

| Designation | | | | Dimensions mm | |
| IEC | ANSI | Size | Voltage | Dia (or L x W) | Height |
|---|---|---|---|---|---|
| SR57 | 1165S0 | 55 | 1.55 | 9.5 | 2.69 |
| SR59 | 1163S0 | 30 | 1.55 | 7.9 | 2.64 |
| SR60 | 1175S0 | 18 | 1.55 | 6.8 | 2.15 |
| SR66 | 1176S0 | 25 | 1.55 | 6.78 | 2.64 |

technical advantages of NiMH. The Batteries Directive is about to be updated and it is likely to propose the following changes:

- EU member states to collect and recycle all batteries, with targets of 75% consumer (disposable or rechargeable) and 95% industrial batteries;
- no less than 55% of all materials recovered from the collection of spent batteries to be recycled.

In the UK in 1999, 654 million consumer batteries were sold, but the rate for recycling consumer rechargeables is a mere 5%, and less than 1% of consumer batteries are collected for recycling. On the other hand, more than 90% of automotive batteries are recycled and 24% of other industrial batteries. Clearly, for consumer batteries at least, a sea change in disposal habits is expected.

### 15.18.5.2 Primary Cells

The most common chemical systems employed in primary, nonrechargeable cells are alkaline manganese dioxide, silver oxide, zinc air and lithium manganese dioxide. Figure 15.41 compares the typical discharge characteristics for lithium and alkaline types of roughly the same volume on various loads.

### Alkaline Manganese Dioxide

The operating voltage range of this type, which uses a highly conductive aqueous solution of potassium hydroxide as its electrolyte, is 1.3 to 0.8 V per cell under normal load conditions, while its nominal voltage is 1.5 V. Recommended end voltage is 0.8 V per cell for up to 6 series cells at room temperature, increasing to 0.9 V when more cells are used. The alkaline battery is well suited to high-current discharge. It can operate between −30 and +80°C, but high relative humidity can cause external corrosion and should be

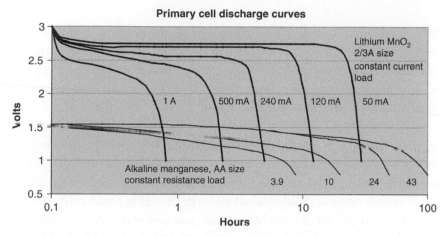

**Figure 15.41: Load discharge characteristics for lithium and alkaline manganese primary cells**

avoided. Shelf life is good, typically 85% of stored energy being retained after 3 years at 20°C. Standard types are now widely and cheaply available in retail outlets and it can therefore be confidently used in most general-purpose applications.

### Silver Oxide

Zinc-silver oxide cells are used as button cells with similar dimensions and energy density to the older and now withdrawn mercury types. Their advantage is that they have a high capacity versus weight, offer a fairly high operating voltage, typically 1.5V, which is stable for some time and then decays gradually, and can provide intermittent high pulse discharge rates and good low temperature operation. They are popular for such applications as watches and photographic equipment. Typical shelf life is two years at room temperature.

### Zinc Air

This type has the highest volumetric energy density, but is very specialized and not widely available. It is activated by atmospheric oxygen and can be stored in the sealed state for several years, but once the seal is broken it should be used within 2 months. It has a comparatively narrow environmental temperature and relative humidity range. Consequently its applications are somewhat limited. Its open circuit voltage is typically 1.45 V, with the majority of its output delivered between 1.3 and 1.1 V. It cannot give sustained high output currents.

### Lithium

Several battery systems are available based on the lithium anode with various electrolyte and cathode compounds. Lithium is the lightest known metal and the most electro-negative element. Their common features are a high terminal voltage, very high energy density, wide operating temperature range, very low self-discharge and hence long shelf life, and relatively high cost. They have been used for military applications for some years. If abused, some types can be potentially very hazardous and may have restrictions on air transport. The lithium manganese dioxide ($LiMnO_2$) couple has become established for a variety of applications, because of its high voltage and "fit-and-forget" lifetime characteristics. Operating voltages range from 2.5 to 3.5 V. Very high pulse discharge rates (up to 30A) are possible. Widely available types are either coin cells, for memory back-up, watches and calculators and other small, low power devices; or cylindrical cells, which offer light weight combined with capacities up to 1.5 Ah and high pulse current capability, together with long shelf life and wide operating temperature range.

Other primary lithium chemistries are lithium thionyl chloride ($Li-SOCl_2$) and lithium sulphur dioxide ($Li-SO_2$). These give higher capacities and pulse capability and wider temperature range but are really only aimed at specialized applications.

### 15.18.5.3 Secondary Cells

There have historically been two common rechargeable types: lead-acid and nickel-cadmium. These have quite different characteristics. Neither of them offer anywhere near the energy density of primary cells. At the same time, their heavy metal content and consequent exposure to environmental legislation (see page 598) have spurred development of other technologies, of which NiMH and Lithium Ion are the frontrunners.

### Lead-Acid

The lead-acid battery is the type that is known and loved by millions all over the world, especially on cold mornings when it fails to start the car. As well as the conventional "wet" automotive version, it is widely available in a valve-regulated "dry" or "maintenance-free" variant in which the sulphuric acid electrolyte is retained in a glass mat and does not need topping-up. This version is of more interest to circuit designers as it is frequently used as the standby battery in mains-powered systems, which must survive a mains failure.

These types have a nominal voltage of 2 V, a typical open circuit voltage of 2.15 V and an end-of-cycle voltage of 1.75 V per cell. They are commonly available in standard

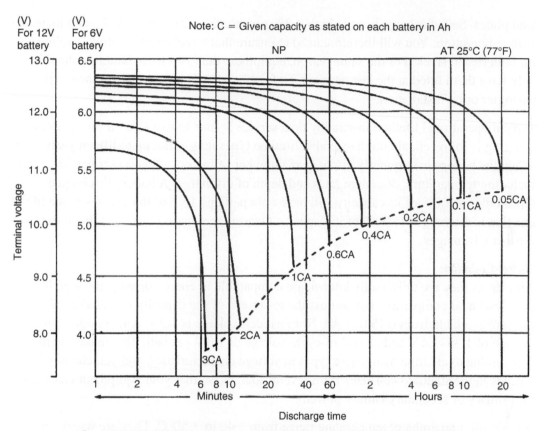

**Figure 15.42: Discharge characteristics for sealed lead-acid batteries.**

Source: Yuasa (dotted line indicates the lowest recommended voltage under load)

case sizes of 6 V or 12 V nominal voltage, with capacities from 1 to 100 Ah. Typical discharge characteristics are as shown in Figure 15.42. The value "C", as noted earlier, is the ampere-hour rating, conventionally quoted at the 20-hour discharge rate (5-hour discharge rate for nickel-cadmium and nickel metal hydride). Ambient temperature range is typically from −30 to +50°C, though capacity is reduced to around 60%, and achievable discharge rate suffers, at the lower extreme.

Valve regulated lead-acid types can be stored for a matter of months at temperatures up to 40°C, but will be damaged, perhaps irreversibly, if they are allowed to spend any length of time fully discharged. This is due to build-up of the sulphur in the electrolyte on the lead plates. Self-discharge is quite high −3% per month at 20°C is typical—and increases with temperature. You will therefore need to ensure that a recharging regime is followed for batteries in stock. For the same reason, equipment that uses these batteries should only have them fitted at the last moment, preferably when it is being dispatched to the customer or on installation.

Typical operational lifetime in standby float service is four to five years if proper float charging is followed, although extended lifetime types now claim up to fifteen years. When the battery is frequently discharged a number of factors affect its service life, including temperature, discharge rates and depth of discharge. A battery discharged repeatedly to 100% of its capacity will have only perhaps 15% of the cyclic service life of one that is discharged to 30% of its capacity. Overrating a battery for this type of duty has distinct advantages.

### Nickel-Cadmium

*NiCads*, as they are universally known, are comparable in energy density and weight to their lead-acid competitors but address the lower end of the capacity range. Typically they are available from 0.15 to 7 Ah. Nominal cell voltage is 1.2 V, with an open circuit voltage of 1.35–1.4 V and an end-of-cycle voltage of 1.0 V per cell. This makes them comparable to alkaline manganese types in voltage characteristics, and you can buy NiCads in the standard cell sizes from several sources, so that your equipment can work off primary or secondary battery power.

NiCads offer an ambient temperature range from −40 to +50°C. They are widely used for memory back-up purposes; batteries of two, three or four cells are available with pcb mounting terminals which can be continuously trickle charged from the logic supply, and can instantly supply a lower back-up voltage when this supply fails. Self-discharge rate is high and a cell that is not trickle charged will only retain its charge for a few months at most. Unlike lead-acid types they are not damaged by long periods of full discharge, and because of their low internal resistance they can offer high discharge rates. On the other hand they suffer from a "memory effect": a cell that is constantly being recharged before it has been completely discharged will lose voltage more quickly, and in fact it is better to recharge a NiCad from its fully discharged condition.

However, NiCads are now frowned upon because of their heavy metal content and hence the environmental consequences of their disposal to landfill. They are largely being superseded by Nickel Metal Hydride.

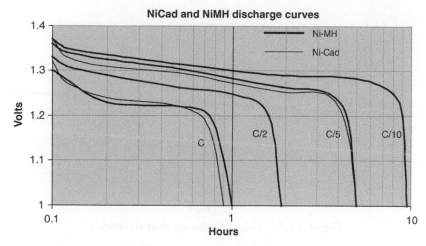

**Figure 15.43: NiCad and NiMH discharge characteristics**

### Nickel Metal Hydride

The discharge characteristics of NiMH are very similar to those of NiCad. The charged open circuit, nominal and end-point voltages are the same. The voltage profile of both types throughout most of the discharge period is flat (Figure 15.43). NiMH cells are generally specified from $-20°C$ to $+50°C$. They are around 20% heavier than their NiCad equivalents, but have about 40% more capacity. Also, they suffer less from the "memory effect" of NiCads (see above). On the other hand, they are less tolerant of trickle charging, and only very low trickle charge rates should be used if at all.

NiMH cells are available in a wide range of standard sizes, including button cells for memory back-up, and are also frequently specified in multi-cell packs for common applications such as mobile phones, camcorders and so on.

### Lithium-Ion

The Lithium-ion cell has considerable advantages over the types described above. Principally, it has a much higher gravimetric energy density (available energy for a given weight)—see Figure 15.45, which compares approximate figures for the three types, drawn from various manufacturers' specifications. But also, its cell voltage is about three times that of nickel batteries, 3.6–3.7 V versus 1.2 V. Its discharge profile with time is reasonably flat with an endpoint of 3 V (Figure 15.44), and it does not suffer from the NiCad memory effect.

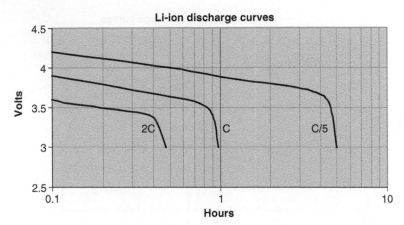

**Figure 15.44: Li-ion discharge characteristics**

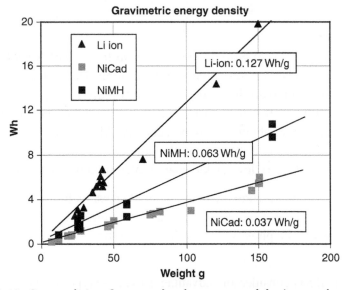

**Figure 15.45: Comparison of energy density versus weight (approximate values)**

These advantages come at a price, and Li-ion batteries are more expensive than the others. Also, they are much more susceptible to abuse in charging and discharging. The battery should be protected from over-charge, over-discharge and over-current at all times and this means that the best way to use it is as a battery pack, purpose designed for a given application, with charging and protection circuits built into the pack. This prevents the user from replacing or accidentally degrading individual cells, and gives the designer greater control over the expected performance of the battery. Since the high cost of a Li-ion battery pack makes it more suited to high value applications such as laptops and mobile phones, the extra cost of the integrated control circuitry is marginal and acceptable.

# CHAPTER 16

# Signals and Signal Processing

**Walt Kester**

## 16.1 Origins of Real-World Signals and their Units of Measurement

In this chapter, we will primarily be dealing with the processing of real-world signals using both analog and digital techniques. Before starting, however, let's look at a few key concepts and definitions required to lay the groundwork for things to come (Figure 16.1).

Webster's New Collegiate Dictionary defines a signal as "a detectable (or measurable) physical quantity or impulse (as voltage, current, or magnetic field strength) by which messages or information can be transmitted." Key to this definition are the words: detectable, physical quantity, and information.

- ■ Signal characteristics
  - ◆ Signals are physical quantities
  - ◆ Signals are measurable
  - ◆ Signals contain information
  - ◆ All signals are analog

- ■ Units of measurement
  - ◆ Temperature: °C
  - ◆ Pressure: Newtons/m²
  - ◆ Mass: kg
  - ◆ Voltage: Volts
  - ◆ Current: Amps
  - ◆ Power: Watts

**Figure 16.1: Signal characteristics**

By their very nature, signals are analog, whether DC, AC, digital levels, or pulses. It is customary, however, to differentiate between *analog* and *digital* signals in the following manner: Analog (or real-world) variables in nature include all measurable physical quantities. In this chapter, *analog* signals are generally limited to electrical variables, their rates of change, and their associated energy or power levels. Sensors are used to convert other physical quantities such as temperature or pressure to electrical signals. The entire subject of signal conditioning deals with preparing real-world signals for processing, and includes such topics as sensors (temperature and pressure, for example), isolation amplifiers, and instrumentation amplifiers.

Some signals result in response to other signals. A good example is the returned signal from a radar or ultrasound imaging system, both of which result from a known transmitted signal.

On the other hand, there is another classification of signals, called *digital*, where the actual signal has been conditioned and formatted into a digit. These digital signals may or may not be related to real-world analog variables. Examples include the data transmitted over local area networks (LANs) or other high speed networks.

In the specific case of digital signal processing (DSP), the analog signal is converted into binary form by a device known as an analog-to-digital converter (ADC). The output of the ADC is a binary representation of the analog signal and is manipulated arithmetically by the digital signal processor. After processing, the information obtained from the signal may be converted back into analog form using a digital-to-analog converter (DAC).

Another key concept embodied in the definition of *signal* is that there is some kind of *information* contained in the signal. This leads us to the key reason for processing real-world analog signals: the *extraction of information* (Figure 16.2).

## 16.2  Reasons for Processing Real-World Signals

The primary reason for processing real-world signals is to extract information from them. This information normally exists in the form of signal amplitude (absolute or relative), frequency or spectral content, phase, or timing relationships with respect to other signals. Once the desired information is extracted from the signal, it may be used in a number of ways.

- Extract Information about the signal (Amplitude, Phase, Frequency, Spectral Content, Timing Relationships)
- Reformat the Signal (FDMA, TDMA, CDMA Telephony)
- Compress data (Modems, Cellular Telephone, HDTV, MPEG)
- Generate feedback control signal (Industrial Process Control)
- Extract signal from noise (Filtering, Autocorrelation, Convolution)
- Capture and store Signal in digital format for analysis (FFT Techniques)

**Figure 16.2: Reasons for signal processing**

In some cases, it may be desirable to reformat the information contained in a signal. This would be the case in the transmission of a voice signal over a frequency division multiple access (FDMA) telephone system. In this case, analog techniques are used to "stack" voice channels in the frequency spectrum for transmission via microwave relay, coaxial cable, or fiber. In the case of a digital transmission link, the analog voice information is first converted into digital using an ADC. The digital information representing the individual voice channels is multiplexed in time (time division multiple access, or TDMA) and transmitted over a serial digital transmission link (as in the T-carrier system).

Another requirement for signal processing is to *compress* the frequency content of the signal (without losing significant information), then format and transmit the information at lower data rates, thereby achieving a reduction in required channel bandwidth. High speed modems and adaptive pulse code modulation systems (ADPCM) make extensive use of data reduction algorithms, as do digital mobile radio systems, MPEG recording and playback, and high-definition television (HDTV).

Industrial data acquisition and control systems make use of information extracted from sensors to develop appropriate feedback signals, which in turn control the process itself. Note that these systems require both ADCs and DACs as well as sensors, signal conditioners, and the DSP (or microcontroller). In some cases, the signal containing the information is buried in noise, and the primary objective is signal recovery. Techniques such as filtering, autocorrelation, and convolution are often used to accomplish this task in both the analog and digital domains.

## 16.3 Generation of Real-World Signals

In most of the previous examples (the ones requiring DSP techniques), both ADCs and DACs are required. In some cases, however, only DACs are required where real-world analog signals may be generated directly using DSP and DACs. Video raster scan display systems are a good example. The digitally generated signal drives a video or RAMDAC. Another example is artificially synthesized music and speech. In reality, however, the real-world analog signals generated using purely digital techniques do rely on information previously derived from the real-world equivalent analog signals. In display systems, the data from the display must convey the appropriate information to the operator. In synthesized audio systems, the statistical properties of the sounds being generated have been previously derived using extensive DSP analysis of the entire signal chain, including sound source, microphone, preamp, and ADC.

## 16.4 Methods and Technologies Available for Processing Real-World Signals

Signals may be processed using analog techniques (analog signal processing, or ASP), digital techniques (digital signal processing, or DSP), or a combination of analog and digital techniques (mixed-signal processing, or MSP). In some cases, the choice of techniques is clear; in others, there is no clear-cut choice, and second-order considerations may be used to make the final decision.

With respect to DSP, the factor that distinguishes it from traditional computer analysis of data is its speed and efficiency in performing sophisticated digital processing functions such as filtering, FFT analysis, and data compression in real time.

The term *mixed-signal processing* implies that *both* analog and digital processing is done as part of the system. The system may be implemented in the form of a printed circuit board, hybrid microcircuit, or a single integrated circuit chip. In the context of this broad definition, ADCs and DACs are considered to be mixed-signal processors, since both analog and digital functions are implemented in each. Recent advances in very large scale integration (VLSI) processing technology allow complex digital processing as well as analog processing to be performed on the same chip. The very nature of DSP itself implies that these functions can be performed in *real time*.

## 16.5 Analog Versus Digital Signal Processing

Today's engineer faces a challenge in selecting the proper mix of analog and digital techniques to solve the signal processing task at hand. It is impossible to process real-world analog signals using purely digital techniques, since all sensors, including microphones, thermocouples, strain gages, piezoelectric crystals, and disk drive heads are analog sensors. Therefore, some sort of signal conditioning circuitry is required in order to prepare the sensor output for further signal processing, whether it be analog or digital. Signal conditioning circuits are, in reality, analog signal processors, performing such functions as multiplication (gain), isolation (instrumentation amplifiers and isolation amplifiers), detection in the presence of noise (high common-mode instrumentation amplifiers, line drivers, and line receivers), dynamic range compression (log amps, LOGDACs, and programmable gain amplifiers), and filtering (both passive and active).

Several methods of accomplishing signal processing are shown in Figure 16.3. The top portion of the figure shows the purely analog approach. The latter parts of the figure show the DSP approach. Note that once the decision has been made to use DSP techniques, the next decision must be where to place the ADC in the signal path.

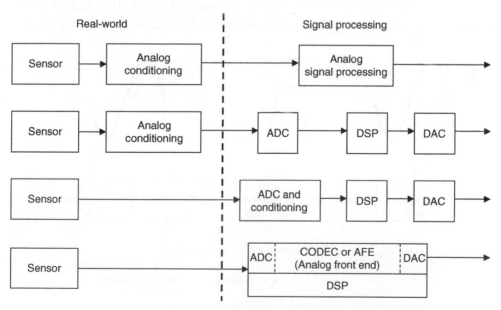

**Figure 16.3: Analog and digital signal processing options**

In general, as the ADC is moved closer to the actual sensor, more of the analog signal conditioning burden is now placed on the ADC. The added ADC complexity may take the form of increased sampling rate, wider dynamic range, higher resolution, input noise rejection, input filtering, programmable gain amplifiers (PGAs), and on-chip voltage references, all of which add functionality and simplify the system. With today's high resolution/high sampling rate data converter technology, significant progress has been made in integrating more and more of the conditioning circuitry within the ADC/DAC itself. In the measurement area, for instance, 24-bit ADCs are available with built-in programmable gain amplifiers (PGAs) that allow full-scale bridge signals of 10 mV to be digitized directly with no further conditioning (e.g., AD773x series). At voice-band and audio frequencies, complete coder/decoders (codecs or analog front ends) are available with sufficient on-chip analog circuitry to minimize the requirements for external conditioning components (AD1819B and AD73322). At video speeds, analog front ends are also available for such applications as CCD image processing and others (e.g., AD9814, AD9816, and the AD984x series).

## 16.6  A Practical Example

As a practical example of the power of DSP, consider the comparison between an analog and a digital low-pass filter, each with a cutoff frequency of 1 kHz. The digital filter is implemented in a typical sampled data system shown in Figure 16.4. Note that there

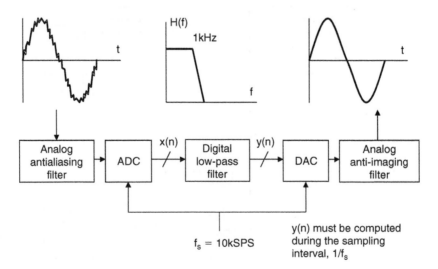

**Figure 16.4: Digital filter**

are several implicit requirements in the diagram. First, it is assumed that an ADC/ DAC combination is available with sufficient sampling frequency, resolution, and dynamic range to accurately process the signal. Second, the DSP must be fast enough to complete all its calculations within the sampling interval, $1/f_s$. Third, analog filters are still required at the ADC input and DAC output for antialiasing and anti-imaging, but the performance demands are not as great. Assuming these conditions have been met, the following offers a comparison between the digital and analog filters.

The required cutoff frequency of both filters is 1 kHz. The analog filter is realized as a 6-pole Chebyshev Type 1 filter (ripple in pass band, no ripple in stop band), and the response is shown in Figure 16.5. In practice, this filter would probably be realized using three 2-pole stages, each of which requires an op amp, and several resistors and capacitors. Modern filter design CAD packages make the 6-pole design relatively straightforward, but maintaining the 0.5 dB ripple specification requires accurate component selection and matching.

On the other hand, the 129-tap digital FIR filter shown has only 0.002 dB pass band ripple, linear phase, and a much sharper roll-off. In fact, it could not be realized using analog techniques. Another obvious advantage is that the digital filter requires no component matching, and it is not sensitive to drift since the clock frequencies are crystal controlled. The 129-tap filter requires 129 multiply-accumulates (MAC) in

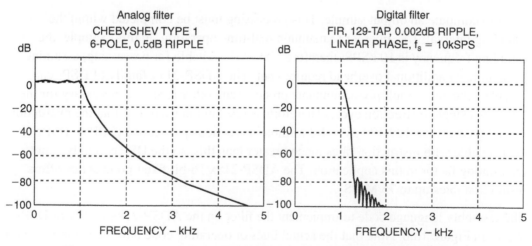

**Figure 16.5: Analog versus digital filter frequency response comparison**

```
        .MODULE          fir_sub;
        {                FIR Filter Subroutine
                         Calling Parameters
                                 I0 --> Oldest input data value in delay line
                                 I4 --> Beginning of filter coefficient table
                                 L0 = Filter length (N)
                                 L4 = Filter length (N)
                                 M1,M5 = 1
                                 CNTR = Filter length - 1 (N-1)
                         Return Values
                                 MR1 = Sum of products (rounded and saturated)
                                 I0 --> Oldest input data value in delay line
                                 I4 --> Beginning of filter coefficient table
                         Altered Registers
                                 MX0,MY0,MR
                         Computation Time
                                 (N - 1) + 6 cycles = N + 5 cycles
                         All coefficients are assumed to be in 1.15 format. }
        .ENTRY           fir;
 ----→  fir:             MR=0, MX0=DM(I0,M1), MY0=PM(I4,M5)
 ----→                   CNTR = N-1;
 ----→                   DO convolution UNTIL CE;
 ----→  convolution:       MR=MR+MX0*MY0(SS), MX0=DM(I0,M1), MY0=PM(I4,M5);
 ----→                   MR=MR+MX0*MY0(RND);
 ----→                   IF MV SAT MR;
 ----→                   RTS;
        .ENDMOD;
```

**Figure 16.6: ADSP-21xx FIR filter assembly code (single precision)**

order to compute an output sample. This processing must be completed within the
sampling interval, $1/f_s$, in order to maintain real-time operation. In this example, the
sampling frequency is 10 kSPS; therefore 100 μs is available for processing, assuming
no significant additional overhead requirement. The ADSP-21xx family of DSPs can
complete the entire multiply-accumulate process (and other functions necessary for the
filter) in a single instruction cycle. Therefore, a 129-tap filter requires that the instruction
rate be greater than 129/100 μs = 1.3 million instructions per second (MIPS). DSPs
are available with instruction rates much greater than this, so the DSP certainly is not
the limiting factor in this application. The ADSP-218x 16-bit fixed-point series offers
instruction rates up to 75 MIPS.

The assembly language code to implement the filter on the ADSP-21xx family of DSPs is
shown in Figure 16.6. Note that the actual lines of operating code have been marked with
arrows; the rest are comments.

- Digital signal processing
  - ADC/DAC Sampling frequency limits signal bandwidth
    - (Don't forget Nyquist)
  - ADC/DAC Resolution/Performance limits signal dynamic range
  - DSP Processor speed limits amount of digital processing available, because:
    - All DSP Computations must be completed during the sampling interval, $1/f_s$, for real-time operation
- Don't forget analog signal processing
  - High frequency/RF filtering, modulation, demodulation
  - Analog antialiasing and reconstruction filters with ADCs and DACs
  - Where common sense and economics Dictate

**Figure 16.7: Real-time signal processing**

In a practical application, there are certainly many other factors to consider when evaluating analog versus digital filters, or analog versus digital signal processing in general. Most modern signal processing systems use a combination of analog and digital techniques in order to accomplish the desired function and take advantage of the best of both the analog and the digital worlds.

## References

Higgins, Daniel, J., *Digital Signal Processing in VLSI*. Prentice-Hall, 1990.

*Practical Design Techniques for Sensor Signal Conditioning*, Analog Devices, 1998.

Sheingold, Daniel, H., (eds). *Transducer Interfacing Handbook.* Analog Devices, Inc, 1972.

# CHAPTER 17

# Filter Design

**Walt Kester**
**Andrew Leven**

## 17.1 Introduction

Electronic filters have many applications in the telecommunications and data communications industry. One such application, which involves a multiple channel communications system employing a technique known as *time-division multiplexing* (TDM), is shown in Figure 17.1. In this system several channels are transmitted through a medium such as an optical fiber, as shown here, or through a coaxial cable or waveguide. Multiplexing means combining several signals into one, and this is accomplished in TDM

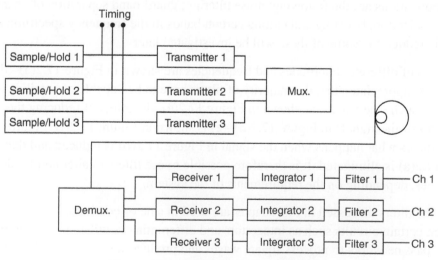

Figure 17.1: Time-division multiplexing

by allocating time slots for each channel so that each channel is transmitted at a particular time. If the signals are synchronized correctly there will be no interference between them. At the transmitter end a multiplexer is used to combine the signals, while at the receiver end a demultiplexer is used to separate the original channels.

However, when the channel signals arrive at the receivers they have deteriorated in shape and amplitude. In order to clean them up they are reconstructed by an integrator that sums up the incoming signal very much as in mathematical integration. Once this has been done a filter is used to pass the wanted channel frequencies while attenuating the unwanted signals such as noise.

The combined functions of the integrator and filter cause the transmitted channels to be reproduced. In this case, where three channels are involved, each filter will be designed to pass the particular channel frequency and its related information, hence a band of frequencies is passed by each filter.

This is an example of where filters are used to pass bands of frequencies such as the voice band (300–3400 Hz). However, filters can also be used to pass frequencies below a certain frequency while attenuating all frequencies above it. Similarly, it is possible to construct a filter which passes all frequencies above a certain frequency while attenuating all frequencies below it.

Other applications are the following: noise filtering; guard band separation of channels; bandpass selection; boosting and cutting certain bands in the frequency spectrum; and harmonic reduction. Some of these will be investigated later.

Sine waves of different amplitudes and frequencies are shown in Figure 17.2(a)–(d). It should be appreciated that the majority of filters have to be capable of handling a mixture of such sine waves, as shown in Figure 17.2(e); the effect of reducing the amplitudes of the signals in Figure 17.2(d)–(e) is shown in Figure 17.2(f). Figure 17.2(g) shows what happens when the signal in Figure 17.2(b) is reduced and that in Figure 17.2(a) is eliminated. It is therefore possible to use filters to alter amplitudes and frequencies, depending on the requirements of the system.

Finally, the filters discussed in this chapter are used in sine or continuous wave circuits. However, certain circuits such as integrators and differentiators utilize passive high-pass and low-pass networks to process square waves and produce wave shaping. When fed through a filter the square wave is modified: the high-frequency edges are rounded when

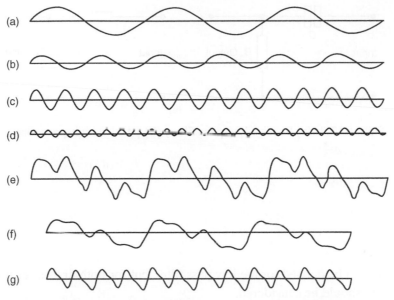

**Figure 17.2: Sine waves of different amplitudes and frequencies**

passing through a low-pass filter, while the flat top and bottom are distorted when passing through a high-pass filter.

## 17.2 Passive Filters

The most elementary types of filters are constructed from $RC$ networks and are known as *passive filters* as they dissipate part of the signal power and pass the rest. Figure 17.3(a) shows a passive low-pass filter, while Figure 17.3(b) shows a passive high-pass filter. These form the basis of more sophisticated filters. Each has a cut-off frequency, which may be derived by considering the high-pass filter as a voltage divider. From Figure 17.3(b) we have:

$$\frac{V_o}{V_i} = \frac{R}{\sqrt{R^2 + X_c^2}} \tag{17.1}$$

and at the cut-off frequency the gain falls by $3\,\text{dB}$ or $1/\sqrt{2}$. Also at this frequency $R = X_c$, which gives:

$$R = X_c = \frac{1}{2\pi f_c C}$$

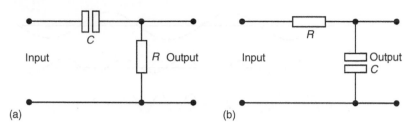

**Figure 17.3: (a) Passive low-pass filter; (b) passive high-pass filter**

Hence,

$$f_c = \frac{1}{2\pi RC} \tag{17.2}$$

A similar result can be derived for the low pass filter, but for both first-order filters the following points should be considered.

(a) Cascading or connecting these networks in series causes the roll-off of the frequency response to increase by 20 dB/decade for each filter, where "decade" refers to a one-to-ten range of frequencies, such as 1–10 Hz, 10–100 Hz, etc.: observe that on a logarithmic scale, such ranges span an equal distance (see Figs 17.42 and 17.43).

(b) A low-pass filter causes a phase lag between the output and input voltages, while a high-pass filter causes a phase lead between the output and input voltages. This has an important bearing on filters used in certain oscillators.

## 17.3 Active Filters

The use of operational amplifiers in active filter devices is now well established in communications systems. Their main advantages over passive filters are:

(a) flexibility in design and construction;

(b) the absence of inductors, which at low frequencies is useful due to their large size and cost;

(c) low-frequency applications down to 1 Hz;

(d) the buffering effect due to the high input impedance and the low output impedance;

(e) with gain setting resistors the op-amp is capable of providing gain; hence, the input signal is not attenuated as it is in passive filters;

(f) they are easier to tune than passive filters.

It is as well at this stage to appreciate that there are many types of filter, such as crystal, acoustical and digital filters, all of which have a specific application. In this chapter we will investigate active filters which are of the analog type but can be used in either digital or analog system applications.

### 17.3.1  Filter Response

Associated with a filter's performance is the frequency response, which involves a plot of frequency against gain or against attenuation. This graph involves a response for all frequencies which the filter is designed to pass. At a particular frequency, known as the *cut-off frequency*, the response starts to decrease in amplitude. This is known as the *roll-off* and is a measure of how sharply the filter responds to attenuate frequencies above or below the cut-off frequency.

The filters in this chapter will have input *RC* networks, and as the signal frequency decreases the capacitive reactance $X_c$ increases. This causes less voltage to be applied across the input impedance of the amplifier because more is dropped across $X_c$. This reduces the overall gain of the filter, and a critical point is reached when the output voltage is 0.707, i.e., $1/\sqrt{2}$, of the input ($V_0 = 0.707V_i$). This condition occurs when $X_c = R$ and is called the –3 dB point of the response as the overall gain is 3 dB down on the pass-band gain. The frequency at which this occurs is the cut-off frequency. This discussion applies to all filter types.

All filters have four basic applications which can be easily understood from the ideal responses shown below. Note that an ideal response is one which has a vertical roll-off at the cut-off frequency. In practice this is not possible, but certain sophisticated filters tend to approach it. The four ideal configurations are shown in Figure 17.4, in which the pass and stop bands are shown.

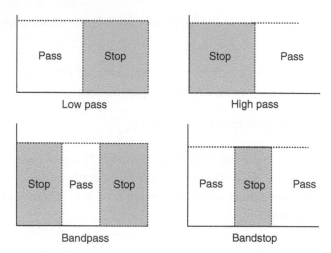

**Figure 17.4: The four ideal filter configurations**

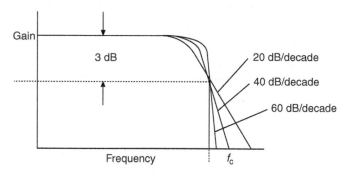

**Figure 17.5: Multiple response diagram**

## 17.3.2  Cut-off Frequency and Roll-off Rate

As has been mentioned, no filter achieves the ideal response shown in Figure 17.4, but the higher the order of the filter the closer it approaches the ideal case. This is shown in Figure 17.5, which shows a multiple response diagram. It can be seen from this diagram that the roll-off rate increases with the order of the filter. This filter order is dependent on the number of *RC* networks (number of poles) included in the filter design. For example, if a single *RC* network is used with a filter it is referred to as a single-pole filter, while two *RC* networks produce a two-pole filter. Correspondingly, the roll-off would be 20 dB/ decade

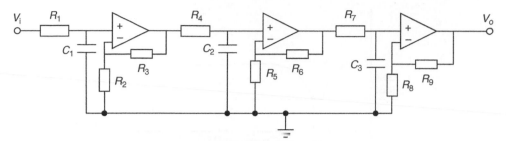

**Figure 17.6: Third-order filter**

and 40 dB/decade, respectively. Hence, increasing the number of *RC* networks increases the order of the filter. A three-pole or third-order filter is shown in Figure 17.6.

It is normally not necessary to go beyond a fourth-order filter, but if this situation arises then it is a simple matter of cascading first and second-order filters to achieve higher orders. We will now examine these two important filters in detail and see how they can be realized in a practical way.

### 17.3.3 Filter Types

There are two fundamental responses generally used in the design of filters; these are referred to as the Butterworth and Chebyshev responses. The low-pass filter responses for these types are shown in Figure 17.7. As can be seen, the two responses are quite different. The Butterworth type has what is called a maximally flat response in the pass band. Hence, there is no ripple in this type of filter and the cut-off frequency is generally taken at the 3 dB level as shown. Note that in Figure 17.7(a) the stop band lies between 0 Hz and $f_c$. In practice this may not be the case, and a minimum gain may be stipulated (say) between point A and $f_c$.

The maximally flat response of the Butterworth is good at frequencies around about zero hertz, but the response is poorer near the edge of the pass band. The Chebyshev filter can solve this problem. The Chebyshev response shown in Figure 17.7(b) contains a ripple in the pass band. However, the attenuation increases more rapidly outside the pass band than the Butterworth. The greater the ripple, the more selective is the filter. The pass band is not so easily defined but is usually taken from the point where the highest-frequency peak ripple occurs. If, for example, the Chebyshev high-pass filter in Figure 17.7(b) has a 0.5 dB ripple as shown and $f_r = 1$ kHz, then its response would be given as $\pm 0.5$ dB from

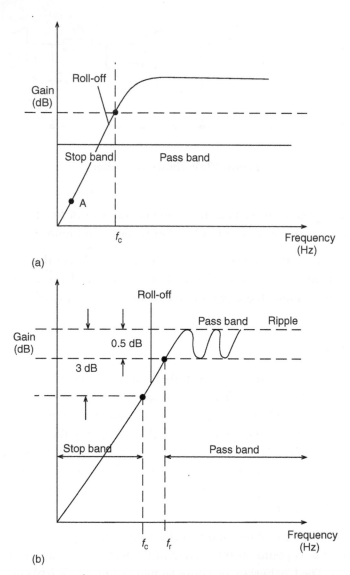

**Figure 17.7: (a) Butterworth filter; (b) Chebyshev filter**

1 kHz onwards with a rapidly increasing attenuation for frequencies less than 1 kHz. However, in some applications the 3 dB bandwidth is required as shown at point *C* on Figure 17.7(b), and this may be calculated using what are called *transfer functions*. These will be discussed later.

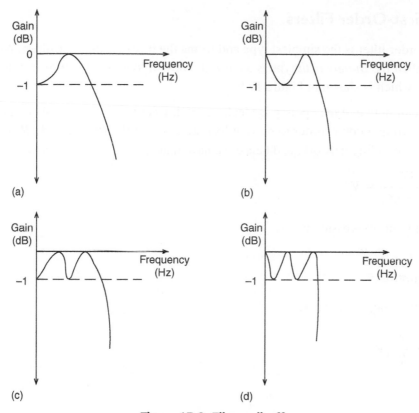

**Figure 17.8: Filter roll-off**

### 17.3.4 Filter Orders

Filter orders have already been mentioned, and it can be seen from Figure 17.4 that the orders would have to be infinitely high in order to achieve ideal responses.

The order of a Chebyshev or Butterworth filter determines the sharpness or roll-off of the response, but the interpretation of order is slightly different because of the ripple pass band in the Chebyshev filter. In this case the number of ripple peaks in the pass band determines the order ($n$) of the filter. This is shown in Figure 17.8. For example, in Figure 17.8(a) $n = 2$ and in Figure 17.8(c) $n = 4$. Note that unlike the Chebyshev filter, the Butterworth low-pass filter will be 3 dB down on its maximum value no matter what the order is. The same points apply to the high-pass filter responses.

## 17.4 First-Order Filters

The first-order filter is the simplest type and forms the basis of all other filters. Normally, what is called the *Butterworth* type is analyzed. We will look at the low-pass filter first, a circuit for which is shown in Figure 17.9.

In this circuit note that the op-amp is ideal, i.e., it draws no current, and also it is used in the noninverting mode in order to prevent loading down of the *RC* network. *R* and *C* act as a voltage-dividing network, and hence we have that:

$$V = -\frac{jX_c}{R - jX_c} = V_i$$

Simplifying this expression gives:

$$V = \frac{V_i}{1 + j2\pi RC}$$

The output voltage is given as:

$$V_o = \left(1 + \frac{R_f}{R_i}\right)V$$

Hence,

$$V_o = \left(1 + \frac{R_f}{R_i}\right)\frac{V_i}{1 + j2\pi RC}$$

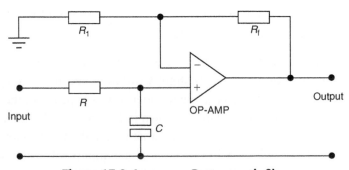

**Figure 17.9: Low-pass Butterworth filter**

or,

$$\frac{V_o}{V_i} = \frac{A}{1 + j(f/f_{3dB})} \tag{17.3}$$

Note that;

$$f_{L(3dB)} = \frac{1}{2\pi RC} \tag{17.4}$$

This has the characteristics of a first-order low-pass filter. When $\omega = 0$ then the pass-band gain is:

$$\frac{V_o}{V_i} = \frac{R_2}{R_1} = K \tag{17.5}$$

This is simply the amplifier gain. Note also that when:

$$\omega = \frac{1}{RC}$$

the gain has dropped by 3 dB after which the gain falls off at the rate of 20 dB/decade. A typical response for this filter is shown in Figure 17.10.

A similar analysis may be carried out for the first-order high-pass filter, which is shown in Figure 17.11. Note that these two filters are identical except that R and C have been interchanged. The output voltage is given by:

$$V_o = \left(1 + \frac{R_f}{R_1}\right)\left(\frac{2j\pi fRC}{1 + j2\pi RC}\right)V_i$$

or,

$$\frac{V_o}{V_i} = A\frac{j(f/f_{3dB})}{1 + j(f/f_{3dB})}$$

Note that:

$$f_{H(3dB)} = \frac{1}{2\pi RC} \tag{17.6}$$

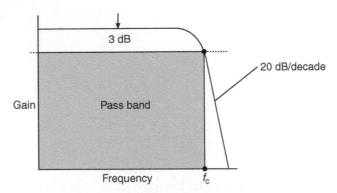

**Figure 17.10: Typical filter response for low-pass**

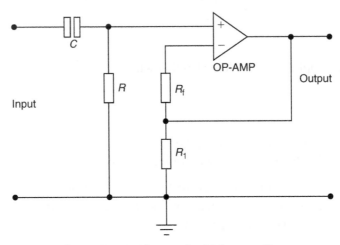

**Figure 17.11: First-order high-pass filter**

The response for this filter is shown below in Figure 17.12.

## 17.5 Design of First-Order Filters

Low- and high-pass first-order filters may be designed very easily if certain steps are followed:

1. The cut-off frequency must be known.

2. A value of C less than $1\,\mu\text{F}$ (say) should be chosen.

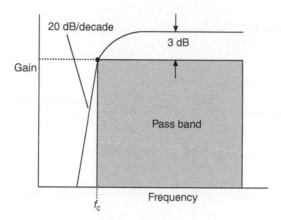

**Figure 17.12: Response for high-pass filter**

3. Then calculate the value of R from equation (17.4) or (17.6), depending on the filter being designed.

4. Determine a value of A and calculate $R_f$ and $R_1$.

**Example 17.1**
Design a low-pass filter at a cut-off frequency of 2.4 kHz with a pass-band gain of 3.

**Solution**
Select a value of C = 0.025 μF. This will give:

$$R = \frac{1}{2\pi \times 2.4 \times 10^3 \times 0.025 \times 10^{-6}} = 2.7\,k\Omega$$

Since the pass-band gain is 3 then:

$$3 = 1 + \frac{R_f}{R_i}$$

Hence, $R_f = 2R_i$ and so various values are possible. If an unusual value is calculated then a potentiometer may be used to set the values. It should also be mentioned at this point that with advanced semiconductor technology a selection of very low

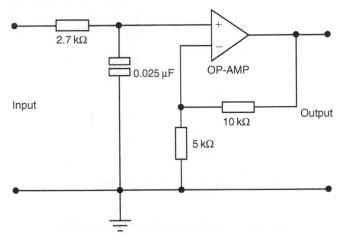

**Figure 17.13: Circuit for Example 17.1**

values of capacitance in the nanofarad range is available from many manufacturers in chip form.

In order to complete the exercise the practical circuit is shown in Figure 17.13 and this can now be set up on a printed circuit board.

**Example 17.2**
Design a high-pass filter at a cut-off frequency of 1 kHz with a passband gain of 2.

**Solution**
Once again select a suitable value of $C$, such as $0.01\,\mu\text{F}$. Hence, since the cut-off frequency is 1 kHz, $R = 15.9\,\text{k}\Omega$. Since $A = 2$, the two feedback resistors are equal. Several solutions are possible, such as $10\,\text{k}\Omega$.

## 17.6 Second-Order Filters

As has already been mentioned, the higher the order of filter the sharper the cut-off. For certain applications, such as radio relay applications and channel separation, it is necessary to have higher-order filters. This chapter only looks at first and second-order filters but many higher orders can be designed by simply cascading these two types; indeed, this is one of the big advantages of using the active filter.

### 17.6.1 Low-Pass Second-Order Filters

Consider two low-pass first-order filters with the same cut-off frequencies, but different pass-band gains:

$$\frac{K_1}{1 + jf/f_{3\text{dB}}} \qquad \frac{K_2}{1 + jf/f_{3\text{dB}}}$$

If these filters are now cascaded, then the overall function will appear as follows,

$$\frac{K_1 K_2}{(1 + jaf)^2} = \frac{K}{(1 + jaf)^2}$$

where $a = 1/f_{3\text{dB}}$ and $K = K_1 K_2$.

Expanding the above expression will give:

$$\frac{V_\text{o}}{V_i} = \frac{K}{a^2(j\omega)^2 + 2a(j\omega) + 1}$$

and in general terms this is stated as:

$$\frac{V_\text{o}}{V_i} = \frac{K}{a_2(j\omega)^2 + a_1(j\omega) + 1} \qquad\qquad (17.7)$$

where $a_1$ and $a_2$ are constants. This expression is the characteristic of a second-order filter, and from it two basic types of filter may be deduced, depending on the values of $a_1$ and $a_2$: Butterworth flat response, where $a_1^2 = 2a_2$; and Chebyshev ripple response, where $a_1^2 < 2a_2$. The responses of both these filters has already been given, but they are combined in Figure 17.14.

The Butterworth response is generally a flatter response than the Chebyshev, but the Chebyshev filter has a faster rate of cut-off immediately after the cut-off frequency. Because of its flat response the Butterworth filter is more popular, but the ripple response of the Chebyshev has applications in satellite transponders where channel separation is tight.

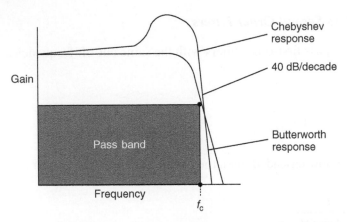

**Figure 17.14: Responses for both low-pass second-order filters**

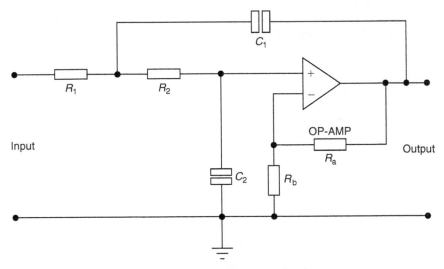

**Figure 17.15: Sallen-Key circuit**

Both these filters can be represented by many circuits, but the easiest configuration is known as the *Sallen-Key circuit* from which most filters may be designed, provided the pass-band gain and cut-off frequency are known. The typical circuit configuration is shown in Figure 17.15.

By using circuit analysis the general transfer function for the circuit in Figure 17.15 may be determined as follows:

$$\frac{V_o}{V_i} = \frac{K/R_1R_2C_1C_2}{s^2 + s\{1/R_1C_1 + 1/R_2C_1 + (1-K)/R_2C_2\} + 1/R_1R_2C_1C_2} \tag{17.8}$$

where $K = 1 + R_a/R_b$ (the DC gain) and $s = j\omega$. A full analysis of the transfer function can be found in standard texts on filters.

The denominator term in equation (17.7) is known as the *polynomial* for the $n$th-order filter. These polynomials may be derived for any filter type or order, but it is more convenient to use polynomial tables. Examples given in this text will use polynomials which are shown in Table 17.1.

This form of the general transfer function is related to Figure 17.15 where $R_1$, $C_1$, etc. are the components used in the Sallen-Key circuit after a multiplication factor has been applied. This is called *denormalization*.

Later in this chapter normalized filter tables, given in Table 17.2, will be used. The term *normalization* is defined usually as the scaling or standardization of a certain parameter. In the case of the filter tables the values are normalized to an angular cut-off frequency of 1 rad/s or 1 Hz. A multiplier is used in order to calculate the actual values of the components which will be used in the printed circuit board design. The operation of this multiplier is known as *denormalization* and will be fully demonstrated by the examples given later.

It is as well to appreciate at this point that filter problems may be solved by using four main methods: the transfer function; normalized tables; identical components; and software. The use of software is widespread and there are many software packages which can be easily used by the novice. The suitability of these packages is a personal matter, but the author has found that the use of spreadsheets gives excellent results. The other three methods of solving active filter problems will be demonstrated by example.

**Table 17.1: Filter polynomials**

| Butterworth polynomials |
| --- |
| $n$ |
| 1  $S + 1$ |
| 2  $S^2 + 1.414 + 1$ |
| 3  $(S + 1)(S^2 + 1.414 + 1)$ |
| 4  $(S^2 + 0.765S + 1)(S^2 + 1.848S + 1)$ |
| 5  $(S + 1)(S^2 + 0.765S + 1)(S^2 + 1.848S + 1)$ |
| 6  $(S^2 + 0.518S + 1)(S^2 + 1.414S + 1)(S^2 + 1.932S + 1)$ |

| Chebyshev polynomials (0.5 dB ripple) |
| --- |
| 1  $S + 2.863$ |
| 2  $S^2 + 1.425S + 1.516$ |
| 3  $(S + 0.626)(S^2 + 0.626S + 1.142)$ |
| 4  $(S^2 + 0.351S + 1.064)(S^2 + 0.845S + 0.356)$ |
| 5  $(S + 0.362)(S^2 + 0.224S + 1.036)(S^2 + 0.586S + 0.477)$ |
| 6  $(S^2 + 0.155S + 1.024)(S^2 + 0.414S + 0.548)(S^2 + 0.580S + 0.157)$ |

| Chebyshev polynomials (1 dB ripple) |
| --- |
| 1  $S + 1.965$ |
| 2  $S^2 + 1.098S + 1.103$ |
| 3  $(S + 0.494) (S^2 + 0.494S + 0.994)$ |
| 4  $(S^2 + 0.279S + 0.987)(S^2 + 0.674S + 0.279)$ |
| 5  $(S + 0.289)(S^2 + 0.179S + 0.989)(S^2 + 0.469S + 0.429)$ |
| 6  $(S^2 + 0.124S + 0.991)(S^2 + 0.340S + 0.558)(S^2 + 0.464S + 0.125)$ |

## 17.7 Using the Transfer Function

*Example 17.3*

Determine suitable values for $R_1$, $R_2$, $C_1$ and $C_2$ for a second-order Butterworth filter with an upper cut-off frequency of 4 kHz and a pass-band gain of 20.

Table 17.2: Normalized filter tables

| Filter type | $R_1$ | $R_2$ | $C_1$ | $C_2$ | Gain $K$ |
|---|---|---|---|---|---|
| **Normalized tables for second-order filters.** | | | | | |
| *(a) Low-pass normalized filter (second order) with cut-off frequency of 1 rad/s* | | | | | |
| Butterworth | 1.000 | 1.000 | 1.000 | 1.000 | 1.585 |
| | 1.000 | 1.000 | 1.414 | 0.707 | 1.000 |
| | 1.000 | 1.000 | 0.874 | 1.144 | 2.000 |
| | 0.707 | 1.414 | 1.000 | 1.000 | 2.000 |
| 0.5 dB ripple Chebyshev | 0.812 | 0.812 | 1.000 | 1.000 | 1.842 |
| | 1.000 | 1.000 | 1.403 | 0.470 | 1.000 |
| | 1.000 | 1.000 | 0.771 | 0.856 | 2.000 |
| | 0.701 | 0.940 | 1.000 | 1.000 | 2.000 |
| 1 dB ripple Chebyshev | 0.952 | 0.952 | 1.000 | 1.000 | 1.954 |
| | 1.000 | 1.000 | 1.822 | 0.498 | 1.000 |
| | 1.000 | 1.000 | 0.938 | 0.967 | 2.000 |
| | 0.911 | 0.996 | 1.000 | 1.000 | 2.000 |
| *(b) High-pass normalized filter (second order) with cut-off frequency of 1 rad/s* | | | | | |
| Butterworth | 1.000 | 1.000 | 1.000 | 1.000 | 1.585 |
| | 0.707 | 1.414 | 1.000 | 1.000 | 1.000 |
| | 1.000 | 1.000 | 1.414 | 0.707 | 2.000 |
| | 1.144 | 0.874 | 1.000 | 1.000 | 2.000 |
| 0.5 dB ripple Chebyshev | 1.231 | 1.231 | 1.000 | 1.000 | 1.842 |
| | 0.713 | 2.127 | 1.000 | 1.000 | 1.000 |
| | 1.000 | 1.000 | 1.426 | 1.064 | 2.000 |
| | 1.247 | 1.169 | 1.000 | 1.000 | 2.000 |
| 1 dB ripple Chebyshev | 1.050 | 1.050 | 1.000 | 1.000 | 1.954 |
| | 0.549 | 2.009 | 1.000 | 1.000 | 1.000 |
| | 1.000 | 1.000 | 1.097 | 1.004 | 2.000 |
| | 1.066 | 1.034 | 1.000 | 1.000 | 2.000 |

### Solution

A problem of this nature requires a normalized response before it can be solved. The second-order Butterworth normalized response in this case will be given as:

$$\frac{H}{(j\omega)^2 + 1.414(j\omega) + 1} \tag{17.9}$$

and the transfer function will be as stated previously in equation (17.8). If we multiply top and bottom of the right-hand side this equation by $R_1 R_2 C_1 C_2$ and substitute $K = 20$, we obtain:

$$\frac{V_o}{V_i} = \frac{20}{R_1 R_2 C_1 C_2 s^2 + s\{R_2 C_2 + (1 - K)R_1 C_1 + R_1 C_2\} + 1/R_1 R_2 C_1 C_2} \tag{17.10}$$

The next step is to equate the coefficients of equations (17.9) and (17.10): for the $s^2$ terms

$$R_1 R_2 C_1 C_2 = 1 \tag{17.11}$$

and for the $s$ terms:

$$R_2 C_2 + (1 - 20) R_1 C_1 + R_1 C_2 = 1.414 \tag{17.12}$$

From (17.11) we may write:

$$R_2 C_2 = 1 \quad \text{and} \quad R_1 C_1 = 1$$

as this will satisfy the right-hand side of the equation. Substituting in (17.12) will give:

$$1 - 19 + R_1 C_2 = 1.414$$

$$\therefore \qquad R_1 C_2 = 19.414$$

Letting $R_1 = 1\,\Omega$ gives $C_2 = 19.414$ F. Since $R_2 C_2 = 1$, we have $R_2 = 1/19.414 = 0.052\,\Omega$. Finally $R_1 C_1 = 1$, hence, $C_1 = 1$F. We now have all the values which will enable us to build the filter, but remember these are normalized values and they have to be denormalized. The method of doing this is shown below.

We will assume a denormalization factor of $10^4$. Note that $10^3$ or $10^5$ could have been used: this is purely arbitrary. Then:

$$R_1' = 1 \times 10^4 = 10\,\text{k}\Omega \tag{17.13}$$

Similarly,

$$R_2' = 1 \times 10^4/19.414 = 515\,\Omega \tag{17.14}$$

The capacitors are treated in a different way, but all you need to know is that the normalized values are divided by the cut-off frequency and the denormalization factor $10^4$ as before:

$$C_1' = \frac{1}{10^4 \times 2\pi \times 5 \times 10^3} = 3.18\,\text{nF} \tag{17.15}$$

$$C_2' = \frac{19.44}{10^4 \times 2\pi \times 5 \times 10^3} = 65\,\text{nF} \tag{17.16}$$

The filter can now be built using the Sallen-Key circuit in Figure 17.16.

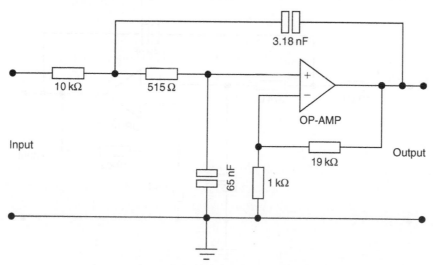

**Figure 17.16: Sallen-Key circuit for Example 17.3**

*Example 17.4*

Design the same filter as in Example 17.3, but with a Chebyshev response given by the following normalized transfer function.

$$\frac{V_o}{V_i} = \frac{H}{1.4125(jw)^2 + 0.9109(jw) + 1}$$

*Solution*

Once again using the procedure adopted in the previous example and equating the coefficients,

$$R_1 = 1\,\Omega, R_2 = \frac{1}{26.74} = 0.0374\,\Omega$$

$$C_1 = 1.4125\,\text{F}, C_2 = 26.74\,\text{F}$$

Denormalizing these values as before gives:

$$C_1' = \frac{1.4125}{10^4 \times 2\pi \times 5 \times 10^3} = 4.49\,\text{nF}$$

$$C_2' = \frac{26.74}{10^4 \times 2\pi \times 5 \times 10^3} = 85.2\,\text{nF}$$

Also, $R_1' = 10\,\text{k}\Omega$ and $R_2' = 374\,\Omega$. The circuit is shown in Figure 17.17.

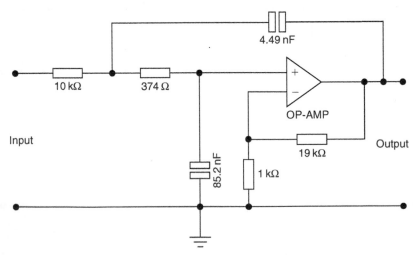

**Figure 17.17: Circuit for Example 17.4**

## 17.8 Using Normalized Tables

If normalized tables are available these can be easily used without much calculation. A set of these tables is shown in Table 17.2. As can be seen, if the pass-band gain (*K*) is known it is simply a matter of selecting the appropriate values. Note that several combinations may be possible, as was the case with the previous method.

Remember these are normalized values, and they have to be denormalized as before. This method is a lot easier than the analytical method discussed previously and where tables are available for a certain pass-band gain this method is by far the easiest to apply.

### Example 17.5
It is required to design a low-pass Butterworth filter with a pass-band gain of 2 and 3 dB cut-off frequency of 17.2 kHz.

### Solution
Consulting the table gives a choice of components in this case, but we will select the following (the choice is purely arbitrary):

$$R_1 = 1.000\,\Omega \qquad R_2 = 1.000\,\Omega$$

$$C_1 = 0.874\,\text{F} \qquad C_2 = 1.144\,\text{F}$$

These are normalized as usual:

$$C_1' = \frac{0.874}{2\pi \times 3.2 \times 10^3 \times 10^4} = 4.35\,\text{nF}$$

$$C_2' = \frac{1.144}{2\pi \times 3.2 \times 10^3 \times 10^4} = 5.7\,\text{nF}$$

Also, $R_1' = R_2' = 10\,\text{k}\Omega$. The gain setting resistors are chosen in the usual way.

## 17.9 Using Identical Components

It is simpler sometimes to use equal components, but it is necessary to adhere to the particular pass-band gain on the normalized tables. In many applications this method should be considered first.

Select a cut-off frequency value and then choose a common value for $C = C_1 = C_2$—some value less than 1 mF, say. Since $R = R_1 = R_2$, $R$ can now be calculated as follows:

$$R = \frac{1}{2\pi f C} \tag{17.17}$$

Note also that the pass-band gain has to be 1.585, this being obtained from the normalized tables.

**Example 17.6**

It is required to design a second-order low-pass filter with a cut-off frequency of 3 kHz.

**Solution**

Let $C_1 = C_2 = 0.047\,\mu\text{F}$. Hence,

$$R_1 = R_2 = \frac{10^6}{2\pi \times 3 \times 10^3 \times 0.047} = 1128.76\,\Omega$$

Selecting the gain setting resistors is once again achieved by using the fact that these have to satisfy the equation:

$$A = 1 + \frac{R_a}{R_b}$$

Hence, $R_a = 0.586\,R_b$ and several combinations are possible.

## 17.10 Second-Order High-Pass Filters

High-pass filters may be designed in a similar manner to low-pass second-order filters, but in this case the normalized response is slightly different. The response for such a filter may be given as:

$$\frac{Hs^2}{a_2 + a_1 s + s^2} \tag{17.18}$$

As before two cases are deduced: the Chebyshev response, where $a_1^2 < 2a_2$ an the Butterworth response, where $a_1^2 < 2a_2$. These responses are shown in Figure 17.18.

As before, a Sallen-Key circuit can be drawn, and this is almost identical to the low-pass circuit except that the components are interchanged. Such a circuit is shown in Figure 17.19.

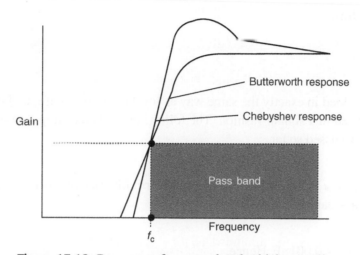

**Figure 17.18: Responses for second-order high-pass filters**

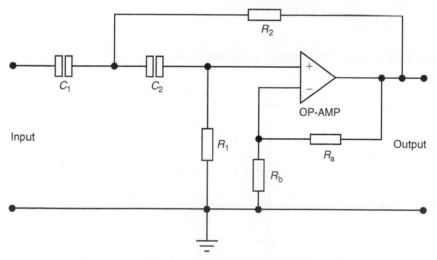

**Figure 17.19: Sallen-Key circuit for high-pass filter**

The transfer function is the same as for the low-pass filter, but it should be remembered that the components have been interchanged and because of this it will now take the form:

$$\frac{V_o}{V_i} = \frac{Ks^2}{s^2 + s\{1/R_2C_1 + 1/R_2C_2 + (1-K)/R_1C_1\} + 1/R_1C_1R_2C_2} \tag{17.19}$$

which is in the form:

$$\frac{Ks^2}{a_2 + sa_1 + s^2}$$

Problems are tackled in exactly the same way as for the low-pass case, and normalized tables may be used in a similar fashion. The following worked examples will now clarify the principles discussed so far.

**Example 17.7**
Draw the circuit of a first-order low-pass Butterworth filter having a cut-off frequency of 10 kHz and a pass-band gain of unity.

**Solution**
Choose a value $C = 0.001\,\mu F$. Hence,

$$R = \frac{10^6}{2\pi \times 10^4 \times 0.001} = 15.9\,k\Omega$$

The circuit for this solution is shown in Figure 17.20.

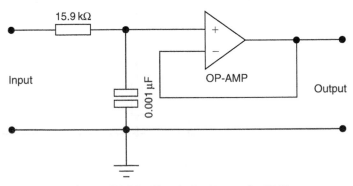

**Figure 17.20: Circuit for Example 17.7**

## Example 17.8
Figure 17.21 represents a first-order filter. Draw the response for this filter showing scaling and relevant points.

### Solution
Gain is given by:

$$\frac{V_o}{V_i} = 1 + \frac{R_2}{R_1} = 1 + \frac{10}{10} = 2$$

Since $R = 15.6\,\text{k}\Omega$ and $C = 0.01\,\mu\text{F}$,

$$f = \frac{1}{2\pi RC} = \frac{1}{2\pi \times 15.6 \times 10^3 \times 0.01 \times 10^{-6}} = 1.020\,\text{kHz}$$

The response for this problem is shown in Figure 17.22.

## Example 17.9
Design a −40 dB/decade low pass filter at a cut-off frequency of 10 krad/s, assuming equal value components.

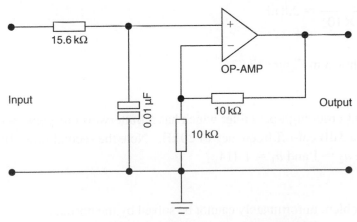

**Figure 17.21: First-order filter for Example 17.8**

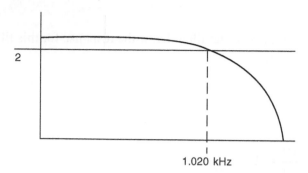

1.020 kHz

**Figure 17.22: Response for Example 17.8**

### Solution

As equal value components are used, from the normalized tables the gain must be 1.585. Hence, as the angular frequency is 10 krad/s,

$$C = \frac{1}{2\pi f R}$$

and selecting a value for R at random, say 36 kΩ, then we simply apply this to the formula as follows:

$$C = \frac{1}{10^4 \times 36 \times 10^3} = 2.8\,\text{nF}$$

The circuit is shown in Figure 17.23.

### Example 17.10

Design a second-order high-pass filter which has a Butterworth response with a pass-band gain of 25 and a 3 dB cut-off frequency of 20 kHz. Note the second-order Butterworth coefficients are $a_2 = 1$ and $a_1 = 1.414$.

### Solution

This type of problem unfortunately cannot be solved by the normalized tables; hence, the analytical method will be used.

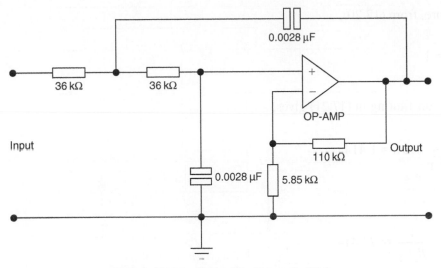

**Figure 17.23: Circuit for Example 17.9**

The second-order Butterworth response is given by:

$$\frac{V_o}{V_i} = \frac{Ks^2}{1 + 1.414s + s^2}$$

$$= \left[ \frac{Ks^2}{1/R_1R_2C_1C_2 + \{(1/R_2C_2) + (1/R_2C_1) - (24/R_1C_1)\}s + s^2} \right]$$

Equating as usual gives:

$$\frac{1}{R_1R_2C_1C_2} = 1 \tag{17.20}$$

$$\frac{1}{R_2C_2} + \frac{1}{R_2C_1} - \frac{24}{R_1C_1} = 1.414 \tag{17.21}$$

Let,

$$\frac{1}{R_2C_2} = 1$$

Therefore, from (17.20),

$$\frac{1}{R_1 C_1} = 1$$

Hence, substituting in (17.21) gives,

$$1 + \frac{1}{R_2 C_1} - 24 = 1.414$$

i.e.,

$$\frac{1}{R_2 C_1} = 24.414$$

Letting $C_1 = 1$ F gives $R_2 = 1/24.414 = 0.0410\ \Omega$; thus $C_2 = 24.414$ F. Also $C_1 = 1/R_1$, therefore $R_1 = 1\Omega$. Assuming a denormalizing factor of 104, we have:

$$C_1' = \frac{1}{2\pi \times 2 \times 10^4 \times 10^4} = 0.79\,\text{nF}$$

$$C_2' = \frac{24.414}{2\pi \times 2 \times 10^4 \times 10^4}$$

$$R_2' = \frac{10^4}{24.414} = 410\,\Omega$$

$$R_1' = 10\,\text{k}\Omega$$

Also, since     $25 = 1 + \dfrac{R_a}{R_b}$

we have $R_a = 1\,\text{k}\Omega$ and $R_b = 24\,\text{k}\Omega$. The circuit is shown in Figure 17.24.

### Example 17.11

Show how a third-order low-pass filter may be designed using a first- and second-order combination in order to achieve a pass-band gain of 2 and a cut-off frequency of 5 kHz.

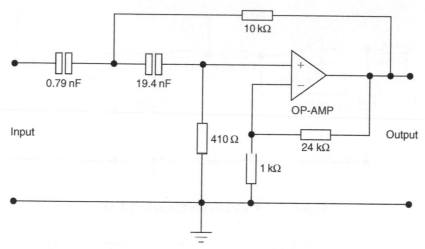

**Figure 17.24: Circuit for Example 17.10**

*Solution*

For the first-order stage we have:

$$R = \frac{1}{2\pi f C}$$

Choosing a value for $C = 0.01\,\mu F$,

$$R = \frac{10^6}{2\pi \times 5 \times 10^3 \times 0.01} = 3.18\,k\Omega\;(\text{use a }5\,k\Omega\text{ pot})$$

For the second-order stage the normalized tables are used for a pass-band gain of 2. Select $R_1 = R_2 = 1$, $C_1 = 0.874$ and $C_2 = 1.414$. Using a denormalizing factor of $10^4$ gives the following values:

$$R_1' = R_2' = 10\,k\Omega$$

$$R_1' = \frac{0.874}{6.28 \times 10^4 \times 5 \times 10^3} = 2.78\,nF$$

$$C_2' = \frac{1.414}{6.28 \times 10^4 \times 5 \times 10^3} = 4.50\,nF$$

$R_a/R_b = 1$, hence, let $R_a = R_b = 10\,k\Omega$. The circuit is shown in Figure 17.25.

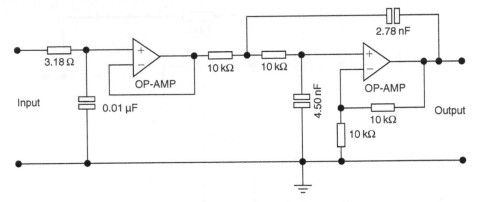

**Figure 17.25: Circuit for Example 17.11**

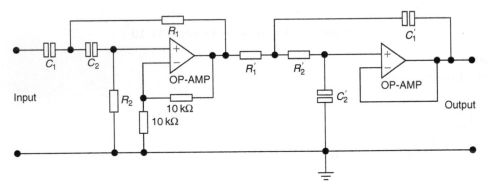

**Figure 17.26: Bandpass filter**

## 17.11  Bandpass Filters

Previously we have looked at single low- or high-pass filters, but a common application of filters is where a band of frequencies has to be passed while all other frequencies are stopped. This is called a *bandpass filter*. Such a filter may be formed from a low- and a high-pass filter in cascade. Generally the low pass is followed by the high pass, but the order of cascade is not important as the same result will be produced.

Consider Figure 17.26. The following points should be noted from this diagram.

1. A second-order low-pass filter is cascaded with a second-order high-pass filter. Note that the labelling of the components should correspond with the normalized tables.

2. The gain of the low-pass filter is unity, while that of the high pass filter is 2. This gives an overall gain of 2.

3. The overall response will give two cut-off frequencies.

4. No buffering is required as op-amps are used.

### Example 17.12

Design a first-order bandpass filter which has a pass-band gain of 4, a lower cut-off frequency $f_l = 200\,\text{Hz}$ and an upper cut-off frequency of $f_h = 1\,\text{kHz}$. Draw the frequency response of this filter.

### Solution

As the gain has to be 4 overall then each filter should have a gain of 2. Hence, if the filter uses op-amps in the noninverting mode, then $R_a$ and $R_b$ are calculated by using:

$$K = 1 + \frac{R_a}{R_b}$$

Let $R1 = 10\,\text{k}\Omega$. So both filter sections will have gain setting resistors of $10\,\text{k}\Omega$. The values for both sections of the filter are calculated as follows. For the high-pass section,

$$f_l = \frac{1}{2\pi RC}$$

Let $C = 0.05\,\mu\text{F}$. Then,

$$R = \frac{1}{200 \times 6.28 \times 0.05 \times 10^{-6}} = 15.9\,\text{k}\Omega$$

For the low-pass section,

$$f_h = \frac{1}{2\pi RC}$$

$$R = \frac{1}{6.28 \times 10^3 \times 0.01 \times 10^{-6}}$$

$$= 15.9\,\text{k}\Omega$$

The response for this filter is shown in Figure 17.27.

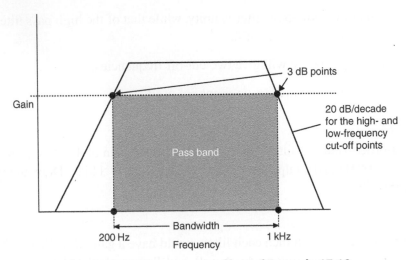

**Figure 17.27: Response for filter of Example 17.12**

### Example 17.13

Design a filter which when cascaded with the high-pass filter in Figure 17.28 will give an overall bandwidth of 35 krad/s and an overall maximum gain of 17.17 at the center frequency. The response should be flat and the roll-off 40 dB/decade.

### Solution

For the high-pass filter in Figure 17.28, normalized values can be calculated by noting that $R_1/R_2 = 1$ and $C_1/C_2 = 2$. Hence, $R_1 = R_2 = 1\Omega$, $C_1 = 1.414$F and $C_2 = 0.707$ F. So from the tables the pass-band gain is 2 for these normalized values. Also,

$$\omega = \frac{1.414}{10^4 \times 28.3 \times 10^{-9}} = 5\,\text{krad/s}$$

Hence a low-pass filter is required with a cut-off frequency of:

$$5 \times 10^3 + 35 \times 10^3 = 40\,\text{krad/s}$$

(this is the upper cut-off frequency). Since the maximum gain at the center frequency has to be $3.17 = 10\,\text{dB}$ then the gain of the second filter is:

$$3.17/2 = 1.585$$

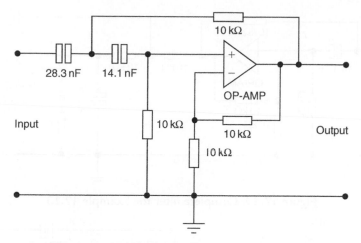

**Figure 17.28: High-pass filter (Example 17.13)**

So the gain of the second filter has to be 1.585, and from the normalized tables for a low-pass Butterworth we have:

$$R_1 = R_2 = 1\Omega, \qquad C_1 = C_2 = 1\,F$$

If a denormalization factor of $10^4$ is used and $\omega = 40$ krad/s, then:

$$R_1' = R_2' = 10\,\text{k}\Omega \qquad C_1' = C_2' = \frac{1}{10^4 \times 4 \times 10^4} = 2.5\,\text{nF}$$

Finally,

$$1 + R_a/R_b = 1.585$$

$$\therefore \qquad R_a/R_b = 0.585$$

Select $R_a = 10\,\text{k}\Omega$ and $R_b = 17\,\text{k}\Omega$. The complete filter is shown in Figure 17.29.

### Example 17.14
It is required to build a third-order low-pass filter with a cut-off frequency of 1 kHz and a pass-band gain of 2. Design such a filter.

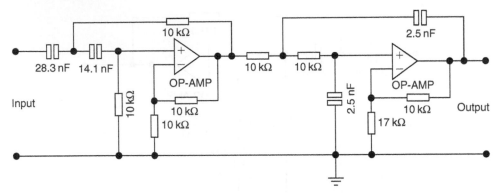

**Figure 17.29: Complete filter for Example 17.13**

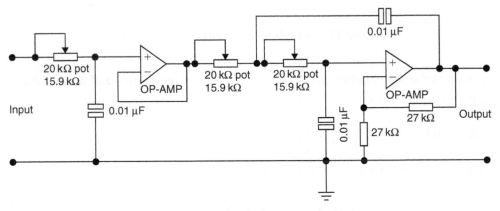

**Figure 17.30: Circuit for Example 17.14**

### Solution

A first-order and second-order filter can be connected in series to satisfy this circuit. In order to guarantee a Butterworth response the gain values of both circuits must be adhered to so for the first order a pass-band gain of 1 will be set, while the second order will have a pass-band gain of 2. The usual calculations are carried out using the normalized tables and the Butterworth low-pass normalized values. The full circuit is given in Figure 17.30.

## 17.12 Switched Capacitor Filter

Switched capacitor filters have become popular mainly because they require no external components such as capacitors or inductors. Besides offering a very sharp cut-off

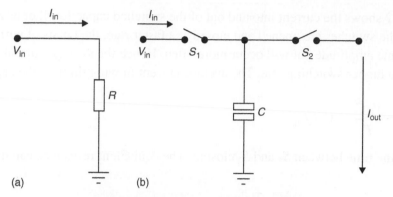

**Figure 17.31: Switched capacitor filter**

frequency, these filters have the following advantages: low cost; high accuracy; good temperature stability; and few external components are required. The main disadvantage is that they generate more noise than standard active filters.

The operation of any RC filter depends on the value of the selected resistors and capacitors. Briefly, the switched capacitor filter simulates the resistance by using a capacitor and a few switches.

In Figure 17.31(a) the value of the simulated resistor is proportional to the rate at which the switches are opened and closed in Figure 17.31(b). If a voltage $V_{in}$ is applied to the resistor then the current through it is given by:

$$I = \frac{V_{in}}{R} \tag{17.22}$$

Figure 17.31(b) consists of a capacitor and two switches, which, in practice, would be MOS transistors etched on the integrated circuit. When $S1$ is open $V_{in}$ is applied to the capacitor $C$ and hence the total charge on the capacitor is:

$$Q = V_{in} C \tag{17.23}$$

When $S_1$ is open and $S_2$ closed, the charge $Q$ flows to ground. Furthermore, if the switches have no resistance, i.e., they are ideal switches, $C$ will charge and discharge instantly.

Figure 17.32 shows the current into and out of the switched capacitor filter as a function of time. If the switches are opened and closed at a faster rate, the bursts of current will have the same amplitude but will occur more often. Hence the average current will be greater for a higher switching rate. The average current flowing through the capacitor is:

$$I_{ave} = \frac{Q}{T} = \frac{V_{in}C}{T} = V_{in}Cf_{clk} \tag{17.24}$$

where $T$ is the time between $S_1$ and $S_2$ closing. The equivalent resistance can now be given by:

$$R = \frac{V_{in}}{I_{ave}} = \frac{V_{in}}{V_{in}Cf_{clk}} = \frac{1}{Cf_{clk}} \tag{17.25}$$

This expression indicates that R is dependent on the clock frequency as C is constant. It should be noted that $V_{in}$ must change at a rate much slower than $f_{clk}$ especially when $V_{in}$ is an AC signal.

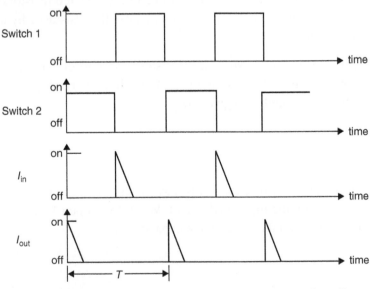

**Figure 17.32: Current into and out of switched capacitor filter**

## 17.13 Monolithic Switched Capacitor Filter

There are many types of switched capacitor chip on the market, and one of the most common is the MF100 universal switched capacitor filter manufactured by National Semiconductor. It can be used as a bandpass, low-pass, high-pass or notch filter simply by connecting the appropriate resistors externally. The values of these resistors determine the shape of the amplitude and phase responses, while the center frequency is set by the external clock. The following points should be noted about the MF100:

1. It is a second-order filter.

2. The maximum recommended clock frequency is 1 MHz.

3. Eight different connecting modes are shown in the data sheets, but for most applications mode 3 is used. This will give low-pass, high-pass and bandpass responses.

4. Mode 3 also allows independent adjustment of gain, $Q$ factor and the clock-to-center frequency ratio. This last feature is particularly advantageous if the only available clock has a frequency other than 50 or 100 times the desired center frequency or if an application requires two or more filters, each with different center or cut-off frequencies.

5. The MF10 chip is a dual version of the MF100.

6. The MF100 can operate with a single or split power supply, but the total supply must be between 8 and 14 V.

7. The $f_{clk}/f_o$ ratio affects the performance of the filter. A ratio of 100 : 1 reduces aliasing and is recommended for wide-band input signals. For noise-sensitive applications a ratio of 50 : 1 is better.

### Example 17.15
It is required to design a second-order Butterworth low-pass filter with a cut-off frequency of 500 Hz and a pass-band gain of $-2$.

*Solution*

Mode 1 is selected as it inverts the signal polarity and also configures for low-pass, bandpass and notch filters. For Mode 1 the following relationships hold:

$$H_{OLP} = -\frac{R_2}{R_1}$$

$$R_2 = -R_1 H_{OLP}$$

Let $R_1 = 10\,k\Omega$. Hence, $R_2 = 20\,k\Omega$. Also,

$$Q = \frac{f_o}{BW} = \frac{R_3}{R_2}$$

Since $Q = 0.707$ for a second-order Butterworth low-pass filter,

$$R_3 = QR_2 = 0.707 \times 20 \times 10^3 = 14.14\ k\Omega$$

(Use $15\,k\Omega$.) Since the cut-off frequency is $500\,Hz$ and $f_{clk}/f_o = 50:1$, the external clock frequency is:

$$50 \times 500 = 25\ kHz$$

L.sh (pin 7) should be connected to ground (pin 11) since the clock is CMOS. Finally, pin 5 should be connected to pin 6. The complete circuit is shown in Figure 17.33.

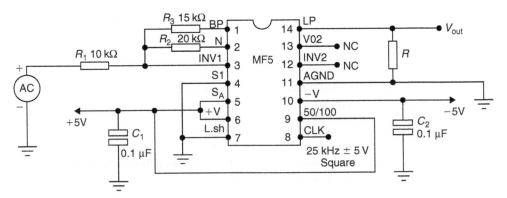

**Figure 17.33: Circuit for Example 17.15**

## 17.14 The Notch Filter

A notch filter is sometimes referred to as a frequency rejection circuit as it functions as a bandstop filter passing all frequencies on either side with a flat response, while filtering out a narrow band of frequencies between these two states. Such filters are commonly used for guard bands in multi-channel systems and to remove mains interference from audio circuits. A typical response for a notch filter is shown in Figure 17.34.

There are two common methods of producing such a filter: using a twin-T network and using a state variable filter. Both methods may be incorporated in an integrated circuit, but a discrete method will be discussed here for the sake of understanding the principles involved.

### 17.14.1  Twin-T Network

Figure 17.35 shows a passive twin-T network. Note the values of the components and their configuration. Frequent problems arise with this circuit because of lack of precision when choosing the components. Also the bandwidth of the notch can be wide. In other words, the *Q* factor is low. This can easily be improved by using an active circuit such as the one shown in Figure 17.36.

The center frequency of the twin-T network may be calculated by using the characteristic expression:

$$f = \frac{1}{2\pi RC} \qquad\qquad (17.26)$$

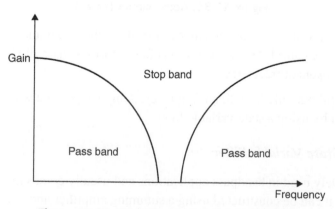

**Figure 17.34: Typical response for notch filter**

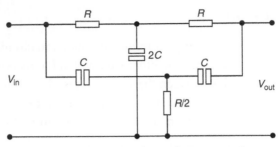

**Figure 17.35: Passive twin-T network**

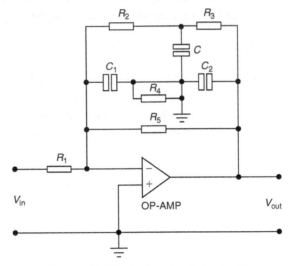

**Figure 17.36: Active circuit (twin-T)**

This is the frequency at which the signals passing along the two branches appear to be in antiphase and hence cancel. This cancellation effect causes a sharp dip in the response at and close to the resonant frequency.

This filter is useful but only for a fixed frequency. A higher $Q$ value with frequency tuning may be achieved by using a state variable filter.

### 17.14.2 The State Variable Filter

This filter is widely used in bandpass applications and usually comes in integrated circuit form. However, it can be constructed using a summing amplifier and two integrators as shown in Figure 17.37.

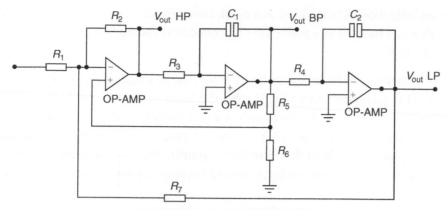

**Figure 17.37: State variable filter**

Note that this filter can be configured as a low-pass and a high-pass circuit as well as a bandpass filter. The center frequency is set by the values of $R$ and $C$ in both the integrators, and when used as a bandpass filter the critical frequencies ($f_c$) of the integrators are usually equal.

At frequencies below the critical frequency the input signal passes through the summing amplifier and integrators and, as can be seen from Figure 17.34, is fed back to the summer amplifiers in antiphase. Hence the feedback and input signals cancel for all frequencies below the critical frequency. This is ideal due to capacitor and resistor tolerances, but the cut-off is sharp in practice. As the low-pass response of the integrators rolls off, the feedback voltage reduces and the input passes through the bandpass output. For signals above the critical frequency the low-pass response disappears and prevents the input signal from passing through the integrators. This results in the bandpass output peaking sharply at the critical frequency.

The $Q$ factor or selectivity of the filter is determined by $R_5$ and $R_6$ in Figure 17.37 and may be calculated from the expression:

$$Q = \frac{1}{3}\left( \frac{R_5}{R_6} + 1 \right)$$  (17.27)

The filter is normally set for a high $Q$ factor, but the high-pass and low-pass filters cannot be simultaneously set for optimum conditions. This is not important, however, when the state variable filter is being used as a notch filter. Figure 17.38 shows how the state variable filter can be used as a notch filter by connecting the high and low-pass outputs to a summer amplifier.

This type of filter can be tuned manually by switching in capacitors or including variable capacitors in the integrator circuits. $RV1$ may also be included to alter the gain of the filter output, while $RV_2$ and $RV_3$ are usually ganged variable resistors used to vary the frequency as they are varied from $1\,k\Omega$ to $10\,k\Omega$. A practical filter using these techniques is shown in Figure 17.39. Note that in order to optimize the low and high-pass outputs a damping circuit would normally be connected between the bandpass output and the

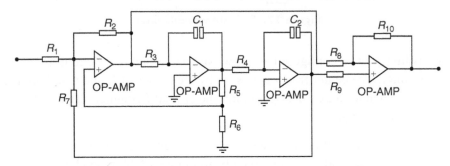

**Figure 17.38: Using state variable filter as notch filter**

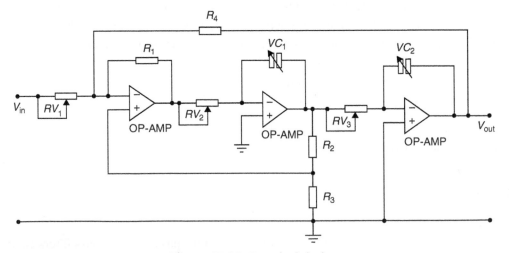

**Figure 17.39: Practical design**

input of the filter. However, as this configuration is being used as a notch filter this is not necessary. It should be appreciated that this filter is manually tuned, but where electronic tuning is required the switched capacitor filter already mentioned is used.

### Example 17.16

A notch filter has to be designed in such a way as to eliminate a 50 Hz hum on a data communications line. In order to achieve this a $Q$ factor of 40 is required. Design a suitable circuit which would practically achieve this

### Solution

The best design for this type of application would be a state variable filter using the summer amplifier. Select a capacitor value of $0.2\,\mu F$ and determine the integrator resistor values.

$$R = \frac{1}{2\pi fC} = \frac{10^6}{2\pi \times 50 \times 0.2} = 15.9 \text{ k}\Omega$$

Also,

$$Q = \frac{1}{3}\left(\frac{R_5}{R_6} + 1\right)$$

$$R_5 = (3Q - 1)\, R_6$$

Select $R_6 = 1\,\text{k}\Omega$. Then:

$$R_5 = \{(3 \times 40) - 1\}1$$

$$= 119 \text{ } k\Omega$$

The complete circuit may now be drawn with a unity gain summer amplifier using $1\,\text{k}\Omega$ resistors. This is shown in Figure 17.40.

## 17.15 Choosing Components for Filters

The selection of components in the construction of filters is more precise than in many electronic circuits as sharp cut-offs and selection bands have to be accommodated. Capacitor selection is perhaps more important as these encompass a large range of materials and tolerances.

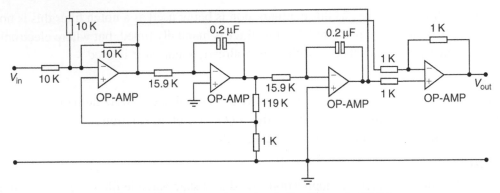

**Figure 17.40: Complete circuit for Example 17.16**

### 17.15.1  Resistor Selection

Generally fixed resistors should have tolerances of $\pm1\%$ or $\pm2\%$, but $\pm5\%$ is adequate for less critical circuit design. Tolerances less than this may be required for notch filters. Carbon track resistors may be suitable if they are properly calibrated on a bridge such as a Wayne-Kerr bridge. An alternative to this would be Cermet track variable resistors, which would give better reliability. However, for greater accuracy a bridge should always be used if available.

### 17.15.2  Capacitor Selection

#### 17.15.2.1  Silvered Mica

These capacitors have the highest tolerance ($\pm1\%$) but the maximum value commonly available may only be 4.7 nF. They have good temperature stability, and this is important if the filter has to operate over a wide range of temperatures.

#### 17.15.2.2  Polystyrene

These capacitors are most suitable for filters because of their close tolerance and large capacitance range. They also have excellent temperature stability.

#### 17.15.2.3  Ceramic

These come in three types, namely metallized, resin-dipped and disc. The metallized type has good tolerance ($\pm2\%$) and temperature stability. The resin-dipped type has tolerances of $\pm5\%$. Disc types have very poor tolerance, making them unsuitable for filter design.

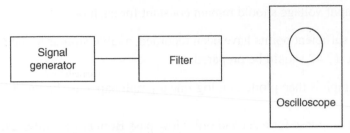

**Figure 17.41: Test set-up**

### 17.15.2.4 Polyester

When a capacitor of larger value is required this may be the choice. Their tolerance is between ±5% and ±10% and their temperature stability is poor.

### 17.15.2.5 Electrolytic

These capacitors have a tolerance of ±20% or more and their capacitance is likely to change more quickly with use. This, together with the fact that they are polarized, makes them unsuitable for filter circuits.

### 17.15.2.6 Tantalum

These capacitors are also unsuitable for filter design for the same reasons as electrolytic capacitors.

## 17.16 Testing Filter Response

There are two basic methods of measuring filter response: the signal generator and oscilloscope method, and the sweep frequency method.

### 17.16.1 Signal Generator and Oscilloscope Method

This method is the one frequently adopted due to the availability of equipment. The test set-up is shown in Figure 17.41. The procedure is as follows:

1. The amplitude of the signal generator is set to a suitable voltage level with no distortion showing on the oscilloscope.

2. The frequency of the signal generator is increased in predetermined steps. Sufficient steps should be selected to give an accurate response when plotted.

3. The input voltage should remain constant for each output.

4. After sufficient points have been recorded, a table similar to the one shown in Figure 17.42 should be prepared.

5. The graph is then plotted on log-linear graph paper as shown in Figure 17.42.

This is a typical response for a second order low-pass Butterworth filter. Note the cut-off frequency, 3 dB point and roll-off which are indicated. Also it is customary to plot decibels vertically on the linear scale while frequency is plotted on the horizontal scale.

A second example is shown in Figure 17.43. In this example, a bandpass filter has been used.

### 17.16.2  The Sweep Frequency Method

This method involves more sophisticated equipment available at the larger telecommunications companies and more sophisticated teaching laboratories. It is a more efficient method and produces very accurate results. A test set-up is shown in Figure 17.44. The sweep frequency generator uses two preset limits sometimes called *markers*; depending on the expected response of the filter, the generator is set between these limits. As the input frequency sweeps through the required range, a response curve is traced out on the spectrum analyzer as shown in Figure 17.44.

## 17.17  Fast Fourier Transforms

### 17.17.1  The Discrete Fourier Transform

In 1807, the French mathematician and physicist Jean Baptiste Joseph Fourier presented a paper to the *Institut de France* on the use of sinusoids to represent temperature distributions. The paper made the controversial claim that *any continuous periodic signal could be represented by the sum of properly chosen sinusoidal waves*. Among the publication review committee were two famous mathematicians: Joseph Louis Lagrange, and Pierre Simon de Laplace. Lagrange objected strongly to publication on the basis that Fourier's approach would not work with signals having discontinuous slopes, such as square waves. Fourier's work was rejected, primarily because of Lagrange's objection, and was not published until the death of Lagrange, some 15 years later. In the meantime, Fourier's time was occupied with political activities, expeditions to Egypt with Napoleon, and trying to avoid the guillotine after the French Revolution. (This bit of history extracted from Reference 1, p. 141.)

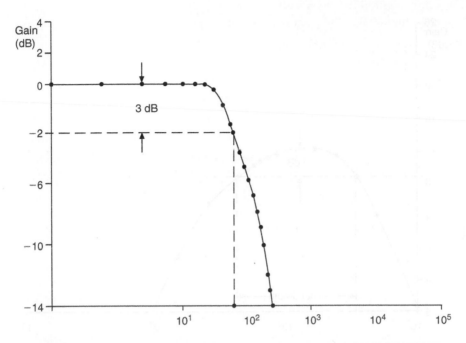

| Frequency | $V_{in}$ | $V_{out}$ | $G = V_{out}/V_{in}$ | dB = 20 log $G$ |
|---|---|---|---|---|
| 100 | 20 | 20 | 1 | 0 |
| 250 | 20 | 20 | 1 | 0 |
| 500 | 20 | 20 | 1 | 0 |
| 750 | 20 | 20 | 1 | 0 |
| 1000 | 20 | 20 | 1 | 0 |
| 1250 | 20 | 20 | 1 | 0 |
| 1500 | 20 | 20 | 1 | 0 |
| 1750 | 20 | 19 | 0.95 | −0.44 |
| 2000 | 20 | 17 | 0.85 | −1.41 |
| 2250 | 20 | 15 | 0.75 | −2.5 |
| 2500 | 20 | 14 | 0.7 | −3 |
| 2750 | 20 | 12 | 0.6 | −4.4 |
| 3000 | 20 | 11 | 0.55 | −5.2 |
| 3250 | 20 | 10 | 0.5 | −6 |
| 3500 | 20 | 9 | 0.45 | −7 |
| 3750 | 20 | 8 | 0.4 | −8 |
| 4000 | 20 | 7 | 0.35 | −9 |
| 4250 | 20 | 6 | 0.3 | −10.5 |
| 4500 | 20 | 5 | 0.25 | −12 |
| 4750 | 20 | 5 | 0.25 | −13 |
| 5000 | 20 | 4 | 0.2 | −14 |

**Figure 17.42: Table and graph showing test data**

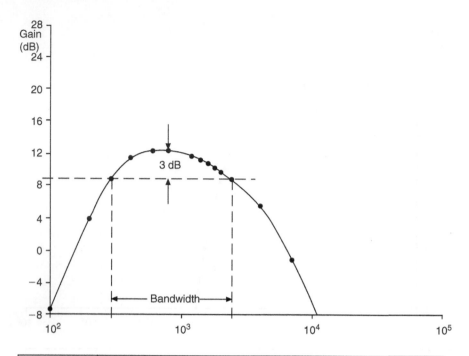

| Frequency (Hz) | $V_{in}$ | $V_{out}$ | $G = V_{out}/V_{in}$ | dB = 20 log $G$ |
|:---:|:---:|:---:|:---:|:---:|
| 100 | 3.6 | 1.5 | 0.42 | −7.6 |
| 200 | 3.6 | 5.8 | 1.61 | 4.14 |
| 400 | 3.6 | 13 | 3.61 | 11.5 |
| 600 | 3.6 | 14.4 | 4 | 12.04 |
| 800 | 3.6 | 14.4 | 4 | 12.04 |
| 1000 | 3.6 | 14.4 | 4 | 12.04 |
| 1200 | 3.6 | 14.2 | 3.94 | 11.91 |
| 1400 | 3.6 | 14 | 3.89 | 11.8 |
| 1600 | 3.6 | 13.6 | 3.78 | 11.5 |
| 1800 | 3.6 | 13.3 | 3.74 | 11.05 |
| 2000 | 3.6 | 12.8 | 3.56 | 11 |
| 2200 | 3.6 | 12.4 | 3.44 | 10.73 |
| 2400 | 3.6 | 11.8 | 3.28 | 10.32 |
| 2600 | 3.6 | 11.2 | 3.11 | 9.86 |
| 2800 | 3.6 | 10.6 | 2.94 | 9.37 |
| 3000 | 3.6 | 10 | 2.78 | 8.88 |
| 3200 | 3.6 | 9.4 | 2.61 | 8.33 |
| 3400 | 3.6 | 8.4 | 2.33 | 7.35 |
| 3600 | 3.6 | 8.2 | 2.28 | 7.16 |
| 3800 | 3.6 | 7.6 | 2.11 | 6.49 |
| 4000 | 3.6 | 7 | 1.99 | 5.76 |
| 7000 | 3.6 | 0.18 | 0.05 | −1 |

**Figure 17.43: Table and graph for second example**

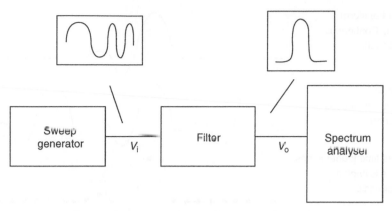

**Figure 17.44: Test set-up for sweep frequency**

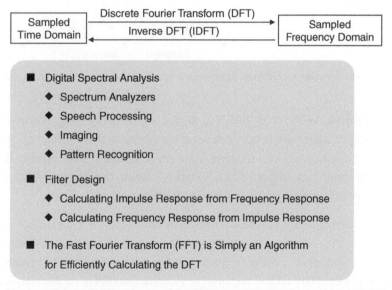

**Figure 17.45: Applications of the Discrete Fourier Transform (DFT)**

It turns out that both Fourier and Lagrange were at least partially correct. Lagrange was correct that a summation of sinusoids cannot *exactly* form a signal with a corner. However, you can get *very* close if enough sinusoids are used. (This is described by the *Gibbs effect,* and is well understood by scientists, engineers, and mathematicians today.)

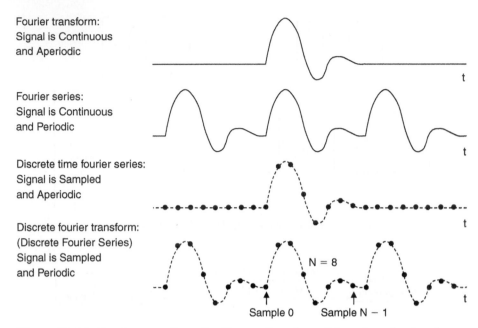

Fourier transform:
Signal is Continuous
and Aperiodic

Fourier series:
Signal is Continuous
and Periodic

Discrete time fourier series:
Signal is Sampled
and Aperiodic

Discrete fourier transform:
(Discrete Fourier Series)
Signal is Sampled
and Periodic

**Figure 17.46: Fourier transform family as a function of time domain signal type**

Fourier analysis forms the basis for much of digital signal processing. Simply stated, the Fourier transform (there are actually several members of this family) allows a time domain signal to be converted into its equivalent representation in the frequency domain. Conversely, if the frequency response of a signal is known, the inverse Fourier transform allows the corresponding time domain signal to be determined.

In addition to frequency analysis, these transforms are useful in filter design, since the frequency response of a filter can be obtained by taking the Fourier transform of its impulse response. Conversely, if the frequency response is specified, the required impulse response can be obtained by taking the inverse Fourier transform of the frequency response. Digital filters can be constructed based on their impulse response, because the coefficients of an FIR filter and its impulse response are identical.

The Fourier transform family (*Fourier Transform, Fourier Series, Discrete Time Fourier Series*, and *Discrete Fourier Transform*) is shown in Figure 17.46. These accepted definitions have evolved (not necessarily logically) over the years and depend upon whether the signal is *continuous-aperiodic, continuous-periodic, sampled-aperiodic, or*

- A periodic signal can be decomposed into the sum of properly chosen cosine and sine waves (Jean Baptiste Joseph Fourier, 1807)

- The DFT operates on a finite number (N) of digitized time samples, x(n). When these samples are repeated and placed "end-to-end," they appear periodic to the transform.

- The complex DFT output spectrum X(k) is the result of correlating the input samples with sine and cosine basis functions.

$$X(k) = \frac{1}{N} \sum_{n=0}^{N-1} x(n)\, e^{\frac{-j2\pi nk}{N}} = \frac{1}{N} \sum_{n=0}^{N-1} x(n) \left[ \cos\frac{2\pi nk}{N} - j\sin\frac{2\pi nk}{N} \right]$$

$$0 \le k \le N - 1$$

**Figure 17.47: The Discrete Fourier Transform (DFT)**

*sampled-periodic.* In this context, the term *sampled* is the same as *discrete* (i.e., a *discrete* number of time samples).

The only member of this family that is relevant to digital signal processing is the *Discrete Fourier Transform (DFT),* which operates on a *sampled* time domain signal that is *periodic.* The signal must be periodic in order to be decomposed into the summation of sinusoids. However, only a finite number of samples (N) are available for inputting into the DFT. This dilemma is overcome by placing an infinite number of groups of the same N samples "end-to-end," thereby forcing mathematical (but not real-world) periodicity as shown in Figure 5.2.

The fundamental analysis equation for obtaining the N-point DFT is as follows:

$$X(k) = \frac{1}{N} \sum_{n=0}^{N-1} x(n)e^{-j2\pi nk/N} = \frac{1}{n} \sum_{n=0}^{N-1} x(n)[\cos(2\pi nk/N) - j\sin(2\pi nk/N)]$$

At this point, some terminology clarifications are in order regarding the above equation (also see Figure 17.47). X(k) (capital letter X) represents the DFT frequency output at the kth spectral point, where k ranges from 0 to N – 1. The quantity N represents the number of sample points in the DFT data frame.

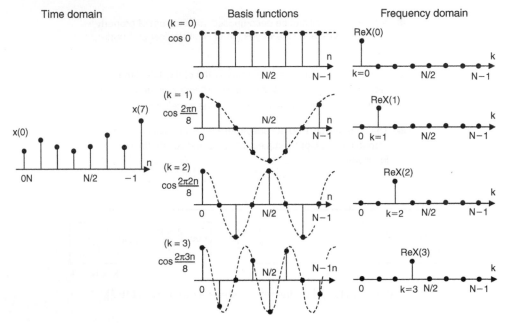

**Figure 17.48: Correlation of time samples with basis functions using the DFT for N = 8**

*Note that "N" should not be confused with ADC or DAC resolution, which is also given by the quantity N in other places in this book.*

The quantity x(n) (lower case letter x) represents the nth time sample, where n also ranges from 0 to N – 1. In the general equation, x(n) can be real or complex. Notice that the cosine and sine terms in the equation can be expressed in either polar or rectangular coordinates using Euler's equation:

$$e^{j\theta} = \cos + j\sin\theta$$

The DFT output spectrum, X(k), is the correlation between the input time samples and N cosine and N sine waves. The concept is best illustrated in Figure 17.48. In this figure, the real part of the first four output frequency points is calculated; therefore, only the cosine waves are shown. A similar procedure is used with sine waves in order to calculate the imaginary part of the output spectrum.

The first point, X(0), is simply the sum of the input time samples, because cos(0) = 1. The scaling factor, 1/N, is not shown, but must be present in the final result. Note that X(0) is the average value of the time samples, or simply the DC offset. The second point, ReX(1), is obtained by multiplying each time sample by each corresponding point on a cosine wave that makes one complete cycle in the interval N and summing the results. The third point, ReX(2), is obtained by multiplying each time sample by each corresponding point of a cosine wave that has two complete cycles in the interval N and then summing the results. Similarly, the fourth point, ReX(3), is obtained by multiplying each time sample by the corresponding point of a cosine wave that has three complete cycles in the interval N and summing the results. This process continues until all N outputs have been computed. A similar procedure is followed using sine waves in order to calculate the imaginary part of the frequency spectrum. The cosine and sine waves are referred to as *basis functions*.

A periodic signal can be decomposed into the sum of properly chosen cosine and sine waves (Jean Baptiste Joseph Fourier, 1807).

The DFT operates on a finite number (N) of digitized time samples, x(n). When these samples are repeated and placed "end-to-end," they appear periodic to the transform.

The complex DFT output spectrum X(k) is the result of correlating the input samples with sine and cosine basis functions:

Assume that the input signal is a cosine wave having a period of N, i.e., it makes one complete cycle during the data window. Also assume its amplitude and phase is identical to the first cosine wave of the basis functions, cos(2pn/8). The output ReX(1) contains a single point, and all the other ReX(k) outputs are zero. Assume that the input cosine wave is now shifted to the right by 90°. The correlation between it and the corresponding basis function is zero. However, there is an additional correlation required with the basis function sin(2pn/8) to yield ImX(1). This shows why both real and imaginary parts of the frequency spectrum need to be calculated in order to determine both the amplitude and phase of the frequency spectrum.

Notice that the correlation of a sine/cosine wave of any frequency other than that of the basis function produces a zero value for both ReX(1) and ImX(1).

A similar procedure is followed when using the *inverse* DFT (IDFT) to reconstruct the time domain samples, x(n), from the frequency domain samples X(k). The synthesis equation is given by:

$$x(n) = \sum_{k=0}^{N-1} X(k)e^{j2\pi nk/N} = \sum_{k=0}^{N-1} X(k)[\cos(2\pi nk/N) + j\sin(2\pi nk/N)]$$

There are two basic types of DFTs: *real* and *complex*. The equations shown in Figure 17.49 are for the complex DFT, where the input and output are both complex numbers. Since time domain input samples are real and have no imaginary part, the imaginary part of the input is always set to zero. The output of the DFT, X(k), contains a real and imaginary component that can be converted into amplitude and phase.

The *real* DFT, although somewhat simpler, is basically a simplification of the *complex* DFT. Most FFT routines are written using the complex DFT format, therefore understanding the complex DFT and how it relates to the real DFT is important. For instance, if you know the real DFT frequency outputs and want to use a complex inverse DFT to calculate the time samples, you need to know how to place the real DFT output points into the complex DFT format before taking the complex inverse DFT.

Frequency Domain ← ← DFT ← ← Time Domain

$$X(k) = \frac{1}{N}\sum_{n=0}^{N-1} x(n)\, e^{\frac{-j2\pi nk}{N}} = \frac{1}{N}\sum_{n=0}^{N-1} x(n)\left[\cos\frac{2\pi nk}{N} - j\sin\frac{2\pi nk}{N}\right]$$

$$W_N = e^{\frac{-j2\pi}{N}}$$

$$= \frac{1}{N}\sum_{n=0}^{N-1} x(n)\, W_N^{nk}, \quad 0 \le k \le N-1$$

Time Domain ← ← INVERSE DFT ← ← Frequency Domain

$$x(n) = \sum_{k=0}^{N-1} X(k)\, e^{\frac{j2\pi nk}{N}} = \sum_{k=0}^{N-1} X(k)\left[\cos\frac{2\pi nk}{N} + j\sin\frac{2\pi nk}{N}\right]$$

$$= \sum_{k=0}^{N-1} X(k)\, W_N^{-nk}, \quad 0 \le n \le N-1$$

**Figure 17.49: The Complex DFT**

Figure 17.50 shows the input and output of a real and a complex FFT. Notice that the output of the real DFT yields real and imaginary X(k) values, where k ranges from only 0 to N/2. Note that the imaginary points ImX(0) and ImX(N/2) are always zero because sin(0) and sin(np) are both always zero.

The frequency domain output X(N/2) corresponds to the frequency output at one-half the sampling frequency, $f_s$. The width of each frequency bin is equal to $f_s/N$.

The complex DFT has real and imaginary values both at its input and output. In practice, the imaginary parts of the time domain samples are set to zero. If you are given the output spectrum for a complex DFT, it is useful to know how to relate them to the real DFT output and vice versa. The crosshatched areas in the diagram correspond to points that are common to both the real and complex DFT.

Figure 17.51 shows the relationship between the real and complex DFT in more detail. The real DFT output points are from 0 to N/2, with ImX(0) and ImX(N/2) always zero.

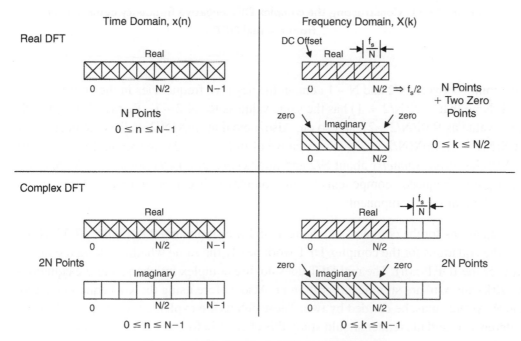

**Figure 17.50: DFT Input/Output Spectrum**

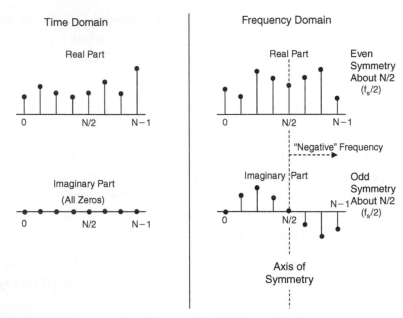

**Figure 17.51: Constructing the complex DFT negative frequency components
from the real DFT**

The points between N/2 and N − 1 contain the negative frequencies in the complex
DFT. Note that ReX(N/2 + 1) has the same value as Re(N/2 − 1), ReX(N/2 + 2) has the
same value as ReX(N/2 − 2), and so on. Also, note that ImX(N/2 + 1) is the negative of
ImX(N/2 − 1), ImX(N/2 + 2) is the negative of ImX(N/2 − 2), and so on. In other words,
ReX(k) has *even symmetry* about N/2 and ImX(k) *odd symmetry* about N/2. In this way,
the negative frequency components for the complex FFT can be generated if you are only
given the real DFT components.

The equations for the complex and the real DFT are summarized in Figure 17.52. Note
that the equations for the complex DFT work nearly the same whether taking the DFT,
X(k) or the IDFT, x(n). The real DFT does not use complex numbers, and the equations
for X(k) and x(n) are significantly different. Also, before using the x(n) equation, ReX(0)
and ReX(N/2) must be divided by two. These details are explained in Chapter 31 of
Reference 1, and the reader should study this chapter before attempting to use these
equations.

COMPLEX TRANSFORM

$$X(k) = \frac{1}{N} \sum_{n=0}^{N-1} x(n)\, e^{\frac{-j2\pi nk}{N}}$$

$$x(n) = \sum_{k=0}^{N-1} X(k)\, e^{\frac{j2\pi nk}{N}}$$

REAL TRANSFORM

$$ReX(k) = \frac{2}{N} \sum_{n=0}^{N-1} x(n)\, \cos(2\pi nk/N)$$

$$ImX(k) = \frac{-2}{N} \sum_{n=0}^{N-1} x(n)\, \sin(2\pi nk/N)$$

$$x(n) = \sum_{k=0}^{N/2} \Big[ ReX(k)\, \cos(2\pi nk/N) - ImX(k)\, \cos(2\pi nk/N) \Big]$$

Time Domain: x(n) is complex, discrete, and periodic. n runs from 0 to N − 1

Frequency Domain: X(k) is complex, discrete, and periodic. k runs from 0 to N − 1
k = 0 to N/2 are positive frequencies.
k = N/2 to N − 1 are negative frequencies

Time Domain: x(n) is real, discrete, and periodic. n runs from 0 to N − 1

Frequency domain:
ReX(k) is real, discrete, and periodic.
ImX(k) is real, discrete, and periodic.
k runs from 0 to N/2

Before using x(n) equation, ReX(0) and ReX(N/2) must be divided by two.

**Figure 17.52: Complex and real DFT equations**

The DFT output spectrum can be represented in either polar form (magnitude and phase) or rectangular form (real and imaginary) as shown in Figure 17.53. The conversion between the two forms is straightforward.

### 17.17.2 The Fast Fourier Transform

In order to understand the development of the FFT, consider first the 8-point DFT expansion shown in Figure 17.54. In order to simplify the diagram, note that the quantity $W_N$ is defined as:

$$W_N = e^{-j2\pi/N}$$

This leads to the definition of the *twiddle factors* as:

$$W_N^{nk} = e^{-j2\pi nk/N}$$

The twiddle factors are simply the sine and cosine basis functions written in polar form. Note that the 8-point DFT shown in the diagram requires 64 complex multiplications.

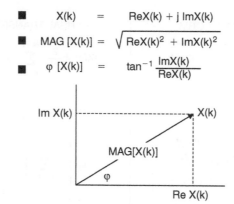

- $X(k) = ReX(k) + j\,ImX(k)$
- $MAG\,[X(k)] = \sqrt{ReX(k)^2 + ImX(k)^2}$
- $\varphi\,[X(k)] = \tan^{-1}\dfrac{ImX(k)}{ReX(k)}$

**Figure 17.53: Converting real and imaginary DFT outputs into magnitude and phase**

$$X(k) = \frac{1}{N}\sum_{n=0}^{N-1} x(n)\, e^{\frac{-j2\pi nk}{N}} \qquad = \frac{1}{N}\sum_{n=0}^{N-1} x(n)\, W_N^{nk} \qquad \boxed{W_N = e^{\frac{-j2\pi}{N}}}$$

| | |
|---|---|
| $X(0) =$ | $x(0)W_8^0 + x(1)W_8^0 + x(2)W_8^0 + x(3)W_8^0 + x(4)W_8^0 + x(5)W_8^0 + x(6)W_8^0 + x(7)W_8^0$ |
| $X(1) =$ | $x(0)W_8^0 + x(1)W_8^1 + x(2)W_8^2 + x(3)W_8^3 + x(4)W_8^4 + x(5)W_8^5 + x(6)W_8^6 + x(7)W_8^7$ |
| $X(2) =$ | $x(0)W_8^0 + x(1)W_8^2 + x(2)W_8^4 + x(3)W_8^6 + x(4)W_8^8 + x(5)W_8^{10} + x(6)W_8^{12} + x(7)W_8^{14}$ |
| $X(3) =$ | $x(0)W_8^0 + x(1)W_8^3 + x(2)W_8^6 + x(3)W_8^9 + x(4)W_8^{12} + x(5)W_8^{15} + x(6)W_8^{18} + x(7)W_8^{21}$ |
| $X(4) =$ | $x(0)W_8^0 + x(1)W_8^4 + x(2)W_8^8 + x(3)W_8^{12} + x(4)W_8^{16} + x(5)W_8^{20} + x(6)W_8^{24} + x(7)W_8^{28}$ |
| $X(5) =$ | $x(0)W_8^0 + x(1)W_8^5 + x(2)W_8^{10} + x(3)W_8^{15} + x(4)W_8^{20} + x(5)W_8^{25} + x(6)W_8^{30} + x(7)W_8^{35}$ |
| $X(6) =$ | $x(0)W_8^0 + x(1)W_8^6 + x(2)W_8^{12} + x(3)W_8^{18} + x(4)W_8^{24} + x(5)W_8^{30} + x(6)W_8^{36} + x(7)W_8^{42}$ |
| $X(7) =$ | $x(0)W_8^0 + x(1)W_8^7 + x(2)W_8^{14} + x(3)W_8^{21} + x(4)W_8^{28} + x(5)W_8^{35} + x(6)W_8^{42} + x(7)W_8^{49}$ |

NOTES: 1. $N^2$ Complex Multiplications

2. $\dfrac{1}{N}$ Scaling Factor Omitted

**Figure 17.54: The 8-Point DFT (N = 8)**

In general, an N-point DFT requires $N^2$ complex multiplications. The number of multiplications required is significant because the multiplication function requires a relatively large amount of DSP processing time. In fact, the total time required to compute the DFT is directly proportional to the number of multiplications plus the required amount of overhead.

The FFT is simply an algorithm to speed up the DFT calculation by reducing the number of multiplications and additions required. It was popularized by J. W. Cooley and J. W. Tukey in the 1960s and was actually a rediscovery of an idea of Runge (1903) and Danielson and Lanczos (1942), first occurring prior to the availability of computers and calculators, when numerical calculation could take many man-hours. In addition, the German mathematician Karl Friedrich Gauss (1777–1855) had used the method more than a century earlier.

In order to understand the basic concepts of the FFT and its derivation, note that the DFT expansion shown in Figure 17.54 can be greatly simplified by taking advantage of the symmetry and periodicity of the twiddle factors as shown in Figure 17.55. If equations are rearranged and factored, the result is the Fast Fourier Transform (FFT) which requires only $(N/2) \log_2(N)$ complex multiplications. The computational efficiency of the FFT versus the DFT becomes highly significant when the FFT point size increases to several

$$\text{Symmetry:} \quad W_N^{r+N/2} = -W_N^r, \quad \text{Periodicity:} \quad W_N^{r+N} = W_N^r$$

$$N = 8$$

$$
\begin{aligned}
W_8^4 &= W_8^{0+4} = -W_8^0 = -1 \\
W_8^5 &= W_8^{1+4} = -W_8^1 \\
W_8^6 &= W_8^{2+4} = -W_8^2 \\
W_8^7 &= W_8^{3+4} = -W_8^3 \\
W_8^8 &= W_8^{0+8} = +W_8^0 = +1 \\
W_8^9 &= W_8^{1+8} = +W_8^1 \\
W_8^{10} &= W_8^{2+8} = +W_8^2 \\
W_8^{11} &= W_8^{3+8} = +W_8^3 \\
\end{aligned}
$$

Figure 17.55: Applying the properties of symmetry and periodicity to $W_N^r$ for $N = 8$

- ■ The FFT is Simply an Algorithm for Efficiently Calculating the DFT
- ■ Computational Efficiency of an N-Point FFT:
  - ◆ DFT:     $N^2$              Complex Multiplications
  - ◆ FFT:     $(N/2) \log_2 (N)$  Complex Multiplications

| N | DFT Multiplications | FFT Multiplications | FFT Efficiency |
|---|---|---|---|
| 256 | 65,536 | 1,024 | 64 : 1 |
| 512 | 262,144 | 2,304 | 114 : 1 |
| 1,024 | 1,048,576 | 5,120 | 205 : 1 |
| 2,048 | 4,194,304 | 11,264 | 372 : 1 |
| 4,096 | 16,777,216 | 24,576 | 683 : 1 |

**Figure 17.56: The Fast Fourier Transform (FFT) vs. the Discrete Fourier Transform (DFT)**

thousand as shown in Figure 17.56. However, notice that the FFT computes *all* the output frequency components (either all or none). If only a few spectral points need to be calculated, the DFT may actually be more efficient. Calculation of a single spectral output using the DFT requires only N complex multiplications.

The radix-2 FFT algorithm breaks the entire DFT calculation down into a number of 2-point DFTs. Each 2-point DFT consists of a multiply-and-accumulate operation called a *butterfly*, as shown in Figure 17.57. Two representations of the butterfly are shown in the diagram: the top diagram is the actual functional representation of the butterfly showing the digital multipliers and adders. In the simplified bottom diagram, the multiplications are indicated by placing the multiplier over an arrow, and addition is indicated whenever two arrows converge at a dot.

The 8-point decimation-in-time (DIT) FFT algorithm computes the final output in three stages as shown in Figure 17.58. The eight input time samples are first divided (or *decimated*) into four groups of 2-point DFTs. The four 2-point DFTs are then combined into two 4-point DFTs. The two 4-point DFTs are then combined to produce the final output X(k). The detailed process is shown in Figure 17.59, where all the multiplications and additions are shown. Note that the basic two-point DFT butterfly operation forms the basis for all computation. The computation is done in three stages. After the first stage

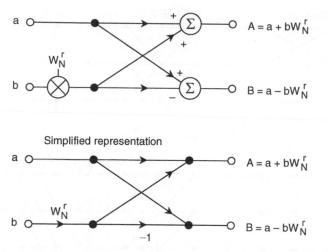

**Figure 17.57: The basic butterfly computation in the decimation-in-time FFT algorithm**

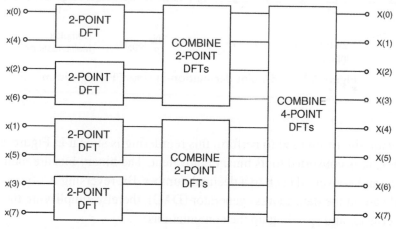

**Figure 17.58: Computation of an 8-point DFT in three stages using decimation-in-time**

computation is complete, there is no need to store any previous results. The first stage outputs can be stored in the same registers that originally held the time samples x(n). Similarly, when the second-stage computation is completed, the results of the first-stage computation can be deleted. In this way, *in-place* computation proceeds to the final stage. Note that in order for the algorithm to work properly, the order of the input time samples, x(n), must be properly reordered using a *bit reversal* algorithm.

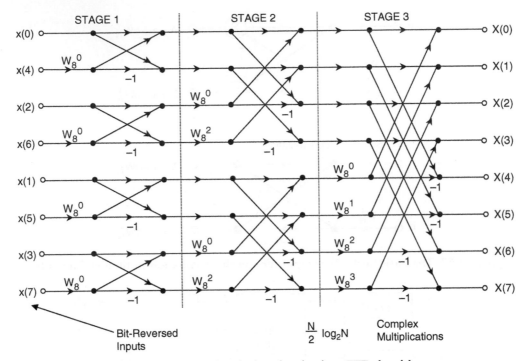

**Figure 17.59: 8-point decimation-in-time FFT algorithm**

The *bit reversal* algorithm used to perform this reordering is shown in Figure 17.60. The decimal index, n, is converted to its binary equivalent. The binary bits are then placed in reverse order, and converted back to a decimal number. Bit reversing is often performed in DSP hardware in the data address generator (DAG), thereby simplifying the software, reducing overhead, and speeding up the computations.

The computation of the FFT using *decimation-in-frequency* (DIF) is shown in Figures 17.61 and 17.62. This method requires that the bit reversal algorithm be applied to the output X(k). Note that the butterfly for the DIF algorithm differs slightly from the decimation-in-time butterfly as shown in Figure 17.63.

The use of decimation-in-time versus decimation-in-frequency algorithms is largely a matter of preference, as either yields the same result. System constraints may make one of the two a more optimal solution.

| Decimal Number : | 0 | 1 | 2 | 3 | 4 | 5 | 6 | 7 |
|---|---|---|---|---|---|---|---|---|
| Binary Equivalent : | 000 | 001 | 010 | 011 | 100 | 101 | 110 | 111 |
| Bit-Reversed Binary : | 000 | 100 | 010 | 110 | 001 | 101 | 011 | 111 |
| Decimal Equivalent : | 0 | 4 | 2 | 6 | 1 | 5 | 3 | 7 |

Figure 17.60: Bit Reversal Example for N = 8

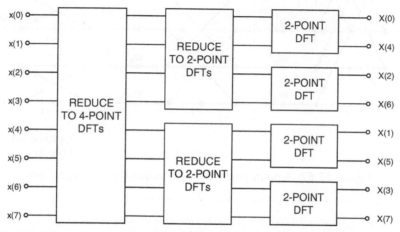

Figure 17.61: Computation of an 8-Point DFT in three stages using decimation-in-frequency

It should be noted that the algorithms required to compute the inverse FFT are nearly identical to those required to compute the FFT, assuming complex FFTs are used. In fact, a useful method for verifying a complex FFT algorithm consists of first taking the FFT of the x(n) time samples and then taking the inverse FFT of the X(k). At the end of this process, the original time samples, ReX(n), should be obtained and the imaginary part, ImX(n), should be zero (within the limits of the mathematical round-off errors).

The FFTs discussed to this point are radix-2 FFTs, i.e., the computations are based on 2-point butterflies. This implies that the number of points in the FFT must be a power of two. If the number of points in an FFT is a power of four, however, the FFT can be broken down into a number of 4-point DFTs as shown in Figure 17.64. This is called a

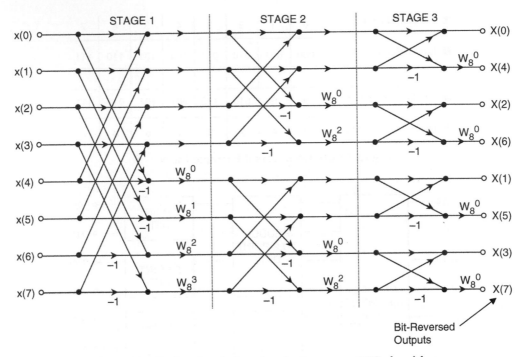

**Figure 17.62: 8-point decimation-in-frequency FFT algorithm**

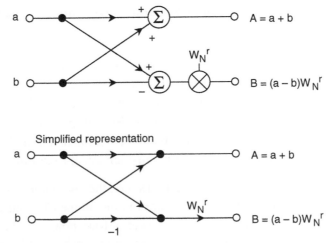

**Figure 17.63: The basic butterfly computation in the decimation-in-frequency FFT algorithm**

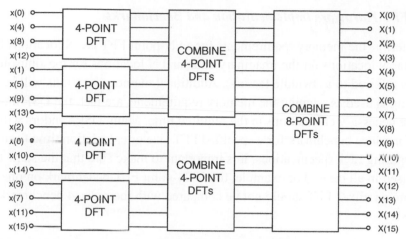

**Figure 17.64: Computation of a 16-point DFT in three stages using radix-4 decimation-in-time algorithm**

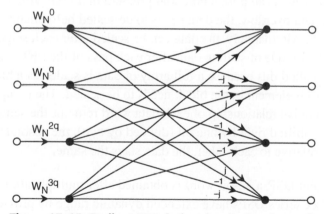

**Figure 17.65: Radix-4 FFT decimation-in-time butterfly**

*radix-4 FFT*. The fundamental decimation-in-time butterfly for the radix-4 FFT is shown in Figure 17.65.

The radix-4 FFT requires fewer complex multiplications but more additions than the radix-2 FFT for the same number of points. Compared to the radix-2 FFT, the radix-4 FFT trades more complex data addressing and twiddle factors with less computation. The resulting savings in computation time varies between different DSPs but a radix-4 FFT can be as much as twice as fast as a radix-2 FFT for DSPs with optimal architectures.

### 17.17.3  FFT Hardware Implementation and Benchmarks

In general terms, the memory requirements for an N-point FFT are N locations for the real data, N locations for the imaginary data, and N locations for the sinusoid data (sometimes referred to as twiddle factors). Additional memory locations will be required if windowing is used. Assuming the memory requirements are met, the DSP must perform the necessary calculations in the required time. Many DSP vendors will either give a performance benchmark for a specified FFT size or calculation time for a butterfly. When comparing FFT specifications, it is important to make sure that the same type of FFT is used in all cases. For example, the 1024-point FFT benchmark on one DSP derived from a radix-2 FFT should not be compared with the radix-4 benchmark from another DSP.

Another consideration regarding FFTs is whether to use a fixed-point or a floating-point processor. The results of a butterfly calculation can be larger than the inputs to the butterfly. This data growth can pose a potential problem in a DSP with a fixed number of bits. To prevent data overflow, the data needs to be scaled beforehand, leaving enough extra bits for growth. Alternatively, the data can be scaled after each stage of the FFT computation. The technique of scaling data after each pass of the FFT is known as *block floating point*. It is called this because a full array of data is scaled as a block, regardless of whether or not each element in the block needs to be scaled. The complete block is scaled so that the relative relationship of each data word remains the same. For example, if each data word is shifted right by one bit (divided by two), the absolute values have been changed but, relative to each other, the data stays the same.

In a 16-bit fixed-point DSP, a 32-bit word is obtained after multiplication. The Analog Devices ADSP-21xx series DSPs have extended dynamic range by providing a 40-bit internal register in the multiply-accumulator (MAC).

The use of a floating-point DSP eliminates the need for data scaling and therefore results in a simpler FFT routine; however, the trade-off is the increased processing time required for the complex floating-point arithmetic. In addition, a 32-bit floating-point DSP will obviously have less round-off noise than a 16-bit fixed-point DSP. Figure 17.66 summarizes the FFT benchmarks for popular Analog Devices DSPs. Notice in particular that the ADSP-TS001 TigerSHARC® DSP offers both fixed-point and floating-point modes, thereby providing an exceptional degree of programming flexibility.

- ADSP-2189M, 16-Bit, Fixed-Point
  - ◆ 453 µs (1024-Point)

- ADSP-21160 SHARC®, 32-Bit, Floating-Point
  - ◆ 90 µs (1024-Point)

- ADSP-TS001 TigerSHARC @ 150 MHz,
  - ◆ 16-Bit, Fixed-Point Mode
    - • 7.3 µs (256-Point FFT)
  - ◆ 32-Bit, Floating-Point Mode
    - • 69 µs (1024-Point)

**Figure 17.66: Radix-2 complex FFT hardware benchmark comparisons**

- Assume 69 µs Execution Time for Radix-2, 1024-Point FFT (TigerSHARC, 32-Bit Mode)

- $f_s$ (maximum) $< \dfrac{1024 \text{ Samples}}{69 \text{ µs}} = 14.8 \text{ MSPS}$

- Input Signal Bandwidth Therefore <7.4 MHz

- This Assumes No Additional FFT Overhead and No Input/Output Data Transfer Limitations

**Figure 17.67: Real-time FFT processing example**

### 17.17.4 DSP Requirements for Real-time FFT Applications

There are two basic ways to acquire data from a real-world signal, either one sample at a time (continuous processing), or one frame at a time (batch processing). Sample-based systems, like a digital filter, acquire data one sample at a time. For each sample clock, a sample comes into the system and a processed sample is sent to the output. Frame-based systems, like an FFT-based digital spectrum analyzer, acquire a frame (or block of samples). Processing occurs on the entire frame of data and results in a frame of transformed output data.

- Signal Bandwidth
- Sampling Frequency, $f_s$
- Number of Points in FFT, N
- Frequency Resolution = $f_s/N$
- Maximum Time to Calculate N-Point FFT = $N/f_s$
- Fixed-Point vs. Floating-Point DSP
- Radix-2 vs. Radix-4 Execution Time
- FFT Processing Gain = $10 \log_{10}(N/2)$
- Windowing Requirements

**Figure 17.68: Real-time FFT considerations**

In order to maintain real-time operation, the entire FFT must therefore be calculated during the frame period. This assumes that the DSP is collecting the data for the next frame while it is calculating the FFT for the current frame of data. Acquiring the data is one area where special architectural features of DSPs come into play. Seamless data acquisition is facilitated by the DSP's flexible data addressing capabilities in conjunction with its direct memory accessing (DMA) channels.

Assume the DSP is the ADSP-TS001 TigerSHARC, which can calculate a 1024-point 32-bit complex floating-point FFT in 69 μs. The maximum sampling frequency is therefore 1024/69 μs = 14.8 MSPS. This implies a signal bandwidth of less than 7.4 MHz. It is also assumed that there is no additional FFT overhead or data transfer limitation.

The above example will give an estimate of the maximum bandwidth signal that can be handled by a given DSP using its FFT benchmarks. Another way to approach the issue is to start with the signal bandwidth and develop the DSP requirements. If the signal bandwidth is known, the required sampling frequency can be estimated by multiplying by a factor between 2 and 2.5 (the increased sampling rate may be required to ease the requirements on the antialiasing filter that precedes the ADC). The next step is to determine the required number of points in the FFT to achieve the desired frequency resolution. The frequency resolution is obtained by dividing the sampling rate $f_s$ by N, the number of points in the FFT. These and other FFT considerations are shown in Figure 17.68.

The number of FFT points also determines the noise floor of the FFT with respect to the broadband noise level, and this may also be a consideration. Figure 17.69 shows

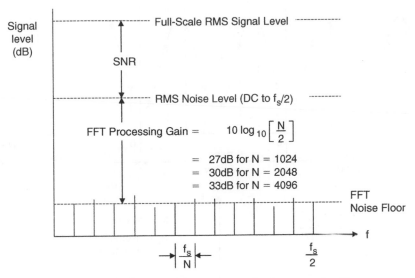

**Figure 17.69: FFT processing gain neglecting round-off error**

the relationships between the system full-scale signal level, the broadband noise level (measured over the bandwidth DC to $f_s/2$), and the FFT noise floor. Notice that the FFT processing gain is determined by the number of points in the FFT. The FFT acts like an analog spectrum analyzer with a sweep bandwidth of $f_s/N$. Increasing the number of points increases the FFT resolution and narrows its bandwidth, thereby reducing the noise floor. This analysis neglects noise caused by the FFT round-off error. In practice, the ADC that is used to digitize the signal produces quantization noise, which is the dominant noise source.

At this point it is time to examine actual DSPs and their FFT processing times to make sure real-time operation can be achieved. This means that the FFT must be calculated during the acquisition time for one frame of data, which is $N/f_s$. Fixed-point versus floating-point, radix-2 versus radix-4, and processor power dissipation and cost may be other considerations.

### 17.17.5 Spectral Leakage and Windowing

Spectral leakage in FFT processing can best be understood by considering the case of performing an N-point FFT on a pure sinusoidal input. Two conditions will be

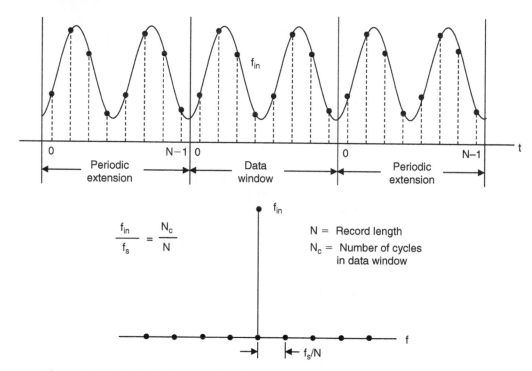

**Figure 17.70: FFT of sine wave having integral number of cycles in data window**

considered. In Figure 17.70, the ratio between the sampling frequency and the input sine wave frequency is such that precisely an integral number of cycles is contained within the data window (frame or record). Recall that the DFT assumes that an infinite number of these windows are placed end-to-end to form a periodic waveform as shown in the diagram as the periodic extensions. Under these conditions, the waveform appears continuous (no discontinuities), and the DFT or FFT output will be a single tone located at the input signal frequency.

Figure 17.71 shows the condition where there is not an integral number of sine wave cycles within the data window. The discontinuities that occur at the endpoints of the data window result in leakage in the frequency domain because of the harmonics that are generated. In addition to the sidelobes, the main lobe of the sine wave is smeared over several frequency bins. This process is equivalent to multiplying the input sine wave by

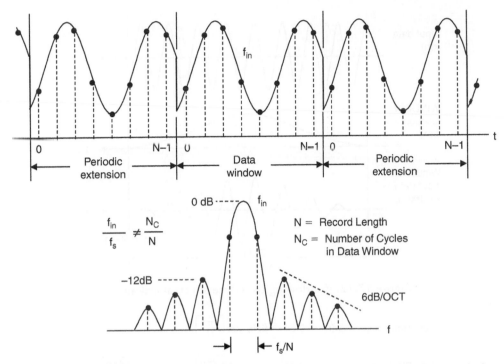

**Figure 17.71: FFT of sine wave having nonintegral number of cycles in data window**

a rectangular window pulse that has the familiar sin(x)/x frequency response and associated smearing and sidelobes.

Notice that the first sidelobe is only 12 dB below the fundamental, and that the sidelobes roll off at only 6 dB/octave. This situation would be unsuitable for most spectral analysis applications. Since in practical FFT spectral analysis applications the exact input frequencies are unknown, something must be done to minimize these sidelobes. This is done by choosing a window function other than the rectangular window. The input time samples are multiplied by an appropriate window function, which brings the signal to zero at the edges of the window as shown in Figure 17.72. The selection of a window function is primarily a trade-off between main lobe spreading and sidelobe roll-off. Reference 7 is highly recommended for an in-depth treatment of windows.

The mathematical functions that describe four popular window functions (Hamming, Blackman, Hanning, and Minimum 4-Term Blackman-Harris) are shown in Figure 17.73.

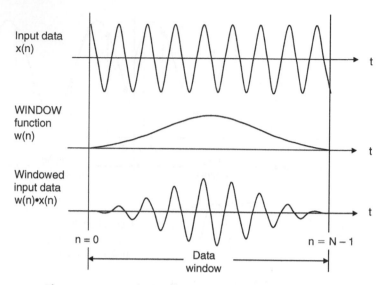

**Figure 17.72: Windowing to reduce spectral leakage**

■ Hamming:     $w(n) = 0.54 - 0.46 \cos\left[\dfrac{2\pi n}{N}\right]$

■ Blackman:     $w(n) = 0.42 - 0.5 \cos\left[\dfrac{2\pi n}{N}\right] + 0.08 \cos\left[\dfrac{4\pi n}{N}\right]$

■ Hanning:     $w(n) = 0.5 - 0.5 \cos\left[\dfrac{2\pi n}{N}\right]$

■ Minimum 4-Term Blackman Harris     $w(n) = 0.35875 - 0.48829 \cos\left[\dfrac{2\pi n}{N}\right]$

$+ 0.14128 \cos\left[\dfrac{4\pi n}{N}\right]$

$\boxed{0 \le n \le N-1}$     $- 0.01168 \cos\left[\dfrac{6\pi n}{N}\right]$

**Figure 17.73: Some popular window functions**

The computations are straightforward, and the window function data points are usually precalculated and stored in the DSP memory to minimize their impact on FFT processing time. The frequency response of the rectangular, Hamming, and Blackman windows are shown in Figure 17.74. Figure 17.75 shows the trade-off between main lobe spreading and sidelobe amplitude and roll-off for the popular window functions.

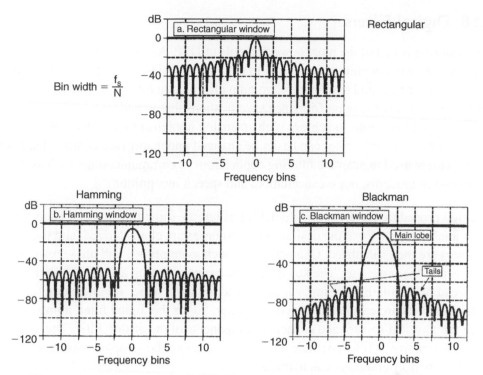

Bin width $= \dfrac{f_s}{N}$

**Figure 17.74: Frequency response of rectangular, Hamming, and Blackman windows for N = 256**

| Window function | 3 dB BW (Bins) | 6 dB BW (Bins) | Highest sidelobe (dB) | SIDELOBE ROLL-OFF (dB/Octave) |
|---|---|---|---|---|
| Rectangle | 0.89 | 1.21 | −12 | 6 |
| Hamming | 1.3 | 1.81 | − 43 | 6 |
| Blackman | 1.68 | 2.35 | −58 | 18 |
| Hanning | 1.44 | 2.00 | −32 | 18 |
| Minimum 4-Term Blackman-Harris | 1.90 | 2.72 | −92 | 6 |

**Figure 17.75: Popular windows and figures of merit**

## 17.18 Digital Filters

Digital filtering is one of the most powerful tools of DSP. Apart from the obvious advantages of virtually eliminating errors in the filter associated with passive component fluctuations over time and temperature, op-amp drift (active filters), and other effects, digital filters are capable of performance specifications that would, at best, be extremely difficult, if not impossible, to achieve with an analog implementation. In addition, the characteristics of a digital filter can easily be changed under software control. Therefore, they are widely used in adaptive filtering applications in communications such as echo cancellation in modems, noise cancellation, and speech recognition.

The actual procedure for designing digital filters has the same fundamental elements as that for analog filters. First, the desired filter responses are characterized, and the filter parameters are then calculated. Characteristics such as amplitude and phase response are derived in the same way. The key difference between analog and digital filters is that instead of calculating resistor, capacitor, and inductor values for an analog filter, coefficient values are calculated for a digital filter. So for the digital filter, numbers replace the physical resistor and capacitor components of the analog filter. These numbers reside in a memory as filter coefficients and are used with the sampled data values from the ADC to perform the filter calculations.

The real-time digital filter, because it is a discrete time function, works with digitized data as opposed to a continuous waveform, and a new data point is acquired each sampling period. Because of this discrete nature, data samples are referenced as numbers such as sample 1, sample 2, and sample 3. Figure 17.76 shows a low frequency signal containing higher frequency noise which must be filtered out. This waveform must be digitized with an ADC to produce samples $x(n)$. These data values are fed to the digital filter, which in this case is a low-pass filter. The output data samples, $y(n)$, are used to reconstruct an analog waveform using a low glitch DAC.

Digital filters, however, are not the answer to all signal processing filtering requirements. In order to maintain real-time operation, the DSP processor must be able to execute all the steps in the filter routine within one sampling clock period, $1/f_s$. A fast general-purpose fixed-point DSP (such as the ADSP-2189M at 75 MIPS) can execute a complete filter tap multiply-accumulate instruction in 13.3 ns. The ADSP-2189M requires $N + 5$ instructions for an N-tap filter. For a 100-tap filter, the total execution time is

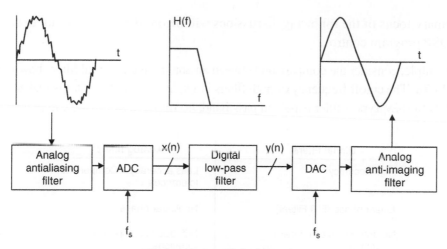

**Figure 17.76: Digital filtering**

approximately 1.4 μs. This corresponds to a maximum possible sampling frequency of 714 kHz, thereby limiting the upper signal bandwidth to a few hundred kHz.

However, it is possible to replace a general-purpose DSP chip and design special hardware digital filters that will operate at video-speed sampling rates. In other cases, the speed limitations can be overcome by first storing the high speed ADC data in a buffer memory. The buffer memory is then read at a rate that is compatible with the speed of the DSP-based digital filter. In this manner, pseudo-real-time operation can be maintained as in a radar system, where signal processing is typically done on bursts of data collected after each transmitted pulse.

Another option is to use a third-party dedicated DSP filter engine like the Systolix PulseDSP filter core. The AD7725 16-bit sigma-delta ADC has an on-chip PulseDSP filter that can do 125 million multiply-accumulates per second.

Even in highly oversampled sampled data systems, an analog antialiasing filter is still required ahead of the ADC and a reconstruction (anti-imaging) filter after the DAC. Finally, as signal frequencies increase sufficiently, they surpass the capabilities of available ADCs, and digital filtering then becomes impossible. Active analog filtering is not possible at extremely high frequencies because of op-amp bandwidth and distortion limitations, and filtering requirements must then be met using purely passive components.

The primary focus of the following discussions will be on filters that can run in real-time under DSP program control.

As an example, consider the comparison between an analog and a digital filter shown in Figure 17.78. The cutoff frequency of both filters is 1 kHz. The analog filter is realized as a 6-pole Chebyshev Type 1 filter (ripple in pass-band, no ripple in stop-band). In practice, this

| Digital filters | Analog filters |
|---|---|
| High accuracy | Less accuracy – Component tolerances |
| Linear phase (FIR Filters) | Nonlinear phase |
| No drift due to component variations | Drift due to component variations |
| Flexible, adaptive filtering possible | Adaptive filters difficult |
| Easy to simulate and design | Difficult to simulate and design |
| Computation must be completed in sampling period–Limits real-time operation | Analog filters required at high frequencies and for antialiasing filters |
| Requires high performance ADC, DAC, and DSP | No ADC, DAC, or DSP required |

**Figure 17.77: Digital vs. analog filtering**

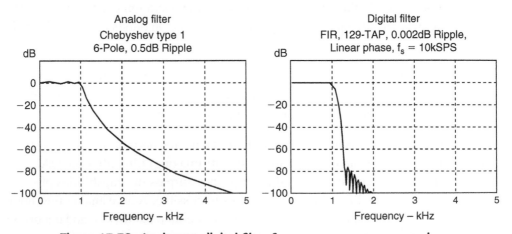

**Figure 17.78: Analog vs. digital filter frequency response comparison**

filter would probably be realized using three 2-pole stages, each of which requires an op-amp, and several resistors and capacitors. The 6-pole design is certainly not trivial, and maintaining the 0.5 dB ripple specification requires accurate component selection and matching.

On the other hand, the digital FIR filter shown has only 0.002 dB pass-band ripple, linear phase, and a much sharper roll-off. In fact, it could not be realized using analog techniques. In a practical application, there are many other factors to consider when evaluating analog versus digital filters. Most modern signal processing systems use a combination of analog and digital techniques in order to accomplish the desired function and take advantage of the best of both the analog and the digital world.

There are many applications where digital filters must operate in real-time. This places specific requirements on the DSP, depending upon the sampling frequency and the

- Signal Bandwidth = $f_a$
- Sampling Frequency $f_s > 2f_a$
- Sampling Period = $1/f_s$
- Filter computational time + Overhead < Sampling period
  - ◆ Depends on number of taps
  - ◆ Speed of DSP Multiplication-Accumulates (MACs)
  - ◆ Efficiency of DSP
    - • Circular buffering
    - • Zero-Overhead looping

**Figure 17.79: Processing requirements for real-time digital filtering**

- Moving average
- Finite impulse response (FIR)
  - ◆ Linear phase
  - ◆ Easy to design
  - ◆ Computationally intensive
- Infinite Impulse Response (IIR)
  - ◆ Based on classical analog filters
  - ◆ Computationally efficient
- Lattice filters (Can be FIR or IIR)
- Adaptive filters

**Figure 17.80: Types of digital filters**

filter complexity. The key point is *that the DSP must finish all computations during the sampling period so it will be ready to process the next data sample.* Assume that the analog signal bandwidth to be processed is $f_a$. This requires the ADC sampling frequency $f_s$ to be at least $2f_a$. The sampling period is $1/f_s$. All DSP filter computations (including overhead) must be completed during this interval. The computation time depends on the number of taps in the filter and the speed and efficiency of the DSP. Each tap on the filter requires one multiplication and one addition (multiply-accumulate). DSPs are generally optimized to perform fast multiply-accumulates, and many DSPs have additional features such as circular buffering and zero-overhead looping to minimize the "overhead" instructions that otherwise would be needed.

### 17.18.1 Finite Impulse Response (FIR) Filters

There are two fundamental types of digital filters: finite impulse response (FIR) and infinite impulse response (IIR). As the terminology suggests, these classifications refer to the filter's impulse response. By varying the weight of the coefficients and the number of filter taps, virtually any frequency response characteristic can be realized with a FIR filter. As has been shown, FIR filters can achieve performance levels that are not possible with analog filter techniques (such as perfect linear phase response). However, high performance FIR filters generally require a large number of multiply-accumulates and therefore require fast and efficient DSPs. On the other hand, IIR filters tend to mimic the performance of traditional analog filters and make use of feedback, so their impulse response extends over an infinite period of time. Because of feedback, IIR filters can be implemented with fewer coefficients than for a FIR filter. Lattice filters are simply another way to implement either FIR or IIR filters and are often used in speech processing applications. Finally, digital filters lend themselves to adaptive filtering applications simply because of the speed and ease with which the filter characteristics can be changed by varying the filter coefficients.

The most elementary form of a FIR filter is a *moving average* filter as shown in Figure 17.81. Moving average filters are popular for smoothing data, such as in the analysis of stock prices. The input samples, x(n) are passed through a series of buffer registers (labeled $z^{-1}$, corresponding to the z-transform representation of a delay element). In the example shown, there are four taps corresponding to a 4-point moving average. Each sample is multiplied by 0.25, and these results are added to yield the final moving average output y(n). The figure also shows the general equation of a moving average filter with

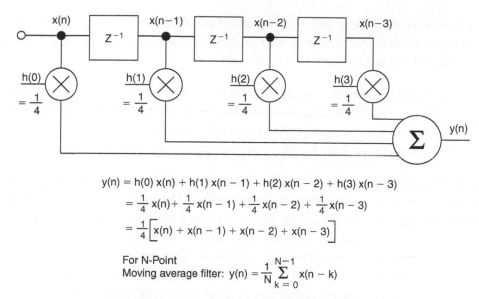

$$y(n) = h(0)\,x(n) + h(1)\,x(n-1) + h(2)\,x(n-2) + h(3)\,x(n-3)$$

$$= \frac{1}{4}\,x(n) + \frac{1}{4}\,x(n-1) + \frac{1}{4}\,x(n-2) + \frac{1}{4}\,x(n-3)$$

$$= \frac{1}{4}\Big[x(n) + x(n-1) + x(n-2) + x(n-3)\Big]$$

For N-Point
Moving average filter: $y(n) = \dfrac{1}{N} \displaystyle\sum_{k=0}^{N-1} x(n-k)$

**Figure 17.81: 4-point moving average filter**

$$y(3) = 0.25\Big[\qquad\qquad\qquad\qquad x(3) + x(2) + x(1) + x(0)\Big]$$

$$y(4) = 0.25\Big[\qquad\qquad\qquad\quad x(4) + x(3) + x(2) + x(1)\qquad\Big]$$

$$y(5) = 0.25\Big[\qquad\qquad\quad x(5) + x(4) + x(3) + x(2)\qquad\qquad\Big]$$

$$y(6) = 0.25\Big[\qquad x(6) + x(5) + x(4) + x(3)\qquad\qquad\qquad\Big]$$

$$y(7) = 0.25\Big[\; x(7) + x(6) + x(5) + x(4)\qquad\qquad\qquad\qquad\Big]$$

•
•  Each output requires:
•  1 Multiplication, 1 Addition, 1 Subtraction

**Figure 17.82: Calculating output of 4-point moving average filter**

N taps. Note again that N refers to the number of filter taps, and not the ADC or DAC resolution as in previous sections.

Since the coefficients are equal, an easier way to perform a moving average filter is shown in Figure 17.82. Note that the first step is to store the first four samples, x(0), x(1), x(2), x(3) in a register. These quantities are added and then multiplied by 0.25 to yield the first output, y(3). Note that the initial outputs y(0), y(1), and y(2) are not valid because all registers are not full until sample x(3) is received.

When sample x(4) is received, it is added to the result, and sample x(0) is subtracted from the result. The new result must then be multiplied by 0.25. Therefore, the calculations required to produce a new output consist of one addition, one subtraction, and one multiplication, regardless of the length of the moving average filter.

The step function response of a 4-point moving average filter is shown in Figure 17.83. Notice that the moving average filter has no overshoot. This makes it useful in signal processing applications where random white noise must be filtered but pulse response preserved. Of all the possible linear filters that could be used, the moving average produces the lowest noise for a given edge sharpness. This is illustrated in Figure 17.84, where the noise level becomes lower as the number of taps are increased. Notice that the 0% to 100% rise time of the pulse response is equal to the total number of taps in the filter multiplied by the sampling period.

The frequency response of the simple moving average filter is SIN (x)/x and is shown on a linear amplitude scale in Figure 17.85. Adding more taps to the filter sharpens the roll-off, but does not significantly reduce the amplitude of the sidelobes which are

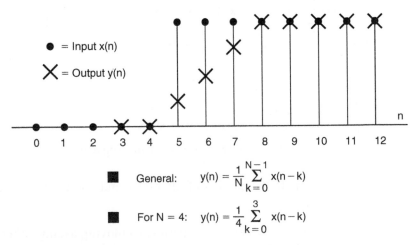

● = Input x(n)

✖ = Output y(n)

■ General: $y(n) = \dfrac{1}{N} \sum\limits_{k=0}^{N-1} x(n-k)$

■ For N = 4: $y(n) = \dfrac{1}{4} \sum\limits_{k=0}^{3} x(n-k)$

**Figure 17.83: 4-tap moving average filter step response**

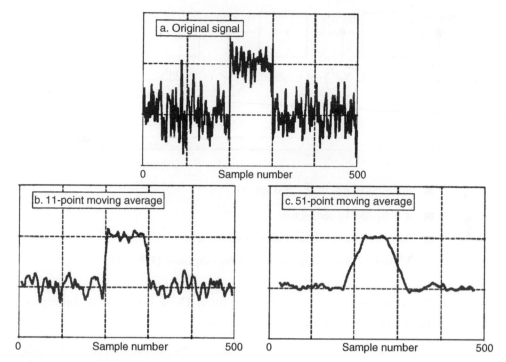

**Figure 17.84: Moving average filter response to noise superimposed on step input**

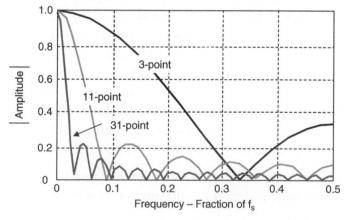

**Figure 17.85: Moving average filter frequency response**

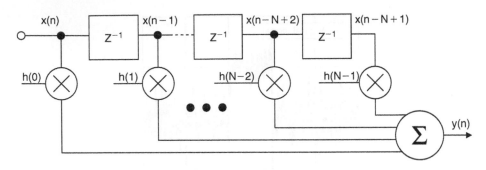

$$y(n) = h(n) * x(n) = \sum_{k=0}^{N-1} h(k)\,x(n-k)$$

■  $*$ = Symbol for convolution

■  Requires N multiply-accumulates for each output

**Figure 17.86: N-tap finite impulse response (FIR) filter**

approximately 14 dB down for the 11- and 31-tap filter. These filters are definitely not suitable where high stop-band attenuation is required.

It is possible to dramatically improve the performance of the simple FIR moving average filter by properly selecting the individual weights or coefficients rather than giving them equal weight. The sharpness of the roll-off can be improved by adding more stages (taps), and the stop-band attenuation characteristics can be improved by properly selecting the filter coefficients. Note that unlike the moving average filter, one multiply-accumulate cycle is now required per tap for the generalized FIR filter. The essence of FIR filter design is the appropriate selection of the filter coefficients and the number of taps to realize the desired transfer function H(f). Various algorithms are available to translate the frequency response H(f) into a set of FIR coefficients. Most of this software is commercially available and can be run on PCs. *The key theorem of FIR filter design is that the coefficients h(n) of the FIR filter are simply the quantized values of the impulse response of the frequency transfer function H(f).* Conversely, the impulse response is the discrete Fourier transform of H(f).

The generalized form of an N-tap FIR filter is shown in Figure 17.86. As has been discussed, an FIR filter must perform the following convolution equation:

$$y(n) = h(k) * x(n) = \sum_{k=0}^{N-1} h(k)x(n-k)$$

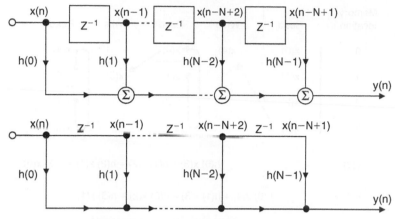

**Figure 17.87: Simplified filter notations**

where h(k) is the filter coefficient array and x(n − k) is the input data array to the filter. The number N, in the equation, represents the number of taps of the filter and relates to the filter performance as has been discussed above. An N-tap FIR filter requires N multiply-accumulate cycles.

FIR filter diagrams are often simplified as shown in Figure 17.87. The summations are represented by arrows pointing into the dots, and the multiplications are indicated by placing the h(k) coefficients next to the arrows on the lines. The $z^{-1}$ delay element is often shown by placing the label above or next to the appropriate line.

### 17.18.2 *FIR Filter Implementation in DSP Hardware Using Circular Buffering*

In the series of FIR filter equations, the N coefficient locations are always accessed sequentially from h(0) to h(N − 1). The associated data points circulate through the memory; new samples are added, replacing the oldest each time a filter output is computed. A fixed-boundary RAM can be used to achieve this circulating buffer effect as shown in Figure 17.88 for a four-tap FIR filter. The oldest data sample is replaced by the newest after each convolution. A "time history" of the four most recent data samples is always stored in RAM.

To facilitate memory addressing, old data values are read from memory starting with the value one location after the value that was just written. For example, x(4) is written into memory location 0, and data values are then read from locations 1, 2, 3, and 0.

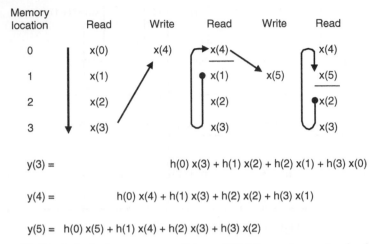

**Figure 17.88: Calculating outputs of 4-tap FIR filter using a circular buffer**

This example can be expanded to accommodate any number of taps. By addressing data memory locations in this manner, the address generator need only supply sequential addresses, regardless of whether the operation is a memory read or write. This data memory buffer is called *circular* because when the last location is reached, the memory pointer is reset to the beginning of the buffer.

The coefficients are fetched simultaneously with the data. Due to the addressing scheme chosen, the oldest data sample is fetched first. Therefore, the last coefficient must be fetched first. The coefficients can be stored backward in memory: $h(N - 1)$ is the first location, and $h(0)$ is the last, with the address generator providing incremental addresses. Alternatively, coefficients can be stored in a normal manner with the accessing of coefficients starting at the end of the buffer, and the address generator being decremented. In the example shown in Figure 17.88, the coefficients are stored in a reverse manner.

A simple summary flowchart for these operations is shown in Figure 17.89. For Analog Devices DSPs, *all operations within the filter loop are completed in one instruction cycle*, thereby greatly increasing efficiency. This is referred to as *zero-overhead looping*. The actual FIR filter assembly code for the ADSP-21xx family of fixed-point DSPs is shown in Figure 17.90. The arrows in the diagram point to the actual executable instructions, and the rest of the code are simply comments added for clarification.

1. Obtain sample from ADC (typically interrupt-driven)
2. Move sample into input signal's circular buffer
3. Update the pointer for the input signal's circular buffer
4. Zero the accumulator
5. Implement filter (control the loop through each of the coefficients)
6. Fetch the coefficient from the coefficient's circular buffer
7. Update the pointer for the coefficient's circular buffer
8. Fetch the sample from the input signal's circular buffer
9. Update the pointer for the input signal's circular buffer
10. Multiply the coefficient by the sample
11. Add the product to the accumulator
12. Move the filtered sample to the DAC

```
ADSP-21xx Example code:

CNTR = N-1;
DO convolution UNTIL CE;
convolution:
   MR = MR + MX0 * MY0(SS), MX0 = DM(I0,M1), MY0 = PM(I4,M5);
```

**Figure 17.89: Pseudocode for FIR filter program using a DSP with circular buffering**

```
.MODULE          fir_sub;
{                FIR Filter Subroutine
                 Calling Parameters
                         I0 --> Oldest input data value in delay line
                         I4 --> Beginning of filter coefficient table
                         L0 = Filter length (N)
                         L4 = Filter length (N)
                         M1,M5 = 1
                         CNTR = Filter length - 1 (N-1)
                 Return Values
                         MR1 = Sum of products (rounded and saturated)
                         I0 --> Oldest input data value in delay line
                         I4 --> Beginning of filter coefficient table
                 Altered Registers
                         MX0,MY0,MR
                 Computation Time
                         (N - 1) + 6 cycles = N + 5 cycles
                 All coefficients are assumed to be in 1.15 format. }
.ENTRY           fir;
fir:             MR=0, MX0=DM(I0,M1), MY0=PM(I4,M5)
                 CNTR = N-1;
                 DO convolution UNTIL CE;
convolution:       MR=MR+MX0*MY0(SS), MX0=DM(I0,M1), MY0=PM(I4,M5);
                 MR=MR+MX0*MY0(RND);
                 IF MV SAT MR;
                 RTS;
.ENDMOD;
```

**Figure 17.90: ADSP-21xx FIR filter assembly code (single precision)**

The first instruction (labeled *fir:*) sets up the computation by clearing the MR register and loading the MX0 and MY0 registers with the first data and coefficient values from data and program memory. The multiply-accumulate with dual data fetch in the *convolution* loop is then executed N − 1 times in N cycles to compute the sum of the first N − 1 products. The final multiply-accumulate instruction is performed with the rounding mode enabled to round the result to the upper 24 bits of the MR register. The MR1 register is then conditionally saturated to its most positive or negative value, based on the status of the overflow flag contained in the MV register. In this manner, results are accumulated to the full 40-bit precision of the MR register, with saturation of the output only if the final result overflowed beyond the least significant 32 bits of the MR register.

The limit on the number of filter taps attainable for a real-time implementation of the FIR filter subroutine is primarily determined by the processor cycle time, the sampling rate, and the number of other computations required. The FIR subroutine presented here requires a total of N + 5 cycles for a filter of length N. For the ADSP-2189M 75 MIPS DSP, one instruction cycle is 13.3 ns, so a 100-tap filter would require 13.3 ns × 100 + 5 × 13.3 ns = 1330 ns + 66.5 ns = 1396.5 ns = 1.4 μs.

### 17.18.3  Designing FIR Filters

FIR filters are relatively easy to design using modern CAD tools. Figure 17.91 summarizes the characteristics of FIR filters as well as the most popular design

- Impulse response has a finite duration (N cycles)
- Linear phase, constant group delay (N must be odd)
- No analog equivalent
- Unconditionally stable
- Can be adaptive
- Computational advantages when decimating output
- Easy to understand and design
  - ◆ Windowed-Sinc method
  - ◆ Fourier series expansion with windowing
  - ◆ Frequency sampling using inverse FFT – Arbitrary frequency response
  - ◆ Parks-McClellan program with Remez exchange algorithm

**Figure 17.91: Characteristics of FIR filters**

techniques. *The fundamental concept of FIR filter design is that the filter frequency response is determined by the impulse response, and the quantized impulse response and the filter coefficients are identical.* This can be understood by examining Figure 17.92. The input to the FIR filter is an impulse, and as the impulse propagates through the delay elements, the filter output is identical to the filter coefficients. The FIR filter design process therefore consists of determining the impulse response from the desired frequency response, and then quantizing the impulse response to generate the filter coefficients.

It is useful to digress for a moment and examine the relationship between the time domain and the frequency domain to better understand the principles behind digital filters such as the FIR filter. In a sampled data system, a convolution operation can be carried out by performing a series of multiply-accumulates. The convolution operation in the time or frequency domain is equivalent to point-by-point multiplication in the opposite domain. For example, convolution in the time domain is equivalent to multiplication in the frequency domain. This is shown graphically in Figure 17.93. It can be seen that filtering in the frequency domain can be accomplished by multiplying all frequency components in the pass band by a 1 and all frequencies in the stop band by 0. Conversely,

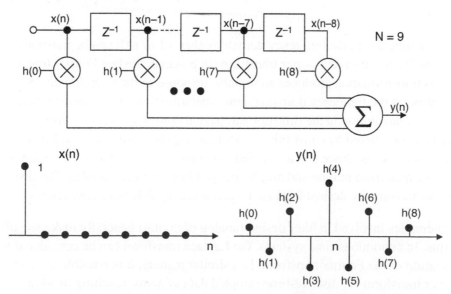

**Figure 17.92: FIR filter impulse response determines the filter coefficients**

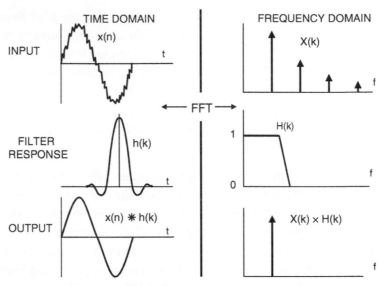

**Figure 17.93: Duality of time and frequency**

convolution in the frequency domain is equivalent to point-by-point multiplication in the time domain.

The transfer function in the frequency domain (either a 1 or a 0) can be translated to the time domain by the discrete Fourier transform (in practice, the fast Fourier transform is used). This transformation produces an impulse response in the time domain. Since the multiplication in the frequency domain (signal spectrum times the transfer function) is equivalent to convolution in the time domain (signal convolved with impulse response), the signal can be filtered by convolving it with the impulse response. The FIR filter is exactly this process. Since it is a sampled data system, the signal and the impulse response are quantized in time and amplitude, yielding discrete samples. The discrete samples comprising the desired impulse response are the FIR filter coefficients.

The mathematics involved in filter design (analog or digital) generally make use of transforms. In continuous-time systems, the Laplace transform can be considered to be a generalization of the Fourier transform. In a similar manner, it is possible to generalize the Fourier transform for discrete-time sampled data systems, resulting in what is commonly referred to as the z-transform. Details describing the use of the z-transform in

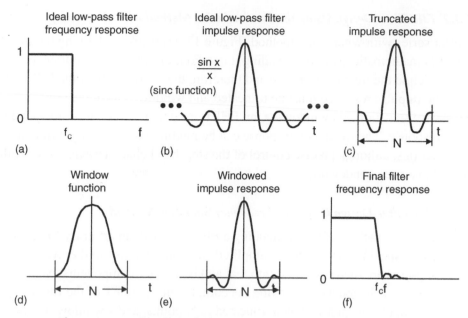

**Figure 17.94: FIR filter design using the windowed-sinc method**

digital filter design are given in References 1 through 6, but the theory is not necessary for the rest of this discussion.

### 17.18.3.1 FIR Filter Design Using the Windowed-sinc Method

An ideal low-pass filter frequency response is shown in Figure 17.94a. The corresponding impulse response in the time domain is shown in Figure 17.94b, and follows the $\sin(x)/x$ (sinc) function. If a FIR filter is used to implement this frequency response, an infinite number of taps are required. The windowed-sinc method is used to implement the filter as follows. First, the impulse response is truncated to a reasonable number of N taps as in Figure 17.94c. As has been discussed previously, the frequency response corresponding to Figure 17.94c has relatively poor sidelobe performance because of the end-point discontinuities in the truncated impulse response. The next step in the design process is to apply an appropriate window function as shown in Figure 17.94d to the truncated impulse. This forces the endpoints to zero. The particular window function chosen determines the roll-off and sidelobe performance of the filter. There are several good choices depending upon the desired frequency response. The frequency response of the truncated and windowed-sinc impulse response of Figure 17.94e is shown in Figure 17.94f.

### 17.18.3.2 FIR Filter Design Using the Fourier Series Method with Windowing

The Fourier series with windowing method (Figure 17.95) starts by defining the transfer function H(f) mathematically and expanding it in a Fourier series. The Fourier series coefficients define the impulse response and therefore the coefficients of the FIR filter. However, the impulse response must be truncated and windowed as in the previous method. After truncation and windowing, an FFT is used to generate the corresponding frequency response. The frequency response can be modified by choosing different window functions, although precise control of the stop-band characteristics is difficult in any method that uses windowing.

### 17.18.3.3 FIR Filter Design Using the Frequency Sampling Method

This method is extremely useful in generating a FIR filter with an arbitrary frequency response. H(f) is specified as a series of amplitude and phase points in the frequency domain. The points are then converted into real and imaginary components. Next, the impulse response is obtained by taking the complex inverse FFT of the frequency response. The impulse response is then truncated to N points, and a window function is applied to minimize the effects of truncation. The filter design should then be tested by taking its FFT and evaluating the frequency response. Several iterations may be required to achieve the desired response.

### 17.18.3.4 FIR Filter Design Using the Parks-McClellan Program

Historically, the design method based on the use of windows to truncate the impulse response and to obtain the desired frequency response was the first method used for

■ Specify H(f)

■ Expand H(f) in a Fourier Series: The Fourier Series Coefficients are the Coefficients of the FIR Filter, h(k), and its Impulse Response

■ Truncate the Impulse Response to N Points (Taps)

■ Apply a Suitable Window Function to h(k) to Smooth the Effects of Truncation

■ Lacks Precise Control of Cutoff Frequency; Highly Dependent on Window Function

**Figure 17.95: FIR filter design using Fourier series method with windowing**

designing FIR filters. The frequency-sampling method was developed in the 1970s and is still popular where the frequency response is an arbitrary function.

Modern CAD programs are available today that greatly simplify the design of low-pass, high-pass, band-pass, or band-stop FIR filters. A popular one was developed by Parks and McClellan and uses the Remez exchange algorithm. The filter design begins by specifying the parameters shown in Figure 17.97: pass-band ripple, stop-band ripple (same as attenuation), and the transition region. For this design example, the QED1000 program from Momentum Data Systems was used (a demo version is free and downloadable from http://www.mds.com).

For this example, we will design an audio low-pass filter that operates at a sampling rate of 44.1 kHz. The filter is specified as shown in Figure 17.97: 18 kHz pass-band frequency, 21 kHz stop-band frequency, 0.01 dB pass-band ripple, 96 dB stop-band ripple (attenuation). We must also specify the word length of the coefficients, which in this case is 16 bits, assuming a 16-bit fixed-point DSP is to be used.

The program allows us to choose between a window-based design or the equiripple Parks-McClellan program. We will choose the latter. The program now estimates the number of taps required to implement the filter based on the above specifications. In this case, it is 69 taps. At this point, we can accept this and proceed with the design or decrease the number of taps and see what degradation in specifications occur.

- Specify H(k) as a Finite Number of Spectral Points Spread Uniformly between 0 and 0.5$f_s$ (512 Usually Sufficient)
- Specify Phase Points (Can Make Equal to Zero)
- Convert Rectangular Form (Real + Imaginary)
- Take the Complex Inverse FFT of H(f) Array to Obtain the Impulse Response
- Truncate the Impulse Response to N Points
- Apply a Suitable Window Function to h(k) to Smooth the Effects of Truncation
- Test Filter Design and Modify if Necessary
- CAD Design Techniques More Suitable for Low-Pass, High-Pass, Band-Pass, or Band-Stop Filters

**Figure 17.96: Frequency sampling method for FIR filters with arbitrary frequency response**

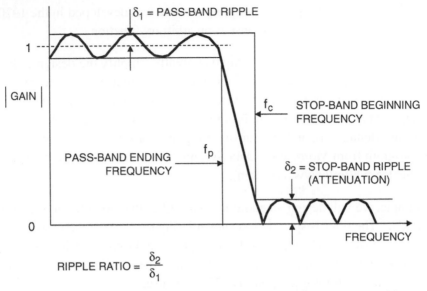

$$\text{RIPPLE RATIO} = \frac{\delta_2}{\delta_1}$$

**Figure 17.97: FIR CAD techniques: Parks McClellan program with Remez Exchange Algorithm**

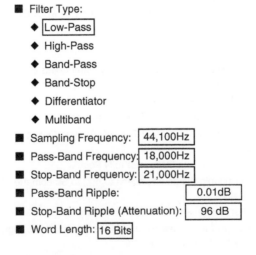

■ Filter Type:
  ◆ Low-Pass
  ◆ High-Pass
  ◆ Band-Pass
  ◆ Band-Stop
  ◆ Differentiator
  ◆ Multiband
■ Sampling Frequency: 44,100Hz
■ Pass-Band Frequency: 18,000Hz
■ Stop-Band Frequency: 21,000Hz
■ Pass-Band Ripple: 0.01dB
■ Stop-Band Ripple (Attenuation): 96 dB
■ Word Length: 16 Bits

**Figure 17.98: Parks McClellan Equiripple FIR filter design: program inputs**

■ Estimated Number of Taps Required: 69
   ◆ Accept? Change? Accept
■ Frequency Response (Linear and Log Scales)
■ Step Response
■ s- and z-Plane Analysis
■ Impulse Response: Filter Coefficients (Quantized)
■ DSP FIR Filter Assembly Code

**Figure 17.99: FIR filter program outputs**

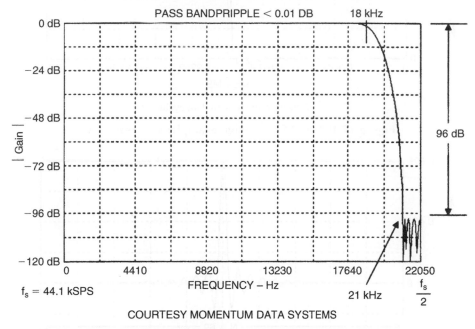

**Figure 17.100: FIR design example: frequency response**

We will accept this number and let the program complete the calculations. The program outputs the frequency response (Figure 17.100), step function response (Figure 17.101), s- and z-plane analysis data, and the impulse response (Figure 17.102). The QED1000 program then outputs the quantized filter coefficients to a program that generates the actual DSP assembly code for a number of popular DSPs, including Analog Devices'.

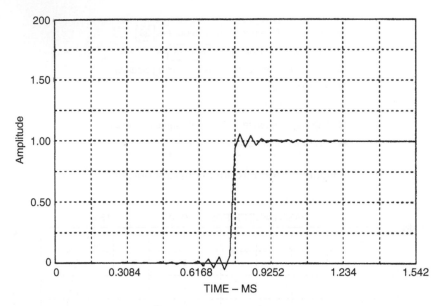

Courtesy momentum data systems

**Figure 17.101: FIR filter design example: step response**

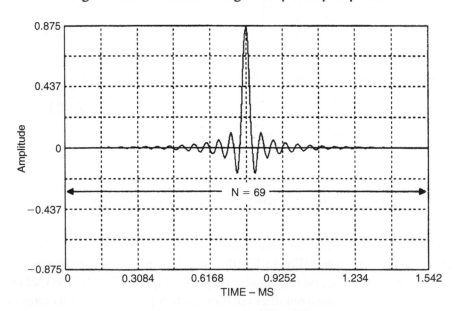

Courtesy momentum data systems

**Figure 17.102: FIR design example: impulse response (filter coefficients)**

The program is quite flexible and allows the user to perform a number of scenarios to optimize the filter design.

The 69-tap FIR filter requires $69 + 5 = 74$ instruction cycles using the ADSP-2189M 75 MIPS processor, which yields a total computation time per sample of $74 \times 13.3 \, \text{ns} = 984 \, \text{ns}$. The sampling interval is $1/44.1 \, \text{kHz}$, or $22.7 \, \mu\text{s}$. This allows $22.7 \, \mu\text{s} - 0.984 \, \mu\text{s} = 21.7 \, \mu\text{s}$ for overhead and other operations.

Other options are to use a slower processor (3.3 MIPS) for this application, a more complex filter that takes more computation time (up to $N = 1700$), or increase the sampling frequency to about 1 MSPS.

### 17.18.3.5 Designing High-Pass, Band-Pass, and Band-Stop Filters Based on Low-Pass Filter Design

Converting a low-pass filter design impulse response into a high-pass filter impulse response can be accomplished in one of two ways. In the *spectral inversion method*, the sign of each filter coefficient in the low-pass filter impulse response is changed. Next, 1 is added to the center coefficient. In the *spectral reversal method*, the sign of every other coefficient is changed. This reverses the frequency domain plot. In other words, if the cutoff of the low-pass filter is $0.2f_s$, the resulting high-pass filter will have a cutoff

- ■ Sampling Frequency $f_s$ = 44.1 kSPS
- ■ Sampling Interval = $1/f_s$ = 22.7 μs
- ■ Number of Filter Taps, N = 69
- ■ Number of Required Instructions = N + 5 = 74
- ■ Processing Time/Instruction = 13.3 ns (75 MIPS) (ADSP-2189M)
- ■ Total Processing Time = 74 × 13.3 ns = 984 ns
- ■ Total Processing Time < Sampling Interval with 22.7 μs − 0.984 μs = 21.7 μs for Other Operations
  - ◆ Increase Sampling Frequency to 1 MHz
  - ◆ Use Slower DSP (3.3 MIPS)
  - ◆ Add More Filter Taps (Up to N = 1700)

**Figure 17.103: Design example using ADSP-2189M: processor time for 69-tap FIR filter**

■ Spectral Inversion Technique:
   ◆ Design Low-Pass Filter (Linear Phase, N Odd)
   ◆ Change the Sign of Each Coefficient in the Impulse Response, h(n)
   ◆ Add 1 to the Coefficient at the Center of Symmetry

■ Spectral Reversal Technique:
   ◆ Design Low-Pass Filter
   ◆ Change the Sign of Every Other Coefficient in the Impulse Response, h(n)
   ◆ This Reverses the Frequency Domain Left-for-Right:
     0 Becomes 0.5, and 0.5 Becomes 0;
     i.e., if the Cut-Off Frequency of the Low-Pass Filter is 0.2, the Cut-Off of the Resulting High-Pass Filter is 0.3

**Figure 17.104: Designing high-pass filters using low-pass filter impulse response**

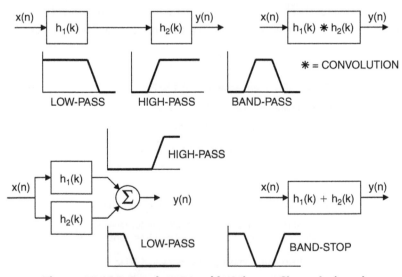

**Figure 17.105: Band-pass and band-stop filters designed from low-pass and high-pass filters**

- Uses Feedback (Recursion)
- Impulse Response has an Infinite Duration
- Potentially Unstable
- Nonlinear Phase
- More Efficient than FIR Filters
- No Computational Advantage when Decimating Output
- Usually Designed to Duplicate Analog Filter Response
- Usually Implemented as Cascaded Second-Order Sections (Biquads)

**Figure 17.106: Infinite impulse response (IIR) filters**

frequency of $0.5f_s – 0.2f_s = 0.3f_s$. This must be considered when doing the original low-pass filter design.

Band-pass and band-stop filters can be designed by combining individual low-pass and high-pass filters in the proper manner. Band-pass filters are designed by placing the low-pass and high-pass filters in cascade. The equivalent impulse response of the cascaded filters is then obtained by *convolving* the two individual impulse responses.

A band-stop filter is designed by connecting the low-pass and high-pass filters in parallel and adding their outputs. The equivalent impulse response is then obtained by *adding* the two individual impulse responses.

### 17.18.4 Infinite Impulse Response (IIR) Filters

As was mentioned previously, FIR filters have no real analog counterparts, the closest analogy being the weighted moving average. In addition, FIR filters have only zeros and no poles. On the other hand, IIR filters have traditional analog counterparts (Butterworth, Chebyshev, Elliptic, and Bessel) and can be analyzed and synthesized using more familiar traditional filter design techniques.

Infinite impulse response filters get their name because their impulse response extends for an infinite period of time. This is because they are recursive, i.e., they utilize feedback. Although they can be implemented with fewer computations than FIR filters, IIR filters do not match the performance achievable with FIR filters, and do not have linear phase.

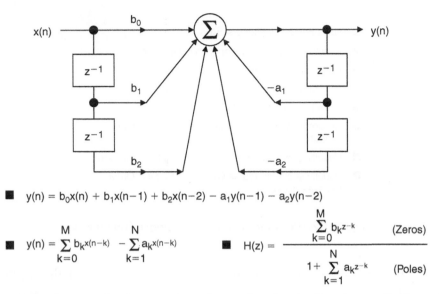

$$y(n) = b_0 x(n) + b_1 x(n-1) + b_2 x(n-2) - a_1 y(n-1) - a_2 y(n-2)$$

$$y(n) = \sum_{k=0}^{M} b_k x(n-k) - \sum_{k=1}^{N} a_k x(n-k)$$

$$H(z) = \frac{\sum_{k=0}^{M} b_k z^{-k}}{1 + \sum_{k=1}^{N} a_k z^{-k}} \quad \begin{array}{l}\text{(Zeros)}\\ \\ \text{(Poles)}\end{array}$$

**Figure 17.107: Hardware implementation of second-order IIR filter (Biquad) Direct Form 1**

Also, there is no computational advantage achieved when the output of an IIR filter is decimated, because each output value must always be calculated.

IIR filters are generally implemented in 2-pole sections called *biquads* because they are described with a biquadratic equation in the z-domain. Higher order filters are designed using cascaded biquad sections, e.g., a 6-pole filter requires three biquad sections.

The basic IIR biquad is shown in Figure 17.107. The zeros are formed by the feedforward coefficients $b_0$, $b_1$, and $b_2$; the poles are formed by the feedback coefficients $a_1$, and $a_2$.

The general digital filter equation is shown in Figure 17.107, which gives rise to the general transfer function H(z), which contains polynomials in both the numerator and the denominator. The roots of the denominator determine the pole locations of the filter, and the roots of the numerator determine the zero locations. Although it is possible to construct a high order IIR filter directly from this equation (called the *direct form* implementation), accumulation errors due to quantization errors (finite word-length arithmetic) may give rise to instability and large errors. For this reason, it is common to cascade several biquad sections with appropriate coefficients rather than use the direct

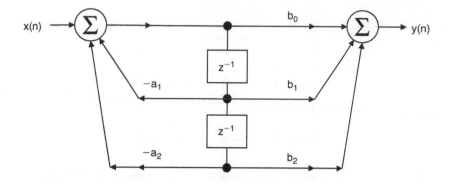

- Reduces to the same equation as Direct Form 1:

$$y(n) = b_0 x(n) + b_1 x(n-1) + b_2 x(n-2) - a_1 y(n-1) - a_2 y(n-2)$$

- Requires only two delay elements (registers)

**Figure 17.108: IIR biquad filter Direct Form 2**

form implementation. The biquads can be scaled separately and then cascaded in order to minimize the coefficient quantization and the recursive accumulation errors. Cascaded biquads execute more slowly than their direct form counterparts, but are more stable and minimize the effects of errors due to finite arithmetic errors.

The Direct Form 1 biquad section shown in Figure 17.107 requires four registers. This configuration can be changed into an equivalent circuit shown in Figure 17.108 that is called the *Direct Form* 2 and requires only two registers. It can be shown that the equations describing the Direct Form 2 IIR biquad filter are the same as those for Direct Form 1. As in the case of FIR filters, the notation for an IIR filter is often simplified as shown in Figure 17.109.

### 17.18.4.1 IIR Filter Design Techniques

A popular method for IIR filter design is to first design the analog-equivalent filter and then mathematically transform the transfer function H(s) into the z-domain, H(z). Multiple pole designs are implemented using cascaded biquad sections. The most popular analog filters are the Butterworth, Chebyshev, Elliptical, and Bessel (see Figure 17.110). There are many CAD programs available to generate the Laplace transform, H(s), for these filters.

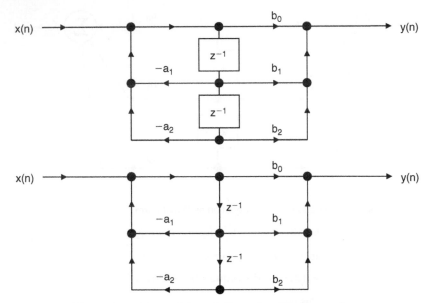

Figure 17.109: IIR biquad filter simplified notations

The all-pole Butterworth (also called *maximally flat*) has no ripple in the pass band or stop band and has monotonic response in both regions. The all-pole Type 1 Chebyshev filter has a faster roll-off than the Butterworth (for the same number of poles) and has ripple in the pass band. The Type 2 Chebyschev filter is rarely used, but has ripple in the stop band rather than the pass band.

The Elliptical (Cauer) filter has poles and zeros and ripple in both the pass band and stop band. This filter has even faster roll-off than the Chebyshev for the same number of poles. The Elliptical filter is often used where degraded phase response can be tolerated.

Finally, the Bessel (Thompson) filter is an all-pole filter optimized for pulse response and linear phase but has the poorest roll-off of any of the types discussed for the same number of poles.

All of the above types of analog filters are covered in the literature, and their Laplace transforms, H(s), are readily available, either from tables or CAD programs. There are three methods used to convert the Laplace transform into the z-transform: *impulse invariant* transformation, *bilinear* transformation, and the *matched z-transform*.

- Butterworth
    - ◆ All Pole, No Ripples in Pass Band or Stop Band
    - ◆ Maximally Flat Response (Fastest Roll-Off with No Ripple)
- Chebyshev (Type 1)
    - ◆ All Pole, Ripple in Pass Band, No Ripple in Stop Band
    - ◆ Shorter Transition Region than Butterworth for Given Number of Poles
    - ◆ Type 2 has Ripple in Stop Band, No Ripple in Pass Band
- Elliptical (Cauer)
    - ◆ Has Poles and Zeros, Ripple in Both Pass Band and Stop Band
    - ◆ Shorter Transition Region than Chebyshev for Given Number of Poles
    - ◆ Degraded Phase Response
- Bessel (Thompson)
    - ◆ All Pole, No Ripples in Pass Band or Stop Band
    - ◆ Optimized for Linear Phase and Pulse Response
    - ◆ Longest Transition Region of All for Given Number of Poles

**Figure 17.110: Review of popular analog filters**

The resulting z-transforms can be converted into the coefficients of the IIR biquad. These techniques are highly mathematically intensive and will not be discussed further.

A CAD approach for IIR filter design is similar to the Parks-McClellan program used for FIR filters. This technique uses the Fletcher-Powell algorithm.

In calculating the throughput time of a particular DSP IIR filter, one should examine the benchmark performance specification for a biquad filter section. For the ADSP-21xx family, seven instruction cycles are required to execute a biquad filter output sample. For the ADSP-2189M, 75 MIPS DSP, this corresponds to $7 \times 13.3\,\text{ns} = 93\,\text{ns}$, corresponding to a maximum possible sampling frequency of 10 MSPS (neglecting overhead).

### 17.18.5 Summary: FIR Versus IIR Filters

Choosing between FIR and IIR filter designs can be somewhat of a challenge, but a few basic guidelines can be given. Typically, IIR filters are more efficient than FIR filters because they require less memory and fewer multiply-accumulates are needed. IIR filters

- Impulse Invariant Transformation Method
  - ◆ Start with H(s) for Analog Filter
  - ◆ Take Inverse Laplace Transform to Get Impulse Response
  - ◆ Obtain z-Transform H(z) from Sampled Impulse Response
  - ◆ z-Transform Yields Filter Coefficients
  - ◆ Aliasing Effects Must Be Considered
- Bilinear Transformation Method
  - ◆ Another Method for Transforming H(s) into H(z)
  - ◆ Performance Determined by the Analog System's Differential Equation
  - ◆ Aliasing Effects Do Not Occur
- Matched z-Transform Method
  - ◆ Maps H(s) into H(z) for Filters with Both Poles and Zeros
- CAD Methods
  - ◆ Fletcher-Powell Algorithm
  - ◆ Implements Cascaded Biquad Sections

**Figure 17.111: IIR filter design techniques**

- Determine How Many Biquad Sections are Required to Realize the Desired Frequency Response
- Multiply This by the Execution Time per Biquad for the DSP (7 Instruction Cycles $\times$ 13.3 ns = 93 ns for the 75 MIPS ADSP-2189M, for example)
- The Result (Plus Overhead) is the Minimum Allowable Sampling Period (1/$f_s$) for Real-Time Operation

**Figure 17.112: Throughput considerations for IIR filters**

can be designed based upon previous experience with analog filter designs. IIR filters may exhibit instability problems, but this is much less likely to occur if higher order filters are designed by cascading second-order systems.

On the other hand, FIR filters require more taps and multiply-accumulates for a given cut off frequency response, but have linear phase characteristics. Since FIR filters operate on

| IIR FILTERS | FIR FILTERS |
| --- | --- |
| More Efficient | Less Efficient |
| Analog Equivalent | No Analog Equivalent |
| May Be Unstable | Always Stable |
| Nonlinear Phase Response | Linear Phase Response |
| More Ringing on Glitches | Less Ringing on Glitches |
| CAD Design Packages Available | CAD Design Packages Available |
| No Effciency Gained by Decimation | Decimation Increases Efficiency |

**Figure 17.113: Comparison between FIR and IIR filters**

a finite history of data, if some data is corrupted (ADC sparkle codes, for example) the FIR filter will ring for only N − 1 samples. Because of the feedback, however, an IIR filter will ring for a considerably longer period of time.

If sharp cut-off filters are needed, and processing time is at a premium, IIR elliptic filters are a good choice. If the number of multiply/accumulates is not prohibitive, and linear phase is a requirement, the FIR should be chosen.

## 17.18.6 Multirate Filters

There are many applications in which it is desirable to change the effective sampling rate in a sampled data system. In many cases, this can be accomplished simply by changing the sampling frequency to the ADC or DAC. However, it is often desirable to accomplish the sample rate conversion after the signal has been digitized. The most common techniques used are *decimation* (reducing the sampling rate by a factor of M), and *interpolation* (increasing the sampling rate by a factor of L). The decimation and interpolation factors (M and L) are normally integer numbers. In a generalized samplerate converter, it may be desirable to change the sampling frequency by a noninteger number. In the case of converting the CD sampling frequency of 44.1 kHz to the digital audio tape (DAT) sampling rate of 48 kHz, interpolating by L = 160 followed by decimation by M = 147 accomplishes the desired result.

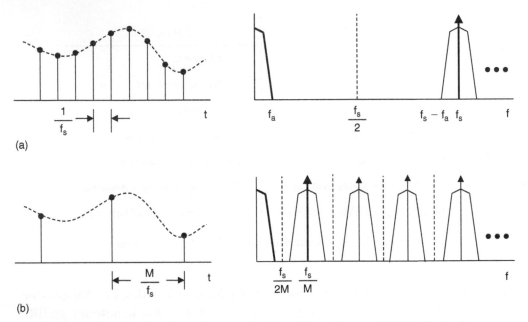

**Figure 17.114: Decimation of a sampled signal by a factor of M (a) Orginal over sampled signal; (b) Signal Decimated by M**

The concept of decimation is illustrated in Figure 17.114. The top diagram shows the original signal, $f_a$, which is sampled at a frequency $f_s$. The corresponding frequency spectrum shows that the sampling frequency is much higher than required to preserve information contained in $f_a$, i.e., $f_a$ is oversampled. Notice that there is no information contained between the frequencies $f_a$ and $f_s - f_a$. The bottom diagram shows the same signal where the sampling frequency has been reduced (decimated) by a factor of M. Notice that even though the sampling rate has been reduced, there is no aliasing and loss of information. Decimation by a larger factor than shown in Figure 17.114 will cause aliasing.

Figure 17.115a shows how to decimate the output of an FIR filter. The filtered data $y(n)$ is stored in a data register that is clocked at the decimated frequency $f_s/M$. This does not change the number of computations required of the digital filter; i.e., it still must calculate each output sample $y(n)$.

Figure 17.115b shows a method for increasing the computational efficiency of the FIR filter by a factor of M. The data from the delay registers are simply stored in N data registers that

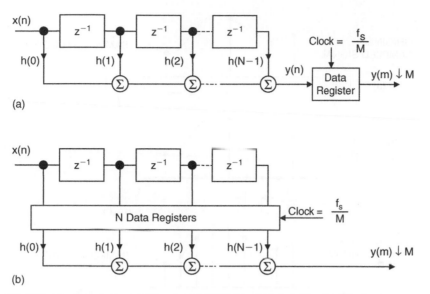

**Figure 17.115: Decimation combined with FIR filtering (a) No change in computational Efficieny; (b) Computational efficiency increased by factor of M**

are clocked at the decimated frequency $f_s/M$. The FIR multiply-accumulates now only have to be done every Mth clock cycle. This increase in efficiency could be utilized by adding more taps to the FIR filter, doing other computations in the extra time, or using a slower DSP.

Figure 17.116 shows the concept of interpolation. The original signal in 17.116a is sampled at a frequency $f_s$. In 17.116b, the sampling frequency has been increased by a factor of L, and zeros have been added to fill in the extra samples. The signal with added zeros is passed through an interpolation filter, which provides the extra data values.

The frequency domain effects of interpolation are shown in Figure 17.117. The original signal is sampled at a frequency $f_s$ and is shown in 17.117a. The interpolated signal in 17.117b is sampled at a frequency $Lf_s$. An example of interpolation is a CD player DAC, where the CD data is generated at a frequency of 44.1 kHz. If this data is passed directly to a DAC, the frequency spectrum shown in Figure 17.117a results, and the requirements on the anti-imaging filter that precedes the DAC are extremely stringent to overcome this. An oversampling interpolating DAC is normally used, and the spectrum shown in Figure 17.117b results. Notice that the requirements on the analog anti-imaging filter are now

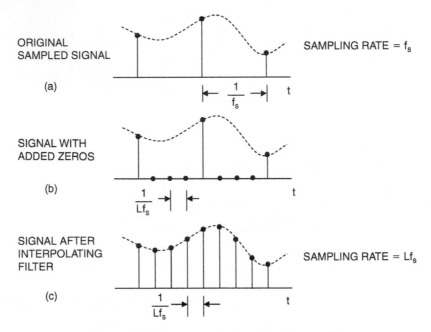

**Figure 17.116: Interpolation by a factor of L**

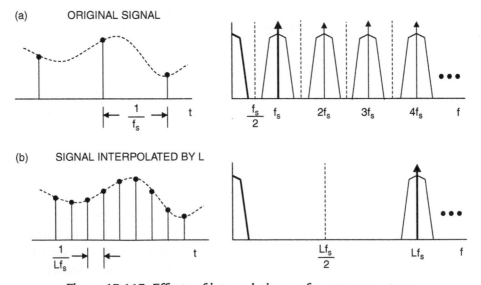

**Figure 17.117: Effects of interpolation on frequency spectrum**

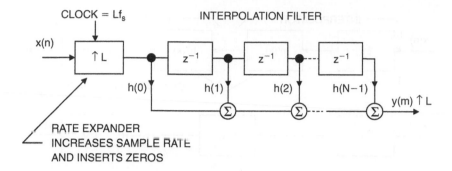

**Figure 17.118: Typical interpolation implementation**

easier to realize. This is important in maintaining relatively linear phase and also reducing the cost of the filter.

The digital implementation of interpolation is shown in Figure 17.118. The original signal x(n) is first passed through a rate expander that increases the sampling frequency by a factor of L and inserts the extra zeros. The data then passes through an interpolation filter that smoothes the data and interpolates between the original data points. The efficiency of this filter can be improved by using a filter algorithm that takes advantage of the fact that the zero-value input samples do not require multiply-accumulates. Using a DSP that allows circular buffering and zero-overhead looping also improves efficiency.

Interpolators and decimators can be combined to perform fractional sample rate conversion as shown in Figure 17.119. The input signal x(n) is first interpolated by a factor of L and then decimated by a factor of M. The resulting output sample rate is $Lf_s/M$. To maintain the maximum possible bandwidth in the intermediate signal, the interpolation must come before the decimation; otherwise, some of the desired frequency content in the original signal would be filtered out by the decimator.

An example is converting from the CD sampling rate of 44.1 kHz to the digital audio tape (DAT) sampling rate of 48.0 kHz. The interpolation factor is 160, and the decimation

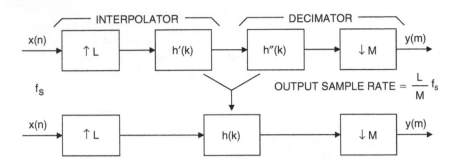

Example: Convert   CD Sampling Rate = 44.1kHz  to
DAT Sampling Rate = 48.0kHz

- Use L = 160, M = 147

- $f_{out} = \dfrac{L}{M} f_s \ = \dfrac{160}{147} \times 44.1\text{kHz} = 48.0\text{kHz}$

- AD189X - Family of Sample Rate Converters

**Figure 17.119: Sample rate converters**

factor, 147. In practice, the interpolating filter h′(k) and the decimating filter h″ (k) are combined into a single filter, h(k).

The entire sample rate conversion function is integrated into the AD1890, AD1891, AD1892, and AD1893 family which operates at frequencies between 8 kHz and 56 kHz (48 kHz for the AD1892). The AD1895 and AD1896 operate at up to 192 kHz.

### 17.18.7 Adaptive Filters

Unlike analog filters, the characteristics of digital filters can easily be changed by modifying the filter coefficients. This makes digital filters attractive in communications applications such as adaptive equalization, echo cancellation, noise reduction, speech analysis, and speech synthesis. The basic concept of an adaptive filter is shown in Figure 17.120. The objective is to filter the input signal, x(n), with an adaptive filter in such a manner that it matches the desired signal, d(n). The desired signal, d(n), is subtracted from the filtered signal, y(n), to generate an error signal. The error signal drives an adaptive algorithm that generates the filter coefficients in a manner that minimizes the error signal. The least-mean-square (LMS) or recursive-least-squares (RLS) algorithms are two of the most popular.

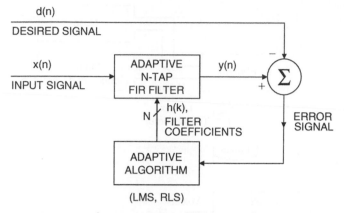

**Figure 17.120: Adaptive filter**

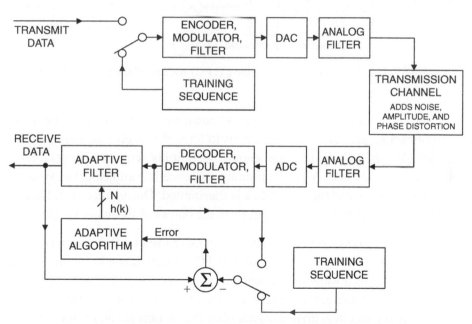

**Figure 17.121: Digital transmission using adaptive equalization**

Adaptive filters are widely used in communications to perform such functions as equalization, echo cancellation, noise cancellation, and speech compression. Figure 17.121 shows an application of an adaptive filter used to compensate for the effects of amplitude and phase distortion in the transmission channel. The filter coefficients are

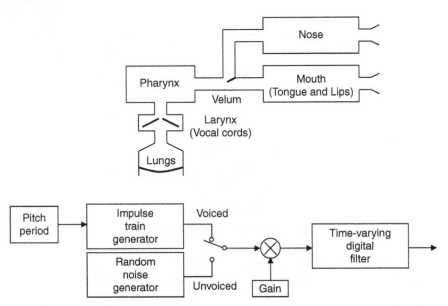

**Figure 17.122: Linear predictive coding (LPC) model of speech production**

determined during a training sequence where a known data pattern is transmitted. The adaptive algorithm adjusts the filter coefficients to force the receive data to match the training sequence data. In a modem application, the training sequence occurs after the initial connection is made. After the training sequence is completed, the switches are put in the other position, and the actual data is transmitted. During this time, the error signal is generated by subtracting the input from the output of the adaptive filter.

Speech compression and synthesis also makes extensive use of adaptive filtering to reduce data rates. The linear predictive coding (LPC) model shown in Figure 17.122 models the vocal tract as a variable frequency impulse generator (for voiced portions of speech) and a random noise generator (for unvoiced portions of speech such as consonant sounds). These two generators drive a digital filter that in turn generates the actual voice signal.

The application of LPC in a communication system such as GSM is shown in Figure 17.123. The speech input is first digitized by a 16-bit ADC at a sampling frequency of 8 kSPS. This produces output data at 128 Kbps, which is much too high to be transmitted directly. The transmitting DSP uses the LPC algorithm to break the speech signal into

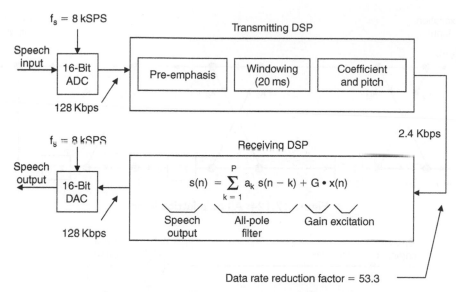

**Figure 17.123: LPC speech companding system**

digital filter coefficients and pitch. This is done in 20 ms windows, which have been found to be optimum for most speech applications. The actual transmitted data is only 2.4 Kbps, which represents a 53.3 compression factor. The receiving DSP uses the LPC model to reconstruct the speech from the coefficients and the excitation data. The final output data rate of 128 Kbps then drives a 16-bit DAC for final reconstruction of the speech data.

The digital filters used in LPC speech applications can either be FIR or IIR, although all-pole IIR filters are the most widely used. Both FIR and IIR filters can be implemented in a lattice structure as shown in Figure 17.122 for a recursive all-pole filter. This structure can be derived from the IIR structure, but the advantage of the lattice filter is that the coefficients are more directly related to the outputs of algorithms that use the vocal tract model shown in Figure 17.122 than the coefficients of the equivalent IIR filter.

The parameters of the all-pole lattice filter model are determined from the speech samples by means of linear prediction as shown in Figure 17.123. Due to the nonstationary nature of speech signals, this model is applied to short segments (typically 20 ms) of the speech signal. A new set of parameters is usually determined for each time segment unless there are sharp discontinuities, in which case the data may be smoothed between segments.

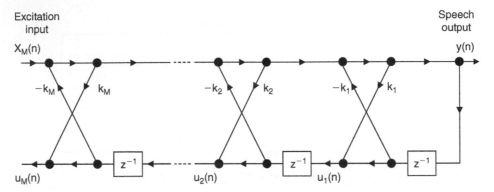

Figure 17.124: All pole lattice filter

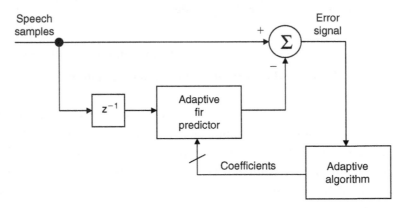

Figure 17.125: Estimation of lattice filter coefficients in transmitting DSP

# References

1. Steven W. Smith, *Digital Signal Processing: A Guide for Engineers and Scientists*, Newnes, 2002.

2. C. Britton Rorabaugh, *DSP Primer*, McGraw-Hill, 1999.

3. Richard J. Higgins, *Digital Signal Processing in VLSI*, Prentice-Hall, 1990.

4. A.V. Oppenheim and R.W. Schafer, *Digital Signal Processing*, Prentice-Hall, 1975.

5. L.R. Rabiner and B. Gold, *Theory and Application of Digital Signal Processing*, Prentice-Hall, 1975.

6.  John G. Proakis and Dimitris G. Manolakis, *Introduction to Digital Signal Processing*, MacMillian, 1988.

7.  J.H. McClellan, T.W. Parks, and L.R. Rabiner, "A Computer Program for Designing Optimum FIR Linear Phase Digital Filters," IEEE Trasactions on Audio and Electroacoustics, Vol. AU-21, No. 6, December, 1973.

8.  Fredrick J. Harris, "On the Use of Windows for Harmonic Analysis with the Discrete Fourier Transform," Proc. IEEE, Vol. 66, No. 1, 1978, pp. 51–83.

9.  Momentum Data Systems, Inc., 17330 Brookhurst St., Suite 140, Fountain Valley, CA 92708, http://www.mds.com.

10. *Digital Signal Processing Applications Using the ADSP-2100 Family*, Vol. 1 and Vol. 2, Analog Devices, Free Download at: http://www.analog.com.

11. *ADSP-21000 Family Application Handbook*, Vol. 1, Analog Devices, Free Download at: http://www.analog.com.

12. B. Widrow and S.D. Stearns, *Adaptive Signal Processing*, Prentice-Hall, 1985.

13. S. Haykin, *Adaptive Filter Theory*, 3rd Edition, Prentice-Hall, 1996.

14. Michael L. Honig and David G. Messerschmitt, *Adaptive Filters – Structures, Algorithms, and Applications*, Kluwer Academic Publishers, Hingham, MA, 1984.

15. J.D. Markel and A.H. Gray, Jr., *Linear Prediction of Speech*, Springer-Verlag, New York, NY, 1976.

16. L.R. Rabiner and R.W. Schafer, *Digital Processing of Speech Signals*, Prentice-Hall, 1978.

17. R.W. Ramirez, *The FFT: Fundamentals and Concepts*, Prentice-Hall, 1985.

18. J.W. Cooley and J.W. Tukey, "An Algorithm for the Machine Computation of Complex Fourier Series," Mathematics Computation, Vol. 19, April 1965, pp. 297–301.

# CHAPTER 18

# Control and Instrumentation Systems

W. Bolton

## 18.1 Introduction

The term *automation* is used to describe the automatic operation or control of a process. In modem manufacturing there is an ever increasing use of automation, e.g., automatically operating machinery, perhaps in a production line with robots, which can be used to produce components with virtually no human intervention. Also, in appliances around the home and in the office there is an ever increasing use of automation. Automation involves carrying out operations in the required sequence and controlling outputs to required values.

The following are some of the key historical points in the development of automation, the first three being concerned with developments in the organization of manufacturing which permitted the development of automated production:

1. Modern manufacturing began in England in the 18th century when the use of water wheels and steam engines meant that it became more efficient to organize work to take place in factories, rather than it occurring in the home of a multitude of small workshops. The impetus was thus provided for the development of machinery.

2. The development of powered machinery in the early 1900s meant improved accuracy in the production of components so that instead of making each individual component to fit a particular product, components were fabricated in identical batches with an accuracy which ensured that they could fit any one of a batch of a product. Think of the problem of a nut and bolt if each nut has to be individually made so that it fitted the bolt and the advantages that are gained

by the accuracy of manufacturing nuts and bolts being high enough for any of a batch of nuts to fit a bolt.

3. The idea of production lines followed from this with Henry Ford, in 1909, developing them for the production of motor cars. In such a line, the production process is broken up into a sequence of set tasks with the potential for automating tasks and so developing an automated production line.

4. In the 1920s developments occurred in the theoretical principles of control systems and the use of feedback for exercising control. A particular task of concern was the development of control systems to steer ships and aircraft automatically.

5. In the 1940s, during the Second World War, developments occurred in the application of control systems to military tasks, such as radar tracking and gun control.

6. The development of the analysis and design of feedback amplifiers, including the paper by Bode in 1945 on Network Analysis and Feedback Amplifier design, was instrumental in further developing control system theory.

7. Numerical control was developed in 1952 whereby tool positioning was achieved by a sequence of instructions provided by a program of punched paper tape, these directing the motion of the motors driving the axes of the machine tool. There was no feedback of positional data in these early control systems to indicate whether the tool was in the correct position, the system being open-loop control.

8. The invention of the transistor in 1948 in the United States led to the development of integrated circuits, and, in the 1970s, microprocessors and computers which enabled control systems to be developed which were cheap and able to be used to control a wide range of processes. As a consequence, automation has spread to common everyday processes such as the domestic washing machine and the automatic focusing, automatic exposure, camera.

This chapter is an introduction to the basic ideas involved in designing control systems with sections 18.2 through 18.7 being an introduction to the basic idea of a control system and the elements used.

## 18.2  Systems

A car gear box can be thought of as a system with an input shaft and an output shaft (Figure 18.l(a)). We supply a rotation to the input shaft and the system then provides a rotation of the output shaft with the rotational speed of the output shaft being related in some way to the rotational speed of the input shaft. Likewise, we can think of an amplifier as a system to which we can supply an input signal and from which we can obtain an output signal which is related in some way to the input signal (Figure 18.1(b)). Thus, we can think of a system as being like a closed box in which the workings of the system are enclosed and to which we can apply an input, or inputs, and obtain an output, or outputs, with the output being related to the input.

A system can be defined as an arrangement of parts within some boundary which work together to provide some form of output from a specified input or inputs (Figure 18.1(c)). The boundary divides the system from the environment and the system interacts with the environment by means of signals crossing the boundary from the environment to the system; that is inputs, and signals crossing the boundary from the system to the environment, i.e., outputs.

With an engineering system an engineer is more interested in the inputs and outputs of a system than the internal workings of the component elements of that system. By considering devices as systems we can concentrate on what they do rather than their internal workings.

Thus if we know the relationship between the output and the input of a system we can work out how it will behave whether it be a mechanical, pneumatic, hydraulic, electrical or electronic system. We can see the overall picture without becoming bogged down by internal detail. An operational amplifier is an example of this approach. We can

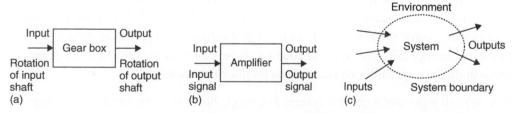

**Figure 18.1: Systems: (a) a gear box; (b) an amplifier; (c) the formal picture defining a system**

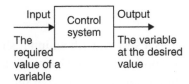

**Figure 18.2: Control system: an input of the required value of some variable and an output of the variable at the desired value**

design circuits involving operational amplifiers by making use of the known relationship between input and output without knowing what is going on inside it.

In this book we are concerned with control systems. *Control systems are systems that are used to maintain a desired result or value* (Figure 18.2). For example, driving a car along a road involves the brain of the driver as a controller comparing the actual position of the car on the road with the desired position and making adjustments to correct any error between the desired and actual position. A room thermostat is another example of a controller, it controlling the heating system to give the required room temperature by switching the heater on or off to reduce the error between the actual temperature and the required temperature.

With a systems approach to control, we express the physical system in terms of a model with the various physical components described as system blocks with inputs and outputs and the relationship between the inputs and outputs expressed by means of a mathematical equation.

### 18.2.1 Block Diagrams

A useful way of representing a system is as a block diagram: within the boundary described by the box outline is the system and inputs to the system are shown by arrows entering the box and outputs by arrows leaving the box. Figure 18.3(a) illustrates this for an electric motor system; there is an input of electrical energy and an output of mechanical energy in the form of the rotation of the motor shaft. We can think of the system in the box operating on the input to produce the output.

While we can represent a control system as a single block with an input and an output, it is generally more useful to consider the system as a series of interconnected system elements with each system element being represented by a block having a particular

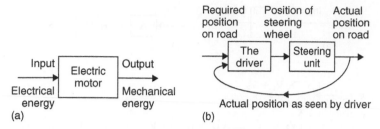

**Figure 18.3: Examples of block diagrams to represent systems: (a) an electric motor; (b) a car driving system involving a number of blocks**

function. Thus, in the case of the driver of a car steering the car along a road we can consider the overall control system to have the elements of: the driver with an input of the actual position he/she sees of the car on the road and also his/her thoughts on where the car should be in relation to the road giving an output of the hands turning the steering wheel; the car steering unit with the input of the steering wheel position and the output of the front wheel positions and hence the positioning of the car on the road. Figure 18.3(b) shows how we might represent these elements.

In drawing formal block diagram models we use a number of conventions to represent the elements and connections:

**1. System element**

A system element is shown as a box with an input shown as an inward directed arrow and an output as an outward directed arrow (Figure 18.4(a)).

**2. Information flows**

A control system will be made up of a number of interconnected systems and we can draw a model of such a system as a series of interconnected blocks. Thus we can have one box giving an output which then becomes the input for another box (Figure 18.4(b)). We draw a line to connect the boxes and indicate a flow of information in the direction indicated by the arrow; the lines does not necessarily represent a physical connection or the form of a physical connection.

**3. Summing junction**

We often have situations with control systems where two signals are perhaps added together or one subtracted from another and the result of such operations then fed on

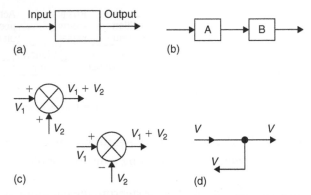

**Figure 18.4: Block diagram elements: (a) system element; (b) information paths; (c) summing junction; (d) takeoff point**

to some system element. This is represented by a circle with the inputs to quadrants of the circle given + or − signs to indicate whether we are summing two positive quantities or summing a positive quantity and a negative quantity and so subtracting signals (Figure 18.4(c)).

4. **Take-off point**

In the case of the car driving system shown in Figure 18.3(b), the overall output is the actual position of the car on the road. But this signal is also tapped off to become an input to the car driver so that he or she can compare the actual position with the required position to adjust the steering wheel accordingly. As another illustration, in the case of a central heating system the overall output is the temperature of a room. But this temperature signal is also tapped off to become an input to the thermostat system where it is compared with the required temperature signal. Such a "tapping-off" point in the system is represented as shown in Figure 18.4(d).

As an illustration of the use of the above elements in drawing a block diagram model for a control system, consider a central heating control system with its input the temperature required in the house and as its output the house at that temperature. Figure 18.5 shows how we can represent such a system with a block diagram.

The required temperature is set on the thermostat and this element gives an output signal which is used to switch on or off the heating furnace and so produce an output affecting

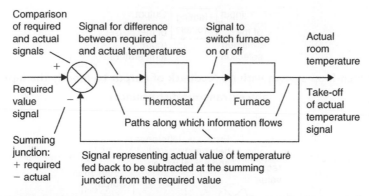

**Figure 18.5: Block diagram for a central heating system employing negative feedback**

the variable which is the room temperature. The room temperature provides a signal which is fed back to the thermostat. This responds to the difference between the required temperature signal and the actual temperature signal.

## 18.3  Control Systems Models

There are two basic types of control systems:

### 1. Open-loop

In an open-loop control system the output from the system has no effect on the input signal to the plant or process. The output is determined solely by the initial setting. Open-loop systems have the advantage of being relatively simple and consequently cheap with generally good reliability. However, they are often inaccurate since there is no correction for errors in the output which might result from extraneous disturbances.

As an illustration of an open-loop system, consider the heating of a room to some required temperature using an electric fire which has a selection switch which allows a 1 kW or a 2 kW heating element to be selected. The decision might be made, as a result of experience, that to obtain the required temperature it is only necessary to switch on the 1 kW element. The room will heat up and reach a temperature which is determined by the fact the 1 kW element is switched on. The temperature of the room is thus controlled by an initial decision and no further adjustments are made. Figure 18.6 illustrates this. If there are changes in the conditions, perhaps someone opening a window, no adjustments are made to the heat output from the fire to

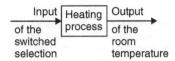

**Figure 18.6: Open-loop system with no feedback of output to modify the input if there are any extraneous disturbances**

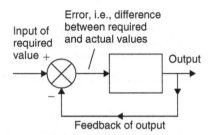

**Figure 18.7: Closed-loop system with feedback of output to modify the input and so adjust for any extraneous disturbances**

compensate for the change. There is no information *fed back* to the fire to adjust it and maintain a constant temperature.

Open-loop control is often used with processes that require the sequencing of events by on-off signals, like washing machines which require the water to be switched on and then, after a suitable time, switched off followed by the heater being switched on and then, after a suitable time, switched off.

## 2. Closed-loop

In a closed-loop control system a signal indicating the state of the output of the system is *fed back* to the input where it is compared with what was required and the difference used to modify the output of the system so that it maintains the output at the required value (Figure 18.7). The term *closed-loop* refers to the loop created by the feedback path. Closed-loop systems have the advantage of being relatively accurate in matching the actual to the required values. They are, however, more complex and so more costly with a greater chance of breakdown as a consequence of the greater number of components.

As an illustration, consider modifications of the open-loop heating system described above to give a closed-loop system. To obtain the required temperature, a person stands

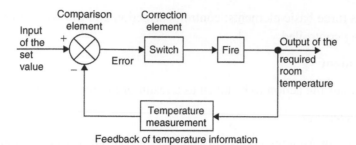

**Figure 18.8: Closed-loop system with feedback being used to modify the input to the controller and so enable the control system to adjust when there are extraneous disturbances**

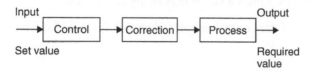

**Figure 18.9: Basic elements of an open-loop control system**

in the room with a thermometer and switches the 1 kW and 2 kW elements on or off, according to the difference between the actual room temperature and the required temperature in order to maintain the temperature of the room at the required temperature. In this situation there is *feedback,* information being fed back from the output to modify the input to the system. Thus if a window is opened and there is a sudden cold blast of air, the feedback signal changes because the room temperature changes and is fed back to modify the input to the system. The input to the heating process depends on the deviation of the actual temperature fed back from the output of the system from the required temperature initially set. Figure 18.8 illustrates this system with the comparison element represented by the summing symbol with a + opposite the set value input and a − opposite the feedback signal to give the sum as + set value − feedback value = error. This error signal is then used to control the process. Because the feedback signal is subtracted from the set value signal, the system is said to have *negative feedback.*

### 18.3.1 Basic Elements of an Open-Loop Control System

The term *open-loop control system* is used for a system where an input to a system is chosen on the basis of previous experience as likely to give the output required. Figure 18.9 shows the basic form of such a system.

The system has three basic elements: control, correction and the process of which a variable is being controlled.

1. **Control element**

   This determines the action to be taken as a result of the input to the system.

2. **Correction element**

   This has an input from the controller and gives an output of some action designed to change the variable being controlled.

3. **Process**

   This is the process of which a variable is being controlled.

There is no changing of the control action to account for any disturbances which change the output variable.

### 18.3.2 Basic Elements of a Closed-Loop System

Figure 18.10 shows the general form of a basic closed-loop system.

The following are the functions of the constituent elements:

1. **Comparison element**

   This element compares the required value of the variable being controlled with the measured value of what is being achieved and produces an error signal:

   error = reference value signal − measured actual value signal

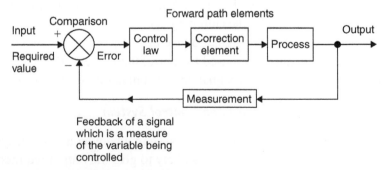

**Figure 18.10: Basic elements of a closed-loop control system**

Thus if the output is the required value then there is no error and so no signal is fed to initiate control. Only when there is a difference between the required value and the actual values of the variable will there be an error signal and so control action initiated.

## 2. Control law implementation element

The control law element determines what action to take when an error signal is received. The control law used by the element may be just to supply a signal which switches on or off when there is an error, as in a room thermostat, or perhaps a signal which is proportional to the size of the error. With a proportional control law implementation, if the error is small a small control signal is produced and if the error is large a large control signal is produced. Other control laws include *integral mode* where a control signal is produced that continues to increase as long as there is an error and *derivative mode* where the control signal is proportional to the rate at which the error is changing.

The term *control unit* or *controller* is often used for the combination of the comparison element, i.e., the error detector, and the control law implementation element. An example of such an element is a differential amplifier which has two inputs, one the set value and one the feedback signal, and any difference between the two is amplified to give the error signal. When there is no difference there is no resulting error signal.

## 3. Correction element

The correction element or, as it is often called, the *final control element,* produces a change in the process which aims to correct or change the controlled condition. The term *actuator* is used for the element of a correction unit that provides the power to carry out the control action. An example is a motor, with an input of a voltage to its armature coils and an output of a rotating shaft which, via possibly a screw, rotates and corrects the position of a workpiece (Figure 18.11(a)). Another example is a hydraulic or pneumatic cylinder (Figure 18.11(b)). The cylinder has a piston which can be moved along the cylinder depending on a pressure signal from the controller.

## 4. Process

The process is the system in which there is a variable that is being controlled; for example, it might be a room in a house with its temperature being controlled.

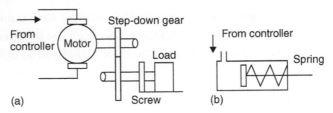

**Figure 18.11: Actuator examples: (a) motor; (b) cylinder**

## 5. Measurement element

The measurement element produces a signal related to the variable condition of the process that is being controlled. For example, it might be a temperature sensor with suitable signal processing.

The following are terms used to describe the various paths through the system taken by signals:

## 6. Feedback path

*Feedback* is a means whereby a signal related to the actual condition being achieved is fed back to modify the input signal to a process. The feedback is said to be *negative* when the signal which is fed back subtracts from the input value. It is negative feedback that is required to control a system. *Positive feedback* occurs when the signal fed back adds to the input signal.

## 7. Forward path

The term *forward path* is used for the path from the error signal to the output. In Figure 18.10 these forward path elements are the control law element, the correction element and the process element.

### 18.3.3 Discrete Event Control

This is often described as *sequential control* and describes control systems where control actions are determined in response to observed time-critical events. For example, the filling of a container with water might have a sensor at the bottom that registers when the container is empty and so gives an input to the controller to switch the water flow on and a sensor at the top that registers when the container is full and so gives an input to the

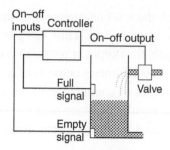

**Figure 18.12: Discrete-event control with the controller switching the valve open when empty signal received and closed when the full signal**

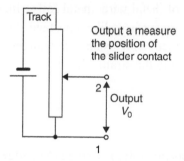

**Figure 18.13: Potentiometer as a sensor of position**

controller to switch off the flow of water. This is a form of closed-loop system since the controller is receiving feedback from the two sensors regarding the state of the variable (Figure 18.12).

## 18.4 Measurement Elements

The following are examples of sensors that are commonly used with the measurement systems of control systems.

### 18.4.1 Potentiometer

A potentiometer consists of a resistance element with a sliding contact which can be moved over the length of the element and connected as shown in Figure 18.13. With a constant supply voltage $V_s$ the output voltage $V_0$ between terminals 1 and 2 is a fraction of the input voltage, the fraction depending on the ratio of the resistance $R_{12}$ between terminals 1 and 2 compared with the total resistance $R$ of the entire length of the track across which the supply voltage is connected. Thus $V_o/V_s = R_{12}/R$. If the track has a constant resistance per unit length, the output is proportional to the displacement of the slider from position 1. A rotary potentiometer consists of a coil of wire wrapped round into a circular track or a circular film of conductive plastic over which a rotatable sliding contact can be rotated, hence an angular displacement can be converted into a potential difference. Linear tracks can be used for linear displacements.

### 18.4.2 Strain-Gauged Element

Figure 18.14(a) shows the basic form of an electrical resistance strain gauge. Strain gauges consist of a fiat length of metal wire, metal foil strip, or a strip of semiconductor material which can be stuck onto surfaces like a postage stamp. When the wire, foil, strip or semiconductor is stretched, its resistance $R$ changes. The fractional change in resistance $\Delta R/R$ is proportional to the strain $\varepsilon$, i.e.:

$$\frac{\Delta R}{R} = G\varepsilon$$

where $G$, the constant of proportionality, is termed the gauge factor. Metal strain gauges typically have gauge factors of the order of 2.0.

When such a strain gauge is stretched, its resistance increases, when compressed its resistance decreases. A displacement sensor might be constructed by attaching strain gauges to a cantilever (Figure 18.14(b)), the free end of the cantilever being moved as a result of the linear displacement being monitored. When the cantilever is bent, the electrical resistance strain gauges mounted on the element are strained and so give a resistance change which can be monitored and which is a measure of the displacement. With strain gauges mounted as shown in Figure 18.14, when the cantilever is deflected downward the gauge on the upper surface is stretched and the gauge on the lower surface compressed. Thus the gauge on the upper surface increases in resistance while that on the lower surface decreases.

### 18.4.3 Linear Variable Differential Transformer

The linear variable differential transformer, generally referred to by the abbreviation LVDT, is a transformer with a primary coil and two secondary coils. Figure 18.15 shows

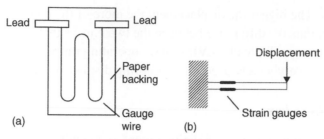

Figure 18.14: (a) Strain gauge; (b) example of use on a cantilever to provide a
displacement sensor

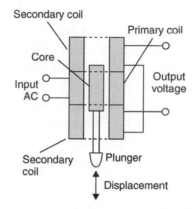

Figure 18.15: LVDT: Giving an output voltage related to the position of the plunger

the arrangement, there being three coils symmetrically spaced along an insulated tube.
The central coil is the primary coil and the other two are identical secondary coils which
are connected in series in such a way that their outputs oppose each other. A magnetic
core is moved through the central tube as a result of the displacement being monitored.
When there is an alternating voltage input to the primary coil, alternating e.m.f.s are
induced in the secondary coils. With the magnetic core in a central position, the amount
of magnetic material in each of the secondary coils is the same and so the e.m.f.s induced
in each coil are the same. Since they are so connected that their outputs oppose each
other, the net result is zero output. However, when the core is displaced from the central
position there is a greater amount of magnetic core in one coil than the other. The result
is that a greater e.m.f. is induced in one coil than the other and then there is a net output

from the two coils. The bigger the displacement the more of the core there is in one coil than the other, thus the difference between the two e.m.f.s increases the greater the displacement of the core. Typically, LVDTs have operating ranges from about $\pm 2\,\text{mm}$ to $\pm 400\,\text{mm}$ and are very widely used for monitoring displacements.

### 18.4.4 Optical Encoders

An encoder is a device that provides a digital output as a result of an angular or linear displacement. Position encoders can be grouped into two categories: incremental encoders, which detect changes in displacement from some datum position, and absolute encoders, which give the actual position.

Figure 18.16 shows the basic form of an *incremental encoder* for the measurement of angular displacement of a shaft. It consists of a disc which rotates along with the shaft. In the form shown, the rotatable disc has a number of windows through which a beam of light can pass and be detected by a suitable light sensor. When the shaft and disc rotates, a pulsed output is produced by the sensor with the number of pulses being proportional to the angle through which the disc rotates. The angular displacement of the disc, and hence the shaft rotating it, can thus be determined by the number of pulses produced in the angular displacement from some datum position. Typically the number of windows on

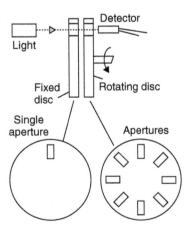

**Figure 18.16: Incremental encoder: angular displacement results in pulses being detected, the number of pulses being proportional to the angular displacement**

the disc varies from 60 to over a thousand with multi-tracks having slightly offset slots in each track. With 60 slots occurring with 1 revolution then, since 1 revolution is a rotation of 360° the minimum angular displacement, i.e., the resolution, that can be detected is 360/60 = 6°. The resolution typically varies from about 6° to 0.3° or better.

With the incremental encoder, the number of pulses counted gives the angular displacement, a displacement of, say, 50° giving the same number of pulses whatever angular position the shaft starts its rotation from. However, the absolute encoder gives an output in the form of a binary number of several digits, each such number representing a particular angular position. Figure 18.17 shows the basic form of an *absolute encoder* for the measurement of angular position. With the one shown in the figure, the rotating disc has four concentric circles of slots and four sensors to detect the light pulses. The slots are arranged in such a way that the sequential output from the sensors is a number in the binary code, each such number corresponding to a particular angular position. A number of forms of binary code are used. Typical encoders tend to have up to 10 or 12 tracks. The number of bits in the binary number will be equal to the number of tracks. Thus with 10 tracks there will be 10 bits and so the number of positions that can be detected is $2^{10}$, i.e., 1024, a resolution of 360/1024 = 0.35°.

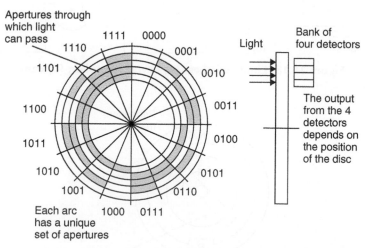

**Figure 18.17: The rotating wheel of the absolute encoder: the binary word output indicates the angular position**

### 18.4.5 Switches

There are many situations where a sensor is required to detect the presence of some object. The sensor used in such situations can be a mechanical switch, giving an on-off output when the switch contacts are opened or closed by the presence of an object. Figure 18.18 illustrates the forms of a number of such switches. An example of switch application is where a work piece closes the switch by pushing against it when it reaches the correct position on a work table, such a switch being referred to as a limit switch. The switch might then be used to switch on a machine tool to carry out some operation on the work piece.

### 18.4.6 Tachogenerator

The basic tachogenerator consists of a coil mounted in a magnetic field (Figure 18.19). When the coil rotates electromagnetic induction results in an alternating e.m.f. being

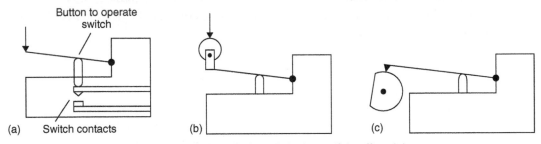

**Figure 18.18: Limit switches: (a) Lever; (b) roller; (c) cam**

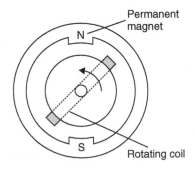

**Figure 18.19: The tachogenerator**

induced in the coil. The faster the coil rotates the greater the size of the alternating e.m.f. Thus the size of the alternating e.m.f. is a measure of the angular speed.

### 18.4.7 Pressure Sensors

The movement of the center of a circular diaphragm as a result of a pressure difference between its two sides is the basis of a pressure gauge (Figure 18.20(a)). For the measurement of the absolute pressure, the opposite side of the diaphragm is a vacuum, for the measurement of pressure difference the pressures are connected to each side of the diaphragm, for the gauge pressure, i.e., the pressure relative to the atmospheric pressure, the opposite side of the diaphragm is open to the atmosphere. The amount of movement with a plane diaphragm is fairly limited; greater movement can, however, be produced with a diaphragm with corrugations (Figure 18.20(b)).

The movement of the center of a diaphragm can be monitored by some form of displacement sensor. Figure 18.21 shows the form that might be taken when strain gauges are used to monitor the displacement, the strain gauges being stuck to the diaphragm

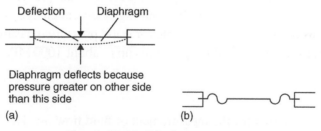

**Figure 18.20: Diaphragm sensors**

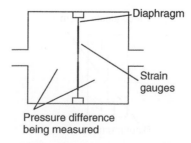

**Figure 18.21: Pressure gauge with strain gauges to sense movement of the diaphragm**

and changing resistance as a result of the diaphragm movement. Typically such sensors are used for pressures over the range 100 kPa to 100 MPa, with an accuracy up to about ±0.1%. Another form of diaphragm pressure gauge uses strain gauge elements integrated within a silicon diaphragm and supplied, together with a resistive network for signal processing, on a single silicon chip as the Motorola MPX pressure sensor. With a voltage supply connected to the sensor, it gives an output voltage directly proportional to the pressure. Such sensors are available for use for the measurement of absolute pressure, differential pressure or gauge pressure; for example, MPX2100 has a pressure range of 100 kPa and with a supply voltage of 16V DC gives a voltage output over the full range of 40 mV.

When certain crystals are stretched or compressed, charges appear on their surfaces. This effect is called *piezo-electricity*. Examples of such crystals are quartz, tourmaline, and zirconate-titanate. A piezoelectric pressure gauge consists essentially of a diaphragm which presses against a piezoelectric crystal (Figure 18.22). Movement of the diaphragm causes the crystal to be compressed and so charges produced on its surface. The crystal can be considered to be a capacitor which becomes charged as a result of the diaphragm movement and so a potential difference appears across it. If the pressure keeps the diaphragm at a particular displacement, the resulting electrical charge is not maintained but leaks away. Thus the sensor is not suitable for static pressure measurements. Typically, such a sensor can be used for pressures up to about 1000 MPa.

### 18.4.8 Fluid Flow

The traditional methods used for the measurement of fluid flow involve devices based on Bernoulli's equation. When a restriction occurs in the path of a flowing fluid, a pressure

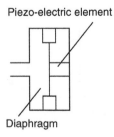

Piezo-electric element

Diaphragm

**Figure 18.22: Basic form of a piezo-electric sensor**

drop is produced with the flow rate being proportional to the square root of the pressure drop. Hence, a measurement of the pressure difference can be used to give a measure of the rate of flow. There are many devices based on this principle. The *Venturi tube* is a tube which gradually tapers from the full pipe diameter to the constricted diameter. Figure 18.23 shows the typical form of such a tube. The pressure difference is measured between the flow prior to the constriction and at the constriction, a diaphragm pressure cell generally being used. The *orifice plate* (Figure 18.24) is simply a disc, with generally a central hole. The orifice plate is placed in the tube through which the fluid is flowing and the pressure difference measured between a point equal to the diameter of the tube upstream and a point equal to half the diameter downstream. Because of the way the fluid flows through the orifice plate, such measurements are equivalent to those taken with the Venturi tube.

The turbine flowmeter (Figure 18.25) consists of a multi-bladed rotor that is supported centrally in the pipe along which the flow occurs. The rotor rotates as a result of the fluid flow, the angular velocity being approximately proportional to the flow rate. The rate of revolution of the rotor can be determined by attaching a small permanent magnet to one of the blades and using a pick-up coil. An induced e.m.f, pulse is produced in the coil

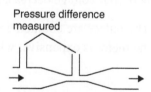

**Figure 18.23: Venturi tube: pressure difference is a measure of the flow rate**

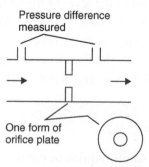

**Figure 18.24: Orifice plate**

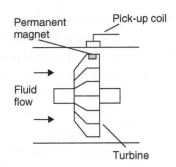

**Figure 18.25: Basic principle of the turbine flowmeter: number of pulses picked-up per second is a measure of flow rate**

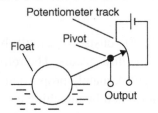

**Figure 18.26: Potentiometer float gauge: as the float rises the output voltage decreases**

every time the magnet passes it. The pulses are counted and so the number of revolutions of the rotor can be determined. The meter is expensive, with an accuracy of typically about ±0.1%.

### 18.4.9 Liquid Level

A commonly used method to measure the level of liquid in a vessel is a float whose position is directly related to the liquid level. Figure 18.26 shows a simple float system. The float is at one end of a pivoted rod with the other end connected to the slider of a potentiometer. Changes in level cause the float to move and hence move the slider over the potentiometer resistance track and so give a potential difference output related to the liquid level.

### 18.4.10 Temperature Sensors

The expansion or contraction of solids, liquids or gases, the change in electrical resistance of conductors and semiconductors, thermoelectric e.m.f.s and the change in the current

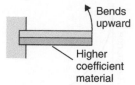

**Figure 18.27: Bimetallic strip; as the temperature increases it bends upward with the higher coefficient material extending more than the lower coefficient material**

across the junction of semiconductor diodes and transistors are all examples of properties that change when the temperature changes and can be used as basis of temperature sensors.

The *bimetallic strip device* consists of two different metal strips of the same length bonded together (Figure 18.27). Because the metals have different coefficients of expansion, when the temperature increases the composite strip bends into a curved strip, with the higher coefficient metal on the outside of the curve. The amount by which the strip curves depends on the two metals used, the length of the composite strip and the change in temperature. If one end of a bimetallic strip is fixed, the amount by which the free end moves is a measure of the temperature. This movement may be used to open or close electric circuits, as in the simple thermostat commonly used with domestic heating systems. Bimetallic strip devices are robust, relatively cheap, have an accuracy of the order of ±1% and are fairly slow reacting to changes in temperature.

*Resistance temperature detectors* (RTDs) are simple resistive elements in the form of coils of wire of such metals as platinum, nickel or copper alloys, the resistance varying as the temperature changes with the change in resistance being reasonably proportional to the change in temperature. Detectors using platinum have high linearity, good repeatability, high long term stability, can give an accuracy of ±0.5% or better, a range of about −200°C to +850°C can be used in a wide range of environments without deterioration, but are more expensive than the other metals. They are, however, very widely used. Nickel and copper alloys are cheaper but have less stability, are more prone to interaction with the environment and cannot be used over such large temperature ranges.

*Thermistors* are semiconductor temperature sensors made from mixtures of metal oxides, such as those of chromium, cobalt, iron, manganese and nickel. The resistance of

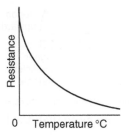

Figure 18.28: Variation of resistance with temperature for thermistors

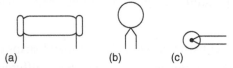

Figure 18.29: Thermistors: (a) rod; (b) disc; (c) bead

thermistors decreases in a very nonlinear manner with an increase in temperature, Figure 18.28 illustrating this. The change in resistance per degree change in temperature is considerably larger than that which occurs with metals. For example, a thermistor might have a resistance of 29 kΩ at −20°C, 9.8 kΩ at 0°C, 3.75 kΩ at 20°C, 1.6 kΩ at 40°C, 0.75 kΩ at 60°C. The material is formed into various forms of element, such as beads, discs and rods (Figure 18.29). Thermistors are rugged and can be very small, so enabling temperatures to be monitored at virtually a point. Because of their small size they have small thermal capacity and so respond very rapidly to changes in temperature. The temperature range over which they can be used will depend on the thermistor concerned, ranges within about −100°C to +300°C being possible. They give very large changes in resistance per degree change in temperature and so are capable, over a small range, of being calibrated to give an accuracy of the order of 0.1°C or better. However, their characteristics tend to drift with time. Their main disadvantage is their nonlinearity.

When two different metals are joined together, a potential difference occurs across the junction. The potential difference depends on the two metals used and the temperature of the junction. A *thermocouple* involves two such junctions, as illustrated in Figure 18.30. If both junctions are at the same temperature, the potential differences across the two

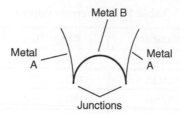

**Figure 18.30: Thermocouple: a temperature difference between the junctions gives a voltage between them**

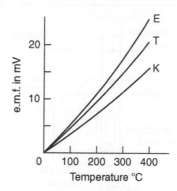

**Figure 18.31: Thermocouples: chromel-constantan (E), chromelalumel (K), copper-constantan (T)**

junctions cancel each other out and there is no net e.m.f. If, however, there is a difference in temperature between the two junctions, there is an e.m.f. The value of this e.m.f. $E$ depends on the two metals concerned and the temperatures $t$ of both junctions. Usually one junction is held at 0°C and then, to a reasonable extent, the following relationship holds:

$$E = at + bt^2$$

where $a$ and $b$ are constants for the metals concerned. Figure 18.31 shows how the e.m.f, varies with temperature for a number of commonly used pairs of metals. Standard tables giving the e.m.f.s at different temperatures are available for the metals usually used for thermocouples. Commonly used thermocouples are listed in Table 18.1, with the temperature ranges over which they are generally used and typical sensitivities. These commonly used thermocouples are given reference letters. The base-metal

**Table 18.1: Thermocouples**

|   | Type Materials | Range °C | Sensitivity μV/°C |
|---|---|---|---|
| E | Chromel-constantan | 0 to 980 | 63 |
| J | Iron-constantan | −180 to 760 | 53 |
| K | Chromel-alumel | −180 to 1260 | 41 |
| R | Platinum-platinum/ rhodium 13% | 0 to 1750 | 8 |
| T | Copper-constantan | −180 to 370 | 43 |

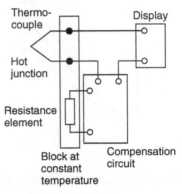

**Figure 18.32: Cold junction compensation to compensate for the cold junction not being at 0°C**

thermocouples, E, J, K and T, have accuracies about ±1 to 3%, are relatively cheap but deteriorate with age. Noblemetal thermocouples, e.g., *R* have accuracies of about ±1% or better, are more expensive but more stable with longer life. Thermocouples are generally mounted in a sheath to give them mechanical and chemical protection. The response time of an unsheathed thermocouple is very fast. With a sheath this may be increased to as much as a few seconds if a large sheath is used.

To maintain one junction of a thermocouple at 0°C it needs to be immersed in a mixture of ice and water. This, however, is often not convenient. A compensation circuit (Figure 18.32) can, however, be used to provide an e.m.f. which varies with the temperature of the cold junction in such a way that when it is added to the thermocouple e.m.f. it generates a combined e.m.f. which is the same as would have been generated if the cold junction had been at 0°C.

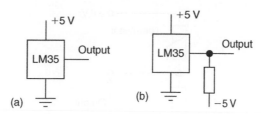

**Figure 18.33: LM35 connections**

There is a change in the current across the junction of *semiconductor diodes* and *transistors* when the temperature changes. For use as temperature sensors they are supplied, together with the necessary signal processing circuitry, as integrated circuits. An integrated circuit temperature sensor using transistors is LM35. This gives an output, which is a linear function of temperature, of 10 mV/°C when the supply voltage is 5 V. Figure 18.33(a) shows the connections for the range 12°C to 110°C and (b) for −40°C to 110°C.

## 18.5  Signal Processing

The output signal from the sensor of a measurement system or the signal from the control unit might have to be processed in some way to make it suitable to operate the next element in the control system. For example, the signal may be too small and have to be amplified, be analog and have to be made digital, be digital and have to be made analog, be a resistance change and have to be made into a current change, be a voltage change and have to be made into a suitable size current change, be a pressure change and have to be made into a current change, etc. All these changes can be referred to as *signal processing*. For example, the output from a thermocouple is a very small voltage, a few millivolts. A signal processing module might then be used to convert this into a larger voltage and provide cold junction compensation (i.e., allow for the cold junction not being at 0°C).

The following are some examples of signal processing commonly encountered in control systems.

### 18.5.1  Resistance to Voltage Converter

Consider how the resistance change produced by a thermistor when subject to a temperature change can be converted into a voltage change. Figure 18.34 shows how a *potential divider circuit* can be used. A constant voltage, of perhaps 6 V, is applied across

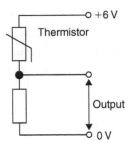

**Figure 18.34: Resistance to voltage conversion for a thermistor**

the thermistor and another resistor in series. With a thermistor with a resistance of $4.7\,\text{k}\Omega$, the series resistor might be $10\,\text{k}\Omega$. The output signal is the voltage across the $10\,\text{k}\Omega$ resistor. When the resistance of the thermistor changes, the fraction of the 6 V across the $10\,\text{k}\Omega$ resistor changes.

The output voltage is proportional to the fraction of the total resistance, which is between the output terminals. Thus:

$$\text{Output} = \frac{R}{R + R_t} V$$

where $V$ is the total voltage applied, in Figure 18.34 this is shown as 6 V, $R$ the value of the resistance between the output terminals ($10\,\text{k}\Omega$) and $R_t$ the resistance of the thermistor at the temperature concerned. The potential divider circuit is thus an example of a simple resistance to voltage converter. Another example of such a converter is the Wheatstone bridge.

### 18.5.2 Protection

An important element that is often required with signal processing is protection against high currents or high voltages. A high current can be protected against by the incorporation in the input line of a series resistor to limit the current to an acceptable level and a fuse to break if the current does exceed a safe level (Figure 18.35).

It is often so vital that high currents or high voltages are not transmitted from the sensor to a microprocessor that it may be necessary to completely isolate circuits so there are

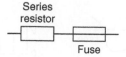

**Figure 18.35: Protection against high currents**

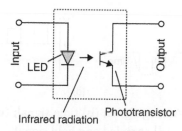

**Figure 18.36: Optoisolator: infrared radiation is used to transmit signal from input to output circuit**

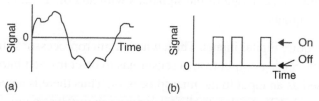

**Figure 18.37: Signals: (a) analog; (b) digital**

no electrical connections between them. This can be done using an *optoisolator* (Figure 18.36). Such a device converts an electrical signal into an optical signal, transmits it to a detector which then converts it back into an electrical signal. The input signal passes through an infrared light-emitting diode (LED) and so produces a beam of infrared radiation which is detected by a phototransistor.

### 18.5.3 Analog-to-Digital Conversion

The electrical output from sensors such as thermocouples, resistance elements used for temperature measurement, strain gauges, diaphragm pressure gauges, LVDTs, etc. is in analog form. An analog signal (Figure 18.37(a)) is one that is continuously variable, changing smoothly over a range of values. The signal is an analog, i.e., a scaled version, of the quantity it represents. A digital signal increases in jumps, being a sequence of pulses, often just on-off signals (Figure 18.37(b)). The value of the quantity instead

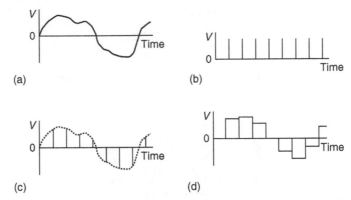

Figure 18.38: (a) Analog signal; (b) time signal; (c) sampled signal;
(d) sampled and held signal

of being represented by the height of the signal, as with analog, is represented by the sequence of on-off signals.

Microprocessors require digital inputs. Thus, where a microprocessor is used as part of a control system, the analog output from a sensor has to be converted into a digital form before it can be used as an input to the microprocessor. Thus there is a need for an *analog-to-digital converter* (ADC). Analog-to-digital conversion involves a number of stages. The first stage is to take samples of the analog signal (Figure 18.38(a)). A clock supplies regular time signal pulses (Figure 18.38(b)) to the analog-to-digital converter and every time it receives a pulse it samples the analog signal. The result is a series of narrow pulses with heights which vary in accord with the variation of the analog signal (Figure 18.38(c)). This sequence of pulses is changed into the signal form shown in Figure 18.38(d) by each sampled value being held until the next pulse occurs. This holding is necessary to allow time for the conversion to take place at an analog-to-digital converter. This converts each sample into a sequence of pulses representing the value. For example, the first sampled value might be represented by 101, the next sample by 011, etc. The 1 represents an "on" or "high" signal, the 0 an "off" or "low" signal. Analog-to-digital conversion thus involves a sample and hold unit followed by an analog-to-digital converter (Figure 18.39).

To illustrate the action of the analog-to-digital converter, consider one that gives an output restricted to three bits. The binary digits of 0 and 1, i.e., the "low" and "high" signals, are referred to as *bits*. A group of bits is called a *word*. Thus the three bits give the *word*

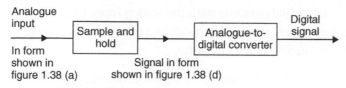

**Figure 18.39: Analog-to-digital conversion**

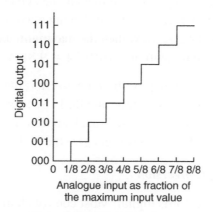

**Figure 18.40: Digital output from an ADC**

*length* for this particular analog-to-digital converter. The word is what represents the digital version of the analog voltage. With three bits in a word we have the possible words of:

000     001     010     011     100     101     110     111

There are eight possible words which can be used to represent the analog input; the number of possible words with a word length of *n* bits is $2^n$. Thus we divide the maximum analog voltage into eight parts and one of the digital words corresponds to each. Each rise in the analog voltage of (1/8) of the maximum analog input then results in a further bit being generated. Thus for word 000 we have 0 V input. To generate the next digital word of 001 the input has to rise to 1/8 of the maximum voltage. To generate the next word of 010 the input has to rise to 2/8 of the maximum voltage. Figure 18.40 illustrates this conversion of the sampled and held input voltage to a digital output.

Thus if we had a sampled analog input of 8 V, the digital output would be 000 for a 0 V input and would remain at that output until the analog voltage had risen to 1 V, i.e., 1/8 of the maximum analog input. It would then remain at 001 until the analog input had risen to

2 V. This value of 001 would continue until the analog input had risen to 3 V. The smallest change in the analog voltage that would result in a change in the digital output is thus 1 V. This is termed the *resolution* of the converter.

The word length possible with an analog-to-digital converter determines its *resolution*. With a word length of $n$ bits the maximum, or full scale, analog input $V_{FS}$ is divided into $2^n$ pieces. The minimum change in input that can be detected, i.e., the *resolution,* is thus $V_{FS}/2^n$. With an analog-to-digital converter having a word length of 10 bits and the maximum analog signal input range 10 V, then the maximum analog voltage is divided into $2^{10} = 1024$ pieces and the resolution is $10/1024 = 9.8 \, \text{mV}$.

There are a number of forms of analog-to-digital converter; the most commonly used being successive approximations, dual-slope and flash. Successive approximations converters are probably the most widely used; dual slope converters have the advantage of excellent noise rejection and flash converters give the highest conversion rates.

### 18.5.4 Digital-to-Analog Conversion

The output from a microprocessor is digital. Most control elements require an analog input and so the digital output from a microprocessor has to be converted into an analog form before it can be used by them. The input to a digital-to-analog converter is a binary word and the output its equivalent analog value. For example, if we have a full scale output of 7 V then a digital input of 000 will give 0 V, 001 give 1 V … and 111 the full scale value of 7 V. Figure 18.41 illustrates this.

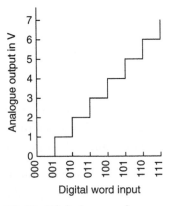

**Figure 18.41: Digital-to-analog conversion**

### 18.5.5 Amplifiers

The *operational amplifier* is the basis of many signal processing elements, the basic amplifier being supplied as an integrated circuit on a silicon chip. It has two inputs, termed the inverting input (−) and the noninverting input (+) and is a high gain DC amplifier, the gain typically being of the order of 100,000 or more. Figure 18.42 shows the pin connections for a 741 operational amplifier with the symbol for the operational amplifier. Pins 4 and 7 are for the connections to the supply voltage for the amplifier, pin 2 for the inverting input, pin 3 for the noninverting input. The output is taken from pin 6. Pins 1 and 5 are for the offset null. These are to enable circuits to be connected to enable corrections to be made for the nonideal behavior of the amplifier.

Consider the amplifier when used as an *inverting amplifier* (Figure 18.43), i.e., an amplifier which gives an output, which is out-of-phase with respect to the input. For the circuit shown in Figure 18.43, the connections for the power supply and the offset null have been omitted. The input is connected to the inverting input, the noninverting

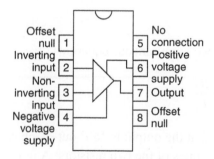

**Figure 18.42: Amplifier**

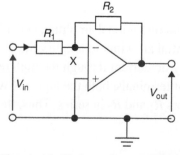

**Figure 18.43: Inverting amplifier**

input being connected to earth. A feedback loop is connected, via the resistor $R_2$, to the inverting input. The output voltage of such an amplifier is limited to about $\pm 10$ V and thus, since the gain is about 100,000, the input voltage to the inverting input at $X$, $V_X$, must be between about $+0.0001$ V and $-0.0001$ V. This is virtually zero and so point $X$ is at virtually earth potential. For this reason it is called a *virtual earth*. The potential difference across the input resistance $R_1$ is $(V_{in} - V_X)$ and thus $(V_{in} - V_X) = I_1 R_1$. But $V_X$ is virtually zero and so we can write:

$$V_{in} = I_1 R_1$$

Operational amplifiers have very high resistance between their input terminals; for example, the resistance with the 741 operational amplifier is about $2 \, \text{M}\Omega$. Thus virtually no current flows from point X through the inverting input and so to earth. Thus the current $I_1$ that flows through $R_1$ must be essentially the current flowing through $R_2$. The potential difference across $R_2$ is $(V_X - V_{out})$. Thus we can write $(V_X - V_{out}). = I_1 R_2$. But as $V_X$ is effectively zero, we can write:

$$-V_{out} = I_2 R_2$$

Eliminating 11 from these two simultaneous equations gives:

$$\text{gain of circuit} = \frac{V_{out}}{V_{in}} = -\frac{R_2}{R_1}$$

The negative sign indicates that the output is 180° out-of-phase with the input. The gain is determined solely by the values of the two resistors. A noninverting amplifier can likewise be produced by taking the input to the noninverting input instead of the inverting input.

As an illustration of the use of an operational amplifier, consider Figure 18.44 which shows how it can be used as a differential amplifier to amplify the difference between two input voltages. Since there is virtually no current through the high resistance in the operational amplifier between the two input terminals, both the inputs X will be at the same potential. The voltage $V_2$ is across resistors $R_1$ and $R_2$ in series. Thus, the potential $V_X$ at X is:

$$\frac{V_X}{V_2} = \frac{R_2}{R_1 + R_2}$$

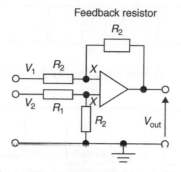

**Figure 18.44: Differential amplifier**

Since the operational amplifier has a very high input resistance, the current through the feedback resistance will be equal to that from $V_1$ through $R_1$. Hence, we have:

$$\frac{V_1 - V_X}{R_1} = \frac{V_X - V_{out}}{R_2}$$

and so:

$$\frac{V_{out}}{R_2} = V_X \left( \frac{1}{R_2} + \frac{1}{R_1} \right) - \frac{V_1}{R_1}$$

Hence, substituting for $V_X$ using the earlier equation, gives:

$$V_{out} = \frac{R_2}{R_1}(V_2 - V_1)$$

The output is a measure of the difference between the two input voltages.

More information on op-amps can be found in Chapter 11.

## 18.6 Correction Elements

The following are examples of correction elements that are commonly encountered in control systems.

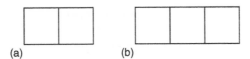

**Figure 18.45: (a) Two position; (b) three position valves**

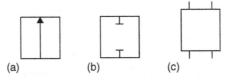

**Figure 18.46: (a) Flow path; (b) shut-off; (c) initial connections**

### 18.6.1 Directional Control Valves

*A directional control valve* on the receipt of some external signal, which might be mechanical, electrical or a pressure signal, change the direction of, or stop, or start the flow of fluid in some part of the pneumatic/ hydraulic circuit. Thus, it might be used to control the direction of fluid flow to a cylinder and so use the movement of its piston to carry out actuation.

The basic symbol for a control valve is a square. With a directional control valve two or more squares are used, with each square representing the positions to which the valve can be switched. Thus, Figure 18.45(a) represents a valve with two switching positions, Figure 18.45(b) a valve with three switching positions. Lines in the boxes are used to show the flow paths with arrows indicating the direction of flow (Figure 18.46(a)) and shut-off positions indicated by terminated lines (Figure 18.46(b)). The pipe connections, i.e., the inlet and outlet ports of the valve, are indicated by lines drawn on the outside of the box and are drawn for just the "rest/initial/neutral position," i.e., when the valve is not actuated (Figure 18.46(c)). You can imagine each of the position boxes to be moved by the action of some actuator so that it connects up with the pipe positions to give the different connections between the ports. Directional control valves are described by the number of ports and the number of positions. Thus, a 2/2 valve has 2 ports and 2 positions, a 3/2 valve 3 ports and 2 positions, a 4/2 valve 4 ports and 2 positions, a 5/3 valve 5 ports and 3 positions. Figure 18.47 shows some commonly used examples and their switching options and Figure 18.48 the means by which valves can be switched between positions.

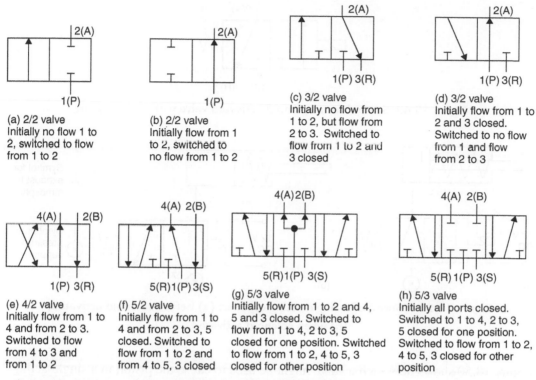

**Figure 18.47: Commonly used direction valves: P or I indicates the pressure supply ports, R and S or 3 and 5 the exhaust ports, A and B or 2 and 4 the signal output ports**

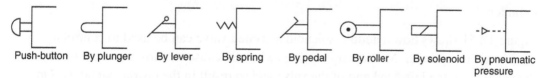

**Figure 18.48: Examples of valve actuation methods**

As an illustration, Figure 18.49 shows the symbol for a 3/2 valve with solenoid activation and return by means of a spring. Thus, when the solenoid is not activated by a current through it, the signal port 2 is connected to the exhaust 3 and so is at atmospheric pressure. When the solenoid is activated, the pressure supply P is connected to the signal port 2 and thus the output is pressurised.

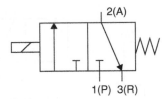

Figure 18.49: Symbol for a solenoid-activated valve with return spring

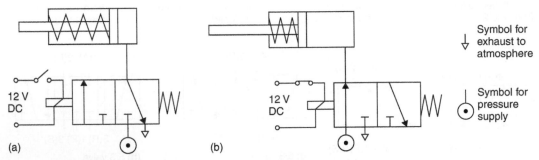

Figure 18.50: Control of a single-acting cylinder: (a) before solenoid activated;
(b) when solenoid activated

Figure 18.50 shows how such a valve might be used to cause the piston in a single-acting cylinder to move; the term *single-acting* is used when a pressure signal is applied to only one side of the piston. When the switch is closed and a current passes through the solenoid, the valve switches position and pressure is applied to extend the piston in the cylinder.

Figure 18.51 shows how a double-solenoid activated valve can be used to control a double-acting cylinder. Momentary closing switch S1 causes a current to flow through the solenoid at the left-hand end of the valve and so result in the piston extending. On opening S1 the valve remains in this extended position until a signal is received by the closure of switch S2 to activate the fight-hand solenoid and return the piston.

### 18.6.2 Flow Control Valves

In many control systems the rate of flow of a fluid along a pipe is controlled by a valve which uses pneumatic action to move the valve stem and hence a plug in the flow path (Figure 18.52), so altering the size of the gap through which the fluid can flow.

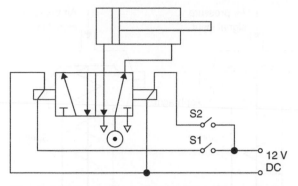

**Figure 18.51: Control of a double-acting cylinder**

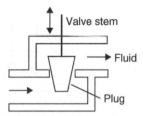

**Figure 18.52: Flow controlled by movement of a plug**

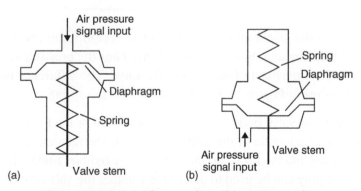

**Figure 18.53: (a) Direct action; (b) reverse action**

The movement of the stem results from the use of a diaphragm moving against a spring and controlled by air pressure (Figure 18.53). The air pressure from the controller exerts a force on one side of the diaphragm, the other side of the diaphragm being at atmospheric pressure, which is opposed by the force due to the spring on the other side. When the air

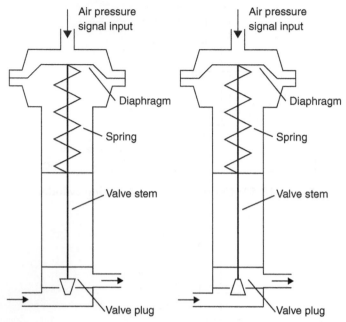

**Figure 18.54: Direct action: (a) air pressure increase to close;
(b) air pressure increase to open**

pressure changes then the diaphragm moves until there is equilibrium between the forces resulting from the pressure and those from the spring. Thus the pressure signals from the controller result in the movement of the stem of the valve. The difference between the direct and reverse forms in Figure 18.53 is the position of the spring.

There are many forms of valve body and plug. The selection of the form of body and plug determine the characteristic of the control valve, i.e., the relationship between the valve stem position and the flow rate through it. For example, Figure 18.54 shows how the selection of plug can be used to determine whether the valve closes when the controller air pressure increases or opens when it increases and Figure 18.55 shows how the shape of the plug determines how the rate of flow is related to the displacement of the valve stem: linear plug—change in flow rate proportional to the change in valve stem displacement; quick-opening plug—a large change in flow rate occurs for a small movement of the valve stem; equal percentage plug—the amount by which the flow rate changes is proportional to the value of the flow rate when the change occurs.

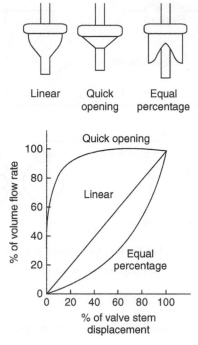

Figure 18.55: Effect of plug shape on flow

### 18.6.3 DC Motors

DC motors are widely used with control systems. In the DC motor, coils of wire are mounted in slots on a cylinder of magnetic material called the *armature*. The armature is mounted on beatings and is free to rotate. It is mounted in the magnetic field produced by *field poles*. This magnetic field might be produced by permanent magnets or an electromagnet with its magnetism produced by a current passing through the, so-termed, *field coils*. Whether permanent magnet or electromagnet, these generally form the outer casing of the motor and are termed the *stator*.

For a DC motor with the field provided by a permanent magnet, the direction of rotation of the motor can be changed by reversing the current in the armature coil. The speed of rotation of such a motor can be changed by changing the size of the current to the armature coil.

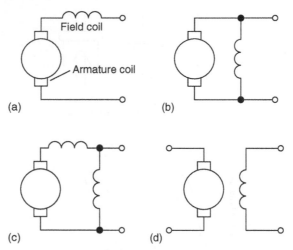

**Figure 18.56: (a) Series; (b) shunt; (c) compound;
(d) separately wound**

DC motors with field coils are classified as series, shunt, compound and separately excited according to how the field windings and armature windings are connected. With the *series-wound motor the* armature and fields coils are in series (Figure 18.56(a)). Such a motor exerts the highest starting torque and has the greatest no-load speed. However, with light loads there is a danger that a series-wound motor might run at too high a speed. Reversing the polarity of the supply to the coils has no effect on the direction of rotation of the motor, since both the current in the armature and the field coils are reversed. With the *shunt-wound motor* (Figure 18.56(b)) the armature and field coils are in parallel. It provides the lowest starting torque, a much lower no-load speed and has good speed regulation. It gives almost constant speed regardless of load and thus shunt wound motors are very widely used. To reverse the direction of rotation, either the armature or field current can be reversed. The *compound motor* (Figure 18.56(c)) has two field windings, one in series with the armature and one in parallel. Compound-wound motors aim to get the best features of the series and shunt-wound motors, namely a high starting torque and good speed regulation. The *separately excited motor* (Figure 18.56(d)) has separate control of the armature and field currents. The direction of rotation of the motor can be obtained by reversing either the armature or the field current. Figure 18.57 indicates the general form of the torque-speed characteristics of the above motors. The separately

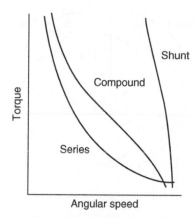

**Figure 18.57: Torque-speed characteristics of DC motors**

excited motor has a torque-speed characteristic similar to the shunt wound motor. The speed of such DC motors can be changed by either changing the armature current or the field current. Generally it is the armature current that is varied.

The choice of DC motor will depend on what it is to be used for. Thus, for example, with a robot manipulator the robot wrist might use a series-wound motor because the speed decreases as the load increases. A shunt-wound motor might be used if a constant speed was required, regardless of the load.

### 18.6.4 Stepper Motor

The *stepper or stepping motor* produces rotation through equal angles, the so-called *steps,* for each digital pulse supplied to its input. For example, if with such a motor 1 input pulse produces a rotation of 1.8° then 20 input pulses will produce a rotation through 36.0°, 200 input pulses a rotation through one complete revolution of 360°. It can thus be used for accurate angular positioning. By using the motor to drive a continuous belt, the angular rotation of the motor is transformed into linear motion of the belt and so accurate linear positioning can be achieved. Such a motor is used with computer printers, *x-y* plotters, robots, machine tools and a wide variety of instruments for accurate positioning.

There are two basic forms of stepper motor, the *permanent magnet* type with a permanent magnet rotor and the *variable reluctance type* with a soft steel rotor. Both form of stepper motor have a stator with a number of diametrically opposite pairs of poles, each wound

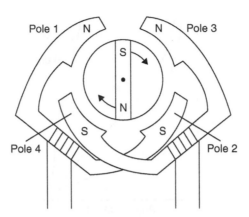

**Figure 18.58: The basic principles of the permanent magnet stepper motor (2-phase) with a rotor giving**

**Table 18.2: Polarities associated with each step**

| Step | Pole 1 | Pole 2 | Pole 3 | Pole 4 |
|------|--------|--------|--------|--------|
| 1 | North | South | North | South |
| 2 | South | North | North | South |
| 3 | South | North | South | North |
| 4 | North | South | South | North |
| 5 | Repeat of steps 1 to 4 | | | |

with a coil. Figure 18.58 shows the permanent magnet type with two pairs of stator poles. Each pole is activated by a current being passed through the appropriate field winding, the coils being such that opposite poles are produced on opposite coils. The current is supplied from a DC source to the windings through switches. With the currents switched through the coils such that the poles are as shown in Figure 18.58, the rotor will move to line up with the next pair of poles and stop there. This would be, for Figure 18.58, an angle of 45°. If the current is then switched so that the polarities are reversed, the rotor will move a step to line up with the next pair of poles, at angle 135° and stop there. The polarities associated with each step are given in Table 18.2.

There are thus, in this case, four possible rotor positions: 45°, 135°, 225° and 315°. Note that the term *phase* is used for the number of independent windings on the stator.

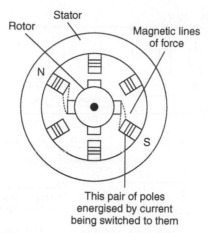

**Figure 18.59: Basic principles of a 3-phase variable reluctance stepper motor**

Figure 18.59 shows the basic form of the *variable reluctance type* of stepper motor. With this form the rotor is made of soft steel and is not a permanent magnet. The rotor has a number of teeth, the number being less than the number of poles on the stator. When an opposite pair of windings on stator poles has current switched to them, a magnetic field is produced with lines of force which pass from the stator poles through the nearest set of teeth on the rotor. Since lines of force can be considered to be rather like elastic thread and always trying to shorten themselves, the rotor will move until the rotor teeth and stator poles line up. This is termed the position of minimum reluctance. Thus by switching the current to successive pairs of stator poles, the rotor can be made to rotate in steps. With the number of poles and rotor teeth shown in Figure 18.59, the angle between each successive step will be 30°. The angle can be made smaller by increasing the number of teeth on the rotor.

To drive a stepper motor, so that it proceeds step-by-step to provide rotation, requires each pair of stator coils to be switched on and off in the required sequence when the input is a sequence of pulses. Driver circuits are available to give the correct sequencing and Figure 18.60 shows an example. The stepper motor will rotate through one step each time the trigger input goes from low to high. The motor runs clockwie when the rotation input is low and anticlockwise when high. When the set pin is made low the output resets. In a control system, these input pulses might be supplied by a microprocessor.

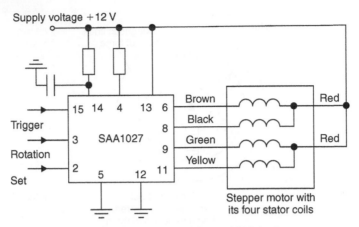

**Figure 18.60: Driver circuit SAA 1027 for a 12 V 4-phase stepper motor**

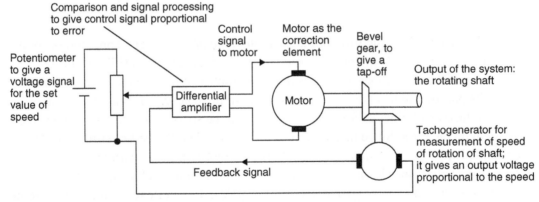

**Figure 18.61: Control of the speed of rotation of a shaft**

## 18.7 Control Systems

The following are examples of closed-loop control systems to illustrate how, despite the different forms of control being exercised, the systems all have the same basic structural elements.

### 18.7.1 Control of the Speed of Rotation of a Motor Shaft

Consider the motor system shown in Figure 18.61 for the control of the speed of rotation of the motor shaft and its block diagram representation in Figure 18.62. The input of the

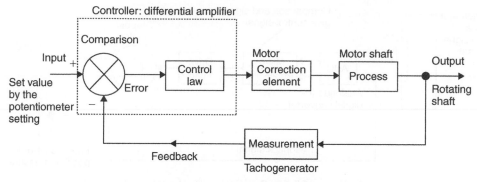

**Figure 18.62: Control of the speed of rotation of a shaft**

required speed value is by means of the setting of the position of the movable contact of the potentiometer. This determines what voltage is supplied to the comparison element, i.e., the differential amplifier, as indicative of the required speed of rotation. The differential amplifier produces an amplified output which is proportional to the difference between its two inputs. When there is no difference then the output is zero. The differential amplifier is thus used to both compare and implement the control law. The resulting control signal is then fed to a motor which adjusts the speed of the rotating shaft according to the size of the control signal. The speed of the rotating shaft is measured using a tachogenerator, this being connected to the rotating shaft by means of a pair of bevel gears. The signal from the tachogenerator gives the feedback signal which is then fed back to the differential amplifier.

### 18.7.2  Control of the Position of a Tool

Figure 18.63 shows a position control system using a belt driven by a stepper motor to control the position of a tool and Figure 18.64 its block diagram representation. The inputs to the controller are the required position voltage and a voltage giving a measure of the position of the workpiece, this being provided by a potentiometer being used as a position sensor. Because a microprocessor is used as the controller, these signals have to be processed to be digital. The output from the controller is an electrical signal which depends on the error between the required and actual positions and is used, via a drive unit, to operate a stepper motor. Input to the stepper motor causes it to rotate its shaft in steps, so rotating the belt and moving the tool.

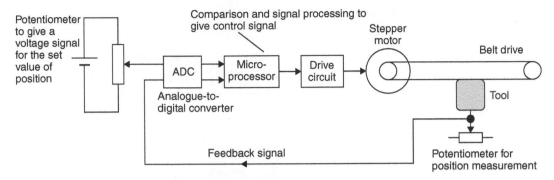

**Figure 18.63: Position control system**

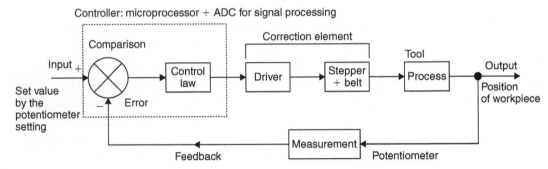

**Figure 18.64: Position control system**

### 18.7.3 Power Steering

Control systems are used to not only maintain some variable constant at a required value but also to control a variable so that it follows the changes required by a variable input signal. An example of such a control system is the power steering system used with a car. This comes into operation whenever the resistance to turning the steering wheel exceeds a predetermined amount and enables the movement of the wheels to follow the dictates of the angular motion of the steering wheel. The input to the system is the angular position of the steering wheel. This mechanical signal is scaled down by gearing and has subtracted from it a feedback signal representing the actual position of the wheels. This feedback is via a mechanical linkage. Thus when the steering wheel is rotated and there is a difference between its position and the required position of the wheels, there is an error signal. The error signal is used to operate a hydraulic valve and so provide a hydraulic

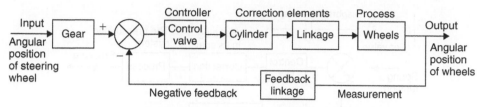

**Figure 18.65: Power assisted steering**

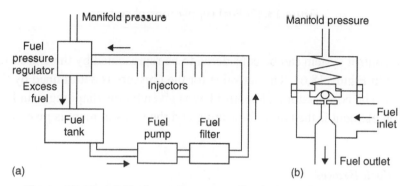

**Figure 18.66: (a) Fuel supply system; (b) fuel pressure regulator**

signal to operate a cylinder. The output from the cylinder is then used, via a linkage, to change the position of the wheels. Figure 18.65 shows a block diagram of the system.

### 18.7.4  Control of Fuel Pressure

The modem car involves many control systems. For example, there is the *engine management system* aimed at controlling the amount of fuel injected into each cylinder and the time at which to fire the spark for ignition. Part of such a system is concerned with delivering a constant pressure of fuel to the ignition system. Figure 18.66(a) shows the elements involved in such a system. The fuel from the fuel tank is pumped through a filter to the injectors, the pressure in the fuel line being controlled to be 2.5 bar (2.5 × 0.1 MPa) above the manifold pressure by a regulator valve. Figure 18.66(b) shows the principles of such a valve. It consists of a diaphragm which presses a ball plug into the flow path of the fuel. The diaphragm has the fuel pressure acting on one side of it and on the other side is the manifold pressure and a spring. If the pressure is too high, the diaphragm moves and opens up the return path to the fuel tank for the excess fuel, so adjusting the fuel pressure to bring it back to the required value.

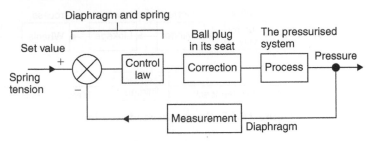

**Figure 18.67: Fuel supply control system**

The pressure control system can be considered to be represented by the closed loop system shown in Figure 18.67. The set value for the pressure is determined by the spring tension. The comparator and control law is given by the diaphragm and spring. The correction element is the ball in its seating and the measurement is given by the diaphragm.

### 18.7.5 Antilock Brakes

Another example of a control system used with a car is the *antilock brake system (ABS)*. If one or more of the vehicle's wheels lock, i.e., begins to skid, during braking, then braking distance increases, steering control is lost and tire wear increases. Antilock brakes are designed to eliminate such locking. The system is essentially a control system which adjusts the pressure applied to the brakes so that locking does not occur. This requires continuous monitoring of the wheels and adjustments to the pressure to ensure that, under the conditions prevailing, locking does not occur. Figure 18.68 shows the principles of such a system.

The two valves used to control the pressure are solenoid-operated valves, generally both valves being combined in a component termed the modulator. When the driver presses the brake pedal, a piston moves in a master cylinder and pressurises the hydraulic fluid. This pressure causes the brake calliper to operate and the brakes to be applied. The speed of the wheel is monitored by means of a sensor. When the wheel locks, its speed changes abruptly and so the feedback signal from the sensor changes. This feedback signal is fed into the controller where it is compared with what signal might be expected on the basis of data stored in the controller memory. The controller can then supply output signals which operate the valves and so adjust the pressure applied to the brake.

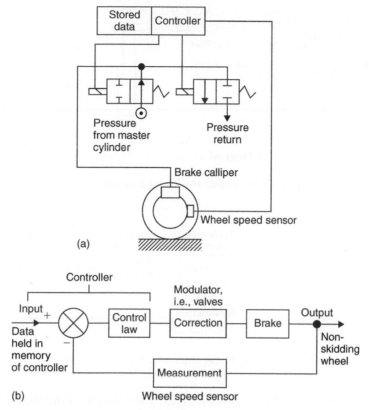

(a)

(b)

**Figure 18.68: Antilock brakes: (a) schematic diagram;
(b) block form of the control system**

### 18.7.6 Thickness Control

As an illustration of a process control system, Figure 18.69 shows the type of system that might be used to control the *thickness of sheet* produced by rollers, Figure 18.70 showing the block diagram description of the system. The thickness of the sheet is monitored by a sensor such as a *linear variable differential transformer* (LVDT). The position of the LVDT probe is set so that when the required thickness sheet is produced, there is no output from the LVDT. The LVDT produces an alternating current output, the amplitude of which is proportional to the error. This is then converted to a DC error signal which is fed to an amplifier. The amplified signal is then used to control the speed of a DC motor,

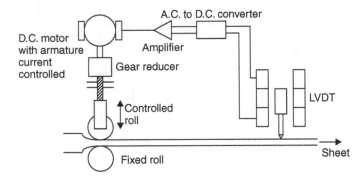

**Figure 18.69: Sheet thickness control system**

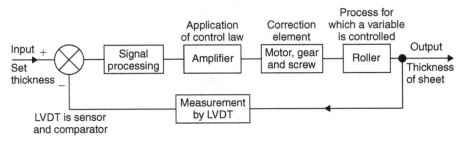

**Figure 18.70: Sheet thickness control system**

generally being used to vary the armature current. The rotation of the shaft of the motor is likely to be geared down and then used to rotate a screw which alters the position of the upper roll, hence changing the thickness of the sheet produced.

### 18.7.7 Control of Liquid Level

Figure 18.71 shows a control system used to control the level of liquid in a tank using a float-operated pneumatic controller, Figure 18.72 showing a block diagram of the system. When the level of the liquid in the tank is at the required level and the inflow and outflows are equal, then the controller valves are both closed. If there is a decrease in the outflow of liquid from the tank, the level rises and so the float rises. This causes point P to move upward. When this happens, the valve connected to the air supply opens and the air pressure in the system increases. This causes a downward movement of the diaphragm in the flow control valve and hence, a downward movement of the valve stem and the valve plug. This then results in the inflow of liquid into the tank being reduced. The increase in

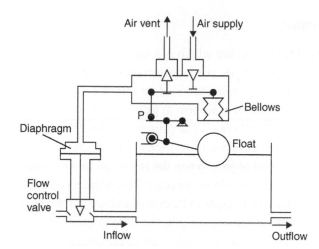

**Figure 18.71: Level control system**

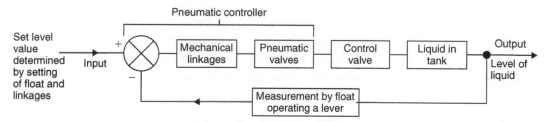

**Figure 18.72: Level control system**

the air pressure in the controller chamber causes the bellows to become compressed and move that end of the linkage downward. This eventually closes off the valve so that the flow control valve is held at the new pressure and hence the new flow rate.

If there is an increase in the outflow of liquid from the tank, the level falls and so the float falls. This causes point P to move downward. When this happens, the valve connected to the vent opens and the air pressure in the system decreases. This causes an upward movement of the diaphragm in the flow control valve and hence an upward movement of the valve stem and the valve plug. This then results in the inflow of liquid into the tank being increased. The bellows react to this new air pressure by moving its end of the linkage, eventually closing off the exhaust and so holding the air pressure at the new value and the flow control valve at its new flow rate setting.

### 18.7.8 Robot Gripper

The term *robot* is used for a machine which is a reprogrammable multi-function manipulator designed to move tools, parts, materials, etc. through variable programmed motions in order to carry out specified tasks. Here just one aspect will be considered, the gripper used by a robot at the end of its arm to grip objects. A common form of gripper is a device which has "fingers" or "jaws." The gripping action then involves these clamping on the object. Figure 18.73 shows one form such a gripper can take if two gripper fingers are to close on a parallel sided object. When the input rod moves toward the fingers they pivot about their pivots and move closer together. When the rod moves outward, the fingers move further apart. Such motion needs to be controlled so that the grip exerted by the fingers on an object is just sufficient to grip it, too little grip and the object will fall out of the grasp of the gripper and too great might result in the object being crushed or otherwise deformed. Thus there needs to be feedback of the forces involved at contact between the gripper and the object. Figure 18.74 shows the type of closed-loop control system involved.

The drive system used to operate the gripper can be electrical, pneumatic or hydraulic. Pneumatic drives are very widely used for grippers because they are cheap to install, the system is easily maintained and the air supply is easily linked to the gripper. Where larger loads are involved, hydraulic drives can be used. Sensors that might be used for

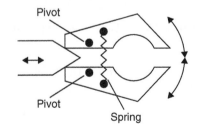

**Figure 18.73: An example of a gripper**

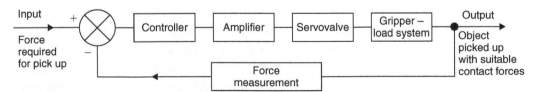

**Figure 18.74: Gripper control system**

measurement of the forces involved are piezoelectric sensors or strain gauges. Thus when strain gauges are stuck to the surface of the gripper and forces applied to a gripper, the strain gauges will be subject to strain and give a resistance change related to the forces experienced by the gripper when in contact with the object being picked up.

The robot arm with gripper is also likely to have further control loops to indicate when it is in the fight position to grip an object. Thus the gripper might have a control loop to indicate when it is in contact with the object being picked up; the gripper can then be actuated and the force control system can come into operation to control the grasp. The sensor used for such a control loop might be a microswitch which is actuated by a lever, roller or probe coming into contact with the object.

### 18.7.9  Machine Tool Control

Machine tool control systems are used to control the position of a tool or workpiece and the operation of the tool during a machining operation. Figure 18.75 shows a block diagram of the basic elements of a closed-loop system involving the continuous monitoring of the movement and position of the work tables on which tools are mounted while the workpiece is being machined. The amount and direction of movement required in order to produce the required size and form of workpiece is the input to the system, this being a program of instructions fed into a memory which then supplies the information as required. The sequence of steps involved is then:

1.  An input signal is fed from the memory store.

2.  The error between this input and the actual movement and position of the work table is the error signal which is used to apply the correction. This may be an electric motor to control the movement of the work table. The work table then moves to reduce the error so that the actual position equals the required position.

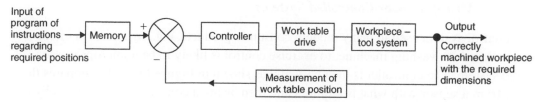

**Figure 18.75: Closed-loop machine tool control system**

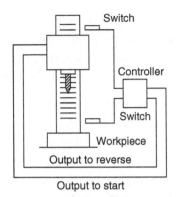

**Figure 18.76: An automatic drill**

3. The next input signal is fed from the memory store.

4. Step 2 is then repeated.

5. The next input signal is fed from the memory store, and so on.

### 18.7.10 An Automatic Drill

As an illustration of the type of control that might be used with a machine consider the system for a drill which is required to automatically drill a hole in a workpiece when it is placed on the work table (Figure 18.76). A switch sensor can be used to detect when the workpiece is on the work table. This then gives an on input signal to the controller and it then gives an output signal to actuate a motor to lower the drill head and commence drilling. When the drill reaches the full extent of its movement in the workpiece, the drill head triggers another switch sensor. This provides an on input to the controller and it then reverses the direction of rotation of the drill head motor and the drill retracts. This is an example of closed-loop discrete-event control.

### 18.7.11 Microprocessor-Controlled Systems

The control system used in many modem consumer products, e.g., in a modem motor car or a modem washing machine, to exercise control is likely to be a microprocessor-based system. The controller is then basically as shown in Figure 18.77. It compares the input from a sensor with what is required and then, using a control law determined by the program stored in its memory, gives an output to a correction element.

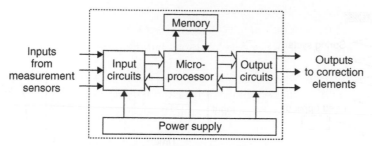

**Figure 18.77: Microprocessor-based controller**

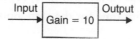

**Figure 18.78: Amplifier system with the output
ten times the input**

## 18.8 System Models

Suppose we have a control system for the temperature in a room. How will the temperature react when the thermostat has its set value increased from, say, 20°C to 22°C? In order to determine how the output of a control system will react to different inputs, we need a mathematical model of the system so that we have an equation describing how the output of the system is related to its input.

Thus, in the case of an amplifier system (Figure 18.78) we might be able to use the simple relationship that the output is always 10 times the input. If we have an input of a 1 V signal we can calculate that the output will be 10V. This is a simple model of a system where the input is just multiplied by a gain of 10 in order to give the output.

However, if we consider a system representing a spring balance with an input of a load signal and an output of a deflection (Figure 18.79) then, when we have an input to the system and put a fixed load on the balance (this type of input is known as a *step input* because the input variation with time looks like a step), it is likely that it will not instantaneously give the weight but the pointer on the spring balance will oscillate for a little time before settling down to the weight value. Thus, we cannot

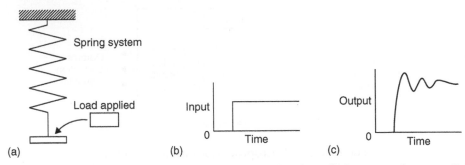

**Figure 18.79: (a) The spring system with a constant load applied at some instant of time; (b) the step showing how theinput varies with time; (c) the output showing how it varies with time for the step input**

just state, for an input of some constant load, that the output is just the input multiplied by some constant number but need some way of describing an output which varies with time. With an electrical system of a circuit with capacitance and resistance, when the voltage to such a circuit is switched on, i.e., there is a constant voltage input to the system, then the current changes with time before eventually settling down to a steady value. With a temperature control system, such as that used for the central heating system for a house, when the thermostat is changed from 20°C to 22°C the output does not immediately become 22°C but there is a change with time and eventually it may become 22°C In general, the mathematical model describing the relationship between input and output for a system is likely to involve terms which give values which change with time and are described by a differential equation (see Appendix B). As we continue through this chapter, we look at how such differential equation relationships arise.

In order to make life simple, what we need is a simple relationship between input and output for a system, even when the output varies with time. It is nice and simple to say that the output is just ten times the input and so describe the system by gain = 10. There is a way we can have such a simple form of relationship where the relationship involves time but it involves writing inputs and outputs in a different form. It is called the *Laplace transform* and it was covered earlier in Chapter 8. Let us consider how we can carry out such transformations; the aim is to enable you to use the transform as a tool to carry out tasks.

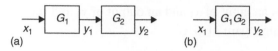

**Figure 18.80: (a) Two systems in series; (b) the equivalent system with a gain equal to the product of the gains of the two constituent systems**

## 18.9 Gain

In the case of an amplifier system we might have the output directly proportional to the input and, with a gain of 10, if we have an input of a 1 V signal we can calculate that the output will be ten times greater and so 10 V. In general, for such a system where the output is directly proportional to the input, we can write:

$$\text{output} = G \times \text{input}$$

with $G$ being the gain.

### Example 18.1
A motor has an output speed which is directly proportional to the voltage applied to its armature. If the output is 5 rev/s when the input voltage is 2 V, what is the system gain?

### Solution
With output $= G \times$ input, then $G = 5/2 = 2.5$ (rev/s)/V.

### 18.9.1 Gain of Systems in Series

Consider two systems, such as amplifiers, in series with the first having a gain $G_1$ and the second a gain $G_2$ (Figure 18.80(a)). The first system has an input of $x_1$ and an output of $y_2$ and thus:

$$y_1 = G_1 x_1$$

The second system has an input of $y_1$ and an output of $y_2$ and thus:

$$y_2 = G_2 y_1 = G_2 \times G_1 x_1$$

The overall system has an input of $x_1$ and an output of $y_2$ and thus, if we represent the overall system as having a gain of $G$:

$$y_2 = Gx_1$$

and so:

$$G = G_1 \times G_2$$

Thus:

For series-connected systems, the overall gain is the product of the gains of the constituent systems.

### Example 18.2

A system consists of an amplifier with a gain of 10 providing the armature voltage for a motor which gives an output speed which is proportional to the armature voltage, the constant of proportionality being 5 (rev/s)/V. What is the relationship between the input voltage to the system and the output motor speed?

### Solution

The overall gain $G = G_1 \times G_2 = 10 \times 5 = 50$ (rev/s)/V.

### 18.9.2 Feedback Loops

Consider a system with negative feedback (Figure 18.81). The output of the system is fed back via a measurement system with a gain $H$ to subtract from the input to a system with gain $G$.

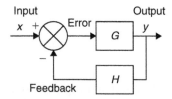

**Figure 18.81: System with negative feedback: the fed back signal subtracts from the input**

The input to the feedback system is y and thus its output, i.e., the feedback signal, is $Hy$. The error is $x - Hy$. Hence, the input to the $G$ system is $x - Hy$ and its output $y$. Thus:

$$y = G(x - Hy)$$

and so:

$$(1 + GH)y = G_x$$

The overall input of the system is $y$ for an input $x$ and so the overall gain $G$ of the system is $y/x$. Hence:

$$\text{system gain} = \frac{y}{x} = \frac{G}{1 + GH}$$

*For a system with a negative feedback, the overall gain is the forward path gain divided by one plus the product of the forward path and feedback path gains.*

For a system with positive feedback (Figure 18.82), i.e., the fed back signal adds to the input signal, the feedback signal is $Hy$ and thus the input to the $G$ system is $x + Hy$. Hence:

$$y = G(x + Hy)$$

and so:

$$(1 - GH)y = Gx$$

$$\text{system gain} = \frac{y}{x} = \frac{G}{1 - GH}$$

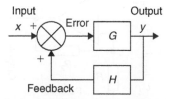

**Figure 18.82: System with positive feedback: the fed back signal adds to the input**

*For a system with a positive feedback, the overall gain is the forward path gain divided by one minus the product of the forward path and feedback path gains.*

### Example 18.3

A negative feedback system has a forward path gain of 12 and a feedback path gain of 0.1. What is the overall gain of the system?

### Solution

$$\text{System gain} = \frac{G}{1+GH} = \frac{12}{1+0.1\times12} = 5.45$$

### 18.9.3  The Feedback Amplifier

Figure 18.83 shows the circuit of a basic feedback amplifier. It consists of an operational amplifier with a potential divider of two resistors $R_1$ and $R_2$ connected across its output. The output from this potential divider is fed back to the inverting input of the amplifier. The input to the amplifier is via its noninverting input. Thus the sum of the inverted feedback input and the noninverted input is the error signal. The op-amp has a very high voltage gain G. Thus *GH, H* being the gain of the feedback loop, is very large compared with 1 and so the overall system gain is:

$$\text{system gain} = \frac{G}{1+GH} \simeq \frac{G}{GH} \simeq \frac{1}{H}$$

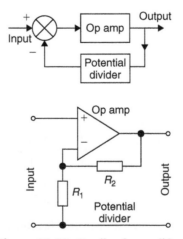

**Figure 18.83: Feedback amplifier**

Since the gain $G$ of the op-amp can be affected by changes in temperature, aging, etc. and thus can vary, the use of the op-amp with a feedback loop means that, since H is just made up of resistances which are likely to be more stable, a more stable amplifier system is produced. The feedback loop gain H is the fraction of the output signal fed back and so is $R_1/(R_1 + R_2)$. Hence, the overall gain of the system is:

$$\text{system gain} = \frac{R_2 + R_1}{R_1}$$

### Example 18.4

What is the overall gain of a noninverting feedback op-amp, connected as in Figure 18.83, if the op-amp has a voltage gain of 200,000, $R_1 = 1\,k\Omega$ and $R_2 = 49\,k\Omega$?

### Solution

The overall system gain is independent of the voltage gain of the op-amp and is given by $(R_1 + R_2)/R_1 = 50/1 = 50$.

## 18.10 Dynamic Systems

The following describes how we can arrive at the input-output relationships for systems by representing them by simple models obtained by considering them to be composed of just a few simple basic elements.

### 18.10.1 Mechanical Systems

Mechanical systems, however complex, have stiffness (or springiness), damping and inertia and can be considered to be composed of basic elements which can be represented by springs, dashpots and masses.

### 1. Spring

The "springiness" or "stiffness" of a system can be represented by a spring. For a linear spring (Figure 18.84(a)), the extension $y$ is proportional to the applied extending force $F$ and we have:

$$F = ky$$

where $k$ is a constant termed the *stiffness*.

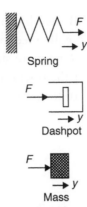

**Figure 18.84: Mechanical system building blocks**

## 2. Dashpot

The "damping" of a mechanical system can be represented by a dashpot. This is a piston moving in a viscous medium in a cylinder (Figure 18.84(b)). Movement of the piston inward requires the trapped fluid to flow out past edges of the piston; movement outward requires fluid to flow past the piston and into the enclosed space. The resistive force F which has to be overcome is proportional to the velocity of the piston and hence the rate of change of displacement y with time, i.e., *dy/dt.* Thus:

$$F = c\frac{dy}{dt}$$

where $c$ is a constant.

## 3. Mass

The "inertia" of a system—i.e., how much it resists being accelerated—can be represented by mass. For a mass $m$ (Figure 18.84(c)), the relationship between the applied force $F$ and its acceleration $a$ is given by Newton's second law as $F = ma$. But acceleration is the rate of change of velocity $v$ with time $t$, i.e., $a = dv/dt$, and velocity is the rate of change of displacement y with time, i.e., $v = dy/dt$. Thus $a = d(dy/dt)/dt$ and so we can write:

$$F = m\frac{d^2y}{dt^2}$$

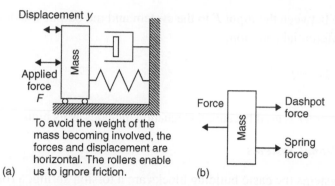

**Figure 18.85: (a) Mechanical system with mass, damping and stiffness; (b) the free-body diagram for the forces acting on the mass**

The following example illustrates how we can arrive at a model for a mechanical system.

*Example 18.5*

Derive a model for the mechanical system given in Figure 18.85(a). The input to the system is the force $F$ and the output is the displacement $y$.

*Solution*

To obtain the system model we draw *free-body diagrams,* these being diagrams of masses showing just the external forces acting on each mass. For the system in Figure 18.84(a) we have just one mass and so just one free-body diagram and that is shown in Figure 18.84(b). As the free-body diagram indicates, the net force acting on the mass is the applied force minus the forces exerted by the spring and by the dashpot:

$$\text{net force} = F - ky - c\frac{dy}{dt}$$

Then applying Newton's second law, this force must be equal to $ma$, where $a$ is the acceleration, and so:

$$m\frac{d^2y}{dt^2} = F - ky - c\frac{dy}{dt}$$

The relationship between the input $F$ to the system and the output $y$ is described by the second-order differential equation:

$$m\frac{d^2y}{dt^2} + c\frac{dy}{dt} + ky = F$$

The term *second-order* is used because the equation includes as its highest derivative $d^2y/dt^2$.

### 18.10.2  Rotational Systems

For rotational systems the basic building blocks are a torsion spring, a rotary damper and the moment of inertia (Figure 18.86).

1. **Torsional spring**

   The "springiness" or "stiffness" of a rotational spring is represented by a torsional spring. For a torsional spring, the angle $\theta$ rotated is proportional to the torque $T$:

   $$T = k\theta$$

   where $k$ is a measure of the stiffness of the spring.

2. **Rotational dashpot**

   The damping inherent in rotational motion is represented by a rotational dashpot. For a rotational dashpot, i.e., effectively a disk rotating in a fluid, the resistive torque $T$ is proportional to the angular velocity $\omega$ and thus:

   $$T = c\omega = c\frac{d\theta}{dt}$$

   where $c$ is the damping constant.

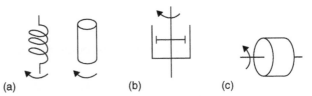

(a)                    (b)              (c)

**Figure 18.86: Rotational system elements: (a) torsional spring or elastic twisting of a shaft; (b) rotational dashpot; (c) moment of inertia**

### 3. Inertia

The inertia of a rotational system is represented by the moment of inertia of a mass. A torque $T$ applied to a mass with a moment of inertia I results in an angular acceleration a and thus, since angular acceleration is the rate of change of angular velocity $\omega$ with time, i.e., $d\omega/dt,$ and angular velocity $\omega$ is the rate of change of angle with time, i.e., $d\theta/dt,$ then the angular acceleration is $d(d\theta/dt)/dt$ and so:

$$T = Ia = I\frac{d^2\theta}{dt^2}$$

The following example illustrates how we can arrive at a model for a rotational system.

### Example 18.6

Develop a model for the system shown in Figure 18.87(a) of the rotation of a disk as a result of twisting a shaft.

### Solution

Figure 18.87(b) shows the free-body diagram for the system. The torques acting on the disk are the applied torque $T$, the spring torque $k\theta$ and the damping torque $c\omega$. Hence:

$$T - k\theta - c\frac{d\theta}{dt} = I\frac{d^2\theta}{dt^2}$$

We thus have the second-order differential equation relating the input of the torque to the output of the angle of twist:

$$I\frac{d^2\theta}{dt^2} + c\frac{d\theta}{dt} + k\theta = T$$

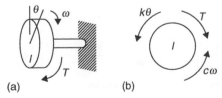

(a)                    (b)

**Figure 18.87: Example**

### 18.10.3 Electrical Systems

The basic elements of electrical systems are the resistor, inductor and capacitor (Figure 18.88).

**1. Resistor**

For a *resistor,* resistance $R$, the potential difference $v$ across it when there is a current $i$ through it is given by:

$$v = Ri$$

**2. Inductor**

For an *inductor,* inductance $L$, the potential difference $v$ across it at any instant depends on the rate of change of current $i$ and is:

$$v = \frac{di}{dt}$$

**3. Capacitor**

For a *capacitor,* the potential difference $v$ across it depends on the charge $q$ on the capacitor plates with $v = q/C$, where $C$ is the capacitance. Thus:

$$v = \frac{1}{C}q$$

$$\frac{dv}{dt} = \frac{1}{C}\frac{dq}{dt}$$

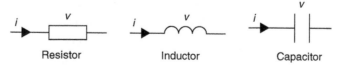

**Figure 18.88: Electrical system building blocks**

Since current $i$ is the rate of movement of charge:

$$\frac{dv}{dt} = \frac{1}{C}\frac{dq}{dt} = \frac{1}{C}i$$

and so we can write:

$$i = C\frac{dv}{dt}$$

To develop the models for electrical circuits we use Kirchhoff's laws. These can be stated as:

### 1. Kirchhoff's current law

The total current flowing into any circuit junction is equal to the total current leaving that junction, i.e., the algebraic sum of the currents at a junction is zero.

### 2. Kirchhoff's voltage law

In a closed circuit path, termed a loop, the algebraic sum of the voltages across the elements that make up the loop is zero. This is the same as saying that for a loop containing a source of e.m.f., the sum of the potential drops across each circuit element is equal to the sum of the applied e.m.f.'s, provided we take account of their directions.

The following examples illustrate the development of models for electrical systems.

*Example 18.7*
Develop a model for the electrical system described by the circuit shown in Figure 18.89. The input is the voltage $v$ when the switch is closed and the output is the voltage $vc$ across the capacitor.

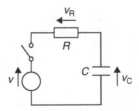

**Figure 18.89: Electrical system with resistance and capacitance**

**Solution**

Using Kirchhoff's voltage law gives:

$$v = v_R + v_C$$

and, since $V_R = Ri$ and $i = C(dv_C/dt)$ we obtain the equation:

$$v = RC\frac{dv_C}{dt} + v_C$$

The relationship between an input $v$ and the output $v_c$ is a first order differential equation. The term *first-order* is used because it includes as its highest derivative $dv_C/dt$.

**Example 18.8**

Develop a model for the circuit shown in Figure 18.90 when we have an input voltage $v$ when the switch is closed and take an output as the voltage $v_C$ across the capacitor.

**Solution**

Applying Kirchhoff's voltage law gives:

$$v = v_R + v_C$$

and so:

$$v = Ri + L\frac{di}{dt} + v_C$$

Since $i = C(dv_C/dt)$, then $di/dt = C(d^2v_C/dt^2)$ and thus, we can write:

$$v = RC\frac{dv_C}{dt} + LC\frac{d^2v_C}{dt^2} + v_C$$

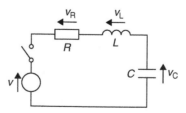

**Figure 18.90: Electrical system with resistance, inductance and capacitance**

The relationship between an input $v$ and output $v_c$ is described by a second-order differential equation.

### 18.10.4 Thermal Systems

Thermal systems have two basic building blocks, resistance and capacitance (Figure 18.91).

### 1. Thermal resistance

The thermal resistance $R$ is the resistance offered to the rate of flow of heat $q$ (Figure 18.91(a)) and is defined by:

$$q = \frac{T_1 - T_2}{R}$$

where $T_1 - T_2$ is the temperature difference through which the heat flows.

For heat conduction through a solid we have the rate of flow of heat proportional to the cross-sectional area and the temperature gradient. Thus for two points at temperatures $T_1$ and $T_2$ and a distance $L$ apart:

$$q = Ak\frac{T_1 - T_2}{L}$$

with $k$ being the thermal conductivity. Thus with this mode of heat transfer, the thermal resistance $R$ is $L/Ak$. For heat transfer by convection between two points, Newton's law of cooling gives:

$$q = Ah(T_2 - T_1)$$

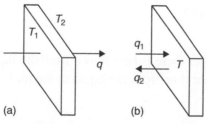

(a)   (b)

**Figure 18.91: (a) Thermal resistance; (b) thermal capacitance**

where $(T_2 - T_1)$ is the temperature difference, $h$ the coefficient of heat transfer and $A$ the surface area across which the temperature difference is. The thermal resistance with this mode of heat transfer is thus *1/Ah*.

## 2. Thermal capacitance

The thermal capacitance (Figure 18.91(b)) is a measure of the store of internal energy in a system. If the rate of flow of heat into a system is $q_1$, and the rate of flow out $q_2$ then the rate of change of internal energy of the system is $q_1 - q_2$. An increase in internal energy can result in a change in temperature:

change in internal energy $= mc \times$ change in temperature

where $m$ is the mass and $c$ the specific heat capacity. Thus the rate of change of internal energy is equal to $mc$ times the rate of change of temperature. Hence:

$$q_1 - q_2 = mc \frac{dT}{dt}$$

This equation can be written as:

$$q_1 - q_2 = C \frac{dT}{dt}$$

where the capacitance $C = mc$.

The following example illustrates the development of models for thermal systems.

### Example 18.9

Develop a model for the simple thermal system of a thermometer at temperature $T$ being used to measure the temperature of a liquid when it suddenly changes to the higher temperature of $T_L$ (Figure 18.92).

### Solution

When the temperature changes there is heat flow $q$ from the liquid to the thermometer. The thermal resistance to heat flow from the liquid to the thermometer is:

$$q = \frac{T_L - T}{R}$$

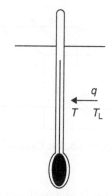

**Figure 18.92: Example**

Since there is only a net flow of heat from the liquid to the thermometer the thermal capacitance of the thermometer is:

$$q = C\frac{dT}{dt}$$

Substituting for $q$ gives:

$$C\frac{dT}{dt} = \frac{T_L - T}{R}$$

which, when rearranged gives:

$$RC\frac{dT}{dt} + T = T_L$$

This is a first-order differential equation.

### Example 18.10

Determine a model for the temperature of a room (Figure 18.93) containing a heater which supplies heat at the rate $q_1$ and the room loses heat at the rate $q_2$.

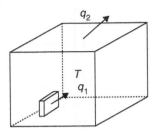

**Figure 18.93: Example**

### Solution

We will assume that the air in the room is at a uniform temperature $T$. If the air and furniture in the room have a combined thermal capacity $C$, since the energy rate to heat the room is $q_1 - q_2$, we have:

$$q_1 - q_2 = C \frac{dT}{dt}$$

If the temperature inside the room is $T$ and that outside the room $T_0$ then

$$q_2 = \frac{T - T_0}{R}$$

where $R$ is the thermal resistance of the walls. Substituting for $q_2$ gives:

$$q_1 - \frac{T - T_0}{R} = C \frac{dT}{dt}$$

Hence:

$$RC \frac{dT}{dt} + T = Rq_1 + T_0$$

This is a first-order differential equation.

### 18.10.5 Hydraulic Systems

For a fluid system the three building blocks are resistance, capacitance and inertance; these are the equivalents of electrical resistance, capacitance and inductance.

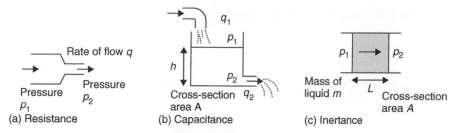

**Figure 18.94: Hydraulic building blocks**

The equivalent of electrical current is the volumetric rate of flow and of potential difference is pressure difference. Hydraulic fluid systems are assumed to involve an incompressible liquid; pneumatic systems, however, involve compressible gases and consequently there will be density changes when the pressure changes. Here we will just consider the simpler case of hydraulic systems. Figure 18.94 shows the basic form of building blocks for hydraulic systems.

1. **Hydraulic resistance**

   Hydraulic resistance $R$ is the resistance to flow which occurs when a liquid flows from one diameter pipe to another (Figure 18.94(a)) and is defined as being given by the hydraulic equivalent of Ohm's law:

   $$p_1 - p_2 = Rq$$

2. **Hydraulic capacitance**

   Hydraulic capacitance $C$ is the term used to describe energy storage where the hydraulic liquid is stored in the form of potential energy (Figure 18.94(b)). The rate of change of volume $V$ of liquid stored is equal to the difference between the volumetric rate at which liquid enters the container $q_1$ and the rate at which it leaves $q_2$, i.e.,

   $$q_1 - q_2 = \frac{dV}{dt}$$

   But $V = Ah$ and so:

   $$q_1 - q_2 = A\frac{dh}{dt}$$

The pressure difference between the input and output is:

$$p_1 - p_2 = p = hpg$$

Hence, substituting for $h$ gives:

$$q_1 - q_2 = \frac{A}{\rho g}\frac{dp}{dt}$$

The hydraulic capacitance $C$ is defined as:

$$C = \frac{A}{\rho g}$$

and thus, we can write:

$$q_1 - q_2 = C\frac{dp}{dt}$$

3. **Hydraulic inertance**

Hydraulic inertance is the equivalent of inductance in electrical systems. To accelerate a fluid a net force is required and this is provided by the pressure difference (Figure 18.93(c)). Thus:

$$(p_1 - p_2)A = ma = m\frac{dv}{dt}$$

where $a$ is the acceleration and so the rate of change of velocity $v$. The mass of fluid being accelerated is $m = ALp$ and the rate of flow $q = Av$ and so:

$$(p_1 - p_2)A = Lp\frac{dq}{dt}$$

$$p_1 - p_2 = I\frac{dq}{dt}$$

where the inertance $I$ is given by $I = Lp/A$.

The following example illustrates the development of a model for a hydraulic system.

**Example 18.11**
Develop a model for the hydraulic system shown in Figure 18.95 where there is a liquid entering a container at one rate $q_1$ and leaving through a valve at another rate $q_2$.

**Solution**
We can neglect the inertance since flow rates can be assumed to change only very slowly. For the capacitance term we have:

$$q_1 - q_2 = C \frac{dp}{dt} = \frac{A}{\rho g} \frac{dp}{dt}$$

For the resistance of the valve we have:

$$p_1 - p_2 = Rq_1$$

Thus, substituting for $q_2$, and recognizing that the pressure difference is $hpg$, gives:

$$q_1 = A \frac{dh}{dt} + \frac{h\rho g}{R}$$

$$A \frac{dh}{dt} + \frac{\rho g}{R} h = q_1$$

This is a first-order differential equation.

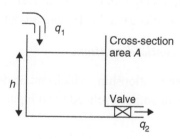

**Figure 18.95: Example**

## 18.11  Differential Equations

As the previous section indicates, the relationship between the input and output for systems is often in the form of a differential equation which shows how, when there is some input, the output varies with time.

### 18.11.1  First-Order Differential Equations

Many systems have input-output relationships which can be described by a first-order differential equation and have an output $y$ related to an input $x$ by an equation of the form:

$$\tau \frac{dy}{dt} + y = kx$$

where $\tau$ and $k$ are constants, $\tau$ being known as the *time constant*.

Consider the response of such a system when subject to a unit step input, i.e., an input which suddenly changes from 0 to a constant value of 1. When we reach the time at which the input $x$ is not changing with time, i.e., we have steady-state conditions, then $dx/dt = 0$ and so we have output $y = kx$ and $k$ is the *steady-state gain*. Thus, with a unit step input the steady-state output is $lk$. Over time, the output is related to the input by an equation of the form:

$$y = \text{steady-state value} \times (1 - e^{-t/\tau})$$

Thus, Figure 18.96 shows how first order systems behave when subject to a unit step input. After a time of $1\tau$ the output has reached $0.63k$, after $2\tau$ is $0.86k$, after $3\tau$ it is $0.95k$, after $4\tau$ it is $0.98k$, and eventually it becomes $1k$.

Examples of first-order systems are an electrical system having capacitance and resistance, an electrical system having inductance and resistance and a thermal system of a room with a heat input from an electrical heater and an output of the room temperature.

### 18.11.2  Second-order Differential Equations

Many systems have input-output relationships which can be described by second-order differential equations and have an output $y$ related to an input $x$ by an equation of the form:

$$\frac{d^2 y}{dt^2} + 2\zeta\omega_n \frac{dy}{dt} + \omega_n^2 y = k\omega_n^2 x$$

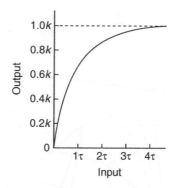

**Figure 18.96: Behavior of a first order system when subject to a unit step input**

where $k$, $\zeta$ and $\omega_n$ are constants for the systems. The constant $\zeta$ is known as the *damping ratio* or *factor* and $\omega_n$ as the *undamped natural angular frequency.*

If the input $y$ is not changing with time, i.e., we have steady-state conditions, then $d^2y/dt^2 = 0$ and $dy/dt = 0$ and so we have output $y = kx$ and $k$ is the *steady-state gain.* Figure 18.97 shows how a second order system behaves when subject to a unit step input.

The general form of the response varies with the damping factor. Systems with damping factors less than 1 are said to be *underdamped,* with damping factors greater than 1 as *overdamped* and for a damping factor of 1 as *critically damped.*

1. With no damping, i.e., $\zeta = 0$, the system output oscillates with a constant amplitude and a frequency of $\omega_n$. (since $\omega_n = 2\pi f_n$, where $f_n$ is the undamped natural frequency, and $f_n - 1/T$, where $T_n$ is the time for one undamped oscillation, then $T_n = 2\pi/\omega_n = 6.3/\omega_n$).

2. With underdamping i.e., $\zeta < 0$, the output oscillates but the closer the damping factor is to 1 the faster the amplitude of the oscillations diminishes.

3. With critical damping, i.e., $\zeta = 1$, there are no oscillations and the output just gradually approaches the steady-state value.

4. With overdamping, i.e., $\zeta > 1$, the output takes longer than critical damping to reach the steady-state value.

A mechanical system that can be modeled by a spring, dashpot and mass is an example of a second order system. When we apply a load to the system then oscillations occur

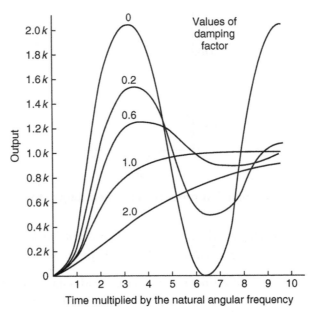

**Figure 18.97: Behavior of a second order system with a unit step input.
With no damping the output is just a continuous oscillation following a step input.
As the damping increases, so the oscillations become damped out and with a damping
factor of 1.0 there are no oscillations and the output just rises over time to the steady
state output value. Further increases in damping mean that the output takes longer to
reach the steady state value.**

which have amplitudes which die away with time. This was illustrated in section 18.8 and Figure 18.79. Likewise, with the second order system of an electrical circuit having resistance, inductance and capacitance: when there is a step voltage input, i.e., a switch is closed and applies a constant voltage to the circuit, then the voltage across the capacitor will be described by a second order differential equation and so can oscillate with amplitudes which die away with time.

The differential equations describe the input/output relationship when we consider the input and output to be functions of time. We can use the model building techniques described in the previous section to arrive at differential equations, alternatively we can find the response of a system to, say, a step input and by examining the response determine the form of the differential equation which described its behavior.

## Example 18.12

An electrical system has an output $v$ related to the input $V$ by the differential equation:

$$RC \frac{dv}{dt} + v = V$$

What are the time constant and the steady state gain of the system?

## Solution

This equation is of the form:

$$\tau \frac{dy}{dt} + y = kx$$

and hence has a time constant of $RC$ and a steady state gain of 1.

## Example 18.13

A mechanical system has an output of a rotation through an angle $\theta$ related to an input torque $T$ by the differential equation:

$$I \frac{d^2\theta}{dt^2} + c \frac{d\theta}{dt} + k\theta = T$$

What are the natural angular frequency and the damping constant of the system?

## Solution

The equation can be put into the form:

$$\frac{d^2y}{dt^2} + 2\zeta\omega_n \frac{dy}{dt} + \omega_n^2 y = k\omega_n^2 x$$

as:

$$\frac{d^2\theta}{dt^2} + \frac{c}{I} \frac{d\theta}{dt} + \frac{k}{I}\theta = \frac{1}{I}T$$

Hence, comparing the terms in front of $y$ and $\theta$ gives $\omega_n = \sqrt{(k/I)}$. Comparing the terms in front of $dy/dt$ and $d\theta/dt$ gives $2\zeta\omega_n = c/I$ and hence, since $\omega_n = \sqrt{(k/I)}$, gives $\zeta = c/[2\sqrt{(kI)}]$.

## 18.12 Transfer Function

In general, when we consider inputs and outputs of systems as functions of time then the relationship between the output and input is given by a differential equation. If we have a system composed of two elements in series with each having its input-output relationships described by a differential equation, it is not easy to see how the output of the system as a whole is related to its input. There is a way we can overcome this problem and that is to transform the differential equations into a more convenient form by using the *Laplace transform*. This form is a much more convenient way of describing the relationship than a differential equation, since it can be easily manipulated by the basic rules of algebra.

To carry out the transformation we follow the following rules:

1. A variable which is a function of time, e.g., the input or output voltage $v$ in a circuit (to emphasise that $v$ is a function of time it might be written as $v(t)$ note that this does not mean that $v$ is multiplied by $t$), becomes a function of $s$. A voltage is thus written as $V(s)$; note that upper case letters are used for the variables when written as functions of $s$ and that this does not mean that $V$ is multiplied by $s$.

2. A constant $k$ which does not vary with time remains a constant. Thus $kv$, where $v$ is a function of time, becomes $kV(s)$. For example, the voltage $3v$ written as an $s$ function is $3V(s)$.

3. If the initial value of the variable $v$ is zero at time $t = 0$, the first derivative of a function of time $dv/dt$ becomes $sV(s)$ *and* $kdv/dt$ becomes $ksV(s)$. For example, with no initial values $4dv/dt$ as an $s$ function is $4sV(s)$.

   Note that if there is an initial value $v_0$ at $t = 0$ then the first derivative of a function of time $dv/dt$ becomes $sV(s) - V_o$, i.e., we subtract any initial value, and $kdv/dt$ becomes $k[sV(s) - v_0]$. For example, if we have $v_0 = 2$ at $t = 0$ then $dv/dt$ becomes $sV(s) - 2$.

4. If the initial value of the variable $v$ and $dv/dt$ is zero at time $t = 0$, the second derivative of a function of time $d^2v/dt^2$ becomes $s^2V(s)$ and $kd^2v/dt^2$ becomes $ks^2V(s)$. For example, with no initial values $4d^2v/dt^2$ as an s function is $4s^2V(s)$.

Note that if there are initial values $v_0$ and $(dv/dt)_0$ then the second derivative of a function of time $d^2v/dt^2$ becomes $s^2V(s) - sv_0 - (dv/dt)_0$ and $kd^2v/dt^2$ becomes $k[s^2V(s) - sv_0 - (dv/dt)_0]$. For example, with initial values of $v_0 = 2$ and $(dv/dt)_0 = 3$ at time $t = 0$, then $4d^2v/dt^2$ as an $s$ function is $4s^2V(s) - 2s - 3$.

With an integral of a function of time:

$$\int_0^t v \, dt \text{ becomes } \frac{1}{s}V(s)$$

$$\int_0^t kv \, dt \text{ becomes } \frac{1}{s}kV(s)$$

Note that, when derivatives are involved, we need to know the initial conditions of a system output prior to the input being applied before we can transform a time function into an s function.

### Example 18.14

Determine the Laplace transform for the following equation where we have $v$ and $v_C$ as functions of time and no initial values.

$$v = RC\frac{dv_C}{dt} + v_C$$

### Solution

The Laplace transform **is:**

$$V(s) = RCsV_C(s) + V_C(s)$$

Thus $V(s)$ is the Laplace transform of the input voltage $v$ and $V_C(s)$ is the Laplace transform of the output voltage $v_C$. Rearranging gives:

$$\frac{V_C(s)}{V(s)} = \frac{1}{RCs + 1}$$

The above equation thus describes the relationship between the input and output of the system when described as s functions.

### 18.12.1 Transfer Function

In section 18.9 we used the term *gain* to relate the input and output of a system with gain $G$ = output/input. When we are working with inputs and outputs described as functions of $s$ we define the *transfer function $G(s)$* as [output $Y(s)$/input $X(s)$] when all initial conditions before we apply the input are zero:

$$G(s) = \frac{Y(s)}{X(s)}$$

A transfer function can be represented as a block diagram (Figure 18.98) with $X(s)$ the input, $Y(s)$ *the* output and the transfer function $G(s)$ as the operator in the box that converts the input to the output. The block represents a multiplication for the input. Thus, by using the Laplace transform of inputs and outputs, we can use the transfer function as a simple multiplication factor, like the gain discussed in section 18.9.

### Example 18.15

Determine the transfer function for an electrical system for which we have the relationship (this equation was derived in the example in the preceding section):

$$\frac{V_C(s)}{V(s)} = \frac{1}{RCs + 1}$$

The transfer function $G(s)$ is thus:

$$G(s) = \frac{V_C(s)}{V(s)} = \frac{1}{RCs + 1}$$

To get the output $V_c(s)$ we multiply the input $V(s)$ by $1/(RCs + 1)$.

**Figure 18.98: Transfer function as the factor that multiplies the input to give the output**

*Example 18.16*
Determine the transfer function for the mechanical system, having mass, stiffness and damping, and input F and output y and described by the differential equation (as in section 18.10.1):

$$F = m\frac{d^2y}{dt^2} + c\frac{dy}{dt} + ky$$

*Solution*
If we now write the equation with the input and output as functions of $s$, with initial conditions zero:

$$F(s) = ms^2Y(s) = csY(s) + kY(s)$$

Hence the transfer function $G(s)$ of the system is:

$$G(s) = \frac{Y(s)}{F(s)} = \frac{1}{ms^2 + cs + k}$$

### 18.12.2 Transfer Functions of Common System Elements

By considering the relationships between the inputs to systems and their outputs we can obtain transfer functions for them and hence describe a control system as a series of interconnected blocks, each having its input-output characteristics defined by a transfer function. The following are transfer functions which are typical of commonly encountered system elements:

1. **Gear train**

   For the relationship between the input speed and output speed with a gear train having a gear ratio $N$:

   transfer function $= N$

2. **Amplifier**

   For the relationship between the output voltage and the input voltage with $G$ as the constant gain:

   transfer function $= G$

### 3. Potentiometer

For the potentiometer acting as a simple potential divider circuit the relationship between the output voltage and the input voltage is the ratio of the resistance across which the output is tapped to the total resistance across which the supply voltage is applied and so is a constant and hence the transfer function is a constant $K$:

transfer function $= K$

### 4. Armature-controlled DC motor

For the relationship between the drive shaft speed and the input voltage to the armature is:

$$\text{transfer function} = \frac{1}{sL + R}$$

where $L$ represents the inductance of the armature circuit and R its resistance.

This was derived by considering armature circuit as effectively inductance in series with resistance and hence:

$$v = L\frac{di}{dt} + Ri$$

and so, with no initial conditions:

$$(V)(s) = sLI(s) + RI(s)$$

and, since the output torque is proportional to the armature current we have a transfer function of the form $1/(sL + R)$.

### 5. Valve controlled hydraulic actuator

The output displacement of the hydraulic cylinder is related to the input displacement of the valve shaft by a transfer function of the form:

$$\text{transfer function} = \frac{K_1}{s(K_2 s + K_3)}$$

where $K_1$, $K_2$ and $K_3$ are constants.

### 6. Heating system

The relationship between the resulting temperature and the input to a heating element is typically of the form:

$$\text{transfer function} = \frac{1}{sC + 1/R}$$

where $C$ is a constant representing the thermal capacity of the system and $R$ a constant representing its thermal resistance.

### 7. Tachogenerator

The relationship between the output voltage and the input rotational speed is likely to be a constant $K$ and so represented by:

$$\text{transfer function} = K$$

### 8. Displacement and rotation

For a system where the input is the rotation of a shaft and the output, as perhaps the result of the rotation of a screw, a displacement, since speed is the rate of displacement we have $v = dy/dt$ and so $V(s) = sY(s)$ and the transfer function is:

$$\text{transfer function} = \frac{1}{s}$$

### 9. Height of liquid level in a container

The height of liquid in a container depends on the rate at which liquid enters the container and the rate at which it is leaving. The relationship between the input of the rate of liquid entering and the height of liquid in the container is of the form:

$$\text{transfer function} = \frac{1}{sA + \rho g/R}$$

where $A$ is the constant cross-sectional area of the container, $p$ the density of the liquid, $g$ the acceleration due to gravity and $R$ the hydraulic resistance offered by the pipe through which the liquid leaves the container.

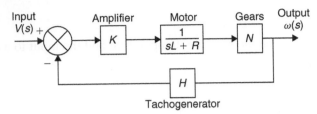

**Figure 18.99: Block diagram for the control system for speed of a shaft with the terms in the boxes being the transfer functions for the elements concerned**

### 18.12.3 Transfer Functions and Systems

Consider a speed control system involving a differential amplifier to amplify the error signal and drive a motor, this then driving a shaft via a gear system. Feedback of the rotation of the shaft is via a tachogenerator.

1.  The differential amplifier might be assumed to give an output directly proportional to the error signal input and so be represented by a constant transfer function $K$, i.e., a gain $K$ which does not change with time.

2.  The error signal is an input to the armature circuit of the motor and results in the motor giving an output torque which is proportional to the armature current. The armature circuit can be assumed to be a circuit having inductance $L$ and resistance $R$ and so a transfer function of $1/(sL + R)$.

3.  The torque output of the motor is transformed to rotation of the drive shaft by a gear system and we might assume that the rotational speed is proportional to the input torque and so represent the transfer function of the gear system by a constant transfer function $N$, i.e., the gear ratio.

4.  The feedback is via a tachogenerator and we might make the assumption that the output of the generator is directly proportional to its input and so represent it by a constant transfer function $H$.

The block diagram of the control system might thus be like that in Figure 18.99.

## 18.13 System Transfer Functions

Consider the overall transfer functions of systems involving series-connected elements and systems with feedback loops.

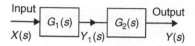

**Figure 18.100: Systems in series**

### 18.13.1 Systems in Series

Consider a system of two subsystems in series (Figure 18.100). The first subsystem has an input of $X(s)$ and an output of $Y_1(s)$; thus, $G_1(s) = Y_1(s)/X(s)$. The second subsystem has an input of $Y_1(s)$ and an output of $Y(s)$; thus, $G_2(s) = Y(s)/Y_1(s)$. We thus have:

$$Y(s) = G_2(s)Y_1(s) = G_2(s)G_1(s)X(s)$$

The overall transfer function $G(s)$ of the system is $Y(s)/X(s)$ and so:

$$G_{\text{overall}}(s) = G_1(s)G_2(s)$$

Thus, in general:

*The overall transfer function for a system composed of elements in series is the product of the transfer functions of the individual series elements.*

### Example 18.17
Determine the overall transfer function for a system which consists of two elements in series, one having a transfer function of $1/(s + 1)$ and the other $1/(s + 2)$.

### Solution
The overall transfer function is thus:

$$G_{\text{overall}}(s) = \frac{1}{s+1} \times \frac{1}{s+2} = \frac{1}{(s+1)(s+2)}$$

### 18.13.2 Systems with Feedback

For systems with a negative feedback loop we can have the situation shown in Figure 18.101 where the output is fed back via a system with a transfer function $H(s)$ to subtract from the input to the system $G(s)$. The feedback system has an input of $Y(s)$ and thus

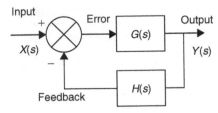

**Figure 18.101: System with negative feedback**

an output of $H(s)Y(s)$. Thus the feedback signal is $H(s)Y(s)$. The error is the difference between the system input signal $X(s)$ and the feedback signal and is thus:

Error $(s) = X(s) - H(s)Y(s)$

This error signal is the input to the $G(s)$ system and gives an output of $Y(s)$. Thus:

$$G(s) = \frac{Y(s)}{X(s) - H(s)Y(s)}$$

and so:

$$[1 + G(s)H(s)]Y(s) = G(s)X(s)$$

which can be rearranged to give:

$$\text{overall transfer function} = \frac{Y(s)}{X(s)} = \frac{G(s)}{1 + G(s)H(s)}$$

*For a system with a negative feedback, the overall transfer function is the forward path transfer function divided by one plus the product of the forward path and feedback path transfer functions.*

For a system with positive feedback (Figure 18.102), the feedback signal is $H(s)Y(s)$ and thus the input to the $G(s)$ system is $X(s) + H(s)Y(s)$. Hence:

$$G(s) = \frac{Y(s)}{X(s) + H(s)Y(s)}$$

and so:

$$[1 - G(s)H(s)]Y(s) = G(s)X(s)$$

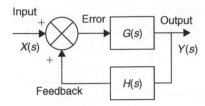

**Figure 18.102: System with positive feedback**

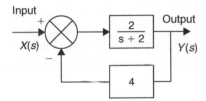

**Figure 18.103: Example**

This can be rearranged to give:

$$\text{overall transfer function} = \frac{Y(s)}{X(s)} = \frac{G(s)}{1 - G(s)H(s)}$$

For a system with a positive feedback, the overall transfer function is the forward path transfer function divided by one minus the product of the forward path and feedback path transfer functions.

### Example 18.18
Determine the overall transfer function for a control system (Figure 18.103) which has a negative feedback loop with a transfer function 4 and a forward path transfer function of $2/(s + 2)$.

### Solution
The overall transfer function of the system is:

$$G_{\text{overall}}(s) = \frac{\dfrac{2}{s+2}}{1 + 4 \times \dfrac{2}{s+2}} = \frac{2}{s+10}$$

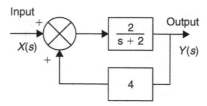

**Figure 18.104: Example**

### Example 18.19

Determine the overall transfer function for a system (Figure 18.104) which has a positive feedback loop with a transfer function 4 and a forward path transfer function of $2/(s + 2)$.

### Solution

The overall transfer function is:

$$G_{overall}(s) = \frac{\dfrac{2}{s+2}}{1 - 4 \times \dfrac{2}{s+2}} = \frac{2}{s-6}$$

## 18.14  Sensitivity

The *sensitivity* of a system is the measure of the amount by which the overall gain of the system is affected by changes in the gain of system elements or particular inputs. In the following, we consider the effects of changing the gain of elements and also the effect of disturbances.

### 18.14.1  Sensitivity to Changes in Parameters

With a control system, the transfer functions of elements may drift with time and thus we need to know how such drift will affect the overall performance of the system.

For a closed-loop system with negative feedback (Figure 18.105):

$$\text{overall transfer function} = \frac{G(s)}{1 + G(s)H(s)}$$

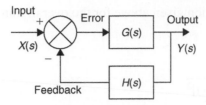

**Figure 18.105: System with negative feedback**

If $G(s)H(s)$ is large, then the above equation reduces to:

$$\text{overall transfer function} \simeq \frac{G(s)}{G(s)H(s)} \simeq \frac{1}{H(s)}$$

Thus, in such a situation, the system is relatively insensitive to variations in the forward path transfer function but is sensitive to variations in the feedback path transfer function. For example, a change in the feedback path transfer function of, say, 10%, i.e., from $H(s)$ to $1.1\,H(s)$, will result in a change in the overall transfer function from $1/H(s)$ to $1/1.1\,H(s)$ or about $0.9/H(s)$ and so a change of about 10%.

This sensitivity is because the feedback transfer function is for the measurement system supplying the signal which is compared with the set value signal to determine the error and so variations in the feedback transfer function directly affect the computation of the error.

If the forward path transfer function G(s) changes then the overall transfer function $G_{overall}$ (s) will change. We can define the sensitivity of the system to changes in the transfer function of the forward element as the fractional change in the overall system transfer function $G_{overall}(s)$ divided by the fractional change in the forward element transfer function G(s), i.e., $(\Delta G_{overall}/G_{overall})/(\Delta G/G)$ where $\Delta G_{overall}$ is the change in overall gain producing a change of $\Delta G$ in the forward element transfer function. Thus, the sensitivity can be written as:

$$\text{sensitivity} = \frac{\Delta G_{overall}(s)}{\Delta G(s)} \frac{G(s)}{G_{overall}(s)}$$

If we differentiate the equation given above for the overall transfer function we obtain:

$$\frac{dG_{overall}(s)}{dG(s)} = \frac{1}{[1 + G(s)H(s)]^2}$$

and since $G_{overall}(s)/G(s) = 1/[1 + G(s)H(s)]$, the sensitivity is:

$$\text{sensitivity} = \frac{1}{1 + G(s)H(s)}$$

Thus the bigger the value of $G(s)H(s)$ the lower the sensitivity of the system to changes in the forward path transfer function. The feedback amplifier discussed in section 18.9.3 is an illustration of this, the forward path transfer function for the op-amp being very large and so giving a system with low sensitivity to changes in the op amp gain and hence a stable system which can have its gain determined by purely changing the feedback loop gain, i.e., the resistors in a potential divider.

### Example 18.20

A closed-loop control system with negative feedback has a feedback transfer function of 0.1 and a forward path transfer function of (a) 50, (b) 5. What will be the effect of a change in the forward path transfer function of an increase by 10%?

### Solution

(a) We have, before the change:

$$\text{overall transfer function} = \frac{G(s)}{1 + G(s)H(s)} = \frac{50}{1 + 50 \times 0.1} = 8.3$$

After the change we have:

$$\text{overall transfer function} = \frac{G(s)}{1 + G(s)H(s)} = \frac{55}{1 + 55 \times 0.1} = 8.5$$

The change is thus about 2%.

(b) We have, before the change:

$$\text{overall transfer function} = \frac{G(s)}{1 + G(s)H(s)} = \frac{5}{1 + 5 \times 0.1} = 3.3$$

After the change we have:

$$\text{overall transfer function} = \frac{G(s)}{1 + G(s)H(s)} = \frac{5.5}{1 + 5.5 \times 0.1} = 3.5$$

The change is thus about 6%.

Thus the sensitivity of the system to changes in the forward path transfer function is reduced as the gain of the forward path is increased.

### 18.14.2 Sensitivity to Disturbances

An important effect of having feedback in a system is the reduction of the effects of disturbance signals on the system. A disturbance signal is an unwanted signal which affects the output signal of a system, such as noise in electronic amplifiers or a door being opened in a room with temperature controlled by a central heating system

Consider the effect of external disturbances on the overall gain of a system. Firstly we consider the effect on a open-loop system and then on a closed-loop system.

Consider the two-element open-loop system shown in Figure 18.106 when there is a disturbance which gives an input between the two elements. For an input $X(s)$ to the system, the first element gives an output of $G_1(s)X(s)$. To this is added the disturbance $D(s)$ to give an input of $G_1(s)X(s) + D(s)$. The overall system output will then be

$$Y(s) = G_2(s)[G_1(s)X(s) + D(s)] = G_1(s)G_2(s)X(s) + G_2(s)D(s)$$

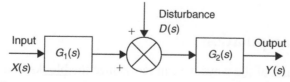

**Figure 18.106: Disturbance with an open-loop system**

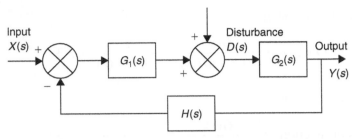

**Figure 18.107: Disturbance with closed-loop system**

The *signal-to-noise ratio* is the ratio of the output due to the signal to that produced by the noise and is thus, for the open-loop system, given by $G_1(s)G_2(s)X(s)/G_2(s)D(s) = G_1(s)X(s)/D(s)$. Thus increasing the gain of the element prior to the disturbance increases the signal-to-noise ratio.

For the system with feedback (Figure 18.107), the input to the first forward element $G_1(s)$ is $X(s) - H(s)Y(s)$ and so its output is $G_1(s)[X(s) - H(s)Y(s)]$. The input to $G_2(s)$ is $G_1(s)[X(s) - H(s)Y(s)] + D(s)$ and so its output is $X(s) = G_2(s)\{G_1(s)[X(s) - H(s)Y(s)] + D(s)\}$. Rearranged this gives:

$$Y(s) = \frac{G_1(s)G_2(s)}{1 + G_1(s)G_2(s)H(s)} X(s) + \frac{G_2(s)}{1 + G_1(s)G_2(s)H(s)} D(s)$$

Comparing this with the equation for the open-loop system of $Y(s) = G_2(s)[G_1(s)X(s) + D(s)] = G_1(s)G_2(s)X(s) + G_2(s)D(s)$ indicates that the effect of the disturbance on the output of the system has been reduced by a factor of $[1 + G_1(s)G_2(s)H(s)]$. This factor is thus a measure of how much the effects of a disturbance are reduced by feedback.

The signal-to-noise ratio is the ratio of the output due to the signal to that produced by the noise and is thus $G_1(s)X(s)/D(s)$ and is the same as when there is no feedback. Thus, as with the open-loop system, the effect of such a disturbance is reduced as the gain of the $G_1(s)$ element is increased.

## 18.15 Block Manipulation

The following are some of the ways we can reorganize the blocks in a block diagram of a system in order to produce simplification and still give the same overall transfer function

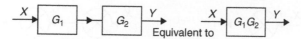

**Figure 18.108: Blocks in series**

**Figure 18.109: Moving a take-off point to beyond a block**

**Figure 18.110: Moving a take-off point to ahead of a block**

for the system. To simplify the diagrams, the (*s*) has been omitted; it should, however, be assumed for all dynamic situations.

### 18.15.1  Blocks in Series

As indicated in section 18.12.1, Figure 18.108 shows the basic rule for simplifying blocks in series.

### 18.15.2  Moving Take-Off Points

As a means of simplifying block diagrams it is often necessary to move take-off points. The following figures (Figures 18.109 and 18.110) give the basic rules for such movements.

### 18.15.3  Moving a Summing Point

As a means of simplifying block diagrams it is often necessary to move summing points. The following figures (Figures 18.111–18.114) give the basic rules for such movements.

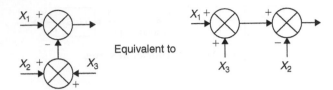

**Figure 18.111: Rearrangement of summing points**

**Figure 18.112: Interchange of summing points**

**Figure 18.113: Moving a summing point ahead of a block**

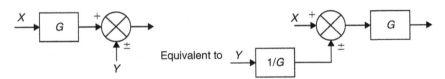

**Figure 18.114: Moving a summing point beyond a block**

### 18.15.4 Changing Feedback and Forward Paths

Figures 18.115 and 18.116 show block simplification techniques when changing feed-forward and feedback paths.

***Example 18.21***
Use block simplification techniques to simplify the system shown in Figure 18.117.

***Solution***
Figures 18.118–18.123 show the various stages in the simplification.

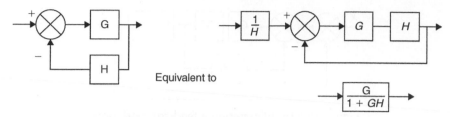

**Figure 18.115: Removing a block from a feedback path**

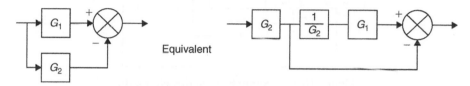

**Figure 18.116: Removing a block from a forward path**

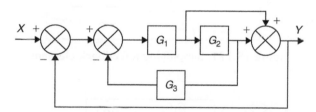

**Figure 18.117: The circuit to be simplified**

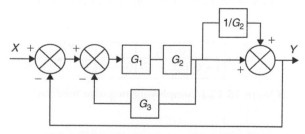

**Figure 18.118: Moving a take-off point**

X
+
−
+
−
G1
G2
1 + 1/G2
Y
G3

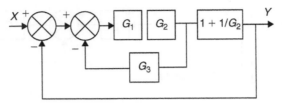

Figure 18.119: Eliminating a feed-forward loop

X
+
−
+
−
G1G2
1 + 1/G2
Y
G3

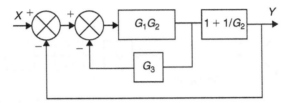

Figure 18.120: Simplifying series elements

X
+
−
G1G2 / 1 + G1G2G3
1 + 1/G2
Y

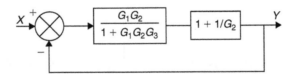

Figure 18.121: Simplifying a feedback element

X
+
−
G1G2 / 1 + G1G2G3 (1 + 1/G2)
Y

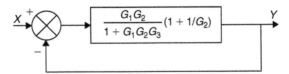

Figure 18.122: Simplifying series elements

X
G1 (G1 + 1) / 1 + G1G2G3 + G1 (G1 + 1)
Y

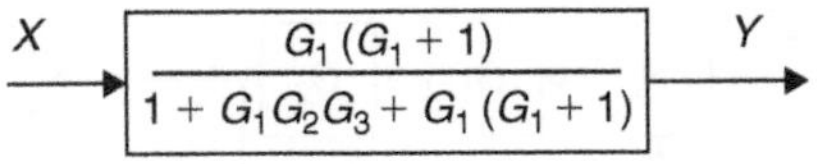

Figure 18.123: Simplifying negative feedback

## 18.16 Multiple Inputs

When there is more than one input to a system, the *superposition principle* can be used. This is *that the response to several inputs simultaneously applied is the sum of the individual responses to each input when applied separately*. Thus, the procedure to be adopted for a multi-input-single output (MISO) system is:

1. Set all but one of the inputs to zero.

2. Determine the output signal due to this one nonzero input.

3. Repeat the above steps for each input in turn.

4. The total output of the system is the algebraic sum of the outputs due to each of the inputs.

*Example 18.22*

Determine the output $Y(s)$ of the system shown in Figure 18.124 when there is an input $X(s)$ to the system as a whole and a disturbance signal $D(s)$ at the point indicated.

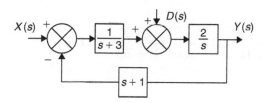

Figure 18.124: System with a disturbance input

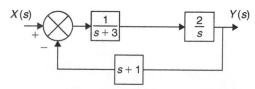

Figure 18.125: System with disturbance put equal to zero

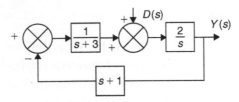

**Figure 18.126: System with input equal to zero**

## Solution

If we set $D(s)$ to zero we have the system shown in Figure 18.125 and the output is given by:

If we now set $X(s)$ to zero we have the system shown in Figure 18.126. This is a system with a forward path transfer function of $2/s$ and a positive feedback of $(1/s + 3)[-(s + 1)]$. This gives an output of:

$$\frac{Y(s)}{D(s)} = \frac{2(s + 3)}{s(s + 3) + 2(s + 1)}$$

The total input is the sum of the outputs due to each of the inputs and so:

$$Y(s) = \frac{2}{s(s + 3) + 2(s + 1)} X(s) + \frac{2(s + 3)}{s(s + 3) + 2(s + 1)} D(s)$$

# CHAPTER 19

# Communications Systems

**Alan Bensky**
**Andrew Leven**

## 19.1 Introduction

Figure 19.1 shows a simplified block diagram of a color television receiver. We will refer to it throughout this chapter. It shows that any receiver must be capable of extracting information from the incoming channels to which it is tuned. The shaded blocks show examples of where demodulation or detection occur for the video and audio signals. Generally the sound uses frequency modulation (FM), while the video signal uses amplitude modulation (AM). The age of digital television and modern data communications uses other techniques. All these methods will be discussed in this chapter.

The requirements for modulation are threefold. First, all channels must be separated from one another to avoid interference in the form of intermodulation distortion and crosstalk. Crosstalk occurs when one channel spills over into an adjacent channel, causing

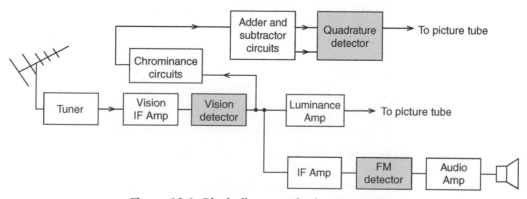

Figure 19.1: Block diagram of color TV receiver

interference. Intermodulation distortion occurs when two signals at frequencies $f_1$ and $f_2$ are amplified by a nonlinear device. Second-order products ($2f_1, f_1 + f_2$ and $f_1 - f_2$) are produced. This might only be troublesome in a broadband system where these products fall within the band. However, third-order components ($2f_1 + f_2$ and $2f_2 - f_1$) usually fall within a system bandwidth, i.e., a particular range of frequencies over which the system operates with good linearity, flat response and minimum distortion, and again cause intermodulation distortion.

In order to achieve good channel separation and avoid interference data, audio and video are generally superimposed on a carrier signal. Each station may have a different carrier or use sophisticated techniques like polarization or frequency sharing, but the point is that frequency translation takes place, with the information signals being shifted to a new frequency.

Second, the physical size of half-wavelength antenna systems would be prohibitive if higher frequencies were not used. In order to understand this, it is convenient to consider the

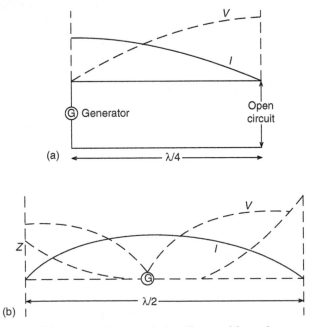

**Figure 19.2: (a) Open-ended transmission line and its voltage and current distributions; (b) Transmission line opened out**

properties of a quarter-wavelength ($\lambda$/4) transmission line. Figure 19.2(a) shows an open-ended $\lambda$/4 transmission line and its voltage and current distributions. If this is opened out as in Figure 19.2(b), then a $\lambda$/2 radiator is produced with the voltage and current distributions as shown. It can be seen that the current is at its maximum at the centre while the voltage is at its minimum. This is equivalent to a low-resistance series resonant circuit which can be tuned to the required transmitted or received channels. However, the point here is that the antenna has an electrical length of half the operating wavelength and is referred to as a $\lambda$/2 dipole. (In practice, it is actually 5% shorter than this theoretical value.)

Consider the case of the speech band being transmitted. This is generally from 300 Hz to 3.4 kHz. The two-dimensional wave equation is used to determine the wavelength:

$$\lambda_1 = \frac{v}{f_1} = \frac{3 \times 10^8}{3 \times 10} = 10^6 \, \text{m}$$

$$\lambda_2 = \frac{v}{f_2} = \frac{3 \times 10^8}{3.4 \times 10^3} = 1.13 \times 10^5 \, \text{m}$$

Here $v$ is the velocity of light ($3 \times 10^8$ m/s). Obviously using a much higher frequency would solve this problem by making the wavelength shorter.

Third, transmitting information in raw form, normally known as the *baseband*, would be impractical due to the low energy content. Losses between transmission and reception would soon attenuate the signals, with a resultant loss in reception. Modulating the signal by analog or digital methods increases the power to the information and gives a higher signal-to-noise ratio.

In this chapter the following modulation techniques will be discussed together with suitable circuits: amplitude modulation (AM); frequency modulation (FM); frequency shift keying (FSK); phase-shift keying (PSK); and quadrature phase-shift keying (QPSK).

## 19.2 Analog Modulation Techniques

### 19.2.1 Amplitude Modulation

When the amplitude of a carrier signal is varied in accordance with the information signal, amplitude modulation is produced. This method is mainly used where large power outputs are required for long-distance communications.

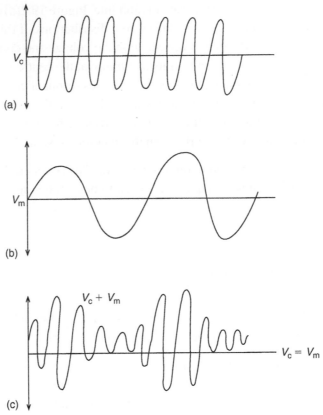

**Figure 19.3: (a) Waveform; (b) Modulating signal; (c) Amplitude-modulated wave**

Figure 19.3 shows a constant-amplitude, constant-frequency carrier being modulated by a single tone. In practice, many modulating signals may be used. The general expression for the waveform in Figure 19.3(a) is:

$$v_c = V_c \sin(\omega_c + \theta) \tag{19.1}$$

where $v_c$ is the instantaneous carrier voltage and $V_c$ is the peak amplitude; and $\omega_c$ is the frequency of the carrier in radians. $\theta$ is the phase of the carrier but this will be ignored in the following analysis.

The modulating signal in Figure 19.3(b) is given by:

$$v_m = V_m \sin \omega_m t \tag{19.2}$$

where $v_m$ is the instantaneous amplitude of the modulating signal and $V_m$ is the peak amplitude.

The amplitude-modulated wave as shown in Figure 19.3(c) is given by:

$$v = (V_c + V_m \sin \omega_m t) \sin \omega_c t \tag{19.3}$$

$$= V_c \sin \omega_c t + V_m \sin \omega_m t \sin \omega_c t \tag{19.4}$$

Using the trigonometric identity:

$$\sin A \sin B = \frac{1}{2} \cos(A - B) - \frac{1}{2} \cos(A + B)$$

equation (19.4) becomes:

$$v = V_c \sin \omega_c t + \frac{V_m}{2} \cos(\omega_c - \omega_m)t - \frac{V_m}{2} \cos(\omega_c + \omega_m)t \tag{19.5}$$

The modulated wave has three frequency components, namely the carrier frequency ($f_c$), the lower sideband ($f_c - f_m$) and the upper sideband ($f_c + f_m$). These components are represented in the form of a line or spectrum diagram as shown in Figure 19.4. If several modulating tones were present as in the speech band they would be as shown in Figure 19.5.

Figure 19.3(c) shows two important factors used in practice: the modulating factor and the depth of modulation. The modulating factor (m) is given by:

$$m = \frac{(V_c + V_m) - (V_c - V_m)}{(V_c + V_m) + (V_c - V_m)} = \frac{V_m}{V_c} \tag{19.6}$$

Expressed as a percentage, this is known as the *depth of modulation*. Hence, the depth to which the carrier is modulated depends on the amplitude of the carrier and the modulating voltage. The maximum modulation factor used is unity. Exceeding this causes overmodulation and break-up of the signal, and hence some figure less than unity is used in practice.

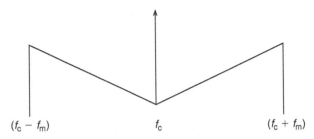

$(f_c - f_m)$                          $f_c$                          $(f_c + f_m)$

**Figure 19.4: Spectrum diagram**

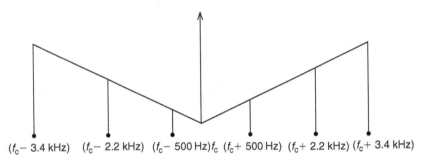

$(f_c - 3.4\ \text{kHz})$   $(f_c - 2.2\ \text{kHz})$   $(f_c - 500\ \text{Hz})f_c$   $(f_c + 500\ \text{Hz})$   $(f_c + 2.2\ \text{kHz})$   $(f_c + 3.4\ \text{kHz})$

**Figure 19.5: With several modulating tones present, as in speech band**

### 19.2.2 Power Distribution in an AM Wave

The power which is coupled to an antenna by an AM wave is developed across its resistance. An antenna must be coupled to a transmitter by means of a transmission line or waveguide in order to be excited and hence produce radiation. The antenna input impedance which the feeder 'sees' must be known in order to achieve efficient coupling, and this requires a knowledge of transmission line theory.

Note, however, that the antenna input impedance generally has a resistive and reactive part. The reactive element originates from the inherent inductance and capacitance in the antenna. The resistive element of the input impedance originates from the numerous losses in the antenna. The radiated loss (radiation resistance) is the actual power transmitted and is a necessary loss caused by the modulated wave generating power in the antenna. However, other losses are present such as ohmic losses and those due to currents lost in the ground. Because of this it is important that the radiation resistance be much greater than all the other losses. The radiation resistance is generally defined as the equivalent resistance that would dissipate an amount of power equal to the total radiated power when the current through the resistance is equal to the current at the antenna input terminals.

Rearranging equation (19.6) as:

$$V_m = mV_c$$

we can rewrite equation (19.5) as:

$$v = V_c \sin \omega_c t + \frac{1}{2} mV_c \left[\cos (\omega_c - \omega_m) t - \cos (\omega_c + \omega_m) t\right] \tag{19.7}$$

The r.m.s. power developed across the antenna resistance ($R_a$) by the carrier and two sidebands is therefore:

$$P_c \left(\frac{V_c}{\sqrt{2}}\right)^2 \frac{1}{R_a} = \frac{V_c^2}{2R_a}$$

$$P_m = \left(\frac{mV_c}{\sqrt{2}}\right)^2 \frac{2}{R_a} = \frac{m^2 V_c^2}{4R_a} \qquad \text{(sidebands)}$$

Total power is:

$$P_T = \frac{V_c^2}{2R_a} + \frac{m^2 V_c^2}{4R_a}$$

$$P_T = \frac{V_c^2}{2R_a}\left[1 + \frac{m^2}{2}\right] \tag{19.8}$$

If several pairs of sidebands are involved, equation (19.8) becomes:

$$P_T = P_c \left[1 + \frac{m_1^2}{2} + \frac{m_2^2}{2} + \cdots\right] \tag{19.9}$$

*Example 19.1*

The carrier of an AM transmitter is 50 W and, when modulated by a sinusoidal tone, the power increases to 59 W. Calculate:

(a) the depth of modulation;

(b) the ratio of maximum to minimum values of the wave envelope.

*Solution*

(a) From equation (19.9):

$$P_T = P_c \left(1 + \frac{m^2}{2}\right)$$

so,

$$m = \sqrt{2\left(\frac{P_T}{P_c} - 1\right)} = \sqrt{2\left(\frac{59}{50} - 1\right)} = 0.6 \quad \text{or} \quad 60\%$$

(b) From the wave envelope:

$$P_c(1 + m) = 50(1 + 0.6) = 80 \, \text{W}$$
$$P_c(1 - m) = 50(1 - 0.6) = 20 \, \text{W}$$

Hence,    $\dfrac{P_{max}}{P_{min}} = 4$

*Example 19.2*

An AM transmitter radiates 2 kW when the carrier is unmodulated and 2.25 kW when the carrier is modulated. When a second modulating signal is applied giving a modulation factor of 0.4, calculate the total radiated power with both signals applied.

*Solution*

$$m = \sqrt{2\left(\frac{P_T}{P_c} - 1\right)} = \sqrt{2\left(\frac{2.25}{2} - 1\right)} = 0.5$$

As the carrier power for the unmodulated wave is unchanged,

$$P_T = 2\left(1 + \frac{0.5^2}{2} + \frac{0.4^2}{2}\right) = 2.4\,\text{kW}$$

### Example 19.3

An AM signal has a 25 V/100 kHz carrier and is modulated by a 5 kHz tone to a modulation depth of 95%.

(a) Sketch the spectrum diagram of this modulated wave, showing all values.

(b) Determine the bandwidth required.

(c) Calculate the power delivered to a 75 Ω load.

### Solution

(a) Using our familiar rearrangement of equation (19.6),

$$V_m = 0.95 \times 25 = 23.75\,\text{V}$$

The amplitude of the sidebands, from (19.7), is

$$\frac{V_m}{2} = \frac{23.75}{2} = 11.875\,\text{V}$$

(See Figure 19.6).

(b) As there is a double sideband, the bandwidth is 10 kHz

(c) We have:

$$P_c = \left(\frac{V_c}{\sqrt{2}}\right)^2 \frac{1}{R_a} = \left(\frac{25}{\sqrt{2}}\right)^2 \frac{1}{75} = 4.17\,\text{W}$$

$$P_m = \frac{m^2 V_c^2}{4R_a} = \frac{(0.95 \times 25)^2}{4 \times 75} = 1.88\,\text{W}$$

Hence, total power is 6.05 W.

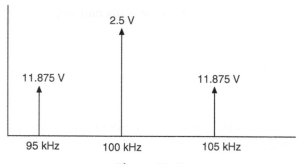

**Figure 19.6:**

### 19.2.3 Amplitude Modulation Techniques

The method of amplitude modulation previously discussed is known as *double sideband modulation (DSB)*. However, this method has a number of disadvantages which can be overcome by filtering out the carrier, one of the sidebands or both. Such a system would have the following advantages:

(a) reduced bandwidth, hence less noise;

(b) more channels available;

(c) increase in efficiency, as power is only transmitted when information is sent;

(d) selective fading is reduced as there is no carrier component to fade below the sideband level and cause sideband beating, which would produce unwanted components;

(e) nonlinearity is reduced as the carrier amplitude is the largest of all the components and this can cause saturation.

Double sideband suppressed carrier (DSBSC) modulation requires the carrier to be reinserted at the receiver with the correct phase and frequency. Single sideband suppressed carrier (SSBSC) modulation only requires the frequency of the reinserted carrier to be correct.

The basic principle of SSBSC is shown in Figure 19.7. The carrier and modulating signal are applied to a balanced modulator (which will be discussed later). The output of the modulator consists of the upper and lower sidebands, but the carrier is suppressed. The band-pass filter then removes one of the sidebands.

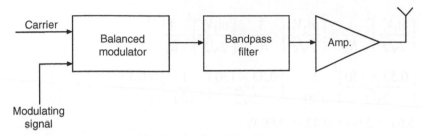

**Figure 19.7: Basic principle of single-sideband suppressed carrier modulation**

### Example 19.4

An AM transmitter is modulated by the audio range 20–15 000 Hz. If the carrier frequency is 820 kHz with a voltage level of 150 V, determine:

    (a) the modulating factor,

    (b) the DSB power, and

    (c) the SSBSC power.

for the frequency components 400 Hz/80 V, 1 kHz/50 V and 10 kHz/20 V if the antenna load is 120 Ω.

### Solution

(a) $\quad m_1 = \dfrac{V_{m1}}{V_c} = \dfrac{80}{150} = 0.53$

$\quad\quad m_2 = \dfrac{V_{m2}}{V_c} = \dfrac{50}{150} = 0.33$

$\quad\quad m_3 = \dfrac{V_{m3}}{V_c} = \dfrac{20}{150} = 0.13$

(b) $\quad P_T = \dfrac{V_c^2}{2R_a} + \dfrac{m_1^2 V_c^2}{4R_a} + \dfrac{m_2 V_c^2}{4R_a} + \dfrac{m_3^2 V_c^2}{4R_a}$

$\quad\quad\quad = \dfrac{150^2}{2 \times 120} + \dfrac{(0.53 \times 150)^2}{4 \times 120} + \dfrac{(0.33 \times 150)^2}{4 \times 120} + \dfrac{(0.13 \times 150)^2}{4 \times 120}$

$\quad\quad\quad = 93.75 + 13.33 + 5.21 + 0.83 = 113.1 \, \text{W}$

(c) $P_T = \left(\dfrac{m_1 V_c}{2\sqrt{2}}\right)^2 \dfrac{1}{R_a} + \left(\dfrac{m_2 V_c}{2\sqrt{2}}\right)^2 \dfrac{1}{R_a} + \left(\dfrac{m_3 V_c}{2\sqrt{2}}\right)^2 \dfrac{1}{R_a}$

$\quad = \left(\dfrac{0.53 \times 150}{2\sqrt{2}}\right)^2 \dfrac{1}{120} + \left(\dfrac{0.33 \times 150}{2\sqrt{2}}\right)^2 \dfrac{1}{120} + \left(\dfrac{0.13 \times 150}{2\sqrt{2}}\right)^2 \dfrac{1}{120}$

$\quad = 6.67 + 2.60 + 0.42 = 9.69\,\text{W}$

## 19.3 The Balanced Modulator/Demodulator

The function of a modulator, as has been shown, is to superimpose the baseband signals on to a carrier while the demodulator provides the reverse role by extracting a carrier, known as the *intermediate frequency*, and leaving the baseband.

There are many modulators and demodulators commercially sold on the market as integrated chips or as part of a front end receiver chip containing other stages such as the tuner and detector. However, it is informative to look at the balanced modulator which is used in the majority of AM and other circuits.

A common method of obtaining a single sideband wave is illustrated in Figure 19.7. The output of this circuit differs from a conventional AM output in that it does not include the original radio frequency signal, but only the two sidebands. The single sideband is obtained by a highly selective filter.

An integrated circuit which is commonly used is the Philips MC 1496. This is a modulator/demodulator chip which uses a monolithic transistor array. It has many applications such as AM and suppressed carrier modulators, AM and FM demodulators and phase detectors. The basic theory of operation is shown in the data sheets at the end of this chapter. Figures 19.8 and 19.9 show the application of this chip as an AM modulator and demodulator, respectively.

The AM modulator shown in Figure 19.8 allows no carrier at the output; by adding a variable offset voltage to the differential pairs at the carrier input the carrier level changes and its amplitude is determined by the AM modulation.

The frequency spectrum is shown in the data sheets; it can be seen that undesired sidebands appear if the modulation or carrier levels are high. These need to be filtered and

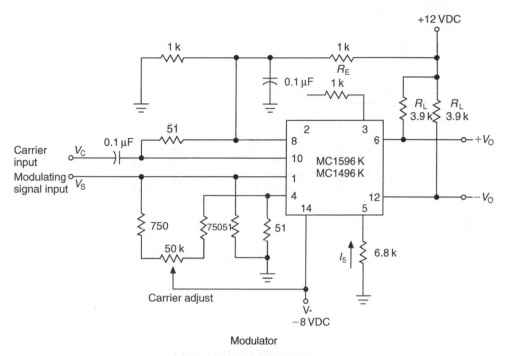

Figure 19.8: AM modulator

a fourth-order Butterworth is ideal. Note also that the modulation levels may be varied by means of $R_E$ connected between pins 2 and 3 in Figure 19.8.

As can be seen from equation (1) in the data sheets, the output of the balanced modulator is a cosine function of the phase between the signal and carrier inputs. If the carrier input is driven hard enough, a switching action occurs and the output becomes a function of the input voltage. The output amplitude is maximized when the phase difference is zero.

A typical demodulator is shown in Figure 19.9. In this case the carrier is amplified by an intermediate frequency chip which provides a limited gain of 55 dB or higher at 400 μV. The carrier is then applied to the demodulator where the carrier frequency is attenuated. Output filtering is required to remove the high-frequency unwanted components.

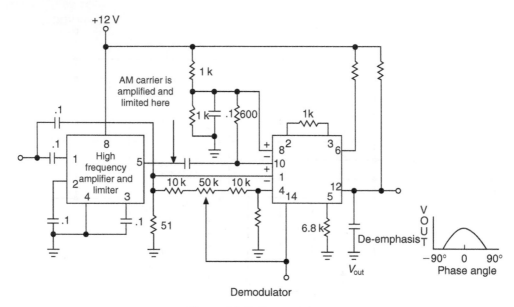

**Figure 19.9: Demodulator**

## 19.4 Frequency Modulation and Demodulation

With frequency modulation the frequency (rather than the amplitude) of a constant-amplitude, constant-frequency sinusoidal carrier is made to vary in proportion to the amplitude of the applied modulating signal. This is shown in Figure 19.10, where a constant-amplitude carrier is frequency-modulated by a single tone. Note how the frequency of the carrier changes.

Frequency modulation can be understood by considering Figure 19.11. This shows that a modulating square or sine wave may be used for this type of modulation. The frequency of the frequency-modulated carrier remains constant, and this indicates that the modulating process does not increase the power of the carrier wave. For FM the instantaneous frequency $\omega$ is made to vary as

$$\omega = \omega_c + kV_m \sin \omega_m t \tag{19.10}$$

In this equation

$$kV_m = \Delta\omega \tag{19.11}$$

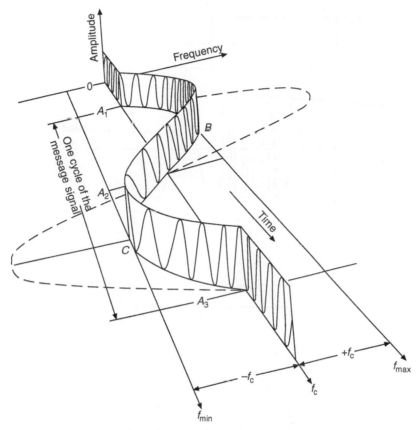

**Figure 19.10: Constant-amplitude carrier is frequency-modulated by single tone**

And we have:

$$\theta = \left( \omega_c t - \frac{\Delta f}{f_m} \cos \omega_m t \right) \tag{19.12}$$

Substituting (19.12) into (19.11) gives:

$$v = V_c \sin \left( \omega_c t - \frac{\Delta f}{f_m} \cos \omega_m t \right) \tag{19.13}$$

for the peak angular frequency shift for the modulating signal.

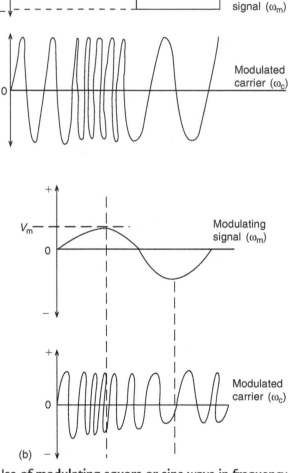

**Figure 19.11: Use of modulating square or sine wave in frequency modulation**

The modulating index is given by:

$$\beta = \frac{\Delta f}{f_m} = \frac{\text{peak frequency shift}}{\text{modulating frequency}} \tag{19.14}$$

for a constant frequency, constant amplitude modulating signal. In practice the modulating signal varies in amplitude and frequency. This leads to two further parameters: a maximum value of the modulating signal ($f_{m(max)}$); and a maximum allowable frequency shift, which is defined as the frequency deviation ($f_d$). The deviation ratio ($\delta$) is then defined as:

$$\delta = \frac{f_d}{f_{m(max)}} \tag{19.15}$$

For any given FM system the frequency will swing to a maximum value of frequency deviation known as the *rated system derivation*. This parameter determines the maximum allowable modulating signal voltage. Equation (19.15) applies for this condition of rated system deviation.

Finally, the frequency-modulated wave can be written as:

$$v = V_c \sin(\omega_c t - \beta \cos \omega_m t) \tag{19.16}$$

for a constant-amplitude, constant-frequency modulating signal such as a square wave, and,

$$v = V_c \sin(\omega_c t - \delta \cos \omega_m t) \tag{19.17}$$

for a variable-amplitude, variable-frequency modulating signal such as a sinusoidal wave.

Expanding (19.17) using the identity:

$$\sin(A + B) = \sin A \cos B - \cos A \sin B$$

gives,

$$\sin \omega_c t \cos(\delta \cos \omega_m t) - \cos \omega_c t \sin(\delta \cos \omega_m t)$$

The second factor of each term expands into an infinite series whose coefficients are a function of $\delta$. These coefficients are called *Bessel functions*, denoted by $J_n(\delta)$, which vary as $\delta$ varies. More specifically, they are Bessel functions of the first kind and of order $n$.

Expanding the second factor gives

$$\cos(\delta \cos \omega_m t) = J_0(\delta) - 2J_2(\delta) \cos 2\omega_m t + 2J_4(\delta) \cos 4\omega_m t - \dots$$

and,

$$\sin(\delta \cos \omega_m t) = 2J_1(\delta)\cos \omega_m t - 2J_3(\delta)\cos 3\omega_m t + \ldots$$

Using the relationships:

$$\sin A \cos B = \frac{1}{2}[\sin(A+B) + \sin(A-B)]$$

$$\cos A \sin B = \frac{1}{2}[\sin(A+B) - \sin(A-B)]$$

we obtain:

$$\begin{aligned}v = V_c\{&J_0(\delta)\sin \omega_c t + J_1(\delta)[\sin(\omega_c+\omega_m)t - \sin(\omega_c-\omega_m)t]\\ &- J_2(\delta)[\sin(\omega_c+2\omega_m)t + \sin(\omega_c-\omega_m)t]\\ &+ J_3(\delta)[\sin(\omega_c+3\omega_m)t - \sin(\omega_c-3\omega_m)t] - \ldots\}\end{aligned}$$

Thus the modulated wave consists of a carrier and an infinite number of upper and lower side-frequencies spaced at intervals equal to the modulation frequency. Also, since the amplitude of the unmodulated and modulated waves are the same, the powers in the unmodulated and modulated waves are equal.

The Bessel coefficients can be determined either from graphs or tables.

### Example 19.5
Determine the values of the amplitudes of the carrier and side frequencies if $f_d$ is 5 kHz, $f_{m(max)}$ is 5 kHz and the carrier amplitude is 10 V.

### Solution
The deviation ratio is unity and the sideband amplitudes and carrier amplitude are:

$$J_0 = 0.77 \quad J_1 = 0.44 \quad J_2 = 0.11 \quad J_3 = 0.02 \quad J_4 = 0$$

As the carrier has an amplitude of 10 V, each component in the spectrum diagram will have the values shown in Figure 19.12.

### Example 19.6
Determine the amplitudes of the side frequencies generated by an FM transmitter having a deviation ratio of 10 kHz, a modulating frequency of 5 kHz and a carrier level of 10 V.

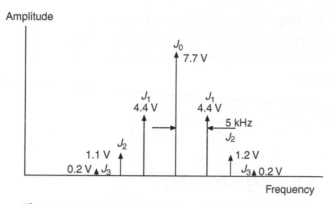

**Figure 19.12: Spectrum diagram for Example 19.5**

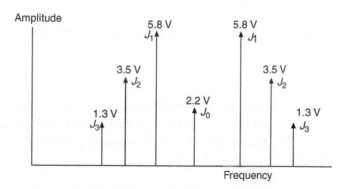

**Figure 19.13: Spectrum diagram for Example 19.6**

### Solution

The deviation ratio is 2 for this system, so once again the following amplitudes are obtained.

$$J_0 = 0.22 \quad J_1 = 0.58 \quad J_2 = 0.35 \quad J_3 = 0.13 \quad J_4 = 0.03 \quad J_5 = 0.01$$

The spectrum diagram is shown in Figure 19.13. There are more side frequencies in this case and hence, the quality of the transmission would be improved. However, not all the side frequencies are relevant, as can be seen from their amplitudes.

### 19.4.1 Bandwidth and Carson's Rule

It has already been mentioned that not all the side frequencies are necessary for satisfactory performance. Generally, an acceptable performance can be obtained with a finite number of side frequencies, and this may be considered satisfactory when not less than 98% of the power is contained in the carrier and its adjacent frequencies. Since the amplitude of the $n$th side frequency is given as $J_n V_c$, the power dissipated in a load ($R$) by the modulated wave is:

$$P = \frac{(J_0 V_c)^2}{R} + 2\left[\frac{(J_1 V_c)^2}{R} + \frac{(J_2 V_c)^2}{R} + \frac{(J_3 V_c)^2}{R}\right]$$

$$\therefore P = \frac{V_c^2}{R} = [J_0^2 + 2(J_1^2 + J_2^2 + J_3^2 + \ldots)] \tag{19.18}$$

$$J_0^2 + (J_1^2 + J_2^2 + J_3^2 + \ldots) > 0.98$$

Thus, for 98% of the power to be contained in the carrier plus the side frequencies, the following applies:

### Example 19.7
An FM transmitter transmits with a rated system deviation of 60 kHz and a maximum modulating frequency of 15 kHz. If the carrier amplitude is 25 V, determine the number of side frequencies required to ensure that 98% of the power is contained in the carrier and side frequencies. Sketch the spectrum diagram.

### Solution
The frequency deviation is given by:

$$\delta = \frac{60}{15} = 4$$

From the Bessel tables,

$$J_0 = -0.3971 \quad J_1 = -0.0660 \quad J_2 = 0.3641 \quad J_3 = 0.4302$$
$$J_4 = 0.2811 \quad J_5 = 0.1321 \quad J_6 = 0.0491 \quad J_7 = 0.0152$$

It will be seen that only $J_2$, $J_3$, $J_4$, $J_5$ and $J_0$ are required:

$$J_0^2 + 2(J_2^2 + J_3^2 + J_4^2 + J_5^2 + J_7^2) > 0.98$$

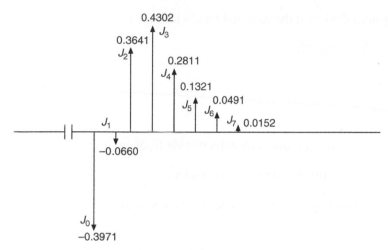

**Figure 19.14: Spectrum diagram for Example 19.7**

The spectrum diagram is sketched in Figure 19.14.

The Bessel tables show negative and positive values for the Bessel functions $J_0$, $J_1$, $J_2$ etc. When the deviation ratio is zero, the carrier is unmodulated and has its maximum value. For any other modulation index the energy levels are distributed between the sidebands and carrier. As the deviation ratio increases the number of sidebands increases, as does the number of negative values. The negative values indicate a 180° phase shift, and this can be seen from the Bessel graphs where the carrier and each sideband behave as sinusoids as frequency deviation takes place. Negative signs are usually ignored in practice, since only the magnitude of the carrier and each sideband is required. Squaring the negative values produces positive or magnitude quantities.

Example 19.7 shows that not all of the side frequencies are necessary for a high-fidelity output in an FM system. The bandwidth can therefore be determined by considering only the useful side frequencies with the higher amplitudes. Note that the required bandwidth, when all the side frequencies are considered, is given as

$$BW = (6 \times 15)2 = 180 \, \text{kHz}$$

This is a considerable saving in bandwidth and hence a reduction in noise in the modulator circuits.

For a rated system deviation the required bandwidth for Example 19.7 is:

$BW = (5 \times 15)2 = 150\,\text{kHz}$

Also since $\delta = 4$,

$\delta + 1 = 5$

This indicates the required number of pairs of side frequencies for the 98% criterion.

Since $BW = (\delta + 1)$ pairs of side frequencies, i.e.,

$$BW = 2f_{\text{m(max)}} \times \text{pairs of side frequencies (Carson's rule)} \qquad (19.19)$$

Since,

$$\delta = \frac{f_{\text{d}}}{f_{\text{m(max)}}}$$

substituting in (19.19) gives:

$$BW = 2f_{\text{m(max)}}\left(\frac{f_{\text{d}}}{f_{\text{m(max)}}} + 1\right)$$

$$\therefore BW = 2(f_{\text{d}} + f_{\text{m(max)}}) \qquad (19.20)$$

Equations (19.19) and (19.20) express a relationship known as *Carson's rule* for determining the bandwidth of an FM system requiring the requisite number of side frequencies to satisfy the 98% criterion. Also $\delta$ indicates the rated system deviation, while $\beta$ is used for values less than this.

### Example 19.8

(a) An FM system uses a carrier frequency of 100 MHz with an amplitude of 100 V. It is modulated by a 10 kHz signal and the rated system deviation is 80 kHz. Determine the amplitude of the center frequency.

(b) An FM station with a maximum modulating frequency of 15 kHz and a deviation ratio of 6 operates at a center frequency ($f_{\text{c}}$) of 10 MHz. Determine the 3 dB bandwidth of the stage following the modulator which would pass 98% of the power in the modulated wave. Also determine the $Q$ factor of this circuit.

**Solution**

(a) $\delta = \dfrac{f_d}{f_{m(max)}} = \dfrac{80}{10} = 8$

For $\delta = 8$, the Bessel tables give $J_0 = 0.1717$. Therefore the amplitude of the center frequency is:

$100 \times 0.1717 = 17.17\,\text{V}$

(b) Assume rated system deviation. Thus:

$$BW = 2(f_d + f_{m(max)})$$

Also, $\qquad \delta = \dfrac{f_d}{f_{m(max)}} = 6$

Hence, $\qquad f_d = 6 \times f_{m(max)} = 6 \times 15 = 90\,\text{kHz}$

$$BW = 2(90 + 15) = 210\,\text{kHz}$$

so,

$$Q = \frac{f}{BW} = \frac{10 \times 10^3}{210} = 47.6$$

**Example 19.9**

An FM system has a rated system deviation of 65 kHz. Determine the maximum permitted value of the modulating signal voltage if the modulator has a sensitivity of 5 kHz/V.

**Solution**

Since the maximum swing is 65 kHz, then:

$65 = 5 \times V_m$

$V_m = \dfrac{65}{5} = 13\,\text{V}$

**Example 19.10**

An FM broadcast station is assigned a channel between 92.1 and 92.34 MHz. If the maximum modulating frequency is 15 kHz determine:

(a) the maximum permissible value of the deviation ratio;

(b) the number of side frequencies.

**Solution**

(a)    $BW = 92.34 - 92.1\,\text{MHz} = 0.24\,\text{MHz} = 240\,\text{kHz}$

$BW = 2(f_\text{d} + f_\text{m(max)})$

$240 = 2(f_\text{d} + 15)$

$\therefore\ f_\text{d} = 105\,\text{kHz}$

Also,    $\delta = \dfrac{f_\text{d}}{f_\text{m(max)}} = \dfrac{105}{15} = 7$

(b) From the Bessel tables, this will give 10 pairs of sidebands.

## 19.5 FM Modulators

The most frequently used modulator in FM systems is the reactance modulator, which incorporates some method of varying the reactance across the oscillator circuit. This can be done by incorporating a device which changes either its inductive or capacitive reactance, depending on the oscillator involved. With a Colpitts oscillator some type of capacitive modulator would be used; a Hartley oscillator would use an inductive modulator.

The capacitance of a simple signal diode depends on the width of its depletion layer when forward or reverse biased. A particular diode, called a *variable reactance* diode (*varactor* for short), is fabricated in such a way that its capacitance is a function of the voltage applied across it.

If a varactor diode is connected across the tuned circuit of an oscillator, and the voltage across the diode is varied, the variation of the diode capacitance will cause a variation in the oscillator's frequency. The circuit virtually functions as a voltage-to-frequency convertor.

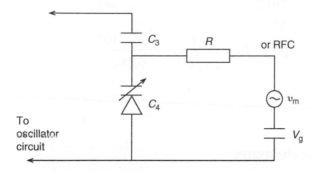

**Figure 19.15: Typical diode FM modulator**

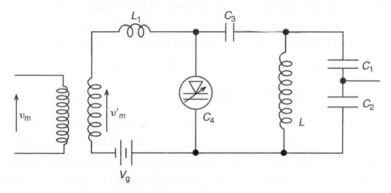

**Figure 19.16: Diode FM modulator**

A typical diode FM modulator is shown in Figure 19.15. In this diagram the modulating signal is fed to a transformer which is coupled to a Colpitts oscillator in this case. The varactor diode $C_4$ is reverse biased by $V_g$, as shown, to a practical point on its characteristics. This reverse bias voltage varies the modulating signal voltage, causing $C_4$ to vary, thus varying the oscillator frequency.

The radio frequency choke is necessary so that the RF voltage across $C_4$ is not shorted out by the sources $v_m$ and $V_g$. $C_3$ is used to block any DC voltage from the oscillator.

### Example 19.11

A reactance modulator is used in an FM transmitter. It consists of a Colpitts oscillator and audio injection circuit as shown in Figure 19.16. The two varactor diodes have a tuneable range from 19.4 to 3 pF. Determine the tuning range of the modulator if the inductance value is $L = 2\,\mu H$.

*Solution*

$$C_T = \frac{2.4 \times 2.4}{2.4 + 2.4} = 1.2\,\text{pF} \quad f = \frac{1}{2\pi\sqrt{LC_T}} = \frac{10^9}{6.28\sqrt{2 \times 1.2}} = 103\,\text{MHz}$$

$$C_T = \frac{3 \times 3}{3 + 3} = 1.5\,\text{pF} \qquad f = \frac{10^9}{6.28\sqrt{2 \times 1.5}} = 92\,\text{MHz}$$

## 19.6  FM Demodulators

The FM demodulator performs the reverse operation to modulation in that it converts variations in frequency into variations in amplitude. The frequency-to-voltage transfer may be nonlinear over the operating range, and several methods are used in practice. However, this text will discuss only two common types of demodulator, namely the phase-looked loop (PLL) demodulator and the ratio detector.

### 19.6.1  The Phase-Locked Loop Demodulator

For the purpose of this particular application, the block diagram shown in Figure 19.17 will be used. This is the simplest type of demodulator and is frequently used in data communications systems. It consists of a phase comparator that has two input signals, one from the voltage controlled oscillator ($f_2$) and the other being the FM signal ($f_1$). The phase comparator compares the phase of the VCO with the incoming FM signal, giving an output proportional to the difference in phase. This is then filtered to remove unwanted high-frequency components and the output from the filter is used to control the frequency of the voltage-controlled oscillator (VCO), locking it to the incoming signal. Hence, the VCO should be capable of tracking the incoming FM signal within the frequency

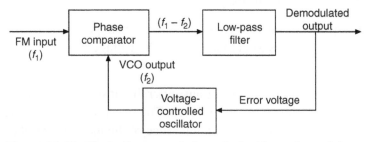

**Figure 19.17: Block diagram of phase-locked loop demodulator**

deviation of the system. The output from the low pass filter, i.e., the error voltage, is used to obtain the demodulated output, and the linearity of the output depends only on the linearity of the voltage-to-frequency characteristics of the VCO.

### 19.6.2 The Ratio Detector

Figure 19.18 shows a circuit which is commonly used in telecommunications applications for FM demodulation. Receivers which use this circuit generally have a bandpass limiter as the previous stage. This improves the filtering before demodulation takes place as ratio detectors have a low-input Q factor which causes the input voltage $V_1$ to change.

The circuit $C_1L_1$ and $C_2L_2$ are tuned to the carrier frequency. The rectifying diodes $D_1$ and $D_2$ are connected such that the DC voltages across $C_a$ and $C_b$ are of the same polarity. $C_3$ electrically connects B to D and must have negligible reactance. The operation of the circuit hinges on the fact that the voltage $V_2$ is 90° out of phase with $V_1$ at resonance.

Consider initially the double tuned circuit shown in Figure 19.19. Since the circuit is in resonance, $V_1$ is in phase with I and IL will lag $V_1$ by 90°. From Figure 19.16, $V_{EC}$ will lead $V_{BA} = V_{DA}$ by 90°. This is shown in the phasor diagram shown in Figure 19.20(a). As the frequency rises above the center frequency ($f_c$), the secondary becomes more inductive and $V_{EC}$ shifts clockwise, as shown in Figure 19.20(b). If the frequency falls below $f_c$ the secondary becomes more capacitive and $V_{EC}$ will shift anticlockwise from the 90° position. This is shown in Figure 19.20(c). Note that in Figure 19.20 $V_{BA} = V_{EC}$ remains constant as the frequency varies.

Consider a DC voltage source $E_X$ replacing the capacitor $C_C$ in Figure 19.18. When the peak-to-peak value of the incoming signal is less than $E_B$, $D_1$ and $D_2$ will not conduct and

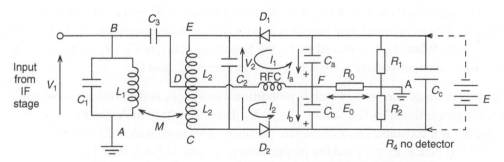

**Figure 19.18: Commonly used circuit for FM demodulation**

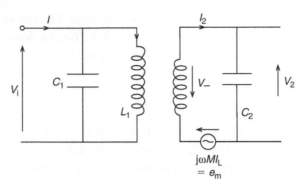

**Figure 19.19: Double tuned circuit**

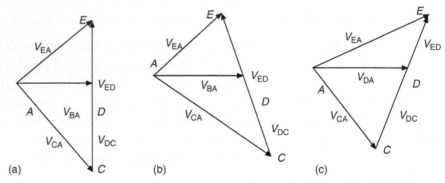

(a)    (b)    (c)

**Figure 19.20: (a) Phasor diagram for circuit in Figure 19.18; (b) $V_{EC}$ shifts clockwise**

the output voltage is zero. An output voltage will only appear if the peak-to-peak value of the input signal is greater than $E_X$. Also the potential across $R_1$ and $R_2$ is clamped to $E_X$ so that $E_X$ acts as an amplitude limiter removing variations in amplitude of the modulated input signal. When $E_X$ is replaced by $C_C$ the large time constant $C_C (R_1 + R_2)$ serves to maintain a constant voltage across $R_1$ and $R_2$ and hence $C_a$ and $C_b$.

If $e_a$ is the voltage across $C_a$ which is proportional to $V_{EA}$, $e_b$ is the voltage across $C_b$ which is proportional to $V_{CA}$ and $e_a + e_b = E_X$ then the voltage at A, the junction of $R_1$ and $R_2$, is a constant, i.e., $E_X/2$. There are three cases to consider. If $f = f_c$ then $V_{EA} = V_{CA}$; hence, $e_a = e_b$ and the output voltage is zero. If $f > f_c$ then $V_{EA} > V_{CA}$; hence, $e_a > e_b$, point $F$ rises in potential above point $A$ and the DC output voltage goes positive.

Finally, if $f < f_c$ then $V_{EA} < V_{CA}$; hence, $e_b > e_c$ and the potential at $F$ falls below the potential of $A$ so that the DC output voltage goes negative. Thus, the value of the output voltage depends on the frequency shift from $f_c$, and the polarity of the output voltage will be determined by whether $f > f_c$ or $f < f_c$.

Note, finally, that only the ratio $e_a : e_b$ changes, which is why the circuit is known as a *ratio detector*.

## 19.7 Digital Modulation Techniques

In the last few sections methods of transmitting analog information using analog signals were explained. This section will consider methods of transmitting digital data using analog signals.

The most familiar use of these methods is in data communications, where modems and telephone networks are used; because integrated circuits are generally used a block diagram approach will be considered.

### 19.7.1 Frequency Shift Keying

In some situations data can be transmitted directly without any modulation technique being necessary. This is applicable over short distances where the baseband signal may be sent in a raw form. However, where distance is involved more sophisticated methods are required.

One of the modulation methods most frequently used is frequency shift keying (FSK). In FSK the transmitted signal is switched between two frequencies every time there is a change in the level of the modulating data stream. The higher frequency may be used to represent a high level (1) and the lower frequency used for the low level (0). This results in a waveform similar to the one shown in Figure 19.21.

Generally the frequencies used in FSK depend on the system application. Most modems traditionally use frequencies within the voice range (300–3400 Hz), while much higher frequencies would be used for satellite or radio relay systems. No matter what system is used there are fundamental blocks which are necessary for successful operation.

A balanced modulator is necessary to generate the required waveforms. This device has been mentioned earlier; it simply multiplies two signals together at its two inputs, the

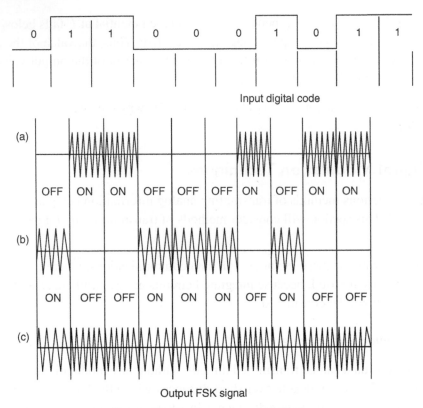

Figure 19.21: Frequency shift keying

output voltage being the product of these two voltages. One of the inputs (the carrier input) is generally AC coupled, while the other (the digital data input) is DC coupled. The block diagram is shown in Figure 19.22. Note that the data stream is inverted in modulator 2 in order to switch to the second carrier frequency.

At the receiver end, the FSK waveform has to be demodulated or, more specifically, decoded. One approach used for this is shown in Figure 19.23. The demodulation of FSK signals can be accomplished by means of a ratio detector or a PLL, but for modem applications the PLL is generally preferred as it can be used for both modulation and demodulation. The data stream consists of marks (1) and spaces (0) which are each allocated a switched frequency. The space is normally allocated the higher frequency. The rate at which the carrier frequency is switched is known as the *baud rate*, and this is the

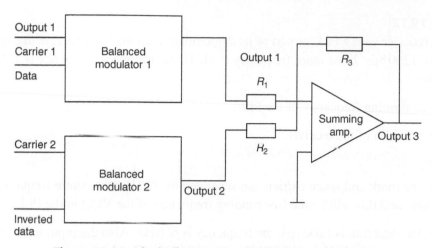

**Figure 19.22: Block diagram showing balanced modulator**

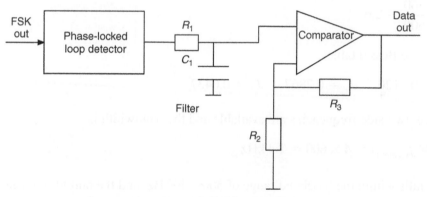

**Figure 19.23: Demodulation of FSK waveform**

same as the digital data rate for FSK. This is not always the case for other demodulation methods. The PLL has a free running frequency of its own and this is normally set between the mark and space carriers when designing such a system.

The output of the PLL in Figure 19.23 contains numerous components due to the interaction of the two frequencies, and hence a low-pass filter is used. However, the output of the filter produces rounded waveforms instead of oblong-shaped pulses, and this is modified by including a comparator.

*Example 19.12*

An FSK receiver uses a PLL as part of its demodulation circuitry, to receive digital data at the rate of 1200 bps. If the mark frequency is 1 kHz and the space frequency is 1.72 kHz, determine:

(a) the free running frequency of the PLL,

(b) the bandwidth of the receiver.

*Solution*

(a) Since the mark and space carriers are separated by 720 Hz, the centre frequency is 1.36 kHz and this will be the free running frequency of the VCO in the PLL.

(b) Since the data rate is 1200 bps, the frequency is 600 Hz. Also the input to the PLL has to swing between 1 kHz and 1.72 kHz, i.e., ±360 Hz. The deviation ratio is thus:

$$\delta = \frac{360}{600} = 0.6$$

From the Bessel tables,

$$J_0 = 0.9120 \quad J_1 = 0.2867 \quad J_2 = 0.0437$$

Hence, two side frequencies are available and the bandwidth is:

$$2(2 \times f_{m(max)}) = 4 \times 600 = 2.4 \, kHz$$

This falls within the baseband range of 300–3400 Hz, and the output will use a filter and comparator as shown in Figure 19.23.

### 19.7.2 Binary Phase-Shift Keying (BPSK)

High-speed modems operating at bit rates of up to 56 kbps require phase-shift keying or quadrature phase-shift keying. It is also the preferred modulation method for satellite and space technology. Unlike FSK, phase-shift keying uses one carrier frequency which is modulated by the data stream. It is a modulation system in which only discrete phase states are allowed. Usually $2^n$ phase states are used, and when $n = 1$ this gives two-phase changes. This is sometimes called *binary phase-shift keying (BPSK)*. When $n = 2$, four phase changes are produced, and this is called *quadrature phase-shift keying (QPSK)*.

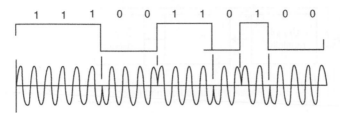

**Figure 19.24: Binary phase-shift keying (BPSK)**

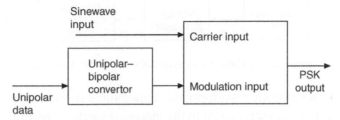

**Figure 19.25: Block diagram showing stages to produce modulator section**

BPSK (Figure 19.24), which will be considered here, is a two-phase modulation method in which a carrier is transmitted to indicate a mark (1) or the phase is reversed (shifted through 180°) to indicate a space (0). Note that the phase shift does not have to be 180°, but this allows for the maximum separation of the digital states between 1 and 0, which is important when noise is prevalent.

The block diagram shown in Figure 19.25 indicates the stages necessary to produce the modulator section. A balanced modulator is used with the carrier applied as shown. The digital input passes through a unipolar-bipolar convertor to ensure that the digital signal passed to the balanced modulator is unipolar.

It can be seen from Figure 19.24 that when the modulation input is positive, the modulator multiplies the carrier input by this constant positive level so that the output signal is simply the carrier sine wave. Note this is in phase with the carrier input. When the digital input data is negative, the modulator multiplies the carrier input by this constant negative level. This causes an output sine wave which is 180° out of phase with the carrier input. The result is that the sine wave at the output is inverted in phase every time the data input changes and produces a transition from 1 to 0 or 0 to 1. The consequence of this action is that the sine wave is inverted each time the modulation input undergoes a transition.

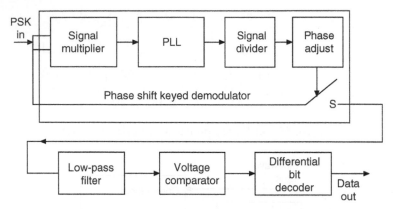

**Figure 19.26: Circuit to demodulate a BPSK waveform**

In order to demodulate a BPSK waveform, the demodulator must have an internal signal whose frequency is exactly equal to the incoming carrier. The PLL on its own is unsuitable in this case because of the sudden phase reversals which cannot produce a discrete carrier component to lock on to. One circuit which overcomes this is shown in Figure 19.26.

The input signal is applied to a signal multiplier which is a square-law device. In this case a balanced demodulator is used, with its inputs tied together. The output from this stage is a signal at twice the original frequency and having phase changes of 0° and 360°. Hence the signal multiplier has removed any phase changes from the original BPSK signal. It now provides a signal which the PLL can lock on to. The output from the PLL is then passed to a divide-by-two network which produces the original BPSK signal. The phase of this signal is then adjusted to the phase of the original BPSK signal. Finally, this output is used to activate an FET switch. When the phase adjust output is high (1) the switch is closed, and the initial BPSK signal is switched through to the demodulator's output. If the phase adjust output is low (0) then the switch is open and the demodulator's output drops to ground potential.

The output is then passed to a low-pass filter to remove unwanted signal components. This is followed by a comparator which squares the output and produces clean positive and negative half-cycles.

One final stage is necessary in order to produce the original data. At the output of the comparator the receiver has to look for level changes, and this has to be done by

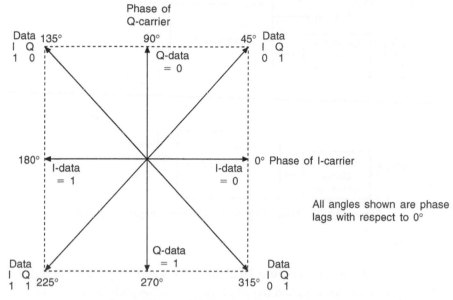

**Figure 19.27: Quadrature phase-shift keying**

a differential decoder block which gives an output (1) when a level change is sensed and no output (0) when no level change takes place. Hence, the original data stream is reproduced.

### 19.7.3 Quadrature Phase-Shift Keying (QPSK)

This type of modulation method has wide application in high-speed data transmission systems. It has two distinct advantages: it produces twice as much data with the same number of phase changes as BPSK, and this also means that the bandwidth is virtually decreased for the same amount of data being transmitted. In order to understand this, it is informative to look at single sideband modulation, which was mentioned in section 19.2.

In quadrature phase-shift keying each pair of consecutive data bits in a data stream is considered a two-bit code called a *dibit*. This is used to switch the carrier at the transmitter between one of four phases, instead of two as was the case with BPSK. The phases selected are 45°, 135°, 225° and 315°, lagging relative to the phase of the original unmodulated carrier. The system is shown in Figure 19.27. This diagram is clarified by looking at the typical QPSK transmitter block diagram shown in Figure 19.28.

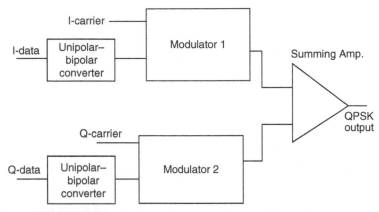

**Figure 19.28: Typical QPSK transmitter block diagram**

The two carrier signals shown in Figure 19.28 have the same frequency but differ in phase by 90°. The 0° phase carrier is called the *in-phase (I) carrier*, while the 90° phase carrier is called the *quadrature (Q) carrier*. The output from the first modulator is a BPSK signal which has phases of 0° and 90° relative to the I carrier while the output of the second modulator is a BPSK signal with phases of 90° and 270° relative to the I carrier. These two signals are then applied to the summing amplifier, but note that there is always ±90° phase difference between the two modulator outputs.

The phase of the summing amplifier's output, relative to the I carrier, can take one of four phase values as shown in Figure 19.26, but this will depend on the dibit code applied to the balanced modulator inputs. When the dibit changes, the phase of the QPSK output changes by 0°, 90°, 180° or 270° from its previous phase position.

It is necessary to include a differentially encoded dibit (DED) sequence at the transmitter in order to avoid phase ambiguity at the receiver. In order to achieve this, two blocks are connected at the input of the unipolar–bipolar converters of Figure 19.28. These blocks cause each pair of consecutive dibits from the data stream to be encoded as a change in the code at the two outputs of the DED. These outputs are then used to drive the modulator inputs and the original dibits cause the appropriate phase changes.

As with the BPSK receiver, the circuitry is fairly complicated but integrated circuitry enables the block diagram of Figure 19.29 to be drawn.

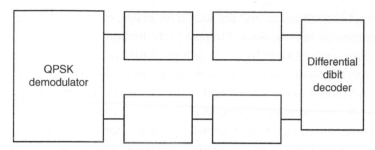

**Figure 19.29: Block diagram of QPSK using integrated circuitry**

The QPSK signal goes to the first stage of the QPSK demodulator, which is a signal squarer. This multiplies the incoming signal by itself causing phase changes of 0° and 180°. (The four original phase changes have been doubled.) The output from this block is then passed on to a second signal squarer and the output from this stage only incorporates a single phase change of 0°, but the frequency is now four times the original. The next stage is a PLL, which locks on to the incoming signal and outputs a clean square wave. This output is then passed to a divide-by-four circuit which outputs the original frequency and passes it on to a phase-changing circuit which generates two square waves at the same frequency but separated by a 90° phase difference. Finally, these outputs are used to operate two FET switches so that when the output is high one switch is closed and the original QPSK signal passes through to the demodulator. If the output is low, one of the FET switches will open and the demodulator input drops to zero.

Figure 19.27 shows how the outputs from the demodulator are arranged with reference to the I and Q signals. Note that the information concerning the original dibit code is incorporated in the average levels of the demodulator's I and Q outputs, hence low-pass filtering is used next to extract the average levels. This is then passed on to two voltage comparators to produce clean square waves. Finally, the change in the dibit code is determined by the differential dibit decoder, which encodes this change as the original dibit pair.

## 19.8 Information Theory

Information theory involves communication in general—on wires, fibers, or over the air—and it's applied to widely varied applications such as information storage on optical disks, radar target identification, and the search for extraterrestrial intelligence. In general, the goal of a communication system is to pass "information" from one place to another through a medium contaminated by noise, at a particular rate, and at a minimum specified level of fidelity to the source. Information theory gives us the means for quantitatively defining our objectives and for achieving them in the most efficient manner. The use of radio as a form of communication presents obstacles and challenges that are more varied and complex than a wired medium. A knowledge of information theory lets us take full advantage of the characteristics of the wireless interface.

In order to understand what information theory is about, you need at least a basic knowledge of probability. We've already encountered uses of probability theory in this book—in comparing different transmission protocols and in determining path loss with random reflections.

The use of coding algorithms is very common today for highly reliable digital transmission even with low signal-to-noise ratios. Error correction is one of the most useful applications of information theory.

Finally, information theory teaches us the ultimate limitations in communication—the highest rate that can be transmitted with a given bandwidth and a given signal-to-noise ratio.

### 19.8.1 Probability

#### 19.8.1.1 Basics

A common use of probability theory in communication is in assessing the reliability of a received message transmitted over a noisy channel. Let's say we send a digital message frame containing 32 data bits. What is the probability that the message will be received in error—that is, that one or more bits will be corrupted? If we are using or are considering using an error-correcting code that can correct one bit, then we will want to know the probability that two or more bits will be in error. Another interesting question: what is the probability of error of a frame of 64 bits, compared to that of 32 bits, when the probability of error of a single bit is the same in both cases?

In all cases we must define what is called an *experiment* in probability theory. This involves defining *outcomes* and *events* and assigning probability measures to them that follow certain rules.

We state here the three axioms of probability, which are the conditions for defining probabilities in an experiment with a finite number of outcomes. We also must describe the concept of field in probability theory. Armed with the three axioms and the conditions of a *field*, we can assign probabilities to the events in our experiments. But, first, let's look at some definitions.

An *outcome* is a basic result of an experiment. For example, throwing a die has six outcomes, each of which is a different number of dots on the upper face of the die when it comes to rest. An *event* is a set of one or more outcomes that has been defined, again according to rules, as a useful observation for a particular experiment. For an example, one event may be getting an odd number from a throw of a die and another event may be getting an even number. The outcomes are assigned probabilities and the events receive probabilities from the outcomes that they are made up of, in accordance with the three axioms. The term *space* refers to the set of all of the outcomes of the experiment. We'll define other terms and concepts as we go along, after we list the three axioms.

### Axioms of probability

I. $P(A) \geq 0$     The probability of an event $A$ is zero or positive.

II. $P(S) = 1$     The probability of space is unity.

III. If $A \cdot B = 0$ then $P(A + B) = P(A) + P(B)$     If two events $A$ and $B$ are mutually exclusive, then the probability of their sum equals the sum of their individual probabilities.

The product of two events, shown in Axiom III as $A \cdot B$ and often called *intersection*, is an event which contains the outcomes that are common to the two events. The sum of two events $A + B$, also called *union*, is an event which contains all of the outcomes in both component events.

*Mutually* exclusive, referred to in Axiom III, means that two events have no outcomes in common. This means that if in an experiment one of the events occurs, the other one doesn't. Returning to the experiment of throwing a die, the odd event and the even event are mutually exclusive.

Now we give the definition of a field, which tells us what events we must include in an experiment. We will use the term *complement* of an event, which is all of the outcomes in the space not included in the event. The complement of $A$ is $A'$. $A \cdot A' = 0$.

*Definition of a field F*

1. If $A \in F$ then $A' \in F$

   If event $A$ is contained in the field $F$, then its complement is also contained in $F$.

2. If $A \in F$ and $B \in F$ then $A + B \in F$

   If the events $A$ and $B$ are each contained in $F$, then the event that is the sum of $A$ and $B$ is also contained in $F$.

Now we need one more definition before we get back down to earth and deal with the questions raised and the uses mentioned at the beginning of this section.

*Definition of independent events*

Two events are called *independent* if the probability of their product (intersection) equals the product of their individual probabilities:

$$P(A \cdot B) = P(A) \times P(B)$$

This definition can be extended to three or more events. For three independent events $A, B, C$:

$$P(A \cdot B \cdot C) = P(A) \times P(B) \times P(C), \text{ and}$$

$$P(A \cdot B) = P(A) \times P(B); \ P(A \cdot C) = P(A) \times P(C); \ P(B \cdot C) = P(B) \times P(C)$$

Similarly, for more than three events the probability of the product of all events equals the product of the probabilities of each of the events, and the probability of the product of any lesser number of events equals the product of the probabilities of those events. If there are $n$ independent events, then the total number of equations like those shown above for three events that are needed to establish their independence is:

$$2^n - (n+1).$$

### 19.8.1.2 Examples

We now can look at some examples of how to use probability theory.

**Example 19.13**

What is the probability of correctly receiving a sequence of 12 bits if the probability of error of a bit is one out of one hundred, or $p_e = .01$? All bits are independent.

### Solution

We look at the problem as an experiment in which we must define space, events according to the conditions of a field, and the probabilities of the events. We'll call the probability of a correct sequence $P_c$ and the probability of an error in the sequence $P_e$.

The space in our experiment contains all conceivable outcomes. Since we have a sequence of 12 bits, we can receive $2^{12} = 4096$ different sequences of bits, or words. The events in our experiment, conforming to the conditions for a field, are:

(1) The reception of the correct word—that is, no bits are in error.

(2) The reception of an erroneous word—a sequence that has 1 or more bits in error.

(3) The reception of any word.

(4) The reception of no word.

The inclusion of event (3) is necessary because of condition 2 in the definition of a field, which says that the event that is the sum of events must also be in the field. The sum of events (1) and (2) is all of the outcomes, which is the space, and this is event (3). Event (4) is needed because of condition 1 of a field—the complement of any event must be included—and event (4) is the complement of event (3). The complements of events (1) and (2) are each other, so the requirements of a field are complied with.

Now we assign probabilities to the events. In the statement of the problem we designated the probability of a bit error as $p_e$. In the field of an individual bit, we have two events: bit error or no bit error. The sum of these two events is the bit space, whose probability is unity. It follows that the probability of no bit error + probability of bit error $(p_e)$ = probability of space (1). Thus, the probability of no bit error = $1 - p_e$.

Now, the bits in the received sequence are independent, so the probability of a particular sequence equals the product of the probabilities of each of its bits. In the case of the errorless sequence, the probability that each bit has no bit error is $1 - p_e$, so this sequence's probability is $(1 - p_e)^{12}$. Using the given bit error probability of .01, we find that the probability of correctly receiving the sequence is $.99^{12}$ or approximately 88.6 percent.

How can we interpret this answer? If the sequence is sent only once for all time, it will either be received correctly or it won't. In this case, the establishment of a probability won't have much significance, except for the purpose of placing bets. However, if sequences are sent repeatedly, we will find that as the number of sequences increases, the percentage of those correctly received approaches 88.6.

Just as we found the probability of no bit error $= 1 - p_e$, we find the probability of a sequence error is $1 - .886 = 0.114$. This is from axioms II and III and the fact that the sum of the two mutually exclusive events—incorrect and correct sequences—is space, whose probability is 1.

Figure 19.30 gives a visual representation of this problem, showing space, the events, and the outcomes. Each outcome, representing one of the 4096 sequences, is assigned a probability $P_i$:

$$P_i = p_1 p_2 p_3 p_4 p_5 p_6 p_7 p_8 p_9 p_{10} p_{11} p_{12}$$

where $p_1$, $p_2$, and so on equals either $p_e$ or $1 - p_e$, depending on whether that specific bit in the sequence is in error or not. Sequences having the same number of error bits have identical probabilities.

For example, an outcome that is a sequence having one bit in error has a probability of $p_e (1 - p_e)^{11}$. There are 12 of these mutually exclusive sequences, so the probability of receiving a sequence having one and only one error bit is:

$$P_1 = 12\, p_e (1 - p_e)^{11} = .107$$

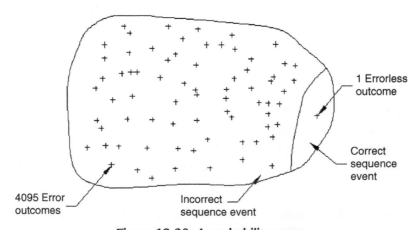

**Figure 19.30: A probability space**

### 19.8.1.3 Conditional Probability

An important concept in probability theory is conditional probability. It is defined as follows:

Given an event $B$ with nonzero probability $P(B) > 0$, then the conditional probability of event $A$ assuming event $B$ is known as:

$$P(A|B) = P(A \cdot B)/P(B) \tag{19.21}$$

A consequence of this definition is that the probability of an event is affected by the assumption of occurrence of another event. We see from the expression above that if $A$ and $B$ are mutually exclusive, the conditional probability is zero because $P(A \cdot B) = 0$. If $A$ and $B$ are independent, then the occurrence of $B$ has no effect on the probability of $A : P(A|B) = P(A)$. Conditional probabilities abide by the three axioms and the definition of a field.

Up to now we have discussed only sets of finite outcomes, but the theory holds when events are defined in terms of a continuous quantity as well. A space can be the set of all real numbers, for example, and subsets, or events, to which we can assign probabilities are intervals in this space. For example, we can talk about the probability of a train arriving between 1 and 2 p.m., or of the probability of a received signal giving a detector output above 1 volt. The axioms and definition of a field still hold but we have to allow for the existence of infinite sums and products of events.

### 19.8.1.4 Density and Cumulative Distribution Functions

In the problem above we found the probability, in a sequence of 12 bits, of getting no errors, of getting an error (one or more errors), and of getting an error in one bit only. We may be interested in knowing the probability of receiving exactly two bits in error, or any other number of error bits. We can find it using a formula called the *binomial distribution*:

$$P_m(n) = \binom{m}{n} p^n q^{m-n} \tag{19.22}$$

where $P_m(n)$ is the probability of receiving exactly $n$ bits in error, $m$ is the total number of bits in the sequence, $p$ is the probability of one individual bit being in error and $q = 1 - p$, the probability that an individual bit is correct. $\binom{m}{n}$ represents the number of different combinations of $n$ objects taken $m$ at a time:

$$\binom{m}{n} = \frac{m!}{n!(m-n)!}$$

The notation ! is factorial—for example, $m! = m(m-1)(m-2)\ldots(1)$.

Each time we send an individual sequence, the received sequence will have a particular number of bits in error—from 0 to 12 in our example. When a large number of sequences are transmitted, the frequency of having exactly $n$ errors will approach the probability given in Eq. (19.22). In probability theory, the quantity that expresses the observed result of a *random process* is called a *random variable*. In our example, we'll call this random variable N. Thus, we could rewrite Eq. (19.22) as:

$$P_m(N = n) = \binom{m}{n} p^n q^{m-n} \tag{19.22a}$$

(We represent uppercase letters as random variables and lowercase letters as real numbers. We may write $P(x)$ which means the probability that random variable $X$ equals real number $x$.)

The random value can be any number that expresses an outcome of an experiment, in this case 0 through 12. The random values are mutually exclusive events.

The probability function given in Eq. (19.22a), which gives the probability that the random variable equals a discrete quantity, is sometimes called the *frequency function*. Another important function is the *cumulative probability distribution function*, or *distribution function*, defined as:

$$F(x) = \text{Prob}(X \leq x) \tag{19.23}$$

defined for any number $x$ from $-\infty$ to $+\infty$ and $X$ is the random value.

In our example of a sequence of bits, the distribution function gives the probability that the sequence will have $n$ or fewer bits in error, and its formula is

$$F(n) = \sum_{n=0}^{m} \binom{m}{n} p^n q^{m-n} \tag{19.24}$$

where $\Sigma$ is the symbol for summation.

The example which we used up to now involves a discrete random variable, but probability functions also relate to continuous random variables, such as time or voltage. The best known of these is probably the Gaussian probability function, which describes, for example, thermal noise in a radio receiver. The analogous type of function to the frequency probability function defined for the discrete variable is called a *density function*. The Gaussian probability density function is:

$$P(x) = \frac{1}{\sqrt{2 \cdot \pi \cdot \sigma^2}} \cdot e^{\frac{-(x-m)^2}{2\sigma^2}} \qquad (19.25)$$

where $\sigma^2$ is the variance and $m$ is the average (defined below).

Plots of the frequency function $P_m(n)$ ($p$ and $q = 1/2$) and the density function $P(x)$, with the same average and variance, are shown in Figure 19.31. While these functions are analogous, as stated above, there are also fundamental differences between them. For one, $P(x)$ is defined on the whole real abscissa, from $-\infty$ (infinity) to $+\infty$ whereas $P_m(n)$ can have only the discrete values 0 to 12. Second, points on $P(x)$ are not probabilities! A continuous random variable can have a probability greater than zero only over a finite interval. Thus, we cannot talk about the probability of an instantaneous noise voltage of 2 V, but we can find the probability of it being between, say, 1.8 and 2.2 V. Probabilities on the curve $P(x)$ are the area under the curve over the interval we are interested in.

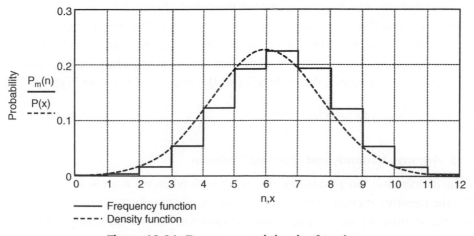

**Figure 19.31: Frequency and density functions**

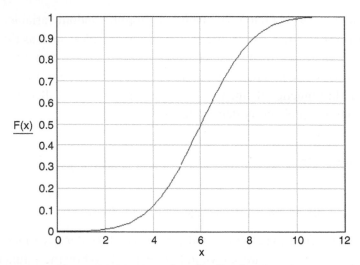

**Figure 19.32: Gaussian distribution function**

We find these areas by integrating the density curve between the endpoints of the interval, which may be plus or minus infinity.

The more useful probability function for finding probabilities directly for continuous random variables is the distribution function. Figure 19.30 shows the Gaussian distribution function, which is the integral of the density function between $-\infty$ and $x$. All distribution functions have the characteristics of a positive slope and values of 0 and 1 at the extremities. To find the probability of a random variable over an interval, we subtract the value of the distribution function evaluated at the lower boundary from its value at the upper boundary. For example, the probability of the interval of 4 to 6 in Figure 19.32 is:

$$F(6) - F(4) = .5 - .124 = .376$$

### 19.8.1.5  Average, Variance, and Standard Deviation

While the distribution function is easier to work with when we want to find probabilities directly, the density function or the frequency function is more convenient to use to compute the statistical properties of a random variable. The two most important of these properties are the *average* and the *variance*.

The statistical average for a discrete variable is defined as:

$$\bar{X} = \sum_i x_i \cdot P(x_i) \tag{19.26}$$

Writing this using the frequency function for the $m$-bit sequence (Eq. (19.2)) we have:

$$\bar{X} = \sum_n n \cdot P_m(n)$$

where $n$ ranges from 0 to $m$.

Calculating the average using $p = .1$, for example, we get $\bar{X} = 1.2$ bits. In other words, the average number of bits in error in a sequence with a bit error probability of $.1$ is just over 1 bit.

The definition of the average of a continuous random variable is:

$$\bar{X} = \int_{-\infty}^{+\infty} x \cdot P(x)dx \tag{19.27}$$

If we apply this to the expression for the Gaussian density function in Eq. (19.5) we get $\bar{X} = m$ as expected.

We can similarly find the average of a function of a random variable, in both the discrete and the continuous cases:

When this function is expressed as $f(x) = x^n$, its average is called the *nth moment of X*. The first moment of $X$ is its average, shown above. The second moment of the continuous random variable $x$ is:

$$\overline{X^2} = \int_{-\infty}^{\infty} x^2 \cdot P(x)dx \tag{19.28}$$

In the case where the random variable has a nonzero average, $a$, a more useful form of the second moment is the second-order moment about a point $a$, also called the *second-order central moment*, defined as:

$$Var(X) = \int_{-\infty}^{\infty} (x - a)^2 \cdot P(x)dx \tag{19.29}$$

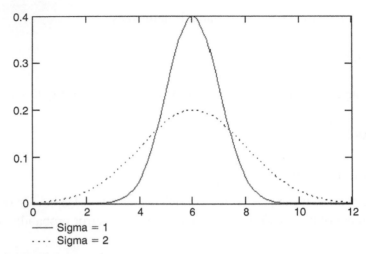

**Figure 19.33: Gaussian density function with different standard deviations**

The second-order central moment of *X* is called the *variance of X*, and its square root is the standard deviation. The standard deviation, commonly represented by the Greek letter sigma ($\sigma$), gives a measure of the form factor of a probability density function. Figure 19.33 shows two Gaussian density functions, one with s = 1 and the other with s = 2. Both have the same average, *m*.

The first and second moments are used all the time by electrical engineers when they are talking about voltages and currents, either steady-state or random. The first moment is the DC level and the second moment is proportional to the power. The variance of a voltage across a unit resistance is its AC power, and the standard deviation is the RMS value of a current or voltage about its DC level.

### 19.8.2 Information Theory

In 1948 C. E. Shannon published his Mathematical Theory of Communication, which was a tremendous breakthrough in the understanding of the possibilities of reliable communication in the presence of interference. A whole new field of study opened up, that of *information theory*, which deals with three basic concepts: the measure of information, the capacity of a communication channel to transfer information, and the use of coding as the means of approaching error-free communication at rates that approach this capacity. Much of the significance of Shannon's work can be realized by considering this statement, which sums up what is called the *fundamental theorem* of information theory:

*It is possible to transmit information through a noisy communication channel at any rate up to the channel capacity with an arbitrarily small probability of error.*

The converse of this statement has also been proven:

*It is not possible to achieve reliable communication through a channel at a rate higher than the channel capacity.*

The key to reliable communication in the presence of noise is *coding*. We will look at an example of error correction coding after a brief review of the basics of information theory.

### 19.8.2.1 Uncertainty, Entropy, and Information

Engineers generally talk about *bit* rate as the number of binary symbols that are transmitted per unit time. In information theory, the bit has a different, deeper meaning. Imagine the transmission of an endless stream of digital ones where each one has a duration of one millisecond. Is the rate of communication 1000 bits/second? According to information theory, the rate is zero. A stream of ones, or any other repetitive pattern of symbols reveals nothing to the receiver, and even in the presence of noise and interference, the "message" is always detected with 100 percent certainty. The more uncertain we are about what is being transmitted, the more information we get by correctly receiving it. In information theory, the term *bit* is a unit used to measure a quantity of information or uncertainty.

Information theory defines mathematically the uncertainty of a message or a symbol. Let's say we want to send a message using a sequence of three binary digits. We know that we can send up to eight different messages using this sequence ($2^3$). Each message has a particular probability of being sent. An example of this situation is a room with 8 patients in a hospital that has a nurse call system. When one of the patients presses a button next to his or her bed, a three-digit message is sent to the nurse's desk where it rings a bell and causes a display to show a number representing that patient. Depending on their condition, some patients may use the call system more than others. We present the probability of a patient pressing the nurse call button in Table 19.1.

The total probability of one of them having pressed the call button when the bell rings is, of course, 1.

**Table 19.1: Data for example**

| Patient's Name | Patient's Number | Probability |
|---|---|---|
| John | 1 | .1 |
| Mary | 2 | .5 |
| Jane | 3 | .2 |
| Mike | 4 | .05 |
| Pete | 5 | .05 |
| Sue | 6 | .03 |
| Tom | 7 | .01 |
| Elaine | 8 | .06 |

When the nurse hears the bell, she won't be surprised to see the number 2 on the display, since Mary requires assistance more than any of the other patients. The display of "7" though will be quite unexpected, since Tom rarely resorts to calling the nurse. The message indicating that Tom pressed his assistance button gives more information than the one triggered by Mary.

If we label the probability of an event, such as a patient pressing the assistance button, by $p_i$, we quantify the self-information of the event $i$ by:

$$I(i) = \log_2(1/p_i) \tag{19.30}$$

With $i$ referring to the number of the patient, we find that the self information of Mary's signal is $I(2) = \log_2(2) = 1$ bit, and Tom's signal self-information is $I(7) = \log_2(100) = 6.64$ bits. The unit of self-information is the bit when the log is taken to the base 2. The self-information of the other patients' signals can be found similarly.

More important than the individual self-information of each of the signals is the average self-information of all the signals. This quantity is commonly labeled $H$ and is called the *uncertainty* of the message, or its *entropy*. The latter term was taken from a value in thermodynamics with similar properties. Taking the statistical average of the self information, we get:

$$H = \sum_i p_i I(i) \tag{19.31}$$

With $i$ ranging from 1 to 8 and using the probabilities in the table above, we get for our present example:

$$H = .1(3.322) + .5(1) + .2(2.322) + .05(2)(4.322) + .03(5.059)$$
$$+ .01(6.644) + .06(4.059)$$

$$H = 2.191 \text{ bits/message}$$

Now let's assume that all of the patients are equally likely to call the nurse. The self-information of each signal $I(i)$ in this case is $\log_2(8) = 3$ bits. The entropy is:

$$H_m = 8 \times 1/8 \times 3 = 3 \text{ bits/message}$$

It turns out that this is the maximum possible entropy of the message—the case where the probability of each of the signals is equal.

Let's stretch our example of the nurse call system to an analogy with a continuous stream of binary digits. We'll assume the patients press their call buttons one after another at a constant rate.

In the case where the signal probabilities are distributed as in the table, the entropy per digit is $H/3 = .73$. If each patient pressed his button with equal probability, the entropy per digit would be the maximum of $H_m/3 = 1$. So with the different probabilities as listed in the table, the system communicates only 73% of the information that is possible to send over the communication channel per unit time. Using coding, discussed below, we can match a source to a channel to approach the channel's capability, or capacity, to any extent that we want it to.

Up to now we have been talking about the entropy, or uncertainty, of a source, and have assumed that what is sent is what is received. At least of equal importance is to measure the entropy, and the information, involved with messages sent and received over a noisy channel. Because of the noise, the probabilities of the received messages might not be the same as those of the source messages, because the receiver will make some wrong decisions as to the identity of the source. We saw above that entropy is a function of probabilities, and in the case of communication over a noisy channel, several sets of probability functions can be defined.

On a discrete, memoryless channel (memoryless because the noise affecting one digit is independent of the noise affecting any other digit) we can present the effect of the noise as a matrix of conditional probabilities. Assume a source transmits symbols having one of four letters $x1$, $x2$, $x3$, and $x4$. $X$ is a random variable expressing the transmitted symbol and $Y$ is a random variable for the received symbol. If a symbol $x1$ happens to be sent, the receiver may interpret it as $y1$, $y2$, $y3$, or $y4$, depending on the effect of the random noise at the moment the symbol is sent. There is a probability of receiving $y1$ when $x1$ is sent, another probability of receiving $y2$ when $x1$ is sent, and so on for a matrix of 16 probabilities as shown below:

$$P(Y|X) = \begin{matrix} p(y1|x1) & p(y2|x1) & p(y3|x1) & p(y4|x1) \\ p(y1|x2) & p(y2|x2) & p(y3|x2) & p(y4|x2) \\ p(y1|x3) & p(y2|x3) & p(y3|x3) & p(y4|x3) \\ p(y1|x4) & p(y2|x4) & p(y3|x4) & p(y4|x4) \end{matrix} \tag{19.32}$$

This example shows a square matrix, which means that the receiver will interpret a signal as being one of those letters that it knows can be transmitted, which is the most common situation. However, in the general case, the receiver may assign a larger or smaller number of letters to the signal so the matrix doesn't have to be square.

The conditional entropy of the output $Y$ when the input $X$ is known is:

$$H(Y|X) = -\sum_i p(x_i)\sum_j p(y_j|x_i)\log p(y_j|x_i) \tag{19.33}$$

where the sums are over the number of source letters $x_i$ and received letters $y_j$, and to get units of bits the log is to the base 2. The expression has a minus sign to make the entropy positive, canceling the sign of the log of a fraction, which is negative.

If we look only at the $Y$s that are received, we get a set of probabilities $p(y1)$ through $p(y4)$, and a corresponding uncertainty $H(Y)$.

Knowing the various probabilities that describe a communication system, expressed through the entropies of the source, the received letters, and the conditional entropy of the channel, we can find the important value of the *mutual information* or *transinformation* of the channel. In terms of the entropies, we defined above:

$$I(X;Y) = H(Y) - H(Y|X) \tag{19.34a}$$

The information associated with the channel is expressed as the reduction in uncertainty of the received letters given by a knowledge of the statistics of the source and the channel. The mutual information can also be expressed as:

$$I(X;Y) = H(X) - H(X|Y) \tag{19.34b}$$

which shows the reduction of the uncertainty of the source by the entropy of the channel from the point of view of the receiver. The two expressions of mutual information are equal.

### 19.8.2.2 Capacity

In the fundamental theorem of information theory, summarized above, the concept of channel capacity is a key attribute. It is connected strongly to the mutual information of the channel. In fact, the capacity is the maximum mutual information that is possible for a channel having a given probability matrix:

$$C = \max I(X;Y) \tag{19.35}$$

where the maximization is taken over all sets of source probabilities. For a channel that has a symmetric noise characteristic, so that the channel probability matrix is symmetric, the capacity is easily shown to be:

$$C = \log_2(m) - h \tag{19.36}$$

where $m$ is the number of different letters for each source symbol and $h = H(Y|X)$, a constant independent of the input distribution when the channel noise is symmetric. This is the case for expression (19–12), when each row of the matrix has the same probabilities except in different orders.

For the noiseless channel, as in the example of the nurse call system, the maximum information that could be transferred was $H_m = 3$ bits/ message, achieved when the source messages all have the same probability. Taking that example into the frame of the definition of capacity, Eq. (19.36), the number of different messages is the number of letters per symbol, so for $m = 8$ ("messages"):

$$C = \log_2 m = \log_2 8 = 3. \tag{19.36a}$$

This is a logical extension of the more general case presented previously. Here the constant $h = 0$ for the situation when there is no noise.

Another situation of interest for finding capacity on a discrete memoryless channel is when the noise is such that the output $Y$ is independent of the input $X$, and the conditional entropy $H(Y|X) = H(Y)$. The mutual information of such a channel $I(X;Y) = H(Y) - H(Y|X) = 0$ and the capacity is also zero.

The notion of channel capacity and the fundamental theorem also hold for continuous, "analog" channels, where signal-to-noise ratio ($S/N$) and bandwidth ($B$) are the characterizing parameters. The capacity in this case is given by the Hartley-Shannon law:

$$C = B \log_2 (1 + S/N) \text{ bits/second.} \tag{19.37}$$

The extension from a discrete system to a continuous one is easy to conceive of when we consider that a continuous signal can be converted to a discrete one by sampling, where the sampling rate is a function of the signal bandwidth (at least twice the highest frequency component in the signal).

From a glance at (19–17) we easily see that bandwidth can be traded off for signal-to-noise ratio, or vice-versa, while keeping a constant capacity. Actually, it's not quite that simple since the signal-to-noise ratio itself depends on the bandwidth. We can show this relationship by rewriting (19.17) using $n = $ the noise density. Substituting $N = n \cdot B$:

$$C = B \log_2 (1 + S/nB) \text{ bits/second} \tag{19.38}$$

In this expression the tradeoff between bandwidth and signal-to-noise ratio (or transmitter power) is tempered somewhat but it still exists. This doesn't mean that increasing the bandwidth automatically lets us send a higher data rate and keep a low probability of errors. We must use coding to keep the error rate down, but the higher bandwidth facilitates the addition of error-correcting bits in accordance with the coding algorithm.

In radio communication, a common aim is the reduction of signal bandwidth. Shannon-Hartley indicates that we can reduce bandwidth if we increase signal power to keep the capacity constant. This is fine, but Figure 19.34 shows that as the bandwidth is reduced more and more below the capacity, large increases in signal power are needed to maintain that capacity.

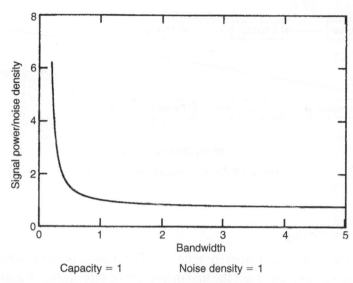

Figure 19.34: Signal power vs. bandwidth at constant capacity

### 19.8.2.3 Coding

We can represent a communication system conveniently by the block diagram in Figure 19.35. Assume the source outputs binary data, although if it were analog, we could make it binary by passing it through an analog-to-digital converter. The modulator and demodulator act as interfaces between the discrete signal parts of the system and the waveform that passes information over the physical channel—the modulated carrier frequency in a wireless system, for example.

We can take the modulator and demodulator blocks together with the channel and its noise input and look at the ensemble as a binary discrete channel. Consider the encoder and decoder blocks as matching networks which process the binary data so as to get the most "power" out of the system, in analogy to a matching network that converts the RF amplifier output impedance to the conjugate impedance of the antenna in order to get maximum power transfer. "Power," in the case of the communication link, may be taken to mean the data rate and the equivalent probability of error. A perfect match of the encoder and decoder to the channel gives a rate of information transfer equaling the capacity of the equivalent binary channel with an error rate approaching zero.

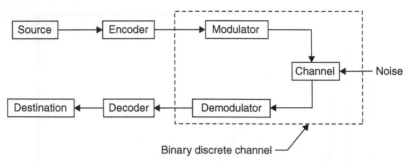

**Figure 19.35: Communication system**

### 19.8.2.4 Noiseless Coding

When there's no noise in the channel (or relatively little) there is no problem of error rate but coding is still needed to get the maximum rate of information transfer. We saw above that the highest source entropy is obtained when each message has the same probability. If source messages are not equi-probable, then the encoder can determine the message lengths that enter the binary channel so that highly probable messages will be short, and less probable messages will be long. On the average, the channel rate will be obtained.

For example, let's say we have four messages to send over a binary channel and that their rates of occurrence (probabilities) are 1/2, 1/4, 1/8, and 1/8. Table 19.2 shows two coding schemes that may be chosen for the messages $m1$, $m2$, $m3$, and $m4$.

One of the possible streams of messages that represents the probabilities is $m1 \cdot m1 \cdot m1 \cdot m1 \cdot m2 \cdot m2 \cdot m3 \cdot m4$. The bit streams that would be produced by each of the two codes is:

CODE A: 0000000001011011

CODE B: 00001010110111

Code A needs 16 binary symbols, or digits, to send the message stream while Code B needs only 14 digits. Thus, using Code B, we can make better use of the binary channel than if we use Code A, since, on the average, it lets us send 14% more messages using the same digit rate.

**Table 19.2: Two coding schemes**

| Message | Probability | Code A | Code B |
|---------|-------------|--------|--------|
| m1 | .5 | 00 | 0 |
| m2 | .25 | 01 | 10 |
| m3 | .125 | 10 | 110 |
| m4 | .125 | 11 | 111 |

We can determine the best possible utilization of the channel by calculating the entropy of the messages, which is (from 19.11 above):

$$H = .5 \times \log_2(1/.5) + .25 \times \log_2(1/.25) + .25 \times \log_2(1/.125)$$
$$+ .25 \times \log_2(1/.125)$$

$H = 1.75$ digits/message

In the example, we sent 8 messages, which can be achieved using a minimum of $8 \times H = 14$ digits, where each digit carries one bit of information. This we get using Code B.

The capacity of the binary symmetric channel, which sends ones and zeros with equal probability, is (from Eq. 19.16):

$$C = \log_2(2) - 0 = 1 \text{ bit/digit.}$$

Each digit on the channel is a bit, so Code B fully utilizes the channel capacity. If each message had equal probability, $H$ would be $\log_2(4) = 2$ bits per message. Then the best code to use would be code A. You can prove it yourself using a message stream with each message occurring an equal number of times and checking how many digits you need for each code. The point is that the chosen code matches the entropy of the source to get the maximum rate of communication.

There are several schemes for determining an optimum code when the message probabilities are known. One of them is Huffman's minimum-redundancy code. This code, like the one in the example above, is easy to decode because each word in the code is distinguishable without a deliminator and decoding is done "on the fly." When input probabilities are not known, such codes are not applicable. An example of a code that works on a bit stream without dependency on source probabilities or knowledge of word lengths is the Lempel-Ziv algorithm, used for file compression. Its basic idea is to look for repeated strings of characters and to specify where a previous version of the string started and how many characters it has.

The various codes used to transfer a given amount of "information" in the shortest time for a given bit rate are often called *compression schemes*, since they also can be viewed as reducing the number of symbols, or bits, needed to represent this information. One should remember that when the symbols in the input message stream are randomly distributed, coding will not make any difference from the point of view of compression or increasing the message transmission rate.

### 19.8.2.5 Error Detection and Correction

A most important use of coding is on noisy channels where you want to reduce the error rate while maintaining a given message transmission rate. We've already learned from Shannon that it's theoretically possible to reduce the error rate to as close to zero as we want, as long as the signal rate is below channel capacity. There's more than one way to look at the advantage of coding for error reduction. For example, if we demand a given maximum error rate in a communication system, we can achieve that goal without coding by increasing transmitter power to raise the signal-to-noise ratio and thereby reduce errors. We can also reduce errors by reducing the transmission rate, allowing us to reduce bandwidth, which will also improve the signal-to-noise ratio (since noise power is proportional to bandwidth). So using coding for error correction lets us either use lower power for a given error rate, or increase transmission speed for the same rate. The reduction in signal-to-noise ratio that can be obtained through the use of coding to achieve a given error rate compared to the signal-to-noise ratio required without coding for the same error rate is defined as the *coding gain*.

A simple and well-known way to increase transmission reliability is to add a parity bit at the end of a block, or sequence, of message bits. The parity bit is chosen to make the number of bits in the message block odd or even. If we send a block of seven message bits, say 0110110, and add an odd parity bit, and the receiver gets the message 01001101, it knows that the message has been corrupted by noise, although it doesn't know which bit is in error. This method of error detection is limited to errors in an odd number of bits, since if, say, two bits were corrupted, the total number of bits is still odd and the errors are not noticed. When the probability of a bit error is relatively low, by far most errors will be in one bit only, so the use of one parity bit may still be useful.

Another simple way to provide error detection is by logically adding together the contents of a number of message blocks and appending the result as an additional block in the message sequence. The receiver performs the same logic summing operation as the transmitter and compares its computation result to the block appended by the transmitter. If the blocks don't match, the receiver knows that one or more bits in the message blocks received are not correct. This method gives a higher probability of detection of errors than does the use of a single parity bit, but if an error is detected, more message blocks must be rejected as having suspected errors.

When highly reliable communication is desired, it's not enough just to know that there is an error in one or more blocks of bits, since the aim of the communication is to receive the whole message reliably. A very common way to achieve this is by an automatic repeat query (ARQ) protocol. After sending a message block, or group of blocks, depending on the method of error detection used, the transmitter stops sending and listens to the channel to get confirmation from the receiver. After the receiver notes that the parity bit or the error detection block indicates no errors, it transmits a short confirmation message to the transmitter. The transmitter waits long enough to receive the confirmation. If the message is confirmed, it transmits the next message. If not, it repeats the previous message.

ARQ can greatly improve communication reliability on a noisy communication link. However, there is a price to pay. First, the receiver must have a transmitter, and the transmitter a receiver. This is OK on a two-way link but may be prohibitive on a one-way link such as exists in most security systems. Second, the transmission rate will be reduced because of the necessity to wait for a response after each short transmission period. If the link is particularly noisy and many retransmissions are required, the repetitions themselves will significantly slow down the communication rate. In spite of its limitations, ARQ is widely used for reliable communications and is particularly effective when combined with a forward error correction scheme as discussed in the next section.

### 19.8.2.6 *Forward Error Correction (FEC)*

Just as adding one odd or even parity bit allows determining if there has been an error in reception, adding additional parity bits can tell the location of the error. For example, if the transmitted message contains 15 bits, including the parity bits, the receiver will need enough information to produce a four-bit word to indicate that there are no errors, (error word 0000), or which bit is in error (one out of 15 error words 0001 through 1111).

The receiver might be able to produce this four-bit error detection and correction word from four parity bits (in any case, no less than four parity bits), and then we could send 11 message information bits plus four parity bits and get a capability of detecting and correcting an error in any one bit.

R.W. Hamming devised a relatively simple method of determining parity bits for correcting single-bit errors. If $n$ is the total number of message bits in a sequence to be transmitted, and $k$ is the number of parity bits, the relationship between these numbers permitting correction of one digit is:

$$2^k \geq n + 1 \tag{19.39}$$

If we represent $m$ as the number of information bits ($n = m + k$) we can find $n$ and $k$ for several values of $m$ in Table 19.3.

As an example, one possible set of rules for finding the four parity bits for a total message length of 12 bits (8 information bits), derived from Hamming's method, is as follows.

We let $x1$, $x2$, $x3$, and so on, represent the position of the bits in the 12-bit word. Bits $x1$, $x2$, $x4$, and $x8$ are chosen in the transmitter to give even parity when the bits are summed up using binary arithmetic as shown in the four equations below (in binary arithmetic, $0 + 0 = 0$, $0 + 1 = 1$, $1 + 0 = 1$, $1 + 1 = 0$). Let $s1$ through $s4$ represent the results of the equations.

| | |
|---|---|
| $x1 + x3 + x5 + x7 + x9 = 0$ | $s1$ |
| $x2 + x3 + x6 + x7 + x10 + x11 = 0$ | $s2$ |
| $x4 + x5 + x6 + x7 + x12 = 0$ | $s3$ |
| $x8 + x9 + x10 + x11 + x12 = 0$ | $s4$ |

In this scheme, $x1$, $x2$, $x4$, and $x8$ designate the location of the parity bits. If we number the information bits appearing, say, in a byte-wide register of a microcomputer, as $m1$

**Table 19.3: Values of *m*, *n*, and *k***

| $m$ | 4 | 8 | 11 | 26 | 57 |
|---|---|---|---|---|---|
| $k$ | 3 | 4 | 4 | 5 | 6 |
| $n$ | 7 | 12 | 15 | 31 | 63 |

through $m8$, and we label the parity bits $p1$ through $p4$, then the transmitted 12-bit code word would look like this:

$$p1 \cdot p2 \cdot m1 \cdot p3 \cdot m2 \cdot m3 \cdot m4 \cdot p4 \cdot m5 \cdot m6 \cdot m7 \cdot m8$$

Those parity bits are calculated by the transmitter for even parity as shown in the four equations.

When the receiver receives a code word, it computes the four equations and produces what is called a *syndrome*—a four-bit word composed of $s4 \cdot s3 \cdot s2 \cdot s1$. $s1$ is 0 if the first equation $= 0$ and 1 otherwise, and so on for $s2$, $s3$, and $s4$. If the syndrome word is 0000, there are no single-bit errors. If there is a single-bit error, the value of the syndrome points to the location of the error in the received code word, and complementing that bit performs the correction. For example, if bit 5 is received in error, the resulting syndrome is 0101, or 5 in decimal notation (you should check this by calculating the equations).

Note that different systems could be used to label the code bits. For example, the four parity bits could be appended after the message bits. In this case a lookup table might be necessary in order to show the correspondence between the syndrome word and the position of the error in the code word.

The one-bit correcting code just described gives an order of magnitude improvement of message error rate compared to transmission of information bytes without parity digits, when the probability of a bit error on the channel is $10^{-2}$. However, one consequence of using an error-correcting code is that, if message throughput is to be maintained, a faster bit rate on the channel is required, which entails reduced signal-to-noise ratio because of the wider bandwidth needed for transmission. This rate, for the above example, is 12/8 times the uncoded rate, an increase of 50%. Even so, there is still an advantage to coding, particularly when coding is applied to larger word blocks.

Error probabilities are usually calculated on the basis of independence of the noise from bit to bit, but on a real channel, this is not likely to be the case. Noise and interference tend to occur in bursts, so several adjacent bits may be corrupted. One way to counter the noise bursts is by *interleaving* blocks of code words. For simplicity of explanation, let's say we are using four-bit code words. Interleaving the code words means that after forming each word with its parity bits in the encoder, the transmitter sends the first bit of the first word, then the first bit of the second word, and so on. The order of transmitted

**Table 19.4: Matrix showing order
of transmitted bits**

| $a1$ | $a2$ | $a3$ | $a4$ |
|------|------|------|------|
| $b1$ | $b2$ | $b3$ | $b4$ |
| $c1$ | $c2$ | $c3$ | $c4$ |
| $d1$ | $d2$ | $d3$ | $d4$ |

bits is best shown as a matrix, as in Table 19.4, where the $a$s are the bits of the first word, and $b$, $c$, and $d$ represent the bits of the second, third, and fourth words.

The order of transmission of bits is $a1 \cdot b1 \cdot c1 \cdot d1 \cdot a2 \cdot b2 \ldots d3 \cdot a4 \cdot b4 \cdot c4 \cdot d4$. In the receiver, the interleaving is decomposed, putting the bits back in their original order, after which the receiver decoder can proceed to perform error detection and correction.

The result of the interleaving is that up to four consecutive bit errors can occur on reception and the decoder can still correct them using a one-bit error correcting scheme. A disadvantage is that there is an additional delay of three word durations until decoded words start appearing at the destination.

Forward error correction schemes which deal with more than one error per block are much more complicated, and more effective, than the Hamming code described here. An important class of codes which are not based on blocks of bits but rather add parity bits based on logic operations on a small number of previous bits is called *convolutional* codes. In general, coding is a very important aspect of radio communication and its use and continued development is a prime reason why radio communication can continue to replace wires while making more effective use of the available spectrum and limited transmitter power.

### 19.8.3 Summary of Probability and Coding Principles

In this chapter we reviewed the basics of probability and coding. Sets of probabilities were applied both to the different signals generated by the source, as well as to the characteristics of a communication channel corrupted by noise. We saw that associated with a noisy communication channel is a maximum data rate called the *capacity* of the channel.

Through the use of coding, which involves inserting a redundancy in the transmitted data, it's possible to approach error-free communication up to the capacity of the channel.

Although effective and efficient coding algorithms are quite complicated, the great advances in integrated circuit miniaturization and logic circuit speed in recent years have allowed incorporating error-correction coding in a wide range of wireless communications applications. Many of these applications are in products that are mass produced and relatively low cost—among them cellular telephones and wireless local area networks. This trend will certainly continue and will give even the most simple short-range wireless products "wired" characteristics from the point of view of communication reliability.

This chapter has discussed matters which are not necessarily limited to radio communication, and where they do apply to radio communication, not necessarily to short-range radio. However, anyone concerned with getting the most out of a short-range radio communication link should have some knowledge of information theory and coding. The continuing advances in the integration of complex logic circuits and digital signal processing are bringing small, battery-powered wireless devices closer to the theoretical bounds of error-free communication predicted by Shannon fifty years ago.

## 19.9  Applications and Technologies

An important factor in the widespread penetration of short-range devices into the office and the home is the basing of the most popular applications on industry standards. In this chapter, we take a look at some of these standards and the applications that have emerged from them. Those covered pertain to HomeRF, Wi-Fi, HIPERLAN/2, Bluetooth, and Zigbee. In order to be successful, a standard has to be built so that it can keep abreast of rapid technological advancements by accommodating modifications that don't obsolete earlier devices that were developed to the original version. A case in point is the competition between the WLAN (wireless local area network) standard that was developed by the HomeRF Working Group based on the SWAP (shared wireless access protocol) specification, and IEEE specification 802.11, commonly known as *Wi-Fi*. The former used frequency-hopping spread-spectrum exclusively, and although some increase of data rate was provided for beyond the original 1 and 2 Mbps, it couldn't keep up with Wi-Fi, which incorporated new bandwidth efficient modulation methods to increase data rates 50-fold while maintaining compatibility with first generation DSSS terminals. Other reasons why HomeRF lost out to Wi-Fi are given below.

Many of the new wireless short-range systems are designed for operation on the 2.4 GHz ISM band, available for license-free operation in North America and Europe, as well as virtually all other regions in the world. Most systems have provisions for handling errors due to interference, but when the density of deployment of one or more systems is high, throughput, voice intelligibility, or quality of service in general is bound to suffer. We will look at some aspects of this problem and methods for solving it in relation to Bluetooth and Wi-Fi.

A relatively new approach to short-range communications with unique technological characteristics is ultra-wideband (UWB) signal generation and detection. UWB promises to add applications and users to short-range communication without impinging on present spectrum use. Additionally, it has other attributes including range finding and high power efficiency that are derived from its basic principles of operation. We present the main features of UWB communication and an introduction to how it works.

### 19.9.1 Wireless Local Area Networks (WLAN)

One of the hottest applications of short-range radio communication is wireless local area networks. While the advantage of a wireless versus wired LAN is obvious, the early versions of WLAN had considerably inferior data rates so conversion to wireless was often not worthwhile, particularly when portability is not an issue. However, advanced modulation techniques have allowed wireless throughputs to approach and even exceed those of wired networks, and the popularity of highly portable laptop and handheld computers, along with the decrease in device prices, have made computer networking a common occurrence in multi-computer offices and homes.

There are still three prime disadvantages to wireless networks as compared to wired: range limitation, susceptibility to electromagnetic interference, and security. Direct links may be expected to perform at a top range of 50 to 100 meters depending on frequency band and surroundings. Longer distances and obstacles will reduce data throughput. Greater distances between network participants are achieved by installing additional access points to bridge remote network nodes. Reception of radio signals may be interfered with by other services operating on the same frequency band and in the same vicinity. Wireless transmissions are subject to eavesdropping, and a standardized security implementation in Wi-Fi called *WEP* (wired equivalent privacy), has been found to be breachable with relative ease by persistent and knowledgeable hackers. More sophisticated encryption techniques can be incorporated, although they may be accompanied by reduction of convenience in setting up connections and possibly in performance.

Various systems of implementation are used in wireless networks. They may be based on an industrial standard, which allows compatibility between devices by different manufacturers, or a proprietary design. The latter would primarily be used in a special purpose network, such as in an industrial application where all devices are made by the same manufacturer and where performance may be improved without the limitations and compromises inherent in a widespread standard.

### 19.9.1.1 The HomeRF Working Group

The HomeRF Working Group was established by prominent computer and wireless companies that joined together to establish an open industry specification for wireless digital communication between personal computers and consumer electronic devices anywhere in and around the home. It developed the SWAP specification—Shared Wireless Access Protocol, whose major application was setting up a wireless home network that connects one or more computers with peripherals for the purposes of sharing files, modems, printers, and other electronic devices, including telephones. In addition to acting as a transparent wire replacement medium, it also permitted integration of portable peripherals into a computer network. The originators expected their system to be accepted in the growing number of homes that have two or more personal computers.

Following are the main system technical parameters:

- Frequency-hopping network:   50 hops per second

- Frequency range:   2.4 GHz ISM band

- Transmitter power:   100 milliwatt

- Data rate:   1 Mbps using 2FSK modulation
  2 Mbps using 4FSK modulation

- Range:   Covers typical home and yard

- Supported stations:   Up to 127 devices per network

- Voice connections:   Up to 6 full-duplex conversations

- Data security:   Blowfish encryption algorithm (over 1 trillion codes)

- Data compression:   LZRW3-A (Lempel-Ziv) algorithm

- 48-bit network ID:   Enables concurrent operation of multiple co-located networks

The HomeRF Working Group ceased activity early in 2003. Several reasons may be cited for its demise. Reduction in prices of its biggest competitor, Wi-Fi, all but eliminated the advantage HomeRF had for home networks—low cost. Incompatibility with Wi-Fi was a liability, since people who used their Wi-Fi equipped laptop computer in the office also needed to use it at home, and a changeover to another terminal accessory after work hours was not an option. If there were some technical advantages to HomeRF, support of voice and connections between peripherals for example, they are becoming insignificant with the development of voice interfaces for Wi-Fi and the introduction of Bluetooth.

### 19.9.1.2 Wi-Fi

Wi-Fi is the generic name for all devices based on the IEEE specification 802.11 and its derivatives. It is promoted by the Wi-Fi Alliance that also certifies devices to ensure their interoperability. The original specification is being continually updated by IEEE working groups to incorporate technical improvements and feature enhancements that are agreed upon by a wide representation of potential users and industry representatives. 802.11 is the predominant industrial standard for WLAN and products adhering to it are acceptable for marketing all over the world.

802.11 covers the data link layer of lower-level software, the physical layer hardware definitions, and the interfaces between them. The connection between application software and the wireless hardware is the MAC (medium access control). The basic specification defines three types of wireless communication techniques: DSSS (direct sequence spread spectrum), FSSS (frequency-hopping spread spectrum) and IR (infra-red). The specification is built so that the upper application software doesn't have to know what wireless technique is being used—the MAC interface firmware takes care of that. In fact, application software doesn't have to know that a wireless connection is being used at all and mixed wired and wireless links can coexist in the same network.

Wireless communication according to 802.11 is conducted on the 2.400 to 2.4835 GHz frequency band that is authorized for unlicensed equipment operation in the United States and Canada and most European and other countries. A few countries allow unlicensed use in only a portion of this band. A supplement to the original document, 802.11b, adds increased data rates and other features while retaining compatibility with equipment using the DSSS physical layer of the basic specification. Supplement 802.11a specifies considerably higher rate operation in bands of frequencies between 5.2 and 5.8 GHz. These data rates were made available on the 2.4 GHz band by 802.11g that has downward compatibility with 802.11b.

### 19.9.1.3 Network Architecture

Wi-Fi architecture is very flexible, allowing considerable mobility of stations and transparent integration with wired IEEE networks. The transparency comes about because upper application software layers (see below) are not dependent on the actual physical nature of the communication links between stations. Also, all IEEE LAN stations, wired or wireless, use the same 48-bit addressing scheme so an application only has to reference source and destination addresses and the underlying lower-level protocols will do the rest.

Three Wi-Fi network configurations are shown in Figures 19.36 through 19.38. Figure 19.36 shows two unattached basic service sets (BSS), each with two stations (STA). The BSS is the basic building block of an 802.11 WLAN. A station can make *ad hoc* connections with other stations within its wireless communication range but not with those in another BSS that is outside of this range. In order to interconnect terminals that are not in direct range one with the other, the distributed system shown in Figure 19.37 is needed. Here, terminals that are in range of a station designated as an access point (AP)

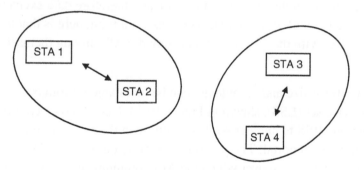

Figure 19.36: Basic service set

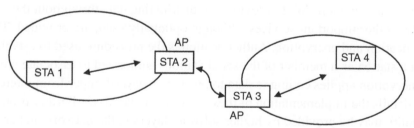

Figure 19.37: Distribution system and access points

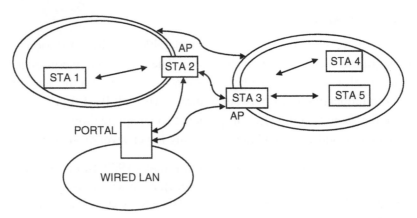

**Figure 19.38: Extended service set**

can communicate with other terminals not in direct range but who are associated with the same or another AP. Two or more such access points communicate between themselves either by a wireless or wired medium, and therefore data exchange between all terminals in the network is supported. The important thing here is that the media connecting the STAs with the APs, and connecting the APs among themselves are totally independent.

A network of arbitrary size and complexity can be maintained through the architecture of the extended service set (ESS), shown in Figure 19.38. Here, STAs have full mobility and may move from one BSS to another while remaining in the network. Figure 19.38 shows another element type—a portal. The portal is a gateway between the WLAN and a wired LAN. It connects the medium over which the APs communicate to the medium of the wired LAN—coaxial cable or twisted pair lines, for example.

In addition to the functions Wi-Fi provides for distributing data throughout the network, two other important services, although optionally used, are provided. They are authentication and encryption. Authentication is the procedure used to establish the identity of a station as a member of the set of stations authorized to associate with another station. Encryption applies coding to data to prevent an eavesdropper from intercepting it. 802.11 details the implementation of these services in the MAC. Further protection of confidentiality may be provided by higher software layers in the network that are not part of 802.11.

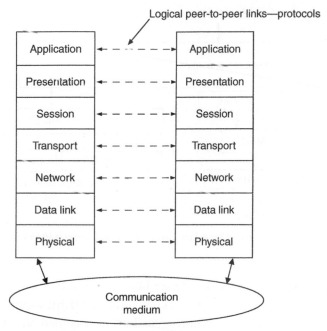

**Figure 19.39: Open System Interconnection reference model**

The operational specifics of WLAN are described in IEEE 802.11 in terms of defined protocols between lower-level software layers. In general, networks may be described by the communication of data and control between adjacent layers of the Open System Interconnection Reference Model (OSI/RM), shown in Figure 19.39, or the peer-to-peer communication between like layers of two or more terminals in the network. The bottom layer, physical, represents the hardware connection with the transmission medium that connects the terminals of the network—cable modem, radio transceiver and antenna, infrared transceiver, or power line transceiver, for example. The software of the upper layers is wholly independent of the transmission medium and in principle may be used unchanged no matter what the nature of the medium and the physical connection to it. IEEE 802.11 is concerned only with the two lowest layers, physical and data link.

IEEE 802.11 prescribes the protocols between the MAC sublayer of the data layer and the physical layer, as well as the electrical specifications of the physical layer. Figure 19.40 illustrates the relationship between the physical and MAC layers of several types of

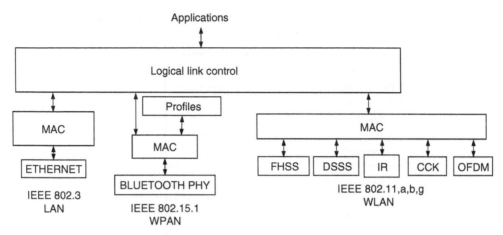

**Figure 19.40: Data Link and Physical Layers (PHY)**

networks with upper-layer application software interfaced through a commonly defined logical link control (LLC) layer. The LLC is common to all IEEE local area networks and is independent of the transmission medium or medium access method. Thus, its protocol is the same for wired local area networks and the various types of wireless networks. It is described in specification ANSI/IEEE standard 802.2.

The Medium Access Control function is the brain of the WLAN. Its implementation may be as high-level digital logic circuits or a combination of logic and a microcontroller or a digital signal processor. IEEE 802.11 and its supplements, (which may be generally designated 802.11x), prescribe various data rates, media (radio waves or infrared), and modulation techniques (FHSS, DSSS, CCK, ODFM) . These are the principle functions of the MAC:

- Frame delimiting and recognition,

- Addressing of destination stations,

- Transparent transfer of data, including fragmentation and defragmentation of packets originating in upper layers,

- Protection against transmission error,

- Control of access to the physical medium,

- Security services—authentication and encryption.

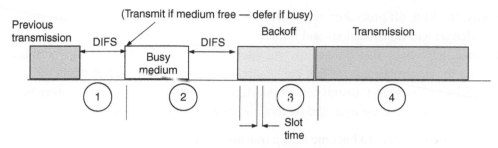

**Figure 19.41: CSMA/CA Access Method**

An important attribute of any communications network is the method of access to the medium. 802.11 prescribes two possibilities: DCF (distributed coordination function) and PCF (point coordination function).

The fundamental access method in IEEE 802.11 is the DCF, more widely known as *CSMA/CA* (carrier sense multiple access with collision avoidance). It is based on a procedure during which a station wanting to transmit may do so only after listening to the channel and determining that it is not busy. If the channel is busy, the station must wait until the channel is idle. In order to minimize the possibility of collisions when more than one station wants to transmit at the same time, each station waits a random time-period, called a *back off interval*, before transmitting, after the channel goes idle. Figure 19.41 shows how this method works.

The figure shows activity on a channel as it appears to a station that is attempting to transmit. The station may start to transmit if the channel is idle for a period of at least a duration of DIFS (distributed coordination function interframe space) since the end of any other transmission (Section 1 of the figure). However, if the channel is busy, as shown in Section 2 of the figure, it must defer access and enter a back off procedure. The station waits until the channel is idle, and then waits an additional period of DIFS. Now it computes a time-period called a *back off window*, which equals a pseudo-random number multiplied by constant called the *slot time*. As long as the channel is idle, as it is in Section 3 of the figure, the station may transmit its frame at the end of the back off window, Section 4. During every slot time of the back off window the station senses the channel, and if it is busy, the counter that holds the remaining time of the back off window is frozen until the channel becomes idle and the back off counter resumes counting down.

Actually, the back off procedure is not used for every access of the channel. For example, acknowledgement transmissions and RTS and CTS transmissions, (see below), do not use it. Instead, they access the channel after an interval called *SIFS* (short interframe space) following the transmission to which they are responding. SIFS is shorter than DIFS, so other stations waiting to transmit cannot interfere since they have to wait a longer time, after the previous transmission, and by then the channel is already occupied.

In waiting for a channel to become idle, a transmission contender doesn't have to listen continuously. When one hears another station access the channel, it can interpret the frame length field that is transmitted on every frame. After taking into account the time of the acknowledgement transmission that replies to a data transmission, the time that the channel will become idle is known even without physically sensing it. This is called a *virtual carrier sense mechanism*.

The procedure shown in Figure 19.41 may not work well under some circumstances. For example, if several stations are trying to transmit to a single access point, two or more of them may be positioned such that they all are in range of the access point but not of each other. In this case, a station sensing the activity of the channel may not hear another station that is transmitting on the same network. A refinement of the described CSMA/SA procedure is for a station thinking the channel is clear to send a short RTS (request to send) control frame to the AP. It will then wait to receive a CTS (clear to send) reply from the AP, which is in range of all contenders for transmission, before sending its data transmission. If the originating station doesn't hear the CTS it assumes the channel was busy and so it must try to access the channel again. This RTS/CTS procedure is also effective when not all stations on the network have compatible modulation facilities for high rate communication and one station may not be able to detect the transmission length field of another. RTS and CTS transmissions are always sent at a basic rate that is common to all participants in the network.

The PCS is an optional access method that uses a master-slave procedure for polling network members. An AP station assumes the role of master and distributes timing and priority information through beacon management transmissions, thus creating a contention free access method. One use of the PCS is for voice communications, which must use regular time slots and will not work in a random access environment.

### Physical Layer

The discussion so far on the services and the organization of the WLAN did not depend on the actual type of wireless connection between the members of the network. 802.11

and its additions specify various bit rates, modulation methods, and operating frequency channels, on two frequency bands, which we discuss in this section.

***IEEE 802.11 Basic*** The original version of the 802.11 specification prescribes three different air interfaces, each having two data rates. One is infrared and the others are based on frequency-hopping spread spectrum (FHSS) and direct-sequence spread-spectrum, each supporting raw data rates of 1 and 2 Mbps. Below is a short description of the IR and FHSS links, and a more detailed review of DSSS.

***Infrared PHY*** Infrared communication links have some advantages over radio wave transmissions. They are completely confined within walled enclosures and therefore eavesdropping concerns are greatly relieved, as are problems from external interference. Also, they are not subject to intentional radiation regulations. The IEEE 802.11 IR physical layer is based on diffused infrared links, and the receiving sensor detects radiation reflected off ceilings and walls, making the system independent of line-of-site. The range limit is on the order of 10 meters. Baseband pulse position modulation is used, with a nominal pulse width of 250 nsec. The IR wavelength is between 850 and 950 nM. The 1 Mbps bit rate is achieved by sending symbols representing 4 bits, each consisting of a pulse in one of 16 consecutive 250 nsec slots. This modulation method is called *16-PPM*. Optional 4-PPM modulation, with four slots per two-bit symbol, gives a bit rate of 2 Mbps.

Although part of the original IEEE 802.11 specification and having what seems to be useful characteristics for some applications, products based on the infrared physical layer for WLAN have generally not been commercially available. However, point-to-point, very short-range infrared links using the IrDA (Infrared Data Association) standard are very widespread (reputed to be in more than 300 million devices). These links work reliably line-of-site at one meter and are found, for example, in desktop and notebook computers, handheld PC's, printers, cameras and toys. Data rates range from 2400 Bps to 16 Mbps. Bluetooth devices will take over some of the applications but for many cases IrDA imbedding will still have an advantage because of its much higher data rate capability.

***Fhss Phy*** While overshadowed by the DSSS PHY, acquaintance with the FHSS option in 802.11 is still useful since products based on it may be available. In FHSS WLAN, transmissions occur on carrier frequencies that hop periodically in pseudo-random order over almost the complete span of the 2.4 GHz ISM band. This span in North America and most European countries is 2.400 to 2.4835 GHz, and in these regions there are 79 hopping carrier frequencies from 2.402 to 2.480 GHz. The dwell on each frequency is a system-determined parameter, but the recommended dwell time is 20 msec, giving a hop rate of 50 hops per second. In order for FHSS network stations to be synchronized, they must all use the same pseudo-random sequence of frequencies, and their synthesizers must be in step, that is, they must all be tuned to the same frequency channel at the same time. Synchronization is achieved in 802.11 by sending the essential parameters— dwell time, frequency sequence number, and present channel number—in a frequency parameter set field that is part of a beacon transmission (and other management frames) sent periodically on the channel. A station wishing to join the network can listen to the beacon and synchronize its hop pattern as part of the network association procedure.

The FHSS physical layer uses GFSK (Gaussian frequency shift keying) modulation, and must restrict transmitted bandwidth to 1 MHz at 20 dB down (from peak carrier). This bandwidth holds for both 1 Mbps and 2 Mbps data rates. For 1 Mbps data rate, nominal frequency deviation is $\pm 160$ kHz. The data entering the modulator is filtered by a Gaussian (constant phase delay) filter with 3 dB bandwidth of 500 kHz. Receiver sensitivity must be better than $-80$ dBm for a 3% frame error rate.

In order to keep the same transmitted bandwidth with a data rate of 2 Mbps, four-level frequency shift-keying is employed. Data bits are grouped into symbols of two bits, so each symbol can have one of four levels. Nominal deviations of the four levels are $\pm 72$ kHz and $\pm 216$ kHz. A 500 kHz Gaussian filter smoothes the four-level 1 Megasymbols per second at the input to the FSK modulator. Minimum required receiver sensitivity is –75 dBm.

Although development of Wi-Fi for significantly increased data rates has been along the lines of DSSS, FHSS does have some advantageous features. Many more independent networks can be collocated with virtually no mutual interference using FHSS than with DSSS. As we will see later, only three independent DSSS networks can be collocated. However, 26 different hopping sequences (North America and Europe) in any of three defined sets can be used in the same area with low probability of collision. Also, the degree of throughput reduction by other 2.4 GHz band users, as well as interference caused to the other users is lower with FHSS. FHSS implementation may at one time also have been less expensive. However, the updated versions of 802.11—specifically 802.11a, 802.11b, and 802.11g—have all based their methods of increasing data rates on the broadband channel characteristics of DSSS in 802.11, while being downward compatible with the 1 and 2 Mbps DSSS modes (except for 802.11a that operates on a different frequency band).

***Dsss Phy*** The channel characteristics of the direct sequence spread spectrum physical layer in 802.11 are retained in the high data rate updates of the specification. This is natural, since systems based on the newer versions of the specification must retain compatibility with the basic 1 and 2 Mbps physical layer. The channel spectral mask is shown in Figure 19.42, superimposed on the simulated spectrum of a filtered 1 Mbps transmission. It is 22 MHz wide at the −30 dB points. Fourteen channels are allocated in the 2.4 GHz ISM band, in which the center frequencies are 5 MHz apart, from 2.412 GHz to 2.484 GHz. The highest channel, number fourteen, is designated for Japan where the allowed band edges are 2.471 GHz and 2.497 GHz. In the US and Canada, the first eleven channels are used. Figure 19.43 shows how channels one, six and eleven may be used by three adjacent independent networks without co-interference. When there are no more than two networks in the same area, they may choose their operating channels to avoid a narrow-band transmission or other interference on the band.

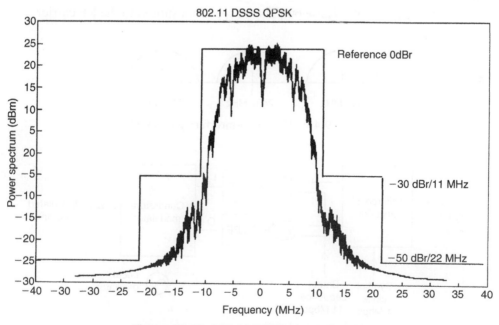

**Figure 19.42: 802.11 DSSS spectral mask**

In 802.11 DSSS, a pseudo-random bit sequence phase modulates the carrier frequency. In this spreading sequence, bits are called *chips*. The chip rate is 11 megachips per second (Mcps). Data is applied by phase modulating the spread carrier. There are eleven chips per data symbol. The chosen pseudo-random sequence is a Barker sequence, represented as 1, −1,1,1, −1,1,1,1, −1, −1, −1. Its redeeming property is that it is optimally detected in a receiver by a matched filter or correlation detector. Figure 19.43 is one possible implementation of the modulator. The DSSS PHY specifies two possible data rates—1 and 2 Mbps. The differential encoder takes the data stream and produces two output streams at 1 Mbps that represent changes in data polarity from one symbol to the next. For a data rate of 1 Mbps, differential binary phase-shift keying is used. The input data rate of 1 Mbps results in two identical output data streams that represent the changes between consecutive input bits. Differential quadrature phase-shift keying handles 2 Mbps of data. Each sequence of two input bits creates four permutations on two outputs. The differential encoder outputs the differences from symbol to symbol on the lines that go to the inputs of the exclusive OR gates shown in Figure 19.44. The outputs on the *I* and *Q* lines are the Barker sequence of 11 Mcps inverted or sent straight through, at a rate of 1 Msps, according to the differentially encoded data at the exclusive OR gate inputs. These outputs are spectrum shifted to the RF carrier

Figure 19.43: DSSS Non-interfering channels

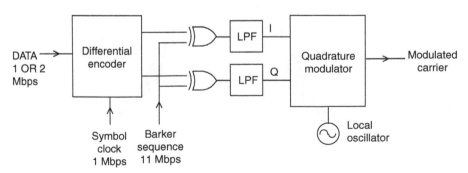

Figure 19.44: DSSS modulation

frequency (or an intermediate frequency for subsequent up-conversion) in the quadrature modulator.

Reception of DSSS signals is represented in Figure 19.45. The downconverted *I* and *Q* signals are applied to matched filters or correlation detectors. These circuits correlate the Barker sequence with the input signal and output an analog signal that represents the degree of correlation. The following differential decoder performs the opposite operation of the differential encoder described above and outputs the 1 or 2 Mbps data.

The process of despreading the input signal by correlating it with the stored spreading sequence requires synchronization of the receiver with transmitter timing and frequency. To facilitate this, the transmitted frame starts with a synchronization field (SYNC), shown at the beginning of the physical layer protocol data unit in Figure 19.46. Then a start frame delimiter (SFD) marks out the commencement of the following information bearing fields. All bits in the indicated preamble are transmitted at a rate of 1 Mbps, no matter what the subsequent data rate will be. The signal field specifies the data rate of the following fields in the frame so that the receiver can adjust itself accordingly. The next field, SERVICE, contains all zeros for devices that are only compliant with the basic version of 802.11, but some of its bits are used in devices conforming with updated versions. The value of the length field is the length, in microseconds, required to transmit

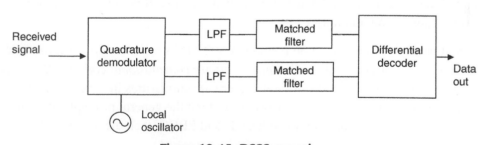

**Figure 19.45: DSSS reception**

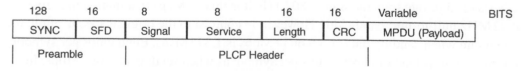

**Figure 19.46: DSSS frame format**

the data-carrying field labeled MPDU (MAC protocol data unit). An error check field, labeled CRC, protects the integrity of the SIGNAL, SERVICE, and LENGTH fields. The last field MPDU (MAC protocol data unit) is the data passed down from the MAC to be sent by the physical layer, or to be passed up to the MAC after reception. All bits in the transmitted frame are pseudo-randomly scrambled to ensure even power distribution over the spectrum. Data is returned to its original form by descrambling in the receiver.

**802.11b**  The "b" supplement to the original 802.11 specification supports a higher rate physical layer for the 2.4 GHz band. It is this 802.11b version that provided the impetus for Wi-Fi proliferation. With it, data rates of 5.5 Mbps and 11 Mbps are enabled, while retaining downward compatibility with the original 1 and 2 Mbps rates. The slower rates may be used not only for compatibility with devices that aren't capable of the extended rates, but also for fall back when interference or range conditions don't provide the required signal-to-noise ratio for communication using the higher rates.

As previously stated, the increased data rates provided for in 802.11b do not entail a larger channel bandwidth. Also, the narrow-band interference rejection, or jammer resisting qualities of direct sequence spread-spectrum are retained. The classical definition of processing gain for DSSS as being the chip rate divided by the data bandwidth doesn't apply here. In fact, the processing gain requirement that for years was part of the FCC Rules paragraph 15.247 definition of direct sequence spread-spectrum was deleted in an update from August 2002, and at the same time reference to DSSS was replaced by "digital modulation."

The mandatory high-rate modulation method of 802.11b is called *complementary code keying (CCK)*. An optional mode called *packet binary convolutional coding (PBCC)* is also described in the specification. Although there are similarities in concept, the two modes differ in implementation and performance. First the general principle of high-rate DSSS is presented below, applying to both CCK and PBCC, then the details of CCK are given.

As in the original 802.11, a pseudo-random noise sequence at the rate of 11 Mcps is the basis of high-rate transmission in 802.11b. It is this 11 Mcps modulation that gives the 22 MHz null-to-null bandwidth. However, in contrast to the original specification, the symbol rate when sending data at 5.5 or 11 Mbps is 1.375 Msps. Eight chips per symbol are transmitted instead of eleven chips per symbol as when sending at 1 or 2 Mbps. In "standard" DSSS as used in 802.11, the modulation, BPSK or QPSK, is applied to the group of eleven chips constituting a symbol. The series of eleven chips in the symbol is always the same (the Barker sequence previously defined). In contrast, high-rate DSSS uses a different 8-chip sequence in each symbol, depending on the sequence of data bits that is applied to each symbol. Quadrature modulation is used, and each chip has an *I* value and a *Q* value which represent a complex number having a normalized amplitude of one and some angle, $\alpha$, where $\alpha = \text{arctangent}\ (Q/I)$. $\alpha$ can assume one of four values divided equally around 360 degrees. Since each complex bit has four possible values, there are a total of $4^8 = 65536$ possible 8-bit complex words. For the 11 Mbps data rate, 256 out of these possibilities are actually used—which one being determined by the sequence of 8 data bits applied to a particular symbol. Only 16-chip sequences are needed for the 5.5 Mbps rate, determined by four data bits per symbol. The high-rate algorithm describes the manner in which the 256 code words, or 16 code words, are chosen from the 65536 possibilities. The chosen 256 or 16 complex words have the very desirable property that when correlation detectors are used on the *I* and *Q* lines of the received signal, downconverted to baseband, the original 8-bit (11 Mbps rate) or 4-bit (5.5 Mbps rate) sequence can be decoded correctly with high probability even when reception is accompanied by noise and other types of channel distortion.

The concept of CCK modulation and demodulation is shown in Figures 19.47 and 19.48. It's explained below in reference to a data rate of 11 Mbps. The multiplexer of Figure 19.47 takes a block of eight serial data bits, entering at 11 Mbps, and outputs them in parallel, with updates at the symbol rate of 1.375 MHz. The six latest data bits determine 1 out of 64 ($2^6$) complex code words. Each code word is a sequence of eight complex

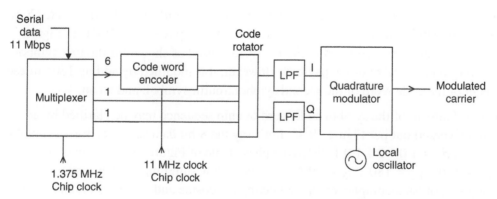

**Figure 19.47: High-rate modulator—11 Mbps**

Data symbol:  $d_0$ $d_1$ $d_2$ $d_3$ $d_4$ $d_5$ $d_6$ $d_7$

| Phase Table | |
|---|---|
| $d_i$   $d_{i+1}$ | $\varphi$ |
| 0    0 | $0^0$ |
| 1    0 | $180^0$ |
| 0    1 | $90^0$ |
| 1    1 | $-90^0$ |

Phase $(d_0,d_1) = \varphi_1$
Phase $(d_2,d_3) = \varphi_2$
Phase $(d_4,d_5) = \varphi_3$
Phase $(d_6,d_7) = \varphi_4$

$\alpha_1 = \varphi_1 + \varphi_2 + \varphi_3 + \varphi_4$
$\alpha_2 = \varphi_1 + \varphi_3 + \varphi_4$
$\alpha_3 = \varphi_1 + \varphi_2 + \varphi_4$
$\alpha_4 = \varphi_1 + \varphi_4 + 180^0$
$\alpha_5 = \varphi_1 + \varphi_2 + \varphi_3$
$\alpha_6 = \varphi_1 + \varphi_3$
$\alpha_7 = \varphi_1 + \varphi_2 + 180^0$
$\alpha_8 = \varphi_1$

$I_i = \cos(\alpha_i)$
$Q_i = \sin(\alpha_i)$
$i = 1...8$

**Figure 19.48: Derivation of code word**

chips, having phase angles $\alpha_1$ through $\alpha_8$ and a magnitude of unity. The first two data bits, $d_0$ and $d_1$, determine an angle, $\alpha'_8$ which, in the code rotator (see Figure 19.47), rotates the whole code word relative to $\alpha_8$ of the previous code word. This angle of rotation becomes the absolute angle $\alpha_8$ of the present code word. The normalized I and Q outputs of the code rotator, which after filtering are input to a quadrature modulator for up-conversion to the carrier (or intermediate) frequency, are:

$$I_i = \cos(\alpha_i), \quad Q_i = \sin(\alpha_i) \quad i = 1 \dots 8.$$

Figure 19.48 is a summary of the development of code words a for 11 Mbps rate CCK modulation. High rate modulation is applied only to the payload—MPDU in Figure 19.47. The code word described in Figure 19.48 is used as shown for the first symbol and then every other symbol of the payload. However, it is modified by adding 180° to each element of the code word of the second symbol, fourth symbol, and so on.

The development of the symbol code word or chip sequence may be clarified by an example worked out per Figure 19.48. Let's say the 8-bit data sequence for a symbol is $d = d_0 \dots d_7 = 1\ 0\ 1\ 0\ 1\ 1\ 0\ 1$. From the phase table of Figure 19.48 we find the angles $\varphi$: $\varphi_1 = 180°$, $\varphi_2 = 180°$, $\varphi_3 = -90°$, $\varphi_4 = 90°$. Now summing up these values to get the angle $\alpha_i$ of each complex chip, then taking the cosine and sine to get $I_i$ and $Q_i$, we summarize the result in Table 19.5.

### Table 19.5: Summarized results

| $i$ | 1 | 2 | 3 | 4 | 5 | 6 | 7 | 8 |
|---|---|---|---|---|---|---|---|---|
| $\alpha$ | 0 | 180 | 90 | 90 | −90 | 90 | 180 | 180 |
| $I$ | 1 | −1 | 0 | 0 | 0 | 0 | −1 | −1 |
| $Q$ | 0 | 0 | 1 | 1 | −1 | 1 | 0 | 0 |

### Table 19.6: 5.5 Mbps CCK decoding

| $d3,d2$ | $\alpha_1$ | $\alpha_2$ | $\alpha_3$ | $\alpha_4$ | $\alpha_5$ | $\alpha_6$ | $\alpha_7$ | $\alpha_8$ |
|---|---|---|---|---|---|---|---|---|
| 00 | 90° | 0° | 90° | 180° | 90° | 0° | −90° | 0° |
| 10 | −90° | 180° | −90° | 0° | 90° | 0° | −90° | 0° |
| 01 | −90° | 0° | −90° | 180° | −90° | 0° | 90° | 0° |

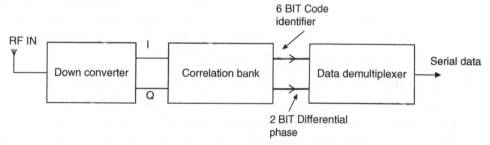

**Figure 19.49: CCK decoding**

The code words for 5.5 Mbps rate CCK modulation are a subset of those for 11 Mbps CCK. In this case, there are four data bits per symbol which determine a total of 16 complex chip sequences. Four 8-element code words (complex chip sequences) are determined using the last two data bits of the symbol, $d2$ and $d3$. The arguments (angles) of these code words are shown in Table 19.6. Bits $d0$ and $d1$ are used to rotate the code words relative to the preceding code word as in 11 Mbps modulation and shown in the phase table of Figure 19.48. Code words are modified by 180° every other symbol, as in 11 Mbps modulation.

The concept of CCK decoding for receiving high rate data is shown in Figure 19.49. For the 11 Mbps data rate, a correlation bank decides which of the 64 possible codes best fits each received 8-bit symbol. It also finds the rotation angle of the whole code relative to the previous symbol (one of four values). There are a total of 256 (64 × 4) possibilities and the chosen one is output as serial data. At the 5.5 Mbps rate there are four code words to choose from and after code rotation a total of 16 choices from which to decide on the output data.

To maintain compatibility with earlier non high-rate systems, the DSSS frame format shown in Figure 19.45 is retained in 802.11b. The 128-bit preamble and the header are transmitted at 1 Mbps while the payload MPDU can be sent at a high rate of 5.5 or 11 Mbps. The long and slow preamble reduces the throughput and cancels some of the advantage of the high data rates. 802.11b defines an optional short preamble and header which differ from the standard frame by sending a preamble with only 72 bits and transmitting the header at 2 Mbps, for a total overhead of 96 μsec instead of 192 μsec for the long preamble and header. Devices using this option can only communicate with other stations having the same capability.

Use of higher data rates entails some loss of sensitivity and hence range. The minimum specified sensitivity at the 11 Mbps rate is −76 dBm for a frame-error rate of 8% when sending a payload of 1024 bytes, as compared to a sensitivity of −80 dBm for the same frame-error rate and payload length at a data rate of 2 Mbps.

***802.11a and OFDM***  In the search for ways to communicate at even higher data rates than those applied in 802.11b, a completely different modulation scheme, OFDM (orthogonal frequency division multiplexing) was adopted for 802.11a. It is not DSSS yet it has a channel bandwidth similar to the DSSS systems already discussed. The 802.11*a* supplement is defined for channel frequencies between 5.2 and 5.85 GHz, obviously not compatible with 802.11b signals in the 2.4 GHz band. However, since the channel occupancy characteristics of its modulation are similar to that of DSSS Wi-Fi, the same system was adopted in IEEE 802.11*g* for enabling the high data rates of 802.11a on the 2.4 GHz band, while allowing downward compatibility with transmissions conforming to 802.11b.

802.11a specifies data rates of 6, 9, 12, 18, 24, 36, 48, and 54 Mbit/s. As transmitted data rates go higher and higher, the problem of multipath interference becomes more severe. Reflections in an indoor environment can result in multipath delays on the order of 100 nsec but may be as long as 250 nsec, and a signal with a bit rate of 10 Mbps (period of 100 nsec) can be completely overlapped by its reflection. When there are several reflections, arriving at the receiver at different times, the signal may be mutilated beyond recognition. The OFDM transmission system goes a long way to solving the problem. It does this by sending the data partitioned into symbols, and the length in time is several times the expected reflected path length time differences. The individual data bits in a symbol are all sent in parallel on separate subcarrier frequencies within the transmission channel. Thus, by sending many bits during the same time, each on a different frequency, the individual transmitted bit can be lengthened so that it won't be affected by the multipath phenomenon. Actually, the higher bit rates are accommodated by representing a group of data bits by the phase and amplitude of a particular transmitted carrier. A carrier modulated using quadrature phase-shift keying (QPSK) can represent two data bits and 64-QAM (quadrature amplitude modulation) can present six data bits as a single data unit on a subcarrier.

Naturally, transmitting many subcarriers on a channel of given width brings up the problem of interference between those subcarriers. There will be no interference between them if all the subcarriers are orthogonal—that is, if the integral of any two different subcarriers over the symbol period is zero. It is easy to show that this condition exists if the frequency difference between adjacent subcarriers is the inverse of the symbol period.

In OFDM, the orthogonal subcarriers are generated mathematically using the inverse Fourier transform (IFT), or rather its discrete equivalent, the inverse discrete Fourier transform (IDFT). The IDFT may be expressed as:

$$x(n) = \frac{1}{N} \sum_{m=0}^{N-1} X(m)[\cos(2\pi mn/N) + j \cdot \sin(2\pi mn/N)]$$

$x(n)$ are complex sample values in the time domain, $n = 0 \ldots N-1$, and $X(m)$ are the given complex values, representing magnitude and phase, for each frequency in the frequency domain. The IDFT expression indicates that the time domain signal is the sum of $N$ harmonically related sine and cosine waves each of which magnitude and phase is given by $X(m)$. We can relate the right side of the expression to absolute frequency by multiplying the arguments $2\pi mn/N$ by $f_s/f_s$ to get:

$$x(n) = \frac{1}{N} \sum_{m=0}^{N-1} X(m)[\cos(2\pi mf_1 nt_s) + j \cdot \sin(2\pi mf_1 nt_s)] \qquad (19.40)$$

where $f_1$ is the fundamental subcarrier and the difference between adjacent subcarriers, and $t_s$ is the sample time $1/f_s$. In 802.11a OFDM, the sampling frequency is 20 MHz and N = 64, so $f_1$ = 312.5 kHz. Symbol time is $Nt_s$ = 64/$f_s$ = 3.2 μsec.

In order to prevent intersymbol interference, 802.11a inserts a guard time of 0.8 μsec in front of each symbol, after the IDFT conversion. During this time, the last 0.8 μsec of the symbol is copied, so the guard time is also called a *circular prefix*. Thus, the extended symbol time that is transmitted is 3.2 + .8 = 4 μsec. The guard time is deleted after reception and before reconstruction of the transmitted data.

Although the previous equation, where $N$ = 64, indicates 64 possible subcarriers, only 48 are used to carry data, and four more for pilot signals to help the receiver phase lock to the transmitted carriers. The remaining carriers that are those at the outside of the occupied bandwidth, and the DC term ($m$ = 0 in Eq. (19.40)), are null. It follows that there are 26 ((48 + 4)/2) carriers on each side of the nulled center frequency. Each channel width is 312.5 kHz, so the occupied channels have a total width of 16.5625 (53 × 312.5 kHz) MHz.

For accommodating a wide range of data rates, four modulation schemes are used— BPSK, QPSK, 16-QAM and 64-QAM, requiring 1, 2, 4, and 6 data bits per symbol, respectively. Forward error correction (FEC) coding is employed with OFDM, which entails adding code bits in each symbol. Three coding rates: 1/2, 2/3, and 3/4, indicate the ratio of data bits to the total number of bits per symbol for different degrees of coding performance. FEC permits reconstruction of the correct message in the receiver, even when one or more of the 48 data channels have selective interference that would otherwise result in a lost symbol. Symbol bits are interleaved so that even if adjacent subcarrier bits are demodulated with errors, the error correction procedure will still reproduce the correct symbol. A block diagram of the OFDM transmitter and receiver is shown in Figure 19.50. Blocks FFT and IFFT indicate the fast Fourier transform and its inverse instead of the mathematically equivalent (in terms of results) discrete Fourier transform and inverse discrete Fourier transform (IFDT) that we used above because it is much faster to implement. Table 19.7 lists the modulation type and coding rate used for each data rate, and the total number of bits per OFDM symbol, which includes data bits and code bits.

The available frequency channels in the 5 GHz band in accordance with FCC paragraphs 15.401–15.407 for unlicensed national information infrastructure (U-NII) devices are

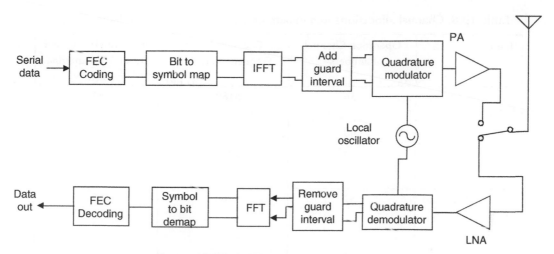

**Figure 19.50: OFDM system block diagram**

**Table 19.7: OFDM characteristics according to data rate**

| Data Rate Mbps | Modulation | Coding Rate | Coded Bits per Subcarrier | Coded Bits per OFDM Symbol | Data Bits per OFDM Symbol |
|---|---|---|---|---|---|
| 6 | BPSK | 1/2 | 1 | 48 | 24 |
| 9 | BPSK | 3/4 | 1 | 48 | 36 |
| 12 | QPSK | 1/2 | 2 | 96 | 48 |
| 18 | QPSK | 3/4 | 2 | 96 | 72 |
| 24 | 16-QAM | 1/2 | 4 | 192 | 96 |
| 36 | 16-QAM | 3/4 | 4 | 192 | 144 |
| 48 | 64-QAM | 2/3 | 6 | 288 | 192 |
| 54 | 64-QAM | 3/4 | 6 | 288 | 216 |

shown in Table 19.8. Channel allocations are 5 MHz apart and 20 MHz spacing is needed to prevent co-channel interference. Twelve simultaneous networks can coexist without mutual interference. Power limits are also shown in Table 19.9.

Extension of the data rates of 802.11b to those of 802.11a, but on the 2.4 GHz band is covered in supplement 802.11 g. The OFDM physical layer defined for the 5 GHz band is applied essentially unchanged to 2.4 GHz. Equipment complying with 802.11 g must also

Table 19.8: Channel allocations and maximum power for 802.11a in United States

| Band | Operation Channel Numbers | Channel Center Frequencies (MHz) | Maximum Power with up to 6 dBi antenna gain (mW) |
|---|---|---|---|
| U-NII lower band (5.15–5.25 GHz) | 36 | 5180 | 40 |
| | 40 | 5200 | |
| | 44 | 5220 | |
| | 48 | 5240 | |
| U-NII middle band (5.25–5.35 GHz) | 52 | 5260 | 200 |
| | 56 | 5280 | |
| | 60 | 5300 | |
| | 64 | 5320 | |
| U-NII upper band (5.725–5.825 GHz) | 149 | 5745 | 800 |
| | 153 | 5765 | |
| | 157 | 5785 | |
| | 161 | 5805 | |

Table 19.9: HIPERLAN/2 Frequency channels and power levels
(Reference 22, ETSI TS 101 475 V1.3.1 (2001–12))

| Center Frequency (MHz) | Radiated Power (mean EIRP) (dBm) |
|---|---|
| Every 20 MHz from 5180 to 5320 | 23 |
| Every 20 MHz from 5500 to 5680 | 30 |
| 5700 | 23 |

have the lower-rate features and the CCK modulation technique of 802.11b so that it will be downward compatible with existing Wi-Fi systems.

### 19.9.1.4  HIPERLAN/2

While 802.11b was designed for compliance with regulations in the European Union and most other regions of the world, 802.11a specifically refers to the regulations of the FCC and the Japanese MPT. ETSI (European Telecommunications Standards Institute) developed a high-speed wireless LAN specification, called *HIPERLAN/2* (high performance local area network), which meets the European regulations and in many ways goes beyond the capabilities of 802.11a. HIPERLAN/2 defines a physical layer essentially identical to that of 802.11a, using coded OFDM and the same data rates up to 54 Mbps. However, its second layer software level is very different from the 802.11 MAC and the two systems are not compatible. Built-in features of HIPERLAN/2 that distinguish it from IEEE 802.11a are the following:

- *Quality of service (QOS)*. Time division multiple access/time division duplex (TDMA/TDD) protocol permits multimedia communication.

- *Dynamic frequency selection (DFS)*. Network channels are selected and changed automatically to maintain communication reliability in the presence of interference and path disturbances.

- *Transmit power control (TPC)*. Transmission power is automatically regulated to reduce interference to other frequency band users and reduce average power supply consumption.

- *High data security*. Strong authentication and encryption procedures.

All of the above features of HIPERLAN/2 are being dealt with by IEEE task groups for implementation in 802.11. Specifically, the features of DFS and TPC are necessary for conformance of 802.11a to European Union regulations.

Frequency channels and power levels of HIPERLAN/2 are shown in Table 19.9.

### 19.9.2 Bluetooth

There are two sources of the Bluetooth specification. One is the Bluetooth Special Interest Group (SIG). The current version at this writing is Version 1.1. It is arranged in two volumes—Core and Profiles. Volume 1, the core, describes the physical, or hardware radio characteristics of Bluetooth, as well as low-level software or firmware which serves as an interface between the radio and higher level specific user software. The profiles in Volume 2 detail protocols and procedures for several widely used applications. The other Bluetooth source specification is IEEE 802.15.1. It is basically a rewriting of the SIG core specification, made to fit the format of IEEE communications specifications in general.

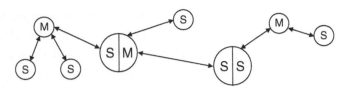

**Figure 19.51: Bluetooth scatternet**

Bluetooth is an example of a wireless personal area network (WPAN), as opposed to a wireless local area network (WLAN). It's based on the creation of *ad hoc*, or temporary, on-the-fly connections between digital devices associated with an individual person and located in the vicinity of around ten meters from him. Bluetooth devices in a network have the function of a master or a slave, and all communication is between a master and one or more slaves, never directly between slaves. The basic Bluetooth network is called a *piconet*. It has one master and from one to seven slaves. A scatternet is an interrelated network of piconets where any member of a piconet may also belong to an adjacent piconet. Thus, conceptually, a Bluetooth network is infinitely expandable. Figure 19.51 shows a scatternet made up of three piconets. In it, a slave in one piconet is a master in another. A device may be a master in one piconet only.

The basic RF communication characteristics of Bluetooth are shown in Table 19.10.

A block diagram of a Bluetooth transceiver is shown in Figure 19.52. It's divided into three basic parts: RF, baseband, and application software. A Bluetooth chip set will usually include the RF and baseband parts, with the application software being contained in the system's computer or controller. The user data stream originates and terminates in the application software. The baseband section manipulates the data and forms frames or data bursts for transmission. It also controls the frequency synthesizer according to the Bluetooth frequency-hopping protocol. The blocks in Figure 19.52 are general and various transmitter and receiver configurations are adopted by different manufacturers. The Gaussian low-pass filter block before the modulator, for example, may be implemented digitally as part of a complex signal *I/Q* modulation unit or it may be a discrete element filter in which the output is applied to the frequency control line of a VCO. Similarly, the receiver may be one of several types. If a superheterodyne configuration is chosen, the filter at the output of the downconverter will be a bandpass type. A direct conversion receiver will use low pass filters in complex *I* and *Q* outputs of the downconverter. While different manufacturers employ a variety of methods to

**Table 19.10: Bluetooth technical parameters**

| Characteristic | Value | Comment |
|---|---|---|
| Frequency Band | 2.4 to 2.483 GHz | May differ in some countries |
| Frequency Hopping Spread Spectrum (FHSS) | 79 1-MHz channels from 2402 to 2480 MHz | May differ in some countries |
| Hop Rate | 1600 hops per second | |
| Channel Bandwidth | 1 MHz | 20 dB down at edges |
| Modulation | Gaussian Frequency Shift Keying (GFSK) | |
| | Filter BT = 0.5 | Gaussian Filter bandwidth = 500 kHz |
| | Nominal modulation index = 0.32 | Nominal deviation = 160 kHz |
| Symbol Rate | 1 Mbps | |
| Transmitter Maximum Power | | |
|    Class 1 | 100 mW | Power control required |
|    Class 2 | 2.5 mW | Must be at least 0.25 mW |
|    Class 3 | 1 mW | No minimum specified |
| Receiver Sensitivity | −70 dBm for BER = 0.1% | |

implement the Bluetooth radio, all must comply with the same strictly defined Bluetooth specification, and therefore the actual configuration used in a particular chipset should be of little concern to the end user.

The Bluetooth protocol has a fixed-time slot of 625 microseconds, which is the inverse of the hop rate given in Table 11.5. A transmission burst may occur within a duration of one, three, or five consecutive slots on one hop channel. As mentioned, transmissions are always between the piconet master and a slave, or several slaves in the case of a broadcast, or point-to-multipoint transmission. All slaves in the piconet have an internal timer synchronized to the master device timer, and the state of this timer determines the transmission hop frequency of the master and that of the response of a designated slave. Figure 19.53 shows a sequence of transmissions between a master and two slaves. Slot are numbered according to the state, or phase, of the master clock, which is copied to each slave when it joins the piconet. Note that master transmissions take place during even numbered clock phases and slave transmissions during odd numbered phases.

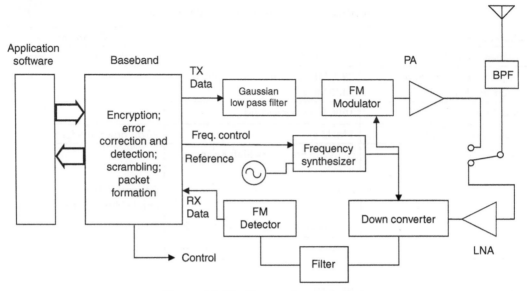

**Figure 19.52: Bluetooth transceiver**

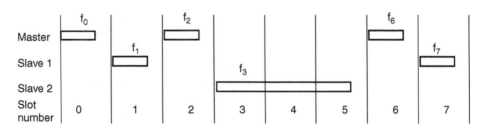

**Figure 19.53: Bluetooth timing**

Transmission frequency depends on the clock phase, and if a device makes a three or five slot transmission (slave two in the diagram), the intermediate frequencies that would have been used if only single slots were transmitted are omitted ($f_4$ and $f_5$ in this case). Note that transmissions do not take up a whole slot. Typically, a single-slot transmission burst lasts 366 microseconds, leaving 259 microseconds for changing the frequency of the synthesizer, phase locked loop settling time, and for switching the transceiver between transmit and receive modes.

There are two different types of wireless links associated with a Bluetooth connection. An asynchronous connectionless link (ACL) is used for packet data transfer while a synchronous connection oriented link (SCO) is primarily for voice. There are two major differences between the two link types. When an SCO link is established between a master and a slave, transmissions take place on dedicated slots with a constant interval between them. Also, unlike an ACL link, transmitted frames are not repeated in the case of an error in reception. Both of these conditions are necessary because voice is a continuous real-time process and the data rate cannot be randomly varied without affecting intelligibility. On the other hand, packet data transmission can use a handshaking protocol to regulate data accumulation and the instantaneous rate is not usually critical. Thus, for ACL links the master has considerable leeway in proportioning data transfer with the slaves in its network. An ARQ (automatic repeat request) protocol is always used, in addition to optional error correction, to ensure the highest reliability of the data transfer.

Bluetooth was conceived for employment in mobile and portable devices, which are more likely than not to be powered by batteries, so power consumption is an important issue. In addition to achieving low-power consumption due to relatively low transmitting power levels, Bluetooth incorporates power saving features in its communication protocol. Low average power is achieved by reducing the transmission duty cycle, and putting the device in a low-power standby mode for as long a period as possible relative to transmit and receive times while still maintaining the minimum data flow requirements. Bluetooth has three modes for achieving different degrees of power consumption during operation: sniff, hold, and park. Even in the normal active mode, some power saving can be achieved, as described below.

### 19.9.2.1 Three Modes

#### Active Mode

During normal operation, a slave can transmit in a particular time slot only if it is specifically addressed by the master in the proceeding slot. As soon as it sees that its address is not contained in the header of the master's message, it can "go to sleep," or enter a low-power state until it's time for the next master transmission. The master also indicates the length of its transmission (one, three, or five slots) in its message header, so the slave can extend its sleep time during a multiple slot interval.

#### Sniff Mode

In this mode, sleep time is increased because the slave knows in advance the time interval between slots during which the master may address the slave. If it's not addressed during the agreed slot, it returns to its low-power state for the same period and then wakes up and listens again. When it is addressed, the slave continues listening during subsequent master transmission slots as long as it is addressed, or for an agreed time-out period.

### Hold Mode

The master can put a slave in the hold mode when data transfer between them is being suspended for a given period of time. The slave is then free to enter a low-power state, or do something else, like participate in another piconet. It still maintains its membership in the original piconet, however. At the end of the agreed time interval, the slave resynchronizes with the traffic on the piconet and waits for instructions from the master.

### Park Mode

Park has the greatest potential for power conservation, but as opposed to hold and sniff, it is not a directly addressable member of the piconet. While it is outside of direct calling, a slave in park mode can continue to be synchronized with the piconet and can rejoin it later, either on its own initiative or that of the master, in a manner that is faster than if it had to join the piconet from scratch. In addition to saving power, park mode can also be considered a way to virtually increase the network's capacity from eight devices to 255, or even more. When entering park mode, a slave gives up its active piconet address and receives an 8-bit parked member address. It goes into low-power mode but wakes up from time to time to listen to the traffic and maintain synchronization. The master sends beacon transmissions periodically to keep the network active. Broadcast transmissions to all parked devices can be used to invite any of them to rejoin the network. Parked units themselves can request re-association with the active network by way of messages sent during an access window that occurs a set time after what is called a *beacon instant*. A polling technique is used to prevent collisions.

### 19.9.2.2 Packet Format

In addition to the data that originates in the high-level application software, Bluetooth packets contain fields of bits that are created in the baseband hardware or firmware for the purpose of acquisition, addressing, and flow control. Packet bits are also subjected to data whitening (randomization), error-correction coding, and encryption as defined for each particular data type. Figure 19.54 shows the standard packet format.

The access code is used for synchronization, DC level compensation, and identification. Each Bluetooth device has a unique address, and it is the address of the device acting as master that is used to identify transmitted packets as belonging to a specific piconet.

| Access code | Header | Payload |
|:---:|:---:|:---:|
| 72 BITS | 54 BITS | 0 TO 2745 BITS |

**Figure 19.54: Bluetooth packet**

A 64-bit synchronization word sandwiched between a four-bit header and four-bit trailer, which provide DC compensation, is based on the master's address. This word has excellent correlation properties so when it is received by any of the piconet members it provides synchronization and positive identification that the packet of which it is a part belongs to their network. All message packets sent by members of the piconet use the same access code.

The header contains six fields with link control information. First, it has a three-bit active member address which identifies to which of the up to seven slaves a master's message is destined. An all zero address signifies a broadcast message to all slaves in the piconet. The next field has four bits that define the type of packet being sent. It specifies, for example, whether one, three, or five slots are occupied, and the level of error correction applied. The remaining fields involve flow control (handshaking), error detection and sequencing. Since the header has prime importance in the packet, it is endowed with forward-error correction having a redundancy of times three.

Following the header in the packet is the payload, which contains the actual application or control data being transferred between Bluetooth devices. The contents of the payload field depend on whether the link is an ACL or SCO. The payload of ACL links has a payload header field that specifies the number of data bytes and also has a handshaking bit for data-buffering control. A CRC (cyclic redundancy check) field is included for data integrity. As stated above, SCO links don't retransmit packets so they don't include a CRC. They don't need a header either because the SCO payload has a constant length.

The previous packet description covers packets used to transfer user data, but other types of packets exist. For example, the minimum length packet contains only the access code, without the four-bit trailer, for a total of 68 bits. It's used in the inquiry and paging procedures for initial frequency-hopping synchronization. There are also NULL and POLL packets that have an access code and header, but no payload. They're sent when slaves are being polled to maintain synchronization or confirm packet reception (in the case of NULL) in the piconet but there is no data to be transferred.

### 19.9.2.3 Error Correction and Encryption

The use of forward error correction (FEC) improves throughput on noisy channels because it reduces the number of bad packets that have to be retransmitted. In the case of SCO links that don't use retransmission, FEC can improve voice quality. However, error correction involves bit redundancy so using it on relatively noiseless links will decrease throughput. Therefore, the application decides whether to use FEC or not.

As already mentioned, there are various types of packets, and the packet type defines whether or not FEC is used. The most redundant FEC method is always used in the packet header, and for the payload in one type of SCO packet. It simply repeats each bit three times, allowing the receiver to decide on the basis of majority rule what data bit to assign to each group of incoming bits.

The other FEC method, applied in certain type ACL and SCO packets, uses what's called a *(15,10)* shortened Hamming code. For every ten data bits, five parity bits are generated. Since out of every 15 transmitted bits only ten are retrieved, the data rate is only two-thirds what it would be without coding. This code can correct all single errors and detect all double errors in each 15-bit code word.

Wireless communication is susceptible to eavesdropping so Bluetooth incorporates optional security measures for authentication and encryption. Authentication is a procedure for verifying that received messages are actually from the party we expect them to be and not from an outsider who is inserting false messages. Encryption prevents an eavesdropper from understanding intercepted communications, since only the intended recipient can decipher them. Both authentication and implementation routines are implemented in the same way. They involve the creation of secret keys that are generated from the unique Bluetooth device address, a PIN (personal identification number) code, and a random number derived from a random or pseudo-random process in the Bluetooth unit. Random numbers and keys are changed frequently. The length of a key is a measure of the difficulty of cracking a code. Authentication in Bluetooth uses a 128-bit key, but the key size for encryption is variable and may range from 8 to 128 bits.

### 19.9.2.4 Inquiry and Paging

A distinguishing feature of Bluetooth is its *ad hoc* protocol and connections are often required between devices that have no previous knowledge of their nature or address. Also, Bluetooth networks are highly volatile, in comparison to WLAN for example, and connections are made and dissolved with relative frequency. To make a new connection, the initiator—the master—must know the address of the new slave, and the slave has to synchronize its clock to the master's in order to align transmit and receive channel hop-timing and frequencies. The inquiry and paging procedures are used to create the connections between devices in the piconet.

By use of the inquiry procedure, a connection initiator creates a list of Bluetooth devices within range. Later, desired units can be summoned into the piconet of which the initiator is master by means of the paging routine.

As mentioned previously, the access code contains a synchronization word based on the address of the master. During inquiry, the access code is a general inquiry access code (GIAC) formed from a reserved address for this purpose. Dedicated inquiry access codes (DIAC) can also be used when the initiator is looking only for certain types of devices. Now a potential slave can lock on to the master, provided it is receiving during the master's transmission time and on the transmission frequency. To facilitate this match-up, the inquiry procedure uses a special frequency hop routine and timing. Only 32 frequency channels are used and the initiator transmits two burst hops per standard time slot instead of one. On the slot following the transmission inquiry bursts, the initiator listens for a response from a potential slave on two consecutive receive channels in which frequencies are dependent on the previously transmitted frequencies.

When a device is making itself available for an inquiring master, it remains tuned to a single frequency for a period of 1.28 seconds and at a defined interval and duration scans the channel for a transmission. At the end of the 1.28-second period, it changes to another channel frequency. Since the master is sending bursts over the whole inquiry frequency range at a fast rate—two bursts per 1250 microsecond interval—there's a high probability the scanning device will catch at least one of the transmissions while it remains on a single frequency. If that channel happens to be blocked by interference, then the slave will receive a transmission after one of its subsequent frequency changes. When the slave does hear a signal, it responds during the next slot with a special packet called *FHS* (frequency hop synchronization) in which is contained the slave's Bluetooth address and state of its internal clock register. The master does not respond but notes the slave's particulars and continues inquiries until it has listed the available devices in its range. The protocol has provisions for avoiding collisions from more than one scanning device that may have detected a master on the same frequency and at the same time.

The master makes the actual connection with a new device appearing in its inquiry list using the page routine. The paging procedure is quite similar to that of the inquiry. However, now the master knows the paged device's address and can use it to form the synchronization word in its access code. The designated slave does its page scan while expecting the access code derived from its own address. The hopping sequence is different during paging than during inquiry, but the master's transmission bursts and the slave's scanning routine are very similar.

A diagram of the page state transmissions is given in Figure 19.55. When the slave detects a transmission from the master (Step 1), it responds with a burst of access code based on its own Bluetooth address. The master then transmits the FHS, giving the slave the access code information (based on the master's address), timing and piconet active member address (between one and seven) needed to participate in the network. The slave acknowledges FHS receipt in Step 4. Steps 5 and 6 show the beginning of the network transmissions which use the normal 79 channel hopping-sequence based on the master's address and timing.

### 19.9.3 Zigbee

Zigbee is the name of a standards-based wireless network technology that addresses remote monitoring and control applications. Its promotion and development is being handled on two levels. A technical specification for the physical and data link layers, IEEE 802.15.4, was drawn up by a working group of the IEEE as a low data rate WPAN (wireless personal area network). An association of committed companies, the Zigbee Alliance, is defining the network, security, and application layers above the 802.15.4 physical and medium access control layers, and will deal with interoperability certification and testing.

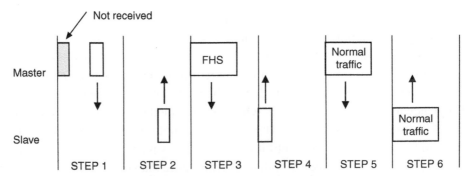

Figure 19.55: Paging transmissions

The distinguishing features of Zigbee to which the IEEE standard addresses itself are P

- Low data rates—throughput between 10 and 115.2 Kbps;
- Low power consumption—several months up to two years on standard primary batteries;
- Network topology appropriate for multisensor monitoring and control applications;
- Low complexity for low cost and ease of use;
- Very high reliability and security.

These will lend themselves to wide-scale use embedded in consumer electronics, home and building automation and security systems, industrial controls, PC peripherals, medical and industrial sensor applications, toys and games and similar applications. It's natural to compare Zigbee with the other WPAN standard, Bluetooth, and there will be some overlap in implementations. However, the two systems are quite different, as is evident from the comparison in Table 19.11.

**Table 19.11: Comparison of Zigbee and Bluetooth**

| | Bluetooth | Zigbee |
|---|---|---|
| Transmission Scheme | FHSS (Frequency Hopping Spread Spectrum) | DSSS (Direct Sequence Spread Spectrum) |
| Modulation | GFSK (Gaussian Frequency Shift Keying) | QPSK (Quadrature Phase Shift Keying) or BPSK (Binary Phase Shift Keying) |
| Frequency Band | 2.4 GHz | 2.4 GHz, 915 MHz, 868 MHz |
| Raw Data Bit Rate | 1 MBPS | 250 KBPS, 40 KBPS or 20 KBPS (depends on frequency band) |
| Power Output | Maximum 100 mW, 2.5 mW, or 1 mW, depending on class | Minimum capability 0.5 mW; maximum as allowed by local regulations |
| Minimum Sensitivity | −70 dBm for 0.1% BER | −85 dBm (2.4 GHz) or −92 dBm (915/868 MHz) for packet error rate <1% |
| Network topology | Master-Slave 8 active nodes | Star or Peer-Peer 255 active nodes |

### 19.9.3.1 Architecture

The basic architecture of Zigbee is similar to that of other IEEE standards, Wi-Fi and Bluetooth for example, a simplified representation of which is shown in Figure 19.56. On the bottom are the physical layers, showing two alternative options for the RF transceiver functions of the specification. Both of these options are never expected to exist in a single device, and indeed their transmission characteristics—frequencies, data rates, modulation system—are quite different. However, the embedded firmware and software layers above them will be essentially the same no matter what physical layer is applied. Just above the physical layers is the data link layer, consisting of two sublayers: medium access control, or MAC, and the logical link control, LLC. The MAC is responsible for management of the physical layer and among its functions are channel access, keeping track of slot times, and message delivery acknowledgement. The LLC is the interface between the MAC and physical layer and the upper-application software.

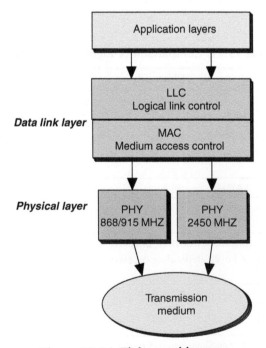

**Figure 19.56: Zigbee architecture**

Application software is not a part of the IEEE 802.15.4 specification and it is expected that the Zigbee Alliance will prepare profiles, or programming guidelines and requirements for various functional classes in order to assure product interoperability and vendor independence. These profiles will define network formation, security, and application requirements while keeping in mind the basic Zigbee features of low power and high reliability.

### 19.9.3.2 Communication Characteristics

In order to achieve high flexibility of adaptation to the range of applications envisioned for Zigbee, operation is being specified for three unlicensed bands—2.4 GHz, 915 MHz and 868 MHz; the latter two being included in the same physical layer. Those two bands are generally mutually exclusive, their use being determined by geographic location and regional regulations. The following 27 transmitting channels are defined, as shown in Table 19.12.

Data rates and modulation types for each of the bands are shown in Table 19.13.

In both physical layers, the modulation is DSSS (direct sequence spread spectrum). The spreading parameters are defined to meet communication authority regulations in the

**Table 19.12: Zigbee transmitting channels**

| Channel Number | Center Frequency Range | Channel Width |
|---|---|---|
| 0 | 868.3 MHz | 600 kHz |
| 1 to 10 | 906 to 924 MHz | 2 MHz |
| 11 to 27 | 2405 to 2480 | 5 MHz |

**Table 19.13: Data rates and modulation**

| PHY (MHz) | Frequency Band (MHz) | Spreading Parameters | | Data Parameters | | |
|---|---|---|---|---|---|---|
| | | Chip Rate (kcps) | Modulation | Bit Rate (kbps) | Symbol Rate (ksps) | Symbols |
| 868/915 | 868–868.6 | 300 | BPSK | 20 | 20 | Binary |
| | 902–928 | 600 | BPSK | 40 | 40 | Binary |
| 2450 | 2400–2483.5 | 2000 | Offset-QPSK | 250 | 62.5 | 16-ary Orthogonal |

various regions as well as desired data rates. For example, the chip rate of 600 Kbps on the 902–928 band allows the transmission to meet the FCC paragraph 15.247 requirement of minimum 500 kHz bandwidth at 6 dB down for digital modulation. However the chip rate, and with it the data rate, has to be reduced on Channel 0 in order to meet the confines of the 868 to 868.6 MHz channel allowed under ERC recommendation 70-03 and ETSI specification EN 300-220. On the 2400 to 2483.5 MHz band, the bit rate of 250 Kbps allows a throughput, after considering the overheads involved in packet transmissions, to attain 115.2 Kbps, a rate used for some PC peripherals for example.

The spreading modulation used on the 2450 MHz physical layer has similarity in principle to that used on IEEE 802.11b (high-rate Wi-Fi) to increase the data bit rate without raising the chip rate, thereby achieving a desired carrier bandwidth. Sixteen different, almost orthogonal 32-bit long spreading sequences are available for transmission at 2 Mchips/second. Each consecutive sequence of four data bits determines which of the sixteen spreading sequences is sent. On reception, the receiver can identify the spreading sequence and thus decode the data bits. The modulation used, O-QPSK (offset quadrature phase-shift keying) with half-sine wave pulse shaping is essentially equivalent to a form of frequency shift keying, MSK (minimum shift keying). It is fairly easy to generate and has a relatively narrow bandwidth for the given chip rate. This latter feature allows a large number of nonoverlapping channels that can be used, with proper upper layer software, on the crowded 2.4 GHz band to avoid interference.

Other physical layer characteristics of Zigbee are output power and receiver sensitivity. The devices must be capable of radiating at least $-3$ dBm although output may be reduced to the minimum necessary in order to limit interference to other users. Maximum power is determined by the regulatory authorities. While much higher powers are allowed, it may not be practical to transmit over, say 10 dBm, because of absolute limits on spurious radiation and the general objective of low-cost and low-power consumption. Minimum receiver sensitivity for the 868/915 MHz physical layer is specified as $-92$ dBm and $-85$ dBm on 2.4 GHz. These limits are for a packet error rate of one percent.

### 19.9.3.3 Device Types and Topologies

Two device types, of different complexities, are defined. A full function device (FFD) will be able to implement the full protocol set and can act as a network coordinator. Devices capable of minimal protocol implementation are reduced function devices (RFD). Due to the distinction between device types, networks in which most members require only minimum functionality, such as switches and sensors, can be made significantly less costly and have lower power consumption than if all devices were constrained to have maximum capability.

Flexibility in network configuration is achieved through two topologies—star and peer-to-peer that are depicted in Figure 19.57. A network may have as many as 255 members, one of which is a PAN (personal area network) coordinator. The function of the PAN coordinator, in addition to any specific application it may have, is to initiate, terminate, or route communication around the network. It also provides synchronization services. In a star network, each device communicates directly with the coordinator. The coordinator must be a FFD, and the others can be FFDs or RFDs. Relatively simple applications, like PC peripherals and toys, would typically use the star topology.

In the peer-to-peer topology, any device can communicate with any other device as long as it is in range. RFDs cannot participate, since an RFD can only communicate with a FFD. More complicated structures can be set up as a combination of peer-to-peer groups and star configurations. There is still just one PAN coordinator in the whole network. One example of such a structure is a cluster-tree network shown in Figure 19.58. In this

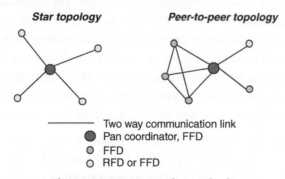

**Figure 19.57: Network topologies**

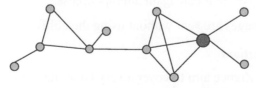

**Figure 19.58: Cluster-tree topology**

| 32 BITS | 8 BITS | 8 BITS | 127 BYTES Maximum |
|---|---|---|---|
| Premable | Start-of-frame deliminator | Frame length | PSDU — PHY Service Data unit |

**Figure 19.59: Transmission packet**

arrangement devices on the network extremities may well be out of radio range of each other, but they can still communicate by relaying messages through the individual clusters.

### 19.9.3.4 Frame Structure, Collision Avoidance, and Reliability

Zigbee frame construction and channel access are similar to those of WLAN 802.11 (Wi-Fi) but are less complex. The transmitted packet has the basic construction shown in Figure 19.59. The purpose of the preamble is to permit acquisition of chip and symbol timing. The PHY header, which is signaled by a delimiter byte, notifies the baseband software in the receiver of the length of the subsequent data. The PSDU (PHY service data unit) is the message that has been passed down through the higher protocol layers. As shown, it can have a maximum of 127 bytes although monitoring and control applications will typically be much shorter. Included in the PSDU are information on the format of the message frame, a sequence number, address information, the data payload itself, and at the end, two bytes that serve as a frame check sequence. Reliability is assured since the receiver performs an independent calculation of this frame check sequence and compares it with the number received. If any bits have been changed by interference or noise, the numbers will not match. Only if a match occurs, the receiving side returns an acknowledgement to the originator of the message. Lacking an acknowledgement, the transmission will be repeated until it is successfully received.

In order to avoid two or more stations trying to transmit at the same time, a carrier sense multiple access with collision avoidance (CSMA-CA) routine is employed, similar to that used in Wi-Fi, IEEE 802.11. The Zigbee receiver monitors the channel and only if it is idle it may initiate a transmission. If the channel is occupied, the terminal must wait a random back off period before it can again attempt access.

Acknowledgement messages are sent without using the collision avoidance mechanism.

### 19.9.3.5 Zigbee Applications

While the promoters of Zigbee aim to cover a very large market for those applications that require relatively low data rates, there will remain applications for which the compromises inherent in a general specification are not acceptable, and producers will continue to develop devices with proprietary specifications and characteristics. However, the open specification and a recognized certification of conformity are an advantage in many situations. For example, a home burglar alarm system would accept wireless sensors produced by different manufacturers, which will facilitate future expansion or allow installers to add sensors of types not available from the original system manufacturer. Use of devices approved according to a recognized standard gives the consumer some security against obsolescence.

Although Zigbee claims to be appropriate for most control applications, it will not fit all of them, and will not necessarily take advantage of all the possibilities of the unlicensed device regulations. Its declared maximum range of some 50 to 75 meters will fall short of the requirements of many systems. Given the meager maximum power allowed, greater range means reduced bandwidth and reduced data rate. In fact, a great many of the applications envisaged by Zigbee can get by very well with data rates of hundreds or a few thousand bits per second, and by matching receiver sensitivity to these rates, ranges of hundreds of meters can be achieved.

One partial answer to the range question is the deployment of the Zigbee network in a cluster-tree configuration, as previously described. Adjacent nodes serve as repeaters so that large areas can be covered, as long as the greatest distance between any two directly communicating nodes does not exceed Zigbee's basic range capability. For example, in a multi-floor building, sensors on the top floor can send alarms to the control box in the basement by passing messages through sensors located on every floor and operating as relay stations.

No doubt that there will be competition between Bluetooth and Zigbee for use in certain applications, but the overall deployment and the reliability of wireless control systems will increase. The proportion of wireless security and automation systems will increase because the new standard will provide a significant boost in reliability, security, and convenience, as compared to most present solutions.

### 19.9.4 Conflict and Compatibility

With the steep rise of Bluetooth product sales and the already large and growing use of wireless local area networks, there is considerable concern about mutual interference between Bluetooth-enabled and Wi-Fi devices. Both occupy the 2.4 to 2.4835 GHz unlicensed band and use wideband spread-spectrum modulating techniques. They will most likely be operating concurrently in the same environments, particularly office/ commercial but also in the home.

Interference can occur when a terminal of one network transmits on or near the receiving frequency of a terminal in another collocated network with enough power to cause an error in the data of the desired received signal. Although they operate on the same frequency band, the nature of Bluetooth and Wi-Fi signals are very different. Bluetooth has a narrowband transmission of 1 MHz bandwidth which hops around pseudo-randomly over an 80 MHz band while Wi-Fi (using DSSS) has a broad, approximately 20 MHz, bandwidth that is constant in some region of the band. The interference phenomenon is apparent in Figure 19.60. Whenever there is a frequency and time coincidence of the transmission of one system and reception of the other, it's possible for an error to occur. Whether it does or not depends on the relative signal strengths of the desired and undesired signals. These in turn depend on the radiated power outputs of the transmitters and the distance between them and the receiver. When two terminals are very close (on the order of centimeters), interference may occur even when the transmitting frequency is outside the bandwidth of the affected receiver.

Bluetooth and Wi-Fi systems are not synchronous and interference between them has to be quantified statistically. We talk about the probability of a packet error of one system caused by the other system. The consequence of a packet error is that the packet will have to be

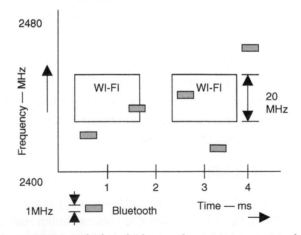

**Figure 19.60: Wi-Fi and Bluetooth spectrum occupation**

retransmitted once or more until it is correctly received, which causes a delay in message throughput. Voice transmissions generally don't allow packet retransmission because throughput cannot be delayed, so interference results in a decrease in message quality.

Following are parameters that affect interference between Bluetooth and Wi-Fi:

- *Frequency and time overlap.* A collision occurs when the interferer transmits at the same time as the desired transmitter and is strong enough to cause a bit or symbol error in the received packet.

- *Packet length.* The longer the packet length of the Wi-Fi system, relative to a constant packet length and hop rate of Bluetooth, the longer the victim may be exposed to interference from one or more collisions and the greater the probability of a packet error.

- *Bit rate.* Generally, the higher the bit rate, the lower the receiver sensitivity and therefore the more susceptible the victim will be to packet error for given desired and interfering signal strengths. On the other hand, higher bit rates usually result in reduced packet length, with the opposite effect.

- *Use factor.* Obviously, the more often the interferer transmits, the higher the probability of packet error. When both communicating terminals of the interferer are in the interfering vicinity of the victim the use factor is higher than if the terminals are further apart and one of them does not have adequate strength to interfere with the victim.

- *Relative distances and powers.* The received power depends on the power of the transmitter and its distance. Generally, Wi-Fi systems use more power than Bluetooth, typically 20 mW compared to 1 mW. Bluetooth Class 1 systems may transmit up to 100 mW, but their output is controlled to have only enough power to give a required signal level at the receiving terminal.

- Signal-to-interference ratio of the victim receiver, SIR, for a specified symbol or frame error ratio.

- Type of modulation, and whether error-correction coding is used.

A general configuration for the location of Wi-Fi and interfering Bluetooth terminals is given in Figure 19.61. In this discussion, only transmissions from the access point to the mobile terminal are considered. We can get an idea of the vicinity around the Wi-Fi

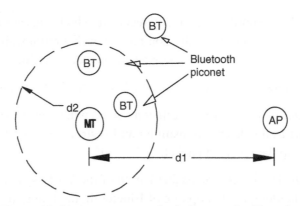

**Figure 19.61: Importance of relative terminal location**

mobile terminal in which operating Bluetooth terminals will affect transmissions from the access point to the mobile terminal by examining the following parameters:

$CI_{cc}$, $CI_{ac}$—Ratio of signal carrier power to co-channel or adjacent channel interfering power for a given bit or packet error rate (probability).

$P_{WF}$, $P_{BT}$—Wi-Fi and Bluetooth radiated power outputs.

$PL = Kd^r$—Path loss which is a function of distance $d$ between transmitting and receiving terminals, and the propagation exponent $r$. $K$ is a constant.

$d1$—Distance between Wi-Fi mobile terminal and access point.

$d2$—Radius of area around mobile terminal within which an interfering Bluetooth transmitter signal will increase the Wi-Fi bit error rate above a certain threshold.

$PR_{WF}$, $PR_{BT}$—Received powers from the access point and from the Bluetooth interfering transmitters.

$d2$ as a function of $d1$ is found as follows, using power in dBm:

1) $PR_{WF} = P_{WF} - 10\log(Kd1^r); PR_{BT} = P_{BT} - 10\log(kd2^r)$

2) $CI_{cc} = PR_{WF} - PR_{BT} = P_{WF} - 10\log(Kd1^r) - P_{BT} + 10\log(Kd2^r)$

3) $(CI_{cc} - P_{WF} + P_{BT})/10r = \log(d2/d1)$

4) $d2 = d1 \cdot 10^{(CI_{CC} - P_{WF} + P_{BT})/(10r)}$ \hfill (19.41)

As an example, the interfering area radius $d2$ is now calculated from equation (19.41) using the following system parameters:

$$CI_{cc} = 10\,\text{dB}, \; P_{WF} = 13\,\text{dBm}, \; P_{DT} = 0\,\text{dBm}, \; r = 2 \; \text{(free space)}$$

Solving equation (19.41): $d2 = d1 \times .71$

In this case, if a Wi-Fi terminal is located 15 meters from an access point, for example, all active Bluetooth devices within a distance of 10.6 meters from it have the potential of interfering. Only co-channel interference is considered. Adjacent channel interference, if significant, would increase packet error probability because many more Bluetooth hop channels would cause symbol errors. However, the adjacent channel $CI_{ac}$ is on the order of 45 dB lower than $CI_{cc}$ and would be noticed only when Bluetooth is several centimeters away from the Wi-Fi terminal.

The effect of an environment where path loss is greater than in free space can be seen by using an exponent $r = 3$. For the same Wi-Fi range of 15 meters, the radius of Bluetooth interference becomes 11.9 meters.

While equation (19.41) does give a useful insight into the range where Bluetooth devices are liable to deteriorate Wi-Fi performance, its development did involve simplifications. It considered that the signal-to-interference ratio that causes the error probability to exceed a threshold is constant for all wanted signal levels, which isn't necessarily so. It also implies a step relationship between signal-to-interference ratio and performance degradation, whereas the effect of changing interference level is continuous. The propagation law used in the development is also an approximation.

### 19.9.4.1 Methods for Improving Bluetooth and Wi-Fi Coexistence

By dynamically modifying one or more system operating parameters according to detected interference levels, coexistence between Bluetooth and Wi-Fi can be improved. Some of these methods are discussed below.

#### Power Control

Limiting transmitter power to the maximum required for a satisfactory level of performance will reduce interference to collocated networks. Power control is mandatory for Class 1 Bluetooth systems, where maximum power is 100 mW. The effect of the power on the interference radius is evident in equation (19.41). For example, in a Bluetooth piconet established between devices located over a spread of distances from the master, the master will use only the power level needed to communicate with each of the slaves in the network. Lack of power control would mean that all devices would communicate at maximum power and the collocated Wi-Fi system would be exposed to a high rate of interfering Bluetooth packets.

### Adaptive Frequency Hopping

Wi-Fi and Bluetooth share approximately 25 percent of the total Bluetooth hop-span of 80 MHz. Probably the most effective way to avoid interference between the two systems is to restrict Bluetooth hopping to the frequency range not used by Wi-Fi. When there is no coordination or cooperation between collocated networks, the Bluetooth piconet master would have to sense the presence of Wi-Fi transmissions and modify the frequency-hopping scheme of the network accordingly. A serious obstacle to this method was lifted by a change to the FCC regulations governing spread-spectrum transmissions in the 2.4 GHz band. Previously, frequency hopping devices were committed to hopping over at least 75 pseudo-randomly selected hop channels. In August 2002, paragraph 15.247, according to which Bluetooth and Wi-Fi devices are regulated, was changed to allow a minimum of 15 nonoverlapping channels in the 2400 to 2483 MHz band. In addition, the regulation allows employing intelligent hopping techniques, when less than 75 hopping frequencies are used, to avoid interference with other transmissions, and also suppression of transmission on an occupied channel provided that there are a minimum of 15 hops. The Bluetooth specification is due to be modified to take advantage of the adaptive frequency hopping method of avoiding interference.

There are situations where adaptive frequency hopping may not be effective or may have a negative effect. When two or more adjacent Wi-Fi networks are operating concurrently, they will utilize different 22 MHz sections of the 2.4 GHz band—three nonoverlapping Wi-Fi channels are possible. In this case, Bluetooth may not be able to avoid collisions while using a minimum of 15 hop frequencies. In addition, if there are several Bluetooth piconets in the same area, collisions among themselves will be much more frequent than when the full 79 channel hopping sequences are used.

### Packet Fragmentation

The two interference-avoiding methods described above are applicable primarily for action by the Bluetooth network. One method that the Wi-Fi network can employ to improve throughput is packet fragmentation. By fragmenting data packets and sending more, but shorter transmission frames, each transmission will have a lower probability of collision with a Bluetooth packet. Although reducing frame size increases the percentage of overhead bits in the transmission, when interference is heavy the overall effect may be higher throughput than if fragmentation was not used. Increasing bit rate for a constant packet length will also result in a shorter transmitted frame and less exposure to interference.

The methods mentioned above for reducing interference presume no coordination between the two different types of collocated wireless networks. However, devices are now being produced, in laptop and notebook computers, for example, that include both Wi-Fi and Bluetooth, sometimes even in the same chipset. In this case collaboration is possible in the device software to prevent inter-network collisions.

### 19.9.5 Ultra-wideband Technology

Ultra-wideband (UWB) technology is based on transmission of very narrow electromagnetic pulses at a low repetition rate. The result is a radio spectrum that is spread over a very wide bandwidth—much wider than the bandwidth used in the spread-spectrum systems previously discussed. Ultra-wideband transmissions are virtually undetectable by ordinary radio receivers and therefore can exist concurrently with existing wireless communications without demanding additional spectrum or exclusive frequency bands.

These are some of the advantages cited for ultra-wideband technology:

- Very low spectral density—Very low probability of interference with other radio signals over its wide bandwidth,

- High immunity to interference from other radio systems,

- Low probability of interception/detection by other than the desired communication link terminals,

- High multipath immunity,

- Many high data rate ultra-wideband channels can operate concurrently,

- Fine range-resolution capability,

- Relatively simple, low-cost construction, based on nearly all digital architectures.

Transmission and reception methods are unique, and are described briefly below.

Differing from conventional radio communication systems, which use up conversion and down conversion to pass information signals between baseband and bandpass frequency channels where wireless propagation occurs, UWB signal generation and detection use baseband techniques. An example of a UWB "carrier" is a Gaussian monopulse, shown in Figure 19.62. Its power spectrum is shown in Figure 19.63. If the time scale in Figure 19.62 is in nanoseconds, then the width of the pulse is 0.5 nanoseconds and the 3 dB bandwidth of the power spectrum is approximately 3.2 GHz with maximum power density at 2 GHz.

In order to pass information over a UWB communication link, trains of pulses must be transmitted with some characteristic of a pulse or group of pulses varied in order to distinguish between "0" and "1." The time between consecutive pulses should be

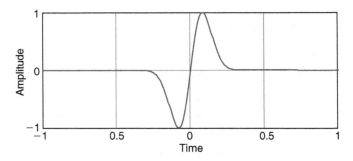

**Figure 19.62: UWB monopulse**

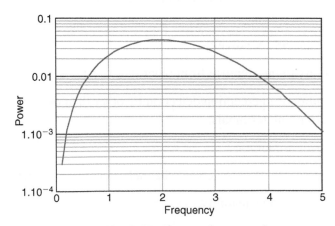

**Figure 19.63: Spectrum of monopulse**

determined in a pseudo-random manner in order to smooth the energy spikes in the frequency spectrum. Reception of the transmitted pulse train is done by correlating the received signal with a similar sequence of pulses generated in the receiver. A large number of communication links can be maintained simultaneously and independently by using different pseudo random sequences for each link.

A pulse similar to that of Figure 19.62 can be generated by applying an impulse, or perhaps more conveniently a step-voltage or current, to a linear band limited network. Figure 19.64 is a simulation of a sequence of UWB pulses created by stimulating a

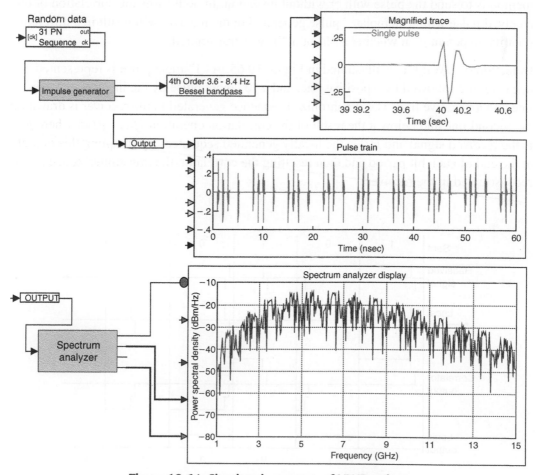

**Figure 19.64: Simulated sequence of UWB pulses**

bandpass filter with a pseudorandomly spaced sequence of impulses. The figure also shows the power spectrum of that sequence. The network that creates the individual UWB pulses includes the transmitter antenna, the propagation channel, and the receiving antenna, the characteristics, in terms of impulse response or amplitude and phase vs. frequency must be known and accounted for in designing the system.

There are several ways of representing a UWB pulse as "1" or "0." One method is to advance or retard the transmitted pulse with respect to the expected time of arrival of the pulse in the receiver according to the agreed pseudo-random time sequence. Another method is to send the pulse with or without inversion. In both cases the correlation of the received pulse with a "template" pulse generated in the receiver will result in a different polarity, depending on whether a "1" or a "0" was transmitted.

Detection of UWB bits is illustrated in Figure 19.65. A "1" monopulse is represented by a negative line followed by a positive line, and a "0" monopulse by the inverse—a positive line and a negative line. The synchronized sequence generated in the receiver is drawn on the second line and below it the result of the correlation operation $\int f(t) \cdot g(t) dt$ where $f(t)$ is the received signal and $g(t)$ is the locally generated sequence. By sampling this output at the end of each bit period and then resetting the correlator, the transmitted sequence is reconstructed in the receiver.

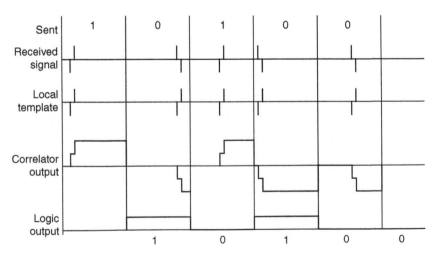

**Figure 19.65: Detection of UWB bit sequence**

As mentioned above, an individual bit can be represented by more than one sequential monopulse. Doing so increases the processing gain by the number of monopulses per bit. Processing gain is also an inverse function of the pulse duty cycle. This is because, for constant average power, the power in the pulses contributing to each bit must be raised by (1/*duty-cycle*). By gating out the noise except during the interval of the expected incoming pulse, the signal-to-noise ratio will only be a function of the power in the pulse, regardless of the duty cycle. An example may make the explanation clearer. Let's say we are sending data a rate of 10 Mbps. A UWB pulse in the transmitter is 200 picoseconds wide; 20 pulses represent one bit. The time between bits is $1/(10*10^6) = 100$ nanoseconds. So the time between pulses is $(100\,ns)/20 = 5\,ns$. The duty cycle is $(200\,ps)/(5\,ns) = 25$. Now the processing gain attributed to the number of UWB pulses per bit is $10\log(20) = 13\,dB$. That due to the duty cycle is $10\log(25) = 14\,dB$. Total processing gain is $13 + 14 = 27\,dB$.

A simplified block diagram of a UWB system is shown in Figure 19.66. A key to the generation of UWB pulses is the ability to create short impulse or step functions with

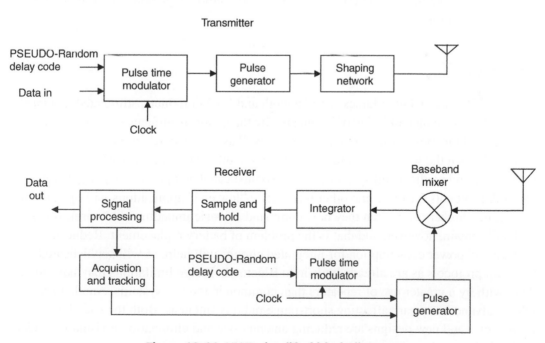

**Figure 19.66: UWB simplified block diagram**

rise times on the order of tens or at the most hundreds of picoseconds, and to detect the UWB pulses that result from their application. High speed integrated circuits can be employed or special circuit elements, such as tunnel diodes or step recovery diodes, can be incorporated.

Conditions for using UWB in Europe are also being considered by the European Union. Due to the fear of interference with vital wireless services in the wide bandwidths covered by UWB radiation, spectral density limits allowed by the FCC and presumably to be permitted in the European Union are relatively low, on the order of spurious radiation limits for conventional unlicensed transmissions. However, as we have seen, high-processing gains can be achieved with UWB and communication ranges on the order of tens or hundreds of meters at highdata rates can be expected. Also, the FCC has indicated its intention to monitor the effects of UWB transmissions on other services, once equipment has been put into service in significant quantities, and the agency may be expected to modify or make its limits more lenient if interference is found not to be a problem. In any case, the unique characteristics of UWB are attractive enough to make this technology an important part of the offerings for short-range wireless communication in the years to come.

### 19.9.6 Summary

The annually increasing volumes for Bluetooth and Wi-Fi products, stimulated in a large part by the acceptance of industrial standards by the major manufacturers, are causing prices to fall on complex integrated circuits as well as the basic RF components. This trend will open the way for the use of these parts in other short-range applications such as security and medical call systems. Use of sophisticated and proven two-way hardware and link protocols for these and other technically "low end" applications will open them up to much higher usage than they now command. A basic impediment to wireless use will still remain, however, and that is the problem of battery replacement. Reduced voltage and power consumption for integrated circuits will help, as will sophisticated wake-up protocols as are already built-in to Bluetooth. Range limitations may have to be dealt with by a greater use of repeaters than common in today's systems. Another area where advancements are affecting short-range radio is antennas. Both the use of higher frequencies and new designs are reducing antenna size and eliminating a visual reminder of the difference between wired and wireless devices.

The unconventional ultra-wideband technology, since its approval by the FCC, is opening up new civilian applications for short-range wireless, notably in the areas of distance measurement, concealed object location, and high precision positioning systems. Because of its high-interference immunity, and its property of not causing interference, it may successfully compete with and complement other technologies used for short-range radio applications such as personal communications systems, security sensors, and RFID tags.

In summary, advances in short-range radio communication developments in one area feeds its expansion in other areas. Overall, short-range radio will continue to play a major part in the ongoing communication revolution.

## Reference

Petroff, Alan and Withington, Paul, "Time Modulated Ultra-Wideband (TM-UWB) Overview," Presented at Wireless Symposium/Portable by Design, Feb 25, 2000, San Jose, California (http://www.time-domain.com).

# CHAPTER 20

# Principles of Electromagnetics

Ron Schmitt

How does electromagnetic theory tie together such broad phenomena as electronics, radio waves, and light? Explaining this question in the context of electronics design is the main goal of this chapter. The basic philosophy is that by developing an understanding of the fundamental physics, you can develop an intuitive feel for how electromagnetic phenomena occur. Learning the physical foundations serves to build the confidence and skills to tackle real-world problems, whether you are an engineer, technician, or physicist. So, in the beginning of this chapter, we will review fundamentals of electrical engineering in the context of electromagnetics.

The many facets of electromagnetics are due to how waves behave at different frequencies and how materials react in different ways to waves of different frequency. Quantum physics states that electromagnetic waves are composed of packets of energy called *photons*. At higher frequencies each photon has more energy. Photons of infrared, visible light, and higher frequencies have enough energy to affect the vibrational and rotational states of molecules and the electrons in orbit of atoms in the material. Photons of radio waves do not have enough energy to affect the bound electrons in a material. Furthermore, at low frequencies, when the wavelengths of the EM waves are very long compared to the dimensions of the circuits we are using, we can make many approximations leaving out many details. These low-frequency approximations give us the familiar world of basic circuit theory.

## 20.1 The Need for Electromagnetics

So why would an electrical engineer need to know all this theory? There are many reasons why any and all electrical engineers need to understand electromagnetics. Electromagnetics is necessary for achieving electromagnetic compatibility of products, for understanding high-speed digital electronics, RF, and wireless, and for optical computer networking.

Certainly any product has some electromagnetic compatibility (EMC) requirements, whether due to government mandated standards or simply for the product to function properly in the intended environment. In most EMC problems, the product can be categorized as either an aggressor or a victim. When a product is acting as an aggressor, it is either radiating energy or creating stray reactive fields at power levels high enough to interfere with other equipment. When a product is acting as a victim, it is malfunctioning due to interference from other equipment or due to ambient fields in its environment. In EMC, victims are not always blameless. Poor circuit design or layout can create products that are very sensitive to ambient fields and susceptible to picking up noise. In addition to aggressor/victim problems, there are other problems in which noise disrupts proper product operation. A common problem is that of cabling, that is, how to bring signals in and out of a product without also bringing in noise and interference. Cabling problems are especially troublesome to designers of analog instrumentation equipment, where accurately measuring an external signal is the goal of the product.

Moreover, with computers and networking equipment of the 21$^{st}$ century running at such high frequencies, digital designs are now in the RF and microwave portion of the spectrum. It is now crucial for digital designers to understand electromagnetic fields, radiation, and transmission lines. This knowledge is necessary for maintaining signal integrity and for achieving EMC compliance. High-speed digital signals radiate more easily, which can cause interference with nearby equipment. High-speed signals also more often cause circuits within the same design to interfere with one another (i.e., crosstalk). Circuit traces can no longer be considered as ideal short circuits. Instead, every trace should be considered as a transmission line because reflections on long traces can distort the digital waveforms. The Internet and the never-ending quest for higher bandwidth are pushing the speed of digital designs higher and higher. Web commerce and applications such as streaming audio and video will continue to increase consumer demand for higher bandwidth. Likewise, data traffic and audio and video conferencing will do the same for businesses. As we enter the realm of higher frequencies, digital designs are no longer a matter of just ones and zeros.

Understanding electromagnetics is vitally important for RF (radio frequency) design, where the approximations of electrical circuit theory start to break down. Traditional viewpoints of electronics (electrons flowing in circuits like water in a pipe) are no longer sufficient for RF designs. RF design has long been considered a "black art," but it is time to put that myth to rest. Although RF design is quite different from low-frequency design, it is not very hard to understand for any electrical engineer. Once you understand the basic concepts and gain an intuition for how electromagnetic waves and fields behave, the mystery disappears.

Optics has become essential to communication networks. Fiber optics are already the backbone of telecommunications and data networks. As we exhaust the speed limits of electronics, optical interconnects and possibly optical computing will start to replace electronic designs. Optical techniques can work at high speeds and are well suited to parallel operations, providing possibilities for computation rates that are orders of magnitude faster than electronic computers. As the digital age progresses, many of us will become "light engineers," working in the world of photonics. Certainly optics is a field that will continue to grow.

## 20.2 The Electromagnetic Spectrum

For electrical engineers the word *electromagnetics* typically conjures up thoughts of antennas, transmission lines, and radio waves, or maybe boring lectures and "all-nighters" studying for exams. However, this electrical word also describes a broad range of phenomena in addition to electronics, ranging from X-rays to optics to thermal radiation. In physics courses, we are taught that all these phenomena concern electromagnetic waves. Even many nontechnical people are familiar with this concept and with the electromagnetic spectrum, which spans from electronics and radio frequencies through infrared, visible light, and then on to ultraviolet and X-rays. We are told that these waves are all the same except for frequency. However, most engineers find that even after taking many physics and engineering courses, it is still difficult to see much commonality across the electromagnetic spectrum other than the fact that all are waves and are governed by the same mathematics (Maxwell's equations). Why is visible light so different from radio waves? I certainly have never encountered electrical circuits or antennas for visible light. The idea seems absurd. Conversely, I have never seen FM radio or TV band lenses for sale. So why do light waves and radio waves behave so differently?

Of course the short answer is that it all depends on frequency, but on its own this statement is of little utility. Here is an analogy. From basic chemistry, we all know that all matter is made of atoms, and that atoms contain a nucleus of protons and neutrons with orbiting electrons. The characteristics of each element just depend on how many protons the atom has. Although this statement is illuminating, just knowing the number of protons in an atom doesn't provide much more than a framework for learning about chemistry. Continuing this analogy, the electromagnetic spectrum as shown in Figure 20.1 provides a basic framework for understanding electromagnetic waves, but there is a lot more to learn.

To truly understand electromagnetics, it is important to view different problems in different ways. For any given frequency of a wave, there is also a corresponding wavelength, time period, and quantum of energy. Their definitions are given below, with their corresponding relationships in free space.

frequency, $f$, the number of oscillations per second wavelength, $\lambda$, the distance between peaks of:

$$\lambda = \frac{c}{f}$$

time period, $T$, the time between peaks of a wave:

$$T = \frac{1}{f}$$

photon energy, $E$, the minimum value of energy that can be transferred at this frequency:

$$E = h \times f$$

where $c$ equals the speed of light and $h$ is Planck's constant.

Depending on the application, one of these four interrelated values is probably more useful than the others. When analyzing digital transmission lines, it helps to compare the signal rise time to the signal transit time down the transmission line. For antennas, it is usually most intuitive to compare the wavelength of the signal to the antenna length. When examining the resonances and relaxation of dielectric materials it helps to compare the frequency of the waves to the resonant frequency of the material's microscopic dipoles. When dealing with infrared, optical, ultraviolet, and x-ray interactions with matter, it is often most useful to talk about the energy of each photon to relate it to the

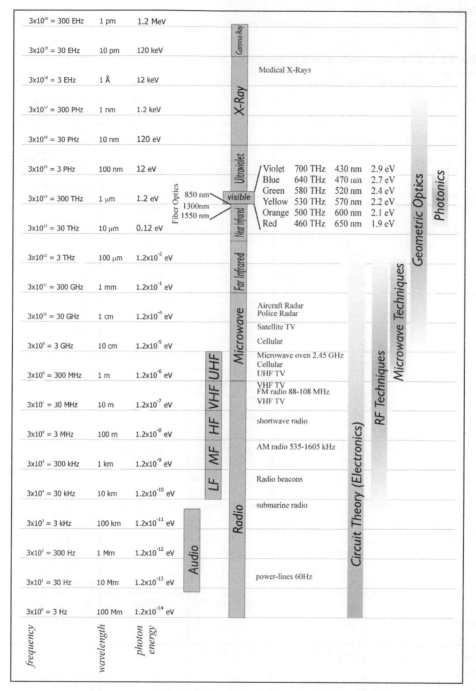

**Figure 20.1: The electromagnetic spectrum**

**Table 20.1: Characteristics of electromagnetic waves at various frequencies**

| Frequency | Wavelength | Photon Energy | Period | Copper Skin Depth | Copper Propagation Phase Angle |
|---|---|---|---|---|---|
| 60 Hz *Power line frequency* | 5000 km | $2.48 \times 10^{-13}$ eV | 16.7 ms | 8.4 mm | 45° (conductor) |
| 440 Hz *audio* | 681 km | $1.82 \times 10^{-12}$ eV | 2.27 ms | 3.1 mm | 45° (conductor) |
| 1 MHz *AM radio* | 300 km | $4.14 \times 10^{-9}$ eV | 1.00 μs | 65 μm | 45° (conductor) |
| 100 MHz *FM radio* | 3.00 m | $4.14 \times 10^{-7}$ eV | 10.0 ns | 6.5 μm | 45° (conductor) |
| 2.45 GHz *Microwave oven* | 12.2 cm | $1.01 \times 10^{-7}$ eV | 40.8 ps | 1.3 μm | 45° (conductor) |
| 160 GHz *Cosmic background radiation ("Big Bang") Peak* | 1.87 mm | $6.62 \times 10^{-4}$ eV | 6.25 ps | 0.16 μm | 46° (conductor) |
| 4.7 THz *Relaxation resonance of Copper* | 63.8 μm | $1.94 \times 10^{-2}$ eV | 213 fs | 27.3 nm | 68° |
| 17.2 THz *Room temperature Blackbody infrared peak* | 17.4 μm | $7.11 \times 10^{-2}$ eV | 5.81 fs | 21.8 nm | 82° |
| 540 THz *Center of visible band* | 555 nm | 2.23 eV | 1.85 fs | 21.8 nm | 90° (reflecting plasma) |
| 5000 THz *Ultraviolet* | 60.0 nm | 20.7 eV | 0.60 fs | 89 μm | 90° (transparent plasma) |
| $1 \times 10^{-7}$ THz *Diagnostic x-ray* | 30 pm | $4.14 \times 10^{-4}$ eV | $1.00 \times 10^{-19}$ s | 400 m | 90° (transparent plasma) |
| $1 \times 10^{-8}$ THz *Gamma ray from $^{198}Hg$ nucleus* | 3.0 pm | $4.15 \times 10^{-5}$ eV | $1.00 \times 10^{-20}$ s | 40 km | 90° (transparent plasma) |

Table 20.1: Continued

| Dipole Radiation Field Border | Blackbody Characteristic Radiation Temperature | Photon Rate for 1 mW Source | Aperture for Human Quality Imaging | Aperture for Minimal Quality Imaging |
|---|---|---|---|---|
| 795 km | <1°K | $2.5 \times 10^{28}$ photons/s | $2.7 \times 10^{10}$ m | $7.0 \times 10^{7}$ m |
| 108 km | <1°K | $3.4 \times 10^{27}$ photons/s | $3.7 \times 10^{9}$ m | $9.5 \times 10^{6}$ m |
| 47.7 m | <1°K | $1.5 \times 10^{24}$ photons/s | $1.6 \times 10^{6}$ m | 4200 m |
| 47.7 m | <1°K | $1.5 \times 10^{22}$ photons/s | 1600 m | 42 m |
| 1.95 cm | <1°K | $6.2 \times 10^{20}$ photons/s | 660 m | 1.7 m |
| 298 μm | 2.72°K (temperature of outer space) | $9.4 \times 10^{18}$ photons/s | 10 m | 2.6 cm |
| 40.2 μm | 80°K | $3.2 \times 10^{17}$ photons/s | 35 cm | 0.89 mm |
| 2.77 μm | 20°C | $8.8 \times 10^{16}$ photons/s | 9.4 cm | 0.24 mm |
| 88.4 nm | 9440°K | $2.8 \times 10^{15}$ photons/s | 3.0 mm | 7.8 μm |
| 9.54 nm | 85,000°K | $3.0 \times 10^{14}$ photons/s | 0.32 mm | 840 nm |
| 4.77 pm | $1.7 \times 10^{8}$°K | $1.5 \times 10^{11}$ photons/s | 160 nm | 420 pm |
| 0.477 pm | $1.7 \times 10^{9}$°K | $1.5 \times 10^{10}$ photons/s | 16 nm | 42 pm |

orbital energy of electrons in atoms. Table 20.1 lists these four values at various parts of the electromagnetic spectrum, and also includes some other relevant information. If some of these terms are unfamiliar to you, don't fret—they'll be explained as you progress through this chapter.

## 20.3 Electrical Length

An important concept to aid understanding of electromagnetics is electrical length. Electrical length is a unitless measure that refers to the length of a wire or device at a certain frequency. It is defined as the ratio of the physical length of the device to the wavelength of the signal frequency:

$$\text{Electrical length} = \frac{L}{\lambda}$$

As an example, consider a 1-meter long antenna. At 1 kHz this antenna has an electrical length of about $3 \times 10^{-16}$. An equivalent way to say this is in units of wavelength; that is, a 1 meter antenna is $3 \times 10^{-16}\lambda$ long at 1 kHz. At 1 kHz this antenna is electrically short. However, at 100 MHz, the frequency of FM radio, this antenna has an electrical length of 0.3 and is considered electrically long. In general, any device in which the electrical length is less than about 1/20 can be considered electrically short. (Beware: When working with wires that have considerable loss or large impedance mismatches, even electrical lengths of 1/50 may not be electrically short.) Circuits that are electrically short can in general be fully described by basic circuit theory without any need to understand electromagnetics. On the other hand, circuits that are electrically long require RF techniques and knowledge of electromagnetics.

At audio frequencies and below (<20 kHz), electromagnetic waves have very long wavelengths. The wavelength is typically much larger than the length of any of the wires in the circuit used. (An exception would be long telephone lines.) *When the wavelength is much* longer *than the wire lengths, the basic rules of electronic circuits apply and electromagnetic theory is not necessary.*

## 20.4 The Finite Speed of Light

Another way of looking at low-frequency circuitry is that the period (the inverse of frequency) of the waves is much larger than the delay through the wires. "What delay in the wires?" you might ask. When we are involved in low-frequency circuit design it is easy to forget that the electrical signals are carried by waves and that they must travel at the speed of light, which is very fast (about 1 foot/ns on open air wires), but not infinite. So, even when you turn on a light switch there is a delay before the light bulb receives the voltage. The same delay occurs between your home stereo and its speakers. This delay is typically too small for humans to perceive, and is ignored whenever you approximate a wire as an ideal short circuit. The speed of light delay also occurs in telephone lines, which can produce noticeable echo (>50 ms) if the connection spans a large portion of the earth or if a satellite feed is used. Long distance carriers use echo-cancellation electronics for international calls to suppress the effects. The speed of light delay becomes very important when RF or high-speed circuits are being designed. For example, when you are designing a digital system with 2 ns rise-times, a couple feet of cable amounts to a large delay.

## 20.5 Electronics

Electronics is the science and engineering of systems and equipment that utilize the flow of electrons. Electrons are small, negatively charged particles that are free to move about inside conductors such as copper and gold. Because the free electrons are so plentiful inside a conductor, we can often approximate electron flow as fluid flow. In fact, most of us are introduced to electronics using the analogy of (laminar) flow of water through a pipe. Water pressure is analogous to electrical voltage, and water flow rate is analogous to electrical current. Frictional losses in the pipe are analogous to electrical resistance. The pressure drop in a pipe is proportional to the flow rate multiplied by the frictional constant of the pipe. In electrical terms, this result is Ohm's law. That is, the voltage drop across a device is equal to the current passing through the device multiplied by the resistance of the device:

*Ohm's law*: $V = I \cdot R$

Now imagine a pump that takes water and forces it through a pipe and then eventually returns the water back to the tank. The water in the tank is considered to be at zero potential—analogous to an electrical ground or common. A pump is connected to the water tank. The pump produces a pressure increase, which causes water to flow. The pump is like a voltage source. The water flows through the pipes, where frictional losses cause the pressure to drop back to the original "pressure potential." The water then returns to the tank. From the perspective of energy flow, the pump sources energy to the water, and then in the pipes all of the energy is lost due to friction, converted to heat in the process. Keep in mind that this analogy is only an approximation, even at DC.

Basic circuit theory can be thought of in the same manner. The current flows in a loop, or circuit, and is governed by Kirchhoff's laws (as shown in Figures 20.2 and 20.3). Kirchhoff's voltage law (KVL) says that the voltages in any loop sum to zero. In other words, for every voltage drop in a circuit there must be a corresponding voltage source. Current flows in a circle, and the total of all the voltage sources in the circle or circuit is always equal to the total of all the voltage sinks (resistors, capacitors, motors, etc.). KVL is basically a consequence of the conservation of energy.

Kirchhoff's current law (KCL) states that when two or more branches of a circuit meet, the total current is equal to zero. This is just conservation of current. For example, if 5 A

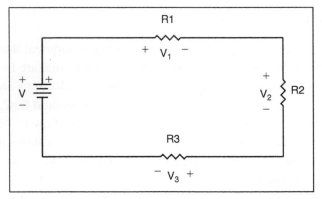

**Figure 20.2: A simple circuit demonstrating Kirchhoff's voltage law ($V = V_1 + V_2 + V_3$)**

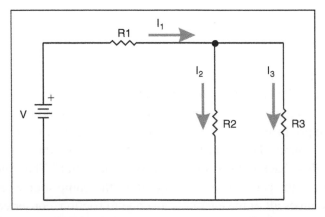

**Figure 20.3: A simple circuit demonstrating Kirchhoff's current law ($I_1 = I_2 + I_3$)**

is coming into a node through a wire, then 5 A must exit the node through another wire(s). In our water tank analogy, this law implies that no water can leave the system. Current can't just appear or disappear.

Additional rules of basic circuit theory are that circuit elements are connected through ideal wires. Wires are considered perfect conductors with no voltage drop or delay. The wires between components are therefore all considered to be at the same voltage potential and are referred to as a node. This concept often confuses the beginning student of electronics. For an example, refer to Figure 20.4. In most schematic diagrams, the wire connections are in fact considered to be ideal. This method of representing electronic circuits is termed "lumped element" design.

The ironic thing about this is that the beginning student is taught to ignore the shape and length of wires, but at RF frequencies the length and shape of the wires become just as important as the components. Engineering and science are filled with similar situations where you must develop a simplified understanding of things before learning all the exceptions and details. Extending the resistance concept to the concept of AC (alternating current) impedance allows you to include capacitors and inductors. That is circuit theory in a nutshell. There are no antennas or transmission lines. We can think of the circuit as electrons flowing through wires like water flowing through a pipe. Electromagnetics is not needed.

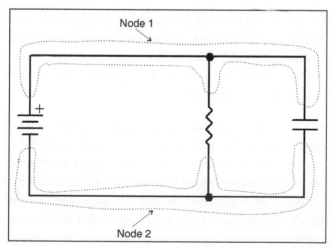

**Figure 20.4: A simple circuit demonstrating the voltage node principle. The voltage is the same everywhere inside each of the dotted outlines**

## 20.6 Analog and Digital Signals

Electronics is typically divided into the categories of analog and digital. Analog signals are continuously varying signals such as audio signals. Analog signals typically occupy a specific bandwidth and can be decomposed in terms of sinusoids using Fourier theory. For example, signals carrying human voice signals through the telephone network occupy the frequency band from about 100 Hz to about 4000 Hz.

Digital signals, on the other hand, are a series of ones and zeroes. A typical method to represent a digital signal is to use 5 V for a one and 0 V for a zero. A digital clock signal is shown as an example in Figure 20.5. Fourier theory allows us to create such a square wave by summing individual sine waves. The individual sine waves are at multiples or harmonics of the clock frequency.* To create a perfectly square signal (signal rise and fall times of zero) requires an infinite number of harmonics, spanning to infinite frequency. Of course, this is impossible in reality, so all real digital signals must have rise and fall times greater than zero. In other words, no real digital signal is perfectly square. *When performing transmission line and radiation analysis for digital designs, the rise and fall times are the crucial parameters.*

## 20.7 RF Techniques

At higher frequencies, basic circuit theory runs into problems. For example, if wires are electrically long, transmission line effects can occur. (Note that rock musicians may find it interesting to know that the signal of an electric guitar with distortion looks very similar to Figure 20.5. The distortion effect for guitars is created by "squaring off" the sine waves from the guitar, using a saturated amplifier.)

The basic theory no longer applies because electromagnetic wave reflections bouncing back and forth along the wires cause problems. These electromagnetic wave reflections can cause constructive or destructive interference resulting in the breakdown of basic circuit theory. In fact, when a transmission line has a length equal to one quarter wavelength of the signal, a short placed at the end will appear as an open circuit at the other end! Certainly, effects like this cannot be ignored. Furthermore, at higher frequencies, circuits can radiate energy much more readily; that is circuits can turn into antennas. Parasitic capacitances and inductances can cause problems too. No component can ever be truly ideal. The small inductance of component leads and wires can cause

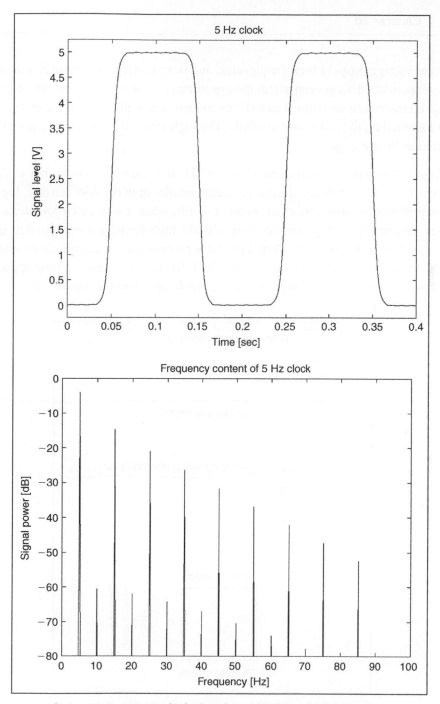

**Figure 20.5: A 5-Hz clock signal and its frequency content**

significant voltage drops at high frequencies, and stray capacitances between the leads of the component packages can affect the operation of a high-frequency circuit. These parasitic elements are sometimes called "the hidden schematic" because they typically are not included on the schematic symbol. (The high-frequency effects just mentioned are illustrated in Figure 20.6.)

How do you define the high-frequency regime? There is no exact border, but when the wavelengths of the signals are similar in size or smaller than the wire lengths, high-frequency effects become important; in other words, when a wire or circuit element becomes electrically long, you are dealing with the high-frequency regime. An equivalent way to state this is that when the signal period is comparable in magnitude or smaller than the delay through the interconnecting wires, high-frequency effects become apparent. *It is important to note that for digital signals, the designer must compare the rise and*

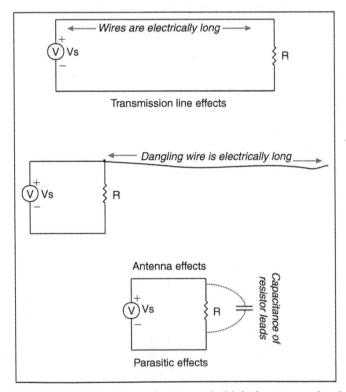

**Figure 20.6: Some effects that occur in high-frequency circuits**

*fall times of the digital signal to the wire delay.* For example, a 10-MHz digital clock signal may only have a signal period of 100 ns, but its rise time may be as low as 5 ns. Hence, the RF regime doesn't signify a specific frequency range, but signifies frequencies where the rules of basic circuit theory breakdown. *A good rule of thumb is that when the electrical length of a circuit element reaches 1/20, RF (or high-speed digital) techniques may need to be used.*

When working with RF and high-frequency electronics it is important to have an understanding of electromagnetics. At these higher frequencies, you must understand that the analogy of electrons acting like water through a pipe is really more of a myth than a reality. In truth, circuits are characterized by metal conductors (wires) that serve to guide electromagnetic energy. The circuit energy (and therefore the signal) is carried between the wires, and not inside the wires. For an example, consider the power transmission lines that deliver the electricity to our homes at 60 Hz. The electrons in the wires do not directly transport the energy from the power plant to our homes. On the contrary, the energy is carried in the electromagnetic field between the wires. This fact is often confusing and hard to accept for circuit designers. The wire electrons are not experiencing any net movement. They just slosh back and forth, and through this movement they propagate the field energy down the wires. A good analogy is a "bucket brigade" that people sometimes use to fight fires. A line of firefighters (analogous to the electrons) is set up between the water source (signal source) and the fire (the load). Buckets of water (the electromagnetic signal) are passed along the line from firefighter to firefighter. The water is what puts out the fire. The people are just there to pass the water along. In a similar manner, the electrons just serve to pass the electromagnetic signal from source to load. This statement is true at all electronic frequencies, DC, low frequency, and RF.

## 20.8 Microwave Techniques

At microwave frequencies in the GHz range, circuit theory is no longer very useful at all. Instead of thinking about circuits as electrons flowing through a pipe, it is more useful to think about circuits as structures to guide and couple waves. At these high frequencies, lumped elements such as resistors, capacitors, and inductors are often not viable. As an example, the free space wavelength of a 30-GHz signal is 1 cm. Therefore, even the components themselves are electrically long and do not behave as intended. Voltage, current,

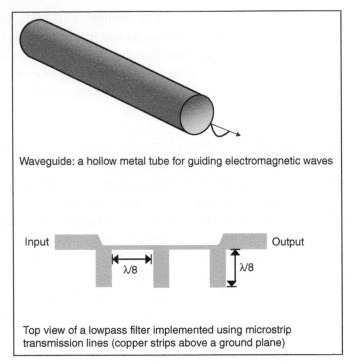

Waveguide: a hollow metal tube for guiding electromagnetic waves

Input                                                          Output

$\lambda/8$                         $\lambda/8$

Top view of a lowpass filter implemented using microstrip transmission lines (copper strips above a ground plane)

**Figure 20.7: Examples of microwave techniques**

and impedance are typically not used. In this realm, electronics starts to become similar to optics in that we often talk of power transmitted and reflected instead of voltage and current. Instead of impedance, reflection/transmission coefficients and S-parameters are used to describe electronic components. Some microwave techniques are shown in Figure 20.7.

## 20.9  Infrared and the Electronic Speed Limit

The infrared region is where the spectrum transitions from electronics to optics. The lower-frequency portion of the infrared is termed the "far infrared," and is the extension of the microwave region. Originally, the edge of the microwave band (300 MHz) was considered the highest viable frequency for electronics. As technology progresses, the limit of electronics extends further into the infrared. Wavelengths in the infrared are under 1 mm, implying that even a 1-mm wire is electrically long, readily radiating energy from electrical currents. Small devices are therefore mandatory.

At the time of writing this chapter, experimental integrated circuit devices of several terahertz (1012 Hz) had been achieved, and 40-GHz digital devices had become commercially available for communications applications. (Terahertz devices were created decades ago using vacuum tube techniques, but these devices are obviously not viable for computing devices.) Certainly digital devices in the hundreds of gigahertz will become commercially viable; in fact, such devices have already been demonstrated by researchers. Making digital devices past terahertz speeds will be a very difficult challenge. To produce digital waveforms, you need an amplifier with a bandwidth of at least 3 to 5 times the clock frequency. Already researchers are pursuing special semiconductors such as Indium Phosphide (InP) electron spin, single-electron, and quantum devices, as well as molecular electronics. Only time will tell what the ultimate "speed limit" for electronics will be.

What is almost certain is that somewhere in the infrared frequencies, electronics will always be impossible to design. There are many problems in the infrared facing electronics designers. The speed of transistors is limited by their size; consequently, to probe higher frequencies, the state of the art in integrated circuit geometries must be pushed to smaller and smaller sizes. Quantum effects, such as tunneling, also cause problems. Quantum tunneling allows electrons to pass through the gate of very small MOSFET transistors. This effect is a major problem facing researchers trying to further shrink CMOS technology. Furthermore, the properties of most materials begin to change in the infrared. The conductive properties of metals begin to change. In addition, most dielectric materials become very lossy. Even dielectrics that are transparent in the visible region, such as water and glass, become opaque in the portions of the infrared. Photons in the infrared are very energetic compared to photons at radio frequencies and below. Consequently, infrared photons can excite resonant frequencies in materials. Another characteristic of the infrared is that the maximum of heat radiation occurs in the infrared for materials between room temperature (20°C) and several thousand degrees Celsius. These characteristics cause materials to readily absorb and emit radiation in the infrared. For these reasons, we can readily feel infrared radiation. The heat we feel from incandescent lamps is mostly infrared radiation. It is absorbed very easily by our bodies.

## 20.10 Visible Light and Beyond

At the frequencies of visible light, many dielectrics become less lossy again. Materials such as water and glass that are virtually lossless with respect to visible light are therefore transparent. Considering that our eyes consist mostly of water, we are very fortunate that water is visibly transparent. Otherwise, our eyes, including the lens, would be opaque and quite useless. A striking fact of nature is that the absorption coefficient of water rises more than 7 decades (a factor of 10 million) in magnitude on either side of the visible band. So it is impossible to create a reasonably sized, water-based eye at any other part of the spectrum. All creatures with vision exploit this narrow region of the spectrum. Nature is quite amazing!

At visible frequencies, the approximations of geometric optics can be used. These approximations become valid when the objects used become much larger than a wavelength. This frequency extreme is the opposite of the circuit theory approximations. The approximation is usually called *ray theory* because light can be approximated by rays or streams of particles. Isaac Newton was instrumental in the development of geometric optics, and he strongly argued that light consisted of particles and not waves. The physicist Huygens developed the wave theory of light and eventually experimental evidence proved that Huygens was correct. However, for geometrical optics, Newton's theory of particle streams works quite well. An example of geometrical optics is the use of a lens to concentrate or focus light. Figure 20.8 provides a lens example. Most visible phenomena, including our vision, can be studied with geometrical optics. The wave theory of light is usually needed only when studying diffraction (bending of light around corners) and coherent light (the basis for lasers). Wave theory is also needed to explain the resolution limits of optical imaging systems. A microscope using visible light can only resolve objects down to about the size of a wavelength.

At the range of ultraviolet frequencies and above (x-rays, etc.) each photon becomes so energetic that it can kick electrons out of their atomic orbit. The electron becomes free and the atom becomes ionized. Molecules that absorb these high-energy photons can lose the electrons that bond the molecules together. Ions and highly reactive molecules

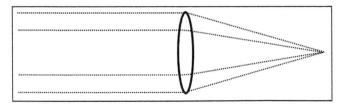

**Figure 20.8: A lens that focuses rays of light**

called *free radicals* are produced. These highly reactive ions and molecules cause cellular changes and lead to biological tissue damage and cancer. Photons of visible and infrared light, on the other hand, are less energetic and only cause molecular heating. We feel the heat of the infrared radiation from the sun. We see the light of visible radiation from the sun. Our skin is burned and damaged by the ultraviolet radiation from the sun.

X-ray photons, being higher in energy, are even more damaging. Most materials are to some degree transparent to x-rays, allowing the use of x-ray photography to "see through" objects. But when x-rays are absorbed, they cause cellular damage. For this reason, limited x-ray exposure is recommended by physicians. The wavelengths of high-energy x-rays are about the same size as the atomic spacing in matter. Therefore, to x-rays, matter cannot be approximated as continuous, but rather is "seen" as lumps of discrete atoms. The small wavelength makes x-rays useful for studying crystals such as silicon, using the effects of diffraction. Above x-rays in energy are gamma rays and cosmic rays. These extremely high-energy waves are produced only in high-energy phenomena such as radioactive decay, particle physics collisions, nuclear power plants, atomic bombs, and stars.

## 20.11 Lasers and Photonics

Electronic circuits can be created to transmit, amplify, and filter signals. These signals can be digital bits or analog signals such as music or voices. The desire to push electronics to higher frequencies is driven by two main applications: computers and communication links. For computers, higher frequencies translate to faster performance. For communication links, higher frequencies translate to higher bandwidth. Oscillator circuits serve as timing for both applications. Computers are in general synchronous and require a clock signal. Communications links need a carrier signal to modulate the information for transmission. Therefore, a basic need to progress electronics is the ability to create oscillators.

In the past few decades, photonics has emerged as an alternative to electronics, mostly in communication systems. Lasers and fiber optic cables are used to create and transmit pulses of a single wavelength (frequency) of light. In the parlance of optics, single-frequency sources are known as *coherent* sources. Lasers produce synchronized or coherent photons; hence, the name photonics. The light that we encounter every day from the sun and lamps is noncoherent light. If we could look at this light on an oscilloscope, it would look like noise. In fact, the visible light that we utilize for our vision is noise—the thermal noise of hot objects such as the sun or the filament in a light bulb. The electrical term *white noise* comes from the fact that optical noise contains all the visible colors (frequencies) and appears white. The white noise of a light bulb extends down to electronic frequencies and is the same white noise produced by resistors and inherent in all circuits. Most imaging devices, like our eyes and cameras, only use the average squared-field amplitude of the light received. (Examination at the quantum level reveals imaging devices to be photon detectors/counters.) Averaging allows us to use "noisy" signals for vision, but because of averaging all phase information is lost. To create sophisticated communication devices, such light is not suitable. Instead the coherent, single-frequency light of lasers is used. Lasers make high-bandwidth fiber optic communication possible.

Until recently, the major limitation of photonics was that the laser pulsed signals eventually had to be converted to electronic signals for any sort of processing. For instance, in data communications equipment, major functions include the switching, multiplexing, and routing of data between cables. In the past, only electronic signals could perform these functions. This requirement limited the bandwidth of a fiber optic cable to the maximum available electronic bandwidth. However, with recent advances in optical multiplexing and switching, many tasks can now be performed completely using photonics. The upshot has been an exponential increase in the data rates that can be achieved with fiber optic technology. The ultimate goal for fiber optics communication is to create equipment that can route Internet protocol (IP) datapackets using only photonics. Such technology would also lead the way for optical computing, which could provide tremendous processing speeds as compared with electronic computers of today.

## 20.12  Summary of General Principles

Different techniques and approximations are used in the various portions of the electromagnetic spectrum. Basic circuit theory is an approximation made for low-frequency electronics. The circuit theory approximations work when circuits are electrically small. In other words, circuit theory is the limit of electromagnetics as the wavelength becomes infinitely larger than the circuit. RF theory takes circuit theory and adds in some concepts and relations from electromagnetics. RF circuit theory accounts for transmission line effects in wires and for antenna radiation. At microwave frequencies it becomes impossible to design circuits with lumped elements like resistors, capacitors, and inductors because the wavelengths are so small. Distributed techniques must be used to guide and process the waves. In the infrared region, we can no longer design circuits. The wavelengths are excessively small, active elements like transistors are not possible, and most materials become lossy, readily absorbing and radiating any electromagnetic energy. At the frequencies of visible light, the wavelengths are typically much smaller than everyday objects, and smaller than the human eye can notice. In this range, the approximations of geometrical optics are used. Geometrical optics is the limit of electromagnetic theory where wavelength becomes infinitely smaller than the devices used. At frequencies above light, the individual photons are highly energetic, able to break molecular bonds and cause tissue damage.

With the arrival of the information age, we rely on networked communications more and more every day, from our cell phones and pagers to our high-speed local-area networks (LANs) and Internet connections. The hunger for more bandwidth consistently pushes the frequency and complexity of designs. The common factor in all these applications is that they require a good understanding of electromagnetics.

## 20.13 The Electric Force Field

To understand high-frequency and RF electronics, you must first have a good grasp of the fundamentals of electromagnetic fields. This section discusses the electric field and is the starting place for understanding electromagnetics. Electric fields are created by charges; that is, charges are the source of electric fields. Charges come in two types, positive (+) and negative (−). Like charges repel each other and opposites attract. In other words, charges produce a force that either pushes or pulls other charges away. Neutral objects are not affected. The force between two charges is proportional to the product of the two charges, and is called *Coulomb's law*. Notice that the charges produce a force on each other without actually being in physical contact. It is a force that acts at a distance. To represent this "force at distance" that is created by charges, the concept of a *force field* is used. Figure 20.9 shows the electrical force fields that surround positive and negative charges.

By convention, the electric field is always drawn from positive to negative. It follows that the force lines emanate from a positive charge and converge to a negative charge. Furthermore, the electric field is a normalized force, a force per charge. The normalization allows the field values to be specified independent of a second charge. In other words, the value of an electric field at any point in space specifies the force that

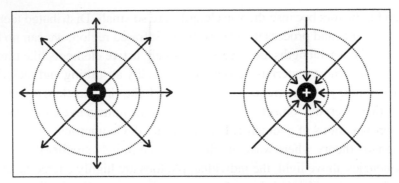

**Figure 20.9: Field lines surrounding a negative and a positive charge. Dotted lines show lines of equal voltage**

would be felt if a unit of charge were to be placed there. (A unit charge has a value of 1 in the chosen system of units.)

*Electric field = Force field as "felt" by a unit charge*

To calculate the force felt by a charge with value, $q$, we just multiply the electric field by the charge:

$$\vec{F} = q \cdot \vec{E}$$

The magnitude of the electric field decreases as you move away from a charge, and increases as you get closer. To be specific, the magnitude of the electric field (and magnitude of the force) is proportional to the inverse of the distance squared. The electric field drops off rather quickly as the distance is increased. Mathematically this relation is expressed as:

$$E = \frac{q}{r^2}$$

where $r$ is the distance from the source and $q$ is the value of the source charge. Putting our two equations together gives us Coulomb's law,

$$F = \frac{q_1 \cdot q_2}{r^2}$$

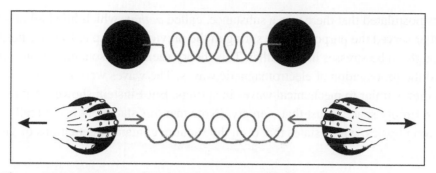

**Figure 20.10: Two balls attached by a spring. The spring exerts an attractive force when the balls are pulled apart**

where $q_1$ and $q_2$ are the charge values and $r$ is the distance that separates them. Electric fields are only one example of fields.

## 20.14 Other Types of Fields

Gravity is another field. The gravitational force is proportional to the product of the masses of the two objects involved and is always attractive. (There is no such thing as negative mass.) The gravitational field is much weaker than the electric field, so the gravitational force is only felt when the mass of one or both of the objects is very large. Therefore, our attraction to the earth is big, while our attraction to other objects like furniture is exceedingly small.

Another example of a field is the stress field that occurs when elastic objects are stretched or compressed. For an example, refer to Figure 20.10. Two balls are connected by a spring. When the spring is stretched, it will exert an attractive force on the balls and try to pull them together. When the spring is compressed, it will exert a repulsive force on the balls and try to push them apart. Now imagine that you stretch the spring and then quickly release the two balls. An oscillating motion occurs. The balls move close together, then far apart and continue back and forth. The motion does not continue forever though, because of friction. Through each cycle of oscillation, the balls lose some energy until they eventually stop moving completely. The causes of fiction are the air surrounding the balls and the internal friction of the spring. The energy lost to friction becomes heat in the air and spring. Before Einstein and his theory of relativity, most scientists thought that the electric field operated in a similar manner. During the 1800s, scientists postulated that there was a substance, called *aether*, which filled all of space. This aether served the purpose of the spring in the previous example. Electric fields were thought to be stresses in the aether. This theory seemed reasonable because it predicted the propagation of electromagnetic waves. The waves were just stress waves in the aether, similar to mechanical waves in springs. But Einstein showed that there was no aether. Empty space is just that—empty. (Note that this statement is not really true in quantum physics, which states that even the vacuum contains fluctuations of virtual particles.)

Without any aether, there is no way to measure absolute velocity. All movement is therefore relative.

## 20.15 Voltage and Potential Energy

A quantity that goes hand in hand with the electric field is *voltage*. Voltage is also called *potential*, which is an accurate description since voltage quantifies potential energy. Voltage, like the electric field, is normalized per unit charge.

*Voltage = Potential energy of a unit charge*

In other words, multiplying voltage by charge gives the potential energy of that charge, just as multiplying the electric field by charge gives the force felt by the charge. Mathematically, we represent this by:

$$U = q \cdot V$$

Potential energy is always a relative term; therefore, voltage is always relative. Gravity provides a great visual analogy for potential. Let's define ground level as zero potential. A ball on the ground has zero potential, but a ball 6 feet in the air has a positive potential energy. If the ball were to be dropped from 6 feet, all of its potential energy will have been converted to kinetic energy (i.e., motion) just before it reaches the ground. Gravity provides a good analogy, but the electric field is more complicated because there are both positive and negative charges, whereas gravity has only positive mass. Furthermore, some particles and objects are electrically neutral, whereas all objects are affected by gravity. For instance, an unconnected wire is electrically neutral, therefore, it will not be subject to movement when placed in an electrical potential. (However, there are the secondary effects of electrostatic induction, which are described later in the chapter.)

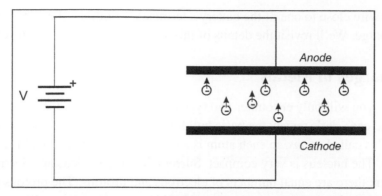

**Figure 20.11: A vacuum tube diode, showing electrons leaving the negatively biased cathode to combine with positive charge at the anode**

Consider another example, a vacuum tube diode, as shown in Figure 20.11. Two metals plates are placed in an evacuated glass tube, and a potential (10 V) is placed across them. The negative electrode is heated. The extra electrons in the negative electrode that constitute the negative charge are attracted to the positive charge in the positive electrode. The force of the electric field pulls electrons from the negative electrode to the positive electrode. (The heating of the electrode serves to "boil off" electrons into the immediate vicinity of the metal.) Once free, the electrons accelerate and then collide with the positive electrode where they are absorbed. Just before each electron collides, it is traveling very fast because of the energy gained from the electric field. Its kinetic energy can be easily calculated, in terms of electron-volts (eV) and in terms of joules (J):

$$U = q \cdot V = e \cdot (10 \text{ volts}) = 10 \text{ eV}$$
$$= 10(1.6 \times 10^{-9} C)V = 1.6 \times 10^{-18} \text{ J}$$

From this example, you can see how natural the unit of electron-volts (eV) is for describing the energy of an electron.

The most important thing for you to remember is that voltage is a relative term. On a 9 V battery, the (+) contact has a voltage of 9 volts relative to the (−) contact. Furthermore, the battery has a net charge of zero, although the charge is separated into negative and positive regions. The charge on the positive side is drawn to the charge on the negative side. Connecting a wire between the terminals allows the charges to recombine. The same result is true for a charged capacitor. What would happen if you brought a neutral unconnected wire close to one of the battery's terminals? Nothing, the wire is neutral. It has no net charge. We'll revisit the details of this situation a little later in the chapter.

## 20.16  Charges in Metals

In electronics you will only encounter two types of charged particles, electrons and ions. To understand each, let's review the basic building blocks of matter. Matter consists of tiny particles called *atoms*. In each atom is a core or nucleus that contains protons and neutrons. The nucleus is very compact. Surrounding the nucleus are electrons. For a neutral atom, there are equal numbers of electrons and protons. The protons possess positive charge, and the electrons possess an equal but opposite charge. The neutrons in the nucleus are neutral. The electrons orbit the nucleus in a special way. You might imagine the electron as a small ball orbiting the nucleus in the same way that planets orbit the sun. However, this analogy is not quite correct. Each electron is smeared out in a three-dimensional cloud called an *orbital*. Atoms can lend out or borrow electrons, which leaves the atom with a net charge. Such atoms are called *ions* and they can be positively charged (missing electrons) or negatively charged (extra electrons). Ions of opposite charge can attract one another and form ionic bonds. These bonded ions are called *molecules*. Table salt, NaCl, is a good example of molecules. Each salt molecule consists of a positive ion ($Na^+$) and a negative ion ($Cl^-$). There also exist other types of molecular bonds. For instance, covalent bonds are established when two atoms share an electron.

Metal materials have special properties that make them good conductors of electricity. First of all, metals are crystals; that is, metals have an orderly construction of atoms. Most people who have not studied solid-state or semiconductor physics find it very surprising that metals are crystals, because we tend to associate crystals with transparent materials like quartz. As with all crystals, the structure of a metal is a three-dimensional lattice like that shown in Figure 20.12. Metals have a rather interesting bonding structure. The positive ions of the metal are held together by a sea of electrons that is shared by the entire crystal. Each atom of the metal typically contributes one or two electrons to this sea. You can picture a metal as closely packed balls (the ions) in a gas of small particles (the electrons). Because the "sea" electrons are free to roam throughout the metal, they serve to conduct electricity quite well. It then makes sense to call them conduction electrons to differentiate them from the electrons that are bound to the ions.

At the microscopic level, a metal has a lot going on, even without any applied electric field. Thermal vibrations cause the lattice of ions to vibrate and cause the conduction

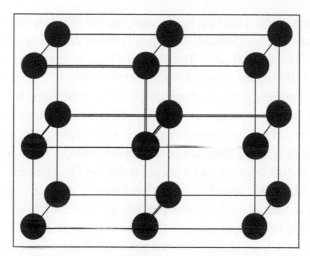

**Figure 20.12: Example of atoms arranged in a crystal lattice structure (simple cubic)**

electrons to move around. With higher temperatures, the vibrations in the metal get larger. This effect stems from the microscopic relationship of temperature. Temperature is proportional to the average kinetic energy of the particles,

$$T = \frac{m}{3k} v^2_{rms}$$

where $k$ is Boltzman's constant, $m$ is the atomic mass, and $v^2_{rms}$ is the root-mean-square particle velocity. (This formula applies directly to ideal gases, but it gives a decent approximation for the temperature of metals.) These thermal vibrations result in the random agitation of electrons and are sources of thermal noise in electronics. Every wire and resistor in a circuit injects white noise (i.e., with the same power level at all frequencies) from its internal thermal vibrations. Now if we place a voltage across a metal wire (or resistor), an electric field is developed through the wire. The electric field causes the conduction electrons to move more toward the positive (+) end of the wire. In other words, a current is produced. This current is a direct current (DC) so it has a constant value.

Current is defined as the amount of charge through a cross section of wire, per second. This definition implies that the charges are moving at a steady, constant value. However, this assumption is incorrect. Only the statistical average of the charge movement is a constant value. Each individual charge's movement is very random. The electrons are constantly colliding with the vibrating lattice ions and the other electrons. On average an electron in copper at room temperature only spends about 25 fs between collisions (1 fs equals 1/1,000,000,000,000,000 sec.). The electrons have a very large velocity between collisions, about 100,000 m/sec. However most (99.999999%) of the velocity is in a random direction. Statistically, the random components of the velocity cancel each other out, leaving a much smaller average velocity. Therefore, while the root-mean-square (RMS) velocity is very high, the average velocity is much lower. The average velocity of the electrons, which is in the direction of the applied field, is very small. This average velocity is called the *drift* velocity, and it accounts for the observed current. For a 1-m long copper wire with 1/10 V across it, the drift velocity is about 0.5 mm/s or 5 feet per hour! It will therefore take an electron about 38 minutes to travel from one side of the wire to the other. Assuming this copper wire is 20 gauge wire, its resistance is 33 mΩ, and the current from 0.1 V is 3 A.Even though the electrons travel very slowly, their effect adds up to large currents because there are so many of them. In a 1-m long 20 gauge copper wire, there are about $4.4 \times 10^{22}$ conduction electrons. Now that is strength in numbers!

## 20.17 The Definition of Resistance

Besides the flow of DC current, something else happens when you apply a voltage across a conductor: it heats up. The temperature of the conductor rises because DC flow causes the electrons to have higher-energy collisions (on average) than they would if the voltage were not present. Since it requires energy to heat something up, energy must be provided by the voltage supply. The energy per second lost to heat is exactly equal to the power calculated from the power law of electric circuits:

$$I^2 \cdot R = \frac{V^2}{R}$$

In a rather roundabout way, we have arrived at the definition of *resistance*. Resistance quantifies the power that is lost to heating when a voltage or current is applied to the conductor.

## 20.18 Electrons and Holes

Imagine that you have a solid metal ball or sphere and you place a charge on it. For a negative charge, this process equates to adding an excess of electrons to the sphere.

For a positive charge, this equates to removing some electrons, leaving an excess of positive ions or *holes*, as they are called in semiconductor parlance. Holes can be thought of as virtual positive particles, which can move around in a material like a bubble moving in water. Whereas negative charges (electrons) move freely about the material, holes must move by means of charge theft. Let's say there is an atom that is missing one electron. This atom is therefore a positive ion (+), or equivalently, this atom is carrying a hole. This ion can "steal" a bound electron from a neighbor. (Remember that not all atoms are ionized at all times.) By stealing an electron from its neighbor, it has in effect given the hole to the neighbor. This way holes can move through the material, and can be thought of as positive particles in their own right.

In semiconductors like silicon, the holes and electrons can move with approximately equal ease. In other words, the hole and electron mobilities are approximately equal in silicon. In most metals, however, the electrons are very mobile, and the holes have negligible mobility. The holes are virtually stationary and only the negatively charged electrons can move. Negative charge in metals is created when electrons move away from a region, leaving positively charged holes behind. Positive charge in metals should be thought of as a lack of free electrons. Most often, when we are not so formal, we do talk about positive charges moving about in a metal; just keep in mind that in most metals, it is always the negatively charged electrons that do the moving.

Back to the problem of the metal ball. When charge is placed onto the ball, the individual charges will immediately spread apart as far as possible because like charges repel each other. The upshot is that all the charge becomes concentrated at the surface. It's like a bunch of people in a large room. If each person tries to avoid the rest, they will migrate to the walls of the room, like a "wallflower" at a high school dance. How quickly charges distribute themselves to the surface is proportional to the *relaxation time* of the material. In the present context, the relaxation time can be approximated as the dielectric constant (discussed later in the chapter) of the material divided by the conductivity of the material. The relaxation time specifies how freely charge can move in a material. For copper, the relaxation constant is:

$$\tau = \frac{\varepsilon}{\sigma} = \frac{8.85 \times 10^{-12}\, \frac{F}{m}}{5.8 \times 10^{7}\, \frac{1}{\Omega \cdot m}} = 1.5 \times 10^{-19}\ \text{sec}$$

Therefore, charge placed on a copper object will very quickly redistribute so that it all resides on the surface. The charge half-life of a material is about 0.7 times the relaxation time. An example will help illustrate this concept. If charge is somehow placed at the center of a metal ball, the charge will immediately start to migrate toward the surface. After a half-life in time $(0.7\tau)$, half of the charge will have migrated away from the center.

## 20.19 Electrostatic Induction and Capacitance

To understand capacitance, you need to first understand the process of *electrostatic induction*. For example, consider that you have a metal ball that is positively charged, near which you bring a neutral metal ball. Even though the second ball has overall neutrality, it still contains many charges. Neutrality arises because the positive and negative charges exist in equal quantities. When placed next to the first ball, the second ball is affected by the electric field of the charged ball. The charges of the second ball separate. Negative charges are attracted, and positive charges are repelled, leaving the second ball polarized, as shown in Figure 20.13. This polarization of charge is called *electrostatic induction*.

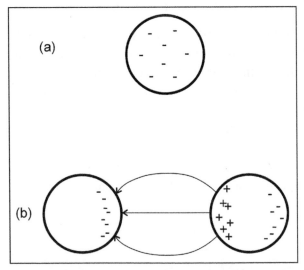

**Figure 20.13: (a) A negatively charged metal sphere; (b) A neutral sphere (right) is brought close to the negative sphere (left)**

A direct consequence of electrostatic induction is that the electric field inside an unconnected conductor is always zero at steady state. When a conductor is first placed in a field, the field permeates the conductor. The charges then separate as described in the preceding paragraph. The separation of charge tends to neutralize the electric field. Charge movement continues until the electric field reaches zero. Another way of stating this is that the voltage inside a conductor is constant at steady state. Placed in an ambient electric field, the conductor quickly adjusts its charge configuration until it has reached the voltage potential of its environment.

Now let's connect a metal object to the second ball using a wire. As shown in Figure 20.14, the charge polarizes even further, with the negative charge of the neutral objects moving as far away as possible. Figure 20.15 takes this one step further by connecting the earth to the second ball. (An earth connection can be achieved by connecting the ball to the third prong of a wall outlet, which typically is "earthed" on the outside of each

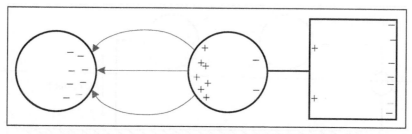

**Figure 20.14: A second neutral object is connected by a wire**

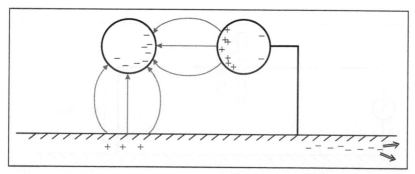

**Figure 20.15: The neutral sphere is connected to ground with a wire**

building using 8 foot or longer copper stakes.) Here the negative charge will move down the wire, into the earth, and go very far away.

Instead of placing a constant charge on the first metal ball, you could connect an oscillating charge, as in Figure 20.16. For example, assume that the AC voltage is at a frequency of 60 Hz; the polarization induced in the second ball will alternate at a frequency of 60 Hz. The alternating polarization will also cause current to flow in the wire that connects the second ball to the ground. What you have created is simply a *capacitor*! Any metal conductors that are separated by an insulator form a capacitor. In Figure 20.17, the two balls have been replaced with metal plates, forming a more familiar and efficient capacitor. Notice that no current actually traverses the gap between the plates, but equal current flows on both sides, as charge rushes to and from the plates of the capacitor. The virtual current that passes between the plates is called *displacement current*. It is really just a changing electric field, but we call it a current.

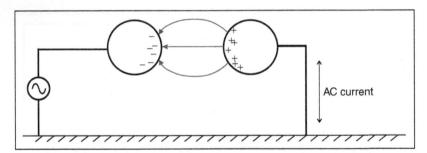

**Figure 20.16: An AC source is connected to the first sphere**

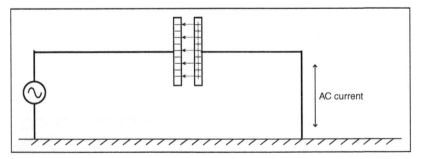

**Figure 20.17: Spheres are replaced by metal plates**

You can get a good feel for electrostatic induction by learning how some simple, but ingenious, inventions work. If you have ever worked on the electric *mains* wiring that runs in the walls of your house and provides the 120V power to your electric outlets and lamps, you have probably purchased what is called a *noncontact field detector*. (Incidentally, the term *mains* refers to the fact that your power is provided by your electric utility company. The same term is also used for the water that enters your house from the water utility.) A noncontact field detector is a device that is about the size and shape of a magic marker. It allows you to determine if a wire is live (is connected to voltage) or not without actually making metallic contact. This nifty invention can be waved near insulated wires, or it can be inserted into an electric outlet. It detects live wires and live outlets using the phenomenon of electrostatic induction that you just learned about. A simple schematic of such a device is shown in Figure 20.18. A metal lead or plate is attached to the input of a high-impedance amplifier. This plate serves one side of a capacitor. When the plate is brought close to an object that has an electric field, charge is induced on the plate. Some of the charge has to pass through the amplifier. Because the amplifier has high input impedance, a reasonable voltage will be created at its output. If the object being tested has a varying electric field, like a wall outlet, an AC current will be induced in the device. The rest of the circuitry serves to light an LED when the induced current varies continuously at around 60 Hz.

A variation of this circuit uses a human as the capacitive plate of the device. Glow tube meters and zero-pressure "touch" buttons work in this manner. A glow tube meter is a device for testing electrical outlets. It looks like a screwdriver, except the handle is made of clear plastic and contains a small glow tube inside. The glow tube is a glass tube filled

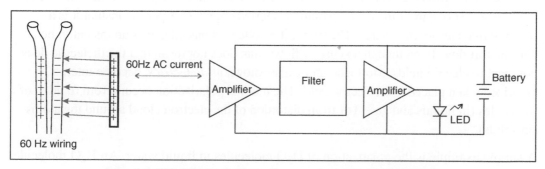

**Figure 20.18: Block diagram for a noncontact field detector**

with neon gas and contains two separated electrodes. One electrode is connected to the screwdriver blade and makes contact inside the electric outlet. The other electrode is connected to a piece of metal at the end of the screwdriver handle. By touching your finger to the handle end, you become part of a circuit. Being a decent conductor, your body serves the function of being one half of a capacitor. A small amount of the 60-Hz displacement current ($-10\,\mu A$) is able to capacitively couple from your body to the ground wires in the wall. This small current causes a voltage drop across you and across the glow tube. The gas inside the glow tube ionizes and conducts electricity. In the process, it gives off a faint orange glow, telling you that the outlet is working. You certainly should use care with such a device, and be sure not to touch the screwdriver shaft directly.

Touch buttons are found in some elevators. These buttons are metal, and if you encounter one, you will notice that the button does not physically depress when you touch it. This button is connected to a high-impedance amplifier. When you touch the button, you again form a capacitor plate. In this application, your body couples 60-Hz energy from the wires in the elevator and the fluorescent lights on the ceiling to the metal touch pad. A high-impedance amplifier amplifies the current and determines that you have, for instance, touched the button for the third floor.

It may be tempting to think of these applications as antennas picking up 60-Hz radiation, but this idea is incorrect. Being electrically small wires, virtually no energy is radiated. In addition, you are standing well within the near field of the source.

## 20.20 Insulators (dielectrics)

To make an electronic circuit work, you not only need conductors, you also need insulators. Otherwise, every part of the circuit would be shorted together! A perfect insulator is a material that has no free charge. Therefore, if a voltage is placed across an insulator, no current will flow. Even though no current flows, that does not mean that the dielectric does not react to electric fields. Most insulators, also known as *dielectrics*, become internally polarized when placed in an electric field. The internal polarization occurs from rotation of molecules (in liquids and gases) or from distortion of the electron cloud around the atoms (in solids).

A simple example is the polarization of $H_2O$ molecules in liquid water. An $H_2O$ molecule is shown in Figure 20.19. Due to the structure of the molecule, the charge is not exactly

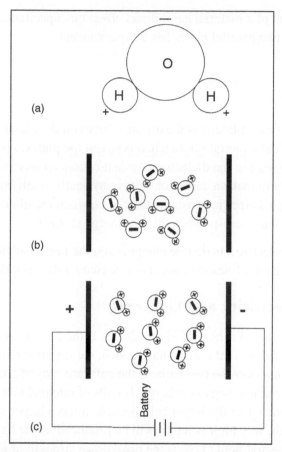

**Figure 20.19: (a) Water molecule; (b) Water molecules in random orientation with no field applied; (c) Water molecules line up when an electric field is applied**

symmetric. Without an external electric field applied, the molecules have random orientations (polarizations) due to thermal motion. When a field is applied, the molecules tend to line up so that their negative sides are facing the positive voltage. The molecules therefore set up a secondary electric field that opposes the direction of the applied field. The result is that inside the dielectric material, the net electric field is reduced in value. The ratio of the applied field to the reduced field is called the *(relative) dielectric constant*, $\varepsilon_r$. This value is relative to the permittivity of a vacuum, $\varepsilon_o = 8.85 \times 10^{-12}$ F/m.

The dielectric constant of a material has a direct effect on capacitance. For example, a capacitor made from two parallel plates has a capacitance,

$$C = \frac{\varepsilon_r \varepsilon_o A}{d}$$

where $A$ is the area of each plate, $d$ is the distance between the plates, and $\varepsilon_r$ is the dielectric constant of the material sandwiched between the plates. As you can see from this formula, increasing the dielectric constant causes an increase in capacitance. Therefore, if you want to make a capacitor that is physically small but has a large capacitance, use a high dielectric constant material between the plates. It follows that a higher dielectric constant corresponds to higher energy storage.

Although a perfect dielectric has no free charge, even the best insulators in the real world will have some free charge. Even air conducts electricity, albeit poorly.

## 20.21 Static Electricity and Lightning

Let's go back to the charged sphere of Figure 20.13. If the second metal ball is placed closer to the charged ball so that they actually touch, the negative charge will now redistribute itself evenly over the two balls. If the balls are moved apart again, each will now have half of the original negative charge. If balls of unequal size were used, then more charge would end up on the larger ball, because it has a larger surface area. If the charge is large enough, very high voltages will be produced as the balls are brought close together. When the electric field exceeds the breakdown strength of air (typically about 5000 volts per centimeter), a spark will occur between the two balls, allowing charge to transfer without physical contact. In this phenomenon, air molecules become ionized, forming a jagged conducting path between the two conductors. In the process, some of the original charge is lost to the air ions. What you have just learned is the process of *static discharge* (illustrated in Figure 20.20). The same effect occurs when you walk across the carpet and then get shocked by a doorknob. Your body becomes charged while walking across the carpet. Called the *triboelectric* effect, it consists of charge separation when certain materials are placed and/or rubbed together and then pulled apart. The outcome is that your body takes up a charge and the carpet stores an equal but opposite charge. Typical static discharges you encounter will be in the range of 5 kV to 15 kV and will produce a peak current of about 1A! Quite a large jolt.

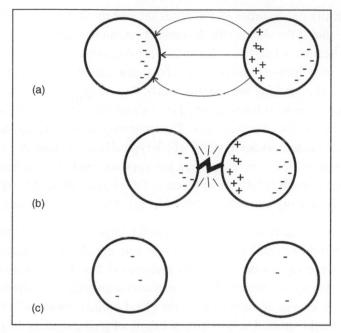

Figure 20.20: (a) A neutral sphere (right) is brought close to the negative sphere (left);
(b) The spheres become very close and a path of ionized molecules forms in the air, causing
a spark; (c) After the spark has dissipated, some of the charge has been carried away by
the ions in the air. The remaining charge is split equally between the spheres

You may wonder why such high voltage and current hurts but is not dangerous. There
are several reasons for this. First of all, the entire discharge only lasts about 1 μsec. For
this same reason, shocks from spark plugs *usually* cause no harm. Second, the current
is mostly concentrated near the discharge (i.e., your finger) and then quickly spreads
and dissipates. Negligible current will traverse your heart, which is the danger zone for
current in the human body. Typically, a person can sense currents of 1 mA or more, with
currents over 40 mA being possibly lethal as they pass through the heart. Therefore, it is
how much current flows through the body and where it flows that is of concern. When a
person becomes part of a circuit, there are four parameters that determine the current: 1)
source voltage, 2) source resistance, 3) contact resistance, and 4) internal body resistance
for the current path. The second two parameters are a function of body physiology. The
contact resistance is mainly caused by the dead skin layer where contact is made. Contact

resistance is typically on the order of 100 ohms for sweaty skin to 100 kohms for very dry skin. Beware that if the skin is cut, the contact resistance becomes negligible. The internal body resistance is fairly low due to the fact that nerves and blood vessels make good conductors. Any limb-to-limb internal resistance can be approximated as about 500 ohms. The effects of electricity are felt differently, depending on frequency. The most dangerous frequency range is from about 5 Hz to about 500 Hz and peaks in danger right about 60 Hz, the frequency of power lines. The frequency sensitivity has to do with the physiology of the human nervous system, which typically communicates via pulse trains in this range of frequencies. A 60 Hz current has approximately two to three times the danger as the same current at DC. As opposed to DC signals, these AC signals can cause muscles to lock up, leaving a person unable to let go of the voltage source.

Lightning is static electricity on a grand scale. In the large circulating winds of a thunderstorm, raindrops and ice crystals collide, causing charge separation. The process of charge separation is not fully understood, but it is known that positive charge collects at the top of the cloud and negative charge collects at the bottom of the cloud. The negative charge at the bottom of the cloud induces an equal but opposite region of charge on the ground below. You can think of it as a localized charge shadow. When the charge builds up to a high enough voltage (typically 10 MV to 100 MV), the air starts to ionize in the form of a jagged "leader" which migrates from the cloud to the ground in discrete jumps of (typically) a few hundred feet. When the ionized leader gets one jump away from the ground, its path is affected by the objects within the immediate vicinity, often connecting to the highest projecting object. Upon contact, a conducting path then connects the cloud with the ground, allowing the cloud to discharge. Peak currents can range from a few kA to 150 kA, and the event typically lasts about several hundred microseconds. With such high currents, lightning is always dangerous to people.

So what happens to all this negative charge that accumulates on the ground? It gradually migrates back to the atmosphere through the small concentration of ions always present in air. During fair weather, the upper atmosphere has a positive charge and the earth has a negative charge, forming a giant capacitor (about 5000 farad) that is discharging an average of 1800 A at any given time. This "fair weather current" is needed to balance out the currents from thunderstorms. We are standing in the middle of it! Consequently, the air of a typical day has a DC electric field in it of about 100 volts/meter. Even though we are in the middle of this high field, we don't experience much of a problem because, being very good conductors when compared to air, we locally short out the field. Approximating head-to-toe human body internal resistance as 700 ohms and the air resistance of a 6 foot long, 3 foot by 3 foot columnar region as about 1014 ohms, an equivalent circuit can be constructed. We end up being a very small resistance in series with a very large resistance (the miles of air between a person and upper atmosphere). Using these assumptions, the voltage across the body is approximated as $3\,\mu V$.

Large electric fields can cause other interesting effects. As mentioned earlier, large electric fields can cause gases to ionize; that is, electrons are freed from the gas atoms, leaving charged ions behind. Ionized gases conduct electricity and produce visible light in the process. Electrical *corona* is the term used to describe the glowing region of ionized gas that can occur around conductors. The glow is visible radiation produced when an atom gives up or accepts an electron. Coronas are created near conductors that have high electric fields emanating from them. High-voltage power lines must be placed far enough apart to avoid creating coronas since a corona consumes energy. Fluorescent light bulbs work via the same phenomena. A high voltage causes the gas inside the bulb to ionize. In fact, an unconnected fluorescent light bulb will glow if held in air near high-voltage wires! The large electric field causes the gas inside the bulb to ionize, which in turn causes the visible light. (To be exact, the gas emits UV radiation, which is absorbed and then reradiated by the fluorescent powder on the inside surface of the bulb. The reradiated light is in the visible region of the spectrum.)

Pointed objects such as flagpoles and ship masts tend to concentrate electric charge and can produce a corona during a strong thunderstorm. Before this effect was understood, sailors thought that it was the sign of a ghost or spirit and called it *St. Elmo's fire*. If you see a corona during a storm, this is a bad sign, since whatever object is producing the corona is acting as a lightning rod and will attract a lightning bolt if one approaches the area. A *lightning rod* is a metal rod that is connected to earth and protects a house or structure by attracting nearby lightning strikes. The lightning rod conducts the lightning current safely into the earth, preventing it from finding alternative paths to ground (like through your roof). It is a common myth believed by many engineers and scientists that a lightning rod prevents lightning by slowly discharging the cloud immediately above. Let me emphasize that this myth has been proven false many times over by lightning researchers. In fact, it would take over 800 hours for a lightning rod to discharge the typical cloud, and most of the ions the lightning rod releases get dispersed by the high storm winds.

The triboelectric effect can cause other, less dramatic problems. Moving cables can cause noise via the triboelectric effect. The movement of the cable causes friction between the insulation and the metal, rubbing charge from the insulation. A similar effect can happen with outdoor cables or antennas that are blown by the wind, especially if the weather is stormy and the air is well ionized.

## 20.22  The Battery Revisited

Earlier in this chapter, during the introduction of voltage, I mentioned that a neutral conductor like a wire or metal plate that is brought close to a battery will not be affected by the battery. Even if the battery or capacitor has a very high voltage, you will not see a spark (static discharge) develop to a third conductor. Why not?

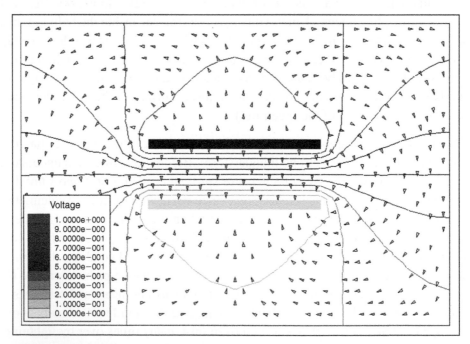

**Figure 20.21: Electric field surrounding a capacitor with DC voltage applied.**
**This figure was created using Ansoft Corporation's Maxwell 2D field solver software**
**(http://www.ansoft.com)**

To avoid discussing the details of the electrochemistry inside a battery, let's assume that we have a capacitor charged to a large voltage. One plate has many positive charges and the other plate has many negative charges. The net charge of the capacitor is neutral. Now bring a neutral, isolated metal ball near to the negative terminal. A slight charge is induced in the wire, but not much because most of the negative charge has a greater attraction to the other plate of the capacitor. Furthermore, we will not see a spark because the capacitor as a whole has a neutral charge. Suppose that charge did start to leave the capacitor to jump to the neural sphere. The capacitor would then have a charge imbalance and the charge would be attracted right back.

## 20.23  Electric Field Examples

Figures 20.21 through 20.23 show examples of the electric fields surrounding various circuit configurations. Figure 20.21 shows the electric field surrounding a simple plate

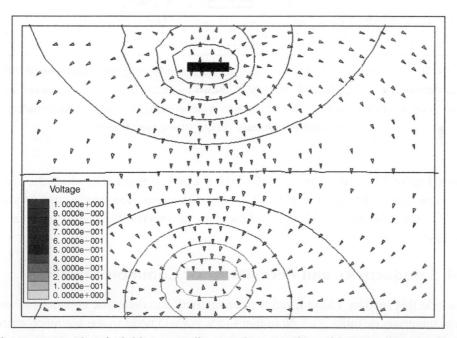

**Figure 20.22: Electric field surrounding another capacitor with DC voltage applied. This figure was created using Ansoft Corporation's Maxwell 2D field solver software (http://www.ansoft.com)**

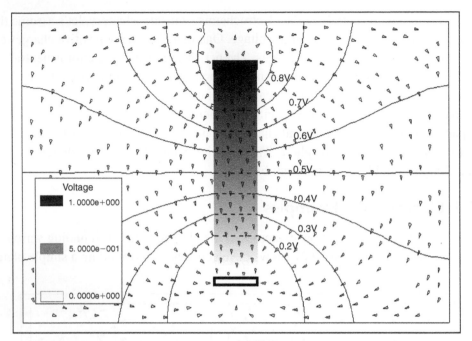

**Figure 20.23: Electric field of DC voltage across a conducting wire. Arrowheads show the direction of the electric field. Voltage is shown by shade inside the conductor and as contours outside the conductor. This figure was created using Ansoft Corporation's Maxwell 2D field solver software (http://www.ansoft.com)**

capacitor. Figure 20.22 shows the electric field of a plate capacitor in which the plates are less wide and farther apart. Figure 20.23 shows the same structure, with a conductor inserted between the plates. This figure shows how the electric field looks inside a current-carrying wire.

## 20.24 Conductivity and Permittivity of Common Materials

To conclude this section, I have listed the conductivity and dielectric constants of several common materials in Table 20.2. Note that salt water and body tissue are much more conductive than distilled water. The conductivity is caused by dissolved chemicals like salt, which dissolves into $Na^+$ and $Cl^-$ when placed in water. These ions act to conduct electricity. Also notice that the typical ground is not a very good conductor.

**Table 20.2: Low-frequency (DC) conductivity and permittivity of various materials**

| Material | Conductivity, $1/[\Omega m]$ | Permittivity (Dielectric Constant $\times$ $\varepsilon0$) |
|---|---|---|
| Copper | $5.7 \times 10^7$ | $1 \times \varepsilon0$ |
| Stainless steel | $10^6$ | $1 \times \varepsilon0$ |
| Salt water | ~4 | $80 \times \varepsilon0$ |
| Fresh water | $\sim 10^{-2}$ | $80 \times \varepsilon0$ |
| Distilled water | $\sim 10^{-4}$ | $80 \times \varepsilon0$ |
| Animal muscle | 0.35 | [no data] |
| Animal fat | $14 \times 10^{-2}$ | [no data] |
| Typical ground (soil) | $\sim 10^{-2}$ to $10^{-4}$ | $1 \times \varepsilon0$ to $14 \times \varepsilon0$ |
| Glass | $\sim 10^{-12}$ | $\sim 5 \times \varepsilon0$ |
| $\varepsilon0 = 8.85 \times 10^{-12}$ F/m | | |

Table 20.2 data adapted from Krauss and Fleisch, *Electromagnetics with Applications*, 5th Edition, Boston: McGraw-Hill, 1999, and from Paul and Nasar, *Introduction to Electromagnetic Fields*, 2nd Edition, New York: McGraw-Hill, 1987.

Keep in mind that these values are for DC and low frequency. All materials, conductors and insulators, change characteristics at different frequencies. Most materials used in electronics have constant properties into the microwave region, and some materials such as metals, glass, and teflon don't change until the infrared. Materials such as water and soil change dramatically in the radio wave frequencies.

# References

## General topics

Button, K. J., Editor, Infrared and Millimeter Waves, Volume I: Sources of Radiation, New York: Academic Press, 1979.

Cogdell, J. R., Foundations of Electrical Engineering, 2nd Edition, Englewood Cliffs, NJ: Prentice-Hall, 1995.

Encyclopedia Britannica Inc, "Electromagnetic Radiation," "Laser," Encyclopedia Britannica, Chicago: Encyclopedia Britannica Inc, 1999.

Feynman, R. P., Leighton, R. B. and Sands, M., The Feynman Lectures on Physics Vol I: Mainly Mechanics, Radiation, and Heat, Reading, Mass.: Addison-Wesley Publishing, 1963.

Feynman, R. P., Leighton, R. B. and Sands, M., The Feynman Lectures on Physics Vol II: Mainly Electromagnetism and Matter, Reading, Mass.: Addison-Wesley Publishing, 1964.

Granatstein, V. L. and Alexeff, I., Editors, High-Power Microwave Sources, Boston: Artech House, 1987.

Halliday, D., Resnick, R. and Walker, J., Fundamentals of Physics, 6th Edition, New York: John Wiley & Sons, 2000.

Halsall, F., Data Communications, Computer Networks and Open Systems, 4th Edition, Reading, Mass.: Addison-Wesley, 1996.

Halsall, F., Multimedia Communications: Applications, Networks, Protocols, and Standards, Reading, Mass.: Addison-Wesley, 2000.

Hecht, E. and Guardino, K., Optics, 3rd Edition, Reading, Mass.: Addison-Wesley, 1997.

Hutchinson, C., Kleinman, J. and Straw, D. R. Editors, The ARRL Handbook for Radio Amateurs, 78th edition, Newington, Conn.: American Radio Relay League, 2001.

Johnson, H. and Graham, M., High-Speed Digital Design: A Handbook of Black Magic, Englewood Cliffs, NJ: Prentice-Hall, 1993.

Kraus, J. D. and Fleisch, D. A., Electromagnetics with Applications, 5th Edition, Boston: McGraw-Hill, 1999.

Montrose, M. I., Printed Circuit Board Design Techniques EMC Compliance—A Handbook for Designers, 2nd Edition, New York: IEEE Press, 2000.

Paul, C. R., Introduction to Electromagnetic Compatibility, New York: John Wiley & Sons, 1992.

Pedrotti, F. L. and Pedrotti, L. S., Introduction to Optics, 2nd Edition, Upper Saddle River, NJ: Prentice Hall, 1993.

Pozar, D. M., Microwave Engineering, 2nd Edition, New York: John Wiley, 1998.

Schmitt, R., "Analyze Transmission Lines with (almost) No Math," EDN, March 18, 1999.

Schmitt, R., "Understanding Electromagnetic Fields and Antenna Radiation Takes (almost) No Math," EDN, March 2, 2000.

Straw, R. D. Editor, The ARRL Antenna Book, 19th Edition, Newington, Conn.: American Radio Relay League, 2000.

Tanenbaum, S., Computer Networks, 3rd Edition, Upper Saddle River, NJ: Prentice Hall, 1996.

## State-of-the-Art Electronics

Brock, D. K., Track, E. K. and Rowell, J. M., "Superconductor ICs: The 100-GHz Second Generation", IEEE Spectrum, December 2000.

Collins, P. G. and Avouris, P., "Nanotubes for Electronics," Scientific American, December 2000.

Cravotta, N., "DWDM: Feeding Our Insatiable Appetite for Bandwidth," EDN, September 1, 2000.

Geppert, L., "Quantum Transistors: Toward Nanoelectronics," IEEE Spectrum, September 2000.

Hopkins, J.-M. and Sibbett, W., "Big Payoffs in a Flash," Scientific American, September 2000.

Israelsohn, J., "Switching the Light Fantastic," EDN, October 26, 2000.

Israelsohn, J., "Pumping Data at Gigabit Rates," EDN, April 12, 2001.

Matsumoto, C. and Wirbel, L., "Vitesse goes with InP process for 40-Gbit devices," EETimes.com, CMP Media Inc, 2000.

Mullins, J., "The Topsy Turvy World of Quantum Computing,", IEEE Spectrum, February 2001.

Nortel Networks, "Pushing the Limits of Real-World Optical Networks," Nortel's Technology Perspectives, October 19, 1998.

Prichett, J., TRW Demonstrates World's Fastest Digital Chip; Indium Phosphide Technology Points To Higher Internet Speeds, Hardware Telecommunications Internet Product Tradeshow, TRW. Inc., 2000.

Raghavan, G., Sokolick, M. and Stanch, W. E., "Indium Phosphide ICs Unleash the High-Frequency Spectrum," IEEE Spectrum, October 2000.

Reed, M. A. and Tour, J. M., "Computing with Molecules," Scientific American, June 2000.

Rodwell, M., "Bipolar Technologies and Optoelectronics," 1999 IEEE MTT-S Symposium Workshop Technologies for the Next Millennium.

Science Wise, "Terahertz Quantum Well Emitters and Detectors," Sciencewise.com, April 14, 2001.

Stix, G., "The Triumph of the Light," Scientific American, January 2001.

Tuschman, R., "Bursting at the Seams," IEEE Spectrum, January 2001.

Zorpette, G., "The Quest for the Spin Transistor," IEEE Spectrum, December 2001.

## General Web Resources

http://www.britannica.com/.

The electromagnetic spectrum

http://imagine.gsfc.nasa.gov/docs/science/know_l1/emspectrum.html

http://observe.ivv.nasa.gov/nasa/education/reference/emspec/emspectrum.html

U.S. Frequency Allocation Chart

http://www.ntia.doc.gov/osmhome/allochrt.html.

Optical Networking News

www.lightreading.com

## Electric Fields and Conduction

Cogdell, J. R., Foundations of Electrical Engineering, 2nd Edition, Englewood Cliffs, NJ: Prentice-Hall, 1995.

Eisberg, R. and Resnick, R., Quantum Physics of Atoms, Molecules, Solids, Nuclei, and Particles, 2nd Edition, New York: John Wiley & Sons, 1985.

Epstein, L. C., Thinking Physics—Is Gedanken Physics; Practical Lessons in Critical Thinking, 2nd Edition, San Francisco: Insight Press, 1989.

Feynman, R. P., Leighton, R. B. and Sands, M., The Feynman Lectures on Physics Vol I: Mainly Mechanics, Radiation, and Heat, Reading, Mass.: Addison-Wesley Publishing, 1963.

Feynman, R. P., Leighton, R. B. and Sands, M., The Feynman Lectures on Physics Vol II: Mainly Electromagnetism and Matter, Reading, Mass.: Addison-Wesley Publishing, 1964.

Glover, J. D. and Sarma, M., Power System Analysis and Design with Personal Computer Applications, Boston: PWS Publishers, 1987.

Griffiths, D. J., Introduction to Electrodynamics, 3rd Edition, Upper Saddle River, NJ: Prentice Hall, 1999.

Halliday, D., Resnick, R. and Walker, J., Fundamentals of Physics, 6th Edition, New York: John Wiley & Sons, 2000.

Heald, M. and Marion, J., Classical Electromagnetic Radiation, 3rd Edition, Fort Worth, Texas: Saunders College Publishing, 1980.

Jackson, J. D., Classical Electrodynamics, 2nd Edition, New York: John Wiley & Sons, 1975.

Kraus, J. D. and Fleisch, D. A., Electromagnetics with Applications, 5th Edition, Boston: McGraw-Hill, 1999.

Pierret, R. F., Semiconductor Device Fundamentals, Reading, Mass.: Addison-Wesley, 1996.

Ramo, S., Whinnery, J. R. and Van Duzer, T., Fields and Waves in Communication Electronics, 2nd Edition, New York: John Wiley, 1989.

Shadowitz, A., The Electromagnetic Field, New York: Dover Publications, 1975.

Ulaby, F. T., Fundamentals of Applied Electromagnetics, Englewood Cliffs, NJ: Prentice-Hall, 1999.

Vanderlinde, J., Classical Electromagnetic Theory, New York: John Wiley & Sons, 1993.

## Static Electricity and Lightning

Adams, J. M., Electrical Safety a Guide to the Causes and Prevention of Electrical Hazards, London: The Institution of Electrical Engineers, 1994.

Anderson, K. Frequently Asked Questions (FAQ) About Lighting, www.nofc.foresty.ca Edmonton: Canadian Forest Service.

Carlson, S., "Detecting the Earth's Electricity," Scientific American, July 1999.

Carlson, S., "Counting Atmospheric Ions," Scientific American, September 1999.

Carpenter, R. B. Jr., Lightning Protection Requirements for Communications Facilities, Lightning Eliminators & Consultants, Inc., Report No. T9408, August 1994.

Carpenter, R. B. Jr. and Tu, Y., The Secondary Effects of Lightning Activity, Lightning Eliminators & Consultants, Inc., January 1997.

Chalmers, J. A., Atmospheric Electricity, Oxford: Pergamon Press, 1967.

Diels, J.-C., Bernstien, R., Stahlkopf, K. E. and Zhao, X. M., "Lightning Control with Lasers," Scientific American, 1997.

Encyclopedia Britannica Inc., "Electricity," Encyclopedia Britannica, Chicago: Encyclopedia Britannica Inc., 1999.

Feynman, R. P., Leighton, R. B. and Sands, M., The Feynman Lectures on Physics Vol II: Mainly Electromagnetism and Matter, Reading, Mass.: Addison-Wesley Publishing, 1964.

Frydenlund, M. M., Lightning Protection for People and Property, New York: Van Nostrand Reinhold, 1993.

Golde, R. H., Lightning Protection, New York: Chemical Publishing, 1973.

Jonassen, N., Electrostatics, New York: Chapman and Hall, 1998.

Kithil, R., Lightning Rod Behavior: A Brief History, National Lightning Safety Institute Facilities Protection, September 18, 2000.

Kraus, J. D. and Fleisch, D. A., Electromagnetics with Applications, 5th Edition, Boston: McGraw-Hill, 1999.

Moore, C. B., "Measurements of Lightning Rod Responses to Nearby Strikes," Geophysical Research Letters, Vol. 27, No. 10: May 15, 2000.

Sorwar, M. G. and Gosling, I. G., "Lightning Radiated Electric Fields and Their Contribution to Induced Voltages," IEEE, 1999.

Uman, M. A., The Lightning Discharge, Orlando, Florida: Academic Press, 1987.

Uman, M. A. and Krider, E. P., "Natural and Artificially Initiated Lightning," Science, Vol. 246, October 27, 1989.

Williams, E. R., "The Electrification of Thunderstorms," Scientific American, November 1988.

# CHAPTER 21

# Magnetic Fields

**Ron Schmitt**

Magnetic fields are inherently different and more difficult to grasp than electric fields. Whereas electric fields emanate directly from individual charges, magnetic fields arise in a subtle manner because there are no magnetic charges. Moreover, because there are no magnetic charges, magnetic field lines can never have a beginning or an end. Magnetic field lines always form closed loops. You may have heard that some particle physicists have been searching for magnetic charges (or "magnetic monopoles," as the particle physicists call them) in high-energy experiments. In fact, many unified theories of physics require such particles. However, at this point such particles have not been found. Even if they were to be found, they would be so rare as to be inconsequential to everyone except the particle physicists and cosmologists. Instead of hoping for magnetic charges to bail us out, you need to just accept the fact that magnetic fields are inherently different from electric fields.

## 21.1 Moving Charges: Source of All Magnetic Fields

Without magnetic charges, magnetic fields can only arise indirectly. In fact, all magnetic fields are generated indirectly by moving electric charges. It is a fundamental fact of nature that moving electrons, as well as any other charges, produce a magnetic field when in motion. Electrical currents in wires also produce magnetic fields because a current is basically the collective movement of a large number of electrons. A steady (DC) current through a wire produces a magnetic field that encircles the wire, as shown in Figure 21.1.

A single charge moving at constant velocity also produces a tubular magnetic field that encircles the charge, as shown in Figure 21.2. However, the field of a single charge decays along the axis of propagation, with the maximum field occurring in the

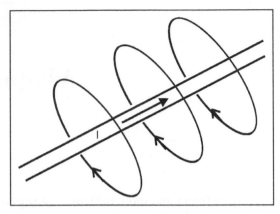

Figure 21.1: Magnetic field lines surrounding a current-carrying wire

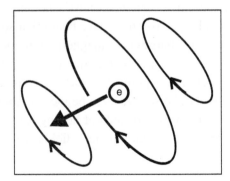

Figure 21.2: Magnetic field lines surround a moving electron

neighborhood of the charge. The law that describes the field is called the *Biot-Savart law*, named after the two French scientists who discovered it.

It is interesting to note that if you were to move along at the same velocity as the charge, the magnetic field would disappear. In that frame of reference, the charge is stationary, producing only an electric field. Therefore, the magnetic field is a relative quantity. This odd situation hints at the deep relationship between Einstein's relativity and electromagnetics.

The magnetic field direction, clockwise or counterclockwise, depends on which direction the current flows. You can use the "right hand rule" for determining the magnetic field

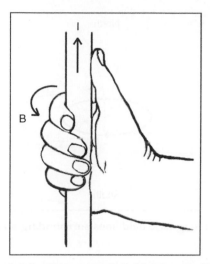

**Figure 21.3: The right hand rule: the magnetic field (*B*) curls like the fingers of the hand around the current (*I*), which points in the direction of the thumb**

direction. Using Figure 21.3 as a guide, extend your hand flat and point your thumb in the direction of the current (i.e., current is defined as the flow of positive charge, which is opposite to the flow of electrons). Now curl the rest of your fingers to form a semicircle. The magnetic field will follow your fingers, flowing from your hand to your fingertips, or in other words, the arrow tips of the field will be at your fingertips.

## 21.2 Magnetic Dipoles

Now, consider a current that travels in a loop, as shown in Figure 21.4. The magnetic field is a toroidal (donut-shaped) form. The magnetic field of this device flows out of one side and back in the other side. Although the field lines still form closed loops, they now have a sense of direction. The side where the field lines emanate is called the *north pole*, and the side they enter is called the *south pole*. Hence, such a structure is called a *magnetic dipole*. Now if a wire is wound in many spiraling loops, a solenoid like that shown in Figure 21.5 is formed. A solenoid concentrates the field into even more of a dipole structure.

Another example of a dipole is the simple bar magnet. The field of such a permanent magnetic is shown in Figure 21.6. This field is just like that of a solenoid, implying that

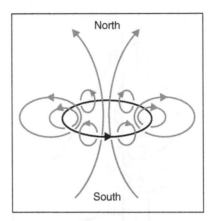

**Figure 21.4: Magnetic field lines surrounding a current loop**

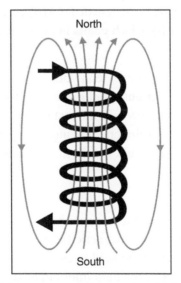

**Figure 21.5: Magnetic field lines surrounding a solenoid that is carrying a DC current**

there must be a net circular current inside the magnetic material. However, in this case the current is due to electron spin. (Spin is an intrinsic quantum property of electrons. Spin describes the angular momentum of an electron. Logically, it follows that if an electron is spinning, then its charge will create a magnetic field like that of the earth.)

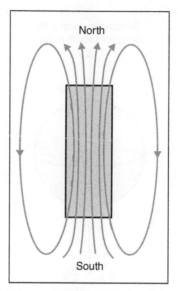

**Figure 21.6: Magnetic field lines surrounding a bar magnet**

The definitions of north pole and south pole come from the natural magnetic field that the earth produces. A sensitive magnetic dipole like a compass needle will rotate itself such that its north pole points toward the Earth's geographic north pole. The Earth's north pole is the side where the global magnetic field enters. The Earth's south pole is therefore the side from which the magnetic field emanates. (The Earth's magnetic poles are therefore opposite to the geographic poles. The geographic north pole is the magnetic south pole and vice versa.) That's right, you guessed it, the Earth's magnetic field (shown in Figure 21.7) also arises from currents. In the case of the Earth, the currents are from charges revolving inside the Earth's molten core.

Even the electron has an inherent dipole magnetic field. An electron has an inherent angular momentum (called *spin*) and it certainly has charge. Although we don't know what an electron is or what really happens inside an electron, we can think of an electron as a spinning ball of charge that creates its own magnetic dipole, just like the rotating currents inside the Earth create its magnetic field. The magnetic dipole of an electron is quite small and we typically can ignore it when we study the movement of a free electron. However, the electron's magnetic field does play an important role when the electron is bound in the atomic structure of materials.

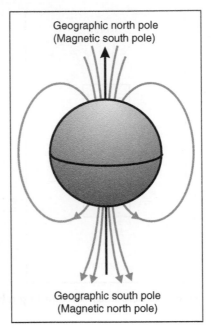

Geographic north pole
(Magnetic south pole)

Geographic south pole
(Magnetic north pole)

**Figure 21.7: Magnetic field lines surrounding the earth**

## 21.3 Effects of the Magnetic Field

### 21.3.1 The Dipole

Now that you understand how magnetic fields are created, you need to understand
how magnetic objects are affected by an external magnetic field. The situation is more
complex than the electric field, where charges just follow the electric field lines. The
effect of the magnetic field is rotational. To analyze how the magnetic field operates, you
need some form of fundamental test particle. For the electric field, we use a point charge
(i.e., a charged, infinitesimally small particle). Since magnetic charges do not exist, some
alternative must be used. One such test particle is an infinitesimally small magnetic
dipole. A magnetic dipole test particle can be thought of as a compass needle made
exceedingly small.

A magnetic dipole has a north pole and a south pole, implying that it has direction
in addition to magnitude. In other words, it is a vector quantity. The property of

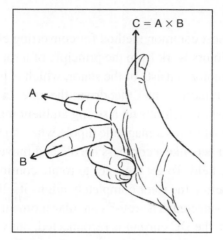

**Figure 21.8: The cross product right hand rule**

direction highlights a fundamental characteristic of the magnetic field that makes it different from the electric field. You know from experience that a compass needle always rotates so that the marked end (north pole) of the needle points north. If we place our conceptual compass in a magnetic field, the needle will likewise rotate until it points along the field lines. Its orientation will be such that its field lines up with those lines of the field in which it is immersed. So instead of a force being transmitted to the test dipole, torque is transmitted. A torque is the rotational analogy to a force. In this instance, the magnetic field acts as a "torque field" in comparison to the electric force field. This relation can be mathematically expressed as the following cross product:

$$\vec{\tau} = \vec{\mu} \times \vec{B},$$

where $\tau$ is the torque in Newton-meters, $\mu$ is the magnetic dipole moment in ampere-meters$^2$, and $B$ is the magnetic field in webers/meter$^2$. All three variables are vector quantities; that is, each has a magnitude and direction. The direction of the torque can be determined by the right hand rule for cross products, as shown in Figure 21.8. The magnitude of the torque is

$$\tau = \mu B \sin(\theta)$$

where $\theta$ is the angle between the dipole, $\mu$, and the magnetic field, $B$.

### 21.3.1.1 Motors

The electric motor is the most common method for converting electromagnetic energy into mechanical energy. Motors work from the principle of a rotating dipole. An example is the DC motor. The DC motor consists of the stator, which is the stationary enclosure, and the rotor, which is the rotating center that drives the axle. In its simplest form, the stator is a permanent magnet, which sets up a strong ambient magnetic field. The rotor is basically a coil of wire that forms a magnetic dipole when a DC current is driven through the wire. The rotor acts like a compass needle and moves to align its dipole moment with the magnetic field. To get the rotor to rotate continuously, some ingenious engineering is used. Just before the rotor completely aligns itself with the field, the DC current in the rotor is disconnected. The rotor's angular momentum causes it to freely rotate past alignment. Then the DC current is reconnected, but with reverse polarity. The rotor's dipole is consequently reversed, and the rotor is now forced to continue rotating another 180 degrees to try to align with the field. The process repeats ad nauseam. This simple example is called a *two-pole motor* because the rotor has two poles, north and south. More than one dipole can be used in a radial pattern on the rotor to produce a more powerful motor. You have now learned another way that electrical energy can be taken from a circuit. In Chapter 20, you learned that a resistor is just a device that converts electrical energy into heat. A motor converts electrical energy into mechanical energy. From the point of view of the circuit, this energy loss also appears as a resistance, although there is no "resistor" involved.

### 21.3.2 The Moving Charge

Another, more fundamental, test particle for the magnetic field is a free charge moving with velocity, *v*. As you learned earlier in this chapter, the magnetic field arises from moving charges. Therefore, a moving charge serves as a good test particle.

You can better understand the effect that a magnetic field has on a moving charge by first understanding a similar mechanical effect, that of the Coriolis force. Without knowing it, you are probably very familiar with the Coriolis force. Imagine that you are standing on a spinning platform, such as a merry-go-round or a giant turntable. You are standing at the center of the platform, and your friend Bob is standing on the other side. Furthermore, the platform is spinning counterclockwise (as seen from above). You are playing catch and you throw a baseball directly at Bob. To your dismay, the ball does not travel in a straight line to Bob, but curves off to the right. From the perspective of you and Bob, it is as if a

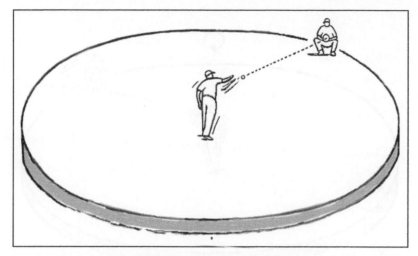

**Figure 21.9: You throw a ball to your friend Bob. The platform is not spinning. The ball, therefore, travels straight to Bob**

force acted on the ball, making it curve to the right. This apparent force is the Coriolis force. (There is also a centrifugal force present, which is discussed briefly later in this chapter.) Although no real force acts on the ball, from the reference frame of the spinning platform, it appears as if a force acts. Figures 21.9 and 21.10 illustrate the situation.

Next, Bob takes a baseball and throws it toward you. This time the ball curves to your left. You have discovered something else. The force depends on the direction of the throw. In fact, it also depends on the speed of the throw. Now what if the platform was spinning at a different rate, or what if the platform changed direction of spin? You can easily convince yourself that both of these changes would affect the path of the ball and the apparent force. The exact mathematical formula for the magnitude of the Coriolis force is

$$F_{Coriolis} = 4\pi m v f \sin(\theta)$$

where $m$ is the mass of the ball, $v$ is the speed of the ball, $f$ is the rotation frequency of the platform, and $\theta$ is the angle between the rotation axis and the direction of the ball's velocity. The rotation axis is the axis on which the platform is rotating. The direction of the force is determined by the cross product right hand rule, and the full formula for the force is

$$\vec{F}_{Coriolis} = 4\pi m(\vec{v} \times \vec{f})$$

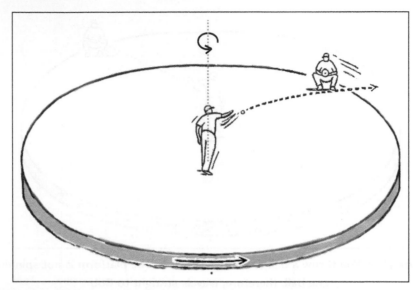

**Figure 21.10: You throw a ball to your friend Bob. The platform is spinning,
and the ball curves to the right of Bob**

where the direction of *f* is defined as upward for a counterclockwise rotating platform,
and downward for a clockwise rotating platform.

Could you throw the ball straight to Bob? No. You could account for the Coriolis force
and aim your throw to the left of Bob. But even though the ball might make it to Bob,
while it was in the air its flight path would still be curved. You could force the ball to
travel straight to Bob if you had a pipe or tube connecting the path between the two of
you. The ball would be forced to travel in a straight line because the pipe would provide
an equal but opposite force to counteract the Coriolis force.

The Coriolis force causes many effects on earth. The spin of the earth about its axis causes
a global Coriolis force, which is responsible for many weather effects on our planet. For
instance, the Coriolis force is what causes hurricanes to rotate counterclockwise in the
Northern Hemisphere and clockwise in the Southern Hemisphere.

The magnetic force acts in the exact same manner as the Coriolis force. Imagine you
now are trying to have the same game of catch with your friend Bob. However, instead of
throwing a baseball, you are now throwing a positively charged metal sphere to him, as

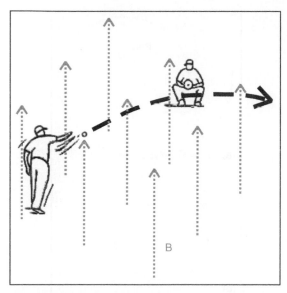

**Figure 21.11: Immersed in a magnetic field, you throw a charged metal ball to your friend Bob. The ball curves to the right of Bob, makes a loop, and eventually returns to you**

shown in Figure 21.11. You also happen to be standing in a constant magnetic field in which the direction is upward. You throw the ball directly at Bob, but to your amazement it curves off to the right. In fact, it behaves the way the baseball did when you were spinning on the merry-go-round. However, in this case neither of you is spinning! Very strange indeed. Continuing the analogy, to a charged particle, the magnetic field makes space seem like it is rotating, with the rotation taking place about an axis in the direction of the magnetic field.

The exact formula for the force is called the *Lorentz force law*, which is expressed as:

$$\vec{F}_{magnetic} = q(\vec{v} \times \vec{B})$$

or, in terms of magnitude only,

$$F_{magnetic} = qvB \sin(\theta)$$

where $q$ is the magnitude of the charge on the moving object, $v$ is the velocity of the object, $B$ is the magnitude of the magnetic field, and $\theta$ is the angle between the magnetic field and the direction of the ball's velocity. The direction of the force is again determined by the

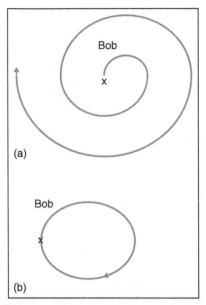

**Figure 21.12: (a) Path of ball thrown in Figure 21.10 (Coriolis force + centrifugal force);
(b) Path of ball thrown in Figure 21.11 (magnetic force)**

cross product right hand rule. The Lorentz force law forms the basis for how the magnetic field transmits its action. All magnetic effects can be ultimately reduced to this law.

There is an interesting side effect to the magnetic force in this example. Assuming that there is no air friction, the charged metal ball will continue to curve forever. Therefore, it will trace out a circular path. Furthermore, the path of the free charge creates a field in which the direction is opposite to that of the applied field. The free charge orients so that its dipole moment opposes that of the field. Therefore, the rotation of objects like compass needles and motor coils, which rotate so as to reinforce the field, must stem from something other than free charges. Although the mechanical example and the magnetic example are very similar, there is one important difference, as illustrated in Figure 21.12. In the mechanical example, there is the additional centrifugal force, which causes the ball to move outward from the center. (The Coriolis and centrifugal forces are actually virtual forces, as your frame of reference, not the ball, is being accelerated.) So the ball thrown from the spinning merry-go-round will appear to spiral away forever. In contrast to this behavior, the ball thrown in the magnetic field behaves like a boomerang, tracing out the same path forever.

### 21.3.2.1 Aurora Borealis: The "Northern Lights"

In Figure 21.10, the charge begins its velocity at a direction exactly perpendicular (90 degrees) to the magnetic field. If the velocity is not exactly perpendicular, the charge will follow a helical path along the magnetic field lines. In other words, it moves in the direction of the field lines, as well as encircling them. This result follows directly from the Lorentz magnetic force law. In addition, this phenomenon is responsible for the aurora borealis or "northern lights," the fantastic natural light show seen in the arctic regions. (In Antarctica it is called the *southern lights*.) Charged particles (protons and electrons) that are part of the solar wind are swept into the vicinity of the Earth's magnetic field, which extends out past the atmosphere. These particles are caught by the earth's magnetic field, and they tend to spiral around the magnetic field lines toward the north and south poles. As the particles get closer to the polar regions, they descend through the atmosphere. In the atmosphere, they collide with gas atoms (mainly oxygen and nitrogen), causing the atoms to ionize. During the ionization process, some of the particles' kinetic energy is converted to light. This visible light is the aurora borealis.

### 21.3.3 Currents

Another test particle that can be used in analyzing magnetic fields is the current segment. Keep in mind that there is a distinct difference between a moving charge and a current. A current consists of a group of moving charge that occupies a length in space, as opposed to a charge, which occupies only a point in space. The different points of a current are also typically rigidly connected, as is the case with a current in a wire. This point is key. In the case of a free electron, the magnetic field acts at one specific point in space. With a current, the magnetic field acts in many places at the same time, acting to move the entire structure. Second, a current always implies the existence of another force with the job of always keeping the current at its same value. In the typical case of a current in an ordinary wire, this second force is the electric field, which is imposed by the source. Another common difference is that most currents occur in wires made of atoms. As opposed to static electricity, the dynamic electricity of electronic circuits involves conductors that have overall neutrality. For every electron flowing in the current, there is a corresponding stationary hole or positive ion (refer back to Chapter 20).

Consider the situation in Figure 21.13. A wire carrying a current is placed in a constant magnetic field that points upward. As in the case of the single charged particle, the charges of the current are initially pulled to the right, dragging the wire with it. However,

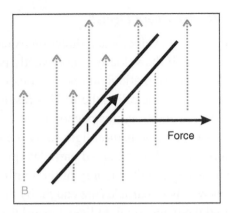

**Figure 21.13: A current-carrying wire is forced to the right by an external field**

with the current, we have a second force—the electric field in the wire. Consequently, the charges also continue to move down the wire. The outcome is that the electric field fights against the tendency for the charges to try to circle back as in the case of the single charge. As long as the source for the current continues to drive a constant current, the wire will continue to move to the right. The energy expended to move the wire comes from the current source. The current source must supply an extra amount of electric field to counteract the magnetic force. Again, this energy loss corresponds to the appearance of a resistance from the circuit point of view.

A similar situation occurs when two parallel, current-carrying wires are placed near each other. From Figure 21.14 you can determine that if the wires are carrying current in the same direction, the wires will be attracted to each other, moving together until their magnetic fields coincide. If the wires carry current in opposite directions, they will be repelled from each other.

For another example, consider a loop of current in a magnetic field as shown in Figure 21.15. By using the cross product right hand rule, you can see that the loop will rotate to align its magnetic dipole with that of the imposed field. You may also note the strange fact that if the dipole is placed exactly opposite to the field, it will not move. The situation is similar to the theoretical fact that a pendulum perfectly balanced at its peak will remain stationary in an inverted position, like a pencil on its point. In reality, any slight deviation from perfect balance will cause the pendulum to fall and eventually settle at its bottom-most position. The same can be said of the magnetic dipole.

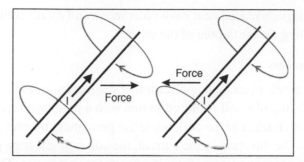

**Figure 21.14: Two wires that carry current in the same direction are attracted to each other**

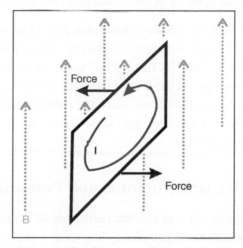

**Figure 21.15: A current-carrying loop experiences a torque (rotational force) causing it to line up with an applied magnetic field**

At this point, I have come full circle. I introduced the magnetic field by describing how dipoles rotate to line up with the field without explaining why. I then introduced the Lorentz law for magnetic forces and the cross product right hand rule to explain the fundamental effects of magnetism. I then proceeded to explain the reason why the dipole orients itself by using the Lorentz force law.

The general law governing magnetics is that currents and magnets will move so that their magnetic fields line up to produce the minimal energy of the total field. This statement explains why two magnets placed near each other move and rotate until the south pole of one magnet is touching the north pole of the other.

### 21.3.3.1 Audio Speakers

Audio speakers are another practical application of magnetics. In the typical speaker, an electromagnet consisting of a coil is placed in line with a permanent magnetic. The cone of the speaker is then attached to the coil, while the permanent magnet is held fixed in place. Depending on the direction of the current, the coil will either be pulled toward the permanent magnetic or repelled away from the permanent magnetic. The AC signals of music or speech cause the speaker coil and cone to vibrate in concert with the signal. The vibrations create sound waves in the surrounding air and you hear the signal. The energy expended to create the sound appears as resistance from the circuit point of view.

Here's a question for you: Why do speakers have plus and minus terminals on them? The speaker is not grounded. The signals are AC, so only the relative motion matters. The answer has nothing to with electronics. Instead this requirement is because of acoustics. The two speakers of a stereo system must have the same polarity so the sound created by the speakers adds constructively. If you mismatch the polarity of one speaker, you will get a dead zone of sound between the two speakers. If you match the polarity of both speakers, your stereo will sound fine.

## 21.4 The Vector Magnetic Potential and Potential Momentum

In the previous chapter, which covered electric fields, one of the first concepts covered was the electric field potential, more commonly known as *voltage*. You may be wondering if a similar potential exists for the magnetic field. If so, you are correct. However, the magnetic potential is a vector quantity. It has both magnitude and direction. The vector potential around a current is shown in Figure 21.16. As you can see, its main characteristic is that it points in a direction parallel to the current, and it decays in magnitude as the distance to the current increases.

The magnetic vector potential is much harder to understand than voltage, the electric potential. However, I will sketch out some of its characteristics. The magnetic field stores energy just as the electric field stores energy. In some situations the vector potential can be interpreted as the potential momentum of a charge. In fact, the units of the vector potential are those of momentum per charge. When Maxwell developed his theory of

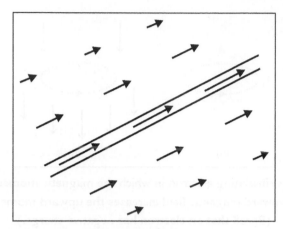

**Figure 21.16: A plot of the magnetic vector potential surrounding
a current-carrying wire**

electromagnetism, he called the vector magnetic potential the *electrodynamic momentum*
because it can be used to calculate the total momentum or total kinetic energy of a system
of charged particles and their electromagnetic fields.

## 21.5 Magnetic Materials

### 21.5.1 Diamagnetism

In Chapter 20, you learned that different materials behave differently in electric fields.
You learned about conductors and dielectrics. Electric fields induce reactions in materials.
In conductors, charges separate and nullify the field within the conductor. In dielectrics,
atoms or molecules rotate or polarize to reduce the field. Magnetic fields also induce
reactions in materials. However, since there are no magnet charges, there is no such
thing as a "magnetic conductor." All materials react to magnetic fields similarly to the
way dielectrics react to electric fields. To be precise, magnetic materials usually interact
with an external magnetic field via dipole rotations at the atomic level. For a simple
explanation, you can think of an atom as a dense positive nucleus with light electrons
orbiting the nucleus, an arrangement reminiscent of the planets orbiting the sun in the
solar system. Another similar situation is that of a person swinging a ball at the end of
a string. In each situation, the object is held in orbit by a force that points toward the
orbit center. This type of force is called a *centripetal* force. The force is conveyed by

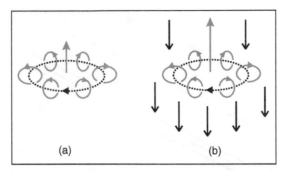

**Figure 21.17:** (a) A circulating electron in which the magnetic moment points upward;
(b) Applying a downward magnetic field increases the upward moment of the electron.
(Recall that an electron has negative charge.)

electricity, gravity, or the string tension, respectively, for the three situations. Referring
to Figure 21.17 and using the cross product right hand rule, you find that the force due
to the external magnetic field points inward, adding to the centripetal force. The increase
in speed increases the electron's magnetic field, which is opposite to the external field.
The net effect is that the orbiting electron tends to cancel part of the external field.
Just as the free electron rotates in opposition to a magnetic field, the orbiting electron
changes to oppose the magnetic field. This effect is called *diamagnetism* and is just like
that of dielectrics, where the dielectrics tend to reduce the applied electric field. The
major difference here is that diamagnetism is an extremely weak effect. Even though all
materials exhibit diamagnetism, the effect is so weak that you can usually ignore it. This
explains the commonly known fact that most materials are not affected by magnets.

### 21.5.2  Paramagnetism

In addition to orbiting the atom, each electron has a spin, which can be thought of in
simple terms as similar to the Earth spinning on its axis. Because the electron has a net
charge, the spin causes a circular current and a magnetic dipole. We learned earlier that
although currents are governed by the same magnetic law as free electrons, they behave
in opposite ways. Therefore, the inherent dipole of the electron will be rotated to line
up with the external magnetic field, thereby increasing the overall field. Paramagnetism,
while stronger than diamagnetism, is another very weak effect and can usually be
ignored. The reason for its weakness is that the electrons in each atom are always grouped
in pairs that spin opposite to one another. Hence, paramagnetism can only occur in atoms

that have an odd number of electrons. For example, aluminum has an atomic number of 13 and thus has an odd number of electrons. It therefore exhibits paramagnetic properties. The random thermal motions of the atoms tend to prevent the dipole moments from lining up well, even when exposed to an external field. (There is a simple mnemonic that can be used to recall the definitions of paramagnetism and diamagnetism. With *para*magnetism, the dipoles line up *para*llel to the external field. With *dia*magnetism, the dipoles line up *dia*metrically opposed to the external field.)

### 21.5.3 Ferromagnetism and Magnets

Diamagnetism and paramagnetism are both rather obscure phenomena to most engineers. The effects of ferromagnetism, however, are quite pronounced and quite well known. Ferromagnetism is responsible for the existence of magnets—that is, permanent magnets like the ones you use to hang up notes on the refrigerator. A ferromagnetic material is like a paramagnetic material with the added feature of "domains." Each domain is a microscopic patch of billions of atoms that have all lined up their dipole moments in the same direction. It so happens that the quantum mechanical properties of certain materials, notably iron, cause these domains to form spontaneously. This is due to the electron spin and to the collective behavior of the outermost electrons of large groups of atoms. Normally, the domains are randomly oriented so that the material still has no overall magnetic dipole. However, when a magnetic field is applied the domains that align to the field grow, while domains of other orientations shrink. In addition, the domains have a tendency to freeze in place after aligning. In other words, ferromagnetic materials have memory. For example, if a bar of iron is placed in a strong magnetic field and then removed, the bar retains a net magnetic dipole moment. It has become a magnet. Some other examples of ferromagnetic materials are nickel, which is used in guitar strings, and cobalt. Several metal alloys are also ferromagnetic.

Incidentally, there is also a ferroelectric effect. Ferroelectric materials tend to retain an electric field after being exposed to a large electric field. Analogous to the term *magnet*, the term *electret* is used for objects that exhibit ferroelectricity. Electrets are used in certain types of microphones.

### 21.5.3.1 Demagnetizing: Erasing a Magnet

There are two common ways to demagnetize (to remove any net magnetic field in) a magnet. First, you can heat the magnetic past its "Curie point." Just as ice melts to water, the frozen magnetic moment will "melt away" because above this temperature the material is no longer ferromagnetic. The second and more practical technique is to expose it to the strong AC magnetic field of an electromagnetic, such as a solenoid. The field is then slowly decreased to zero. By the end of the process, the object will have a negligible magnetic field. This technique is known as *degaussing*.

### 21.5.4 Summary of Magnetic Materials

In summary, some magnetic materials line up with an external field and some materials line up opposite to the field. This result is similar to the way free electrons line up to oppose a field, whereas controlled currents move and/or rotate to reinforce a field. Table 21.1 summarizes some of the types of magnetic materials.

Table 21.2 gives the properties of a few magnetic materials. The relative permeability of each material is given. Permeability quantifies how a material responds to magnetic fields in a manner analogous to how permittivity quantifies the material response to an electric field. It follows that permeability is a measure of the magnetic energy storage capabilities of material. A material with a relative permeability of 1 is magnetically identical to a vacuum, and therefore stores no magnetic energy. Paramagnetic and ferromagnetic materials have relative permeability greater than 1, which implies that the material aligns its dipole moments to an induced field and therefore stores energy. Higher permeability translates to a larger reaction and higher energy storage. Diamagnetic materials are characterized by relative permeabilities less than 1. This fact implies that the material aligns its dipole moments opposite to an induced field. Because the material reacts to the field, it also stores energy. Hence a lower permeability for a diamagnetic material translates to higher energy storage.

## 21.6 Magnetism and Quantum Physics

This chapter's descriptions of diamagnetism, paramagnetism, and ferromagnetism are only approximations. To truly explain these effects, quantum mechanics and solid state physics (quantum theory of solids) are necessary. In quantum physics, the electron, due to its inherent wave nature, acts more like a spread-out glob engulfing the nucleus rather than a miniature planet. Furthermore, in quantum mechanics the electron's spin is more of a theoretical quantity and can occur in only one of two quantum states, up or down.

**Table 21.1: Magnetic classification of materials**

| Material Type | Description |
|---|---|
| Nonmagnetic | No magnetic reaction. |
| Diamagnetic | Induced dipole moment *opposes* applied field. |
| | Repelled by bar magnet. |
| | Very weakly magnetic. |
| Paramagnetic | Induced dipole moment aligns to applied field. |
| | Attracted by bar magnet. |
| | Weakly magnetic. |
| Ferromagnetic | Induced dipole moment aligns to applied field. |
| | Attracted by bar magnet. |
| | Very strongly magnetic. |
| | Has *memory* and so can be used to create permanent magnets. |
| | High electrical conductivity. |
| Ferrimagnetic | Type of ferromagnetic material. |
| | Induced dipole moment aligns to applied field. |
| | Attracted by bar magnet. |
| | Very strongly magnetic. |
| Ferrites | Type of ferrimagnetic material. |
| | Induced dipole moment aligns to applied field. |
| | Attracted by bar magnet. |
| | Very strongly magnetic. |
| | Low electrical *conductivity*. |
| Superparamagnetic | Material mixture: ferromagnetic particles suspended in a plastic binder. |
| | Induced dipole moment aligns to applied field. |
| | Very strongly magnetic. |
| | Has memory, which allows for uses in audio, video, and data recording. |

Data adapted from Krauss and Fleisch, *Electromagnetics with Applications*, 5th Edition, McGraw-Hill, 1999.

#### Table 21.2: Low-frequency permeability of various materials

| Material | Type | Permeability |
|---|---|---|
| Copper | Diamagnetic | $0.999991 \times \mu_0$ |
| Water | Diamagnetic | $0.999991 \times \mu_0$ |
| Vacuum | Nonmagnetic | $1 \times \mu_0$ |
| Air | Paramagnetic | $1.0000004 \times \mu_0$ |
| Aluminum | Paramagnetic | $1.00002 \times \mu_0$ |
| Nickel | Ferromagnetic | 600 |
| Ferroxcube 3, Mn-An Ferrite Powder | Ferromagnetic (Ferrite) | 1500 |
| Iron (0.2% impurities) | Ferromagnetic | 5000 |
| Iron (0.05% impurities) | Ferromagnetic | 200,000 |

$\mu_0 = 4\pi \times 10^{-7} H/m$

Data adapted from Krauss and Fleisch, *Electromagnetics with Applications*, 5th Edition, McGraw-Hill, 1999.

# References

Blatt, F. J., Principles of Physics, 3rd Edition, Boston, Mass.: Allyn and Bacon, 1989.

Cogdell, J. R., Foundations of Electrical Engineering, 2nd Edition, Englewood Cliffs, N.J.: Prentice-Hall, 1995.

Encyclopedia Britannica Inc., "Magnetism"; "Aurora," Encyclopedia Britannica, Chicago: Encyclopedia Britannica Inc., 1999.

Epstein, L. C., Thinking Physics—Is Gedanken Physics; Practical Lessons in Critical Thinking, 2nd Edition, San Francisco, Calif.: Insight Press, 1989.

Feynman, R. P., Leighton, R. B. and Sands, M., The Feynman Lectures on Physics Vol II: Mainly Electromagnetism and Matter, Reading, Mass.: Addison-Wesley Publishing, 1964.

Fowles, G. R. and Cassiday, G. L., Analytical Mechanics, 6th Edition, Fort Worth, Texas: Saunders College Publishing, 1999.

Griffiths, D. J., Introduction to Electrodynamics, 3rd Edition, Upper Saddle River, NJ: Prentice Hall, 1999.

Halliday, D., Resnick, R. and Walker, J., Fundamentals of Physics, 6th Edition, New York: John Wiley & Sons, 2000.

Kittel, C., Introduction to Solid State Physics, 7th Edition, New York: John Wiley, 1996.

Kraus, J. D. and Fleisch, D. A., Electromagnetics with Applications, 5th Edition, Boston: McGraw-Hill, 1999.

Purcell, E. M., Electricity and Magnetism, Boston, Mass.: McGraw-Hill, 1985.

# CHAPTER 22

# Electromagnetic Transients and EMI

Keith H. Sueker

Some aspects of transient voltages have been mentioned previously, but a more detailed examination will be made in this chapter, with regards to power electronics design techniques. The long-term health of a power electronics system often hinges on its ability to withstand transient voltages arising both inside and outside of the equipment itself.

## 22.1 Line Disturbances

Lightning strikes and switching transients on power lines will propagate down the line and eventually arrive at a substation. There, the voltage will be clamped by lightning arresters to a level the substation equipment can handle without damage. This voltage will be passed on to the distribution lines in two forms: differential and common modes. The differential-mode voltage is the voltage between the power line conductors themselves, and it does not directly involve voltages to ground. The common-mode voltages are the voltages of the several conductors to ground.

Differential-mode voltages are passed directly through transformers and appear on the secondaries as transformed by the turns ratio. Some attenuation may result from intrawinding capacitances, but interwinding capacitances may actually increase the voltage. Common-mode voltages are transferred to the secondary through the interwinding capacitances and can be effectively stopped by an electrostatic shield between the windings. Absent the shield, however, they can appear on the secondaries with a magnitude close to that on the primary. This can be a severe problem on medium-voltage systems where there is the possibility of 10 kV or more being developed on secondaries to ground.

The best protection from line-induced transients of all types on secondaries is a set of MOVs. Line-to-line MOVs on the secondary are best for differential-mode voltages, and line-to-ground MOVs will provide the best protection from common-mode voltages if a shield is not used. An electrostatic shield (Faraday screen) is a relatively low-cost addition to a transformer, and it is a good practice to specify a shield on transformers with medium-voltage primaries. With a shield, line-to-line MOVs are likely to provide sufficient transient protection. Dry-type transformers should be equipped with at least distribution-class lightning arresters on medium-voltage circuits.

## 22.2 Circuit Transients

Most gate drives for SCRs or insulated gate bipolar transistors (IGBTs) are supplied in one way or another through transformers with primaries at control circuit potential and secondaries at the cathode or emitter voltage, which may be far above ground. The transformers have interwinding capacitances that will couple the power voltage transients on the semiconductors to the low-voltage control circuits. Here, they may flow through printed circuit board traces and cause improper operation or even component failure. The best protection is to equip the gate drive transformers with electrostatic shields and to be sure their primaries are tied directly to ground whenever possible.

The use of multilayer PC boards has resulted in much lower-inductance ground planes than earlier traces could provide. Still, it is not wise to run grounds in on one end of a board and out on the other. Ground and common leads entering and exiting a PC board should be on the same end if possible, and the same is true for power leads.

The operating coils on contactors and power relays will generate transient voltages when they are deenergized. R/C snubbers on the coils will reduce the effect on other circuit elements, but it is good practice to separate low-power control signals from higher-power switching circuits by running them in separate wiring troughs. Power circuits should never be run in control or signal circuit troughs.

Signal circuits should be further protected by using shielded wire. The shields should be continuously connected, but they should be grounded only at a single point that serves as an earth ground for all signal commons. Figure 22.1 shows a problem that can arise with stray pickup, even with shielded wire. The shielded signal lead between boards A and B at top is run directly, while the control commons and signal shield are connected to a common ground point at a distance. This results in a large area that stray flux can penetrate and induce spurious voltages into the signal. If the signal lead is routed along

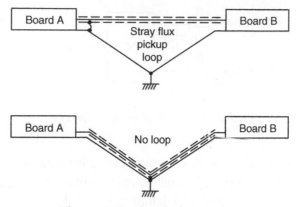

**Figure 22.1: Signal wire routing**

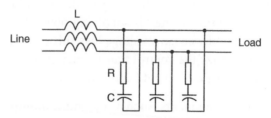

**Figure 22.2: R/C notch reduction filter**

the ground paths, the loop is eliminated. Shielding can eliminate electrostatic effects, but it does little to prevent electromagnetic couplings. Twisted pairs are the answer here.

Switching currents in semiconductors can cause very high *di/dt* levels, particularly in IGBTs, and these transients can couple to other circuitry. Gate drives and low-level circuitry should be kept away from power switching circuits.

The commutation period in SCRs will drop the source voltage to zero in a microsecond or so and hold it there until commutation is completed. The effect is called *notching*, and this disturbance can propagate to other equipment in the area through common power lines. One way to reduce this effect is to install R/C snubbers on the power lines. Such filtering is also useful to control the maximum *dv/dt* on SCRs from power line disturbances. Figure 22.2 shows a simple input circuit where the source inductance per phase is shown as *L*. The equivalent value for each R/C component as seen by a line-to-line transient

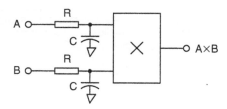

**Figure 22.3: Multiplier input filtering**

is $Req = 2R/3$ and $Ceq = 3C/2$ for the capacitor. For an instantaneous line-to-line disturbance of V, the *di/dt* will be $V/2L$, and the maximum *dv/dt* passed on to the power electronics will be $Req(di/dt)$. For critical damping, $Ceq = Req^2/8L$. Any SCR turn-on current limit must also be observed in sizing such a filter.

Although most transients are associated with rising voltages, falling voltages can also interrupt operation. If a control transformer is supplying power to large contactor coils or fan motors, the inrush current may cause a voltage loss sufficient to drop relays or disrupt operations. The cure is to oversize the control transformer or use a low-leakage reactance design so that the control voltage can be held above the required minimum.

Signal processing electronic components, such as op-amps, multipliers, digital processors, and the like, have a limited ability to reject common-mode noise on differential inputs. If noisy signals are expected, it is good practice to attenuate the noise before it arrives at the sensitive circuitry. Once inside, it is too late, and noise overloads can cause anomalies that are difficult to analyze. Figure 22.3 shows a multiplier chip as an example. The circuit is presumably immune to high-frequency noise because of the common-mode rejection of the multiplier. But if the noise level is not really known, it is a good idea to install a pole of rolloff ahead of each multiplier input. The pole should be located about a decade above the highest frequency that the multiplier is expected to pass.

## 22.3 Electromagnetic Interference

Electromagnetic interference (EMI) is a double-acting problem. Especially in military usage, equipments must not radiate interference beyond levels allowed by specifications nor be affected by defined levels of external interference. These levels are defined for both radiated and conducted levels. Radiated interference into and out of equipment

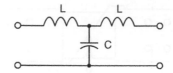

**Figure 22.4: T-section filter**

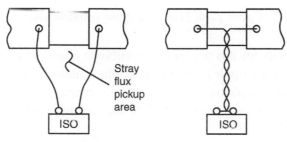

**Figure 22.5: Shunt wiring**

can usually be thwarted by good grounding techniques and shielding of critical circuits. Interference conducted by supply, control, or load wiring may require the installation of low-pass traps such as shown in Figure 22.4. A "T" section of this sort will reduce conducted noise in both directions, into and out of the equipment.

Local EMI can cause problems with instrumentation. Shunts are often used to measure DC current levels and occasionally for AC. Isolation amplifiers are used to amplify the shunt output voltage that is usually in the 50 to 100 mV range. Difficulties may arise if care is not taken with the lead dress to the shunt, especially if the power circuit has a high ripple level. Figure 22.5 shows the nature of the problem. The lead dress at left provides a large loop that magnetic fields can penetrate and induce spurious voltages. At right, the leads have been twisted to minimize the loop.

Even with twisted leads, there is a less obvious loop caused by the thickness of the shunt itself. The shunt should be made such that the terminals are symmetric about the geometric center of the shunt, and this requires an even number of leaves and location of the terminals on opposite sides of the shunt body and opposite ends of the leaves. Figure 22.6 shows a recommended shunt construction that several vendors are willing to make at a small premium in price.

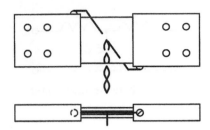

**Figure 22.6: Preferred shunt construction**

Even with these precautions, it is not always possible to eliminate all stray pickup in circuits with AC or high-ripple currents. However, contaminating signals can be neutralized with an air core pickup coil of a few turns connected in series with the shunt output. The location and orientation of the pickup coil must be determined experimentally.

The precautions on shunt metering apply also to oscilloscope measurements of low-level signals in the presence of interference. The ground lead of the scope should be looped back and forth along the probe to minimize the loop area and the resultant induced voltages.

# CHAPTER 23

# Traveling Wave Effects

Keith H. Sueker

Traveling wave effects were once the nearly exclusive domain of communications engineers and electric utility transmission line engineers. However, as variable frequency, pulse width modulated (PWM) motor drives have gone to higher and higher PWM frequencies, standing-wave phenomena have appeared in motor circuits. Traveling waves are now of concern to many power electronics engineers, and an understanding of their effects is necessary for motor protection in installations with long cables or high PWM frequencies. In modern terms, a transmission line is any set of parallel or coaxial conductors of finite length, and they may be printed circuit conductors or miles of overhead wires.

## 23.1 Basics

Standing waves appear when a length of line is excited at a frequency for which the electrical line length is a significant part of an electrical wavelength. They result from the constructive and destructive interference of forward and reflected waves on the line. The behavior of the line can be determined by solving the applicable differential equations relating the line parameters to the exciting frequency. The solution of the equations for a line with losses is rather complex and adds little to the practical considerations, so the lossless line will be analyzed instead.

In the lossless line, $L$ is the series inductance per unit length, and $C$ is the shunt capacitance. If a differential length, $dx$, is considered, the inductance for that length is $L\,dx$, and the voltage in that length is $e = -L\,dx(di/dt)$. Since $e = (de/dx)dx$, the equation can be written as $dx(de/dx) = -L\,dx(di/dt)$. Similarly, $dx(di/dx) = -C\,dx(de/dt)$. Dividing out the $dx$ terms and substituting partial derivatives, the fundamental forms of transmission line equations result:

$$-\partial e/\partial x = \mathrm{L}\,\partial i/\partial t \quad \text{and} \quad -\partial i/\partial x = \mathrm{C}\,\partial e/\partial t$$

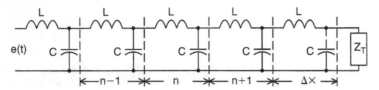

**Figure 23.1: Transmission line difference equations**

By differentiating with respect to $x$ and then with respect to $t$, these equations can be solved simultaneously to yield second-order, elliptical, partial differential equations for both $e$ and $i$ individually with respect to $t$ and $x$. The classical forms then result:

$$LC\ \partial^2 i/\partial t^2 - \partial^2 i/\partial x^2 = 0 \quad \text{and} \quad LC\ \partial^2 e/\partial t^2 - \partial^2 e/\partial x^2 = 0$$

These equations can now be solved by transforms or classical methods. Explicit solutions can be developed with hyperbolic functions in the complex plane, and these solutions were the only practical means of line analysis until the digital computer was developed. Fortunately, the computer offers an easier method of analysis by numerical integration, and line losses can be incorporated with relative ease. The difference equations can be solved by simple Euler integration, so the whole process is not nearly as daunting as in earlier years.

The difference equations for differential sections of line are shown in Figure 23.1. These equations allow numerical solutions for the voltages and currents on the line as functions of distance and time. Although it may not be immediately apparent, these difference equations, in the limit, replicate the differential equations.

Before proceeding to typical solutions, several derived parameters should be defined. First, the line has a surge, or characteristic, impedance defined as $Z_0 = (L/C)^{1/2}$ and, second, a velocity of propagation $v = 1/(LC)^{1/2}$. The characteristic impedance defines the relationship between the line and its attached load, and the velocity of propagation defines the speed of signal transmission along the line and consequently its electrical length. The electrical length of the line, in terms of wavelengths for any given exciting frequency, is $\lambda_p/\lambda_e = v/c$, where $\lambda_p$ is the physical line length, $\lambda_e$ is the exciting frequency wavelength in free space, $v$ is the velocity of propagation, and $c$ is the speed of light. The parameters vary widely among the various types of transmission lines and cables typically encountered in power electronics. Figure 23.2 shows two examples of such lines.

Overhead line of 1/0 conductors with 6-ft spacing:

$$Z_0 = 325\ \Omega,\ v = 83\%$$

Shielded coaxial cable, 15 kV with 500 kcm conductor:

$$Z_0 = 21\ \Omega,\ v = 29\%$$

Velocities are shown as percentage of $c = 3 \times 10^8$ m/s

**Figure 23.2. Transmission line parameters**

The overhead line has a high series inductance and relatively low shunt capacitance that leads to a high surge impedance. It also has a relatively high velocity of propagation because of the low capacitance. In the cable, things are reversed. Shielded cable has a very high capacitance that makes the surge impedance low, and the velocity of propagation is also low. Note that the physical wavelength of a signal in such shielded cable is less than one-third of the wavelength in free space.

## 23.2 Transient Effects

The problems that can arise from traveling waves in motor circuits can best be illustrated by examining the response of a transmission line to applied transient voltages. Figure 23.3 shows the extreme case of a transmission line with a zero source impedance and no load termination. The reflection coefficient, $RC$, is given by:

$$RC = \frac{Z_T - Z_0}{Z_T + Z_0}$$

where $Z_T$ is the terminating impedance and $Z_0$ the surge impedance. This parameter defines the reflected waves that are generated with a given terminating impedance. $RC = 1$ for the unterminated line.

When a traveling wave reaches the end of a line, a reflected wave will be developed unless the line is terminated in its surge impedance. If $RC$ is positive, the reflected wave will be positive and will add to the incident wave. If $RC$ is negative, the reflected wave is negative and subtracts from the incident wave. In the limiting cases of open and short circuit terminations, the open circuit termination doubles the incident voltage and the short circuit brings it to zero. In either case, however, the reflected wave continues in the

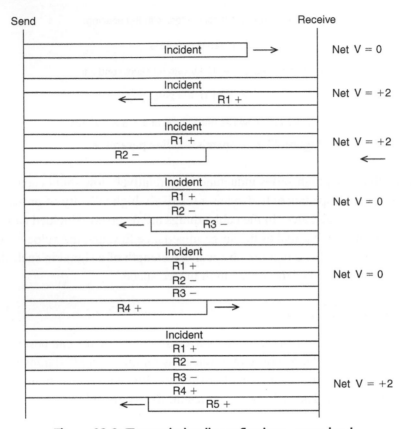

**Figure 23.3: Transmission line reflections—open load**

reverse direction, and the incident wave continues, in effect, in the forward direction. The process is shown in Figure 23.3 with a unit applied voltage. At any point on the line, the voltage is the sum of the incident and reflected waves.

The receiving end voltage will continue the sequence of $+2, +2, 0, 0, +2\dots$ as long as the voltage is supplied at the sending end. The sending end will continue to invert each reflected wave, and the receiving end will return it with the same polarity. Although this little exercise examined a lossless line with pure reflections, much the same process obtains with typical shielded cable or conductors in conduit.

Figure 23.4 shows the effects of shaping the voltage wavefront to reduce the rate of rise. A unit voltage wave with various rise times is applied and the receiving end voltage

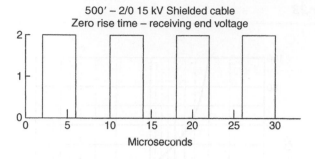

500′ – 2/0 15 kV Shielded cable
Zero rise time – receiving end voltage

Microseconds

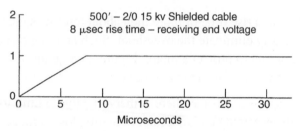

4 μsec rise time – receiving end voltage

Microseconds

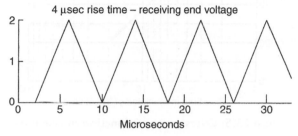

500′ – 2/0 15 kv Shielded cable
8 μsec rise time – receiving end voltage

Microseconds

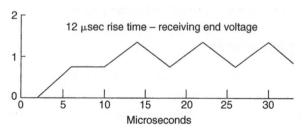

12 μsec rise time – receiving end voltage

Microseconds

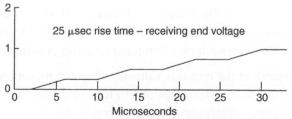

25 μsec rise time – receiving end voltage

Microseconds

**Figure 23.4: Front-of-wave shaping**

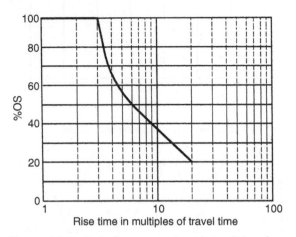

**Figure 23.5: Overshoot as a function of rise time**

shown for each. Note that a rise time of more then about ten times the transit time, 2 μs in this example, will nearly eliminate the overshoot. Such shaping can be done with low-pass filters on a drive output to protect a motor. Figure 23.5 quantifies the needed shaping for a given overshoot. The curve is the maximum envelope of an oscillatory function.

When the exciting voltage is repetitive, there is theoretically no limit to the magnitude of voltage overshoot at the receiving end of a line. In reality, losses reduce the overshoot, but it can ratchet up well beyond what a single pulse can produce.

## 23.3 Mitigating Measures

Reducing the load voltage overshoot means reducing the slope of the wavefront at either the source end or at the load to control the reflected waves. In principle, it is possible to match the line impedance with a $Z_0$ termination, but this is not often practical because of the unknown high-frequency characteristics of the load. The usual measures are to install an inductor or low-pass filter on the source. If an inductor alone is used, it should have an inductance on the order of five times the line inductance to the load. A low-pass filter should be simulated with the system on a computer to avoid possible resonance problems.

Installing a surge capacitor at the motor is valuable when the motor terminals are accessible to the drive designer. The capacitor will tend to reduce the end turn effects that can damage motor windings. However, the capacitor may need a series damping resistor and, again, a computer simulation is suggested.

# CHAPTER 24

# Transformers

Mike Tooley

Transformers provide us with a means of coupling AC power or signals from one circuit to another. Voltage may be *stepped-up* (secondary voltage greater than primary voltage) or *stepped-down* (secondary voltage less than primary voltage). Since no increase in power is possible (transformers are passive components like resistors, capacitors and inductors) an increase in secondary voltage can only be achieved at the expense of a corresponding reduction in secondary current, and vice versa (in fact, the secondary power will be very slightly less than the primary power due to losses within the transformer). Typical applications for transformers include stepping-up or stepping-down mains voltages in power supplies, coupling signals in AF amplifiers to achieve impedance matching and to isolate DC potentials associated with active components. Figure 24.1 shows a selection of various transformers.

**Figure 24.1: A selection of transformers with power ratings from 0.1 VA to 100 VA**

The electrical characteristics of a transformer are determined by a number of factors including the core material and physical dimensions. The specifications for a transformer usually include the rated primary and secondary voltages and current the required power rating (i.e., the maximum power, usually expressed in volt-amperes, VA) which can be continuously delivered by the transformer under a given set of conditions, the frequency range for the component (usually stated as upper and lower working frequency limits), and the *regulation* of a transformer (usually expressed as a percentage of full-load). This last specification is a measure of the ability of a transformer to maintain its rated output voltage under load.

Table 24.1 summarizes the properties of three common types of transformer. The photo in Figure 24.2 shows the parts of a typical iron-cored power transformer, and Figure 24.3 shows the construction of a typical iron-cored power transformer.

## 24.1 Voltage and Turns Ratio

The principle of the transformer is illustrated in Figure 24.4. The primary and secondary windings are wound on a common low-reluctance magnetic core. The alternating flux

**Table 24.1: Characteristics of common types of transformer**

| Property | Transformer core type | | | |
|---|---|---|---|---|
| | **Air cored** | **Ferrite cored** | **Iron cored (audio)** | **Iron cored (power)** |
| Core material/ construction | Air | Ferrite ring or pot | Laminated steel | Laminated steel |
| Typical frequency range (Hz) | 30M to 1G | 10k to 10M | 20 to 20k | 50 to 400 |
| Typical power rating (VA) | (see note) | 1 to 200 | 0.1 to 50 | 3 to 500 |
| Typical regulation | (see note) | (see note) | (see note) | 5 to 15% |
| Typical applications | RF tuned circuits and filters | Filters and HF transformers, switched mode power supplies | Smoothing chokes and filters, audio matching | Power supplies |
| Note: Not usually important for this type of transformer | | | | |

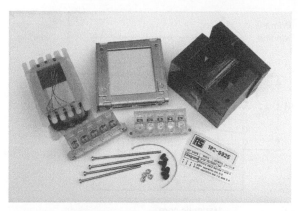

Figure 24.2: Parts of a typical iron-cored power transformer prior to assembly

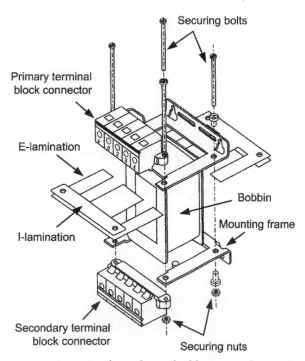

Figure 24.3: Construction of a typical iron-cored transformer

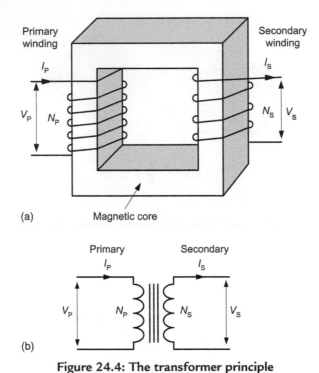

(a)            Magnetic core

(b)

**Figure 24.4: The transformer principle**

generated by the primary winding is therefore coupled into the secondary winding (very little flux escapes due to leakage). A sinusoidal current flowing in the primary winding produces a sinusoidal flux. At any instant the flux in the transformer is given by the equation:

$$\phi = \phi_{max} \sin(2\pi f t)$$

where $\phi_{max}$ is the maximum value of flux (in webers), $f$ is the frequency of the applied current (in hertz), and $t$ is the time in seconds.

The r.m.s. value of the primary voltage, $V_P$, is given by:

$$V_P = 4.44 \ f N_P \phi_{max}$$

Similarly, the r.m.s. value of the secondary voltage, $V_S$, is given by:

$$V_S = 4.44 \ fN_P\phi_{max}$$

Now,

$$\frac{V_P}{V_S} = \frac{N_P}{N_S}$$

where $N_P/N_S$ is the *turns ratio* of the transformer.

Assuming that the transformer is loss-free, primary and secondary powers ($P_P$ and $P_S$, respectively) will be identical. Hence:

$$P_P = P_S \text{ thus, } V_P \times I_P = V_S \times I_P$$

Hence

$$\frac{V_P}{V_S} = \frac{I_S}{I_P} \quad \text{and} \quad \frac{I_S}{I_P} = \frac{N_P}{N_S}$$

Finally, it is sometimes convenient to refer to a *turns-per-volt* rating for a transformer. This rating is given by:

$$\text{turns-per-volt} = \frac{N_P}{V_P} = \frac{N_P}{V_S}$$

### Example 24.1
A transformer has 2,000 primary turns and 120 secondary turns. If the primary is connected to a 220V r.m.s. AC mains supply, determine the secondary voltage.

### Solution
Rearranging $\dfrac{V_P}{V_S} = \dfrac{N_P}{N_S}$ gives:

$$V_S = \frac{N_S \times V_P}{N_P} = \frac{120 \times 220}{2,000} = 13.2 \text{ V}$$

Figure 24.5: Resonant air-cored transformer arrangement

Figure 24.6: A 1:1 ratio toroidal transformer that forms part of a noise filter connected in the input circuit of a switched mode power supply

**Example 24.2**
A transformer has 1,200 primary turns and is designed to operate with a 200 V AC supply. If the transformer is required to produce an output of 10 V, determine the number of secondary turns required. Assuming that the transformer is lossfree, determine the input (primary) current for a load current of 2.5 A.

**Solution**
Rearranging $\dfrac{V_P}{V_S} = \dfrac{N_P}{N_S}$ gives:

$$N_S = \frac{N_P \times V_S}{V_P} = \frac{1,200 \times 10}{200} = 60 \text{ turns}$$

Rearranging $\dfrac{I_S}{I_P} = \dfrac{N_P}{N_S}$ gives:

$$N_S = \frac{N_S \times I_S}{N_P} = \frac{200 \times 2.5}{1,200} = 0.42 \text{ A}$$

Figure 24.5 shows a resonant air-cored transformer arrangement. The two inductors are tuned to resonance at the operating frequency (145 MHz) by means of the two small capacitors.

Figure 24.6 shows a small 1:1 ratio toroidal transformer that forms part of a noise filter connected in the input circuit of a switched mode power supply. The transformer is wound on a ferrite core and acts as a choke, reducing the high-frequency noise that would otherwise be radiated from the mains supply wiring.

# CHAPTER 25

# Electromagnetic Compatibility (EMC)

M.A. Laughton
A. Maddocks
D.F. Warne

## 25.1 Introduction

Electromagnetic compatibility (EMC) has now become a major consideration on any project involving the design, construction, manufacture and installation of electrical and electronic equipment and systems. Electrical equipment must be designed not only to meet a functional technical performance specification but due consideration must also be given to the interaction the equipment has with the electromagnetic environment in its intended operating location. If the equipment is expected to operate reliably in a steel works, for example, it is imperative that the designers, purchasers, installers and operators are aware of the nature of the electromagnetic environment and the potential for unwanted coupling to the equipment which could cause equipment mis-operation or malfunction. For example, for safety-related equipment, any interference to the operation of the system could have serious consequences. Equally, electromagnetic disturbance generated by the equipment itself could cause interference to radio reception. In the case of domestic radio and TV reception this may adversely affect the quality of reception, but can also block emergency channels and in some cases; for example, a radio controlled crane, could cause malfunction with potential reliability and safety implications. Generally, both aspects of controlling emissions from the equipment and providing adequate immunity to the expected electromagnetic disturbances in the intended operating environment are key features of any equipment or system design.

The need for emission control has been recognized for many years and most countries have introduced legal regulations to support the efficient utilization of the electro magnetic spectrum. This is manifest in the form of requirements to comply with EMC emission standards in product certification. In some territories such as Europe and Australia, the legal requirements also address immunity aspects. For most manufacturers of good quality equipment this is not an added burden because they are keen to demonstrate to their customers that the equipment will prove to be reliable in the field, and moreover, recognize that any actual incidents of malfunction could be damaging and costly to the business.

Fortunately there are advantages for both equipment suppliers, purchasers and operators in the availability of nationally or internationally approved high quality EMC standards for both emissions and immunity which can be referenced in contractual agreements as well as providing an excellent foundation for the designers of the equipment. Manufacturers and users alike can apply these standards to their mutual advantage, particularly in the planning of major projects.

EMC considerations need to be addressed at the outset of the development. It is well known that the costs of achieving EMC conformity rise almost exponentially with the delay in first considering the requirements. The key issue in addressing EMC matters is to adopt a strategic approach where the EMC requirements are recognized and clearly understood at the design concept stage, through the product development to in service use, and throughout the lifetime of the equipment.

## 25.2 Common Terms

There are a number of common terms used in the science of EMC. EMC itself can be defined as the ability of equipment to operate satisfactorily within its intended electromagnetic environment without contributing to the disturbance level in that environment such that radio communication is not adversely affected. Other related terms are RFI and EMI: RFI is radio frequency interference, which is usually defined as interference to radio services in the radio bands, 9 kHz to 300 GHz; EMI, electromagnetic interference, is generally accepted as interference both in the radio frequency bands and in the low frequency region DC to 9 kHz.

Because of the large ranges of values that are dealt with in EMC it is customary to express emission limits and system performance in logarithmic ratios, i.e., in dBs. For a voltage or current value, the value in dBs is given by $20. \log_{10}$ (ratio) and for power ratios, $10. \log_{10}$ (ratio). It is important to recognize that 1 dB in voltage is equivalent to 1 dB in power, although the linear ratios are different, 1.12 and 1.26, respectively.

Electric and plane wave fields are expressed in volts/meter or dB($\mu$V/m) and magnetic fields are expressed in amps/meter or dB($\mu$A/m). (1 amp/meter is equivalent to 1.25 micro tesla of magnetic induction in free space.)

## 25.3  The EMC Model

An example of electromagnetic coupling between two items of electrical equipment is illustrated in Figure 25.1. Electrical disturbances generated by equipment #1 may be coupled to equipment #2 by a variety of means: conduction via a common connection to the mains supply, inductive and capacitive coupling between interface and power conductors, and by direct radiation. Conducted coupling tends to be the dominant coupling mechanism at lower frequencies, e.g., below 1 MHz, where conductor impedances are low; capacitive and inductive coupling is more important at higher frequencies where the capacitive impedance between long parallel runs of conductors is relatively low. Radiation coupling dominates at frequencies where the length of the radiating conductor is comparable with a wavelength. For a small computer system, radiation from cables will be prominent in the range 30–300 MHz but at higher frequencies, direct radiation from circuit board tracks dominates. Due consideration must be given to these effects in designing and installing equipment and systems.

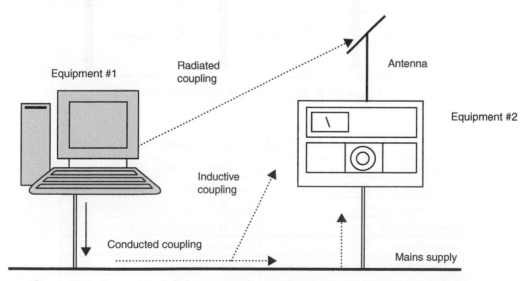

**Figure 25.1: Example of electromagnetic coupling between two items of equipment**

There are many types of coupling that may occur in any particular application and the key factor for the designer is to recognize and understand the various mechanisms. The simple but effective model source-path-receptor model shown can be applied effectively in dealing with overall requirements (Figure 25.2).

For the equipment designer, his product must be considered both as a source of disturbance and as a potential receptor to disturbances in the intended environment. Emission control is achieved by recognising the potential sources of disturbance within the equipment and the paths by which they may couple to the outside world and cause interference in radio communications. Examples of significant sources of internally generated electromagnetic disturbance are given in Table 25.1.

Emission control is achieved by effective design, filtering and suppression measures. The level of control that is normally required is that sufficient for interference free reception of radio communication and radio and TV broadcast services.

But the designer of electronic equipment must also consider his product as a receptor of electromagnetic disturbance in the intended operating environment. For the equipment

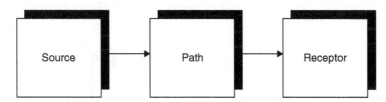

**Figure 25.2: Source-Path-Receptor model**

**Table 25.1: Man-made sources in equipment**

| Source or device | Electromagnetic disturbance |
|---|---|
| Digital electronic circuits | Harmonics of clock oscillators |
| Commutator motors | High frequency switching transients |
| Contact devices | Showering arc discharges |
| RF oscillators | RF fields and voltages |
| Luminaires and Lighting equipment | Arc discharges |
| Switch mode power supplies | Harmonics of the switching frequency |

to work reliably in the field the designer must consider the types and nature of electromagnetic disturbances which are likely to be present, their magnitude, and even their probability of occurrence if it is a safety related system. Generally there are two categories, man made and naturally occurring as shown in Tables 25.2 and 25.3.

For radio transmitters and overhead power lines the level experienced by the receptor circuit will depend on the separation distance from the source. Some VHF and UHF rf transmitters are designed so that the field strength at ground level does not exceed 1 volt/meter but mobile and portable transmitters such as mobile phones can generate field levels of up to 100 volt/meter close to the antenna. In cases where the equipment is intended to be operated close to one of these sources, compliance with commercial EMC standards may be insufficient and additional measures may be required, either in the equipment design or the installation, to ensure that the equipment will operate reliably in practice.

The electrostatic discharge is a high voltage disturbance, typically 8 kV, with a very short rise time, about 1 ns and generates high level spectral components extending to high frequencies and careful equipment design is often required to provide adequate protection. The residual effects of a lightning event are high voltage surges of about 2 kV but with longer durations and much greater energy.

The degree to which electronic equipment may be affected by the electromagnetic disturbances in the environment depends on many factors but is primarily dependent,

**Table 25.2: Man-made sources in the environment**

| Source or device | Electromagnetic disturbance |
|---|---|
| Radio transmitters | Rf fields |
| Power distribution | Surges, fast transients, dips and interruptions |
| Overhead power lines and railway traction | Magnetic and fields, corona discharges |

**Table 25.3: Naturally occurring sources in the environment**

| Source or device | Electromagnetic disturbance |
|---|---|
| Human body electrostatic charging | Electrostatic discharges |
| Lightning | Fields and power surges |

on: a) the coupling between the equipment b) the magnitude of the external source, and c) the sensitivity of the internal electronics. In broad terms any semiconductor device in the equipment must be considered as a potential receptor.

Most digital electronic devices generally require a signal of at least 0 5 V to change state and for protection against ESD with a peak amplitude of 5 kV, the designer must provide a circuit isolation in excess of 10,000 to one or 80 dB. The first 60 dB may be relatively easy to achieve through natural circuit losses but the next 20 dB will inevitably require careful design. Analog semiconductor devices have higher sensitivities with devices in the control application requiring only millivolts of disturbance to cause interference. Here the main concern is the continuously applied or long duration disturbance such as from radio transmitters. An applied field of 3 volt/meter could induce a voltage of 3 V into a conductor connected to a remote analog sensor causing an rf current of up to 10 mA to flow in the wiring. To reduce this level of disturbance to acceptable proportions high degrees of isolation, possibly in the order of 60–70 dB are required. Most of the coupling at high frequencies is in common mode and there will be an inefficient coupling to a differential voltage that could be confused with the wanted signal, but inevitably some form of circuit protection, often in the form of filtering will be required.

## 25.4 EMC Requirements

The overriding requirement in the supply of equipment to the customer, or in placing the equipment on the market, is to meet the accepted or agreed conformity assessment requirements. For supply of defense equipment, for example, the manufacturer will be required to demonstrate that the equipment complies with the EMC specification for the project. The manufacturer ensures that the equipment is tested to the standard and submits the test report to the project office for approval. In the case of equipment for residential, commercial or industrial application, emission control regulations apply in most territories, and compliance with a relevant EMC standard must be demonstrated.

The relevant EMC standards are of three types, high frequency emissions standards for the protection of the radio spectrum, low frequency emissions standards for the protection of the power distribution network, and immunity standards for demonstrating the equipment's robustness in the presence of electromagnetic disturbances. Typical emission limits, in this case for Information Technology Equipment are presented in Figure 25.3(a) and (b).

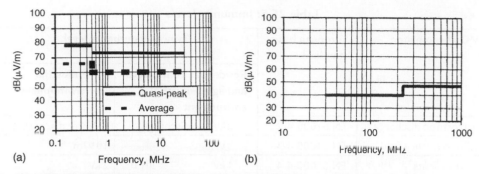

**Figure 25.3: Harmonized European conducted and radiated emissions limits for ITE (EN 55022) Class A (a) Conducted emissions limits; (b) Padiated emissions limits at 10 m**

**Table 25.4: Low frequency emissions**

| EMC Phenomenon | European Basic Standard |
|---|---|
| Main harmonics | EN 61000-3-2 |
| Voltage fluctuations (flicker) | EN 61000-3-3 |

Limits for equipment intended for residential application are generally more severe by 10 dB because of the shorter separation distances and tighter coupling that can occur in domestic premises.

Standards are available for the protection of the power distribution network. Mains harmonics such as generated by high power semiconductor switching devices can cause problems in the supply of electricity for which limits apply for the frequency range 0–2 kHz. Voltage fluctuations due to switching of heavy loads, which can cause lighting flicker is another low frequency phenomenon for which relevant standards exist. For example, equipment containing switch mode power supplies or switching heating loads as in air handling units will be subject to the standards listed in Table 25.4.

The European immunity requirements are primarily based on international IEC standards. Both product specific and generic standards (to be applied where no product standard exists), are based on a raft of basic standards addressing fundamental EMC phenomena These are listed in Table 25.5 together with typical levels for a commercial or light industrial application.

Table 25.5: Immunity levels

| EMC phenomenon | European basic standard | Typical disturbance level AC power port | |
| --- | --- | --- | --- |
| | | Residential, commercial and light industrial environment | Industrial environment |
| Electrostatic discharge | EN 61000-4-2 | 8 kV (air discharge) | 8 kV (air discharge) |
| Radio frequency field | EN 61000-4-3 | 3 V/m | 10 V/m |
| Fast transients | EN 61000-4-4 | 1 kV | 2 kV |
| Surge | EN 61000-4-5 | 1 kV | 2 kV |
| Common mode rf voltage | EN 61000-4-6 | 3 V | 10 V |
| Power frequency magnetic field | EN 61000-4-8 | 3 A/m | 30 A/m |
| Dips and interruptions | EN 61000-4-11 | 30% for 10 ms 60% for 100 ms | 30% for 10 ms 60% for 100, 1000 ms |

The descriptions for these two environments may be insufficient in some applications. For example if the equipment is to be installed in an airfield environment, immunity may be required against radar transmitter fields of up to 100 V/m and these requirements would need to be incorporated in the technical specification for the project.

## 25.5 Product design

In designing equipment for EMC conformity it is important to identify the primary paths whereby electromagnetic energy may be transferred. These are conduction and induction via the power lead and interface cables, radiation from the PCB circuit tracks and radiation leakage through apertures and discontinuities in any shielded enclosure. The aim of designing for EMC conformity is to provide sufficient attenuation in each of the relevant paths such that emission control and provision of adequate immunity is achieved at an economic cost to the product. For many protection measures, e.g., shielding and filtering there is reciprocity between measures for emission control and for immunity, although the required degree of attenuation in a particular path may be different. Experience with previous generations of similar products and some exploratory work in

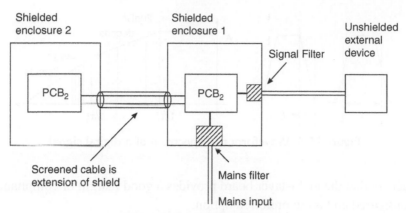

**Figure 25.4: Design concepts**

determining the immunity margins by testing to the level of equipment mis-operation, can provide valuable information as what is the more critical, aspect emissions or immunity. Further development work can then concentrate on testing to only the critical EMC phenomenon.

For equipment contained in metal enclosures which can provide a reasonable degree of shielding, there are a number of important considerations. As shown in Figure 25.4, the barrier for the power cable is a filter mounted at the point of entry for the cable.

Similarly, unshielded signals lines to inputs and outputs entering the enclosure should also be filtered at the point of entry. Where screened cables are employed, usually for the protection of sensitive circuits, or for emission control for high frequency signals, the cable screen should be regarded as an extension of the shield of the enclosure.

Where metal enclosures are not employed, the transfer of high frequency electromagnetic energy must be controlled by good design at the circuit board level. The use of multiplayer PCBs offers such a facility but sometimes there appears to be a cost penalty associated with the more complex structure and fabrication. These costs should be compared with the costs of retrospective remedial measures should the equipment fail to comply with radiated emission and radiated immunity requirements. Generally, most

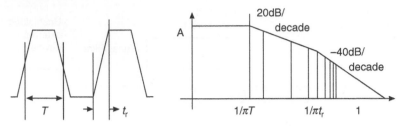

**Figure 25.5: Waveform and spectrum of a digital signal**

designers accept that the multi-layer board provides a good margin of compliance if properly configured and is the preferred option.

## 25.6  Device Selection

The selection of the type of semiconductor device or integrated circuit technology can have an effect on overall EMC performance. Although the trend is for faster devices for higher processing speeds, emission control is easier with slower devices. Figure 25.5 shows the waveform of a digital signal and its associated spectrum. The spectrum starts to decay at 20 dB/decade at just below $1/(\pi \cdot T)$ where $1/T$ is the repetition rate, but decays even faster at frequencies above $1/(\pi \cdot t_r)$ where $t_r$ is the rise-time. Lower clock rates and longer rise times thus reduce high frequency emissions.

For improved immunity, circuit bandwidth is important, the narrower the bandwidth, the lower the energy from impulsive disturbance and there is a greater probability of discriminating against high frequency single frequency disturbances. Although device selection may not be a prime consideration in designing for EMC conformity because of performance constraints, knowledge on the basic principles can be extremely valuable.

## 25.7  Printed Circuit Boards

Tracks on circuit boards can act as antennas and can radiate signals efficiently at high frequencies. There are essentially two basic mechanisms, monopole and loop type.

Where, for example, the signal from a high frequency clock oscillator is taken to another device on the circuit board and the return conductor is not immediately adjacent to the signal line, a loop is formed which will radiate the signal and generate a field at a remote

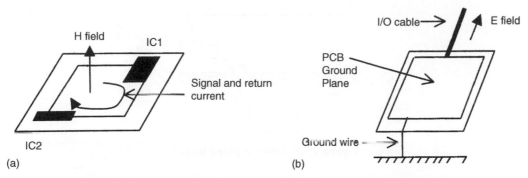

**Figure 25.6: Radiation from printed circuit board tracks: (a) differential mode radiation (loop type); (b) common mode radiation (monopole type)**

point (Figure 25.6(a)). Taking the example of TTL logic for a 1 mA signal at 200 MHz in a 3 cm$^2$ loop the field at 10 m would equal the limits of EN 55022 Class B. The areas of all such loops are to be minimized for lowest radiated field. Similarly, immunity to RF radiation and possibly other types of disturbance such as indirect electrostatic discharges (ESD) would also be improved by this measure.

However, a more important mechanism is where a voltage is generated across the length of the ground return on the board where the voltage drives into an attached wire or cable causing it to radiate as a monopole (Figure 25.6(b)). This is a far more efficient radiation mechanism and can give rise to high emission levels. The key factor is to ensure that the cable shield termination is connected to the metalwork of the enclosure and not to the zero volt conductor/plane of the printed circuit board.

Multilayer PCBs are effective at achieving the design aim of reduced loop area and close tracking. The ground plane acts as the high frequency earth return for all signals and all the high frequency energy is effectively contained within the layer of insulations. Compared with a single-sided board, EMC performance is improved by up to 30 dB by this measure. (See Figure 25.7.)

## 25.8 Interfaces

The multilayer PCB circuitry communicates with the outside world through attached shielded interface cables and it is important to minimize the amount of high frequency energy such as high order clock harmonics from exiting the equipment via the interface.

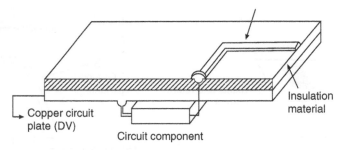

**Figure 25.7: Ground plane board**

One method of control (Reference 1) is to isolate the interface connectors on a piece of "clean" earth ground plane having only a single point of connection to the main part of the board.

The clean interface ground is bonded to the equipment chassis at the cable port; the decoupling capacitors are thus far more effective than they would be if terminated to the noisy ground of the main part of the PCB.

Digital signal interfaces may require additional protection against incoming transient disturbances particularly where the interface cable is unshielded or the pins are exposed to ESD. Here transient suppression in the form of gas tubes and Transorbs may be applied to good effect, provided a good low impedance earth is available. (See Figure 25.9.)

## 25.9  Power Supplies and Power-Line Filters

Many power supplies are of the switch mode types. These are well-known sources of high levels of high frequency disturbance, comprising harmonics of the switching frequency and cover the spectrum up to 30 MHz. These power supplies usually contain well designed high performance filters which attenuate the internally generated disturbance to levels below the most stringent emission limits. Any disturbance generated by the electronic circuit powered by the SMPSU is also attenuated strongly and there is usually no requirement for additional power line filtering for either emissions control or immunity.

The power supplies are usually supplied compliant to national and international EMC standards and the selection of the power supply can then be based solely on functional requirements.

Where a separate power line filter is installed in equipment, a single stage filter can often be selected by choosing the largest line inductance at the rated current. However, overall EMC performance is generally determined by EMC tests on the complete system because the practical terminating impedances are usually unknown for all the modes of propagation, and calculations or estimates based on manufacturers' data can often be accompanied by high uncertainties.

The performance of the filter will be critically dependent on the filter installation method. Figure 25.10 shows an example of bad installation practice. The performance is severely degraded by cross coupling via the bundled input and output wires and the earth is provided by a wire. The impedance of the wire is in series with the suppression capacitors between the power conductors and the equipment chassis and the high frequency performance of the filter will be degraded. The metal body of the filter should be bonded directly to the chassis via a shake-proof washer and not by a wire and the input and output wires should be separated and aligned 180 degrees in opposition to one another.

## 25.10  Signal Line Filters

Where there is no shield for the signal cable a signal line filter may need to be employed and should be installed at the point of entry to the equipment enclosure. It is important to ensure that the introduction of the filter does not cause too much attenuation of the wanted signal. Generally low pass filters are employed and for digital data circuits the half power point of the filter response should be no lower than the 9th harmonic of the fundamental data rate in order to preserve the quality of the waveform. Installation is also important for signal line filters. The decoupling capacitors should be connected to a good chassis ground. Various signal line filters are shown in Figure 25.8.

The simplest form of signal line filter is the capacitive filter comprising a series of 1 nF capacitors from each line to ground at the point of entry of say an RS232 cable to the equipment enclosure. These may be obtained incorporated within the D connector.

In most cases an L filter can be employed using series inductance (denoted "L" in the figures), but where extra stages are required, e.g., for additional protection, a T filter is usually preferred to a Pi filter because it is less dependent on a good earth.

For some high integrity systems, the signal line filters are used in conjunction with screened cables, the filter components often being incorporated within the plug or socket

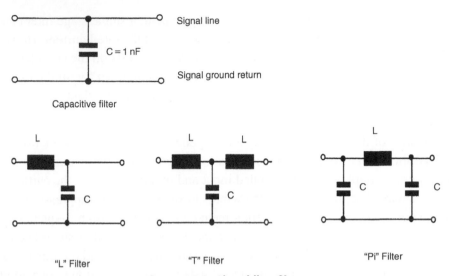

Figure 25.8: Signal line filters

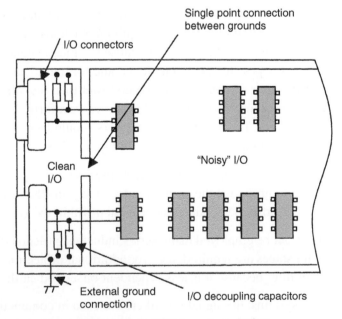

Figure 25.9: PCB interface design

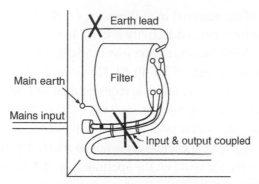

**Figure 25.10: Example of very poor filter installation**

at the end of the cable. However, the small size available will limit the value of series inductance that can be employed and the series elements often comprise a ferrite sleeve or series of beads that give good performance, but only at the higher frequencies.

Additional protection for analog circuits such as operational amplifiers can be achieved by the use of a balanced (equal value) pair of series resistors at the input pins and by 1 nF decoupling capacitors at the input. It is important to decouple the supply rails using capacitors having a good RF performance to minimize unwanted potential differences and to reduce the size of loops to avoid coupling with applied electromagnetic fields.

## 25.11  Enclosure Design

Metal equipment enclosures can be configured to provide an effective barrier to electromagnetic waves. Basic shielding theory states that an incident wave is partly reflected at the surface and is then attenuated in its passage though the medium. The total shielding loss is the sum of the reflection and absorption losses. There is another factor known as the *secondary reflection loss*, but this is only taken into account for very thin shields where the absorption is less than 10 dB.

For most practical low frequency applications, a reliable minimum measure of shield performance can be obtained by calculating the absorption loss factor.

$$A = 131 \cdot t \cdot (f \mu_r \sigma_r)^{1/2} \text{ dB}$$

where $t$ is the thickness of the material in millimeters, $f$ is the frequency in MHz and $\mu_r$ and $\sigma_r$ are the permeability and conductivity relative to copper, respectively. For a 0.5 mm copper sheet at 1 MHz the absorption loss is 65 dB, more than adequate for most commercial/industrial requirements of 20–40 dB. However the same copper sheet provides only 0 5 dB at 50 Hz. If effective protection is required against power frequency magnetic fields, higher permeability materials such as steel ($\mu_r = 300$–$1000$) will be required. For a 20 dB attenuation at 50 Hz the thickness of the steel will need to be about 5 mm. The high frequency performance of practical enclosures is not so much dependent on the inherent properties of the material but far more on the apertures and discontinuities in the surface.

Electromagnetic energy will leak through apertures to a degree dependent on the ratio of the aperture diameter to wavelength. Where the aperture diameter is one half wave-length it is virtually transparent. The degree of attenuation at lower frequencies can be easily derived using the expression $SE = 20 \log_{10}(2D/\lambda)$, where $\lambda$ is the wavelength at the frequency of interest and D is the largest dimension of the aperture. For high attenuation D must be as small as possible. For large apertures for ventilation etc., a mesh may be used to cover the area thus providing an effective shield. In general terms, a large number of small apertures is preferred to a small number of large apertures.

Discontinuities in the surface can also degrade shielding performance, especially where the length of the slot is one half wavelength. The length of the slot should be reduced by providing more points of contact between the sections. Ideally there should be a good surface to surface bond at the connections between modules in a rack enclosure, and the additional expense of modules with rf spring fingers along the edge will pay dividends both for reducing the transfer of electromagnetic wave energy but also in protecting against ESD. Gaskets can also be used to achieve good high frequency electrical bonding between metal panels and enclosure sections. However, the enclosure must be designed from the outset to accept the gasket.

Many enclosures are not metal, but plastic or other non-conducting material. These would provide no shielding to electromagnetic waves but it is possible to introduce a shielding effect by various means. These include painted, sprayed sputtered or plated metallic coatings to the inside surface of the plastic enclosure or by using metallic loaded plastic material. Generally these measures produce only low levels of shielding; for example, less than 30 dB, but they can be extremely effective in improving the situation for marginally noncompliant equipment. (See Figures 25.11 and 25.12.)

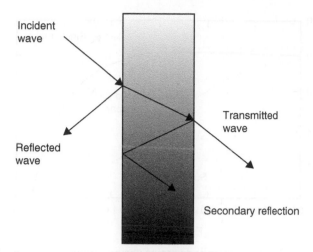

**Figure 25.11: Shielding theory**

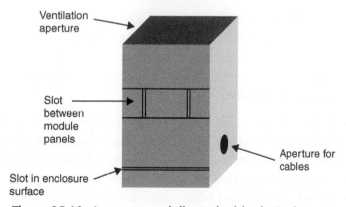

**Figure 25.12: Apertures and discontinuities in Enclosures**

## 25.12  Interface Cable Connections

Cable characteristics can have an impact on the EMC performance of a system. Unshielded twisted pair cable is often used for low and medium speed data links. In the presence of an incident electromagnetic wave, voltages will be induced in the cable both in common mode and differential mode. Although the differential voltage is minimized at low frequencies by the twists in the cable, the common mode voltage remains and will be applied to the circuit to which the cable is connected. The balance of the circuit is

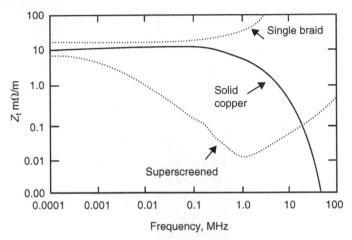

**Figure 25.13: Transfer impedance of various types of screened cable**

therefore crucial and if the impedances to ground at the circuit are equal the conversion to unwanted differential mode noise is minimized. The difficulty is that parasitic components such as stray capacitance tend to unbalance the circuit at high frequencies and the ability to reject common mode disturbance reduces rapidly with increasing frequency, thus causing interference.

Additional protection can be provided by the addition of an outer screen usually of copper braid or aluminum foil. At high frequencies, the electromagnetic energy is confined to the inner and outer surface layers of the screen by skin effect and isolation between the inner conductors carrying the wanted signal and the external environment can be readily achieved through good cable screen design. Screened cable performance can be characterized by its surface transfer impedance ($Z_t$) which is the ratio of the voltage induced on the inner conductor to a current in the outer surface of the outer conductor. The shielding effectiveness of coaxial cables is given approximately by $SE = 20 \cdot \log_{10}(50/Z_t)$. The transfer impedances of various types of screened cable are shown in Figure 25.13.

At low frequencies performance is determined by the self impedance of the screen conductor. For the single copper braid for example, the weave allows energy to pass through the shield and high frequency performance is reduced, although the level of performance is usually adequate for most purposes. For special applications such as for

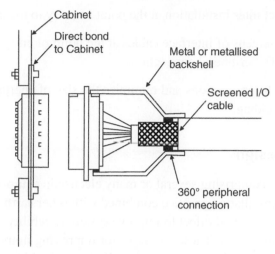

**Figure 25.14: Correct 360° peripheral termination of screened cable**

cables carrying very low level signals in an intense electromagnetic environment such as a nuclear power station the outer screen is composed of many layers including inductive materials as in super screened cables, and excellent performance is achieved.

Having the correct termination for the cable shield is an important consideration in achieving the maximum performance of the cable. Ideally the cable screen should be terminated in a 360° peripheral glanded connection as shown in Figure 25.14.

If the cable screen is made up into a 'pigtail' connection, performance at high frequencies will be severely degraded and pigtail screen connections should be avoided. At high frequencies the impedance of the pigtail connection becomes significant and is a point of common coupling between the inner conductors and the external environment increasing, emissions and reducing immunity to external disturbances.

## 25.13 Golden Rules for Effective Design for EMC

(1)  Consider the location and design of EMC "barriers" at the concept stage.

(2)  Consider the use of multilayer printed circuit boards with good layout and interface design.

(3)  Select filters for unscreened power and/or signal cables entering the enclosure.

(4) Ensure correct filter installation at the point of entry to the enclosure.

(5) Connect the screens of interface cables at the point of entry to the enclosure by a glanded 360° peripheral connection.

(6) Bond the enclosure sections and minimize the size of apertures for best enclosure shielding.

## 25.14 System Design

For distributed systems comprising several or many electronic equipment interconnected by cables good cable, installation practice combined with other earthing and grounding measures can be applied to good effect to improve system reliability in the presence of electromagnetic disturbances. These measures for improving immunity are equally effective in reducing emission levels from an installation and will ease the task of operators and system installers in ensuring compliance with the essential requirements of the EMC Directive. The main methods of control are: cable segregation where cables of various types are routed in groups so that cross talk or disturbance pick up is minimized cable separation where cables with high levels of disturbance are spatially separated from the cables carrying sensitive signals and isolation by the use of earthed conductors which reduce coupling between cable types.

In a first step toward good EMC engineering practice the following general guidelines should be observed:

- *Cable segregation:* Sensitive cables such as signal cables may be grouped together; mains cables including power feeds and lighting circuits' carrying up to 250 V may also be grouped together but the cable types should not be mixed.

- *Cable separation:* In addition to the groupings, data, telecoms or sensitive cabling should be separated from three-phase cables used for heavy electrical inductive load switching; for example, air conditioning, welding equipment and motors by the largest practicable distance.

- *Isolation:* Metal cable trays, if not already in use, should be implemented Having an adequate low impedance for the frequencies in use, and with good earthing, the tray will effectively become a partial screen for the cables.

- *Shield termination:* Cable shield termination is also a key factor in controlling EMC but the best practice is often dependent on the particular circumstances. For low frequency applications the shield may be terminated at only one end in order to mitigate against ground noise currents but this will reduce its effectiveness, particularly against magnetic fields.

For cables carrying low frequency signals, cable terminations have to be designed carefully to avoid coupling with noise currents flowing in the ground. If the sensor is not directly connected to ground as in Figure 25.15(a) above it may be possible to terminate the screen at both ends, thus providing maximum protection against inductively coupled disturbances. If the sensor is grounded as in Figure 25.15(b), the voltage drop across the ground impedance to noise currents in the ground will give rise to high currents in the shield and noise voltages may be present at the input to the amplifier. This can be overcome as shown by terminating the cable at one end only, thus avoiding the ground loop, but the performance of the shield may be reduced. If high performance is required under all conditions, e.g., with the sensor grounded it may be necessary to introduce transformer coupling or opt-isolation in order to minimize unwanted coupling.

Cables transporting similar signals can often be bundled together. With cables transporting different signals it is possible to differentiate between cables as shown in Table 25.6.

The five types of cables listed above should be separated in sequence from each other by at least 150 mm on cable trays or racks. That means that a very sensitive cable should he

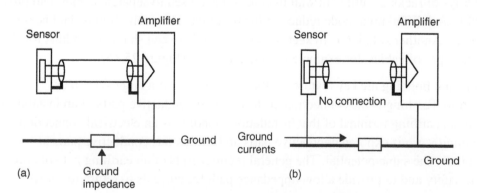

**Figure 25.15: Screened cable termination methods (a) Sensor not grounded; (b) Sensor grounded**

**Table 25.6: Cable type classification**

| Cable type | Characteristics |
|---|---|
| Very sensitive | Low-voltage or low-currents as from sensors |
| Sensitive | Signalling cables. 24V flat cables for parallel data transfer |
| Indifferent | AC power between 100V and 250V, depending on the EMC properties of the apparatus connected |
| Noisy | AC and DC relay feed line without protection measures such as filters or diodes |
| Very noisy | Leads to DC motors with brushes, switched power lines cables and earth wires in high voltage switchyards etc, |

separated from a sensitive cable by at least 150, mm and from a very noisy cable by at least 600 mm. In this latter case a minimum separation distance of 1 m is recommended if cable racks are used.

The use of parallel earthing conductors (PEC) can reduce common mode currents in signal leads by reducing the system common mode impedance and the loop area. Routing cables through trays, conduit or ducting has the effect of introducing a parallel path for disturbance current, which not only is capable of diverting and withstanding high currents but also of providing the necessary low impedance thus protecting the signal cable. Additional connections to earth should be made at regular intervals for very long cable runs.

Using trays or racks of sufficient wall thickness when used to separate cables, can provide both PEC and differential mode reduction in cross-talk. They can often be laid next to each other. Another solution is to keep some distance between shallow conduits for the different types of cables, for example by stacking them (Figure 25.16).

Earthing and bonding are key features in the EMC design of a large system or installation. Earthing is the connection of the exposed conductive parts of an installation to the main earthing terminal of that installation; bonding is an electrical connection maintaining various exposed conductive parts and extraneous conductive parts at substantially the same potential. The general requirements of an earthing network are to provide safety and to provide a low impedance path for currents to return to source.

The current preferences are for a multiple bonded earthing network which is compatible with the measures for lightning protection (Figure 25.17). The frame of the building is

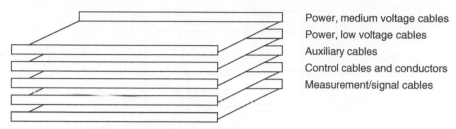

Power, medium voltage cables
Power, low voltage cables
Auxiliary cables
Control cables and conductors
Measurement/signal cables

**Figure 25.16: Stacking conduits to avoid cross-talk**

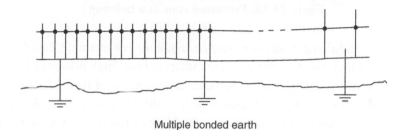

Multiple bonded earth

**Figure 25.17: Preferred method of earthing network**

bonded to the lightning conductors to avoid flash-over with multiple connections to earth electrodes.

Internal to a building, the recommended approach is a three-dimensional network, one earthing network per floor, each network connected to one another and to the earth electrode. The mass of inter-connections is called a *common bonding network (CBN)*. In some cases, e.g., telephone exchanges, an earthing mat may be installed but only connected to the building frame at one point to provide a mesh isolated bonding network (IBN). The main feature of the CBN is that each floor of a building should have a meshed ground network beneath the floor with multiple connections between the floors. The mesh network can take the form of not just the dedicated ground structure, but installed cable trays (metallic), water pipes and ducts. The meshed earthing scheme will produce an earthing system which should have low impedance, and a high probability of remaining so, even if a number of earth connections become corroded and fail.

## 25.15 Buildings

In most practical applications, electrical and electronic equipment is installed and used inside a building. Where the building is exposed to intense sources of electromagnetic

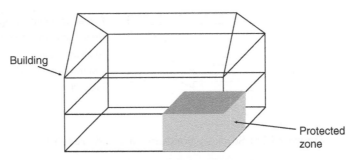

**Figure 25.18: Protected zone in a building**

disturbance, some additional protection may be required. For example, buildings on airfields are often illuminated by the incident radiation from high power radio and radar transmitters in the near vicinity, where the fields are in excess of the equipment immunity test levels of 3 and 10 volts/meter. Measures such as shielding on the windows may be a minimum requirement but in some cases shielding of the room containing the electronic equipment may be required. This introduces the concept of protected zones which can also be developed for protection against other threats such as lightning strikes (Figure 25.18).

Such protection is usually considered at the design stage but can be introduced retrospectively in some cases on a small scale. The need for additional measures is usually determined by a site survey of the electromagnetic environment at the building location. Other sources that often need to be considered are overhead power lines and overhead railway AC electrified lines. Here the magnetic field at the power frequency can cause distortion of the display on computer monitors. The threshold for distortion is about 1 Amp/meter and provision for some measure of control, usually by re-location or local shielding for the monitors, should be made if the field is likely to exceed this level.

## 25.16 Conformity Assessment

In Europe, products placed on the market must comply with the essential requirements of the EMC Directive. (The exceptions are military equipment, which must meet defense standards, and equipment covered by other directives that address EMC; for example, medical devices.) The manufacturer is required by law to make a declaration of conformity stating that the product complies with relevant harmonized standards referenced for application under the EMC Directive. He may then apply the CE mark to

**Table 25.7: European directives and EMC standards**

| Product | Directive | EMC Standards |
|---|---|---|
| Household equipment | EMC Directive 89/336/EEC | EN 55014-1 and –2 |
| Maritime navigation equipment | EMC Directive 89/336/EEC | EN 60945 |
| Lighting equipment | EMC Directive 89/336/EEC | EN 55015 and EN 61547 |
| Information technology equipment | EMC Directive 89/336/EEC | EN 55022 and EN 55024 |
| Railway equipment | EMC Directive 89/336/EEC | prEN 50121 parts 2–5 |
| Low voltage switchgear | EMC Directive 89/336/EEC | EN 60947 |
| Radio and TV receivers | EMC Directive 89/336/EEC | EN 55013 and EN 55020 |
| Radio and radio-telecommunications equipment | RTTE Directive 99/5/EC | Applicable TBR, EN or ETS standards |
| Medical equipment | Medical Devices Directive 93/42 EEC | EN 60601-1-2 |
| Motor vehicles | Automotive EMC Directive 95/54/EEC | The directive contains the technical requirements |

his product and will obtain the benefit of free circulation of the product within the Single European Market without experiencing any barriers to trade. It should be noted that the CE Mark denotes compliance with *all* applicable directives which for most products, will also include the electrical safety requirements of the Low Voltage Directive 73/23/EEC.

For finished products, complex components and systems, the manufacturer has the option of either applying appropriate harmonized EMC standards, i.e., those approved by the European Commission, or can prepare a Technical Construction File where the harmonized standards are not available or are not applied in full. Generally most manufacturers apply the harmonized standards as the most cost-effective route and this usually means that the product must be tested to EMC standards to determine compliance.

Table 25.7 presents examples of the requirements for common types of electrical and electronic equipment.

It is the responsibility of the manufacturer to determine which directives apply to his product but information is usually generally available from standards bodies such as the

BSI, from EMC test laboratories and Competent Authorities such as the UK's Department of Trade and Industry.

The European assessment requirements for fixed installations are somewhat different. The installation may comprise compliant (CE marked) equipment and/or non-compliant equipment. The installation must meet the 'essential requirement' of the EMC Directive, i.e., not to cause interference and not to be affected by electromagnetic disturbance. There is no necessity for declaring compliance nor for CE marking. However the installation designer should ensure that the equipment is installed according to good engineering practice and is advised to maintain a file containing a description of the EMC measures taken, together with any test data etc. This file should be available for inspection by the authorities if challenged at some later date. The key issue is the acceptance by the designer or operator of noncompliant equipment. This is permissible but becomes the responsibility of the operator if interference results and therefore the operator should ensure that this equipment is compatible with its environment either by imposing some contractual requirements for EMC characterisation on the supplier or by in-situ confidence testing.

## 25.17  EMC Testing and Measurements

For the vast majority of electrical and electronic equipment, compliance with the relevant technical specification is achieved by testing to EMC standards. Emissions measurements can comprise two types, a) tests for low frequency phenomena such as power frequency harmonics and voltage fluctuations to protect the power distribution network, and b) high frequency emissions to protect the radio spectrum. For commercial and industrial products, high frequency emission measurements are normally made over the frequency range 150 kHz to 1 GHz but some standards call for measurements down to 9 kHz. It should be noted that it is becoming increasingly necessary to measure emissions above 1 GHz to protect cellular and other radio systems.

The low frequency phenomena are usually measured with proprietary test instrumentation dedicated for that purpose comprising a harmonics analyzer and flicker meter. Most high frequency tests comprise measurements of conducted disturbance over the frequency range 150 kHz to 30 MHz and radiated disturbance at higher frequencies, 30 MHz to 1000 MHz. The different test methods effectively reflect the propagation mechanisms that dominate in practice.

Conducted disturbance measurements on the mains power input comprise a measurement of rf voltage across a passive network (50 uH/50 ohms) having an input impedance representative of the RF impedance of the mains at high frequencies. The voltage is measured using a calibrated EMC measuring receiver or spectrum analyzer. Most modern instrumentation systems are computer controlled for maximum efficiency.

Radiated measurements are performed on a test range meeting particular requirements for path loss, or in a facility giving results which can be correlated to those on the test range. The standard test range comprises on open area site which is flat and free from reflecting objects and typically having the dimensions shown in Figure 25.19. A significant proportion of the area should be covered by a conducting ground plane in order to achieve the site attenuation calibration requirements.

Shielded enclosures lined with radio-wave absorbing material or ferrite tiles on all internal sides except the floor can be constructed to achieve site attenuation performance characteristics comparable with an open area test site. Due to cost limitations, most chambers are built to accommodate a 3 meter range and some additional calibration work may be required to relate the results to the 10 meter range, particularly for physically large items of equipment.

Measurements of field strength are made using a receiver or spectrum analyzer and a calibrated antenna situated at a fixed distance from the equipment under test. A key factor in emissions measurements is to configure and operate the product in a manner which is both representative of practical use, yet maximizes the emissions from the product. For products and small systems which contain cables some pre-testing is often required to determine the optimum configuration for test, particularly for radiated emissions tests.

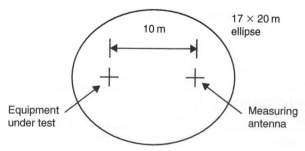

**Figure 25.19: Dimensions of the CISPR open area test for radiated emissions**

Immunity tests are performed by subjecting the equipment to electromagnetic disturbances of the type and maximum level that the product is likely to encounter in its intended operating environment. For many of the tests, specialized test instrumentation is required, providing the well-defined disturbance characteristics described in the standards. The instrumentation is commonly available from a number of specialist suppliers and the tests which involve the direct application of disturbance voltages are relatively straightforward.

The facilities for subjecting the product to disturbance fields are more complex and costly. The RF field immunity test is performed inside a shielded room lined with radio-wave absorbing material or ferrite tiles and the costs of purchasing and installing the facility are often quite high. The test method requires that the test volume is precalibrated to achieve the desired test level. There is an additional field uniformity requirement for the maximum level of field over 75% of a 15 m square area to be no greater than twice the required test level. RF power over the frequency range 80–1000 MHz is fed to a dedicated radiating antenna in the room from a signal generator and high power amplifier. The system is usually computer controlled to achieve the calibrated level in an efficient and reproducible manner.

To simulate coupling with RF fields at frequencies below 80 MHz, a common mode voltage test is applied to the cables attached to the product, i.e., the mains leads and signal and interface cables using coupling/decoupling networks (CDNs). The measurements are usually made inside the shielded enclosure for convenience and to avoid causing interference to radio services.

Immunity to power frequency magnetic fields is performed by setting up current flow in a loop around the product or preferably by placing the product within the confines of a Helmholtz coil.

**Table 25.8: Test equipment**

| Test | Equipment |
|---|---|
| ESD | ESD Gun |
| Surges | Surge generator |
| Dips and interruptions | Dip generator |
| Rf fields | Signal generator, power amplifiers, antennas |
| Rf voltages | Signal generator, power amplifiers, coupling networks |

In immunity testing, the configuration and operating conditions of the equipment under test can be very important. These conditions are usually specified in the EMC standard applicable to the product but in the absence of specific information the test engineer must attempt to adjust the conditions for maximum sensitivity to the applied disturbance. This can be difficult because immunity testing does not give a resultant variable, only an attribute of compliance or non-compliance, and care is required to follow the specific instructions of the standard and the advice and guidance of the manufacturer. Monitoring of the product's performance should be performed in a non-invasive method that does not influence the coupling of the applied disturbance. In addition the criteria for assessing the product's performance need to be clearly established prior to testing. For example for an analog instrument such as a temperature sensor, the margin of acceptable error, say 1%, needs to be stated in the User Manual.

## 25.18 Management Plans

For all product, system or installation developments, compliance with EMC requirements can be made much easier by the adoption and adherence to an EMC management plan. This would probably be integrated within a broader conformity assessment program which would include safety and other approvals. The plan would be prepared at the inception of the product development with an intention of achieving complete compliance at the lowest cost and in the shortest timeframe both at the initial certification stage and also through the lifetime of the product. The plan would include Control Plans which would specify the intended operating electromagnetic environment, any particular EMC elements in the Technical Specification, a Compatibility Matrix to identify couplings with other systems, EMC Test Plans, and applicable codes of practice to aid the design team. Table 25.9 lists the EMC inputs required at the various stages of a product development.

**Table 25.9: Product development process**

| Product development stage | EMC input |
|---|---|
| Design Concept | Apply EMC control principles |
| Design Process | EMC design measures |
| Prototype | Pre-compliance EMC tests |
| Production Prototype | EMC certification |
| Production | Conformity in production tests |
| Upgrades/modifications | EMC re-assessment |

Procedures can be set up for regular EMC design reviews and formal acceptance of supplied subassemblies or equipment from subcontractors. In some cases, particularly for large systems and installations, it may be necessary to establish procedures for dealing with concession requests and dispensations where supplied items are non-compliant.

## References

*Cabling Installations: User Friendly Guide*, ERA Report No 98-0668

Degauque, P and Hamelin, J. *Electromagnetic Compatibility*. Oxford University Press, 1993.

Design for EMC. *ERA Report 90-006*, ERA Technology Ltd (1990)GOEDBLOED, J. J., *Electromagnetic Compatibility*, Prentice Hall (1992).

GREEN, M., *Optimised and Superscreened cables*, Raychem Ltd.

Information Technology—*Cabling Installation*, Part 2 Installation Planning and Practices Inside Buildings, prEN 50174-2 June 1998.

Ott HW., *Noise Reduction Techniques in Electronic Systems*, John Wiley Interscience, NY, 1988.

Paul CR. *Introduction to Electromagnetic Compatibility*, John Wiley and Sons Inc, 1992.

# General Reference

### John Bird

## A.1 Standard Electrical Quantities—Their Symbols and Units

| Quantity | Quantity symbol | Unit | Unit symbol |
|---|---|---|---|
| Admittance | $Y$ | siemen | S |
| Angular frequency | $\omega$ | radians per second | rad/s |
| Area | $A$ | square meters | $m^2$ |
| Attenuation coefficient (or constant) | $\alpha$ | neper | Np |
| Capacitance | $C$ | farad | F |
| Charge | $Q$ | coulomb | C |
| Charge density | $\sigma$ | coulomb per square meter | $C/m^2$ |
| Conductance | $G$ | siemen | S |
| Current | $I$ | ampere | A |
| Current density | $J$ | ampere per square meter | $A/m^2$ |
| Efficiency | $\eta$ | per-unit or per cent | p.u. or % |
| Electric field strength | $E$ | volt per meter | V/m |
| Electric flux | $\psi$ | coulomb | C |
| Electric flux density | $D$ | coulomb per square meter | $C/m^2$ |
| Electromotive force | $E$ | volt | V |
| Energy | $W$ | joule | J |
| Field strength, electric | $E$ | volt per meter | V/m |
| Field strength, magnetic | $H$ | ampere per meter | A/m |

(*Continued*)

| Quantity | Quantity symbol | Unit | Unit symbol |
|---|---|---|---|
| Flux, electric | $\psi$ | coulomb | C |
| Flux, magnetic | $\Phi$ | weber | Wb |
| Flux density, electric | $D$ | coulomb per square meter | $C/m^2$ |
| Flux density, magnetic | $B$ | tesla | T |
| Force | $F$ | newton | N |
| Frequency | $f$ | hertz | Hz |
| Frequency, angular | $\omega$ | radians per second | rad/s |
| Frequency, rotational | $n$ | revolutions per second | rev/s |
| Impedance | $Z$ | ohm | $\Omega$ |
| Inductance, self | $L$ | henry | H |
| Inductance, mutual | $M$ | henry | H |
| length | $l$ | meter | m |
| Loss angle | $\delta$ | radian or degrees | rad or ° |
| Magnetic field strength | $H$ | ampere per meter | A/m |
| Magnetic flux | $\Phi$ | weber | Wb |
| Magnetic flux density | $B$ | tesla | T |
| Magnetic flux linkage | $\Psi$ | weber | Wb |
| Magnetising force | $H$ | ampere per meter | A/m |
| Magnetomotive force | $F_m$ | ampere | A |
| Mutual inductance | $M$ | henry | H |
| Number of phases | $m$ | – | – |
| Number of pole-pairs | $p$ | – | – |
| Number of turns (of a winding) | $N$ | – | – |
| Period, Periodic time | $T$ | second | s |
| Permeability, absolute | $\mu$ | henry per meter | H/m |
| Permeability of free space | $\mu_0$ | henry per meter | H/m |
| Permeability, relative | $\mu_r$ | – | – |
| Permeance | $\Lambda$ | weber per ampere or per henry | Wb/A or /H |

*(Continued)*

| Quantity | Quantity Symbol | Unit | Unit symbol |
|---|---|---|---|
| Permittivity, absolute | $\varepsilon$ | farad per meter | F/m |
| Permittivity of free space | $\varepsilon_0$ | farad per meter | F/m |
| Permittivity, relative | $\varepsilon_r$ | – | – |
| Phase-change coefficient | $\beta$ | radian | rad |
| Potential, Potential difference | $V$ | volt | V |
| Power, active | $\rho$ | watt | W |
| Power, apparent | $S$ | volt ampere | VA |
| Power, reactive | $Q$ | volt ampere reactive | Var |
| Propagation coefficient (or constant) | $\gamma$ | – | – |
| Quality factor, magnification | $Q$ | – | – |
| Quantity of electricity | $Q$ | coulomb | C |
| Reactance | $X$ | ohm | $\Omega$ |
| Reflection coefficient | $P$ | – | – |
| Relative permeability | $\mu_r$ | – | – |
| Relative permittivity | $\varepsilon_r$ | – | – |
| Reluctance | $S\ or\ R_m$ | ampere per weber or per henry | A/Wb or /H |
| Resistance | $R$ | ohm | $\Omega$ |
| Resistance, temperature coefficient of | $A$ | per degree Celsius or per kelvin | /°C or /K |
| Resistivity | $\rho$ | ohm meter | $\Omega$m |
| Slip | $s$ | per unit or per cent | p.u. or % |
| Standing wave ratio | $S$ | – | – |
| Susceptance | $B$ | siemen | S |
| Temperature coefficient of resistance | $\alpha$ | per degree Celsius or per kelvin | /°C or /K |
| Temperature, thermodynamic | $T$ | kelvin | K |
| Time | $t$ | second | S |

(*Continued*)

| Quantity | Quantity Symbol | Unit | Unit symbol |
|----------|-----------------|------|-------------|
| Torque | $T$ | newton meter | Nm |
| Velocity | $\nu$ | meter per second | m/s |
| Velocity, angular | $\omega$ | radian per second | rad/s |
| Volume | $V$ | cubic meters | $m^3$ |
| Wavelength | $\lambda$ | meter | m |
| (Note that m/s may also be written as $ms^{-1}$, $C/m^2$ as $Cm^{-2}$, /K as $K^{-1}$, and so on.) | | | |

## B.1 Differential Equations

As will be evident from the models discussed earlier, many systems have input-output relationships which have to be described by differential equations. A *differential equation* is an equation involving derivatives of a function, i.e., terms such as $dy/dt$ and $d^2y/dt^2$. Thus:

$$\tau \frac{dy}{dt} + y = 0$$

is a differential equation. The term *ordinary differential equation* is used when there are only derivatives of one variable, e.g., we only have terms such as $dy/dt$ and $d^2y/dt^2$ and not additionally $dx/dt$ or $dx^2/dt^2$. The *order* of a differential equation is equal to the order of the highest derivative that appears in the equation. For example,

$$\tau \frac{dy}{dt} + y = 0$$

and,

$$\tau \frac{dy}{dt} + y = kx$$

are first-order ordinary differential equations since the highest derivative is $dy/dt$ and there are derivatives of only one variable. The first of the above two equations is said to be *homogeneous* since it only contains terms involving $y$. Such an equation is given by a system which has no forcing input; one with a forcing input gives a non-homogeneous equation. For example, an electrical circuit containing just a charged capacitor in series with a resistor will give a homogeneous differential equation describing how the potential difference across the capacitor changes with time as the charge leaks off the capacitor;

there is no external source of voltage. However, if we have a voltage source which is switched into the circuit with the capacitor and resistor then the voltage source gives a forcing input and a non-homogeneous equation results.

The equations:

$$m\frac{d^2y}{dt^2} + c\frac{dy}{dt} + ky = 0$$

and

$$m\frac{d^2y}{dt^2} + c\frac{dy}{dt} + ky = F$$

are examples of a second-order differential equation since the highest derivative is $d^2y/dt^2$. The first of the two is homogeneous since there is no forcing input and the second is non-homogeneous with a forcing input $F$.

### B.1.1  Solving a First-Order Differential Equation

With a first-order differential equation, if the variables are separable, i.e., it is of the form:

$$\frac{dy}{dt} = f(t)$$

then we can solve such equations by integrating both sides of the equation with respect to $x$:

$$\int \frac{dy}{dt}\,dt = \int f(t)dt$$

This is equivalent to separating the variables and writing:

$$\int dt = \int f(t)\,dt$$

Consider the response of a first-order system to a step input, e.g., the response of a thermometer when inserted suddenly into a hot liquid. This sudden change is an example of a step input. In Chapter 18 the differential equation for such a change was determined as:

$$RC\frac{dT}{dt} + T = T_L$$

where $T$ is the temperature indicated by the thermometer, $T_L$ the temperature of the hot liquid, $R$ the thermal resistance and $C$ the thermal capacitance. We can solve such an equation by the technique of 'separation of variables'. Separating the variables gives:

$$\frac{1}{T_L - T} dT = \frac{1}{RC} dt$$

Integrating then gives:

$$-\ln(T_L - T) = (1/RC)t + A$$

where $A$ is a constant. This equation can be written as:

$$T_L - T = e^A e^{-t/\tau} = B e^{-t/\tau}$$

with $B$ being a constant and $\tau = RC$. $\tau$ is termed the *time constant* and can be considered to be the time that makes the exponential term $e^{-1}$. If we consider the thermometer to have been inserted into the hot liquid at time $t = 0$ and to have been indicating the temperature $T_0$ at that time, then, since $e^0 = 1$, we have $B = T_L - T$. Hence the equation can be written as:

$$T = (T_0 - T_L)e^{-t/\tau} + T_L$$

The exponential term will die away as $t$ increases and so gives the transient part of the response. $T_L$ is the steady-state value that will be attained eventually.

### B.1.2 Complementary Function and Particular Integral

Suppose we have the first-order differential equation $dy/dt + y = 0$. Such an equation is homogeneous and, if we apply the technique of separation of the variables, has the solution $y = C e^{-t}$. Now suppose we have the nonhomogeneous equation $dy/dt + y = 2$. If we now use the separation of variable technique we obtain the solution $y = C e^{-t} + 2$. Thus, its solution is the sum of the solution for the homogeneous equation plus another term. The solution of the homogeneous differential equation is called the *complementary function* and, when there is a forcing input with that differential equation, the term added to it for the non-homogeneous solution is called the *particular integral*. We can, easily, obtain a particular integral by assuming it will be of the same form as the forcing input.

Thus if this is a constant then we try $y = A$, if of the form $a + bx + cx^2 + \ldots$ then $y = A + Bx + Cx^2 + \ldots$ is tried, if an exponential then $y = A\,e^{kx}$ is tried, if a sine or cosine then $y = A \sin \omega x + B \cos \omega x$ is tried.

As an illustration, consider the system involving solved in the previous section of a thermometer being inserted into a hot liquid and the relationship being given by:

$$RC\frac{dT}{dt} + T = T_{\text{L}}$$

The homogeneous form of this equation is:

$$RC\frac{dT}{dt} + T = 0$$

We can solve this equation by the technique of the separation of variables to give the complementary function:

$$\frac{1}{T}\,dT = -\frac{1}{RC}\,dt$$

Integrating then gives:

$$-\ln T = -(1/RC)t + A$$

where $A$ is a constant. This equation can be written as:

$$-T = e^A\,e^{-t/\tau} = B\,e^{-t/\tau}$$

$$T = -B\,e^{-t/\tau}$$

Now consider the particular integral for the non-homogeneous equation and, because the forcing input is a constant, we try $T = C$. Since $dT/dt = 0$ then the substituting values in the differential equation gives:

$$0 + C = T_{\text{L}}$$

Thus the particular solution is $T = T_{\text{L}}$. Hence, the full solution is the sum of the complementary solution and the particular integral and so:

$$T = -B\,e^{-t/\tau} + T_{\text{L}}$$

If we consider the thermometer to have been inserted into the hot liquid at time $t = 0$ and to have been indicating the temperature $T_0$ at that time, then, since $e^0 = 1$, we have $B = T_L - T$. Hence, as before, the equation can be written as:

$$T = (T_0 - T_L)\,e^{-t/\tau} + T_L$$

As a further illustration, consider the thermometer, in equilibrium in a liquid at temperature $T_0$, when the temperature of the liquid is increased at a constant rate $a$, i.e., a so-called ramp input. The temperature will vary with time so that after a time $t$ it is $at + T_0$. This is then the forcing input to the system and so we have the differential equation:

$$RC\frac{dT}{dt} + T = at + T_L$$

The homogeneous form of this equation is:

$$RC\frac{dT}{dt} + T = 0$$

and, as before, the complementary solution is:

$$T = -B\,e^{-t/\tau}$$

For the particular integral we try a solution of the form $T = D + Et$. Since $dT/dt = E$, substituting in the non-homogeneous differential equation gives:

$$RCE + D + Et = aT + T_0$$

Equating all the coefficients of the $t$ terms gives $E = a$ and equating all the non-$t$ terms gives $RCE + D = T_0$ and so $D = T_0 - RCa$. Thus the particular integral is:

$$T = T_0 - RCa + at$$

and so the full solution, with $\tau = RC$, is:

$$T = -B\,e^{-t/\tau} + T_0 - \tau a + at$$

When $t = 0$ we have $T = T_0$ and so $B = -\tau a$. Hence we have:

$$T = \tau a\,e^{-t/\tau} + T_0 - \tau a + at$$

### B.1.3 Solving a Second-Order Differential Equation

We can use the technique of finding the complementary function and the particular integral to obtain the solution of a second-order differential equation. Consider a spring—damper—mass system. The differential equation for the displacement $y$ of the mass when subject to step input at time $t = 0$ of a force $F$ is:

$$m\frac{d^2 y}{dt^2} + c\frac{dy}{dt} + ky = F$$

In the absence of damping and the force $F$ we have the homogeneous differential equation:

$$m\frac{d^2 y}{dt^2} + ky = 0$$

This describes an oscillation which has an acceleration $d^2 y/dt^2$ which is proportional to $-y$ and is a description of simple harmonic motion; we have a mass on a spring allowed to freely oscillate without any damping. Simple harmonic motion has a displacement $y = A \sin \omega t$. If we substitute this into the differential equation we obtain:

$$-mA\omega^2 \sin \omega t + kA\omega \sin \omega t = 0$$

and so $\omega = \sqrt{(k/m)}$. This is termed the *natural angular frequency* $\omega_n$. If we define a constant called the *damping ratio* of:

$$\zeta = \frac{c}{2\sqrt{mk}}$$

then we can write the differential equation as:

$$\frac{1}{\omega_n^2}\frac{d^2 y}{dt^2} + \frac{2\zeta}{\omega_n}\frac{dy}{dt} + y = \frac{F}{k}$$

This differential equation can be solved by the method of determining the complementary function and the particular integral. For the homogeneous form of the differential equation, i.e., the equation with zero input, we have:

$$\frac{1}{\omega_n^2}\frac{d^2 y}{dt^2} + \frac{2\zeta}{\omega_n}\frac{dy}{dt} + y = 0$$

We can try a solution of the form $y = A\, e^{st}$. This, when substituted, gives:

$$\frac{1}{\omega_n^2} S^2 + \frac{2\zeta}{\omega_n} s + 1 = 0$$

$$s^2 + 2\omega_n \zeta s + \omega_n^2 = 0$$

This equation has the roots:

$$s = \frac{-2\omega_n \zeta \pm \sqrt{4\omega_n^2 \zeta^2 - 4\omega_n^2}}{2} = -\omega_n \zeta \pm \omega_n \sqrt{\zeta^2 - 1}$$

When we have:

1. Damping ratio between 0 and 1

   There are two complex roots:

   $$s = -\zeta\omega_n \pm j\omega_n \sqrt{1 - \zeta^2}$$

   If we let:

   $$\omega = \omega_n \sqrt{1 - \zeta^2}$$

   we can write:

   $$s = -\zeta\omega_n \pm j\omega$$

   Thus:

   $$y = A\, e^{(-\zeta\omega_n + j\omega)t} + B\, e^{-\zeta(\omega_n - j\omega)t}$$
   $$= e^{-\zeta\omega_n t}(A\, e^{j\Psi t} + B\, e^{-j\omega t})$$

   Using Euler's equation, we can write this as:

   $$y = e^{-\zeta\omega_n t}(P \cos \omega t + Q \sin \omega t)$$

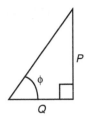

**Figure B.1: Angle $\phi$**

This can be written in an alternative form. If we consider a right-angled triangle with angle $\phi$ with $P$ and $Q$ being opposite sides of the triangle (Figure ApB. 1) then $\sin \phi = P/\sqrt{(P^2 + Q^2)}$ and $\cos \phi = P/\sqrt{(P^2 + Q^2)}$. Hence, using the relationship $\sin(\omega t + \phi) = \sin \omega t \cos \phi + \cos \omega t \sin \phi$, we can write:

$$y = C e^{-\zeta \omega_n t} \sin(\omega t + \phi)$$

where $C$ is a constant and $\phi$ a phase difference. This describes a damped sinusoidal oscillation. Such a motion is said to be *under damped*.

2.  Damping ratio equal to 1

    This gives two equal roots $s_1 = s_2 = -\omega_n$. and the solution:

$$y = (At + B) e^{-\omega_n t}$$

where $A$ and $B$ are constants. This describes an exponential decay with no oscillations. Such a motion is said to be *critically damped*.

3.  Damping ratio greater than 1

    This gives two real roots:

$$s_1 = -\omega_n \zeta + \omega_n \sqrt{\zeta^2 - 1}$$
$$s_2 = -\omega_n \zeta - \omega_n \sqrt{\zeta^2 - 1}$$

and hence:

$$y = A e^{s_1 t} + B e^{s_2 t}$$

where $A$ and $B$ are constants. This describes an exponential decay taking longer to reach the steady-state value than the critically damped case. It is said to be *over damped*.

The above analysis has given the complementary functions for the second-order differential equation. For the particular integral, in this case where we have a step input of size $F$, we can try the particular integral $x = A$. Substituting this in the differential equation gives $A = F/k$ and thus the particular integral is $y = F/k$. Thus, the solutions to the differential equation are:

1. Damping ratio between 0 and 1, i.e., under damped

$$y = C\,e^{-\zeta\omega_n t}\sin\left(\omega t + \phi\right) + F/k$$

2. Damping ratio equal to 1, i.e., critically damped

$$y = (At + B)\,e^{-\omega_n t} + F/k$$

3. Damping ratio greater than 1, i.e., over damped

$$y = A\,e^{S_1 t} + B\,e^{S_2 t} + F/k$$

In all cases, as $t$ tends to infinite then $y$ tends to the value $F/k$. Thus, the steady-state value is $F/k$.

# *Index*

Printed and bound by CPI Group (UK) Ltd, Croydon, CR0 4YY

03/10/2024

01040342-0013